In accordance with the latest syllabus prescribed by the Central Board of Secondary Education, New Delhi

MOST LIKELY CBSE QUESTION BANK

CLASS XII

PHYSICS

Category wise & Chapter wise

By

A PANEL OF AUTHORS

DISCLAIMER

ISBN : 978-93-91184-26-1

PUBLISHED BY

GURUKUL BOOKS & PACKAGING
(Unit of Oswal Printers & Publishers Pvt. Ltd.)

 1/12, Sahitya Kunj, M.G. Road, Agra - 282002

 (0562) 2527771-4, +91 7534077222

 info@oswalpublishers.in

 www.oswalpublishers.com

The cover of this book has been designed using resources from Freepik.com

PREFACE

Gurukul's CBSE Most Likely Question Bank series is updated as per the latest syllabus given by the Central Board of Secondary Education.

This title highlights the knowledge-based and skill-based goals of the Bloom's Taxonomy by making the students acquainted with relevant facts and concepts and also teaching them ways to apply their subjective knowledge into practise for desired academic results.

This book provides ample practice questions arranged section-wise so that the students can revise the whole syllabus in a comparatively lesser time and develop the ability of prioritising and categorising topics for effective learning. It covers all probable types of questions asked in the exams. The solutions have been prepared by eminent subject experts with explanations. Solutions follow the standard Marking Scheme issued by the CBSE Board.

In addition, questions from compartment paper, foreign paper and latest board paper have also been incorporated in the chapters.

This series is an attempt to instil confidence in students to face the board examination. The language used is simple, to the point and questions cover all the important topics as per the weightage given to them by the board.

We hope this book will be a valuable asset for the students. All further suggestions towards improving the series are welcome and would be incorporated in the future editions.

Note : Questions marked with :

'★' are board exam questions from previous years
'★ ★' are frequently asked board exam questions

—Publisher

How to Prepare for Board Exams Using Gurukul's Most Likely Question Bank

The Most Likely Question Bank is a great learning tool for CBSE Examinations as it is the most comprehensive and concisely composed set of questions for students. In order to prepare well for their exams, students may follow the following approach.

- **Step 1 : Prepare Your Theory**

  Students must go through the theory of the chapters, from their textbooks covering the syllabus as per the Board Guidelines.

- **Step 2 : Understand the Categories**

  The contents have been arranged as per different 'Category' of Questions. Each category in turn covers topics from all the chapters of the syllabus. e.g., Short Answers category will cover the short answer type questions from all the chapters of the syllabus in that subject.

- **Step 3 : Complete Syllabus Revision**

  When students revise questions from any single category, they are revising the entire syllabus in different assemblages.

- **Step 4 : Customised Learning**

  Students with different learning calibers can customize their preparations. They can refer the objective questions first to get a basic understanding of the chapters and topics, and then they can take up the most complex categories.

- **Step 5 : Check Your Progress**

  Students can check their answers against the solutions given next to the questions. This will save time and help in faster revisions. Students can benefit tremendously from this series by using them optimally for their exam preparations.

FINDING THE RIGHT BIN FOR MASKS?

Masks are a powerful shield against virus. The purpose behind wearing is to stop the spread of saliva and nasal discharge from infecting the other person.

Global use of disposable face masks and gloves is estimated to have been 100 billion pieces every month.

Masks are made up of a synthetic material called, TNT which takes more than 400 years to decompose.

Also masks come in a myriad of shades. The first step towards its healthy disposal is to choose which one to wear at the first place.

For casual use, choose tight cloth masks which can be reused over and over again. However, wearing it in vicinity of an infected person can be a wrong choice.

When visiting hospitals and other contaminated zones, choose loose fitting surgical masks which blocks droplets and splashes of saliva. However, they are ineffective against very small particles that come off sneezing and coughing in close vicinity.

N95 Masks are not for routine use and must be spared for medical experts who are in direct contact with the patient.

When it comes to disposing a mask, the very first step is to wash your hands before removing it from your face.

Wash your hands for atleast 20 seconds, using either an alcohol based sanitizer (having 60% alcohol) or using soap and water.

Remove your mask either using the mask strings or downside up. Don't touch the center of mask.

If you are wearing a cloth mask, put it separately into wash without mixing it with other clothes.

In case of a surgical mask, fold it inwards such that the droplets are not exposed to the air. Make it into a roll and wrap it up in two polythene bags or tissue bag, one inside the other.

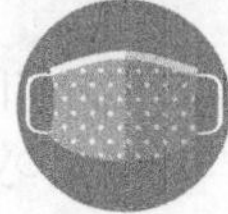

Prefer cloth masks for casual use and avoid one time used masks.

CONTENTS

SYLLABUS

CLASS XII

(THEORY)

Time : 3 Hours **Max. Marks : 70**

		No. of Periods	Marks
Unit – I	**Electrostatics** Chapter - 1 : Electric Charges and Fields Chapter - 2 : Electrostatic Potential and Capacitance	**24**	16
Unit – II	**Current Electricity** Chapter - 3 : Current Electricity	18	
Unit – III	**Magnetic Effects of Current and Magnetism** Chapter - 4 : Moving Charges and Magnetism Chapter - 5 : Magnetism and Matter	22	17
Unit – IV	**Electromagnetic Induction and Alternating Currents** Chapter - 6 : Electromagnetic Induction Chapter - 7 : Alternating Current	20	
Unit – V	**Electromangnetic Waves** Chapter - 8 : Electromagnetic Waves	04	18
Unit – VI	**Optics** Chapter - 9 : Ray Optics and Optical Instruments Chapter - 10 : Wave Optics	**27**	
Unit – VII	**Dual Nature of Radiation and Matter** Chapter - 11 : Dual Nature of Radiation and Matter	**08**	12
Unit – VIII	**Atoms and Nuclei** Chapter - 12 : Atoms Chapter - 13 : Nuclei	**15**	
Unit – IX	**Electronic Devices** Chapter 14 : Semiconductor Electronics : Materials, Devices and Simple Circuits	**12**	**07**
	Total	**150**	**70**

Unit I : **Electrostatics** **24 Periods**

Chapter-1 : Electric Charges and Fields

Electric charges; Conservation of charge, Coulomb's law-force between two point charges, forces between multiple charges; superposition principle and continuous charge distribution.

Electric field, electric field due to a point charge, electric field lines, electric dipole, electric field due to a dipole, torque on a dipole in uniform electric field.

Electric flux, statement of Gauss's theorem and its applications to find field due to infinitely long straight wire, uniformly charged infinite plane sheet and uniformly charged thin spherical shell (field inside and outside).

Chapter-2 : Electrostatic Potential and Capacitance

Electric potential, potential difference, electric potential due to a point charge, a dipole and system of charges; equipotential surfaces, electrical potential energy of a system of two point charges and of electric dipole in an electrostatic field.

Conductors and insulators, free charges and bound charges inside a conductor. Dielectrics and electric polarisation, capacitors and capacitance, combination of capacitors in series and in parallel, capacitance of a parallel plate capacitor with and without dielectric medium between the plates, energy stored in a capacitor.

Unit II : **Current Electricity** **18 Periods**

Chapter-3 : Current Electricity

Electric current, flow of electric charges in a metallic conductor, drift velocity, mobility and their relation with electric current; Ohm's law, electrical resistance, V-I characteristics (linear and non-linear), electrical energy and power, electrical resistivity and conductivity, Carbon resistors, colour code for carbon resistors; series and parallel combinations of resistors; temperature dependence of resistance.

Internal resistance of a cell, potential difference and emf of a cell, combination of cells in series and in parallel, Kirchhoff's laws and simple applications, Wheatstone bridge, metre bridge.

Potentiometer - principle and its applications to measure potential difference and for comparing EMF of two cells, measurement of internal resistance of a cell.

Unit III : **Magnetic Effects of Current and Magnetism** **22 Periods**

Chapter-4 : Moving Charges and Magnetism

Concept of magnetic field, Oersted's experiment.

Biot - Savart law and its application to current carrying circular loop.

Ampere's law and its applications to infinitely long straight wire. Straight and toroidal solenoids (only qualitative treatment), force on a moving charge in uniform magnetic and electric fields, Cyclotron.

Force on a current-carrying conductor in a uniform magnetic field, force between two parallel current-carrying conductors-definition of ampere, torque experienced by a current loop in uniform magnetic field; moving coil galvanometer-its current sensitivity and conversion to ammeter and voltmeter.

Chapter-5 : Magnetism and Matter

Current loop as a magnetic dipole and its magnetic dipole moment, magnetic dipole moment of a revolving electron, magnetic field intensity due to a magnetic dipole (bar magnet) along its axis and perpendicular to its axis, torque on a magnetic dipole (bar magnet) in a uniform magnetic field; bar magnet as an equivalent solenoid, magnetic field lines; earth's magnetic field and magnetic elements.

Para-, dia- and ferro - magnetic substances, with examples. Electromagnets and factors affecting their strengths, permanent magnets.

Unit IV : **Electromagnetic Induction and Alternating Currents** **20 Periods**

Chapter-6 : Electromagnetic Induction

Electromagnetic induction; Faraday's laws, induced EMF and current; Lenz's law, Eddy currents. Self and mutual induction.

Chapter-7 : Alternating Current

Alternating currents, peak and RMS value of alternating current/voltage; reactance and impedance; LC oscillations (qualitative treatment only), LCR series circuit, resonance; power in AC circuits, power factor, wattless current.

AC generator and transformer.

Unit V : **Electromagnetic Waves** **04 Periods**

Chapter-8 : Electromagnetic Waves

Basic idea of displacement current, Electromagnetic waves, their characteristics, their Transverse nature (qualitative ideas only).

Electromagnetic spectrum (radio waves, microwaves, infrared, visible, ultraviolet, X-rays, gamma rays) including elementary facts about their uses.

Unit VI : **Optics** **27 Periods**

Chapter-9 : Ray Optics and Optical Instruments

Ray Optics : Reflection of light, spherical mirrors, mirror formula, refraction of light, total internal reflection and its applications, optical fibres, refraction at spherical surfaces, lenses, thin lens formula, lensmaker's formula, magnification, power of a lens, combination of thin lenses in contact, refraction of light through a prism.

Scattering of light - blue colour of sky and reddish appearance of the sun at sunrise and sunset.

Optical instruments : Microscopes and astronomical telescopes (reflecting and refracting) and their magnifying powers.

Chapter-10 : Wave Optics

Wave Optics : Wave front and Huygen's principle, reflection and refraction of plane wave at a plane surface using wave fronts. Proof of laws of reflection and refraction using Huygen's principle. Interference, Young's double slit experiment and expression for fringe width, coherent sources and sustained interference of light, diffraction due to a single slit, width of central maximum, resolving power of microscope and astronomical telescope, polarisation, plane polarised light, Brewster's law, uses of plane polarised light and Polaroids.

Unit VII : **Dual Nature of Radiation and Matter** **08 Periods**

Chapter-11 : Dual Nature of Radiation and Matter

Dual nature of radiation, Photoelectric effect, Hertz and Lenard's observations; Einstein's photoelectric equation-particle nature of light.

Experimental Study of Photoelectric effect.

Matter waves-wave nature of particles, de-Broglie relation, Davisson-Germer experiment (experimental details should be omitted; only conclusion should be explained).

Unit VIII : **Atoms and Nuclei** **15 Periods**

Chapter-12 : Atoms

Alpha-particle scattering experiment; Rutherford's model of atom; Bohr model, energy levels, hydrogen spectrum.

Chapter-13 : Nuclei

Composition and size of nucleus, Radioactivity, alpha, beta and gamma particles/rays and their properties; radioactive decay law, half life and mean life.

Mass-energy relation, mass defect; binding energy per nucleon and its variation with mass number; nuclear fission, nuclear fusion.

Unit IX : **Electronic Devices** **12 Periods**

Chapter-14 : Semiconductor Electronics : Materials, Devices and Simple Circuits

Energy bands in conductors, semiconductors and insulators (qualitative ideas only).

Semiconductor diode- I-V characteristics in forward and reverse bias, diode as a rectifier.

Special purpose p-n junction diodes : LED, photodiode, solar cell and Zener diode and their characteristics, zener diode as a voltage regulator.

QUESTION PAPER DESIGN
THEORY (Class : XI/XII)

Maximum Marks: 70 **Duration: 3 hrs.**

S. No.	Typology of Question	Total Marks	Approximate Percentage
1.	**Remembering:** Exhibit memory of previously learned material by recalling facts, terms, basic concepts, and answers. **Understanding:** Demonstrate understanding of facts and ideas by organizing, comparing, translating, interpreting, giving descriptions, and stating main ideas.	27	38%
2.	**Applying:** Solve problems to new situations by applying acquired knowledge, facts, techniques and rules in a different way.	22	32%
3.	**Analysing:** Examine and break information into parts by identifying motives or causes. Make inferences and find evidence to support generalizations. **Evaluating:** Present and defend opinions by making judgements about information, validity of ideas, or quality of work based on a set of criteria. **Creating:** Compile information together in a different way by combining elements in a new pattern or proposing alternative solutions.	21	30%
	Total Marks	**70**	**100**

Practical: 30 Marks.

Note:

1. **Internal Choice:** *There is no overall choice in the paper. However, there will be at least 33% internal choice.*
2. *The above template is only a sample. Suitable internal variations may be made for generating similar templates keeping the overall weightage to different form of questions and typology of questions same.*

Question Set 1

Multiple Choice Questions

Chapter 1. Electric Charges and Fields

1. A rectangular frame of area 10 m^2 is placed in a uniform electric field of 20 NC^{-1}, with normal drawn on the surface of the frame making 60° angle with the direction of field. The electric flux through the frame is :
   (a) 100 Vm (b) 200 Vm (c) $50\sqrt{3}$ Vm (d) $100\sqrt{3}$ Vm
2. A charge Q is enclosed by a Gaussian spherical surface of radius R. If the radius is doubled then the outward electric flux will :
   (a) be doubled (b) increase four times
   (c) be reduced half (d) remains the same
3. The work done in taking a charge q one round a circle of radius r having a charge Q at the centre is :
   (a) $\frac{qQ}{4\pi\varepsilon_0 r}$ (b) $\frac{qQ}{4\pi\varepsilon_0 r^2}$ (c) $\frac{qQ}{4\pi\varepsilon_0}(2\pi r)$ (d) Zero.
4. A charge placed at a distance from an electric dipole in the end-on position experiences a force F. If the distance be doubled, the force will become :
   (a) 2 F (b) F/2 (c) F/4 (d) F/8
5. Two point charges (+e) and (–e) are kept inside a large metallic cube without touching its sides. Electric flux emerging out of the cube is :
   (a) $\frac{e}{\varepsilon_0}$ (b) $\frac{-e}{\varepsilon_0}$ (c) Zero (d) $\frac{2e}{\varepsilon_0}$
6. A soap bubble is given a negative charge. Then its radius :
   (a) decreases (b) increases (c) remains unchanged (d) None of the above
7. Relative permittivity of water is 81. If ε_w and ε_0 are permittivities of water and vacuum respectively, then :
   (a) $\varepsilon_0 = 9\varepsilon_w$ (b) $\varepsilon_0 = 81\varepsilon_w$ (c) $\varepsilon_w = 9\varepsilon_0$ (d) $\varepsilon_w = 81\varepsilon_0$
8. A body has a positive charge of 8×10^{-19} C. It has :
   (a) an excess of 5 electrons (b) a deficiency of 5 electrons
   (c) an excess of 8 electrons (d) a deficiency of 8 electrons
9. Electric field intensity, 'E' at a point P (Figure) at a perpendicular distance 'r' from an infinitely long line X'X having linear charge density λ is given by :

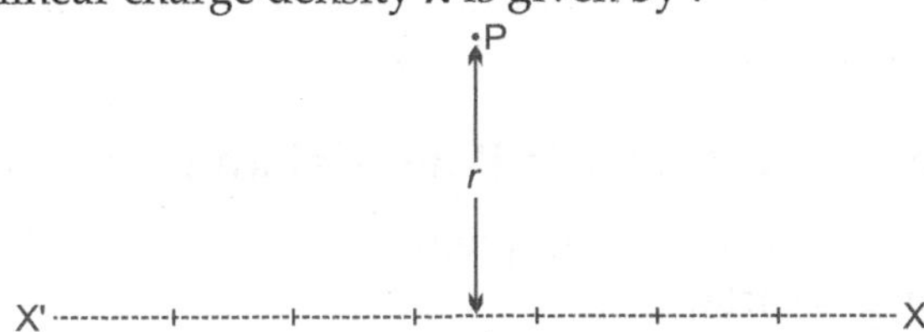

   (a) $E = \left(\frac{1}{4\pi\varepsilon_0}\right)\frac{2\lambda}{r^2}$ (b) $E = \left(\frac{1}{4\pi\varepsilon_0}\right)\frac{2\lambda}{r}$
   (c) $E = \left(\frac{1}{4\pi\varepsilon_0}\right)\frac{\lambda}{r^2}$ (d) $E = \left(\frac{1}{4\pi\varepsilon_0}\right)\frac{\lambda}{r}$
10. Electric charges q_1, $-q_1$, $+q_2$ and $-q_2$ are placed in free space and S is a spherical Gaussian surface. The electric flux passing over the surface S is :

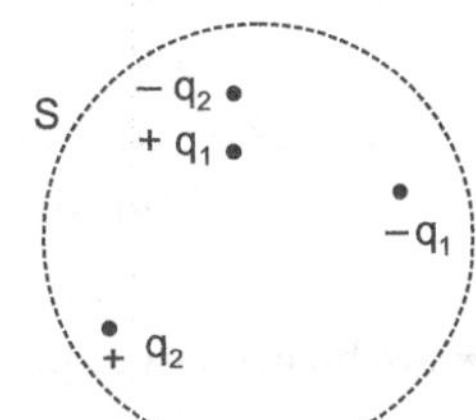

   (a) due to $+q_2$ only
   (b) due to all charges
   (c) zero
   (d) due to positive charge only

11. What is the locus of an electron projected perpendicular to an uniform electric field ?
(a) Circle (b) Parabola (c) Ellipse (d) Hyperbola

12. Gauss's theorem states that :
(a) Electric flux = $\frac{q}{\varepsilon_0}$ (b) Electric potential = εp
(c) Electric field intensity = $\varepsilon_0 q$ (d) None of these

13. When 10^{14} electrons are removed from a neutral metal sphere, the charge on the sphere becomes :
(a) 16 μC (b) 32 μC (c) – 32 μC (d) – 16 μC

14. Two positive point charges are 3 m apart and their combined charge is 20 μC. If the force between them is 0.075 N, then the charges are :
(a) 10 μC, 10 μC (b) 15 μC, 5 μC (c) 12 μC, 8 μC (d) 14 μC, 6μC

15. If an electron has an initial velocity in a direction different from that of an electric field, the path of the electron is :
(a) a straight line (b) a circle (c) an ellipse (d) a parabola

16. If a dipole of dipole moment $\vec{p}$ is placed in a uniform electric field $\vec{E}$, then torque acting on it is given by :
(a) $\vec{\tau} = \vec{p}\,\vec{E}$ (b) $\vec{\tau} = \vec{p} + \vec{E}$ (c) $\vec{\tau} = \vec{p} \times \vec{E}$ (d) $\vec{\tau} = \vec{p} - \vec{E}$

17. The angle between the dipole moment and electric field at any point on the equatorial plane is :
(a) 180° (b) 0° (c) 45° (d) 90°

18. The surface considered for Gauss's law is called :
(a) Closed surface (b) Spherical surface (c) Gaussion surface (d) Plane surface

19. Debye is the unit of :
(a) Electric flux (b) Torque
(c) Electric dipole moment (d) None of these

20. If the charge on an object is halved then electric field becomes :
(a) half (b) double (c) unchanged (d) thrice

21. The electric flux through a closed Gaussian surface depends upon :★
(a) Net charge enclosed and permittivity of the medium
(b) Net charge enclosed, permittivity of the medium and the size of the Gaussian surface
(c) Net charge enclosed only
(d) Permittivity of the medium only

Ans. 1. (a), 2. (d), 3. (d), 4. (d), 5. (c), 6. (b), 7. (d), 8. (b), 9. (b), 10. (c), 11. (d), 12. (a), 13. (a), 14. (b), 15. (d), 16. (c), 17. (a), 18. (c), 19. (c), 20. (a), 21. (a)

Chapter 2. Electrostatic Potential and Capacitance

1. The electric potential inside a conducting sphere :
(a) increases from centre to surface
(b) decreases from centre to surface
(c) remains constant from centre to surface
(d) is zero at every point inside

2. The potential at a point due to a charge of 4×10^{-7} C located 10 cm away is :
(a) 3.6×10^5 V (b) 3.6×10^4 V (c) 4.5×10^4 V (d) 4.5×10^5 V

3. What is the electric potential at a distance of 9 cm from 3 μC ?
(a) 3 V (b) 300 V (c) 270 V (d) 30 V

4. The potential energy of a system of two charges is negative when :
(a) both the charges are positive

★ are board based questions from previous year

(b) both the charges are negative
(c) one charge is positive and other is negative
(d) both the charges are separated by infinite distance

5. An electric dipole of moment $\vec{p}$ is placed normal to the lines of force of electric intensity $\vec{E}$, then the work done in deflecting it through an angle of 180° is :
(a) pE (b) $2pE$ (c) $-2pE$ (d) Zero

6. The air between the plates of a capacitor is replaced by a medium of permittivity K. The potential difference between the plates :
(a) increases K times (b) decreases K times
(c) increases $\sqrt{K}$ times (d) decreases $\sqrt{K}$ times.

7. A conductor is charged to a potential V by importing a charge q, the variations of q with V is represented by figures :

(a)
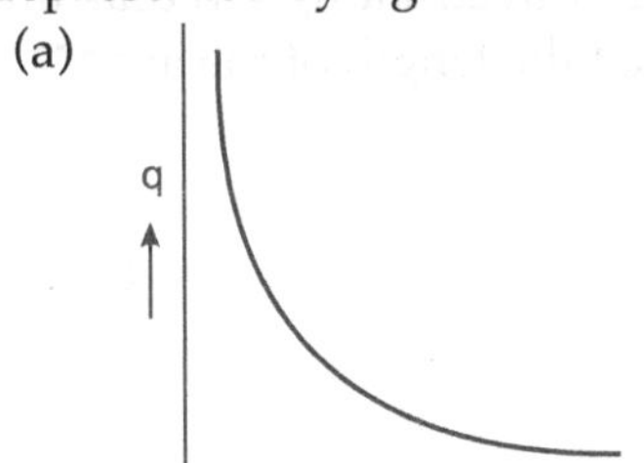

(b)
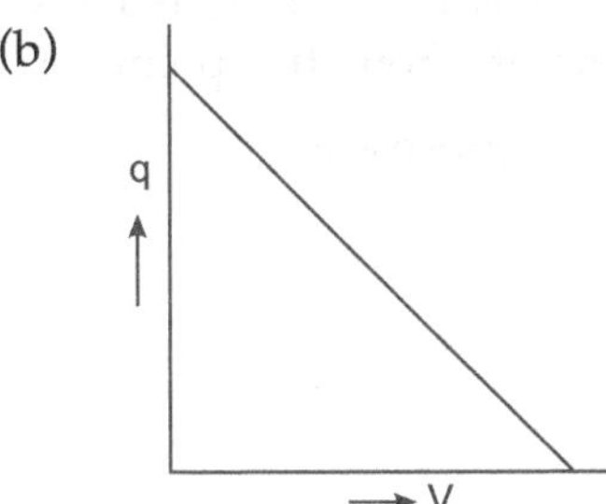

(c)
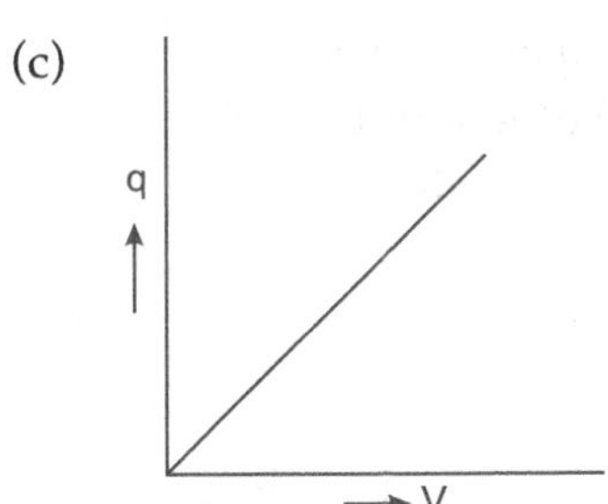

(d)
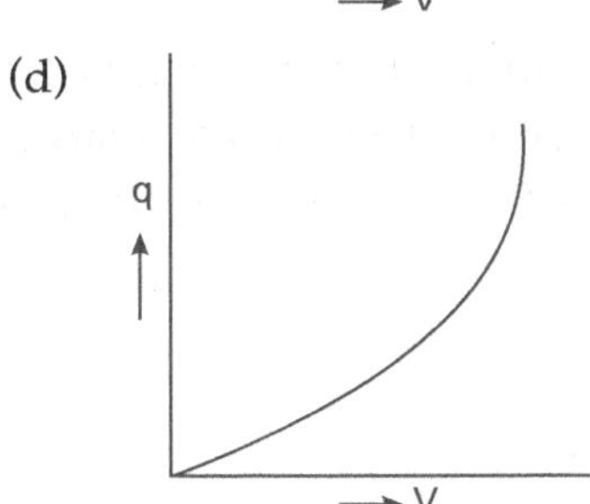

8. 'n' identical capacitors joined in parallel are charged to a common potential V. The battery is disconnected. Now, the capacitors are separated and joined in series. For the new combination :
(a) energy and p.d. both will remain unchanged
(b) energy will remain same, p.d. will become nV
(c) energy and p.d. both will become n times
(d) energy will become n times, p.d. will remain V.

9. Potential gradient 'k' remains constant until :
(a) current in potentiometer wire remains constant
(b) potential difference across the potentiometer wire remains constant
(c) e.m.f. of the testing cell remains constant
(d) none

10. An electric dipole of moment $\vec{p}$ placed in a uniform electric field $\vec{E}$ has minimum (negative) potential energy when the angle between $\vec{p}$ and $\vec{E}$ is :
(a) $\pi/2$ (b) zero (c) π (d) $3\pi/2$.

11. When a test-charge is brought from infinity along the perpendicular bisector of an electric dipole, then the work done is :
(a) Positive (b) Negative (c) Zero (d) None

12. Four capacitors of capacitance 2 μF each are connected in series. The capacitance of the equivalent capacitors will be :

(a) 4 μF (b) 1 μF (c) 0.5 μF (d) 2 μF

13. The equivalent capacitance between points A and B in the given figure would be :

(a) 3C (b) $\frac{C}{3}$

(c) 3C/2 (d) None.

14. The capacitance of a capacitor does not depend upon :

(a) size of the plates (b) shape of the plates

(c) separation between the plates (d) charge on the plates

15. In the given figure, a charge Q is fixed. Another charge q is moved along circular arc MN of radius r around it, from the point M to the point N such that the length of the arc MN = l. The work done in this process is :

(a) zero (b) $\frac{1}{4\pi\varepsilon_0} \cdot \frac{Qq}{r^2} l$

(c) $\frac{Qq}{2\varepsilon_0 r^2} l$ (d) $\frac{Qq}{2\pi\varepsilon_0 r^2}$

16. A short electric dipole (which consists of two point charges, $+q$ and $-q$) is placed at the centre O and inside a large cube (ABCDEFGH) of length L, as shown in the figure. The electric flux emanating through the cube is :

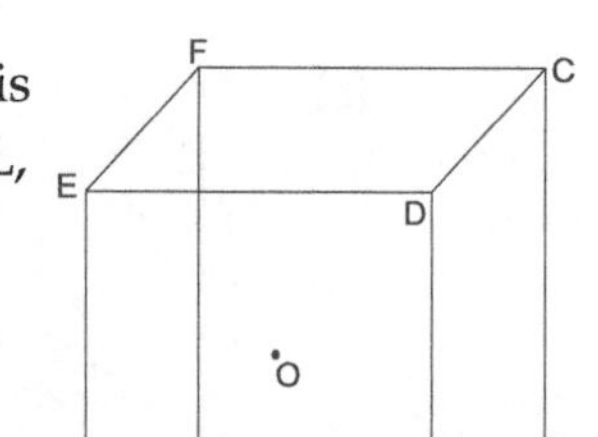

(a) $\frac{q}{4\pi\varepsilon_0 L}$ (b) zero

(c) $\frac{q}{2\pi\varepsilon_0 L}$ (d) $\frac{q}{3\pi\varepsilon_0 L}$

17. An electric dipole is placed at the centre of a hollow conducting sphere. Which of the following is correct ?

(a) Electric field is zero at every point of the sphere
(b) Electric field is not zero anywhere on the sphere
(c) The flux of electric field is not zero through the sphere
(d) All of these

18. The electrostatic potential energy of a charge of 5 C at a point in the electrostatic field is 50 J. The potential at that point is :

(a) 0.1 V (b) 5 V (c) 10 V (d) 250 V

19. A conductor with a positive charge :

(a) is always at positive potential (b) is always at zero potential

(c) is always at negative potential (d) may be at positive, zero or negative potential

20. Dielectric constant of a medium is also known as :

(a) relative permeability (b) permeability

(c) permittivity (d) relative permittivity

21. The electrostatic potential energy of two point charges, 1 μC each, placed 1 m apart in air is :

(a) 9×10^3 J (b) 9×10^9 J (c) 9×10^{-3} J (d) 9×10^{-3} eV

Ans. 1. (c), 2. (b), 3. (b), 4. (c), 5. (d), 6. (b), 7. (c), 8. (b), 9. (a), 10. (b), 11. (c), 12. (c), 13. (a), 14. (d), 15. (a), 16. (b), 17. (b), 18. (c), 19. (a), 20. (d), 21. (c)

Chapter 3. Current Electricity

1. The filaments of 60 W and 100 W bulbs are of the same length. Then :
   (a) 60 W filament is thicker (b) 100 W filament is thicker
   (c) both are of same thickness (d) both cannot have same length
2. The order of coloured rings in a carbon resistor is red, yellow, blue and silver. The resistance of the carbon resistor is :
   (a) $24 \times 10^6\ \Omega \pm 5\%$ (b) $24 \times 10^6\ \Omega \pm 10\%$
   (c) $34 \times 10^4\ \Omega \pm 10\%$ (d) $26 \times 10^4\ \Omega \pm 5\%$
3. The time-constant of a circuit having a capacitor C and a resistor R in series is :
   (a) CR (b) $\frac{C}{R}$ (c) $\frac{R}{C}$ (d) $\sqrt{CR}$
4. The Kirchhoff's first law ($\Sigma i = 0$) and second law ($\Sigma iR = \Sigma E$), where the symbols have their usual meanings, are respectively based on :
   (a) Conservation of charge, conservation of momentum
   (b) Conservation of energy, conservation of charge
   (c) Conservation of momentum, conservation of charge
   (d) Conservation of charge, conservation of energy
5. The drift velocity of the free electrons in a conducting wire carrying a current i is v. If in a wire of the same metal, but of double the radius, the current be $2i$ then the drift velocity of the electrons will be :
   (a) $v/4$ (b) $v/2$ (c) v (d) $4v$
6. A student connects four cells, each of e.m.f. 1.5 V and internal resistance 0.25 Ω in series but one cell has its terminals reversed. This battery sends current in a 2 Ω resistor. The current is :
   (a) Zero (b) 1 A (c) 1.5 A (d) 2.5 A.
7. In current electricity, Ohm's law is obeyed by all :
   (a) solids (b) metals (c) liquids (d) gases
8. Following are the graphs between the current i drawn from a cell and the terminal voltage V of the cell. Which one is correct ?

(a)
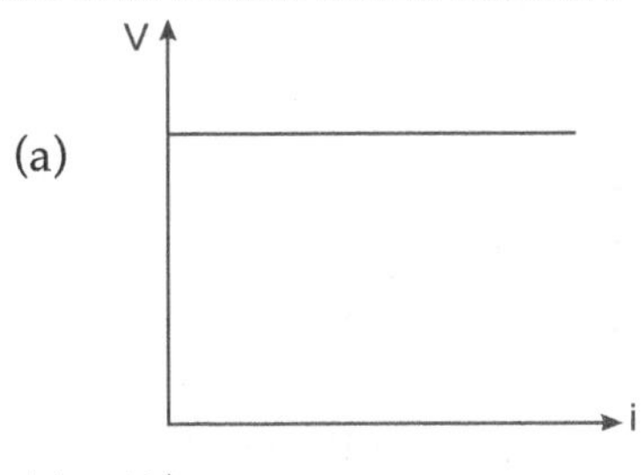

(b)
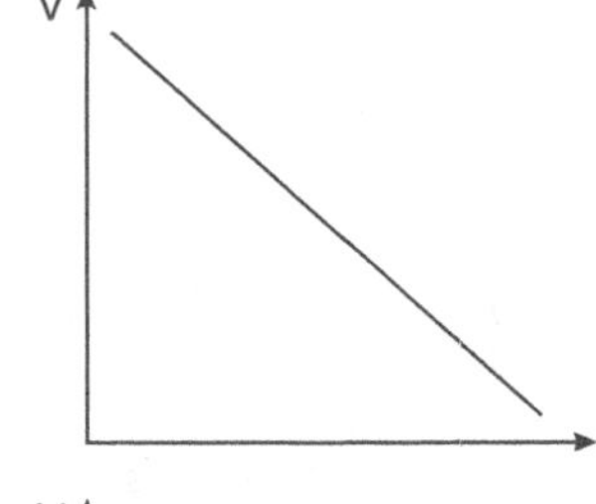

(c) V, i

(d) V, i

9. Two unequal resistors joined in parallel are connected to a cell. Which of the following statement is true ?
   (a) current is same in both
   (b) potential drop is same in both
   (c) current is more in greater resistance
   (d) potential drop is smaller in smaller resistance

10. The current through a 1 mH coil drops from 5 A to 3 A in 10^{-3} s. The e.m.f. developed in the coil is :

(a) 2×10^{-6} V (b) 8×10^{-6} V (c) 2 V (d) 8 V

11. The ratio of the resistances of 100 W and 40 W bulbs of the same rated voltage is :

(a) 2 : 5 (b) 5 : 2 (c) 25 : 4 (d) 4 : 25

12. Five resistors are connected as shown in figure :

The effective resistance *i.e.,* equivalent resistance between the points A and B is :

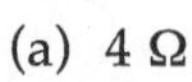

(a) 4 Ω (b) 5 Ω

(c) 15 Ω (d) 20 Ω

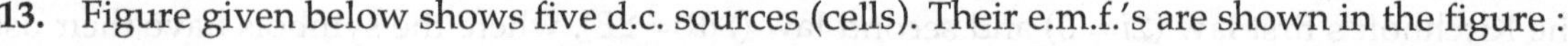

13. Figure given below shows five d.c. sources (cells). Their e.m.f.'s are shown in the figure :

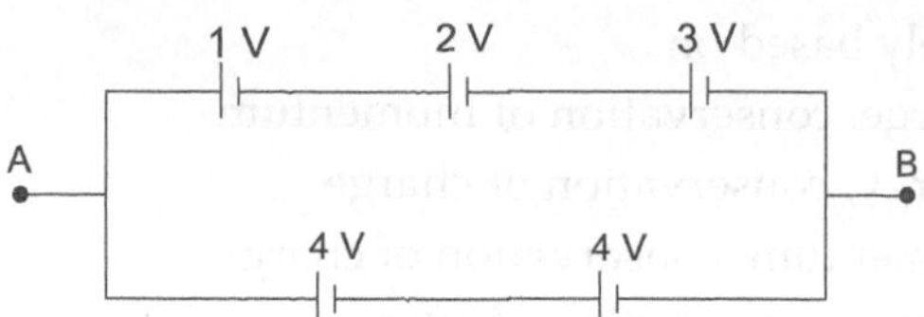

E.m.f. of the battery AB is :

(a) 8 V (b) 16 V (c) 4 V (d) 2 V

14. If a wire is stretched to make it 0.1% longer, its resistance will :

(a) decrease by 0.05% (b) increase by 0.05%

(c) increase by 0.2% (d) decrease by 0.2%

15. A wire of resistance 'R' is cut into '*n*' equal parts. These parts are then connected in parallel with each other. The equivalent resistance of the combination is :

(a) *n*R (b) R/*n* (c) n/R^2 (d) R/n^2

16. The equivalent resistance between points *a* and *f* of the network shown in figure is :

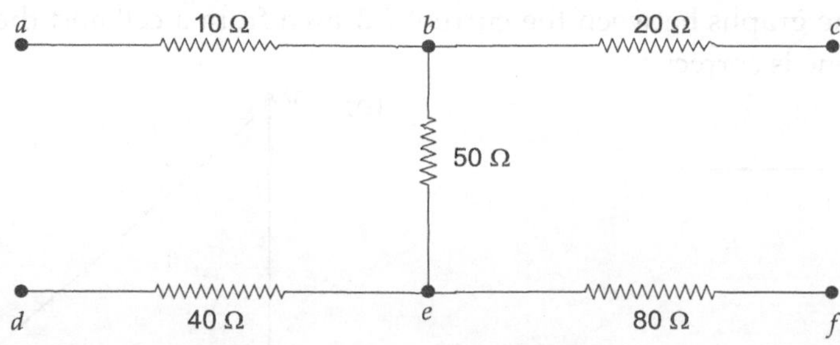

(a) 24 Ω (b) 110 Ω (c) 140 Ω (d) 200 Ω

17. If R_1 and R_2 are filament resistances of a 200 W and a 100 W bulb respectively, designed to operate on the same voltage, then :

(a) $R_1 = R_2$ (b) $R_2 = 2 R_1$ (c) $R_2 = 4 R_1$ (d) $R_1 = 4 R_2$

18. What resistance should be added in the arm RS of Wheatstone bridge to balance the bridge ?

(a) 90 Ω in parallel

(b) 30 Ω in parallel

(c) 1 Ω in series

(d) None of these

19. In a meter bridge experiment the ratio of left gap resistance to right gap resistance is 2 : 3. The balance point from left is :

(a) 20 cm (b) 40 cm (c) 50 cm (d) 60 cm

20. In the given figure, when the galvanometer shows no deflection, the current flowing through the 5 Ω resistance will be :

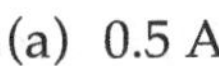
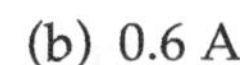
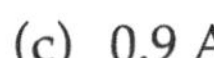

(a) 0.5 A (b) 0.6 A

(c) 0.9 A (d) 1.5 A

21. A cell of e.m.f. (E) and internal resistance r is connected acros a variable external resistance R. The graphy of terminal potential difference V as function of R is :★

(a) 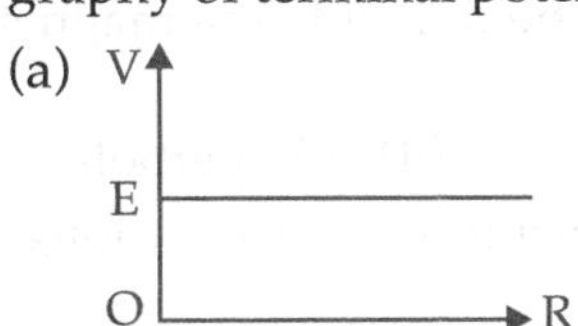

(b) 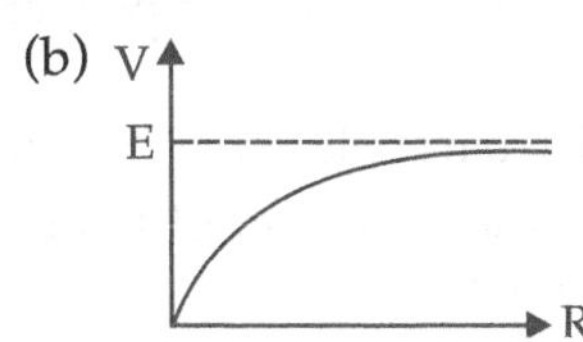

(c) 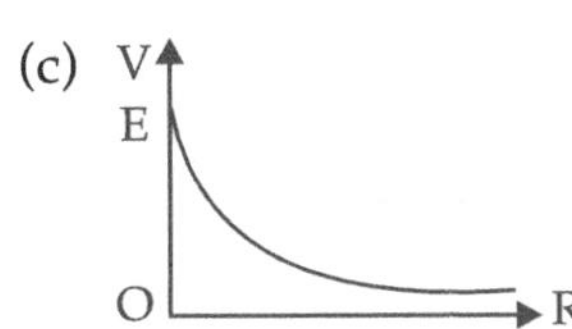

(d) 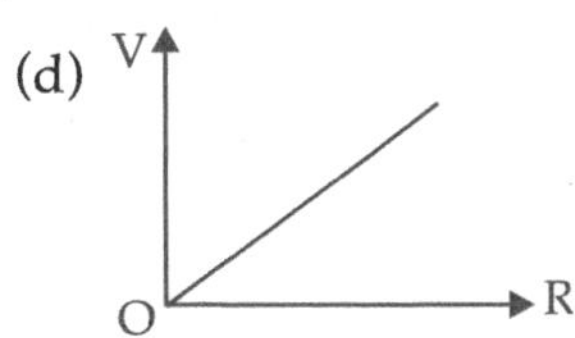

22. A uniform wire of resistance 2R is bent in the form of a circle. The effective resistance between the ends of any diameter of the circle is :★

(a) 2R (b) R (c) $\frac{R}{2}$ (d) $\frac{R}{4}$

Ans. 1. (b), 2. (a), 3. (a), 4. (d), 5. (b), 6. (b), 7. (b), 8. (b), 9. (b), 10. (c), 11. (a), 12. (a), 13. (a), 14. (c), 15. (d), 16. (c), 17. (b), 18. (a), 19. (b), 20. (b), 21. (b), 22. (c).

Chapter 4. Moving Charges and Magnetism

1. A charged particle moves through a magnetic field in a direction perpendicular to it. Then the :

(a) direction of the particle remains unchanged

(b) velocity remains unchanged

(c) acceleration remains unchanged

(d) speed of the particle remains unchanged

2. A charge q is moving in a magnetic field then the magnetic force does not depend upon :

(a) mass (b) velocity (c) charge (d) magnetic field

3. A 2 MeV proton is moving perpendicular to a uniform magnetic field of 2.5 T. The force on the proton is :

(a) 2.5×10^{-10} N (b) 2.5×10^{-11} N (c) 7.6×10^{-11} N (d) 7.6×10^{-12} N

4. A charge of 1 C is moving in a magnetic field of 0.5 T with velocity of 10 m/s. Force experienced is :

(a) 0.5 N (b) 5 N (c) 10 N (d) 0 N

5. A circular coil carrying a current I has radius R and number of turns N. If all the three, *i.e.,* the current I, radius R and number of turns N are doubled, then, magnetic field at its centre becomes:

(a) Double (b) Half (c) Four times (d) One-fourth

6. A current flows along the length of a infinitely long, straight thin-walled pipe. Then :

(a) the magnetic field at all points inside the pipe is same, but not zero

(b) the magnetic field at any point inside the pipe is zero

★ are board based questions from previous year

(c) the magnetic field is zero only on the axis of the pipe

(d) the magnetic field is different at different points inside the pipe

7. When a charged particle is projected perpendicular to a uniform magnetic field, it describes a circular path in which :

(a) its speed remains constant (b) its velocity remains constant

(c) its momentum remains constant (d) its kinetic energy increases

8. A moving electron enters a uniform and perpendicular magnetic field. Inside the magnetic field, the electron travels along :

(a) a straight line (b) parabola (c) a circle (d) a hyperbola

9. Which graph represents the variation of magnetic field B with distance '*a*' from a long straight current carrying conductor ?

(a)

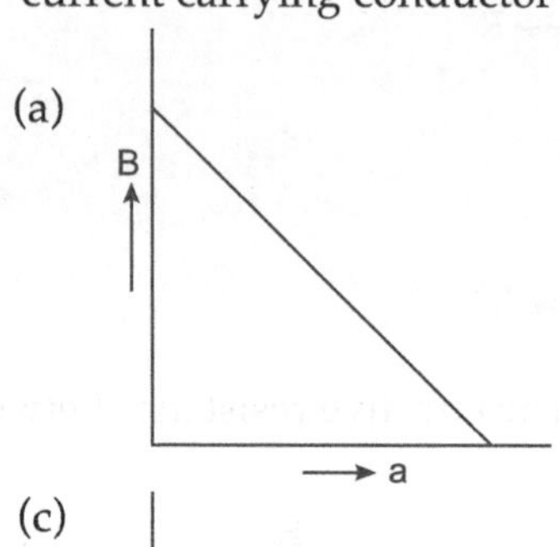

(b)

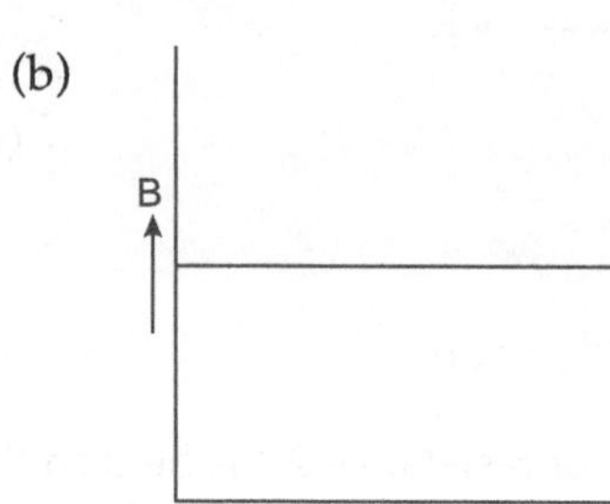

(c)

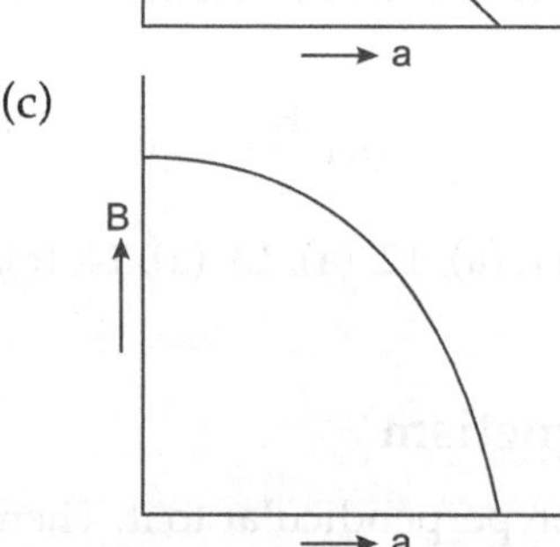

(d)

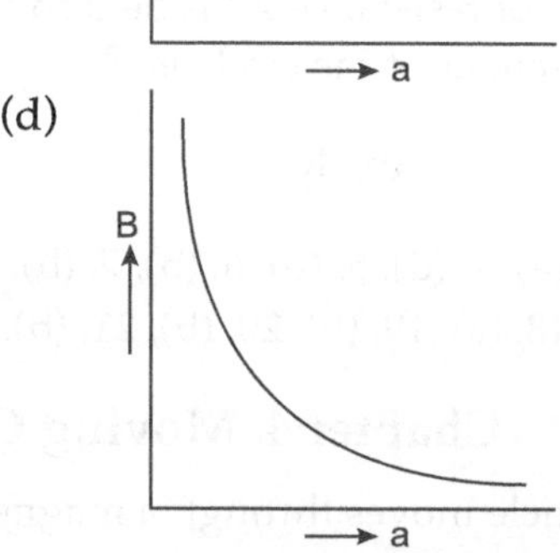

10. A 1 m metallic rod is being revolved in a vertical plane about one of its ends with an angular velocity of 5 radian sec^{-1}. The earth's horizontal magnetic field is 0.2×10^4 T. The e.m.f. induced across the ends of the rod will be :

(a) 5 mV (b) 5×10^{-4} V (c) 50 mV (d) 50 μV

11. A uniform electric field and a uniform magnetic field are acting along the same direction in a certain region. If an electron is projected along the direction of the field with a certain velocity then :

(a) it will turn towards right of the direction of motion

(b) it will turn towards left of the direction of motion

(c) its velocity will decrease

(d) its velocity will increase

12. Magnetic susceptibility of platinum is 0.0001. Its relative permeability is :

(a) 1.0000 (b) 0.9999 (c) 1.0001 (d) zero

13. The Biot-Savart's law in vector form is :

(a) $\vec{dB} = \frac{\mu_0}{4\pi}\frac{dl\,(\vec{I} \times \vec{r})}{r^3}$ (b) $\vec{dB} = \frac{\mu_0}{4\pi}\frac{I\,(\vec{dl} \times \vec{r})}{r^3}$

(c) $\vec{dB} = \frac{\mu_0}{4\pi}\frac{I\,(\vec{r} \times \vec{dl})}{r^3}$ (d) $\vec{dB} = \frac{\mu_0}{4\pi}\frac{I\,(\vec{dl} \times \vec{r})}{r^2}$

14. When a charged particle is projected perpendicular to a uniform magnetic field, it describes a circular path in which :

(a) its speed remains constant (b) its velocity remains constant

(c) its momentum remains constant (d) its kinetic energy increases

15. A vertical wire carries a current in upward direction. An electron beam sent horizontally towards the wire will be deflected :

(a) towards right (b) towards left (c) upwards (d) downwards

16. A charged particle moves through a magnetic field perpendicular to its direction. Then :

(a) both momentum and kinetic energy of the particle are not constant

(b) both momentum and kinetic energy of the particle are constant

(c) kinetic energy changes but the momentum is constant

(d) the momentum changes but the kinetic energy is constant

17. A moving electron enters a uniform and perpendicular magnetic field. Inside the magnetic field, the electron travels along :

(a) a straight line (b) a parabola (c) a circle (d) a hyperbola

18. A moving coil galvanometer can be converted into a voltmeter by connecting :

(a) a low resistance in series with its coil (b) a low resistance in parallel with its coil

(c) a high resistance in parallel with its coil (d) a high resistance in series with its coil

19. A current I flows through a long straight conductor which is bent into a circular loop of radius R in the middle as shown in the figure.★

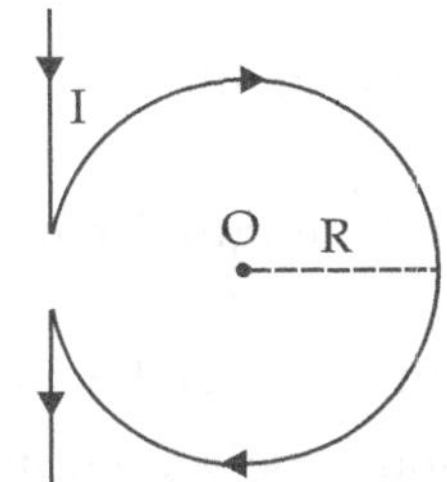

The magnitude of the net megnetic field at point O will be :

(a) Zero (b) $\frac{\mu_0 I}{2R}(1+\pi)$ (c) $\frac{\mu_0 I}{4\pi R}$ (d) $\frac{\mu_0 I}{2R}\left(1-\frac{1}{\pi}\right)$

20. A current of 10A is flowing from east to west in a long straight wire kept on a horizontal table. The magnetic field developed at a distance of 10 cm due north on the table is :

(a) 2×10^{-5} T, acting downwards (b) 2×10^{-5} T, acting upwards

(c) 4×10^{-5} T, acting downwards (d) 4×10^{-5} T, acting upwards

21. An electron and a proton are moving along the same direction with the same kinetic energy. They enter a uniform magnetic field acting perpendicular to their velocities. The dependence of radius of their path on their masses is :★

(a) $r \propto m$ (b) $r \propto \sqrt{m}$ (c) $r \propto \frac{1}{m}$ (d) $\frac{1}{\sqrt{m}}$

22. A charge particle after being accelerated through a potential difference 'V' enters in a uniform magnetic field and moves in a circle of radius r. If V is doubled, the radius of the circle will become :★

(a) $2r$ (b) $\sqrt{2}r$ (c) $4r$ (d) $\frac{r}{\sqrt{2}}$

Ans. 1. (d), 2. (a), 3. (d), 4. (b), 5. (a), 6. (b), 7. (a), 8. (c), 9. (d), 10. (d), 11. (c), 12. (c), 13. (b), 14. (a), 15. (c), 16. (d), 17. (c), 18. (d), 19. (d), (20. (a), 21. (a), 22. (b).

★ are board based questions from previous year

Chapter 5. Magnetism and Matter

1. The magnetic dipole moment of a solenoid having N turns is given as :
(a) N*i*A (b) N*i*A^2 (c) N*i*^{2}A (d) N*i*^{2}A^2
2. Gauss's law should be invalid if :
(a) the inverse square law were not exactly true
(b) the velocity of light were not a universal constant
(c) there were magnetic monopoles
(d) none of these
3. The intensity of magnetisation of a bar magnet is 5.0×10^4 Am^{-1}. The magnetic length and the area of cross-section of the magnet are 12 cm and 1 cm^2 respectively. The magnitude of magnetic moment of this bar magnet is (in SI unit) :
(a) 2.4 (b) 1.3 (c) 1.24 (d) 0.6
4. Which of the following is responsible for the earth's magnetic field ?
(a) Rotational motion of earth (b) Translational motion of earth
(c) Convective currents in earth's core (d) Diversive current in earth's core
5. The north pole of a magnet is brought near a stationary negatively charged conductor. Will the pole experience any force ?
(a) Yes
(b) No
(c) Depends on the magnitude of pole strength
(d) Can't say
6. The magnetisation of a bar magnet of length 5 cm, cross-sectional area 2 cm^2 and magnetic moment 1 Am^2 is :
(a) 2×10^5 A/m (b) 4×10^5 A/m (c) 1×10^5 A/m (d) 3×10^5 A/m
7. If a bar magnet is placed with its south pole pointing towards geographic south in magnetic meridian, then the neutral points are located :
(a) at the middle point of the magnet
(b) on the line perpendicular to the axis of the magnet
(c) at the angle of 45° with the magnetic axis
(d) along the axis of the magnet
8. The magnetic induction B and the force F on a pole m are related by :
(a) $F = mB$ (b) $F = \frac{m}{B}$ (c) $B = mF$ (d) $F = \frac{B}{m}$
9. A north pole of strength 50 Am and south pole of strength 100 Am are separated by a distance of 10 cm in air. Find the force between them.
(a) 20×10^{-6} N (b) 25×10^{-3} N (c) 30×10^{-18} N (d) 50×10^{-3} N
10. A beam of electrons passes through crossed electric and magnetic fields of magnitudes 7.2×10^6 NC^{-1} and 2.0 T respectively. The electrons which pass undeviated have velocity :
(a) 14.4×10^6 ms^{-1} (b) 3.6×10^6 ms^{-1} (c) 3.0×10^8 ms^{-1} (d) 0.28×10^{-6} ms^{-1}
11. The lines of force due to horizontal component of earth's magnetic field are :
(a) straight and parallel (b) concentric circles
(c) parabolic (d) elliptical
12. When a ferromagnetic substance is heated above Curie temperature, then :
(a) it is demagnetised (b) it becomes diamagnetic
(c) it becomes paramagnetic (d) it remains unaffected
13. In a plane perpendicular to magnetic meridian the dip needle will be :
(a) vertical

(b) horizontal
(c) inclined equal to the angle of dip at that place
(d) pointing in any direction

14. If a diamagnetic solution is poured into a U-tube and one arm of this U-tube is placed between the poles of a strong magnet with the meniscus in a line with the field, then the level of the solution will :
(a) rise (b) fall (c) remain same (d) oscillate slowly

15. Earth's magnetic field inside a closed iron box, as compared to outside is :
(a) more (b) less (c) same (d) zero

16. If the current (I) flowing through a circular coil, its radius (R) and number of turns (N) in it are each doubled, magnetic flux density at its centre becomes :
(a) two times (b) four times (c) eight times (d) sixteen times

17. The lines of force due to horizontal component of earth's magnetic field are :
(a) straight and parallel (b) concentric circles
(c) parabolic (d) elliptical

18. A paramagnetic substance is placed in a magnetic field which is increased till the magnetisation of the substance becomes constant. Now, on lowering the temperature, the magnetisation :
(a) will increase (b) will decrease
(c) will remain unchanged (d) may increase or decrease

Ans. 1. (a), 2. (c), 3. (d), 4. (c), 5. (b), 6. (c), 7. (b), 8. (a), 9. (d), 10. (b), 11. (a), 12. (c), 13. (a), 14. (b), 15. (b), 16. (a), 17. (a), 18. (c).

Chapter 6. Electromagnetic Induction

1. To induce an emf in a coil, the linking magnetic flux :
(a) must remain constant (b) can either increase or decrease
(c) must increase (d) must decrease

2. The magnetic flux (ϕ) linked with a coil due to its own magnetic field is related to the number (N) of turns of the coil as :
(a) $\phi \propto N^2$ (b) $\phi \propto N^{-1}$ (c) $\phi \propto N$ (d) $\phi \propto N^{-2}$

3. When current changes from 13 A to 7 A in 0.5 s through a coil, the emf induced is 3×10^{-4} V. The coefficient of self induction is :
(a) 25×10^{-6} H (b) 25×10^{-5} H (c) 25×10^{-4} H (d) 25×10^{-3} H

4. A coil of area 10 cm^2 has 200 turns. Magnetic field of 0.1 Wb/m^2 is perpendicular to the coil. The field is reduced to zero in 0.1 s, the induced emf in the coil is :
(a) Zero (b) 0.2 V (c) 1 V (d) 2 V

5. A small coil of N_1 turns, l_1 length is tightly wound over the centre of a long solenoid of length l_2, area of cross-section A and number of turns N_2. If a current I flows in the small coil, then the flux through the long solenoid is :
(a) Zero (b) $\frac{\mu_0 N_1 N_2 AI}{l_2}$ (c) $\frac{\mu_0 N_1^2 AI}{l_1}$ (d) Infinite

6. Lenz's law gives :
(a) the direction of the induced current
(b) the magnitude of the induced emf
(c) the magnitude of the induced current
(d) both the magnitude and direction of the induced current

7. A moving conductor coil in a magnetic field produces an induced emf. This is in accordance with :
(a) Lenz's Law (b) Coulomb's Law (c) Faraday's Law (d) Ampere's Law

8. A rectangular coil of 100 turns and size 0.1 m × 0.05 m is placed perpendicular to a magnetic field of 0.1 T. If the field drops to 0.05 T in 0.05 s, the magnitude of the emf induced in the coil is :
(a) 0.5 (b) 3 (c) 2 (d) 6

9. The magnetic flux through a circuit of resistance R changes by an amount $\Delta\phi$ in a time Δt. Then the total quantity of electric charge Q that passes any point in the circuit during the time Δt is represented by :

(a) $Q = \frac{\Delta\phi}{R}$ (b) $Q = R\frac{\Delta\phi}{\Delta t}$ (c) $Q = \frac{1}{R}\cdot\frac{\Delta\phi}{\Delta t}$ (d) $Q = \frac{\Delta\phi}{\Delta t}$

10. The cause of induced e.m.f. is :

(a) magnetic flux (b) magnetic field
(c) area (d) change in magnetic flux

11. The property of coil by which a counter e.m.f. is induced in it when the current through the coil changes is known as :

(a) self induction (b) mutual induction (c) capacitance (d) None of these

12. The direction of induced e.m.f. can be found by :

(a) Kirchhoff's voltage law (b) Kirchhoff's current law
(c) Lenz's law (d) Fleming's right hand rule

13. Which of the following is the unit of inductance ?

(a) ohm (b) henry (c) ampere turns (d) weber/meter

14. Faraday's laws are consequences of conservation of :

(a) energy and magnetic field (b) energy
(c) Magnetic field (d) charge

15. The force between two current carrying parallel wires is due to :

(a) magnetic field of electric current (b) electrostatic interaction
(c) electromagnetic induction (d) electric effect of magnetic field

16. In electromagnetic induction, the induced charge is independent of :

(a) resistance of the coil (b) change of flux
(c) time (d) None of these

17. A circular loop of radius r, carrying a current I lies in y-z plane with its centre at the origin. The net magnetic flux through the loop is :★

(a) directly proportional to r (b) zero
(c) inversely proportional to r (d) directly proportional to I

Ans. 1. (b), 2. (a), 3. (a), 4. (b), 5. (b), 6. (a), 7. (c), 8. (a), 9. (a), 10. (d), 11. (a), 12. (c), 13. (b), 14. (b), 15. (a), 16. (c), 17. (b)

Chapter 7. Alternating Current

1. Alternating current cannot be measured by DC ammeter because :

(a) AC is virtual (b) AC changes its direction
(c) AC cannot pass through DC ammeter (d) Average value of complete cycle is zero

2. The alternating current of equivalent value of $\frac{I_0}{\sqrt{2}}$ is :

(a) rms current (b) DC current (c) Peak current (d) all of these

3. In a series resonant circuit, having L, C, and R as its elements, the resonance current is i. The power dissipated in circuit at resonance is :

(a) Zero (b) i^2R (c) $i^2\omega L$ (d) $\frac{i^2R}{\left(\omega L - \frac{1}{\omega C}\right)}$

4. In an LCR circuit, the sharpness of resonance depends on :

(a) Resistance (R) (b) Capacitance (C) (c) Inductance (L) (d) All of these

★ are board based questions from previous year

5. The values of resistance and inductive reactance of a choke coil are 8 Ω and 6 Ω respectively. What is the power factor of the coil ?
(a) 0.3 (b) 0.4 (c) 0.6 (d) 0.8

6. In a circuit, the reactance X and resistance R are equal. What is the power factor ?
(a) $\frac{1}{2}$ (b) 1 (c) $\frac{1}{\sqrt{2}}$ (d) $\sqrt{2}$

7. The ratio of mean value over half cycle to rms value of AC is :
(a) $\sqrt{2}:1$ (b) $2:\pi$ (c) $2\sqrt{2}:\pi$ (d) $\sqrt{2}:\pi$

8. The power factor of LCR circuit at resonance is :
(a) Zero (b) 0.5 (c) 0.707 (d) 1

9. Average value of an a.c. for a complete cycle is :
(a) $\frac{2i_0}{\pi}$ (b) $\frac{-2i_0}{\pi}$ (c) $\frac{i_0}{\pi}$ (d) Zero

10. The value of impedance in a series resonant circuit is :
(a) Zero (b) equal to the resistance
(c) equal to the capacitive reactance (d) equal to the inductive reactance

11. In a series AC circuit, $R = 100\ \Omega$, $X_L = 300\ \Omega$ and $X_C = 200\ \Omega$. The phase difference between the applied e.m.f. and the current will be :
(a) 0 (b) 37° (c) 45° (d) 90°

12. In an a.c. circuit, $R = 3\ \Omega$, $Z = 4\ \Omega$, $V_{rms} = 200$ V and $I_{rms} = 2.0$ A. The average power consumed over a full cycle is :
(a) 30 W (b) 300 W (c) 200 W (d) 3 kW

13. The loss of power in a transformer can be reduced by :
(a) increasing the number of turns in primary
(b) increasing a.c. voltage applied to primary
(c) using a solid core made of steel
(d) using a laminated core of soft iron

14. Which one of the following graphs in figure represents variation of reactance 'X_C' of a capacitor with frequency 'f' of an a.c. supply :

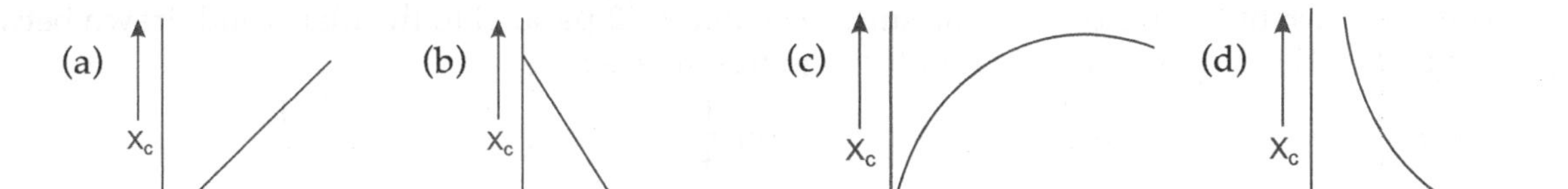

15. In an a.c. circuit, an alternating voltage $e = 200\sqrt{2}\sin 100t$ volt is connected to a capacitor of capacity 1 μF. The r.m.s. value of current in the circuit is :
(a) 20 mA (b) 10 mA (c) 100 mA (d) 200 mA

16. An a.c. voltage is applied to a resistance R and inductor L in a series. If R and inductive reactance are both equal to 3 Ω. The phase difference between the applied voltage and the current in the circuit is :
(a) zero (b) $\frac{\pi}{6}$ (c) $\frac{\pi}{4}$ (d) $\frac{\pi}{2}$

17. In an a.c. circuit, the e.m.f. (E) and the current (i) at any instant are given respectively $E = E_0 \sin \omega t$ and $i = i_0 \sin(\omega t - \phi)$. The average power in the circuit over one cycle of a.c. is :
(a) $\frac{E_0 i_0}{2}$ (b) $\frac{E_0 i_0}{2}\sin\phi$ (c) $\frac{E_0 i_0}{2}\cos\phi$ (d) $E_0 i_0$

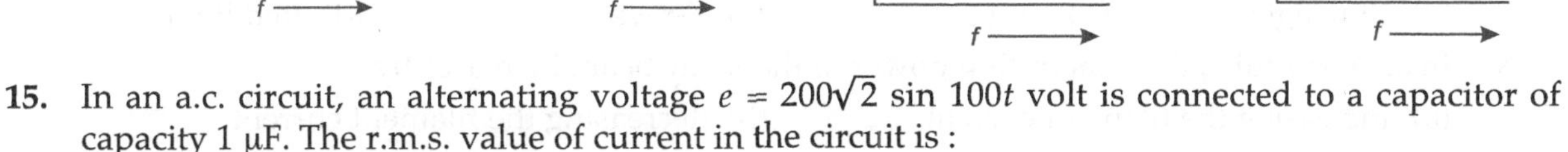

18. An a.c. voltage source of variable angular frequency ω and fixed amplitude V_0 is connected in series with a capacitance C and an electric bulb of resistance R (inductance zero). When ω is increased :

(a) the bulb glows dimmer.
(b) the bulb glows brighter.
(c) total impedance of the circuit is unchanged.
(d) total impedance of the circuit increases.

19. A generator of 220 V having internal resistance $r = 10\ \Omega$ and external resistance $R = 100\ \Omega$. What is the power developed in the external circuit ?

(a) 369 W (b) 400 W (c) 441 W (d) 484 W

Ans. 1. (d), 2. (a), 3. (b), 4. (d), 5. (d), 6. (c), 7. (c), 8. (d), 9. (d), 10. (b), 11. (c), 12. (b), 13. (d), 14. (d), 15. (a), 16. (c), 17. (c), 18. (b), 19. (b).

Chapter 8. Electromagnetic Waves

1. When the sun shines on your hands, your hands get warm because :

(a) electromagnetic waves transfer large amount of radiation pressure
(b) electromagnetic waves transfer large amount of momentum
(c) electromagnetic waves transfer energy
(d) all of the above

2. Fundamental particle in an electromagnetic wave is :

(a) proton (b) electron (c) phonon (d) photon

3. Which of the following is not transported by the electromagnetic waves ?

(a) Charge (b) Energy (c) Information (d) None of these

4. Which of the following is the unit of displacement current ?

(a) Cs^{-1} (b) Vs^{-1} (c) Vm^{-1} (d) Cm^{-1}

5. Radio wave diffract around building although light wave do not. The reason is that radio waves :

(a) Travel with speed larger than *c*
(b) Have much larger wavelength than light
(c) Carry news
(d) Are not electromagnetic waves

6. Outer layer of atmosphere called ionosphere plays important role in transmission of :

(a) optical waves (b) micro waves (c) radio waves (d) infra-red waves

7. A parallel plate capacitor with plate area A and separation between the plates *d*, is charged by a constant current I. Consider a plane surface of area A/2 parallel to the plates and drawn between the plates. The displacement current through this area is :

(a) I (b) $\frac{I}{8}$ (c) $\frac{I}{4}$ (d) $\frac{I}{2}$

8. Pick out the longest wavelength from the following types of radiations :

(a) Blue light (b) γ rays (c) X-rays (d) Red light

9. In an X-ray tube, the penetrating power of the X-ray beam increases by :

(a) increasing the filament current
(b) decreasing the filament current
(c) increasing the p.d. across the tube
(d) decreasing the p.d. across the tube.

10. The X-ray region lies between :

(a) short radio waves and visible region
(b) visible and ultraviolet region
(c) gamma rays and the ultraviolet region
(d) short radio waves and microwaves

11. The displacement current is the current which comes into play in the region in which :

(a) electric field changing with time
(b) electric flux changing with time
(c) electric field and electric flux changing with time
(d) None of the above

12. The expression of displacement current is :
(a) $I_D = \frac{\varepsilon_0}{2}\frac{d\phi_E}{dt}$ (b) $I_D = \varepsilon_0 \cdot \frac{d\phi_E}{dt}$ (c) $I_D = 2\varepsilon_0 \frac{d\phi_E}{dt}$ (d) $I_D = \frac{d\phi_E}{dt}$

13. Electromagnetic waves carry :
(a) negative charge (b) positive charge
(c) no charge (d) both positive and negative charge

14. Which of the following have lowest frequency ?
(a) Radio waves (b) Infra-red (c) Ultraviolet (d) Gamma rays

15. All electromagnetic waves travel through a vacuum at :
(a) the same speed
(b) speed that are proportional to their frequency
(c) speed that are inversely proportional to their frequency
(d) None of these

16. The E and B fields in electromagnetic waves are oriented :
(a) parallel to the wave's direction of travel, as well as to each other
(b) parallel to the wave's direction of travel and perpendicular to each other
(c) perpendicular to the wave's direction of travel and parallel to each other
(d) perpendicular to the wave's direction of travel and also to each other

17. Which of the following correctly lists electromagnetic waves in order from longest to shortest wavelength ?
(a) Gamma rays, Ultraviolet, Infra-red, Microwaves
(b) Microwaves, Ultraviolet, Visible light, Gamma rays
(c) Radio waves, Infra-red, Gamma rays, Ultraviolet
(d) Radio waves, Infra-red, visible light, X-rays

18. Which of the following electromagnetic waves has smaller wavelength ?
(a) X-rays (b) Microwaves
(c) γ-rays (d) Radio waves

19. Which waves are used in Radar system for aircraft navigation ?
(a) Infra-red waves (b) Radio waves
(c) Microwaves (d) Visible light

20. Visible light is used :
(a) in electric bulbs and tube lights (b) in movie screen in cinema hall.
(c) in lasers (d) All of the above

Ans. 1. (c), 2. (d), 3. (a), 4. (a), 5. (b), 6. (c), 7. (d), 8. (d), 9. (c), 10. (c), 11. (c), 12. (b), 13. (c), 14. (a), 15. (a), 16. (d), 17. (d), 18. (c), 19. (c), 20. (d)

Chapter 9. Ray Optics and Optical Instruments

1. An object is kept on the principal axis of a concave mirror of focal length 10 cm, at a distance of 15 cm from its pole. The image formed by the mirror is :
(a) Virtual and magnified (b) Virtual and diminished
(c) Real and magnified (d) Real and diminished

2. A light ray passes from vacuum into a medium of refractive index n. The angle of refraction is found to be half of the angle of incidence. The angle of incidence is :
(a) $\cos^{-1}(n/2)$ (b) $2\cos^{-1}(n/2)$ (c) $2\sin^{-1}(n)$ (d) $2\sin^{-1}(n/2)$

3. When an extended white object is placed before a convex lens, coloured images are formed. The image of the smallest size is of the colour :
(a) violet (b) green (c) yellow (d) red

4. A piece of glass immersed in a colourless liquid is not visible. It is so because :
(a) glass and liquid are both colourless
(b) both are of same density
(c) both are of same refractive index
(d) glass reflects the light transmitted by the liquid.

5. A fish which is at a depth of 12 cm in water $\left(\mu = \frac{4}{3}\right)$ is viewed by an observer on the bank of a lake. Its apparent depth as observed by the observer is :
(a) 3 cm (b) 9 cm (c) 12 cm (d) 16 cm

6. The focal length of a convex lens of glass of refractive index 1.5 is 2 cm. The focal length of the lens when immersed in a liquid of refractive index 1.25 will be :
(a) 1 cm (b) 2.5 cm (c) 5 cm (d) 3 cm

7. Refractive index of a transparent material is :
(a) same for all colours (b) maximum for violet colour
(c) minimum for violet colour (d) maximum for red colour

8. On increasing the focal length of the objective, the magnifying power :
(a) of microscope will increase, of telescope will decrease
(b) of both will increase
(c) of both will decrease
(d) of microscope will decrease, of telescope will increase

9. An object is placed at a distance of 5 cm more than the focal length f of the convex lens. Its real image is formed at a distance of $3f$ beyond the focus. The focal length of the lens is :
(a) 5 cm (b) 10 cm (c) 15 cm (d) 20 cm

10. Two thin lenses, one convex of focal length 30 cm and the other concave of focal length 10 cm, are in contact. The combination will behave like :
(a) a converging lens (b) a diverging lens
(c) a plane glass slab (d) a lens having no power

11. The critical angle of light passing from glass to air is minimum for :
(a) red light (b) yellow light (c) green light (d) violet light

12. A double convex lens of glass (n = 1.5) has a focal length of 10 cm in air. When immersed in a liquid of refractive index 3.0, the lens will behave as :
(a) converging lens of focal length 10 cm. (b) converging lens of focal length 30 cm.
(c) diverging lens of focal length 10 cm. (d) converging lens of focal length $\frac{10}{3}$ cm.

13. It is possible to observe total internal reflection when light travels from :
(a) glass to water (b) air to glass (c) air to water (d) water to glass

14. A physical quantity in optics without unit :
(a) Power (b) Refractive index (c) Wavelength (d) Frequency

15. A lens which can form virtual as well as real images :
(a) concave lens (b) convex lens (c) plano convex lens (d) concavo convex

16. Which mirror is used as a rear-view mirror in vehicles ?
(a) Plane mirror (b) Convex lens (c) Convex mirror (d) All of the above

17. In an astronomical telescope of refracting type :
(a) Eyepiece has greater focal length.
(b) Objective has greater focal length.
(c) Objective and eyepiece have equal focal length.
(d) Eyepiece has greater aperture than the objective.

18. A biconvex lens of focal length 15 cm is in front of a plane mirror. The distance between the lens and the mirror is 10 cm. A small object is kept at a distance of 30 cm from the lens. The final image is :
(a) virtual and at a distance of 16 cm from the mirror.
(b) real and at a distance of 16 cm from the mirror.
(c) virtual and at a distance of 20 cm from the mirror.
(d) real and at a distance of 20 cm from the mirror.

19. Two thin lenses of focal lengths f_1 and f_2 are in contact and coaxial. The power of combination is :
(a) $\sqrt{\frac{f_1}{f_2}}$ (b) $\sqrt{\frac{f_2}{f_1}}$ (c) $\frac{f_1+f_2}{f_1}$ (d) $\frac{f_1+f_2}{f_1 f_2}$

20. A biconcave lens of power P vertically splits into two identical plano concave parts. The power of each part will be :★
(a) 2P (b) $\frac{P}{2}$ (c) P (d) $\frac{P}{\sqrt{2}}$

Ans. 1. (c), 2. (b), 3. (a), 4. (c), 5. (b), 6. (c), 7. (b), 8. (d), 9. (c), 10. (b), 11. (d), 12. (c), 13. (a), 14. (b), 15. (b), 16. (c), 17. (b), 18. (b), 19. (d), 20. (b).

Chapter 10. Wave Optics

1. The spectral line for a given element of light received from a distant star is shifted towards longer wavelength side by 0.025%. The velocity of star in the line of light is :
(a) -7.5×10^3 ms^{-1} (b) -7.5×10^4 ms^{-1} (c) -8.5×10^4 ms^{-1} (d) -7.5×10^5 ms^{-1}

2. During reflection of a plane wavefront from plane reflecting surface, the secondary wavelets are drawn with :
(a) decreasing radii (b) increasing radii (c) same radii (d) any radii

3. When light suffers reflection at the interface between water and glass, the change of phase in the reflected wave is :
(a) Zero (b) π (c) 2π (d) $\pi/2$

4. The refractive index of glass is 1.5 for light waves of $\lambda = 6000$ Å in vacuum. Its wavelength in glass is :
(a) 2000 Å (b) 4000 Å (c) 1000 Å (d) 3000 Å

5. Monochromatic light is refracted from air into a medium of refractive index n. The ratio of the wavelengths of the incident and the refracted waves is :
(a) 1 : 1 (b) 1 : n (c) n : 1 (d) n^2 : 1

6. In Young's double-slit experiment, sodium lamp is replaced by a blue-light lamp. Then :
(a) The fringes will become brighter. (b) The fringes will become fainter .
(c) The fringe width will increase. (d) The fringe width will decrease.

7. The interference phenomenon can take place :
(a) in all waves (b) in transverse waves only
(c) in longitudinal waves only (d) in standing waves only

8. Young's double slit arrangement is contained in a transparent chamber. If the chamber is evacuated, then :
(a) fringes will disappear (b) fringe width will slightly increase
(c) fringe width will slightly decrease (d) there will be no effect

9. The first diffraction minima due to a single slit are observed at 30° for light of wavelength 6500 Å. The width of the slit is :

(a) 3250 Å (b) 1.3 micron (c) 6.5×10^{-4} mm (d) 2.6×10^{-4} cm.

10. The angle of polarisation for water of refractive index 1.33 is :

(a) 42° (b) 5.36° (c) 60° (d) 53.06°

11. An astronomical telescope set for normal adjustment has a magnifying power of 15. If the focal length of the object is 1.05 m then the focal length of the eyepiece will be :

(a) 7 cm (b) 70 cm (c) 7 m (d) 0.7 cm

12. Two waves of equal amplitude and wavelength but differing in phase are superimposed. The amplitude of the resultant wave is maximum when phase difference is :

(a) zero (b) $\frac{\pi}{2}$ (c) π (d) $\frac{3\pi}{2}$

13. When light passes from air to glass then :

(a) the wavelength increases (b) the wavelength decreases

(c) the frequency increases (d) nothing is changed

14. The resolving power of a telescope depends upon :

(a) the focal length of the objective (b) the focal length of the eyepiece

(c) the length of the telescope (d) the aperture of the objective

15. A beam of electrons is used in Young's double slit experiment. If the speed of electrons is increased, then the fringe width will :

(a) increase (b) decrease

(c) remain same (d) Fringes will not be seen

16. In Young's double slit experiment, the slit separation is made 3-fold. The fringe width becomes :

(a) 1/3 times (b) 1/9 times (c) 3 times (d) 9 times

17. When a polaroid placed in the path of light is rotated, the intensity of light appears to vary but never reduces to zero. The light is :

(a) unpolarised (b) plane polarised

(c) partially plane polarised (d) None of these

18. A transparent thin plate of polaroid is placed on another similar plate such that the angle between their axes is 30°. The ratio of intensities of the emergent light and the unpolarised light will be :

(a) 1 : 4 (b) 1 : 3 (c) 3 : 4 (d) 3 : 8

19. Which of the following does not support wave nature of light ?

(a) Interference (b) Diffraction (c) Polarisation (d) Photolectric effect

20. When a wave undergoes reflection at an interface from rarer to denser medium, adhoc change in its phases is :★

(a) $\frac{\pi}{2}$ (b) 0 (c) π (d) $\frac{\pi}{4}$

21. The relationship between Brewester angle 'θ' and the speed of light 'v' in the denser medium is :★

(a) $v \tan\theta = c$ (b) $c \tan\theta = v$ (c) $v \sin\theta = c$ (d) $c \sin\theta = v$

Ans. 1. (b), 2. (a), 3. (b), 4. (b), 5. (c), 6. (d), 7. (a), 8. (d), 9. (b), 10. (d), 11. (a), 12. (a), 13. (b), 14. (d), 15. (b), 16. (a), 17. (c), 18. (d), 19. (d), 20. (c), 21. (a).

★ are board based questions from previous year

Chapter 11. Dual Nature of Radiation and Matter

1. A hydrogen atom moving at a speed v absorbs a photon of wavelength 122 nm and stops. Find the value of v. (Mass of hydrogen atom = 1.67×10^{-27} kg) :
(a) 3.25 m/s (b) 3.05 m/s (c) 32.5 m/s (d) 3.5 m/s
2. Energy of an α-particle, having de-broglie wavelength of 0.004 Å is approximately :
(a) 1200 GeV (b) 1200 KeV (c) 1200 MeV (d) 1275 eV
3. Planck's constant :
(a) depends upon medium (b) depends upon wavelength of light
(c) depends upon frequency of light (d) is universal constant
4. According to De-Broglie, wave is associated with matter :
(a) when it is in motion with any velocity.
(b) when it is in motion with the velocity of light only.
(c) when it is stationary.
(d) None of the above.
5. The stopping potential is directly related to :
(a) intensity of incident radiation.
(b) the saturation current for the given frequency.
(c) the kinetic energy gained by the photoelectrons.
(d) the work function of the metal.
6. Matter waves are :
(a) electromagnetic waves.
(b) mechanical waves.
(c) neither mechanical nor electromagnetic waves.
(d) either mechanical or electromagnetic waves.
7. Which of the following proves particle nature of light ?
(a) Refraction (b) Interference (c) Polarisation (d) Photoelectric effect
8. Einstein's theory of photoelectric effect based on the equation :
(a) $E = mc^2$ (b) $E = h\nu$ (c) $E = \frac{1}{2}mv^2$ (d) $E = qV$
9. Which of the following particles will experience the maximum force when projected perpendicular to a magnetic field with same velocity ?
(a) Electron (b) Proton (c) He^+ (d) Li^{++}
10. Einstein's photoelectric equation is :
(a) $E_{max} = h\lambda - \phi_0$ (b) $E_{max} = \frac{hc}{\lambda} - \phi_0$ (c) $E_{max} = h\nu + \phi_0$ (d) $E_{max} = \frac{hc}{\lambda} + \phi_0$
11. Dimensions of Planck's constant are same as of :
(a) force × time (b) energy × distance (c) energy / time (d) energy × time
12. Highest energy photoelectrons will be produced by :
(a) visible light (b) X-rays (c) ultraviolet light (d) γ-rays
13. The de-broglie wavelength of an electron beam accelerated through a p.d. of 60 V will be :
(a) 2.0 Å (b) 1.58 Å (c) 3.2 Å (d) 158 Å
14. The work function of a surface of photosensitive material is 6.2 eV. The wavelength of the incident radiation for which the stopping potential is 5 V lies in the :
(a) UV region (b) Visible region (c) Infra-red region (d) X-ray region
15. If a source of power 4 kW produces 10^{20} photons per second, the radiation belongs to a part of the spectrum called :
(a) microwaves (b) γ-rays (c) X-rays (d) UV rays

16. Light of two different frequencies whose photons have energies 1 eV and 2.5 eV respectively illuminate a metallic surface whose work function is 0.5 eV, successively. Ratio of maximum speeds of emitted electrons will be :

(a) 1 : 5 (b) 1 : 4 (c) 1 : 2 (d) 1 : 1

17. The kinetic energy of a proton is equal to the energy E of a photon. If λ_1 be the de-broglie wavelength of the proton and λ_2 that of the photon, then the ratio λ_1/λ_2 is proportional to :

(a) E (b) $E^{1/2}$ (c) $E^{-1/2}$ (d) E^{-1}

18. In photoelectric emission process from a metal of work function 1.8 eV, the kinetic energy of most energetic electrons is 0.5 eV. The corresponding stopping potential is :

(a) 2.3 V (b) 1.8 V (c) 1.3 V (d) 0.5 V

19. In Einstein's photoelectric equation, the difference of the energy of the incident photon and the work function is equal to :

(a) kinetic energy of the photoelectron.
(b) mean kinetic energy of the photoelectron.
(c) minimum kinetic energy of the photoelectron.
(d) maximum kinetic energyof the photoelectron.

20. Work function of a photoelectric material is 3.3 eV. What will be its threshold frequency ?

(a) 8×10^{10} Hz (b) 8×10^{14} Hz (c) 8×10^{20} Hz (d) 8×10^{25} Hz

21. The kinetic energy of a proton and that of an α-particle are 4 eV and 1 eV, respectively. The ratio of the de-Broglie wavelengths associated with them, will be :★

(a) 2 : 1 (b) 1 : 1 (c) 1 : 2 (d) 4 : 1

22. A photocell connected in an electrical circuit is placed at a distance *'d'* from a source of light. As a result, current I flows in the circuit. What will be the current in the circuit when the distance is reduced to *'d/2'* ?★

(a) 1 (b) 21 (c) 41 (d) 1/2

23. The graph showing the correct variation of linear momentum (p) of a charge particle with its de-Broglie wavelength (λ) is :★

(a) p vs λ (b) p vs λ (c) p vs λ (d) p vs λ

24. If photons of frequency v are incident on the surfaces of metals. A and B of threshold frequencies $v/2$ and $v/3$ respectively, the ratio of the maximum kinetic energy of electrons emitted from A to that from B is :★

(a) 2 : 3 (b) 3 : 4 (c) 1 : 3 (d) $\sqrt{3} : \sqrt{2}$

Ans. 1. (a), 2. (d), 3. (d), 4. (a), 5. (c), 6. (d), 7. (d), 8. (b), 9. (d), 10. (b), 11. (d), 12. (d), 13. (b), 14. (a), 15. (c), 16. (c), 17. (b), 18. (d), 19. (d), 20. (b), 21. (b), 22. (c), 23. (b), 24. (b).

Chapter 12. Atoms

1. Rutherford's experiment on scattering of α-particles proved that :

(a) atom contains positrons
(b) number of positive charges is not equal to the number of negative charges
(c) positive charge is uniformly distributed in the atom
(d) atom is mostly empty

2. In Geiger-Marsden scattering experiment, the trajectory traced by an α-particle depends on :

(a) impact parameter (b) number of scattered α-particles
(c) number of collisions (d) none of these

★ are board based questions from previous year

3. The relationship between kinetic energy (K) and potential energy (U) of electron moving in an orbit around the nucleus is :
(a) $U = -\frac{1}{2}K$ (b) U = – 3 K (c) U = – 2 K (d) U = – K

4. When an electron jumps from the fourth orbit to the second orbit, one gets the :
(a) second line of Lyman series (b) first line of Pfund series
(c) second line of Balmer series (d) second line of Paschen series

5. The energy of hydrogen atom in its ground state is – 13.6 eV. The energy of the level corresponding to the quantum number n equals to 5 is :
(a) – 0.54 eV (b) – 0.85 eV (c) – 2.72 eV (d) – 5.40 eV

6. The shortest wavelength in the Balmer series is :
(a) 364.6 nm (b) 300 nm (c) 256.8 nm (d) 200 nm

7. Bohr's atom model assumes :
(a) mass of electron remains constant
(b) electrons in a quantized orbit will not radiate energy
(c) the nucleus is of infinite mass and is at rest
(d) all the above conditions

8. The first model of atom in 1898 was proposed by :
(a) Niels Bohr (b) J.J. Thomson (c) Albert Einstein (d) Ernst Rutherford

9. In Bohr's model of hydrogen atom, radius of the first orbit of an electron is r_0. Then, radius of the third orbit is :
(a) $\frac{r_0}{9}$ (b) r_0 (c) $3r_0$ (d) $9r_0$

10. In Bohr's hydrogen atom the electron around the nucleus is in a circular orbit of radius 5×10^{-11} m, with a time-period of 1.5×10^{-16} s. The associated current is :
(a) zero (b) 1.6×10^{-19} A (c) 0.17 A (d) 1.07×10^{-3} A.

11. If the radius of second electron orbit in hydrogen atom be r then the radius of the third orbit will be :
(a) $3r$ (b) $2.25\ r$ (c) $9r$ (d) $\frac{r}{3}$

12. If E_p and E_k represent potential energy and kinetic energy respectively, of an orbital electron, then according to Bohr's theory :
(a) $E_k = -\frac{E_p}{2}$ (b) $E_k = -E_p$ (c) $E_k = -2\ E_p$ (d) $E_k = 2\ E_p$

13. The energy of hydrogen atom in its nth orbit is $E_n = \left(\frac{13.6}{n^2}\right)$ eV. The energy required for sending electron from 1st orbit to 2nd orbit will be :
(a) 10.2 eV (b) 12.1 eV (c) 13.6 eV (d) 3.4 eV

14. A hydrogen atom (ionisation potential 13.6 eV) makes a transition from third excited state to second excited state. The energy of the photon emitted in the process is :
(a) 1.89 eV (b) 2.55 eV (c) 12.09 eV (d) 12.75 eV

15. Total energy of electron in n^{th} stationary orbit of hydrogen atom is :
(a) $\frac{13.6}{n^2}$ J (b) $\frac{13.6}{n^2}$ eV (c) $-\frac{13.6}{n^2}$ eV (d) $\frac{-13.6}{n^2}$ J

16. The series of hydrogen spectrum which lies in the visible region is :
(a) Lyman series (b) Balmer series (c) Paschen series (d) None of these

17. The wavelength of the first spectral line in the Balmer series of hydrogen atom is 6561 Å. The wavelength of the second spectral line in the Balmer series of singly ionized helium atom is :
(a) 1215 Å (b) 1640 Å (c) 2430 Å (d) 4687 Å

18. The atomic model based on quantum theory was first proposed by :
(a) Plank (b) Sommer field (c) Bohr (d) Thomson

19. Paschen series of atomic spectrum of hydrogen gas lies in :★
(a) Infrared region (b) Ultraviolet region
(c) Visible region (d) Partly in ultraviolet and partly in visible region

20. In the α-particle scattering experiment, the shape of the trajectory of the scattered α-particles depend upon :★
(a) only on impact parameter
(b) only on the source of α-particles
(c) both impact parameter and source of α-particles
(d) impact parameter and the screen material of the detector

Ans. 1. (d), 2. (a), 3. (c), 4. (c), 5. (a), 6. (a), 7. (d), 8. (b), 9. (d), 10. (d), 11. (b), 12. (a), 13. (a), 14. (a), 15. (c), 16. (b), 17. (a), 18. (c), 19. (a), 20. (a).

Chapter 13. Nuclei

1. Consider α and β particles and γ-rays each having energy of 0.5 MeV. In the increasing order of penetrating power, the radiation is respectively.
(a) γ, β, α (b) β, γ, α (c) α, γ, β (d) α, β, γ

2. Nuclear binding energy is equivalent to :
(a) Mass of proton (b) Mass of neutron
(c) Mass of nucleus (d) Mass defect of nucleus

3. If the binding energy of the deuterium is 2.23 MeV. The mass defect given in amu is :
(a) 0.00024 (b) 0.00012 (c) 0.0012 (d) 0.0024

4. Mass equivalent of 931 MeV energy is :
(a) 1.66×10^{-19} kg (b) 6.6×10^{-34} kg (c) 6.6×10^{-27} kg (d) 1.66×10^{-27} kg

5. The natural boron of atomic weight 10.81 is found to have two isotopes ^{10}B and ^{11}B. The ratio of abundance of isotopes of natural boron should be :
(a) 10 : 11 (b) 81 : 19 (c) 19 : 81 (d) 11 : 10

6. When a microgram of matter is converted to energy, the amount of energy released will be :
(a) 9×10^{10} J (b) 9×10^{7} J (c) 3×10^{4} J (d) 9×10^{17} J

7. Which of the following is true for a sample of isotope containing U^{235} and U^{238} ?
(a) U^{238} contains 3 less neutron then U^{235}
(b) Contain same number of protons and electrons but U^{238} contains 3 more neutrons than U^{235}
(c) Number of protons, electrons and neutrons are same in both
(d) Number of neutrons are same in both

8. The mass and energy equivalent to 1 amu respectively :
(a) 1.67×10^{-27} gm, 9.30 MeV (b) 1.67×10^{-27} kg, 930 MeV
(c) 1.67×10^{-27} kg, 1 MeV (d) 1.67×10^{-34} kg, 1 MeV

9. Complete the series $^{6}He \rightarrow e^{+} + {}^{6}Li +$:
(a) Neutron (b) proton (c) antineutrino (d) neutrino

10. As a result of radioactive decay, $_{92}U^{238}$ nucleus is converted into $_{91}Pa^{234}$. The particles emitted during this decay are :
(a) a proton and a neutron (b) an α-particle and an β-particle
(c) two β-particles and a neutron (d) two β particles and a proton

★ are board based questions from previous year

11. What is the difference between $_{92}U^{238}$ and $_{92}U^{235}$ atoms ?
    (a) $_{92}U^{238}$ has 3 protons more
    (b) $_{92}U^{238}$has 3 neutrons more
    (c) $_{92}U^{238}$has 3 protons and 3 electrons more
    (d) $_{92}U^{238}$has 3 neutrons and 3 electrons more.
12. A 280 days old radioactive sample shows an activity of 6000 disintegrations per second (dps). The activity decreases to 3000 dps in the next 140 days. The original activity of the sample was :
    (a) 9,000 dps (b) 12,000 dps (c) 20,000 dps (d) 24,000 dps.
13. The stability of a nucleus is decided by its :
    (a) number of protons (b) number of nucleons
    (c) binding energy per nucleon (d) binding energy
14. The half life of a radioactive substance is 20 minute. The approximate time interval $(t_2 - t_1)$ between the time t_2 when 2/3 of it has decayed and time t_1 when 1/3 of it has decayed is :
    (a) 28 min (b) 7 min (c) 14 min (d) 20 min
15. A radiactive substance decays to $\frac{1}{16}$ th of its initial mass in 40 days. The half life of the substance, in days is :
    (a) 20 (b) 10 (c) 5 (d) 2.5
16. Binding energy of a nucleus is of the order of :
    (a) electron volt (eV) (b) kilo electron volt (keV)
    (c) mega electron volt (MeV) (d) a joule (J)
17. Which equation represents the emission of a beta particle by a radioactive nucleus :
    (a) ${}_{0}^{1}n \rightarrow {}_{1}^{1}\text{H} + {}_{-1}^{0}e + v$ (b) $\gamma \rightarrow {}_{1}^{0}e + {}_{-1}^{0}e$
    (c) $4\,{}_{1}^{1}\text{H} \rightarrow {}_{2}^{4}\text{He} + 2\,{}_{1}^{0}e$ (d) ${}_{2}^{4}\text{He} + {}_{7}^{14}\text{N} \rightarrow {}_{8}^{17}\text{O} + {}_{1}^{1}\text{H}$
18. The mass defect for the nucleus of lithium is 0·058135 amu. What is the binding energy per nucleon for lithium in MeV ?
    (a) 7.736 (b) 54.153 (c) 8.790 (d) 9.848
19. The relation between decay constant (λ) and half life (T) of a radioactive element is :
    (a) $T = \frac{1}{\lambda}$ (b) $\lambda = \frac{1}{T^2}$ (c) $T = \frac{0.693}{\lambda}$ (d) $\lambda = 0.693\,T$
20. Which material is used as a fuel in nuclear reactor ?
    (a) Helium (b) Hydrogen (c) Uranium (d) Boron

Ans. 1. (d), 2. (d), 3. (d), 4. (d), 5. (c), 6. (b), 7. (b), 8. (b), 9. (d), 10. (b), 11. (b), 12. (d), 13. (b), 14. (d), 15. (b), 16. (c), 17. (a), 18. (a), 19. (c), 20. (c).

Chapter 14. Semiconductor Electronics : Materials, Devices and Simple Circuits

1. A solar cell is a *p-n* junction operating in :
   (a) reverse bias condition (b) unbiased condition
   (c) forward bias condition (d) in both forward and reverse bias condition
2. A *p-n* photodiode is fabricated from a semiconductor with a band gap of 2.5 eV. The signal wavelength is :
   (a) 496 Å (b) 5000 Å (c) 5000 nm (d) 500 Å
3. In a semiconductor :
   (a) there are no free electrons at 0 K
   (b) there are no free electrons at any temperature
   (c) the number of free electrons increases with pressure
   (d) the number of free electrons is more than that in a conductor

4. The resistivity of a semiconductor at room temperature is in between :
(a) 10^{-2} to 10^{-5} Ω cm (b) 10^{-3} to 10^{6} Ω cm
(c) 10^{6} to 10^{8} Ω cm (d) 10^{10} to 10^{12} Ω cm

5. A semiconductor has equal electron and hole concentration of 6×10^8 m^{-3}. On doping with certain impurity, electron concentration increases to 9×10^{12} m^{-3}. The new hole concentration is :
(a) 4×10^3 m^{-3} (b) 4×10^4 m^{-3} (c) 4×10^5 m^{-3} (d) 4×10^6 m^{-3}

6. LED is the abbreviation of :
(a) Light Editing Diode (b) Light Editing Display
(c) Light Emitting Display (d) Light Emitting Diode

7. In *n*-type semiconductor when all donor states are filled, then the net charge density in the donor states becomes :
(a) 1 (b) < 1, but not zero (c) > 1 (d) zero

8. The forbidden energy band gap in conductors, semi-conductors and insulators are E_{g1}, E_{g2}, and E_{g3} respectively. The relation among them is :
(a) $E_{g1} = E_{g2} = E_{g3}$ (b) $E_{g1} < E_{g2} < E_{g3}$ (c) $E_{g1} > E_{g2} > E_{g3}$ (d) $E_{g1} < E_{g2} > E_{g3}$

9. If the forward bias on *p-n* junction is increased from zero to 0.045 V, then no current flows in the circuit. The contact potential of junction *i.e.*, V_B is :
(a) zero (b) 0.045 V (c) more than 0.045 V (d) less than 0.045 V

10. In an unbaised *p-n* junction, holes diffuse from the *p*-region to *n* region because :
(a) free electrons in the *n* region attract them
(b) hole concentration in *p*-region is more as compared to *n* region
(c) they move across the junction with same potential
(d) all of these

11. In photodiode, the condition for emission of diet electrons is :
(a) $h\nu = E_g$ (b) $h\nu > E_g$ (c) $h\nu < E_g$ (d) $h\nu \leq E_g$

12. A *p-n* junction diode cannot be used :
(a) as rectifier (b) for detecting light intensity
(c) for obtaining light radiation (d) as an amplifier

13. When an impurity is doped into an intrinsic semiconductor, the resistance of the semiconductor :
(a) increase (b) decrease (c) remains unchanged (d) becomes zero

14. An *n*-type and a *p*-type Si can be obtained by doping pure Si with :
(a) sodium and magnesium respectively. (b) phosphorous and boron respectively.
(c) boron and phosphorus respectively. (d) indium and sodium respectively.

15. If a *p-n* junction diode is not connected in any circuit :
(a) the potential is same everywhere
(b) the *p*-type side is at a higher potential than the *n*-type side
(c) there is an electric field at the junction directed from *n*-type side to *p*-type side
(d) there is an electric field at the junction directed from *p*-type side to *n*-type side

16. The distinction between conductors, insulators and semiconductors is largely connected with :
(a) the type of crystal lattice (b) binding energy of their electrons
(c) relative width of their energy gap (d) None of these

17. A single diode can be used :
(a) only as full wave rectifier (b) only as half wave rectifier
(c) both half wave and full wave rectifier (d) neither half wave nor full wave rectifier

18. The heavily and lightly doped regions of a bipolar junction transistor are respectively.

(a) Base and emitter (b) Base collector

(c) Emitter and base (d) Collector and emitter

19. Logic gate is shown in figure. The equivalent circuit is :

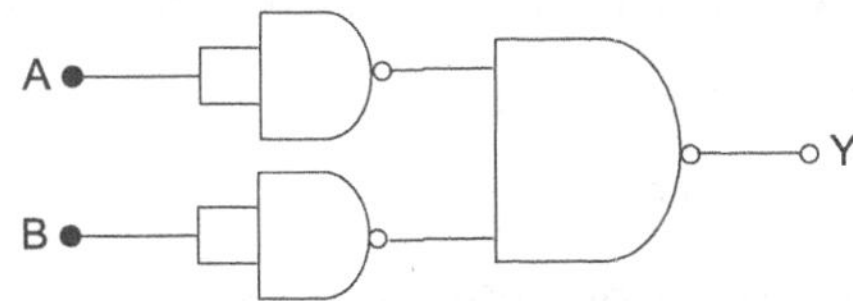

(a) AND gate (b) NOR gate (c) OR gate (d) NOT gate

20. Photo diodes are used to detect :★

(a) radio waves (b) gamma rays (c) IR rays (d) optical signals

21. The wavelength and intensity of light emitted by a LED depend upon :★

(a) forward bias and energy gap of the semiconductor

(b) energy gap of the semiconductor and reverse bias

(c) energy gap only

(d) forward bias only

Ans. 1. (b), 2. (b), 3. (a), 4. (b), 5. (b), 6. (d), 7. (c), 8. (b), 9. (c), 10. (b), 11. (b), 12. (d), 13. (b), 14. (b), 15. (c), 16. (c), 17. (b), 18. (c), 19. (c), 20. (d), 21. (a).

★ are board based questions from previous year

Question Set 2

Very Short Answers

Chapter 1. Electric Charges and Fields

Q. 1. If there were only one type of charge in the universe, then :

(i) $\phi = \oint E \cdot \partial A \neq 0$ on any surface.

(ii) $\phi = \oint E \cdot \partial A = 0$ if the charge is outside a surface.

(iii) $\phi = \oint E \cdot \partial A$ will not be defined.

(iv) $\phi = \oint E \cdot \partial A = \frac{q}{\varepsilon_0}$, if the charge is inside the surface.

Ans. (ii) and (iv) are correct answers.

Q. 2. A certain region has cylindrical symmetry of electric field. Name the charge distribution producing such a field.

Ans. Linear charge distribution.

Q. 3. State Coulomb's law.

Ans. It states that the electrostatic force of attraction or repulsion between two point charges is directly proportional to the product of the magnitude of charges and inversely proportional to the square of the distance between them. The direction of this force is along the line joining the two charges.

Q. 4. Name the physical quantity whose S.I. unit is volt/metre. Is it a scalar or a vector quantity.

Ans. Electric field is the quantity having S.I. unit volt/metre. It is a vector quantity.

Q. 5. A metallic sphere is placed in a uniform electric field as shown in the figure. Which path is followed by electric field lines ? why ?

Ans. Path d is followed by electric field lines. Electric field intensity inside the metallic sphere will be zero, therefore, electric lines of force exist inside the sphere, also lines falls normally on the surface. Electric field lines are always perpendicular to the surface of the conductor.

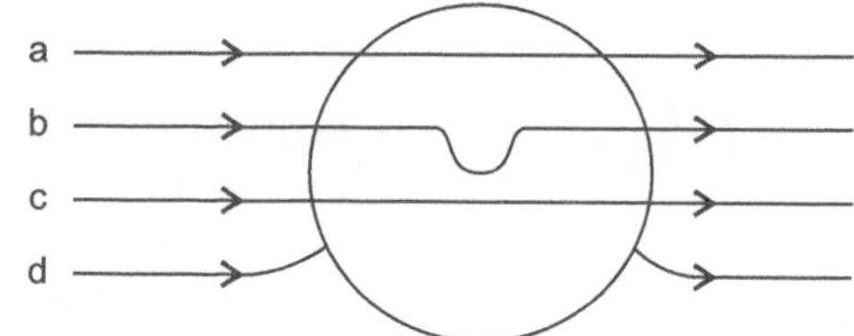

Q. 6. Does the charge given to a metallic sphere depend on whether it is hollow or solid ? Give reason for your answer.

Ans. No, because all the charge resides on the surface of the sphere only.

Q. 7. Why do the electrostatic field lines not form closed loops ?★

Ans. Electrostatic field lines never form loops because they do not converge at only one point as in the case with magnetic field lines. They depend on +ve and – ve charges that can extend to infinity in any particular direction.

Q. 8. What is the electric flux through a cube of side 1 cm which encloses an electric dipole ?★

Ans. The electric flux through a cube of side 1 cm which encloses an electric dipole will be zero, as net charge enclosed by a cube is zero.

Q. 9. An electrostatic field line is a continuous curve. That is, a field line can not have sudden breaks. Why not ? **[NCERT]**

Ans. Electrostatic field line is a continuous curve because a test charge experience continuous force when traced along locus in electrostatic field. The field lines cannot have sudden break as charges moves continuously and doesn't jump from one point to another.

★ are board based questions from previous year

Q. 10. Explain why two field lines never cross each other ? **[NCERT]**

Ans. If two field lines crosses each other at single point, then electric field intensity will have two directions at that point, which is not possible. Hence, two field lines never crosses each other.

Q. 11. State Gauss's law of electric flux.

Ans. The total electric flux associated with any surface is equal to the ratio of electric charge enclosed in surface to ε_0.

$$\phi = \oint \vec{E} \cdot \vec{dS} = \frac{\Sigma q}{\varepsilon_0}$$

Here ε_0 is the permittivity of free space.

Q. 12. Depict the orientation of the dipole in (i) stable, (ii) unstable equilibrium in electric field.

Ans.

(i)

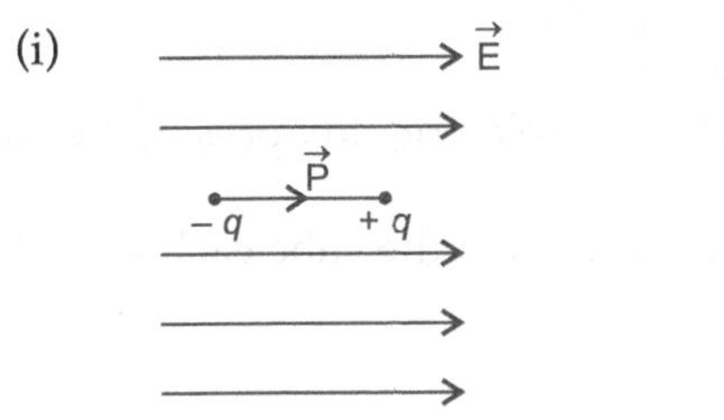

Stable equilibrium ($\theta = 0°$)

(ii)

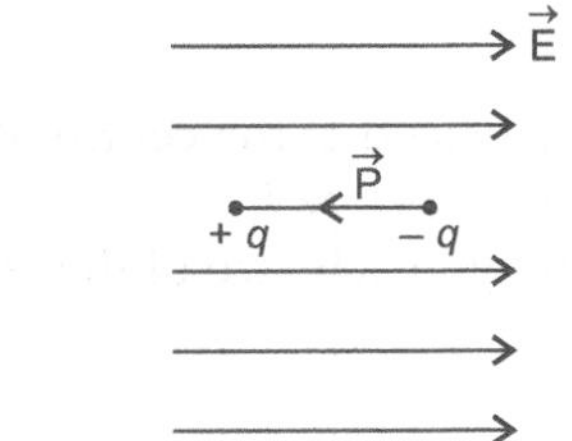

Unstable equilibrium ($\theta = 180°$)

Q. 13. Write properties/characteristics of electric field lines.

Ans. Characteristics of electric field lines are as follows :

(i) The tangent drawn at any point on electric field line shows direction of electric field intensity at that point.

(ii) Two electric field lines can never intersect each other.

(iii) Electric field lines originate from positive charge and end at negative charge.

Q. 14. What are electric field lines ?

Ans. It is a path, straight or curved in the electric field, such that tangent to it at any point gives the direction of electric field intensity at that point.

Q. 15. Arrange the following forces in order of their magnitudes (a) Electrostatic force, (b) Gravitational force, (c) Nuclear force.

Ans. Nuclear forces > Electrostatic forces > Gravitational forces.

Q. 16. Two charges of magnitude – 2 Q and Q are located at point (a, 0) and ($4a$, 0) respectively. What is the electric flux due to these charges through a sphere of radius '$3a$' with its centre at origin.

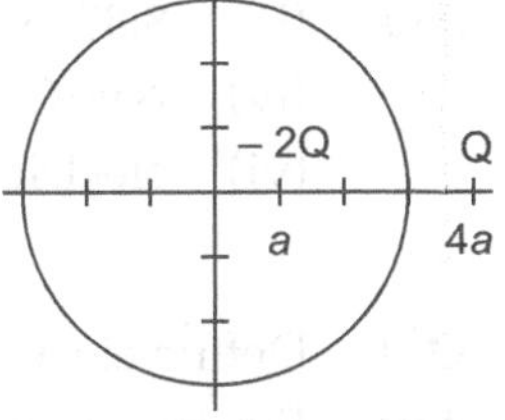

Ans. While drawing as per details given, it has been found that only – 2Q charge will be inside the sphere of radius $3a$. Thus, from Gauss's law,

$$\phi = \frac{q}{\varepsilon_0} = \frac{-2Q}{\varepsilon_0} \text{ Wb}$$

Q. 17. What is conservative field. Give example.

Ans. A conservative field is a field in which the work done on an object by the force due to the field depends only on initial and final positions of objects and not on actual path followed by the object. *e.g.*, Gravitational field, electrostatic field.

Q. 18. A charge q is placed at the centre of an imaginary spherical surface. What will be the electric flux due to this charge through any half of the sphere ?

Ans. The flux through the sphere is given by $\frac{q}{\varepsilon_0}$.

Hence, the flux through the half of the sphere will be half of the total flux *i.e.*, $\frac{q}{2\varepsilon_0}$.

Q. 19. Figure shows five charged lumps of plastic and an electrically neutral coin. The cross-section of a Gaussian surface S is indicated. What is the net electric flux through the surface ?

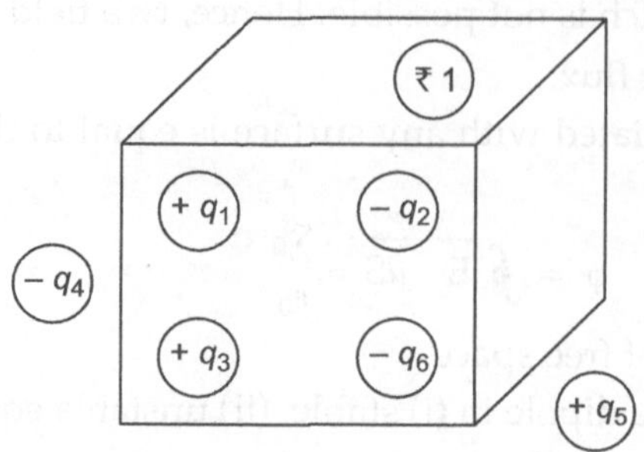

Ans.
$$\phi = \frac{q_{enc}}{\varepsilon_0} = \frac{(q_1 + q_3 - q_2 - q_6)}{\varepsilon_0}$$

Q. 20. A charge q is placed at the centre of a cube of side l. What is the electric flux passing through any one surface of the cube ?

Ans. The flux through each face of the cube is same. There are six faces in the cube each with the same flux 'ϕ''.

Now,
$$\phi = 6\phi'$$

or
$$\phi' = \frac{\phi}{6} = \frac{q}{6\varepsilon_0}$$

Q. 21. The work done in moving a charge particle between two points in an uniform electric field, does not depend on the path followed by the particle. Why ?★

Ans. Because electrostatic force is conservative in nature.

Q. 22. Fill in the blanks with appropriate answer :

(i) The value of permittivity of free space ε_0 is ....

(ii) The S.I. unit of .... is coulomb-metre (C– m).

(iii) The process of protecting certain field from external electric field is called ....

(iv) The S.I. unit of electric flux is ....

(v) Torque acting on an electric dipole placed in an electric field is maximum when the angle between the electric field and the dipole moment is ....★

(vi) The physical quantity having SI unit NC^{-1} m is ....

(vii) A point charge is palced at the centre of a hollow conducting sphere of internal radius 'r' and outer radius '$2r$'. The ratio of the surface charge density of the inner surface to that of the outer surface will be ....★

Ans. (i) $8.854 \times 10^{-12}\ C^2\ N^{-1}\ m^{-2}$ (ii) electric dipole moment (iii) electrostatic shielding

(iv) Nm^2C^{-1} (v) $\theta = 90°$

(vi) electric potential (vii) 4 : 1

Chapter 2. Electrostatic Potential and Capacitance

Q. 1. Define electric potential and give the formula corresponding to it. Write its units and dimensions.

Ans. Electric potential is a physical quantity which determines the flow of charges from one body to another. It is a physical quantity that gives the degree of electrification of a body. Electric potential at a point in the electric field is defined as the amount of work done in moving (without any acceleration) a unit positive charge from infinity to that point against the electrostatic force irrespective of the path followed.

If potential at infinity is chosen to be zero : potential at a point with position vector r due to a point charge Q placed at the origin is given by

$$V(r) = \frac{1}{4\pi\varepsilon_0}\frac{Q}{r}$$

Its unit is Volts and dimensional formula is $[ML^2\ T^{-3}\ A^{-1}]$.

★ are board based questions from previous year

Q. 2. Explain relation between electric field and potential difference.

Ans. Electric field is defined as the negative potential gradient

$$E = \frac{-dV}{dr}$$

The negative sign of potential gradient shows that the rate of change of potential with distance is always against the electric field intensity.

Q. 3. What are dielectrics ?

Ans. Usually, a non-conducting medium or insulator is known as dielectric, more specifically. The non-conducting materials in which induced charges are produced on their faces on the application of electric fields are called dielectrics. Like, glass, air, oil etc.

Q. 4. For any charge configuration, equipotential surface through a point is normal to the electric field. Justify.

Ans. If the electric field were not normal to equipotential surface, it would have non-zero component along the surface. To move a charge against this component, work would have to be done. But no work is needed to move a test charge on an equipotential surface. Hence, electric field must be normal to the equipotential surface at every point.

Q. 5. A charge 'q' is moved from a point A above a dipole of dipole moment 'p' to a point B below the dipole in equatorial plane without acceleration. Find the work done in the process.★

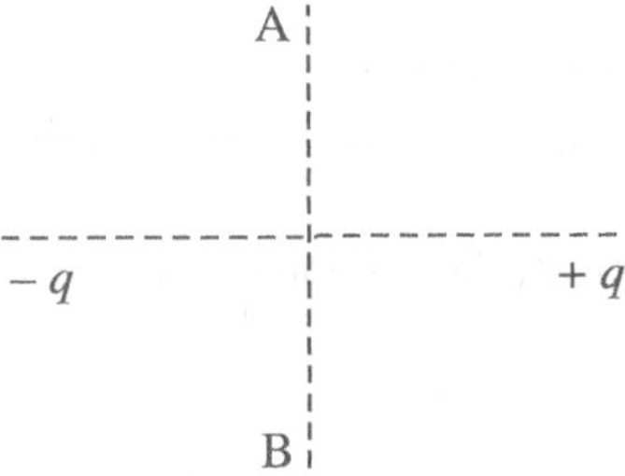

Ans. Work done, $W = q\,(V_B - V_A) = q \times 0 = 0$

Q. 6. For any charge configuration, equipotential surface through a point is normal to the electric field. Justify.★

Ans. If the electric field were not normal to equipotential surface, it would have non-zero component along the surface. To move a charge against this component, work would have to be done. But no work is needed to move a test charge on an equipotential surface. Hence, electric field must be normal to the equipotential surface at every point.

Q. 7. Differentiate between polar and non-polar dielectric.

Ans.

	Polar dielectric	Non-polar dielectric (Insulators)
1.	Each atom of such dielectrics has two centres of charges *i.e.*, positive and negative and they don't coincide.	In insulators, two centre of charges coincide with each other.
2.	It has permanent dipole moment. In external electric field, it experiences torque. *eg.*, H_2O, NH_3 etc.	It doesn't have dipole moment. In an external electric field, it experiences induced dipole moment. *e.g.*, H_2, N_2 etc.

Q. 8. What is electric polarisation ?

Ans. Induction of dipole moment in non polar dielectric in an electric field is known as electric polarisation. When a non-polar dielectric placed in external electric field, two centres (+ and –) gets separated by small distance which gives development of dipole moment for such molecules and it is electric polarisation.

★ are board based questions from previous year

Q. 9. Fill in the blanks with appropriate answer :

(i) Capacitance of an isolated spherical conductor is given by ....

(ii) Work done in moving a charge over an equipotential surface is ....

(iii) For safety reasons, electric field applied is only .... of the dielectric strength of the dielectric material.

(iv) A proton released from rest in an electric field, will start moving towards a region of .... potential in the field.★

(v) The potential difference between any two points of equipotential surface is ....★

Ans. (i) $C = 4\pi\varepsilon_0 r$ (ii) zero (iii) 10% (iv) decreasing

(v) Zero

Q. 10. Without referring to the formula $C = \varepsilon_0 A/d$. Explain why the capacitance of a parallel plate capacitor decreases on increasing the separation between the plates ?

Ans. A capacitor consists of two conductors separated by an insulator. The capacitance is the charge stored divided by the voltage. If the conductors are further apart, the amount of work done to move a charge from one to the other increases so, the voltage increases. Hence, the capacitance decreases.

Chapter 3. Current Electricity

Q. 1. Define the term 'mobility' of charge carriers.

Ans. Mobility of charge carriers in a conductor is defined as the magnitude of their drift velocity per unit applied electric field.

$$\text{Mobility } (\mu) = \text{Drift of electric field}$$

$$\mu = \frac{v_d}{E}$$

Q. 2. Under what condition the emf of a cell will be equal to potential difference ?

Ans. When cell is in open circuit, its internal resistance will not come in picture.

Q. 3. Define internal resistance of cell.

Ans. The resistance offered by the electrolyte of cell is called internal resistance of cell.

Q. 4. Why electrical conductance of electrolytes is less than that of metals ?

Ans. As electrolytes has ions and it have lesser mobility compared to free electrons.

Q. 5. (i) What are limitations of ohm's law for conductors ?

(ii) State conditions under which ohm's law is not obeyed conductor.

Ans. (i) When temperature changes.

(ii) When conductor is under strain.

Q. 6. What happens during death due to electric shock ?

Ans. Electric shock causes electrolysis of blood, which takes place even with small currents.

Q. 7. Kirchoff's current and voltage laws based on which universal laws.

Ans. Kirchoff's current law (KCL) based on conservation of electric charge and Kirchoff's voltage law based on conservation of energy.

Q. 8. What are applications of potentiometer ?

Ans. Potentiometer is used for :

(i) to find internal resistance of cells.

(ii) to compare emf of cells.

(iii) to compare resistances.

★ are board based questions from previous year

Q. 9. Draw V-I (Voltage-current) graph for conductors and semi-conductors.

Ans.

V-I Graph for conductors :

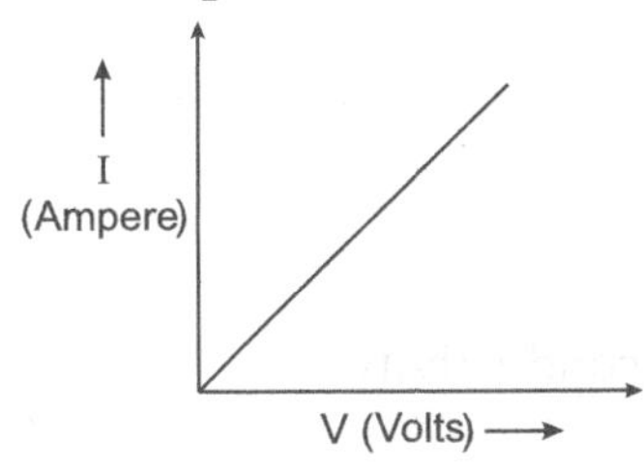

V-I Graph for semi-conductors :

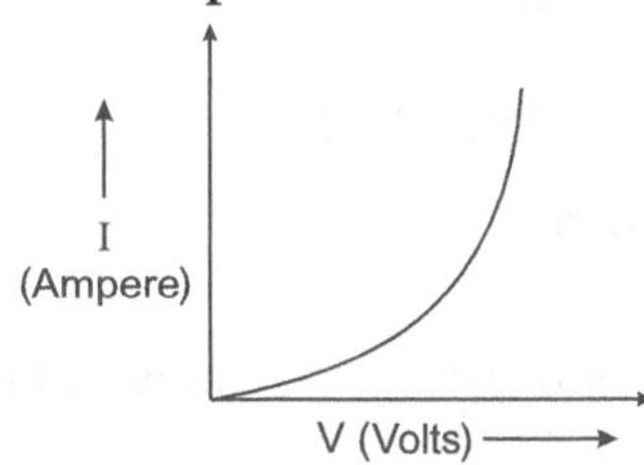

Q. 10. A 4 Ω resistance wire is bent in the middle by 180° and both the halves are twisted with each other. Find new resistance.

Ans. When 4 Ω resistance bent in the middle by 180° then 2 Ω resistance will come in parallel. Thus resultant of this resistance will be 1 Ω.

Q. 11. Does the charge given to a metallic sphere depend on whether it is hollow or solid ? Give reason for your answer.

Ans. No, because all the charge reside on the surface of the sphere only.

Q. 12. Nichrome and copper wires of same length and same radius are connected in a series. Current I is passed through them. Which wire gets heated up more ? Justify your answer.★

Ans. We know that,

$$R = r$$

$$\frac{R_{Cu}}{R_{Ni}} = \frac{\rho_{Cu}}{\rho_{Ni}}$$

$\because$ $\rho_{Cu} < \rho_{Ni}$

$\therefore$ $R_{Cu} < R_{Ni}$

$\because$ $H = I^2Rt$

$H \propto R$

$\therefore$ Nichrome wire will have more heat.

Q. 13. I-V graph for a metallic wire at two different temperatures, T_1 and T_2 is as shown in the figure. Which of the two temperature is lower and why ?★

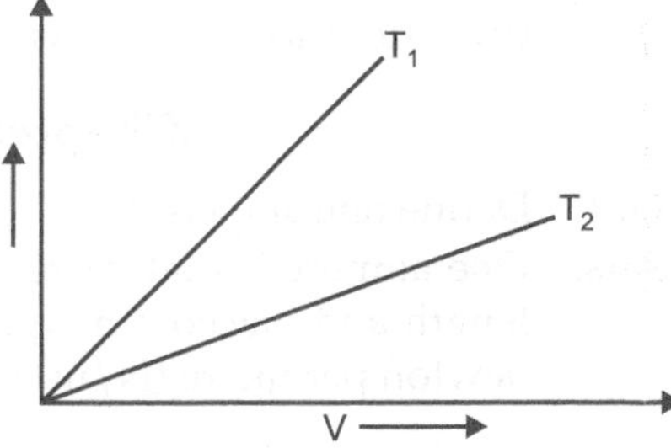

Ans. The slope of a V-I graph is given by the formula $\frac{1}{V} = \frac{1}{R}$. Thus, the smaller the slope larger is the resistance. As the resistance of a metal increases with the increase in temperature, so resistance at T_2 is higher and T_1 is lower.

Q. 14. Show variation of resistivity of copper as a function of temperature in a graph.★

Ans. The variation of resistivity of copper with temperature is parabolic in nature. This is shown in the following graph :

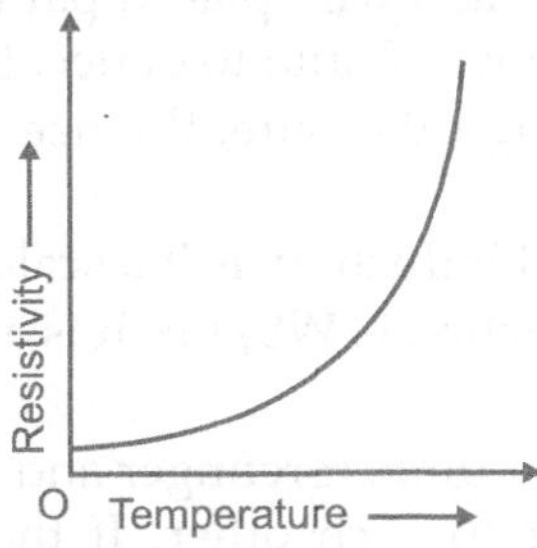

★ are board based questions from previous year

Q. 15. Define the term 'electrical conductivity' of a metallic wire. Write its S.I. unit.★

Ans. The electrical conductivity of a metallic wire is defined as the ratio of the current density to the electric field it creates.

Electrical conductivity, $\sigma = \frac{J}{E}$

S.I. unit = $(\text{ohm m})^{-1}$

OR

The reciprocal of resistivity of a material is called its electrical conductivity.

$$\text{Conductivity} = \frac{1}{\text{Resistivity}}$$

or
$$\sigma = \frac{1}{\rho}$$

Q. 16. Why alloys like constantan and manganin are used for making standard resistors ?★

Ans. They are used to make standard resistors because :

(i) They have high value of resistivity.
(ii) Temperature coefficient of resistance is less.
(iii) They are least affected by temperature.

Q. 17. Fill in the blanks with appropriate answer :

(i) The S.I. unit of mobility is ....
(ii) The average time interval between two successive collisions of a free electron with the positive ion in metal is called ....
(iii) The unit of electric energy is ....
(iv) To minimize the percentage error in the determination of unknown resistance of a conductor in meter bridge experiment, the balance point is adjusted near .... of the wire.★
(v) In potentiometer, uniform wire is used to .... potential gradient along the wire.★
(vi) A copper wire of non-uniform area of cross-section is connected to a d.c. battery. The physical quantity which remains constant along the wire is ....★

Ans. (i) $m^2 v^{-1} s^{-1}$ (ii) relaxation time (iii) 1 kWh (iv) mid-point (middle)
(v) decrease (vi) electric current

Chapter 4. Moving Charges and Magnetism

Q. 1. Define one ampere ?

Ans. One ampere is that current which, it passed through each of two parallel conductors of infinite length and placed 1 m apart in vacuum causes each conductor to experience a force of 2×10^{-7} newton per metre (N/m) of length of the conductor.

$$\frac{F}{l} = \frac{\mu_0 I_1 I_2}{2\pi r} \text{N/m}$$

When $I_1 = I_2 = 1$ ampere and $r = 1$ m, then $F = 2 \times 10^{-7}$ N/m.

Q. 2. What is the Oersted's experiment ?

Ans. When current was allowed to flow in a wire placed parallel to the axis of a magnetic needle kept directly below the wire, the needle was found to deflect from its normal position.

When current was reversed through the wire, the needle was found to deflect in the opposite direction to the earlier case.

Q. 3. Write the S.I. unit of magnetic field induction. Is it a scalar or a vector quantity ?

Ans. The S.I. unit of magnetic field induction is Wb/m^2. It is a vector quantity.

Q. 4. Explain Fleming's left hand rule.

Ans. According to this rule, stretch the thumb, forefinger and middle finger of your left hand such that they are mutually perpendicular to each other. If the first finger points in the direction of

★ are board based questions from previous year

magnetic field and the second finger in the direction of current, then the thumb will point in the direction of motion or the force acting on the conductor.

Q. 5. State Maxwell's cork screw rule or right hand screw rule ?

Ans. If the forward motion of a right handed imaginary conductive screw is in the direction of the current through a linear conductor, then the direction of rotation of the screw gives the direction of the magnetic lines of force around the conductor.

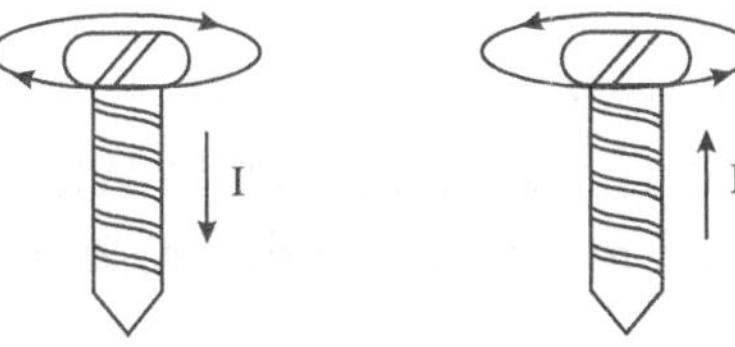

Q. 6. When will the Lorentz force be zero ?

Ans. The Lorentz force on the charge q is 'Zero' when,

(i) $\vec{v}$ and $\vec{B}$ are parallel or antiparallel.
i.e., $\theta = 0°$ or $\theta = 180°$ as $\sin 0° \sin 180° = 0$

(ii) $\vec{v} = 0$

(iii) The particle is neutral.

Q. 7. Write the features of Lorentz force.

Ans. The features of Lorentz force are :

(i) Magnetic Lorentz force does not work on the charged particle, because it is prependicular to $\vec{v}$ and $\vec{B}$.

(ii) Magnetic Lorentz force does not change the kinetic energy of the charged particle.

(iii) Magnetic Lorentz force changes the momentum of the charged particle.

Q. 8. What is limitation of cyclotron ?

Ans. Due to the relativistic effect, the mass of the particle being accelerated increases with speed. Therefore, cyclotron frequency constantly changes and hence it is difficult to synchronise with the frequency of the a.c. oscillator.

Q. 9. A straight wire of mass 200 g and length 1.5 m carries a current of 2 A. It is suspended in mid-air by a uniform horizontal magnetic field B (Fig.). What is the magnitude of the magnetic field ?

[NCERT]

Ans. In the given figure, there is an upward force F, of magnitude I*l*B. For mid-air suspension, this must be balanced by the force due to gravity :

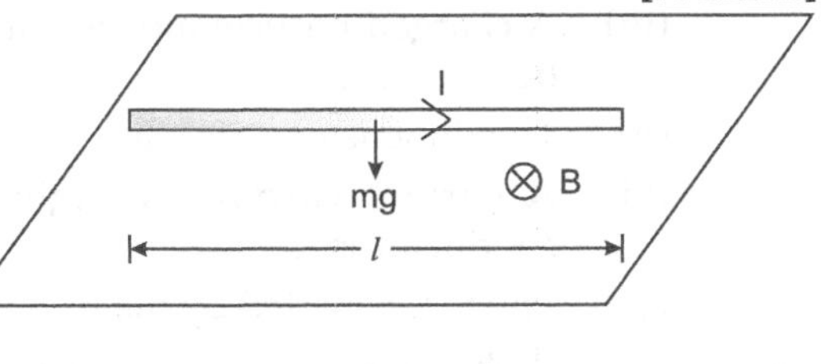

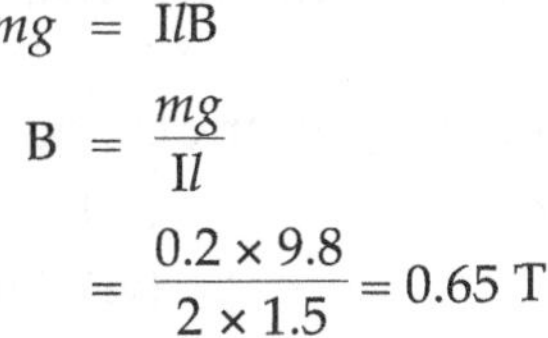

$$mg = IlB$$

$$B = \frac{mg}{Il}$$

$$= \frac{0.2 \times 9.8}{2 \times 1.5} = 0.65 \text{ T}$$

Q. 10. What is the effect of increasing the number of turns on magnetic field produced due to a circular coil ?

Ans. Magnetic field increases.

Q. 11. In moving coil galvanometer, what is meant by radial magnetic field ?

Ans. **Radial magnetic field :** It means that the plane of the coil is always parallel to the magnetic lines of forces, so in every position $\theta = 90°$.

Q. 12. The net charge in a current carrying conductor is zero, even then it experiences force in a magnetic field why ?

Ans. In a current carrying conductor free electrons move with the drift velocity and hence magnetic force acts on them.

Q. 13. (i) Does magnetic field exert a force on a static charge ?
(ii) Does it exert a force on a moving charge in the same direction as the direction of the field itself ?

Ans. (i) No, magnetic field doesn't exert a force on a static charge.
(ii) No, the direction of the magnetic force is always perpendicular to the magnetic field.

Q. 14. The north pole of a magnet is brought near stationary positively charged conductor, will the pole experience any force ?

Ans. No, a stationary charge doesn't produce magnetic effect.

Q. 15. Define magnetic dipole moment. Is it scalar or vector ? What are its units ?

Ans. The product of pole strength and magnetic length (*i.e.*, $2l$) is called the magnetic dipole moment.

It is denoted by $\vec{M}$.

$$\vec{M} = m(2l)$$

It is a vector quantity and its unit is Am^2.

Q. 16. Write the underlying principle of a moving coil galvanometer.★

Ans. When a current carrying coil is placed in magnetic field then it experiences a torque.

$$NIAB = k\alpha$$

$$\Rightarrow \qquad I = \frac{k}{NAB}\alpha$$

where N = Number of turns.
I = Current.
A = Area of the loop.
B = Magnetic field.
k = Torsional constant of the wire
α = Angle of deflection

Q. 17. Fill in the blanks with appropriate answer :
(i) The S.I. unit of magnetic permeability of free space is ....
(ii) The value of constant $\frac{\mu_0}{4\pi}$ is ....
(iii) A current carrying straight wire is aligned in north-south direction. The force on the conductor is ....
(iv) A charged particle moves parallel to a uniform magnetic field. The path of the particle will be ....
(v) The dimensions of μ_0 are ....
(vi) A current carrying straight wire is aligned in east-west direction. The force on the conductor is ....
(vii) The force exerted by magnetic field on moving electric charge is known as ....

Ans. (i) Weber per ampere (ii) $10^{-7}TA^{-1}m$ (iii) zero
(iv) straight line (v) $MLT^{-2}A^{-2}$ (vi) Nil
(vii) Lorentz force

Chapter 5. Magnetism and Matter

Q. 1. Give relation between angle of dip (θ), B_V and B_H.

Ans.
$$\tan\theta = \frac{B_V}{B_H} \qquad \left[\because B_H = B\cos\theta \text{ and } B_V = B\sin\theta\right]$$

Where θ is angle of dip, B_V and B_H are vertical and horizontal components of earth's magnetic field.

★ are board based questions from previous year

Q. 2. What types of magnetic material used in making permanent magnets ?

Ans. Ferromagnetic materials are used in making permanent magnets.

Q. 3. What is magnetic domains ?

Ans. A magnetic domain is a region in which the magnetic fields of atoms are grouped together and aligned. In unmagnetised objects, all of the magnetic domains point in different directions. When it becomes magnetised all magnetic domains lined up and align in same direction.

Q. 4. Define coercive force.

Ans. **Coercive force :** It is the intensity of the applied magnetic field required to reduce the magnetisation of a given material to zero.

Q. 5. State how magnetic susceptibility of a ferromagnetic material changes when it is heated.

Ans. Magnetic susceptibility of a ferromagnetic material decreases with increase in temperature.

Q. 6. How value of magnetic permeability (μ) varies for dia, para and ferromagnetic materials.

Ans. For diamagnetic material $\mu < 1$

For paramagnetic material $\mu > 1$

For ferromagnetic material $\mu >> 1$

Q. 7. Can two magnetic lines of forces intersect ? Justify your answers.

Ans. No, because if two magnetic lines of force intersect, then there will be two tangents and hence there will be two directions of magnetic field at the point of intersection, which is impossible.

Q. 8. State difference between the atoms of diamagnetic and paramagnetic materials.

Ans. Atoms of paramagnetic material have a permanent magnetic moment while in atoms of diamagnetic material net magnetic moment is zero.

Q. 9. Compare magnetic properties of soft iron and steel.

Ans. Soft iron has smaller retentivity but has smaller coercivity than steel and have low hysteresis loss.

Q. 10. What is relation between permeability and magnetic susceptibility of medium.

Ans. If χ_m is the magnetic susceptibility of a medium, the permeability of a medium is $\mu = \mu_0 (1 + \chi_m)$ where μ_0 is permeability of free space drift velocity, hence magnetic force acts on them. Since positive ions are stationary they don't experience any force.

Q. 11. An electric current is flowing along a power line. What is direction of magnetic field :

(a) above it, (b) below it.

Ans. From right hand thumb rule direction of magnetic field is towards west above the wire and towards east below the wire.

Q. 12. What happens if bar magnet is cut into two pieces ?

(i) Perpendicular to its length.

(ii) Along its length.

Ans. (i)Each piece is magnet with north and south poles of same pole strength as before but reduced length.

(ii) Each piece is a magnet with same length but reduced pole strength.

Q. 13. Does a magnetic monopole exist ?

Ans. No, it doesn't exist.

Q. 14. Why does the magnetisation of a paramagnetic salt increases on cooling ?

Ans. This is because tendency to disrupt alignment of dipoles due to random thermal agitation which reduces at lower temperature *i.e.*, on cooling paramagnetic salt.

Q. 15. In what way is the behaviour of a diamagnetic material different from that of a paramagnetic, when kept in an external magnetic field ?★

Ans. The magnetic field lines pass through the paramagnetic material while the magnetic field lines move away from the diamagnetic material.

★ are board based questions from previous year

Q. 16. Define 'magnetic declination' at a place earth.★

Ans. The angle between the geographical meridian and the magnetic meridian at a place is called the magnetic declination (α) at a place.

Q. 17. Fill in the blanks with appropriate answer :`

(i) When a current carrying coil is placed in magnetic field then it experiences a ....

(ii) The net magnetic moment of an atom of a diamagnetic substance is ....

(iii) 0° angle of dip means that earth's magnetic field is .... as on equator.

(iv) The ability of a material to retain a certain amount of residual magnetic field when the magnetising force is removed after achieving saturation is called ....

(v) The neon gas is .... as there are no unpaired electrons in neon atom.

(vi) When the dipole is placed parallel to the magnetic field. The dipole has minimum ....

Ans. (i) torque (ii) zero (iii) horizontal

(iv) retentivity (v) dimagnetic (vi) potential energy

Chapter 6. Electromagnetic Induction

Q. 1. A copper loop and an aluminium loop are removed from a magnetic field in the same time interval. In which loop will the induced emf and induced current be greater ?

Ans. The induced *emf* will be same in both, but the induced current will be greater in copper loop because its resistance is smaller than that of aluminium loop.

Q. 2. What are two applications of Eddy currents ?

Ans. Eddy currents are widely used in electric breaks, energy meters, induction furnaces, speedometer etc.

Q. 3. Does a transformer change frequency of a.c. ?

Ans. No, it doesn't.

Q. 4. Can a d.c. dynamo be used as d.c. motor ?

Ans. Yes, by replacing the load resistance of a d.c. dynamo by a d.c. source, we get a d.c. motor.

Q. 5. A transformer is used to step-down a.c. voltage. Which appliance will you use to step-down d.c. voltage.

Ans. Pure resistance/potential divider can be used to step-down d.c. voltage.

Q. 6. Why can't transformer be used to step-up d.c. voltage ?

Ans. The d.c. supply doesn't produce a changing magnetic flux in the primary coil and hence no emf is set-up in the secondary coil of the transformer by d.c. supply.

Q. 7. What is the average value of a.c. during a half cycle and during full cycle ?

Ans. Average value of a.c. during half cycle

$$i_{avg} = \frac{2i_0}{\pi}$$

$$e_{avg} = \frac{2e_0}{\pi}$$

average value of d.c. during full cycle = zero.

Q. 8. Predict the polarity of the capacitor in the situation described below :★

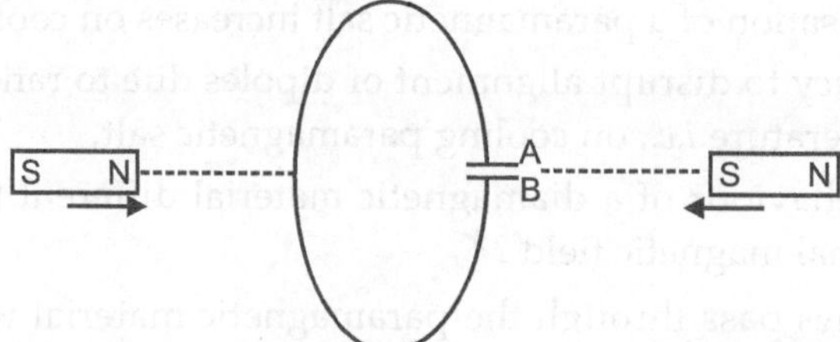

Ans. A will be positive and B will be negative.

★ are board based questions from previous year

Q. 9. A bar magnet is moved in the direction indicated by the arrow between two coils PQ and CD. Predict the direction of the induced current in each coil.★

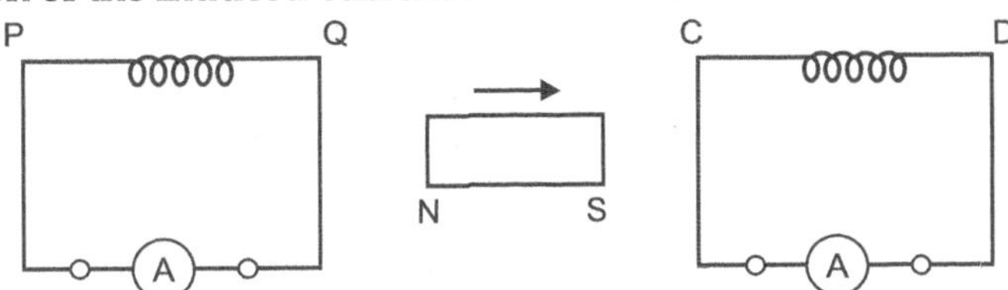

Ans. From Q to P (*i.e.*, anticlockwise) as seen from the left end.

From C to D (*i.e.*, clockwise) as seen from the left end.

Q. 10. What is the direction of induced currents in metal rings 1 and 2 when current I in the wire is increasing steadily ?★

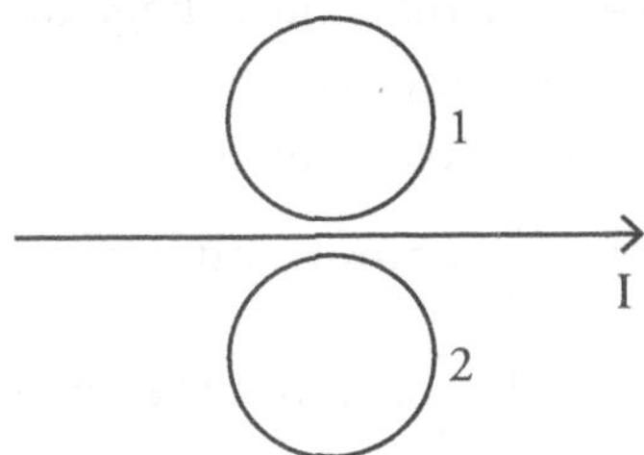

Ans. Ring - 1 → Clockwise

Ring - 2 → Anticlockwise

Q. 11. A long straight current carrying wire passes normally through the centre of circular loop. If the current through the wire increases, will there be an induced emf in the loop ? Justify.★

Ans. No, the emf will not be induced as the magnetic field lines are parallel to the plane of the circular loop.

$$\phi = BA\cos\theta = BA\cos 90° = 0$$

Hence, induced emf,
$$\varepsilon = -\frac{d\phi}{dt} = 0$$

Q. 12. Define the term 'self-inductance' of a coil. Write its S.I. unit.★

Ans. Self-inductance of a coil is numerically equal to the amount of magnetic flux linked with the coil when unit current flows through the coil.

The S.I. unit of self inductance is henry (H) or weber per ampere.

$$1\text{ H} = 1\text{ Wb/A}$$

Q. 13. The electric current flowing in a wire in the direction from B to A is decreasing. Find out the direction of the induced current in the metallic loop kept above the wire as shown.★

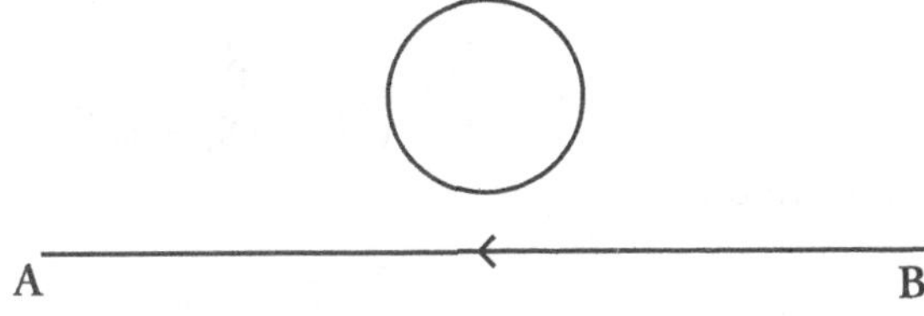

Ans. The direction of the current in the loop will be such as to oppose the decrease of this field (Lenz's Law).

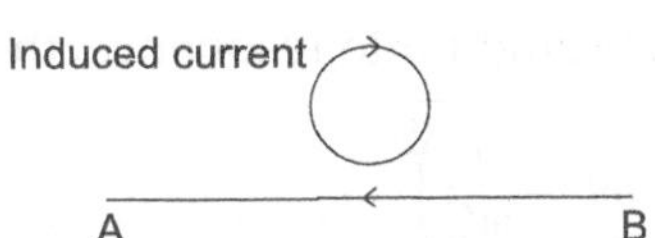

Q. 14. A conducting loop is held above a current carrying wire 'PQ' as shown in the figure. Depict the direction of the current induced in the loop when the current in the wire PQ is constantly increasing.★

★ are board based questions from previous year

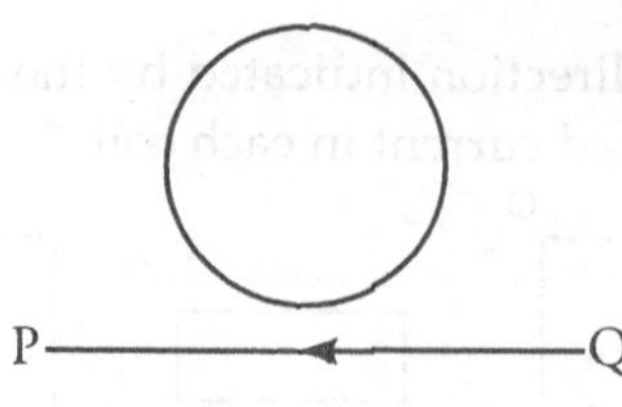

Ans. Anticlockwise

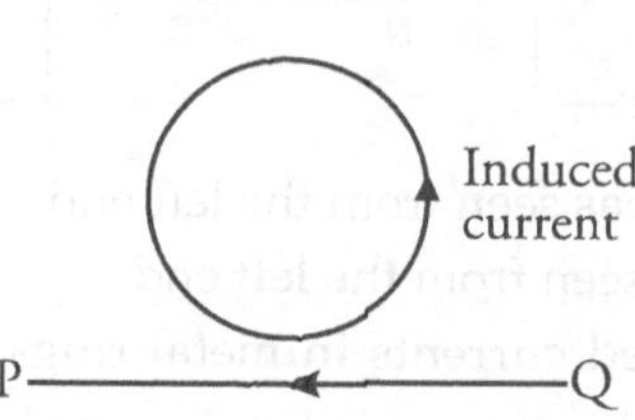

Q. 15. What is phase relationship between current and voltage in following conditions :
(i) Pure resistor (ii) Pure inductor (iii) Pure capacitors **[NCERT]**

Ans. (i) Voltage (V) and current (I) are in phase always.

(ii) Voltage (V) leads current (I) by $\frac{\pi}{2}$ in purely inductive circuit.

(iii) Voltage (V) lags behind current by $\frac{\pi}{2}$ in purely capacitive circuit.

Q. 16. Define mutual inductance.★

Ans. Mutual induction is the phenomenon of production of induced emf in one coil due to change of current in the neighbouring coil. The coil in which the current changes is called primary coil and the coil in which emf is induced is called the secondary coil.

Q. 17. Fill in the blanks with appropriate answer :

(i) The transformer works on the principle of ....

(ii) The electromagnetic radiations used for water purification is ....

(iii) The electromagnetic radiations used for eye-surgery is ....

(iv) a.c. generator can be converted into d.c. generator by using ....

Ans. (i) mutual induction (ii) ultraviolet radiation (iii) infrared radiation
(iv) split ring commutator

Chapter 7. Alternating Current

Q. 1. What is the power factor of a series LCR circuit ?

Ans. If ϕ is the phase angle by which emf lead or lag the current in an LCR circuit, then $\cos\phi$ is called the power factor of circuit

$$\cos\phi = \frac{R}{Z} = \frac{R}{\sqrt{R^2 + (X_L - X_C)^2}}$$

R = Resistance of the circuit
Z = Impedance
X_L = Inductive reactance
X_C = Capacitive reactance

Q. 2. When a series LCR a.c. circuit is brought into resonance, the current has a large value, why ?

Ans. The impedance is $Z = \sqrt{R^2 + \left(\omega L - \frac{1}{\omega C}\right)^2}$

At resonance $\omega L = \frac{1}{\omega C}$, the impedance decreases and hence the current increases.

★ are board based questions from previous year

Q. 3. In LCR series circuit, when voltage and current are in phase.

Ans. At resonance.

Q. 4. Can we use capacitors instead of choke coil for reducing current in an a.c. circuit ? Why ?

Ans. Yes, because the average power dissipated per cycle in an ideal capacitor is also zero.

Q. 5. Sketch the graph showing variation of :

(i) A capacitor with frequency of applied voltage.

(ii) An inductor with frequency of applied voltage.

Ans.

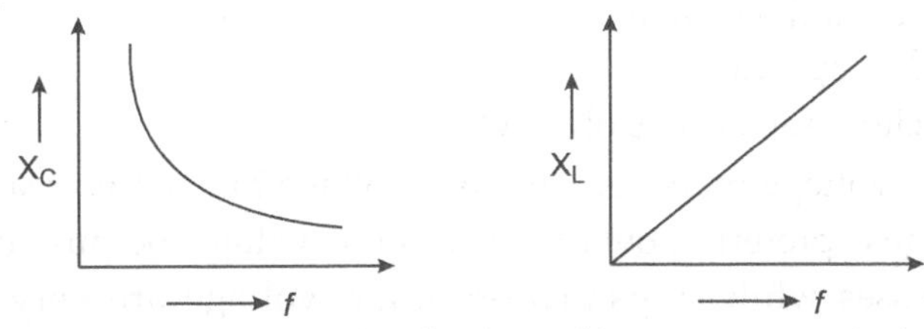

Q. 6. Which current do we required a.c. or d.c. for long distance transmission ?

Ans. The a.c. current is required for long distance transmission.

Q. 7. Variation of alternating current I with time (t) is shown in graph below.

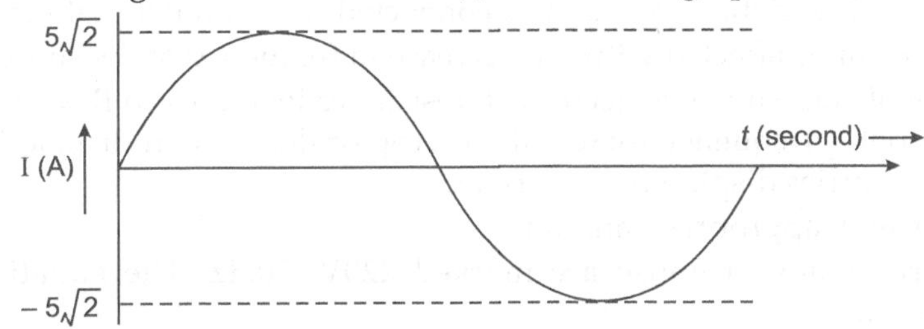

What is the r.m.s. value of this current ?

Ans. The r.m.s. value of current is given by

$$i_{r.m.s.} = \frac{i_0}{\sqrt{2}} = \frac{5\sqrt{2}}{\sqrt{2}} = 5 \text{ Amp.}$$

Q. 8. An ideal inductor doesn't consume any power even though both V and I are non-zero. Explain in brief.

Ans. For an ideal inductor the phase angle is 90°.

We know,

$$P_{av} = E_{r.m.s.} \times i_{r.m.s.} \times \cos\phi$$

$$P_{av} = 0 \qquad [\cos\phi \Rightarrow \cos 90° = 0]$$

Q. 9. Give the expression for the power factor in an LR circuit in terms of the resistance (R) and the inductive reactance X_L.

Ans.

$$\text{Power factor} = \frac{\text{True power}}{\text{Apparent power}} = \frac{\text{Resistance}}{\text{Impedance}} = \frac{R}{\sqrt{R^2 + X_L^2}}$$

Q. 10. Can an ordinary ammeter or moving coil galvanometer be used for measuring a.c. why ?

Ans. No, it can't be used to measure a.c. in a circuit because it measures the average value of current which is zero over a cycle.

Q. 11. Define 'quality factor' of resonance in series LCR circuit. What is its S.I. unit ?★

Ans. The Q factor of series resonance circuit is defined as the ratio of the voltage developed across the inductor or capacitor at resonance to the impressed voltage, which is the voltage across R.

$$Q = \frac{1}{R}\sqrt{\frac{L}{C}}$$

It is dimensionless. Hence, it has no unit.

★ are board based questions from previous year

Q. 12. Define capacitor reactance. Write its S.I. units.★

Ans. Capacitor reactance is the resistance offered by a capacitor to the flow of a.c. It is given by

$$X_C = \frac{1}{\omega C}$$

Where $\omega = 2\pi f$,

f = frequency of the source

$$X_C = \frac{1}{2\pi f C}$$

ω = Angular frequency of the source

C = Capacitance of the capacitor.

The S.I. unit of capacitor reactance is ohm (Ω).

Q. 13. Why is the use of a.c. voltage preferred over d.c. voltage ? Give two reasons.★

Ans. The use of a.c. voltage is preferred over the use of d.c. voltage because of the following reasons :

(i) The energy losses while transmission of a.c. voltage are very less as compared to d.c. voltage.

(ii) a.c. voltage can be controlled as required by using a transformer (*i.e.*, stepped up or stepped down).

Q. 14. An A.C. source with variable frequency is connected to a parallel plate capacitor. How will the displacement current be affected with the decrease in frequency of the source ?★

Ans. When frequency of A.C. source is decreased, resistance increases so flow of current decreases. By dropping the variable frequency there will be drop or decrease in displacement current through the capacitor as it carries displacement current.

Q. 15. Fill in the blanks with appropriate answer :

(i) The electric mains in house are marked 220V, 50Hz. The equation for instantaneous volatage is ....

(ii) The current which does not dissipate any energy is called ....

(iii) The core of transformer is laminated so as to reduce the energy loss due to ....

Ans. (i) $311.12 \sin(314t + \phi)$ (ii) wattless current (iii) eddy current

Chapter 8. Electromagnetic Waves

Q. 1. Which waves are known as heat waves ?

Ans. Infrared waves.

Q. 2. Identify the part of the electromagnetic spectrum having largest penetrating power.

Ans. Gamma rays.

Q. 3. Name the most harmful radiation entering the atmosphere of earth from outer space. What stops this radiation from coming down on to earth ?

Ans. Ultraviolet radiations. Ozone layer stops ultraviolet radiations from coming down to earth.

Q. 4. Arrange the following radiations in descending order of their frequencies. X-rays, radio waves, U.V. rays, Gamma rays, Infrared rays, visible light, Microwaves.

Ans. Gamma rays, X-rays, UV rays, visible light, Infrared rays, microwaves, radio waves.

Q. 5. What is displacement current ? Explain its cause.

Ans. It is a current, which is produced in region where the electric field and hence the electric flux is changing with time. The displacement current is given by

$$I_D = \varepsilon_0 \frac{d\phi_E}{dt}$$

Q. 6. What physical quantity is the same for X-rays of wavelength 10^{-10} m, red light of wavelength 6800Å and radiowaves of wavelength 500 ? **[NCERT]**

Ans. The speed of light (3×10^8 m/s) in a vacuum is the same for all wavelengths. It is independent of the wavelength in the vacuum.

★ are board based questions from previous year

Q. 7. A charged particle oscillates about its mean equilibrium position with a frequency of 10^9 Hz. What is the frequency of the electromagnetic waves produced by the oscillator ? **[NCERT]**

Ans. The frequency of an electromagnetic wave produced by the oscillator is the same as that of a charged particle oscillating about its mean position *i.e.,* 10^9 Hz.

Q. 8. Prove that the unit of $\frac{1}{\sqrt{\mu_0 \varepsilon_0}}$ is that of velocity, using unit of μ_0 and ε_0.

Ans. The unit of $\mu_0 = \frac{N}{A^2}$, and the unit of $\varepsilon_0 = \frac{C^2}{Nm^2}$.

$$\therefore \quad \left[\frac{1}{\sqrt{\mu_0 \varepsilon_0}}\right] = \frac{1}{\sqrt{\frac{N}{A^2}\frac{C^2}{Nm^2}}}$$

$$= \frac{1}{\sqrt{\frac{A^2 s^2}{A^2 m^2}}} = ms^{-1}.$$

Q. 9. Do electromagnetic waves carry energy and momentum ?★

Ans. Yes, the electromagnetic waves carry energy and momentum because electromagnetic waves contains both electric and magnetic field, there is a non-zero energy density associated with it.

Q. 10. In which directions do the electric and magnetic field vectors oscillate in an electro-magnetic wave propagating along the X-axis ?★

Ans. Electric component → Y-axis

Magnetic component → Z-axis.

Q. 11. Why are microwaves considered suitable for radar systems used in aircraft navigation ?★

Ans. Microwaves of frequency 1 GHz to 300 GHz bounces from even the smallest aircraft so that they are suitable to avoid getting bombed. Microwaves can penetrate through clouds also.

Q. 12. To which part of the electromagnetic spectrum does a wave of frequency 3×10^{13} Hz belong ?★

Ans. The frequency 3×10^{13} Hz belongs to infrared region of electromagnetic spectrum.

Q. 13. To which part of the electromagnetic spectrum does a wave of frequency 5×10^{11} Hz belong ?★

Ans. A wave of frequency 5×10^{11} Hz will belong to the microwaves of electromagnetic spectrum.

Q. 14. Identify the electromagnetic waves whose wavelengths vary as :

(i) 10^{-12} m $< \lambda < 10^{-8}$ m (ii) 10^{-3} m $< \lambda < 10^{-1}$ m

Write one use for each.★

Ans. (i) **X-rays :** To detect fractures in the human body.

(ii) **Microwaves :** For aircraft navigation in RADAR systems.

Q. 15. Write the relation for the speed of electromagnetic waves in terms of the amplitudes of electric and magnetic fields.★

Ans.
$$c = \frac{E_0}{B_0}$$

Q. 16. Identify the electromagnetic waves whose wavelength lie in the range :

(i) $10^{-11} < \lambda < 10^{-14}$ m (ii) 10^{-4} m $< \lambda < 10^{-6}$ m

Write one use of each.★

Ans. (i) γ-rays.

Use : For treatment of cancer.

(ii) Infrared rays.

Use : In remote control of TV, VCR, etc.

★ are board based questions from previous year

Q. 17. Depict the fields diagram of an electromagnetic wave propagating along positive X-axis with its electric field along Y-axis.★

Ans.

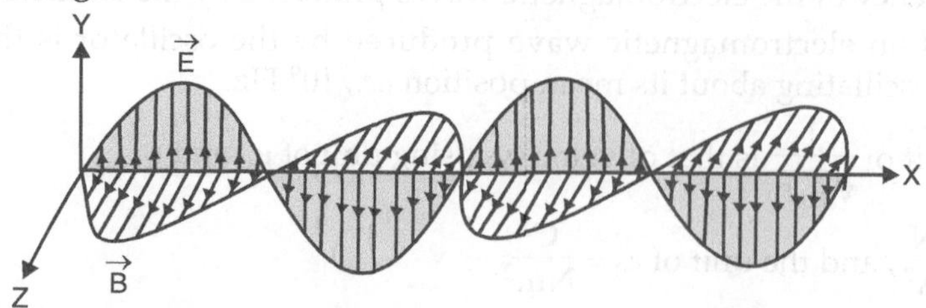

Q. 18. Fill in the blanks with appropriate answer :

(i) The frequency 5×10^{19} Hz lies in the .... region of the electromagnetic spectrum.

(ii) The electromagnetic radiation used for viewing the objects through fog is ....

(iii) The speed of electromagnetic waves in vacuum in terms of electric and magnetic fields is expressed by ....

Ans. (i) gamma region (ii) infrared rays (iii) $c = \frac{E_0}{B_0}$

Chapter 9. Ray Optics and Optical Instruments

Q. 1. Is it possible to photograph a virtual image ?

Ans. Yes, the rays which appears to diverge from the virtual image are real. Hence, they can be focussed on screen of the camera to form real image.

Q. 2. For which light colour, the refractive index of a transparent material is maximum and for which minimum.

Ans. Maximum for violet colour and minimum for red colour.

Q. 3. Does the dispersive power of a prism depend upon the angle of prism ?

Ans. No, its independent of prism.

Q. 4. A distant object is being seen by a telescope. If a fly sits on objective lens, what will be its effect on the image of obejct ?

Ans. The fly will not be seen in the image, while intensity of image will be slightly decreased, since the fly will stop some light from entering the telescope.

Q. 5. Define refractive index of the medium.

Ans. Refractive index of the medium defined as the ratio of the velocity of light waves in vacuum to velocity in medium. It is denoted by μ

$$\mu = \frac{\text{Velocity of light in vacuum}}{\text{Velocity of light in the medium}}$$

Q. 6. Do the frequency and wavelength change when light passes from rarer to denser medium ?

Ans. On change of medium only the wavelength changes and the frequency remains unchanged.

Q. 7. Define power of lens. Name the unit in which it is expressed.

Ans. The power of lens is its ability to deviate the rays towards its principal axis and is defined as the reciprocal of its focal length (expressed in metres)

$$P = \frac{1}{f} \text{ (in meter)}$$

The unit of power of lens is m^{-1} also known as dioptre (D).

Q. 8. Define the angular magnification of an optical instrument.

Ans. Angular magnification is the ratio of the angle subtended at an unaided eye by the object at the least distance of distinct vision ($M = \beta/\alpha$).

★ are board based questions from previous year

Q. 9. What is the magnification produced by the single convex lens used as a simple microscope in the normal use.

Ans. The magnification produced by single convex lens, as a simple microscope in the normal use is given by

$$M = 1 + \frac{D}{f}$$

Q. 10. Why are halos (rings) observed around the sun and the moon.

Ans. The halos are due to the refraction of light from the ice crystals present in the clouds.

Q. 11. A concave mirror is held in water. What are changes that you expect in its focal length ?

Ans. The focal length of mirror will remain same.

Q. 12. Why does sun appear red at sunrise and sunset ?★

Ans. Sun appears red at sunrise and sunset due to the least scattering of red light as it has the longest wavelength.

Q. 13. A concave lens of refractive index 1.5 is immersed in a medium of refractive index 1.65. What is the nature of the lens ?★

Ans. Since, μ_g lens < μ_m surroundings.
It behaves like a converging lens.

Q. 14. A biconvex lens made of a transparent material of refractive index 1.25 is immersed in water of refractive index 1.33. Will the lens behave as a converging or a diverging lens ? Give reason.★

Ans. In this case the biconvex lens will behave as a diverging lens because the refractive index of water (1.33) is more than that of the material (1.25) of the lens.

Q. 15. Explain two advantages of a reflecting telescope over a refracting telescope.★

Ans. Advantages of reflecting telescope over a refracting telescope are :

(a) Reflecting telescope is free from chromatic and spherical aberrations unlike refracting telescope. Thus, image formed is sharp and bright.

(b) It has a larger light gathering power so that a bright image of even far off object is obtained.

Q. 16. A section of sphere has a radius of curvature of 0.80 m. Both inside and outside surfaces have a mirror like polish. What are the focal lengths of the inside and outside surfaces ?

Ans. When inside surface has mirror like polish, the mirror is convex,

$$f = \frac{R}{2} = \frac{+0.8}{2} = 0.4 \text{ m}$$

When outside surface has mirror like polish, the mirror is concave,

$$f = \frac{R}{2} = \frac{-0.8}{2} = -0.4 \text{ m}$$

Q. 17. Why should the objective of a telescope have large focal length and large aperture ? Justify your answer.★

Ans. Magnifying power of telescope, $m = -\frac{f_o}{f_e}$

$\therefore f_o$ should be large.

Also, R.P. of telescope $= \frac{A}{1.22\,\lambda}$

$\therefore$ Aperture should also be large.

Q. 18. What is meant by short sightedness ?

Ans. The short sightedness is the defect in which a person cannot see a distant object clearly but can see nearer objects clearly. Image of a distant object formed infront of retina.

Q. 19. What is meant by long sightedness ? How can this defect be corrected ?

Ans. Long sightedness (hypermetropia) : It is the defect of eye in which a defective person cannot nearer object clearly but can see far object clearly, is called hypermetropia. To correct this defect we use convex lens. The correct diagram is given below :

★ are board based questions from previous year

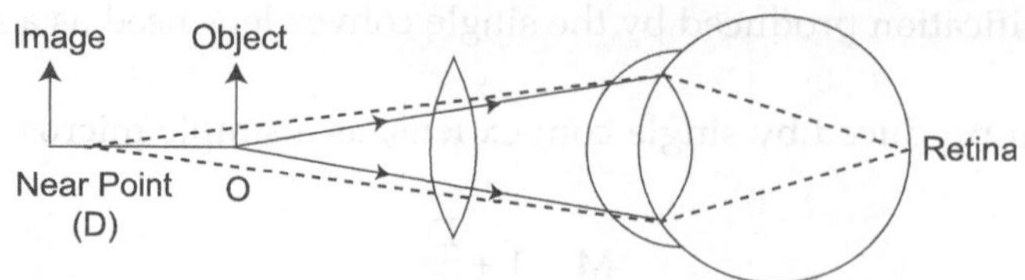

Q. 20. Fill in the blanks with appropriate answer :

(i) The dark lines seen in the solar spectrum is called ....

(ii) The branch of physics that deals with measurement of light-emitted by or reflected from objects is called ....

(iii) When a wave is reflected into rarer medium from the surface of denser medium, it undergoes a phase change of ....

(iv) The refractive index of material of lens (1.5) is greater than the refractive index of water (1.33) so it will behave as ....

Ans. (i) Fraunhofer line (ii) photometry (iii) π-radian (iv) converging lens

Chapter 10. Wave Optics

Q. 1. Sketch the wavefronts corresponding to : (i) Converging rays and (ii) Diverging rays.

Ans. (i)

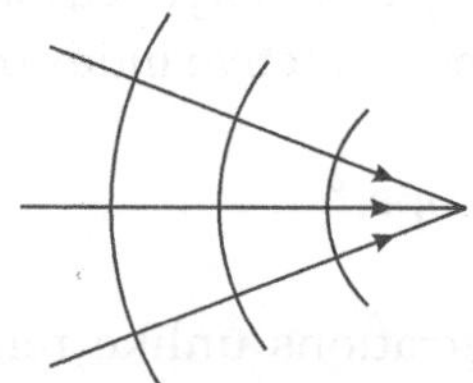

(ii)

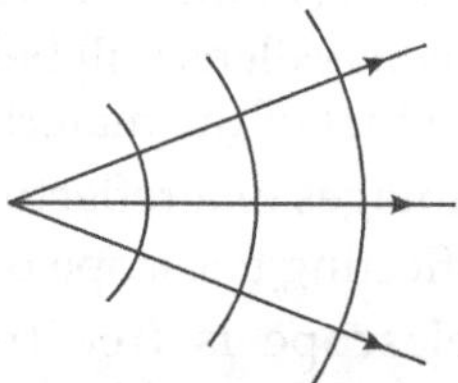

Q. 2. What are the drawbacks of wave theory of light ?

Ans. The wave theory of light is based on the existence of ether medium. However, Michelson and Morley's experiment proved that the ether medium did not exist at all. It led to the wave theory of light.

Q. 3. In the wave picture of light intensity of light is determined by the square of amplitude of wave. What determines the intensity of light in the photon picture of light ?

Ans. In the wave picture of light, intensity of light at a point is determined by the number of photons incident per unit area around that point.

Q. 4. What are coherent sources of light ?

Ans. Two sources are said to be coherent if they have exactly same frequency and have zero or constant phase difference.

Q. 5. Can two independent sources of light be coherent ?

Ans. No.

Q. 6. State the essential conditions, which must be satisfied for two light sources to be coherent.

Ans. (i) The two light waves should be of same wavelength.

(ii) The two light waves should either be in phase or should have a constant phase difference.

Q. 7. State the reason, why two independent sources of light cannot be considered as coherent sources.

Ans. No. It is because, the phase difference between the light waves from the two independent sources keeps on changing continuously.

Q. 8. What happens to fringe width, when the separation between the sources is increased ?

Ans. We know that

$$\beta = \frac{D\lambda}{d}.$$

The fringe width decreases, when the separation d between the two sources in the Young's double slit experiment is increased.

Q. 9. State the condition for destructive interference.

Ans. Path difference,

$$x = (2n+1)\frac{\lambda}{2}, \text{ where } n = 0, 1, 2, \ldots\ldots$$

Q. 10. Can white light produce interference ? What is its nature ?

Ans. Yes, white light can produce interference. However, only first few coloured fringes can be detected and beyond that the fringes of different colours overlap to produce general illumination, white in colour.

Q. 11. What is diffraction of light ?

Ans. The phenomenon of bending of light round the sharp corners and its spreading into the regions of the geometrical shadow is called diffraction.

Q. 12. Does phenomenon of diffraction help us to decide about the longitudinal or transverse nature of light ?

Ans. No.

Q. 13. State the condition for diffraction of light to occur.

Ans. The size of the obstacle should be of the order of the wavelength of light used.

Q. 14. What is unpolarised light ?

Ans. The light, which has vibrations in all the planes perpendicular to the direction of its propagation, is called unpolarised light.

Q. 15. What do you mean by polarisation of light ?

Ans. The phenomenon, due to which the vibrations of light are restricted in a particular plane, is called the polarisation of light.

Q. 16. Write simple uses of polaroids.

Ans. Polaroids are used in sun glare, in window planes, in wind shields of automobiles.

Q. 17. Unpolarised light is incident on a plane glass surface. What should be the angle of incidence so that the reflected and refracted rays are perpendicular to each other ? **[NCERT]**

Ans. When $i + r$ equal to $\pi/2$, $\tan i_B = \mu = 1.5$. This gives $i_B = 57°$. This is the Brewster's angle for air to glass interface.

Q. 18. Fill in the blanks with appropriate answer :

(i) The relation between refractive index and angle of polarisation is ....

(ii) A ray of light is incident on a medium at polarising angle. The angle between the reflected and refracted rays is ....

(iii) Sound wave cannot be polarised because they are .... in nature.

(iv) The reciprocal of the limit of resolution is called its ....

(v) The relation between path difference and phase difference is ....

(vi) The phase difference between any two points on a wavefront is ....

(vii) The wavefront diverging from a point of source is .... in shape.

(viii) Unpolarised light of intensity I_0 is incident on two crossed polaroids. the intensity of light transmitted by the combination will be ....★

(ix) In a linearly polarized light, the electric vector oscillates along a direction perpendicular to the aligned molecules of the polaroid, called .... axis.★

Ans. (i) $\mu = \tan i_p$ (ii) 90° (iii) longitudinal

(iv) resolving power (v) Phase difference $= \frac{2\pi}{\lambda} \times$ path difference

(vi) zero (vii) spherical (viii) zero

(ix) Pass

★ are board based questions from previous year

Chapter 11. Dual Nature of Radiation and Matter

Q. 1. Does each and every incident photon essentially eject an electron ?

Ans. No, it may be absorbed in some other way. Only about 1% of incident photons eject out electrons.

Q. 2. Does the X-rays shows the phenomenon of photoelectric effect ?

Ans. Yes.

Q. 3. Name a phenomenon or an experiment which proves :
(i) Particle nature of electromagnetic radiations.
(ii) Wave nature of particles.
(Description of the phenomenon/experiment is not required)

Ans. (i) Photoelectric effect.
(ii) Davisson and Germer experiment.

Q. 4. What is de-Broglie wavelength ?

Ans. The wavelength of the waves associated with a beam of moving particles is called de-Broglie wavelength.

$$\text{de-Broglie wavelength } (\lambda) = \frac{h}{p} = \frac{h}{mv}$$

Q. 5. What is the threshold frequency for a metal whose work function is ϕ_0 ?

Ans. ϕ_0/h, where h is the Planck's constant.

Q. 6. Is photoelectric emission a spontaneous process or it is instantaneous process.

Ans. No, it is not a spontaneous process, it is an instantaneous process.

Q. 7. Write the expression for the de-Broglie wavelength associated with a charged particle having charge 'q' and mass 'm', when it is accelerated by a potential V.

Ans. Let v be the velocity gained by the given charge particle when it is accelerated through a potential difference of 'V' volts kinetic energy of the particle $= \frac{1}{2}mv^2$.

Kinetic energy of the particle = Work done on the particle by electric field.

$$\therefore \quad \frac{1}{2}mv^2 = qV$$

or
$$\frac{P^2}{2m} = qV$$

$$P = \sqrt{2mqV}$$

Q. 8. Define intensity of radiation on the basis of photon picture of light. Write its S.I. unit.

Ans. The intensity of radiation can be defined as the energy related with photons emitted from a unit surface area in unit time. Its S.I. units is joule/metre^2 second (J/m^{2}s).

Q. 9. The historic experiment on the diffraction of electrons confirmed a new nature of electron. What is this new nature of electron ? Who had proposed it ?

Ans. de-Broglie proposed wave nature of the electron.

According to wave particle duality, an electron (a charged particle) can also behave as a wave of wavelength λ.

$$\therefore \quad \lambda = \frac{h}{p}$$

where h is called Planck's constant and p is the momentum.

Q. 10. What do you understand by the work function of a metal ? Name the units in which the work function of a metal is commonly expressed.

Ans. The work function of a metal is defined as the energy required to eject an electron away from the surface of the metal, without imparting any kinetic energy to it. It is quite large for heavier elements like platinum whereas for other elements like alkali metals, it is quite small. It is generally denoted by W_0 or ϕ_0 and usually measured in eV.

Q. 11. Plot a graph between stopping potential and frequency.

Ans.

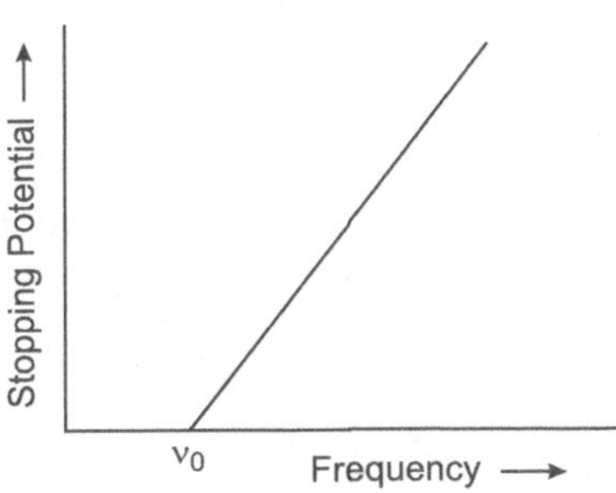

Q. 12. Plot a graph between de-Broglie wavelength and the momentum of a particle.

Ans.

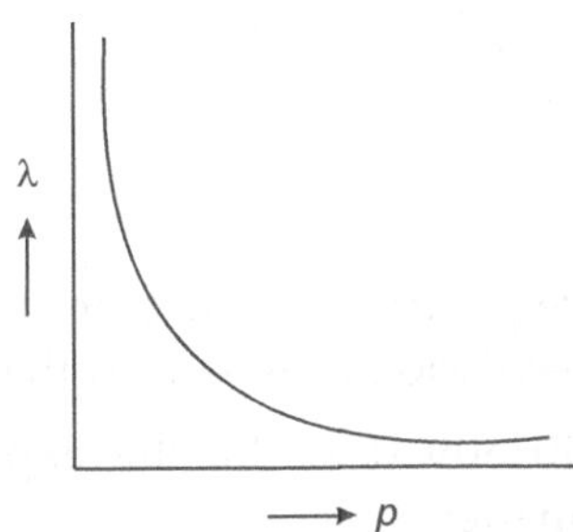

Q. 13. Every metal has a definite work function. Why do photoelectrons not come out with the same energy if incident radiation is monochromatic ? Why is there an energy distribution of photo-electrons ?

Ans. Electrons are present in a continuous band of levels, different electrons require different energy to get out of the atom. For this reason, the electrons knocked off by a monochromatic radiation possesses different energies.

Q. 14. Why are alkali metals most suitable for photoelectric emission ?

Ans. Alkali metals exhibit photoelectric effect with visible light because the work function of an alkali metal is quite low. Therefore, they emit photoelectrons even when visible light falls on them.

Q. 15. It is harder to remove a free electron from copper than from sodium. Which metal has greater work function ? Which metal has higher threshold wavelength ?

Ans. Since, it is difficult to remove an electron from copper than sodium, thus work function of copper is greater. Threshold wavelength will be more for sodium.

Q. 16. An electron and a proton possess the same kinetic energy. Which of the two has greater de-Broglie wavelength ? Write down the formula used.

Ans. Since,

$$\lambda = \frac{h}{\sqrt{2mK}}$$

or

$$\lambda \propto \frac{1}{\sqrt{m}}$$

The mass of an electron is less than that of a proton. Hence, the de Broglie wavelength of electron will be greater than that of proton.

Q. 17. What is 'threshold wavelength' in photoelectric effect ?

Ans. If the wavelength of light incident on a metal exceeds a certain value, no photoelectrons are emitted from the metal. The maximum wavelength of light capable of ejecting photoelectrons from a metal is called the 'threshold wavelength' for that metal.

Q. 18. How does the stopping potential depends upon the frequency of the incident light ?

Ans. The stopping potential V_0 is the measure of the maximum kinetic energy E_k of the photoelectrons ($E_k = eV_0$). Hence, according to Einstein's equation $E_k = h(\nu - \nu_0)$, the stopping potential increases linearly with increasing frequency of light.

Q. 19. The graph shows variation of stopping potential V_0 versus frequency of incident radiation ν for two photosensitive metals A and B. Which of the two metals has higher threshold frequency and why ?★

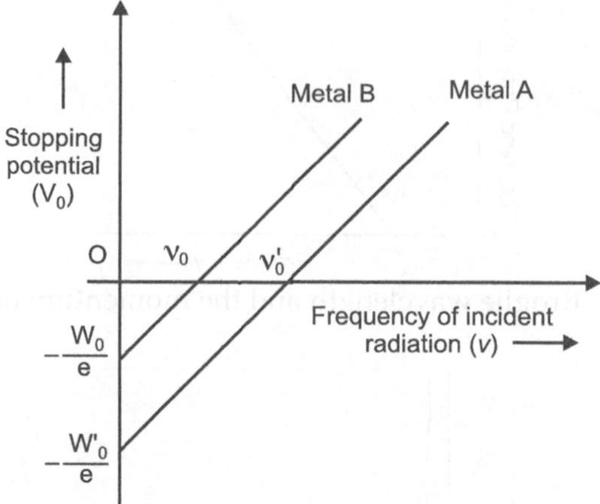

Ans. Metal A has higher threshold frequency because from the graph it is clear that the minimum frequency required to start photoemission is more in A than that of B.

Q. 20. What are de-Broglie waves ? Briefly describe the experiment of Davisson and Germer to demonstrate the wave nature of matter. **[NCERT]**

Ans. In 1924, de-Broglie introduced a bold hypothesis which states that, like radiation, matter should also have a dual nature. According to him, all material particles in motion have wave nature also. The waves associated with moving particles of matter are called 'de Broglie waves' or 'matter waves'.

Q. 21. Draw a plot showing the variation of photoelectric current versus the intensity of incident radiation on a given photosensitive surface.★

Ans.

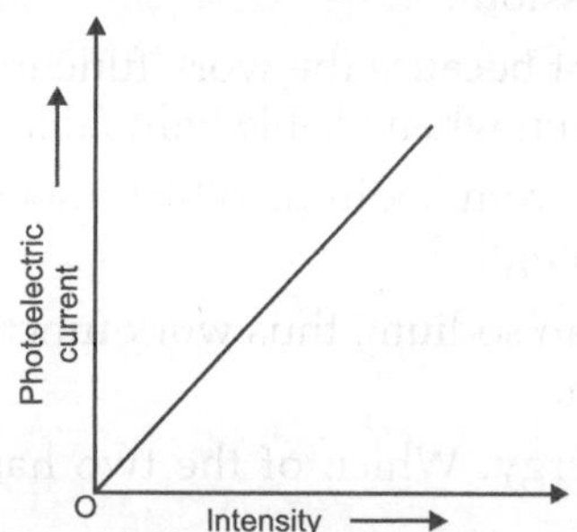

Q. 22. State two important features of Einstein's photoelectric equation.★

Ans. (i) By Einstein's photoelectric equation we can explain the laws of photoelectric emission.

(ii) By this we can find the value of Planck's constant and work function.

Q. 23. Fill in the blanks with appropriate answer :

(i) The relation between wavelength and momentum of moving particles is ....

(ii) The phenomenon which shows the quantum nature of electromagnetic radiation is ....

(iii) For the emission of photoelectrons from the cesium metal the minimum energy of light photon incident on the metal surface should be ....

(iv) .... is that minimum negative potential applied to the anode which completely stops the photoelectric current.

Ans. (i) $\lambda = \frac{h}{p}$ (ii) photoelectric effect (iii) 2eV (iv) cut off or stopping potential

★ are board based questions from previous year

Chapter 12. Atoms

Q. 1. On the basis of Bohr's theory, draw energy level diagram in hydrogen spectrum.

Ans. **Energy level diagram :**

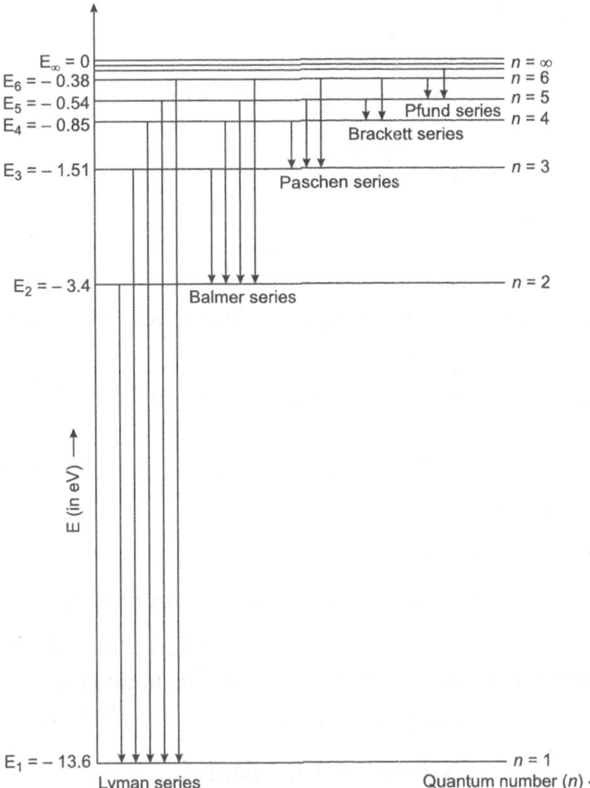

Note : Energy of electron in the infinite energy level is maximum *i.e.*, 0 eV because below this value every energy level has negative value.

Q. 2. State any two postulates of Bohr's theory of hydrogen atom.

Ans. Postulates of Bohr's theory of hydrogen atom :

(i) In an atom, the electrons revolve around the nucleus in certain definite circular paths called orbits or shells.

(ii) Each shell or orbit corresponds to a definite energy. Therefore, these circular orbits are also known as energy levels or energy shells.

Q. 3. Write the expression for Bohr's radius in hydrogen atom ?

Ans. The radius of the innermost orbit ($n = 1$) of the H-atom is called Bohr's radius. It is given by

$$r_0 = 4\pi\varepsilon_0 \cdot \frac{h^2}{4\pi^2 m_e e^2}$$

Q. 4. What is the ratio of radii of the orbits corresponding to first excited state and ground state of the hydrogen atom ?

Ans. Radius of nth orbit, $r_n \propto n^2$.

The ground state corresponds to $n = 1$ and the first excited state corresponds to $n = 2$.

$$\therefore \quad \frac{r_2}{r_1} = \frac{2^2}{1^2}$$

or

$$\frac{r_2}{r_1} = 4$$

Q. 5. The wavelength of some of the spectral lines obtained in hydrogen spectrum are 9,546 Å, 6,463 Å and 1,216 Å, which one of these wavelengths belongs to Lyman series ?

Ans. Lyman series lies in ultraviolet region. Therefore, wavelength of the spectral lines in Lyman series must be less than the wavelengths in visible region *i.e.*, 3,900 Å. Therefore, spectral line of wavelength 1,216 Å lies in Lyman series.

Q. 6. Name the spectral series of hydrogen atom lying in visible region.

Ans. Balmer series lies in visible region.

Q. 7. Name the spectral series of hydrogen atom lying in infra red region.

Ans. Paschen series lies in infrared region.

Q. 8. What is the energy possessed by an electron in $n = \infty$?

Ans. Energy of an electron in the nth orbit hydrogen atom is given by

$$E_n = -\frac{13.6}{n^2} \text{ eV}$$

Here, $n = \infty$

$$\therefore \quad E_\infty = -\frac{13.6}{\infty^2} = 0$$

Q. 9. Write two important limitations of Rutherford nuclear model of the atom.

Ans. Limitations :

(i) It is not in accordance with the Maxwell's theory and could not explain the stability of an atom.

(ii) It did not say anything about the arrangement of electrons in an atom.

Q. 10. Fill in the blanks with appropriate answer :

(i) The ionisation potential of hydrogen atom is ....

(ii) The series of hydrogen spectrum lying in ultra-violet region is ....

(iii) The total energy of an electron in the first excited state of the hydrogen atom is about – 3.4 eV. The potential energy of the electron in this state is ....

Ans. (i) 13.6 eV (ii) Lyman series (iii) – 6.8 eV

Chapter 13. Nuclei

Q. 1. What percentage of a radioactive substance will be left after four half-life periods ?

Ans. Given, $n = 4$

We have,

$$N = N_0 (1/2)^n$$

$$\frac{N}{N_0} = \left(\frac{1}{2}\right)^4 \text{ or } \frac{N}{N_0} \times 100 = \frac{1}{16} \times 100$$

$N = 6.25\%$ of numbers of radioactive substance.

Q. 2. Among alpha, beta and gamma radiations, which are affected by a magnetic field ?

Ans. Among alpha, beta and gamma radiation, alpha and beta radiations are affected by the magnetic field, as they are charged particles. Alpha particles are positively charged and beta particles are negatively charged. Gamma radiations suffer no deflection in the magnetic field as they are neutral in nature.

Q. 3. What do you understand by the terms radioactivity and radioactive substances ? Name two radioactive elements.

Ans. The phenomenon of emission of invisible radiations by certain heavy elements like uranium, radium etc., is called radioactivity and substances which emit such radiation are called radioactive substances. The invisible rays emitted by the radioactive substances consist of three different types of radiation. These are :

(i) Alpha-particles (ii) Beta-particles (iii) Gamma-rays

Two radioactive elements are :

(i) Thorium (ii) Plutonium

Q. 4. What is the significance of binding energy per nucleon of a nucleus ?

Ans. Binding energy per nucleon : It is the average energy required to remove a nucleon from the nucleus to infinite distance.

Higher the average binding energy/nucleon, greater is the stability of the nucleus.

Significance of binding energy : The stability of the nucleus is determined by the binding energy per nucleon.

Q. 5. What is meant by the binding energy of the nucleus ? How is it related to mass defect ?

Ans. The binding energy of a nucleus is the energy required to take its nucleons away from one another. Thus, the binding energy of the nucleus is the measure of the stability of the nucleus.

We know,

$$\text{Mass defect, } \Delta m = [Zm_p + (A - Z)m_n] - M$$

Where A, Z, m_p, m_n and M are the atomic number, mass number, mass of the proton, mass of the neutron and mass of the nucleus respectively.

From Einstein's mass energy equivalence,

The binding energy in terms of mass defect is given by

$$\text{B.E.} = \Delta mc^2$$

$$\text{B.E.} = \Delta m \times c^2 = [Zm_p + (A - Z)\, m_n - M]c^2 \text{ J.}$$

Q. 6. Why is the neutron so effective as bombarding particle ?

Ans. If we use positively charged particles such as proton, α-particle, they experience large repulsion due to atomic nuclei. On the other hand, the negatively charged particles such as electron will be repelled by the electron cloud around the nucleus. However, neutrons having no charge can penetrate the target without being repelled by the atomic nucleus or the electron cloud. The neutron increases the neutron to proton ratio of the nucleus and tends to make it unstable and hence, causes the nuclear reaction.

Q. 7. How is radioactivity taken into the body ?

Ans. Radioactivity is taken into the body in three ways :

(i) Inhalation (breathing it in if it is in gaseous form)

(ii) Ingestion (through the mouth, or if it administered as a drug by injection).

(iii) Absorption (through the skin).

Q. 8. Show graphically the variation in the mass of radioactive substance with time. Mark the values of time on the X-axis in terms of half-life (T, 2T, 3T etc.) and the corresponding values of mass on the Y-axis.

Ans. Graph showing the variation in mass of a radioactive substance with time is shown alongside M_0 shows original mass.

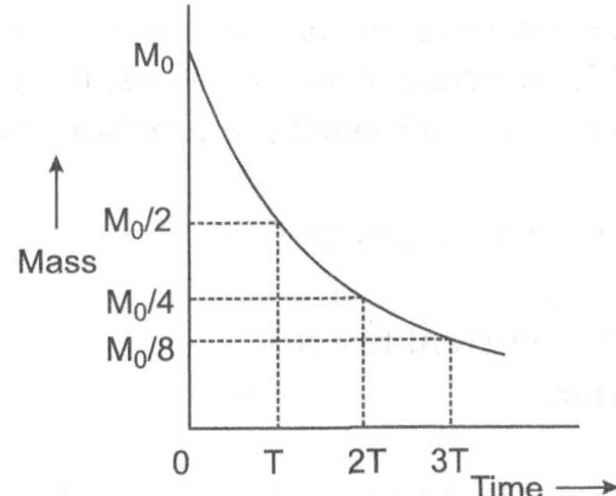

Q. 9. What do you mean by packing fraction ?

Ans. **Packing fraction :** The mass defect per nucleon is known as "packing fraction."

$$\text{Packing fraction} = \frac{\Delta m}{A}$$

Q. 10. How is the principle of conservation of momentum obeyed in the process of β-emission ?

Ans. Before the emission of β-particle, the momentum of the nucleus is zero. Together with β-particle, anti-neutron is emitted with a momentum such that the vector sum of momentum of anti-neutron, β-particle and the new nucleus is zero.

Q. 11. Write down the laws of radioactive decay.

Ans. There are two important radioactive decay laws :

(a) In any radioactive transformation, either an α-particle or a β-particle is emitted but it is never, that particles are ejected simultaneously.

(b) At any instant, the rate of decay of radioactive atoms is proportional to the number of atoms present at that instant.

Q. 12. Fill in the blanks with appropriate answer :

(i) α-particles carry .... units of positive charge.

(ii) A nuclear reaction which produces particles that started the reaction originally is called ....

(iii) Name the particle emitted spontaneously in the following nuclear reaction :

$$^{32}_{15}P \longrightarrow {}^{32}_{16}S + \bar{v} + ... \bigstar$$

(iv) In β-decay, a neutron is converted into a proton, e^- and ....★

Ans. (i) 2 (ii) chain reaction (iii) β^- - particle (– 1°e) (iv) $n \longrightarrow p + e^- + \bar{v}_e$

Chapter 14. Semiconductor Electronics : Materials, Devices and Simple Circuits

Q. 1. In an *n*-type silicon, which of the following statement is true :

(i) Electrons are majority carriers and trivalent atoms are the dopants.

(ii) Electrons are minority carriers and pentavalent atoms are the dopants.

(iii) Holes are minority carriers and pentavalent atoms are the dopants.

(iv) Holes are majority carriers and trivalent atoms are the dopants. **[NCERT]**

Ans. (iii) The electrons are the majority carriers in *n*-type silicon while the holes are the minority carriers. An *n*-type semiconductor is obtained when pentavalent atoms, such as phosphorus, are doped in silicon atoms.

Q. 2. Which of the statements given in previous question is true for *p*-type semiconductors ? **[NCERT]**

Ans. (iv) The holes are the majority carriers in *p*-type semiconductor while the electrons are the minority carriers. A *p*-type semiconductor is obtained when trivalent atoms, such as aluminium, are doped in silicon atoms.

Q. 3. In an unbiased *p-n* junction, holes diffuse from the *p*-region to *n*-region because :

(i) free electrons in the *n*-region attract them.

(ii) they move across the junction by the potential difference.

(iii) hole concentration in *p*-region is more as compared to *n*-region.

(iv) All the above. **[NCERT]**

Ans. (iii) The diffusion of charge carriers across a junction occurs from the region of higher concentration to the region of lower concentration. Thus, the *p*-region has greater concentration of holes than the *n*-region. Hence, in an unbiased *p-n* junction, holes diffuse from the *p*-region to the *n*-region.

Q. 4. When a forward bias is applied to a *p-n* junction, it

(i) raises the potential barrier.

(ii) reduces the majority carrier current to zero.

(iii) lowers the potential barrier.

(iv) None of the above. **[NCERT]**

Ans. (iii) As soon as forward bias is applied to a *p-n* junction, it lowers the value of potential barrier. In the case of a forward bias, the potential barrier opposes the applied voltage. Hence, the potential barrier across the junction gets reduced.

Q. 5. Define Fermi energy.

Ans. The highest energy level occupied by electrons at absolute zero temperature in a partially filled conduction band is called the 'Fermi level' and the corresponding energy is called the Fermi energy.

★ are board based questions from previous year

Q. 6. Define the V-I characteristics of a junction diode.

Ans. The V-I characteristic of the diode is the graph drawn between the voltage V and current I in forward bias and reverse bias of a junction diode.

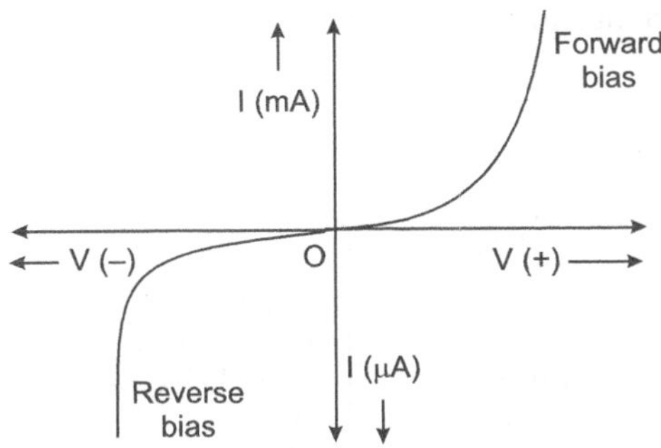

Q. 7. What is the Zener diode ?

Ans. A zener diode is a reverse biased heavily doped *p-n* junction diode, which is operated in breakdown region, where the current is limited by both external resistance and power dissipation of the diode.

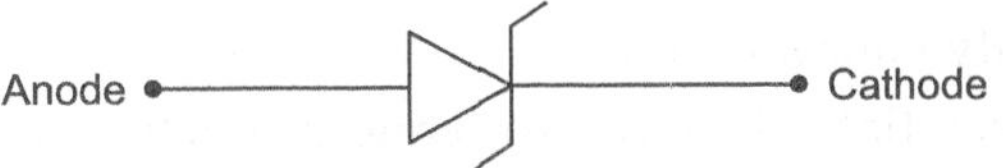

Q. 8. What is the principle of zener diode ?

Ans. The principle behind the working a zener diode lies in the cause of breakdown of a diode in reverse biased condition.

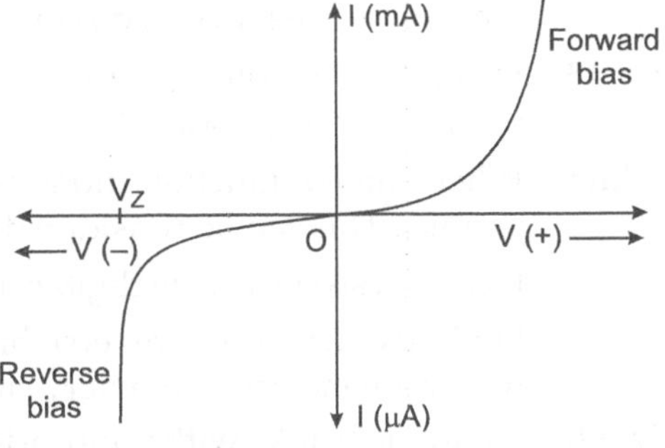

We know that reverse current is due to the flow of electrons (minority carriers) from p to n and holes from n to p. As the reverse bias voltage is increased, the electric field at the junction becomes significant. When the reverse bias voltage $V = V_Z$, then the electric field strength is high enough to pull valence electrons from the host atoms on the p-side which are accelerated to n-side. These electrons causes high current at breakdown.

Q. 9. Name the junction diode whose I-V characteristics are drawn below :★

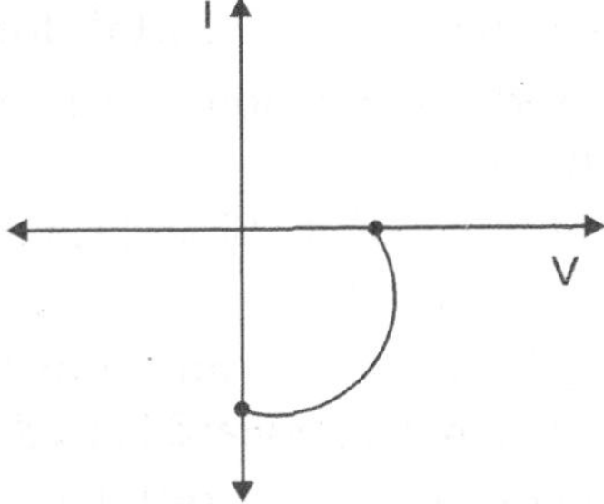

Ans. This is the characteristic of solar cell.

Q. 10. Graph showing the variation of current versus voltage for a material GaAs is shown in the figure. Identify the region of :★

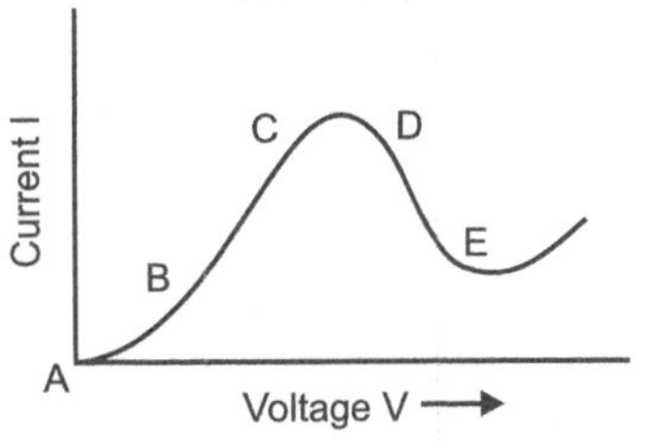

(i) negative resistance

(ii) where Ohm's law is obeyed.

Ans. (i) DE is the region of negative resistance because the slope of curve in this part is negative.

★ are board based questions from previous year

(ii) BC is the region where Ohm's law is obeyed because in this part, the current varies linearly with the voltage.

Q. 11. Plot a graph showing variation of current versus voltage for the material Ga.

Ans. Current-Voltage characteristics graph for Ga :

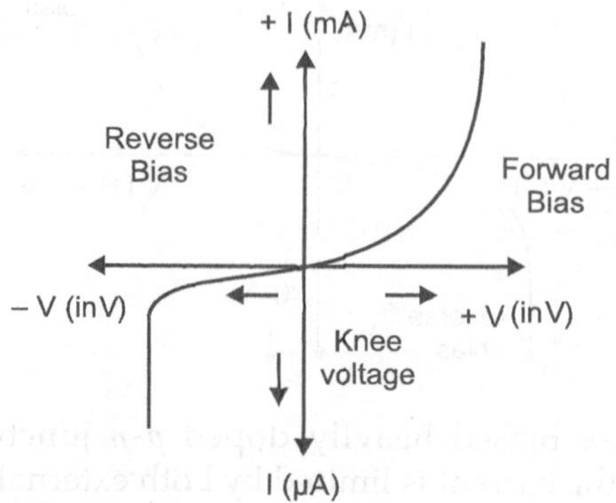

Q. 12. Can a slab of *p*-type semiconductor be physically joined to another *n*-type semiconductor slab to form p-n junction ? Justify your answer.★ **[NCERT]**

Ans. No, any slab, howsoever flat will have roughness much larger than the interatomic crystal spacing (2Å to 3Å) and hence continuous contact at the atomic junction will not be possible. The junction will behave as discontinuous for the following charger carriers.

Q. 13. In a *p-n* junction diode the forward bias resistance is low as compared to the reverse bias resistance. Give reason.★

Ans. When the *p-n* junction diode is forward biased, the diode's depletion layer or potential barrier shrinks. This can be considered as a practical short circuit.

Hence, resistance is negligible and current flow in forward bias is maximum.

Similarly, when it is reverse bias, depletion layer width (or potential barrier) increases. Hence, resistance increases and very small amount of current flow in reverse bias.

Q. 14. Fill in the blanks with appropriate answer :

(i) When semiconductors are doped with pentavalent impurities then donor level is created and .... decreases.

(ii) The forbidden energy gap remains unchanged with .... in temperature.

(iii) The energy band gap of Si is about .... while for GaAs, it is about ....

(iv) .... is a forward biased *p-n* junction which converts electrical energy into optical energy of infrared and visible light.

(v) The ...., a property of materials C, Si and Ge depends upon the energy gap between their conduction and valence bands.★

(vi) The ability of a junction diode to .... an alternating voltage, is based on the fact that it allows current to pass only when it is forward biased.★

Ans. (i) energy gap (ii) increase (iii) 1.1 eV, 1.53 eV (iv) LED

(v) conductivity (vi) rectify

★ are board based questions from previous year

Question Set 3

Short Answers

Chapter 1. Electric Charges and Fields

Q. 1. A spherical conducting shell of inner radius r_1 and outer radius r_2 has a charge Q. A charge q is placed at centre of the shell.

(i) What is the surface charge density on :

(a) inner surface ?

(b) outer surface of the shell ?

(ii) Write expression for the electric field at a point $X > r_2$ from the centre of shell.

Ans. (i) As per question, if we draw spherical shell it will be like fig.

(a) The $+q$ charge at the centre of a shell induces $-q$ charge at the inner surface of the shell. Therefore, total charge on the inner surface of the shell is $-q$.

Surface charge density at inner surface of the shell,

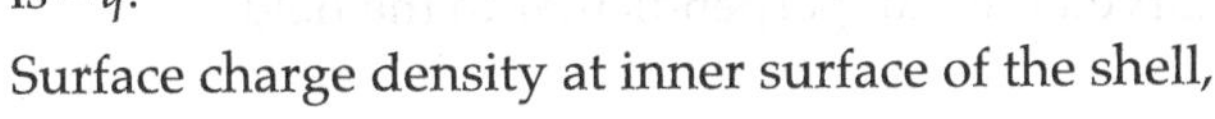

$$\sigma_1 = \frac{-q}{4\pi r_1^2}$$

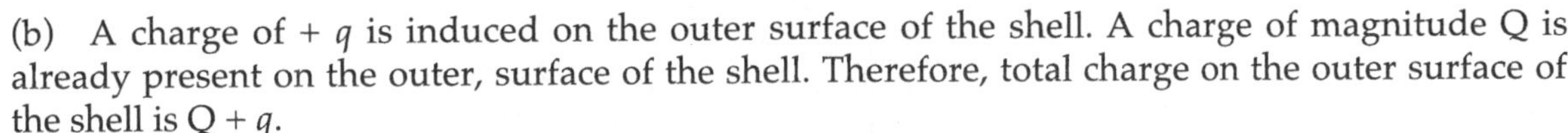

(b) A charge of $+q$ is induced on the outer surface of the shell. A charge of magnitude Q is already present on the outer, surface of the shell. Therefore, total charge on the outer surface of the shell is $Q + q$.

Surface charge density on the outer surface,

$$\sigma_2 = \frac{Q+q}{4\pi r_2^2}$$

(ii) For $X > r_2$,

$$E = \frac{1}{4\pi\varepsilon_0}\frac{Q+q}{x^2}$$

Q. 2. Explain, using suitable diagrams, the difference in the behaviour of a (i) conductor and (ii) dielectric in the presence of external electric field. Define the terms polarization of a dielectric and write its relation with susceptibility.★

Ans. (i) **Conductor :**

Let E_0 = external field, and

E_m = internal field created by the redistribution of electrons inside the metal.

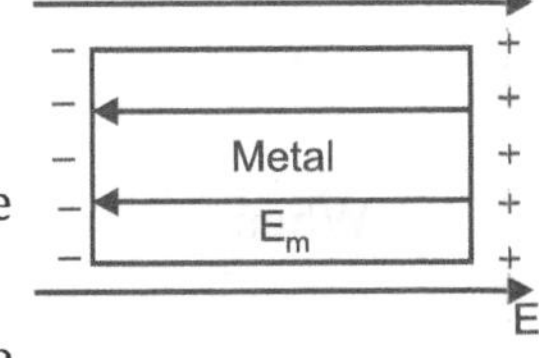

When a conductor like a metal is subjected to external electric field, the electrons experience a force in the opposite direction collecting on the left hand side.

A positive charge is therefore induced on the right hand side. This creates an opposite electric field (E_m) that balances out (E_0).

∴ The net electric field inside the conductor becomes zero.

Here, E_0 is the external field and E_m is the internal field created by the re-distribution of electrons inside the metal.

(ii) **Dielectric** : When external electric field is applied, dipoles are created for non-polar dielectrics and in case of polar dielectrics dipoles are aligned. An internal field is created such that it reduces the external field. The placement of dipoles is as shown in the given figure.

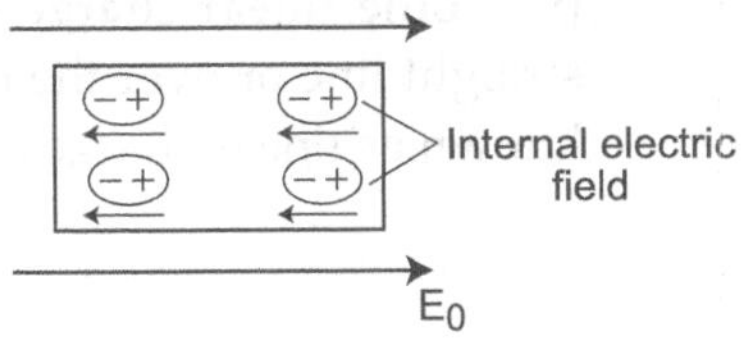

★ are board based questions from previous year

The dipole moment per unit volume of the polarised dielectric is known as the polarisation of a dielectric. It is denoted by P.

The formula for polarization is given by,

$$P = \chi_0 \varepsilon_0 E; \chi_0 \text{ is susceptibility, and}$$

E is the electric field.

Q. 3. Differentiate between uniform electric field and non-uniform electric field. Also draw the diagram.

Ans. **Uniform electric field :** In case of uniform electric field, the field lines are straight and equally spaced. Therefore, equipotential surfaces will be parallel planes at right angles to field lines.

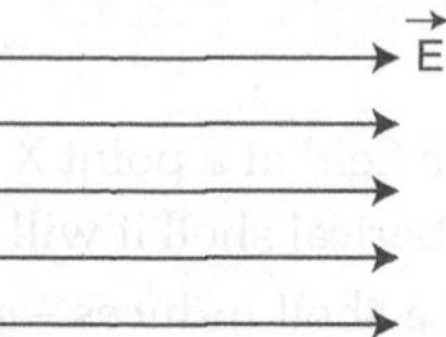

Non-uniform electric field : In case of non-uniform electric field, the field lines are not straight. In that case, equipotential surfaces are curved but still perpendicular to the field.

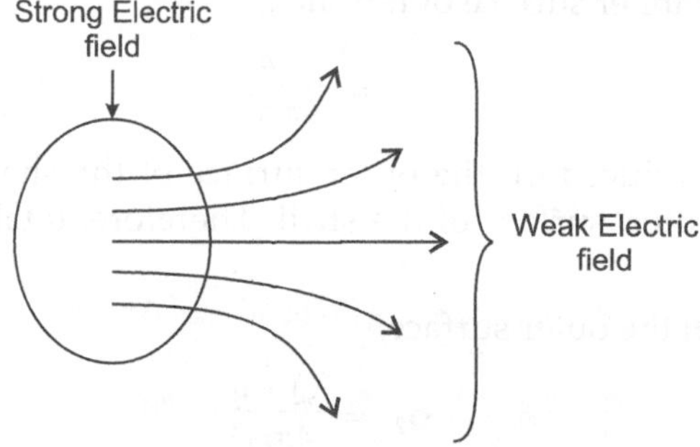

Q. 4. Define one coulomb, what is SI-CGS unit of charge. Give dimensional formula of Coulomb as well.

Ans. One coulomb is the charge transferred in one second across a section of wire carrying current of one ampere.

As
$$i = \frac{dq}{dt}$$
$$\int dq = \int i\, dt$$
$$q = it$$
When
$$i = 1A$$
$$t = I \text{ sec}$$
Then
$$q = I\,C$$

Coulomb is SI unit of charge. In CGS, charge is measured in terms of stat coulomb (esu). 1 coulomb = 3 × 10^9 stat coulomb. Dimensional formula for coulomb is [AT].

Q. 5. What are different types of charge distributions.

Ans. Other than discrete distribution of charges, continuous charges are distributed over length, area and volume which gives following distributions.

(i) **Line/linear charge distribution :** If the charge distributed along straight line or over the circumference of circle then the distribution is known as linear charge distribution.

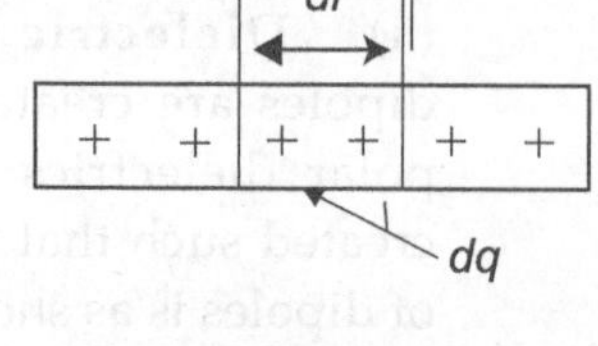

$$\lambda = \frac{dq}{dl}$$
$$q = \int \lambda \cdot dl$$

(ii) **Surface charge distribution :** If charge is distributed over surface area then the distribution is known as surface charge distribution.

$$\sigma = \frac{dq}{ds}$$

$$q = \int_s \sigma \cdot ds$$

(iii) **Volume charge distribution :** If the charge is distributed over a volume, then the distribution is called, volume charge distribution.

$$\rho = \frac{dq}{dV}$$

$$q = \int_V \rho \cdot dV$$

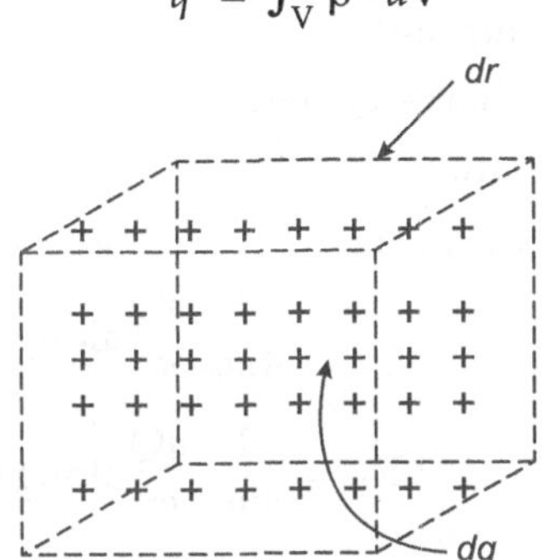

Q. 6. A charge Q is distributed over a metal sphere of radius R. What is the electric field and electric potential at the centre ?

Ans. According to the question, the charge distributed on the metal sphere is on the surface of metal sphere and from property of electric field we know that there is no electric field inside the metal conducting sphere. Hence, electric field inside the charge conducting sphere is zero *i.e.*, E = 0, and the electric field at centre is also zero.

Now, at inner point inside the sphere it is considered that the whole charge is concentrated at one point. Hence, the electric potential inside sphere at any point is

$$V = V_{surface} = \frac{1}{4\pi\varepsilon_0} \cdot \frac{Q}{R} \text{ volt.}$$

Q. 7. An electric dipole is held in a uniform electric field :

(i) Show that net force acting on it is zero.

(ii) The dipole is aligned parallel to the field. Find the work done is rotating it through the angle of 180°.

Ans. (i) Consider an electric dipole consisting of two equal and opposite point charges $-q$ and $+q$ at positions A and B respectively, separated by a distance $2a$.

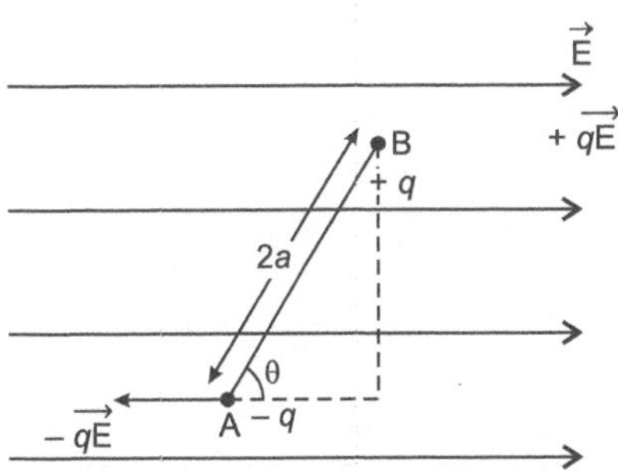

Dipole moment of a dipole is given by $|\vec{P}| = q \cdot 2a$ consider that this dipole is placed in uniform external electric field $\vec{E}$ at an angle with the direction of $\vec{E}$) θ.

Force on charge $-q$ at A $= -q\,\vec{E}$, in a direction opposite to $\vec{E}$.

Force on charge $+q$ at B $= +q\,\vec{E}$, along the direction of $\vec{E}$.

Net force on dipole $= -q\vec{E} + q\vec{E} = 0$

(ii) Net work done on dipole,

$$W = \Delta U = pE(\cos\theta_1 - \cos\theta_2)$$
$$= pE(\cos 0° - \cos 180°)$$
$$= 2pE$$

Q. 8. If a body contains n_1 electrons and n_2 protons then what is the net charge on the body ?

Ans. The charge on the electron and proton is same in magnitude but opposite in sign. Electron has negative charge and proton has positive charge.

Charge on a proton $= +1.6 \times 10^{-19}$ C

Charge on an electron $= -1.6 \times 10^{-19}$ C

Total charge on the body = (No. of proton × Charge on a single proton) + (No. of electrons × Charge on a single electron)

$= n_2 \times 1.6 \times 10^{-19}\text{ C} - n_1 \times 1.6 \times 10^{-19}\text{ C}$

$= 1.6 \times 10^{-19}(n_2 - n_1)\text{ C} = (n_2 - n_1)\text{ C}$

Q. 9. Four point charges Q, q, Q and q are placed at the corners of a square of side 'a' as shown in the figure. Find the★

(i) resultant electric force on a charge Q, and

(ii) potential energy of this system.

Ans. (i) Force on charge Q at B due to charge q

$$F_1 = \frac{1}{4\pi\varepsilon_0}\frac{qQ}{a^2} \text{ along } \vec{BY}$$

$$F_2 = \frac{1}{4\pi\varepsilon_0}\frac{qQ}{a^2} \text{ along } \vec{BX}$$

Resultant force, $F = \sqrt{F_1^2 + F_2^2}$ along $\vec{BO}$

$$= \sqrt{2\left(\frac{1}{4\pi\varepsilon_0}\frac{qQ}{a^2}\right)^2} = \frac{qQ\sqrt{2}}{4\pi\varepsilon_0 a^2}$$

Force due to charge Q $= F_3 = \frac{1}{4\pi\varepsilon_0}\frac{QQ}{(\sqrt{2}a)^2}$ along $\vec{BO}$

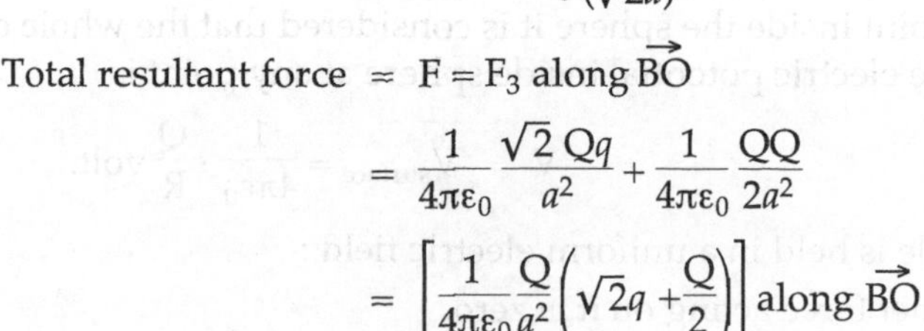

Total resultant force $= F = F_3$ along $\vec{BO}$

$$= \frac{1}{4\pi\varepsilon_0}\frac{\sqrt{2}Qq}{a^2} + \frac{1}{4\pi\varepsilon_0}\frac{QQ}{2a^2}$$

$$= \left[\frac{1}{4\pi\varepsilon_0}\frac{Q}{a^2}\left(\sqrt{2}q + \frac{Q}{2}\right)\right] \text{ along } \vec{BO}$$

(ii) Total potential energy of the system.

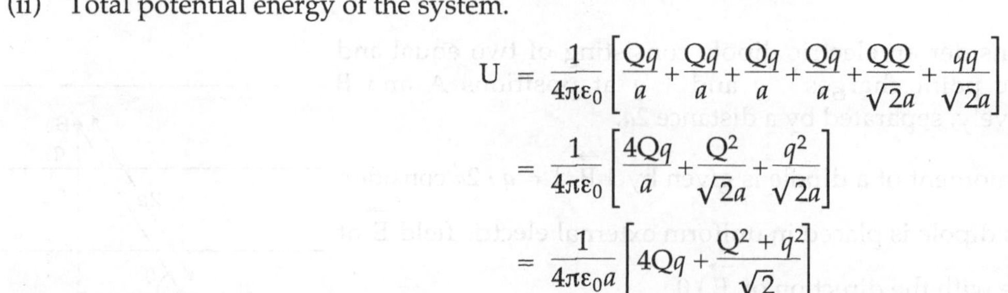

$$U = \frac{1}{4\pi\varepsilon_0}\left[\frac{Qq}{a} + \frac{Qq}{a} + \frac{Qq}{a} + \frac{Qq}{a} + \frac{QQ}{\sqrt{2}a} + \frac{qq}{\sqrt{2}a}\right]$$

$$= \frac{1}{4\pi\varepsilon_0}\left[\frac{4Qq}{a} + \frac{Q^2}{\sqrt{2}a} + \frac{q^2}{\sqrt{2}a}\right]$$

$$= \frac{1}{4\pi\varepsilon_0 a}\left[4Qq + \frac{Q^2 + q^2}{\sqrt{2}}\right]$$

★ are board based questions from previous year

Q. 10. The distance of the field point on the equatorial plane of small electric dipole is halved. By what factor will the electric field change for the dipole ?

Ans. The electric field due to an electric dipole on the equatorial plane is given by

$$E = \frac{1}{4\pi\varepsilon_0}\frac{-p}{r^3}$$

Now,
$$r = \frac{r}{2}$$
$$E' = 8\,E$$

The electric field will increase and becomes 8 times of the previous value.

Q. 11. What is the total positive or negative charge present in 1 molecule of water ?

Ans. A water molecule has 10 protons and 10 electrons (8 from oxygen and 2 from hydrogen atoms).
1 proton or electron has charge equal to 1.6×10^{-19} C in magnitude
Total positive charge = $10 \times (1.6 \times 10^{-19})$ C = 1.6×10^{-18} C
Total negative charge = $10 \times (-1.6 \times 10^{-19})$ C = -1.6×10^{-18} C
The water molecule is neutral as a whole.

Q. 12. Consider two hollow concentric spheres, S_1 and S_2, enclosing charges 2Q and 4Q respectively as shown in the figure. (i) Find out the ratio of the electric flux through them. (ii) How will the electric flux through the sphere S_1 change if a medium of dielectric constant 'ε_r' is introduced in the space inside S_1 in place of air ?★

Deduce the necessary expression.

4Q
2Q
S_2
S_1

Ans. (i) Charge enclosed by sphere S_1 = 2Q

By Gauss' law, electric flux through sphere S_1 is

$$\phi_1 = \frac{2Q}{\varepsilon_0}$$

Charge enclosed by sphere

$$S_2 = 2Q + 4Q = 6Q$$

$$\therefore \quad \phi_2 = \frac{6Q}{\varepsilon_0}$$

The ratio of the electric flux is

$$\frac{\phi_1}{\phi_2} = \frac{\frac{2Q}{\varepsilon_0}}{\frac{6Q}{\varepsilon_0}} = \frac{2}{6} = \frac{1}{3}$$

(ii) For sphere S_1, the electric flux is

$$\phi' = \frac{2Q}{\varepsilon_r}$$

$$\therefore \quad \frac{\phi'}{\phi_1} = \frac{\varepsilon_0}{\varepsilon_r}$$

$$\Rightarrow \quad \phi' = \phi_1 \times \frac{\varepsilon_0}{\varepsilon_r}$$

$$\because \quad \varepsilon_r > \varepsilon_0$$

$$\therefore \quad \phi' < \phi_1$$

Therefore, the electric flux through the sphere S_1 decreases with the introduction of the dielectric inside it.

★ are board based questions from previous year

Chapter 2. Electrostatic Potential and Capacitance

Q. 1. Two charges 2 μC and – 2 μC are placed at points A and B 6 cm apart.

(i) Identify an equipotential surface of the system.

(ii) What is the direction of the electric field at every point on this surface ? **[NCERT]**

Ans. The situation is represented in the given figure.

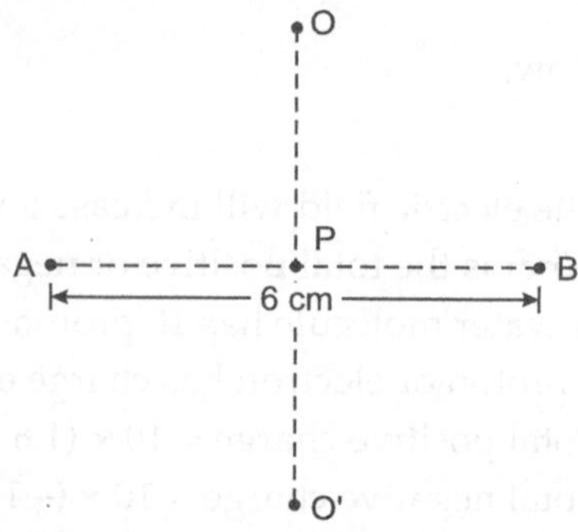

(i) An equipotential surface is the plane on which total potential is zero everywhere. This plane is normal to line AB. The plane is located at the mid-point of line AB because the magnitude of charges is the same.

(ii) The direction of the electric field at every point on this surface is normal to the plane in the direction of AB.

Q. 2. Define (i) Polarisation vector. (ii) Dielectric strength. State their S.I. units also.

Ans. (i) **Polarisation vector :** It is a measure of degree of polarisation of dielectric . It is dipole moment of unit volume of polarised dielectric.

If n is the number of atoms or molecules per unit volume of the dielectric, then polarisation vector is given by

$$\vec{P} = n\,\vec{p}$$

The S.I. unit of polarisation vector is C/m^2.

(ii) **Dielectric strength :** It is the maximum value of the electric field intensity that can be applied to the dielectric without its breakdown.

Its SI unit is V/m and practical unit is kV/mm.

Q. 3. Give the definition and units of capacitance. On which factors does the value of capacitance depend ? Give the symbol of capacitor.

Ans. The measure of the ability of a conductor to store charges is known as capacitance

$$Q \propto V$$

$$Q = CV$$

if V = 1 Then, $$C = Q$$

Capacitance of a conductor is defined as the charge required to raise its potential through one unit.

The S.I. unit of capacitance is Farad or Coulomb per volt. A Farad is very large therefore Pico and Microfarads are common units of capacitance.

The capacitance of a capacitor depends on the following factors :

(i) Geometry of the plates.

(ii) Distance between the capacitor plates and

(iii) Nature of dielectric medium separating the plates.

The symbol of capacitor is ——| |—— .

Q. 4. Differentiate between electric field and potential difference due to point charge ?

Ans. As we know, the electric potential at a point at r distance from point charge is given by

$$V(r) = \frac{Q}{4\pi\varepsilon_0 r}$$

Electric field at a point at r distance from point charge is given by

$$E(r) = \frac{Q}{4\pi\varepsilon_0 r^2}$$

Electric field is a vector quantity while potential difference is a scalar quantity.

Following fig. shows the variation in electric field versus distance and potential difference versus distance.

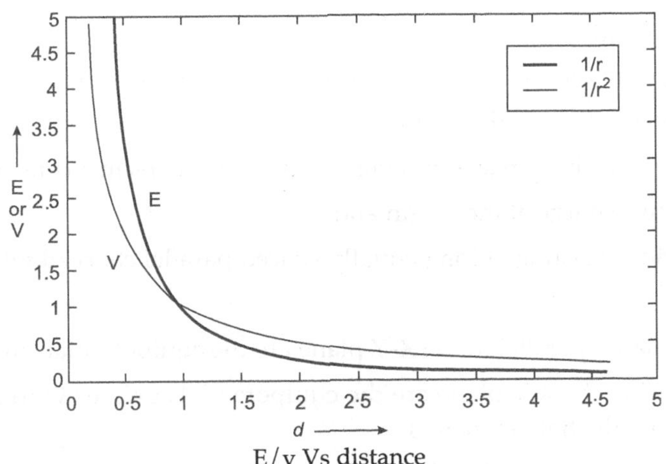

E/v Vs distance

The electrostatic potential ($\propto 1/r$) and the electrostatic field ($\propto 1/r^2$) varies with r.

Q. 5. Explain polarisation of dielectrics.

Ans. When a non-polar dielectric slab is subjected to an electric field, dipoles are induced due to separation of effective positive and negative centers.

If E_o is applied field and Ep is induced field in the dielectric.

The net field is

$$E_N = E_o - E_P$$

Hence, the electric field get reduced when dielectric gets inserted.

The dielectric constant is given by

$$K = \frac{E_o}{E_o - E_P}$$

Q. 6. What are properties of equipotential surfaces ?

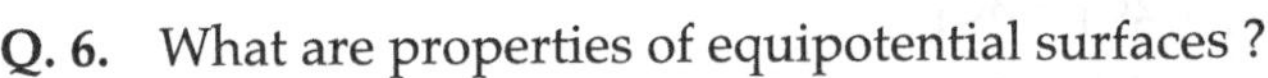

Ans. Equipotential surfaces have following properties :

(i) The electric field is always perpendicular to the element dl of the equipotential surface.

(ii) No work is done in moving a test charge from one point to another on an equipotential surface.

(iii) Equipotential surfaces indicate regions of strong or weak electric fields.

(iv) Two equipotential surfaces cannot intersect. If two equipotential surfaces intersect, then at the points of intersection, there will be two values of the electric potential which is not possible.

Q. 7. A point charge +Q is placed at point O as shown in the figure. Is the potential difference $V_A - V_B$ is positive, negative or zero ?★

Ans. Potential at a distance r from a given point charge +Q is given by

$$V = \frac{1}{4\pi\varepsilon_0} \cdot \frac{Q}{r}$$

$$V_A = \frac{Q}{4\pi\varepsilon_0 r_A}$$

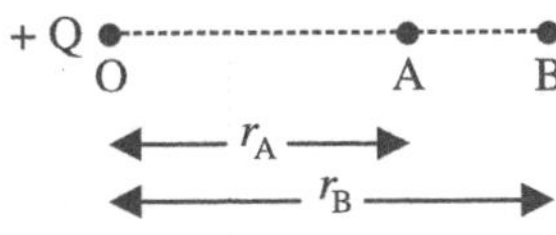

★ are board based questions from previous year

$$V_B = \frac{Q}{4\pi\varepsilon_0 r_B}$$

Since $r_A < r_B$

$\Rightarrow$ $V_A > V_B$

$\therefore$ $V_A - V_B > 0$

Hence, $V_A - V_B$ is positive.

Q. 8. Describe schematically the equipotential surfaces corresponding to :

(i) a constant electric field in the *z*-direction,

(ii) a field that uniformly increases in magnitude but remains in a constant. (say *z*) direction.

(iii) a single positive charge at the origin and

(iv) a uniform grid consisting of long equally spaced parallel charged wires in a plane.

[NCERT]

Ans. (i) Equidistant planes parallel to the X-Y plane are the equipotential surfaces.

(ii) Planes parallel to the X-Y plane are the equipotential surfaces with the exception that when the planes get closer, the field increases.

(iii) Concentric spheres centered at the origin are equipotential surfaces.

(iv) A periodically varying shape near the given grid is the equipotential surface. This shape gradually reaches the shape of planes parallel to the grid at a larger distance.

Q. 9. Find the ratio of the potential differences that must be applied across the parallel and series combination of two capacitors C_1 and C_2 with their capacitances in the ratio 1 : 2 so that the energy stored in the two cases becomes the same.★

Ans. Given,

$$C_1 : C_2 = 1 : 2$$

$\Rightarrow$ $$C_2 = 2C_1$$

For parallel combination of capacitor,

$$C_p = C_1 + C_2$$

$$= C_1 + 2C_1 = 3C_1$$

The energy stored in capacitor

$$E_p = \frac{1}{2}C_p V_p^2$$

$$= \frac{1}{2}3C_1 V_p^2 = \frac{3}{2}C_1 V_p^2 \quad \text{...(i)}$$

For series combination of capacitor,

$$\frac{1}{C_s} = \frac{1}{C_1} + \frac{1}{C_2}$$

$$C_s = \frac{2}{3}C_1$$

The energy stored in capacitor

$$E = \frac{1}{2}C_s V_s^2$$

$$E = \frac{C_1 V_s^2}{3} \quad \text{...(ii)}$$

★ are board based questions from previous year

Equating equation (i) and (ii), we get

$$\frac{3}{2}C_1V_p^2 = \frac{C_1V_s^2}{3}$$

$$\frac{V_p}{V_s} = \frac{\sqrt{2}}{3}$$

Q. 10. Calculate capacitance of an isolated spherical conductor or show that capacitance of earth is 711 μF.

Ans. Consider a point charge q concentrated at the centre of sphere of radius r. The potential at any point on the surface of sphere (*i.e.,*. at r distance from point charge) is given by

$$V = \frac{q}{4\pi\varepsilon_0 r}$$

But $$C = \frac{q}{V}$$

Thus, $$C = \frac{q}{q/4\pi\varepsilon_0 r} = 4\pi\varepsilon_0 r$$

Thus, capacitance (C) of spherical conductor is directly proportional to its radius.

If we substitute value of radius of earth then the value of

$$C = 711 \times 10^{-6}\ \mu F$$
$$= 711\ \mu F$$

Q. 11. Draw the equipotential surfaces due to an electric dipole. Locate the points where the potential due to the dipole is zero.★

Ans.

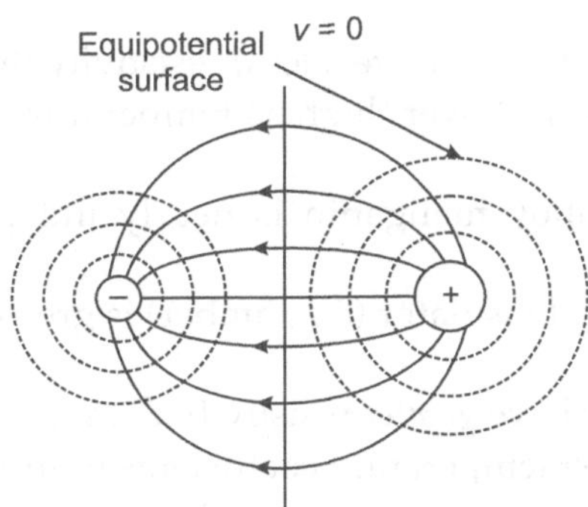

The potential will be zero at equatorial plane of electric dipole going through the centre of dipole.

Chapter 3. Current Electricity

Q. 1. A cell of emf E and internal resistance r is connected to two external resistance R_1 and R_2 and a perfect ammeter. The current in the circuit is measured in four different situations :

(i) Without any external resistance in the circuit.

(ii) With resistance R_1 only

(iii) With R_1 and R_2 in series combination

(iv) With R_1 and R_2 in parallel combination

The currents measured in the four cases are 0.42 A, 1.05 A, 1.4 A and 4.2 A, but not necessarily in that order. Identify the current corresponding to the four cases mentioned above.

Ans. (i) Current in case I :

$$i_1 = \frac{E}{r}$$

★ are board based questions from previous year

(ii) Current in case II :

$$i_2 = \frac{E}{r + R_1}$$

(iii) Current in case III :

$$i_3 = \frac{E}{r + R_1 + R_2}$$

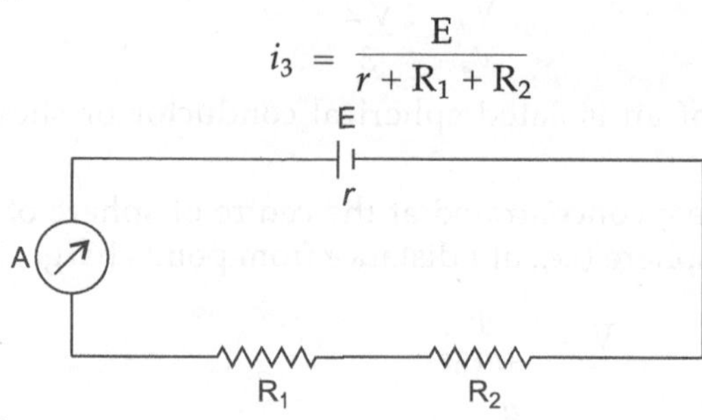

(iv) Current in case IV :

$$i_4 = \frac{E}{r + \frac{R_1R_2}{R_1 + R_2}}$$

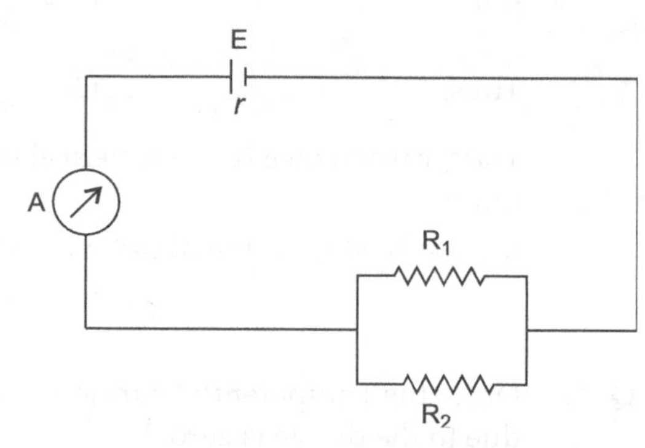

Logically from the above expressions,

$$i_3 < i_2 < i_4 < i_1$$

So,

$$i_1 = 4.2 \text{ A}$$
$$i_4 = 1.4 \text{ A}$$
$$i_2 = 1.05 \text{ A}$$
$$i_3 = 0.42 \text{ A}$$

Q. 2. Choose the correct alternative :

(i) Alloys of metals usually have (greater/less) resistivity than that of their constituent metals.

(ii) Alloys usually have much (lower/higher) temperature coefficients of resistance than pure metals.

(iii) The resistivity of the alloy manganin is nearly independent of/increases rapidly with increase of temperature.

(iv) The resistivity of a typical insulator (*e.g.*, amber) is greater than that of a metal by a factor of the order of ($10^{22}/10^3$). **[NCERT]**

Ans. (i) Alloys of metals usually have *greater* resistivity than that of their constituent metals.

(ii) Alloys usually have *lower* temperature coefficients of resistance than pure metals.

(iii) The resistivity of the alloy, manganin, is nearly *independent of* increase of temperature.

(iv) The resistivity of a typical insulator is greater than that of a metal by a factor of the order of 10^{22}.

Q. 3. A battery of emf 12 V and internal resistance 2 Ω is connected to a 4 Ω resistor as shown in the figure. ★

(i) Show that a voltmeter when placed across the cell and across the resistor, in turn, gives the same reading.

(ii) To record the voltage and the current in the circuit, why is voltmeter placed in parallel and ammeter in series in the circuit ?

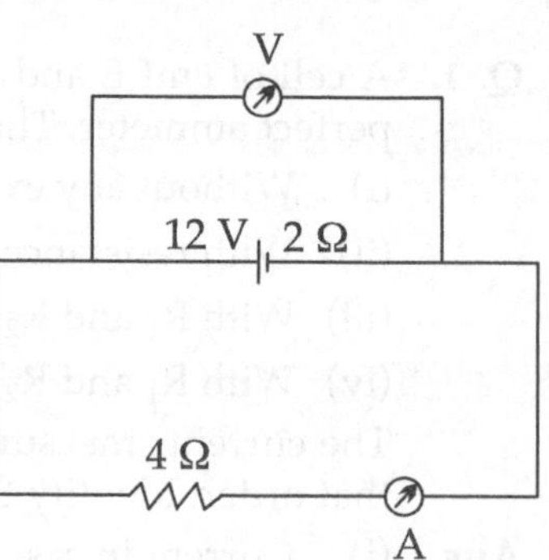

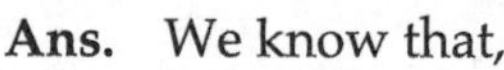

Ans. We know that,

$$E = V + Ir$$
$$V = E - Ir$$

(i) Current,

$$I = \frac{E}{(R + r)}$$

★ are board based questions from previous year

$$= \frac{12}{(4+2)} = \frac{12}{6} = 2 \text{ A}$$

Terminal potential difference across the cell can be calculated as,

$$V = E - Ir = 12 - 2 \times 2 = 12 - 4 = 8 \text{ V}$$

Also, potential difference across 4 Ω resistor can be calculated as

$$V = IR = 2 \times 4 = 8 \text{ V}.$$

So, a voltmeter when placed across the cell and across the resistor, gives the same reading.

(ii) An ammeter is connected in series because it has very low resistance. So, when, an ammeter is connected in series, then there is not much increase in the resistance of the circuit and hence the current through the circuit is unchanged.

A voltmeter is connected in parallel because it has very high resistance. So, it draws a very small current from the circuit.

Q. 4. Use Kirchhoff's rules to obtain conditions for the balance condition in a Wheatstone bridge.★

Ans. Let us consider a Wheatstone bridge arrangement as shown below :

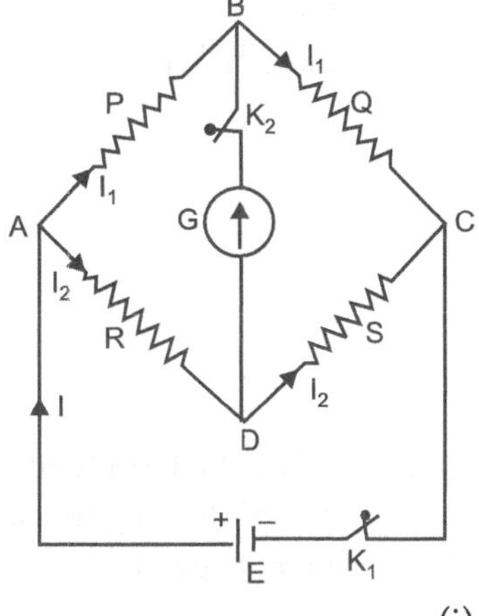

Wheatstone bridge is a special bridge type circuit which consists of four resistances, a galvanometer and a battery. It is used to determine unknown resistance.

In figure four resistance P, Q, R and S are connected in the form of four arms of a quadrilateral. Let the current given by battery in the balanced position be I. This current on reaching point A is divided into two parts I_1 and I_2. As there is no current in galvanometer in balanced state, therefore, current in resistances P and Q is I_1 and in resistances R and S it is I_2.

Applying Kirchhoff's first law at point A

$$I - I_1 - I_2 = 0 \text{ or } I = I_1 + I_2 \quad \text{...(i)}$$

Applying Kirchhoff's second law to closed mesh ABDA

$$-I_1P + I_2R = 0 \text{ or } I_1P = I_2R \quad \text{...(ii)}$$

Applying Kirchhoff's second law to mesh BCDB

$$-I_1Q + I_2S = 0 \text{ or } I_1Q = I_2S \quad \text{...(iii)}$$

Dividing equation (ii) by (iii), we get

$$\frac{I_1P}{I_1Q} = \frac{I_2R}{I_2S} \text{ or } \frac{P}{Q} = \frac{R}{S}$$

This is condition of balance Wheatstone's bridge.

Q. 5. State Kirchhoff 's rules. Explain briefly how these rules are justified ?★

Ans. **Kirchhoff 's first Law–Junction Rule :** In an electrical circuit, the algebraic sum of the currents meeting at a junction is always zero.

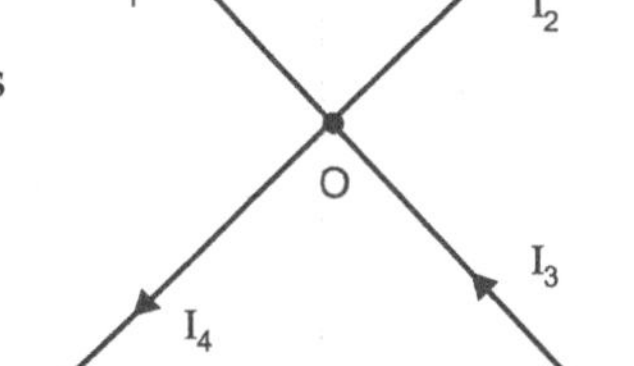

I_1, I_2, I_3 and I_4 are the currents flowing through the respective wires.

Convention : The current flowing towards the junction is taken as positive.

The current flowing away from the junction is taken as negative.

$$I_3 + (-I_1) + (-I_2) + (-I_4) = 0$$

This law is based on the law of conservation of charge.

★ are board based questions from previous year

Kirchhoff 's Second Law – Loop rule : In a closed loop, the algebraic sum of the emf's is equal to the algebraic sum of the products of the resistances and the currents flowing through them.

For the closed loop BACB :

$$E_1 - E_2 = I_1R_1 + I_2R_2 - I_3R_3$$

For the closed loop CADC :

$$E_2 = I_3R_3 + I_4R_4 + I_5R_5$$

This law is based on the law of conservation of energy.

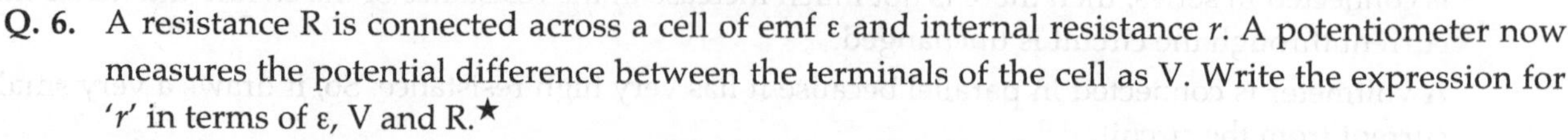

Q. 6. A resistance R is connected across a cell of emf ε and internal resistance r. A potentiometer now measures the potential difference between the terminals of the cell as V. Write the expression for 'r' in terms of ε, V and R.★

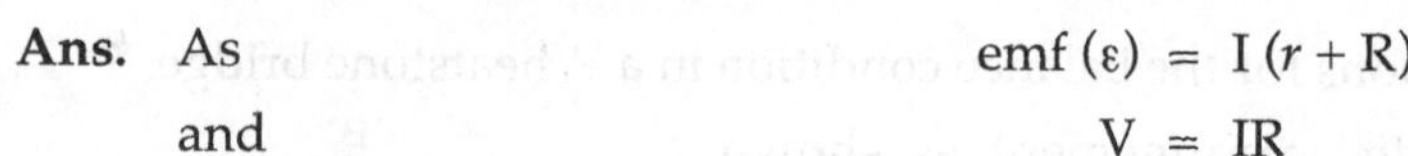

Ans. As $\quad \text{emf } (\varepsilon) = I(r + R)$

and $\quad V = IR$

Taking the ratio,

$$\frac{\varepsilon}{V} = \frac{(r+R)}{R}$$

$$\varepsilon = V\left(\frac{1+r}{R}\right)$$

$$r = R\left(\frac{E}{V-1}\right)$$

Q. 7. Define the term 'drift velocity' of charge carriers in a conductor and write its relationship with the current flowing through it.★

Ans. The net speed achieved by an electron in a current carrying conductor is called as drift velocity. The average velocity acquired by the free electrons along the length of a metallic conductor under a potential difference applied across the conductor is called drift velocity of the electrons.

$$v_d = \frac{I}{neA}$$

Here :

I is the current flowing through the conductor.

n is the number density of an electron.

A is the area of the conductor,

e is the charge of the electron.

Q. 8. Three bulbs of specifications 60 W/220 V, 100 W/220 V, 1000 W/220 V are connected in series, which one will glow the brightest ?

Ans. We know that, $\quad P = \frac{V^2}{R}$

or $\quad R = \frac{V^2}{P}$

$$\therefore \quad \frac{(220)^2}{60} = R_{60\,W}$$

$$\frac{(200)^2}{100} = R_{100\,W}$$

$$\frac{(220)^2}{1000} = R_{1000\,W}$$

★ are board based questions from previous year

$$P \propto \frac{1}{R}$$

Bulb with higher resistance will glow brighter.

As resistance of 60 W > Resistance of 100 W > Resistance of 1000 W so, brightness of 60 W bulb will be more.

Q. 9. Two electric bulbs P and Q have their resistances in the ratio of 1 : 2. They are connected in series across a battery. Find the ratio of the power dissipation in these bulbs.★

Ans. Let the resistances of bulbs P and Q be R and 2R respectively.

As they are connected in series, so current through each bulb is same.

$$\therefore \quad \frac{P_1}{P_2} = \frac{I^2R_1}{I^2R_2} = \frac{R_1}{R_2} = \frac{1}{2}$$

i.e., $\quad P_1 : P_2 = 1 : 2$

Q. 10. A wire of length L_o has a resistance R_o. It is gradually stretched till its length becomes 2 L_o.★

(a) Plot a graph showing variation of its resistance R with its length *l* during stretching.

(b) What will be its resistance when its length becomes 2 L_o?

Ans. (a)

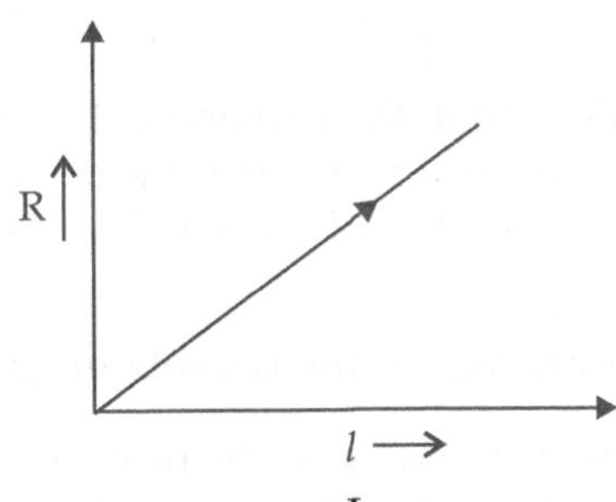

(b) Since, $\quad R_o = \frac{\rho L_o}{A}$

Here, $\quad L_{new} = 2\,L_o$

$$A_{new} = \frac{A}{2}$$

ρ remains same (Property of material)

$$\therefore \quad R_{new} = \frac{\rho \times 2\,L_o}{A/2}$$

$$R_{new} = \frac{4\,\rho\,L_o}{A} = 4\,R_o$$

New resistance will become 4 times of the original.

Chapter 4. Moving Charges and Magnetism

Q. 1. How galvanometer can be converted to ammeter and voltmeter ?

Ans. Conversion of galvanometer to ammeter : Galvanometer can be converted into ammeter by shunting it with a very small resistance. Potential differences across the galvanometer and shunt resistance are equal.

$$(I - I_g)\,S = I_g\,G$$

$$\Rightarrow \quad S = \frac{I_g G}{I - I_g}$$

★ are board based questions from previous year

Conversion of galvanometer to voltmeter : Galvanometer can be converted into voltmeter by connecting a very high resistance in series with the galvanometer. Potential difference across the given load resistance is the sum of p.d. across galvanometer and p.d. across the high resistance.

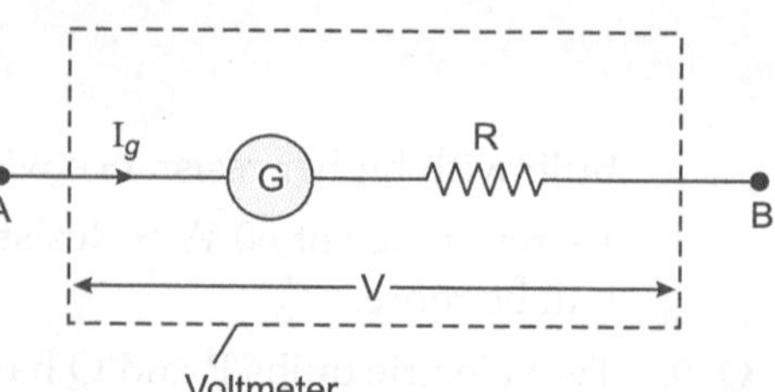

$$V = I_g (G + R)$$

$$R = \frac{V}{I_g} - G$$

Q. 2. A current is induced in coil C_1 due to the motion of current carrying coil C_2.

(i) Write any two ways by which a large deflection can be obtained in the galvanometer G.

(ii) Suggest an alternative device to demonstrate the induced current in place of a galvanometer.

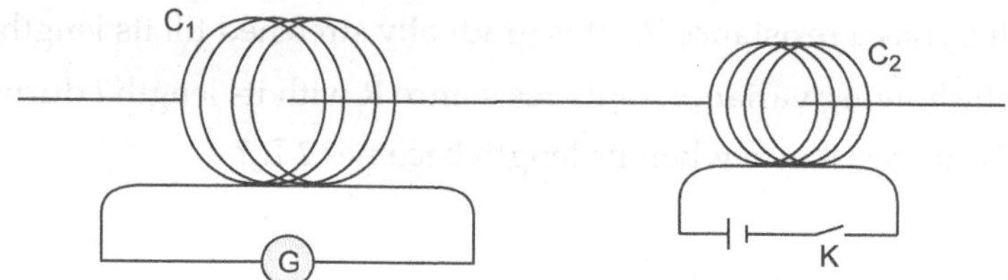

Ans. (i) Two ways to get large deflection in the galvanometer are :

(a) By increasing the relative motion between the coils.

(b) By inserting an iron rod into the coils along their axes.

(ii) An LED (Bulb).

Q. 3. Write the expression, in a vector form, for the Lorentz magnetic force $\vec{F}$ due to a charge moving with velocity $\vec{v}$ in a magnetic field $\vec{B}$. What is the direction of the magnetic force ?★

Ans. The Lorentz magnetic force is given by the following relation :

$$\vec{F} = q(\vec{v} \times \vec{B})$$

Here q, is the magnitude of the moving charge.

The direction of the magnetic force is perpendicular to the plane containing the velocity vector $\vec{v}$ and the magnetic field vector $\vec{B}$.

Q. 4. (i) Write the expression for the force $\vec{F}$ acting on a particle of mass m and charge q moving with velocity $\vec{v}$ in a magnetic field $\vec{B}$. Under what conditions will it move in (a) a circular path and (b) a helical path ?★

(ii) Show that the kinetic energy of the particle moving in magnetic field remains constant.★

Ans. (i)

$$\vec{F} = q(\vec{v} \times \vec{B})$$

$$= qvB \sin \theta$$

If the angle between v and B is 90° then it will move in circular path.

If the angle is other than, 0°, 90° and 180° the path will be helical.

(ii) Since the work done on the charged particle moving in the magnetic field is zero. Hence, according to work energy theorem the change in kinetic energy is zero.

Q. 5. The magnetic field at a point near the centre but outside a current carrying solenoid is zero. Why ?

Ans. Consider a vertical plane through the centre of the solenoid and perpendicular to its length. The plane divides the solenoid into two equal parts. At a point outside the solenoid on this plane, corresponding turns of the two halves of the solenoid produce equal and opposite magnetic field. Hence, the magnetic field at such a point is zero.

★ are board based questions from previous year

Q. 6. Two identical circular wires P and Q each of radius R and carrying current 'I' are kept in perpendicular planes such that they have a common centre as shown in the figure. Find the magnitude and direction of the net magnetic field at the common centre of the two coils.

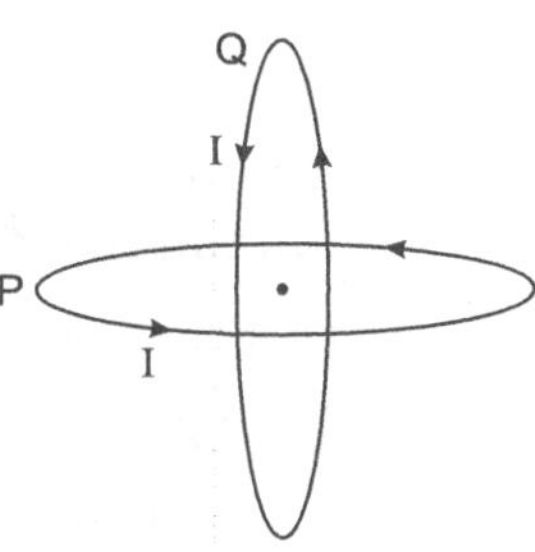

Ans. Magnetic field produced by the two coils at their common centre having currents I_1 and I_2, radius a_1 and a_2, number of turns N_1 and N_2 respectively are given by :

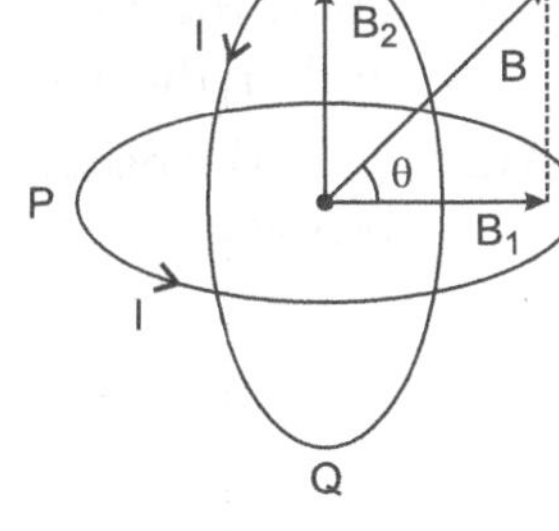

$$B_1 = \frac{\mu_0 N_1 I_1}{2a_1}$$

and

$$B_2 = \frac{\mu_0 N_2 I_2}{2a_2}$$

The resultant field at the common centre is :

$$B = \sqrt{B_1^2 + B_2^2}$$

$$= \sqrt{\left(\frac{\mu_0 N_1 I_1}{2a_1}\right)^2 + \left(\frac{\mu_0 N_2 I_2}{2a_2}\right)^2}$$

Here, $N_1 = N_2 = 1, I_1 = I_2 = I, a_1 = a_2 = R$

$$B = \sqrt{\left(\frac{\mu_0 \times 1 \times I}{2R}\right)^2 + \left(\frac{\mu_0 \times 1 \times I}{2R}\right)^2}$$

$$= \frac{\mu_0 I}{2R}\sqrt{1+1} = \sqrt{2}\left(\frac{\mu_0 I}{2R}\right)$$

Direction of B, $\tan\theta = \frac{B_1}{B_2} = 1$

$\therefore \qquad \theta = 45°$

Hence, the net magnetic field is directed at an angle of 45° with either of the fields.

Q. 7. Write the expression for the magnetic moment $\vec{m}$ due to a planer square loop of side 'l' carrying a steady current I in a vector form.

In the given figure this loop is placed in a horizontal plane near a long straight conductor carrying a steady current I_1 at a distance l as shown. Give reasons to explain that the loop will experience a net force but no torque. Write the expression for this force acting on the loop.★

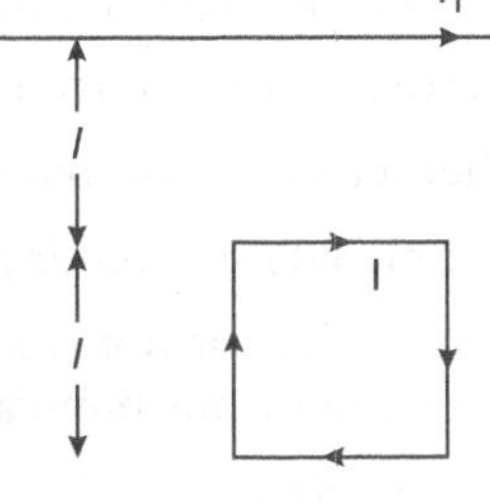

Ans. Magnetic moment of current carrying loop :

$$m = IA = I \times L \times L = IL^2$$

The net torque on the loop will be zero as $\vec{\tau} = \vec{m} \times \vec{B}$, here the angle between vector m and B is zero. ($\sin\theta = 0$).

Force experienced by the nearer parallel side of the loop = $\frac{\mu_0 I I_1}{2\pi L}$, nature of force will be attractive.

Force experienced by the farther parallel side of the loop = $\frac{\mu_0 I I_1}{2\pi\, 2L}$, Nature of force will be repulsive.

Net force on the loop towards the wire : $\frac{\mu_0 I I_1}{2\pi L} - \frac{\mu_0 I I_1}{2\pi\, 2L} = \frac{\mu_0 I I_1}{2\pi\, 2L}$

★ are board based questions from previous year

Chapter 5. Magnetism and Matter

Q. 1. Write two properties of a material suitable for making (i) a permanent magnet, and (ii) an electromagnet.★

Ans. (i) Properties of a material suitable for making permanent magnet :

(a) High retentivity (b) High coercivity

(ii) Properties of a material suitable for making electromagnet :

(a) High permeability (b) Low retentivity

Q. 2. Show diagrammatically the behaviour of magnetic field lines in the presence of (i) paramagnetic and (ii) diamagnetic substances. How does one explain this distinguishing feature ?★

Ans. (i) The behaviour of magnetic field lines in the presence of a paramagnetic substance is shown below :

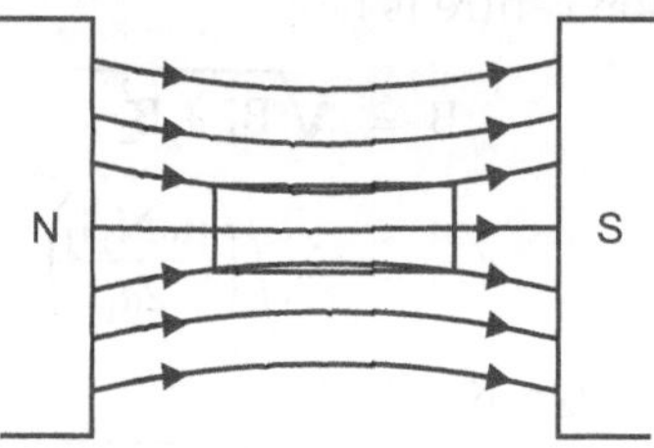

(ii) The behaviour of magnetic field lines in the presence of a diamagnetic substance is shown below :

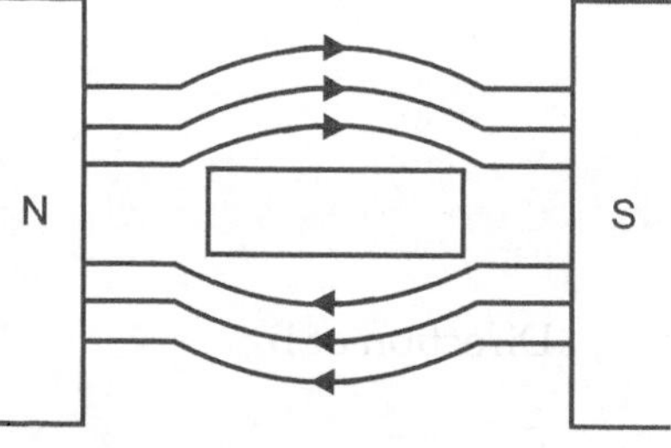

This distinguishing feature is because of the difference in their relative permeability. The relative permeability of diamagnetic substance is less than 1, so the magnetic lines of force do not prefer passing through the substance whereas the relative permeability of a paramagnetic substance is greater than 1, so the magnetic lines of force prefer passing through the substance.

Q. 3. (i) Write two characteristics of a material used for making permanent magnets.★

(ii) Why is core of an electromagnet made of ferromagnetic materials ?★

Ans. (i) Two characteristics of materials used for making permanent magnets :

(a) High retentivity for making strong magnet .

(b) High coercivity for non-removal of magnetisation due to stray magnetic fields, temperature fluctuations or minor mechanical damage.

(ii) Core of electromagnets are made of ferromagnetic materials because they have high permeability and low retentivity. This gives minimum heating losses because of narrow hysteresis curve.

Q. 4. What do we infer from the large value of the susceptibility of the material ? For example, susceptibility of iron is more than that of copper. Explain.

Ans. The susceptibility of a material is defined as the ratio of the intensity of magnetisation (I) to the magnetic intensity (H)

$$\therefore \quad \chi_m = \frac{I}{H}$$

Thus, greater the value of the susceptibility of a material, greater will be the value of intensity of magnetisation for it *i.e.*, more easily it can be magnetised. Thus, greater value of susceptibility for iron means that iron can be magnetised more easily than copper.

★ are board based questions from previous year

Q. 5. What are permanent magnets ? What is an efficient way of preparing a permanent magnet ? Write two characteristic properties of materials which are required to select them for making permanent magnets.

Ans. **Permanent Magnets :** The magnets prepared from ferromagnetic materials which retain their magnetic properties for a long time are called permanent magnets.

An efficient way to make a permanent magnet is to place a ferromagnetic rod in a solenoid and pass a current. The magnetic field of the solenoid magnetises the rod.

The materials used for permanent magnet must have the following characteristic properties :

(i) High retentivity so that the magnet may cause strong magnetic field.

(ii) High coercivity so that the magnetisation is not wiped out by strong external fields, mechanical ill-treatment and temperature changes. The loss due to hysteresis is immaterial because the magnet in this case is never put to cyclic changes.

Q. 6. The susceptibility of a magnetic material is 2.6×10^{-5}. Identify the type of magnetic material and state its two properties.

Ans. The material having positive and small susceptibility is paramagnetic material.

Properties :

(i) They have tendency to move from a region of weak magnetic field to strong magnetic field, *i.e.*, they get weakly attracted to a magnet.

(ii) When a paramagnetic material is placed in an external field the field lines get concentrated inside the material, and the field inside is enhanced.

Q. 7. Explain the following :

(i) Why do magnetic field lines form continuous closed loops ?

(ii) Why are the field lines repelled (expelled) when a diamagnetic material is placed in an external uniform magnetic field ?

Ans. (i) Magnetic lines of force form continuous closed loops because a magnet is always a dipole and as a result, the net magnetic flux of a magnet is always zero.

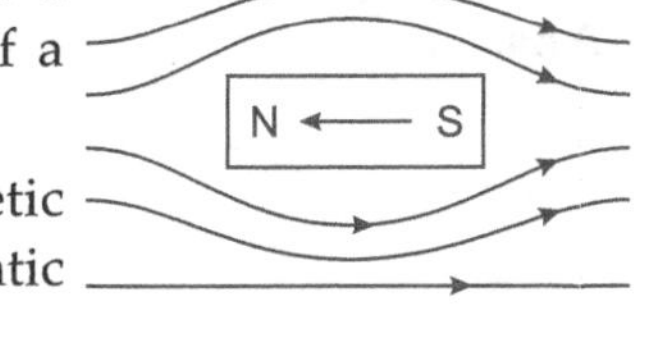

(ii) When a diamagnetic substance is placed in an external magnetic field, a feeble magnetism is induced in opposite direction. So, magentic lines of force are repelled.

Q. 8. Why does a magnetic dipole possess potential energy, when placed at some inclination with the direction of the field ?

Ans. In equilibrium, a magnetic dipole always aligns itself along the direction of the magnetic field. When the magnetic dipole is displaced from the equilibrium position, a restoring torque acts on the dipole to bring it back. Therefore, to place the dipole at some inclination with the field, work has to be done against the restoring torque. This work done is stored in the dipole as its potential energy.

Chapter 6. Electromagnetic Induction

Q. 1. (i) What would you do to obtain a large deflection of the galvanometer ?

(ii) How would you demonstrate the presence of an induced current in the absence of a galvanometer ? **[NCERT]**

Ans. (i) To obtain a large deflection, one or more of the following steps can be taken :

(a) Use a rod made of soft iron inside the coil C_2.

(b) Connect the coil to a powerful battery, and

(c) Move the arrangement rapidly towards the test coil C_1.

(ii) Replace the galvanometer by a small bulb, the one of the same kind, found in a small torch light. The relative motion between the two coils will cause the bulb to glow and thus demonstrate the presence of an induced current.

Q. 2. State Lenz's law. Explain how in the phenomenon of electromagnetic induction, the law of conservation of energy is not violated (or) show it is consistent to law of conservation of energy.

Ans. Lenz's law states that the induced current produced in circuit always flows in such a direction that it opposes the change or cause that produces it. This law is in accordance with the principle of conservation of energy.

The electrical energy is produced at a cost of mechanical energy. To produce induced current equivalent work must be done by the agent in pushing the magnet towards the solenoid. Thus magnet must be moving against some forces as warranted by lenz's law and the work done in moving the magnet against this force, result in the form of induced current in the circuit.

Q. 3. Differentiate between self induction and mutual induction. **[NCERT]**

Ans.

	Self Induction	Mutual Induction
(i)	When the current flowing through a coil changes then induced emf is produced in the coil itself. This phenomenon is called self induction.	When current in one of the two closely lying coils is changed, the emf is induced in the second coil. This phenomenon is called mutual induction.
(ii)	Phenomenon of self induction is characteristic of single coil.	It is characteristic of two coil.
(iii)	The induced current is produced in the same coil in which main current is flowing. Hence, the induced current affects the main current.	Main current flows in one coil and induced current is produced in the second coil. Hence, main current is not directly affected.

Q. 4. Obtain the expression for the mutual inductance of two long co-axial solenoids S_1 and S_2 wound one over the other, each of length L and radii r_1 and r_2 and n_1 and n_2 number of turns per unit length, when a current I is set-up in the outer solenoid S_2.

Ans.

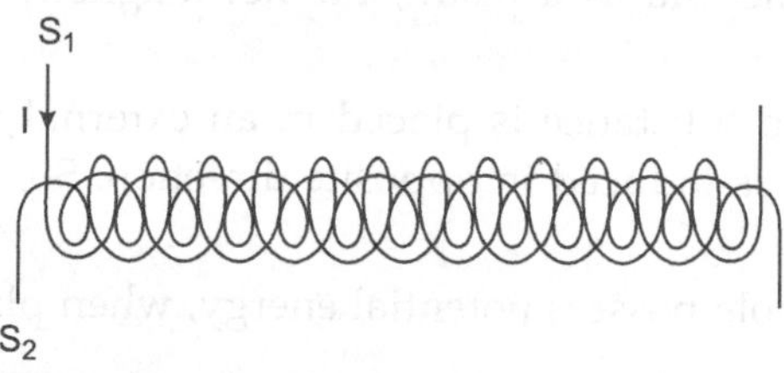

Magnetic field produced, (B) = $\mu_0 n_2 I$.

The flux produced in it, $\phi_1 = \mu_0 n_2 I \pi r_1^2$.

This flux is linked with all the turns of S_2,

$$\phi_1 = \mu_0 n_2 I \pi r_1^2 \cdot n_1 L \quad \text{...(i)}$$

Also

$$\phi = MI \quad \text{...(ii)}$$

Comparing (i) and (ii), we get

$$M = \mu_0 n_1 n_2 L \pi r_1^2$$

Q. 5. The electric current in a wire in the direction from B to A is increasing. What is the direction of induced current in the metallic loop kept above the wire as shown in figure ?

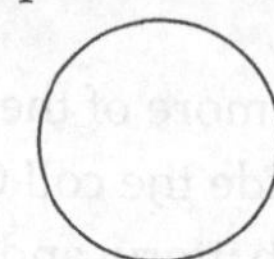

Ans. When the increasing current flows through the wire in the direction from point B to A, the increasing magnetic field is produced which is directed perpendicular to the plane of the loop (or the plane of paper) and in inward direction. Due to this, induced e.m.f. is produced in the loop which opposes the magnetic field produced due to the current flowing through the wire *i.e.*, that it produces magnetic field perpendicular to the plane of the loop and in outward direction, Maxwell's cork screw rule tells that induced current in the loop will flow in *anticlockwise direction.*

Q. 6. An induced current has no direction of its own. Explain, why ?

Ans. The induced current always flows in a direction so that it opposes the cause or the change which produces it. For example, as shown in figure, when the magnet is dropped through the copper ring, the induced current flows in the ring in clockwise direction (when seen from above), till the magnet does not come out of the ring from the other side. But as soon as the magnet leaves the ring, the direction of the current will become anticlockwise in accordance with lenz's law. It shows that an induced current has no direction of its own.

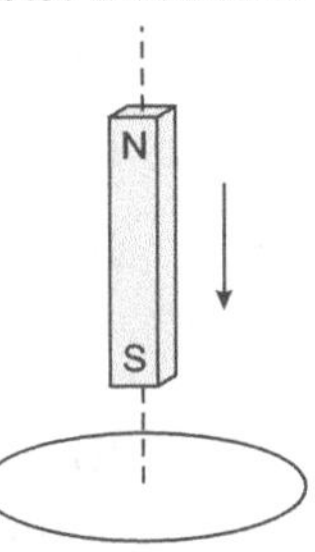

Q. 7. Use Lenz's law to determine the direction of induced current in the situations described by the figures below :

(i) A wire of irregular shape turning circular.

(ii) A circular loop being deformed into a narrow straight wire.

The cross (×) indicates a magnetic field into the paper and the dot (·) a magnetic field out of the paper.

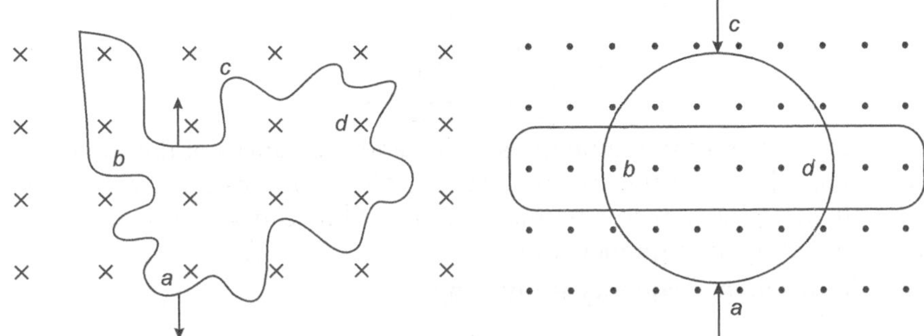

Ans. (i) The current should be along *adcb* (anticlockwise).

(ii) The current should be along *abcd* (clockwise).

Chapter 7. Alternating Current

Q. 1. Explain the importance of power factor.

Ans. The average power, $P = E_v I_v \cos \phi$.

For operating an electric appliance, the power factor of the electric circuit is kept as low as possible. It is done so as to make the power consumption in the circuit low. On the other hand, when power is to be transmitted, the power factor is made as large as possible.

Since,
$$I_v = \frac{P}{E_v \cos \phi}$$

the large value of power factor will reduce current and hence dissipation of electric energy in the form of heat along the line wires will be low.

Q. 2. The core of a transformer is made of a material whose hysteresis loop is narrow. Why ?

Ans. A transformer works on a.c. supply. Therefore, during its operation, the iron core of the transformer is taken over the complete cycle of magnetisation and demagnetisation again and again. In taking the core over a complete cycle of magnetisation, the energy spent per unit volume of the core is equal to area of the hysteresis loop. Therefore, to minimise the loss of energy, the core of the transformer is made of a material, whose hysteresis loop is narrow.

Q. 3. As shown in fig. an electric lamp having coil of negligible inductance connected in series with a capacitor and an a.c. source is glowing with certain brightness.

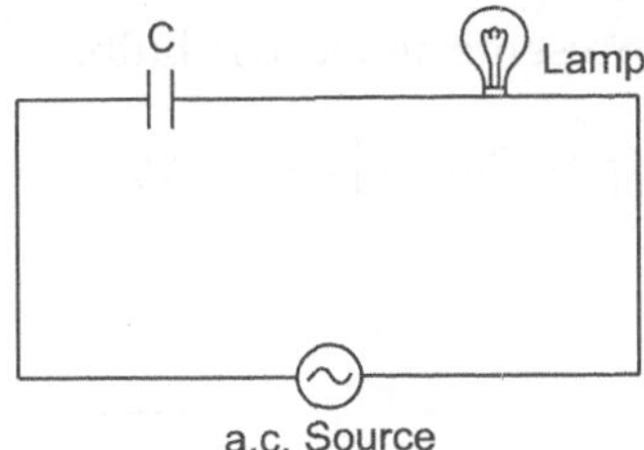

How does the brightness of the lamp change on reducing.
(i) the capacitance and (ii) the frequency ?
Justify your answer.

Ans. Now, capacitive reactance, $X_C = \frac{1}{2\pi fC}$

If E is e.m.f. of the source, then current through the lamp,

$$I = \frac{E}{X_C} = \frac{E}{1/2\,\pi fC} = 2\pi fCE$$

(i) When the capacitance (C) is decreased, it follows that current through the lamp will decrease and hence the brightness of the lamp will *decrease*.

(ii) When the frequency of the source (*f*) is decreased, it will also decrease the current through the lamp and hence the brightness of the lamp will *decrease*.

Q. 4. Three students X, Y and Z performed an experiment for studying the variation of alternating currents with angular frequency in a series LCR-circuit and obtained the graphs shown in fig.

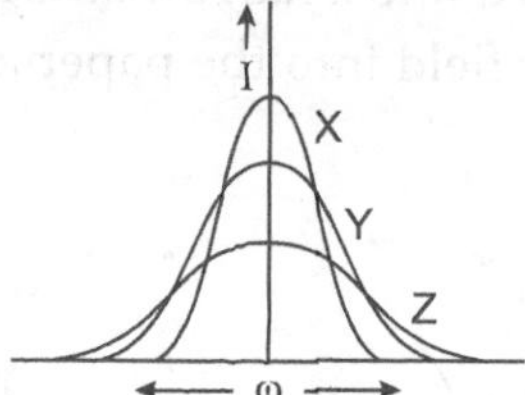

They all used a.c. sources of the same r.m.s. value and inuctances of the same value. What can we (qualitative) conclude about the (i) capacitance value (ii) resistance values used by them ? In which case will the quality factor be maximum ? What can we conclude about nature of the impedance of the set up at frequency ω_0 ?

Ans. (i) The angular resonant frequency is given by

$$\omega_0 = \frac{1}{\sqrt{LC}}$$

Since all the three students have used inductances of the same value, the capacitors used by them must be of the same values of capacitance.

(ii) Since peak values of the alternating currents obtained by them are different, the resistors used by them are of different values of resistance. As the peak value of the currents differ as shown in the figure, $R_Z > R_Y > R_X$.

Now, quality factor, $Q = \frac{\omega_0 L}{R}$

It follows that the quality factor will be maximum for the observation made by the student X.

At the resonant frequency, the impedance of the circuit is equal to the resistance of the circuit. Therefore, in all three cases, impedance will be different *i.e.,* $Z_Z > Z_Y > Z_X$.

Q. 5. Distinguish between resistance, reactance and impedance for an a.c. circuit.

Ans. **Resistance :** The resistance of an a.c. circuit is the ohmic resistance offered by a conductor connected in the circuit. It is due to the nature of the material (resistivity) of the conductor.

Reactance : The reactance of an a.c. circuit is the resistance offered by an inductor or a capacitor connected in the circuit. It arises because of the fact that the alternating e.m.f. and the current differ in phase by $\pi/2$, when a.c. flows through an inductor or a capacitor.

Impedance : The impedance of an a.c. circuit is the effective resistance offered by the LR-circuit or CR-circuit or LCR-circuit.

Q. 6. In the circuit shown in fig., R represents an electric bulb.

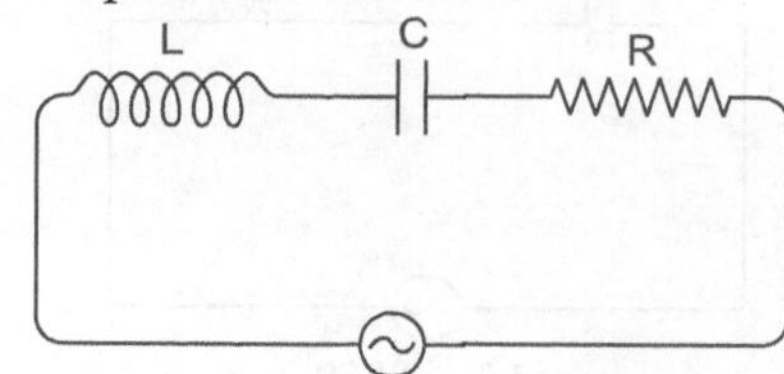

If the frequency of the supply is doubled, how should the values of C and L be changed so that glow in the bulb remains unchanged ?

Ans. The glow of the bulb will remain unchanged, if the current through the bulb does not change. It will happen so, if the impedance of the circuit remains unchanged.

The impedance of the LCR circuit is given by

$$Z = \sqrt{R^2 + \left(2\pi f L - \frac{1}{2\pi f C}\right)^2}$$

It follows that when f is doubled, Z will remain unchanged, if the values of both L and C are halved.

Q. 7. Show that the rms value of current is given by $I_0/\sqrt{2}$ where I_0 is the peak value of current. **[NCERT]**

Ans. The rms value of current is given by

$$I_{rms}^2 = \frac{\int_0^T I^2\,dt}{\int_0^T dt}$$

But $I = I_0 \sin \omega t$

$$I_{rms}^2 = \frac{I_0^2}{2T}\int_0^T \sin^2 \omega t\, dt$$

$$= \frac{I_0^2}{2T}\int_0^T (1 - \cos 2\omega t)\, dt = \frac{I_0^2}{2T}\left(\int_0^T dt - \int_0^T \cos 2\omega t\, dt\right)$$

$$= \frac{I_0^2}{2T} \times T = \frac{I_0^2}{2}$$

$$I_{rms} = \frac{I_0}{\sqrt{2}}$$

Chapter 8. Electromagnetic Waves

Q. 1. (i) Why are infra red waves often called heat waves ? Explain. ★

(ii) What do you understand by the statement, "Electromagnetic waves transport momentum" ?★

Ans. (i) Infra red waves are called heat waves because they raise the temperature of the object on which they fall. They also affect the photographic plate and are readily absorbed by most materials.

(ii) Electromagnetic waves transport momentum. This means that when an EM wave travels through space with energy U and speed c, then it transports linear momentum $p = \frac{U}{c}$. If a surface absorbs the waves completely, then momentum 'p' is delivered to the surface. If the surface reflects the wave, then momentum delivered by both incident and reflected wave adds on to give '$2p$' momentum.

Q. 2. How does Ampere-Maxwell law explain the flow of current through a capacitor when it is being charged by a battery ? Write the expression for the displacement current in terms of the rate of change of electric flux.★

Ans. During the charging of capacitor, displacement current between the plates is produced. Hence circuit becomes complete and current flow through the circuit.

$$I_D = \varepsilon_0 \frac{d\phi_E}{dt}$$

★ are board based questions from previous year

Q. 3. Considering the case of a parallel plate capacitor being charged, show how one is required to generalize Ampere's circuital law to include the term due to displacement current.★

Ans. Gauss' law states that the electric flux ϕ_E of a parallel plate capacitor having an area A, and a total charge Q is given by

$$\phi_E = EA = \frac{Q}{\varepsilon_0 A} \times A \qquad \left[\because E = \frac{Q}{A\varepsilon_0}\right]$$

$$= \frac{Q}{\varepsilon_0}$$

As the charge Q on the capacitor plates change with time, so current is given by

$$i = \frac{dQ}{dt}$$

$$\therefore \qquad \frac{d\phi_E}{dt} = \frac{d}{dt}\left(\frac{Q}{\varepsilon_0}\right) = \frac{1}{\varepsilon_0}\frac{dQ}{dt}$$

$$\Rightarrow \qquad \varepsilon_0 \frac{d\phi_E}{dt} = \frac{dQ}{dt} = i$$

This is the missing term in Ampere's circuital law.
So, the total current through the conductor is
i = Conduction current (i_c) + Displacement current (i_d)

$$i = i_c + \varepsilon_0 \frac{d\phi_E}{dt} \qquad \text{...(i)}$$

Ampere's circuital law states that,

$$\oint \vec{B} \cdot \vec{dl} = \mu_0 i \qquad \text{...(ii)}$$

Putting equation (i) in (ii), we get

$$\oint \vec{B} \cdot \vec{dl} = \mu_0 i_c + \mu_0 \varepsilon_0 \frac{d\phi_E}{dt}$$

This is the required generalised form of Ampere's circuital law.

Q. 4. Name the constituent radiation of Electromagnetic spectrum which

(i) is used in ovens for cooking purposes.
(ii) is used in scientific research.
(iii) is used in treatments of cancer and tumours.
(iv) is used for sterlizing the surgical instruments.
(v) is used in solar water heaters.
(vi) is used to broadcast radio programmes to long distance.
(vii) is used in electric bulb.

Ans. (i) Microwaves (ii) X-rays (iii) Gamma rays (iv) Ultraviolet rays
(v) Infrared waves (vi) Radio waves (vii) Visible light

Q. 5. (i) Identify the part of the electromagnetic spectrum which is :★

(a) suitable for radar system used in aircraft navigation,
(b) produced by bombarding a metal target by high speed electrons.

(ii) Why does a galvanometer show a momentary deflection at the time of charging or discharging a capacitor ? Write the necessary expression to explain this observation.★

Ans. (i) (a) Microwaves, (b) X-rays

(ii) During the charging and discharging of a capacitor, a flow of charges take place from the battery to the plates of the capacitor. This produces a conduction current in the circuit and a displacement current between plates. Hence, the galvanometer shows a momentary deflection.

★ are board based questions from previous year

$$\oint B.dl = \mu_0 I + I_d$$

Where $$I_d = \frac{\varepsilon_0 d\phi_E}{dt}$$

Q. 6. How are e.m. waves produced by oscillating charges ?

Draw a sketch of linearly polarized e.m. waves propagating in the Z-direction. Indicate the directions of the oscillating electric and magnetic fields.★

Ans. A charge oscillating with some frequency, produces an oscillating electric field in space, which in turn produces an oscillating magnetic field perpendicular to the electric field, this process goes on repeating, producing e.m. waves in space perpendicular to both the fields.

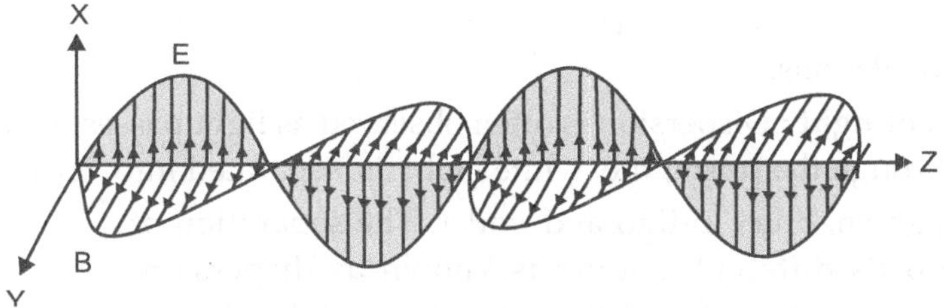

The direction of electric and magnetic fields are perpendicular to each other and are also perpendicular to the direction of propagation of the wave.

Chapter 9. Ray Optics and Optical Instruments

Q. 1. (i) Write the necessary conditions for the phenomenon of total internal reflection to occur.★

(ii) Write the relation between the refractive index and critical angle for a given pair of optical media.★

Ans. (i) **Necessary conditions for total internal reflection to occur are :**

(a) The incident ray on the interface should travel in optically denser medium.

(b) The angle of incidence should be greater than the critical angle for the given pair of optical media.

(ii) $${}^a\mu_b = \frac{1}{\sin C}$$

where a and b are the rarer and denser media respectively. C is the critical angle for the given pair of optical media.

Q. 2. A ray of light falls on a triangular glass prism in such a way that the deviation of the emergent ray is minimum for the prism.

Draw the ray diagram for this case and write the relation between the angle of incidence and angle of emergence.

Ans. The ray diagram is shown in fig. For minimum deviation angle of incidence (i) = angle of emergence (e).

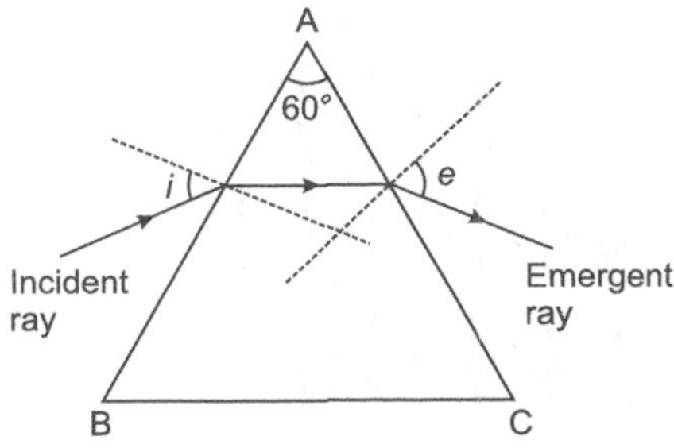

★ are board based questions from previous year

Q. 3. (i) What is dispersion of light ? What is its cause ?

(ii) A ray of light incident normally on one face of a right isosceles prism is totally reflected as shown in figure. What must be the minimum value of refractive index of glass ?★

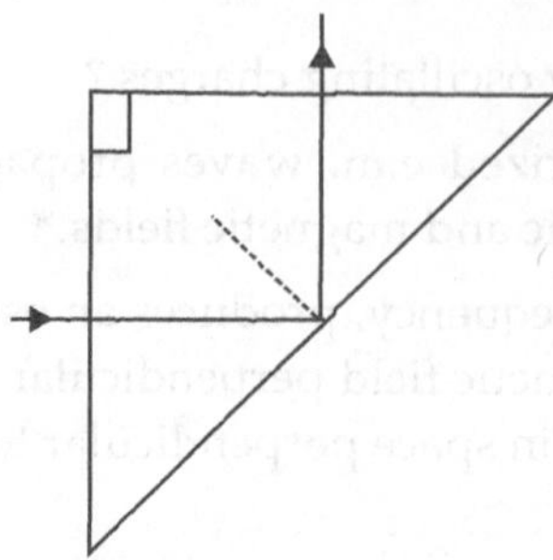

Give relevant calculations.

Ans. (i) **Dispersion of light :** Dispersion is often observed as light passes through a triangular prism. Upon passing through the prism, the white light is separated into its component colours : red, orange, yellow, green, blue, Indigo and voilet. The separation of visible light into its different colours is known as dispersion. Dispersion occurs because for different colour of light in a transparent medium will have different refractive indices (μ).

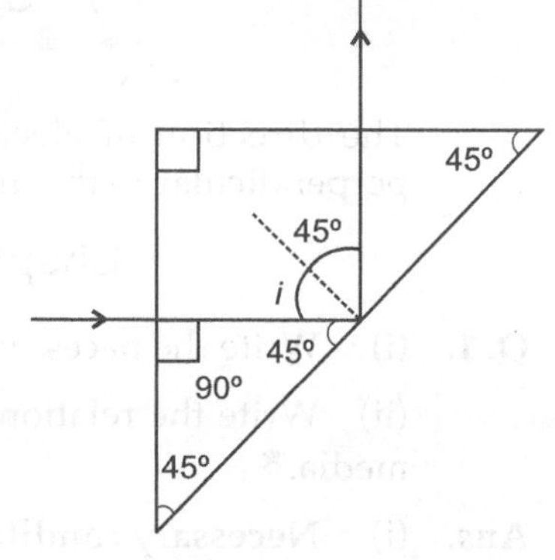

(ii) **For total internal reflection :**

$$i \geq \theta_c$$
$$\sin i \geq \sin \theta_c$$
$$\sin 45° \geq \frac{1}{\mu}$$
$$\frac{1}{\sqrt{2}} \geq \frac{1}{\mu}$$
$$\mu \geq \sqrt{2}$$
$$\mu_{min} = \sqrt{2}$$

Q. 4. Use the mirror equation to show that an object placed between f and $2f$ of a concave mirror produces a real image beyond $2f$.★

Ans. Mirror equation is

$$\frac{1}{f} = \frac{1}{u} + \frac{1}{v}$$

Where u is distance of object from the mirror,

v is the distance of image from the mirror and

f is the focal length of the mirror.

For a concave mirror f is negative *i.e.* $f < 0$, $u < 0$

For a real object (on the left of mirror)

$\therefore$
$$2f < u < f$$
$$\frac{1}{2f} > \frac{1}{u} > \frac{1}{f}$$
$$\frac{1}{2f} - \frac{1}{f} > \frac{1}{u} - \frac{1}{f} > \frac{1}{f} - \frac{1}{f}$$
$$\frac{1}{f} - \frac{1}{2f} < \frac{1}{f} - \frac{1}{u} < \frac{1}{f} - \frac{1}{f}$$

★ are board based questions from previous year

$$\Rightarrow \qquad \frac{1}{2f} < \frac{1}{v} < 0$$

$$\Rightarrow \qquad v < 0. \text{ Also } v > 2f$$

∴ Image is real.

This implies that v is negative and greater than $2f$. This means that the image lies beyond $2f$.

Q. 5. A person standing before a concave mirror can not see his image, unless he is beyond the centre of curvature. Explain.

Ans. In case of a concave mirror, the real and inverted image is formed, when the object is placed beyond its focus, for the position of the man between the focus and the centre of curvature, the image will be formed beyond the centre of curvature *i.e.,* behind him. Obviously, the man will be unable to see his image. But when the man is standing beyond the centre of curvature, his real and inverted image will be formed between the focus and the centre of curvature *i.e.,* just in front of him.

Therefore, it follows that a man can not see his inverted image in a convex mirror, unless he is standing beyond the centre of curvature of the mirror.

Q. 6. What is the advantage of using a concave mirror (in place of convex lens) as objective in a telescope ?

Ans. When a convex lens is used as objective in a telescope, it forms image of different colours due to its prismatic action. This defect in lenses is known as chromatic aberration. However, a concave mirror (used as objective) is free from the defect of chromatic aberration. It is because, the image is formed by a concave mirror due to reflection of light (and not refraction).

In order to view even faint objects, the objective of the telescope should be of large aperture so as to gather a large amount of light from the object. For this, if a convex lens of large aperture is used, it will suffer from spherical aberration. However, a concave mirror of large aperture and free from spherical aberration can be made by giving it a parabolic shape.

Q. 7. A monochromatic ray of light is made to fall on a normal 60° prism under minimum deviation condition. What is the relation between the angle of incidence and the angle of emergence ?

Ans. When light is incident on a prism placed in the minimum deviation position, the refracted ray travels parallel to the base of the prism. Consequently, the prism lies symmetrically w.r.t. the incident and emergent rays. Hence, in minimum deviation position, the angle of incident and angle of emergence are equal.

Q. 8. Can a convergent lens in one medium behave as a divergent lens in some other medium ?

Ans. Yes, a convex lens made of glass behaves as a convergent lens when placed in air or water. However, when the same lens is immersed in carbon disulphide (μ = 1.63), it behaves as a divergent lens. In other words, when a convergent lens is placed inside transparent medium of refractive index greater than that of the material of the lens, it behaves as a divergent lens.

Q. 9. Why is a reflecting type telescope preferred in astronomy ?

Ans. A lens of large aperture and at the same time free from lens—aberrations cannot be manufactured. Therefore, the lens used cannot be of very large aperture. Obviously, light gathering power of such a telescope will be very low and hence one will fail to see faint stars.

The spherical aberrations can be eliminated by making use of a parabolic mirror. Since mirrors of large apertures can be easily designed, reflecting type telescope will have large light gathering power.

Q. 10. A hypermetropic person prefers to remove his spectacles, while driving. Explain, why ?

Ans. Wearing the spectacles (used for near vision), when the person looks at the incoming distant objects, the parallel rays from the objects get further converged. As a result, the image is formed in front of retina and it appears blurred. So that it does not happen, he prefers to remove his spectacles.

Q. 11. When are two objects just resolved ? Explain. How can the resolving power of a compound microscope be increased ? Use relevent formula to support your answer.★

★ are board based questions from previous year

Ans. When the maxima of diffraction pattern from one object coincide with the minima of second object then they are just resolved.

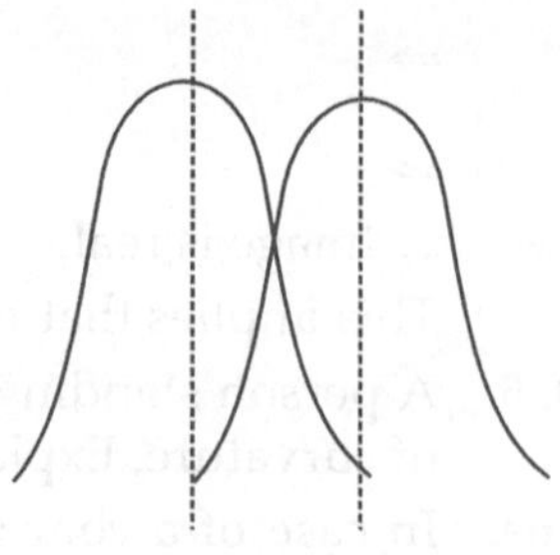

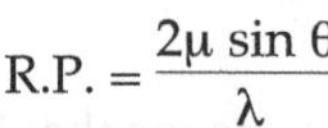

$$\text{R.P.} = \frac{2\mu \sin \theta}{\lambda}$$

The resolving power of a compound microscope can be increased by increasing μ and by decreasing λ.

Chapter 10. Wave Optics

Q. 1. Let us list some of the factors, which could possibly influence the speed of wave propagation :

(i) Nature of the source.

(ii) Direction of propagation.

(iii) Motion of the source and observer.

(iv) Wavelength.

(v) Intensity of the wave.

On which of these factors, if any, does :

(a) The speed of light in vacuum.

(b) The speed of light in a medium (say, glass or water) depend ? **[NCERT]**

Ans. (a) The speed of light in a vacuum *i.e.*, 3×10^8 m/s (approximately) is a universal constant. It is not affected by the motion of the source, the observer, or both. Hence, the given factor does not affect the speed of light in a vacuum.

(b) Out of the listed factors, the speed of light in a medium depends on the wavelength of light in that medium.

Q. 2. In dividing the single slit diffraction pattern, it was stated that the intensity is zero at angles of $n\lambda/a$. Justify this by suitable dividing the slit to bring out the cancellation. **[NCERT]**

Ans. Consider that a single slit of width d is divided into n smaller slits.

$\therefore$ Width of each slit, $d' = \dfrac{d}{n}$

Angle of diffraction is given by the relation,

$$\theta = \frac{\frac{d}{d'}\lambda}{d} = \frac{\lambda}{d'}$$

Now, each of these infinitesimally small slit sends zero intensity in direction θ. Hence, the combination of these slits will give zero intensity.

Q. 3. Draw the intensity pattern for single slit diffraction and double slit interference. Hence, state two differences between interference and diffraction patterns.★

Ans. Intensity pattern for single slit diffraction :

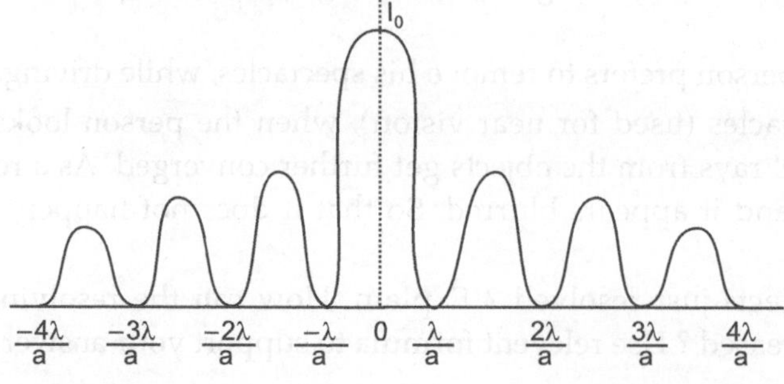

★ are board based questions from previous year

Intensity pattern for double slit interference :

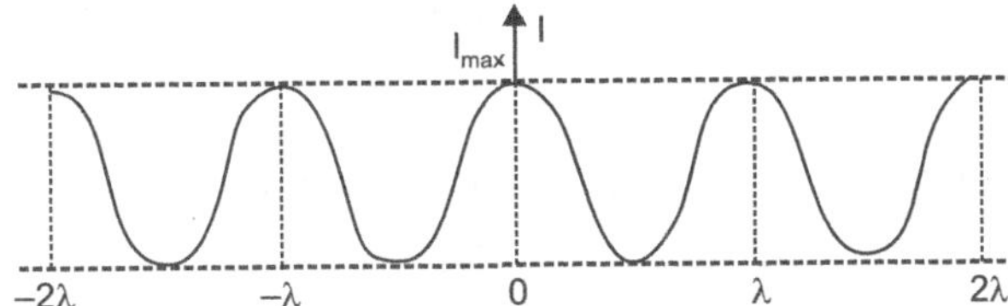

Difference between interference and diffraction patterns :

(a) Interference fringes are of the same width while diffraction fringes are not of the same width.

(b) In interference pattern, all bright bands are of same intensity while in diffraction pattern all bright bands are not of same intensity.

Q. 4. Unpolarised light is passed through a polaroid P_1. When this polarised beam passes through another polaroid P_2 and if the pass axis of P_2 makes angle θ with the pass axis of P_1, then write the expression for the polarised beam passing through P_2. Draw a plot showing the variation of intensity when θ varies from 0 to 2π.★

Ans.

$$I = \frac{I_0}{2}\cos^2\theta$$

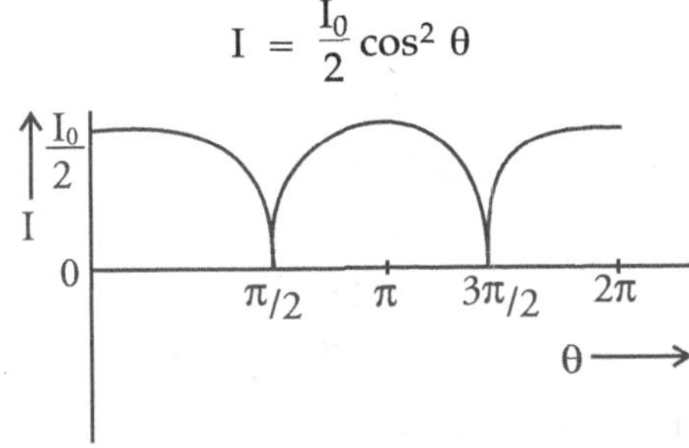

Q. 5. State Brewster's law. The value of Brewster angle for a transparent medium is different for light of different colour. Give reason.★

Ans. **Brewster's law** : The law states that the tangent of the polarising angle of incidence of a transparent medium is equal to its refractive index. The light incident at this angle when reflects back is perfectly polarised.

$$\mu = \tan i_p$$

The refractive index of a material depends on the colour or wavelength of light. As the polarising angle depends on refractive index ($\mu = \tan i_p$), so it also depends on wavelength of light.

Q. 6. What kind of fringes do you expect to observe if white light is used instead of monochromatic light ?★

Ans. If instead of monochromatic light, white light is used, then the central fringe will be white and the fringes on either side will be coloured. Blue colour will be nearer to central fringe and red will be farther away. The path difference at the centre on perpendicular bisector of slits will be zero for all colours and each colour produces a bright fringe thus resulting in white fringe. Farther, the shortest visible wave, blue, produces a bright fringe first.

Q. 7. (i) State law of Malus.★

(ii) Draw a graph showing the variation of intensity (I) of polarised light transmitted by an analyser with angle (θ) between polariser and analyser.★

(iii) What is the value of refractive index of a medium of polarising angle 60°?★

Ans. (i) Malus discovered that when a beam of completely plane polarised light is passed through the analyser, the intensity 'I' of transmitted light changes directly as the square of the cosine of the angle (θ) between the transmission directions of polariser and analyser.

★ are board based questions from previous year

This is known as the law of Malus.

$$I \propto \cos^2\theta$$

or

$$I = I_0 \cos^2 \theta$$

Where, I_0 is the maximum intensity of the transmitted light.

(ii)

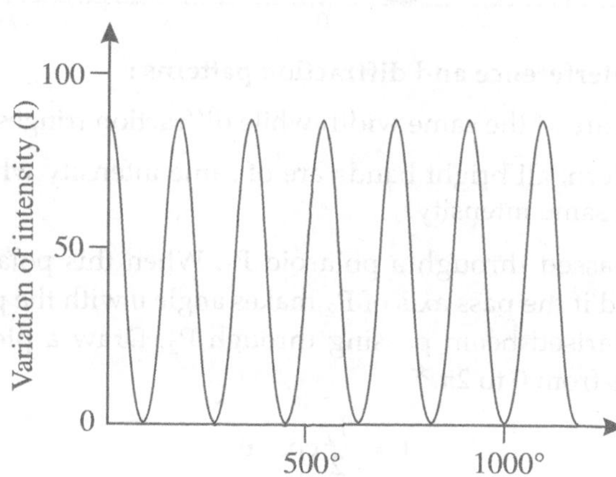

(iii) From the Brewster's law of polarisation :

$$\mu = \tan i_p$$
$$= \tan 60°$$
$$\approx \sqrt{3} = 1.7320$$

Thus, the refractive index of the material is 1.73.

Chapter 11. Dual Nature of Radiation and Matter

Q. 1. A proton and a deuteron are accelerated through the same accelerating potential. Which one of the two has

(i) greater value of de-Broglie wavelength associated with it, and

(ii) less momentum ?

Give reasons to justify your Answer.★

Ans. (i) de-Broglie wavelength,

$\lambda \propto 1/\text{mass}$ (for same accelerating potential)

Mass of a proton is less as compared to a deuteron. So, proton will have greater value of de-Broglie wavelength associated with it.

(ii) Momentum, $p \propto$ mass (for same accelerating potential)

Mass of deuteron is more as compared to a proton. So, it will have a greater value of momentum.

Q. 2. A deuteron and an alpha particle are accelerated with the same accelerating potential. Which one of the two has

(i) greater value of de-Broglie wavelength, associated with it, and

(ii) less kinetic energy ? Explain.

Ans. (i) de-Broglie wavelength of a particle is dependent on its mass and charge for same accelerating potential, such that

$$\lambda \propto \frac{1}{\sqrt{(\text{Mass})(\text{charge})}}$$

Mass and charge of a deuteron are $2m_p$ and e respectively and, mass and charge of an alpha particle are $4m_p$ and $2e$ respectively.

★ are board based questions from previous year

Where,

$$m_p = \text{the mass of a proton.}$$

and

$$e = \text{the charge of an electron.}$$

$$\frac{\lambda_D}{\lambda_\alpha} = \sqrt{\frac{m_\alpha q_\alpha}{m_D q_D}} = \sqrt{\frac{(4m_p)(2e)}{(2m_p)(e)}} = \frac{2}{1}.$$

Thus, de-Broglie wavelength related with deuteron is twice of the de-Broglie wavelength of alpha particle.

(ii) $$\text{K.E.} \propto q$$

(for same accelerating potential)

Charge of a deuteron is less as compared to an alpha particle.

So, deuteron will have less kinetic energy.

Q. 3. An electron, an α-particle and a proton have the same kinetic energy. Which of these particles has the shortest de-Broglie wavelength ? **[NCERT]**

Ans. For a particle,

$$\text{de-Broglie wavelength } (\lambda) = \frac{h}{p}$$

$$\text{Kinetic energy (K)} = \frac{p^2}{2m}$$

Then

$$\lambda = \frac{h}{\sqrt{2mK}}$$

For the same kinetic energy K, the de-Broglie wavelength associated with the particle is inversely proportional to the square root of their masses. A proton (^1_1H) is 1836 times massive than an electron and an α-particle (^4_2He) four times that of a proton.

Hence, α-particle has the shortest de-Broglie wavelength.

Q. 4. An electron is accelerated from rest through a potential difference V volt and it acquires a final velocity v m/s. Obtain an expression for $\frac{e}{m}$ of electron in terms of v and V.

Ans. Let the mass and charge of an electron be m and e respectively. The work done in accelerating a charge through a potential difference V is eV. If the final velocity acquired by the electron is v then the kinetic energy acquired is $\frac{1}{2}mv^2$.

But

$$\frac{1}{2}mv^2 = eV$$

or

$$2\,eV = mv^2$$

$$\frac{e}{m} = \frac{v^2}{2V}$$

Q. 5. Plot a graph showing variation of de-Broglie wavelength λ versus $\frac{1}{\sqrt{V}}$, where V is accelerating potential for two particles A and B carrying same charge but of masses m_1, m_2 ($m_1 > m_2$). Which one of the two represents a particle of smaller mass and why ?

Ans.

$$qV = \frac{1}{2}mv^2$$

$$qV = \frac{p^2}{2m}$$

$$\Rightarrow \quad p = \sqrt{2mqV} = \frac{h}{\lambda}$$

$$\Rightarrow \quad \lambda = \frac{h}{\sqrt{2mqV}}$$

$$\Rightarrow \quad \text{Slope} \propto \frac{1}{\sqrt{m}}$$

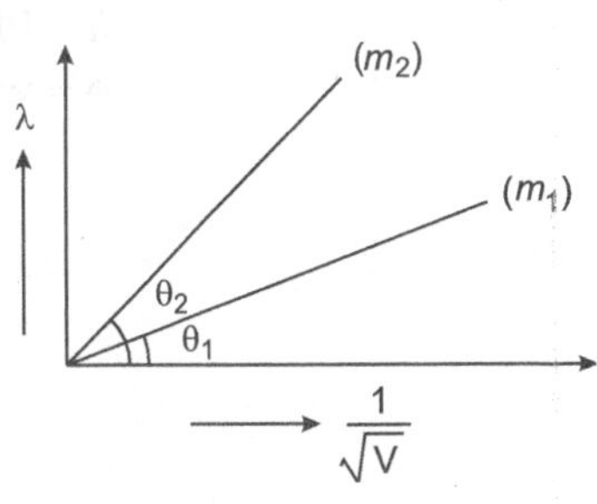

Q. 6. Is it essential that each incident photon should eject a photoelectron ? Explain.

Ans. The photoelectric emission takes place, when the incident photon is absorbed by the electron in the atom. The energy equal to work function of metal is used up in ejecting the photoelectron and the difference appears as the kinetic energy of the electron. Since, one photon can be absorbed by one electron, a photon can eject only one electron.

Q. 7. If the intensity of light falling on a metal plate is doubled, what will be the effect or photocurrent and maximum kinetic energy of emitted photoelectrons ?

Ans. As photocurrent is proportional to intensity of incident light, the photocurrent will be doubled; but the maximum kinetic energy of emitted photoelectrons does not depend upon the intensity of incident light. Hence, the maximum kinetic energy remains unchanged.

Q. 8. There are two sources of light A and B. The wavelenth of light emitted from A is from 8000 Å to 11000 Å, while that from B is from 3000 Å to 6000 Å. The intensity of A is 4 times of B. But when light of A falls on metal, photoelectrons are not emitted whereas light of B can eject photoelectrons from the same metal. Explain its reason.

Ans. The photoelectrons from a metallic surface are emitted only when the energy of incident photons is equal to or more than work function of metal. In other words, the wavelength of incident photon must be equal to or less than the threshold wavelength. In this case the threshold wavelength is less than 8000 Å. Hence, source A cannot eject photoelectrons while B can eject photoelectrons.

Q. 9. What is ionisation potential ? Calculate it for hydrogen atom.

Ans. The minimum accelerating potential required to knock out the electron from the outermost orbit of an atom is called its ionisation potential.

In case of hydrogen atom, the energy of electron in the first orbit of the atom is – 13.6 eV. Since to knock out this electron, an energy equal to 13.6 eV is required, the ionisation potential of H-atom is 13.6 V.

Q. 10. Figure shows the stopping potential (V_0) for the photoelectron versus ($1/\lambda$) graph, for two metals A and B, λ being the wavelength of incident light.★

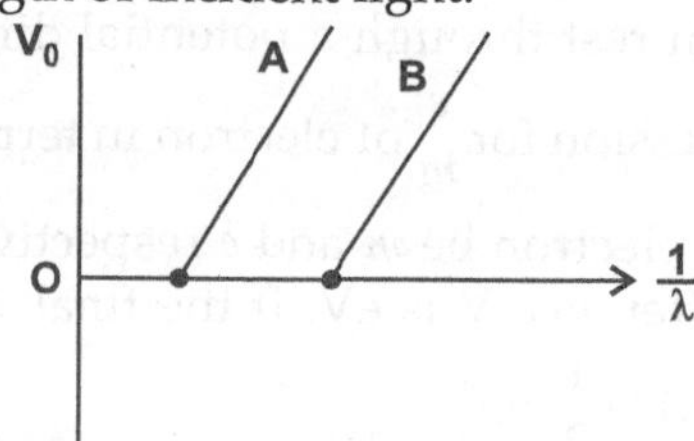

(a) How is the value of Planck's constant determined from the graph?

(b) If the distance between the light source and the surface of metal A is increased, how will the stopping potential for the electrons emitted from it be effected? Justify your answer.

Ans. (a) We know that

$$eV_0 = \frac{hc}{\lambda} - \phi$$

or

$$V_0 = \frac{hc}{e}\left(\frac{1}{\lambda}\right) - \phi \qquad \text{...(i)}$$

where, e = Charge on electron

h = Planck's contant

ϕ = work function of metal surface

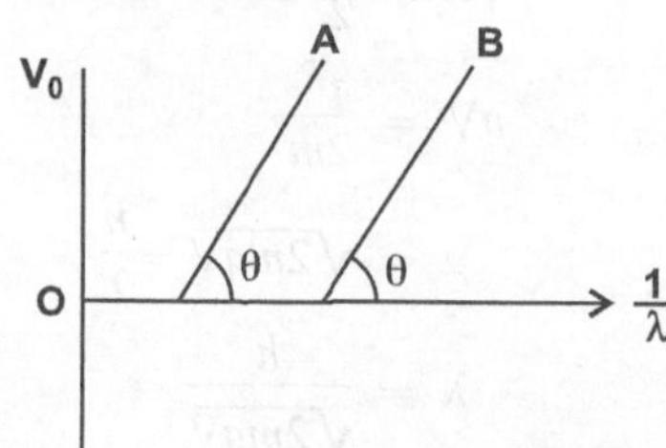

★ are board based questions from previous year

Equation (i) is a straight line as shown in the graph.

Here, slope of line $= \tan\theta = \dfrac{hc}{e}$

or $$h = \frac{e\tan\theta}{c} \quad \text{...(ii)}$$

Planck's constant can easily be determined by substituting the values of the slope of graph, speed of light and electronic charge in equation (ii).

(b) Stopping potential only depends on frequency of incident light. If the distance is increased, intensity will decrease but no change takes place in stopping potential.

Q. 11. Write three characteristic features in photoelectric effect which cannot be explained on the basis of wave theory of light, but can be explained only using Einstein's equation.★

Ans. (i) **Existence of threshold frequency** : According to wave theory, there should not exist any threshold frequency but Einstein's theory explains the existence of threshold frequency.

(ii) **Dependence of kinetic energy on frequency of incident light :** According to wave theory, the maximum kinetic energy of emitted electrons should depend on intensity of incident light and not on frequency whereas Einstein's equation explains that it depends on frequency and not on intensity.

(iii) **Instantaneous emission of electrons :** According to wave theory there should be time lag between emission of electrons and incident of light whereas Einstein's equation explains why there is no time lag between incident of light and emission of electrons.

Chapter 12. Atoms

Q. 1. State Bohr's postulate to define stable orbits in hydrogen atom. How does de Broglie's hypothesis explain the stability of these orbits ?★

Ans. **Bohr's postulate for stable orbits in hydrogen atom :** An electron can revolve only in those circular orbits in which its angular momentum is an integral multiple of $\dfrac{h}{2\pi}$.

If n is the principal quantum number of orbit, then an electron can revolve only in certain orbits or definite radii. These are called stable orbits.

De-Broglie explanation of stability of orbits : According to de Broglie, orbiting electron around the nucleus is associated with a stationary wave. Electron wave is a circular standing wave. Since destructive interference will occur if a standing wave does not close upon itself, only those de-Broglie waves exist for which the circumference of circular orbit contains a whole number of wavelengths *i.e.*, for orbit circumference of nth orbit as $2\pi r_n$,

$$2\pi r_n = n\lambda$$

$$2\pi r_n = n\left(\frac{h}{mv}\right)$$

or $$mvr_n = n\left(\frac{h}{2\pi}\right)$$

which is quantum condition proposed by Bohr.

Q. 2. Define the distance of closest approach. An α-particle of kinetic energy 'K' is bombarded on a thin gold foil. The distance of the closest approach is 'r'. What will be the distance of closest approach for an α-particle of double the kinetic energy ?★

Ans. It is defined as the minimum distance of the charged particle from the nucleus at which initial kinetic energy of the particle is equal to the potential energy due to the charged nucleus.

At this distance, K.E = P.E

★ are board based questions from previous year

$$K = \frac{1}{4\pi\varepsilon_0}\frac{Ze\,(2e)}{r}$$

$$r = \frac{1}{4\pi\varepsilon_0}\frac{2Ze^2}{K} \quad \text{...(i)}$$

If K is doubled,

$$r' = \frac{1}{4\pi\varepsilon_0}\frac{2Ze^2}{2K} \quad \text{...(ii)}$$

Dividing (ii) by (i), we get

$$\frac{r'}{r} = \frac{1}{2}$$

$$\Rightarrow \quad r' = \frac{r}{2}$$

Q. 3. In an experiment on α-particle scattering by a thin foil of gold, draw a plot showing the number of particles scattered versus the scattering angle θ.

Why is it that a very small fraction of the particles are scattered at θ > 90° ?

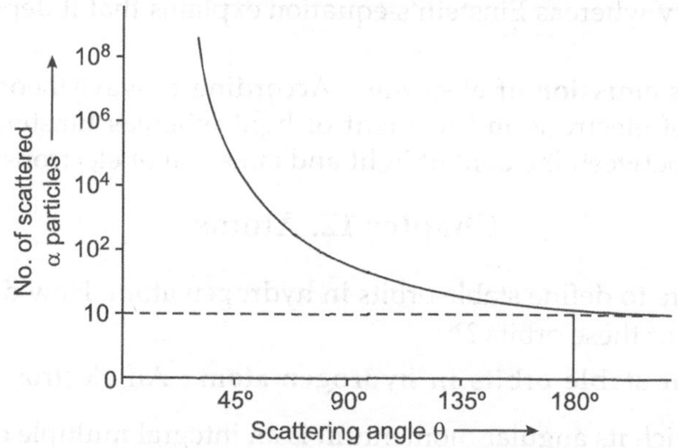

Ans. A small fraction of the alpha particles scattered at angle θ > 90° is due to the reason, that if impact parameter '*b*' reduces to zero, coulomb force increase, hence alpha particles are scattered at angle θ > 90°, and only one alpha particle is scattered at angle 180°.

Q. 4. The electron in hydrogen atom is initially in the third excited state. What is the maximum number of spectral lines which can be emitted, when it finally moves to the ground state ?

Ans. Total number of spectral lines emitted between the energy levels $n = 4$ (third excited state) and $n = 1$ (ground state) is given by

$$N = \frac{4\,(4-1)}{2} = 6$$

Fig. shows the schematic emission of these spectral lines.

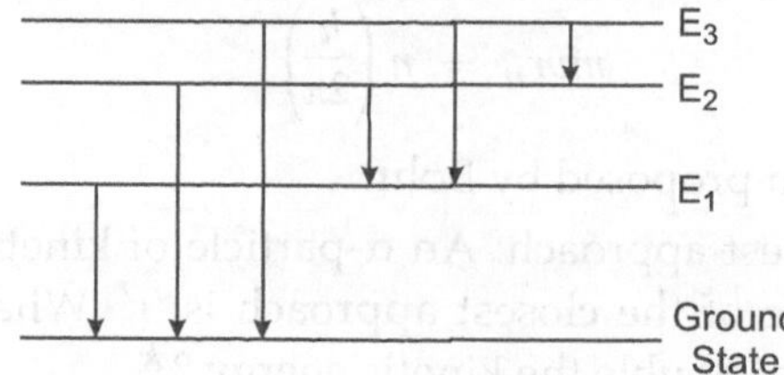

Q. 5. Explain, how Rutherford's experiment on scattering of α-particles led to the estimation of the size of the nucleus ?

Ans. As most of the α-particles passed straight through the gold foil and a very few were scattered back, it indicated that the central part of the atom must be positive and massive. As only 1 in about 8,000 α-particles retraced its path, it indicated that this central part should be only a very small portion of the whole atom. The size of this central part, called nucleus, was estimated to be about $\frac{1}{10,000}$ th of the size of the atom.

Q. 6. Use Bohr's model of hydrogen atom to obtain the relationship between the angular momentum and the magnetic moment of the revolving electron.★

Ans. Circulating electron around a heavy nucleus constitutes a current given by :

$$I = \frac{e}{T} = \frac{e}{\frac{2\pi R}{v}} = \frac{ev}{2\pi R}$$

So, there will be magnetic moment associated with this circulating current is given by :

$$\mu_e = I.A.$$

$$= \frac{ev}{2\pi R} \times R^2$$

$$= \frac{evR}{2} \quad \text{...(i)}$$

Angular momentum of circulating electron is given by :

$$l = mvR$$

So, $$vR = \frac{l}{m} \quad \text{...(ii)}$$

Put the value of equation (ii) in equation (i), we get

$$\mu_e = \frac{el}{2m}$$

which is the required expression.

Q. 7. Nuclear atom model is required to explain the results of Rutherford experiment. Why ?

Ans. In Rutherford experiment, a few α-particles are scattered through large angles. Therefore, massive and positive part of atom should be concentrated in a small portion of the atom. The small positive part having almost the entire mass of the atom was called nucleus. Therefore, the main feature of any atom model should require the presence of the central positive and massive nucleus.

Chapter 13. Nuclei

Q. 1. Explain pair production and pair annihilation.

Ans. **Pair production :** When an energetic γ-ray photon falls on a heavy substance, it is absorbed by some nucleus of the substance, and its energy gives rise to the production of an electron and a positron (Fig.). This phenomenon, in which energy is converted into mass, is called 'pair-production' and may be represented by the following equation :

$$\underset{(\gamma\text{-Photon})}{h\nu} = \underset{(\text{Electron})}{{}_{-1}\beta^0} + \underset{(\text{Positron})}{{}_{+1}\beta^0}$$

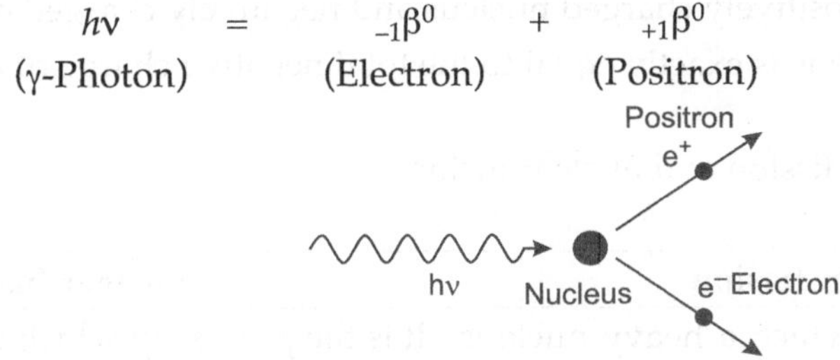

Pair Annihilation : Whenever an electron and a positron come very close to each other, they annihilate each other by combining together and two γ-photons (energy) are produced (Fig.). This

★ are board based questions from previous year

phenomenon in which mass is converted into energy, is called 'pair-annihilation, and is represented by the following equation :

$$_{-1}\beta^0 + _{+1}\beta^0 = h\nu + h\nu$$

(Electron) (Positron) (γ-Photon) (γ-Photon)

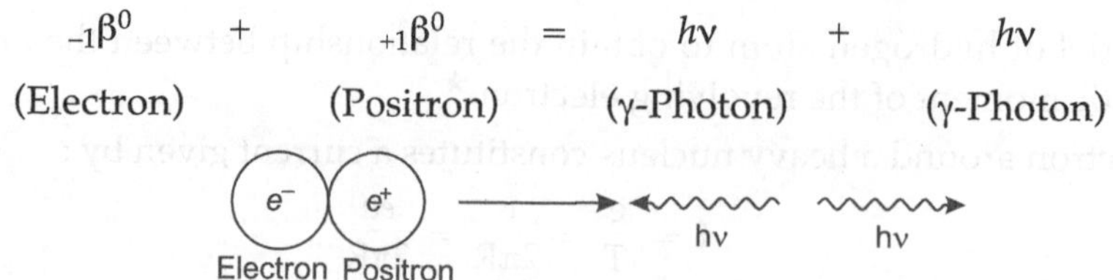

Q. 2. Draw a rough sketch showing the variation of binding energy per nucleon with mass number of various nuclei. Label the graph and state the region where the nuclei are most stable.

Ans. A graph between binding energy per nucleon and the mass number of different nuclei is shown in figure. It is clear from the graph that binding energy per nucleon for the nuclei having atomic numbers less than 20 and more than 80 is relatively lower than the nuclei having atomic number nearly 50.

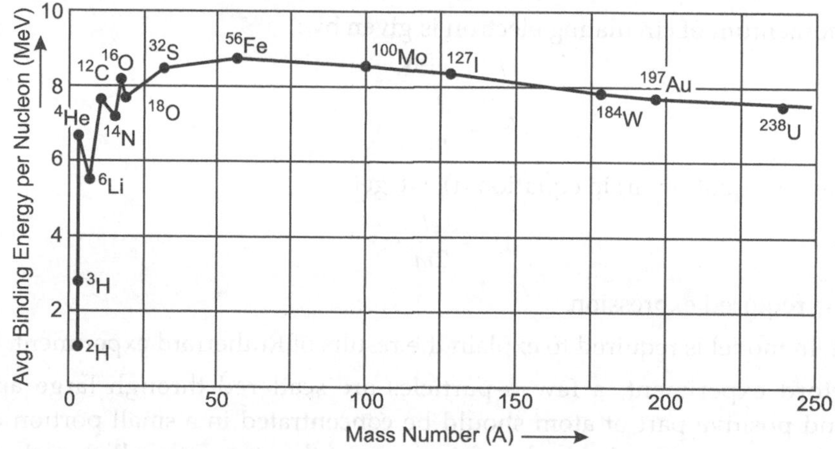

Hence, Fe (A = 56) like substances are most stable. It is easy to break nuclei for A < 20 or A > 80.

Q. 3. State the main assumptions of Rutherford's model of atom.

Ans. Assumptions of Rutherford's model of atom :

(i) An atom consists of a very small central core called the nucleus.

(ii) The nucleus carries all the positive charge and most of the mass of the atom (99.9%).

(iii) The size of the nucleus is of the order of 10^{-15} m which is very small in comparison to the size of the atom which is of the order of 10^{-10} m. Thus, major portion of atom is empty space.

(iv) The negatively charged electrons revolve around the nucleus in circular orbits with nucleus at the centre of the orbit. The necessary centripetal force for circular motion is provided by the force of attraction between positively charged nucleus and negatively charged electron.

(v) Positive charge on nucleus is exactly equal to the total negative charge of an electron so that atom is electrically neutral.

Q. 4. Distinguish between nuclear fission and nuclear fusion.

Ans.

	Nuclear fission	Nuclear fusion
(i)	It is the process in which a heavy nucleus split up into two lighter nuclei of nearly equal masses.	It is the process in which two lighter nuclei combine together to form a heavy nucleus.
(ii)	Comparatively less amount of energy is released.	Enormous amount of energy is released.
(iii)	Nuclear fission may take place at ordinary temperature.	A very high temperature of the order of million of degree is required.

(iv)	The sources of fissionable materials is limited.	The sources of fusion reaction *i.e.*, hydrogen is more plentiful (air and water).
(v)	The products of nuclear fission are in general radioactive and hence pose a radiation hazard.	The products of fusion are non-radioactive and pose no radiation hazard.

Q. 5. Define half-life of a radioactive substance. Establish a relationship between half-life (T) and decay constant ?

Ans. Half life of a radioactive substance is defined as the time taken by half the number of the atoms of the radioactive element, to disintegrate completely.

$$N = N_0 e^{-\lambda t}$$

When, $$t = T, N = \frac{N_0}{2}$$

$\Rightarrow$ $$\frac{N_0}{2} = N_0 e^{-\lambda T}$$

$\Rightarrow$ $$\frac{1}{2} = e^{-\lambda T}$$

or $$\frac{2}{1} = e^{+\lambda T}$$

or $$\log_e 2 = \lambda T$$

$\therefore$ $$T = \frac{\log_e 2}{\lambda} = \frac{0.6931}{\lambda}$$

Q. 6. What is nuclear fusion ? How energy is obtained in this process ? Give a suitable example.

Ans. When two lighter nuclei moving at very high speeds fuse together to form a single heavier nucleus then this phenomenon is called 'nuclear fusion'. The mass of the nucleus obtained after fusion is less than the sum of the nuclei which are fused together. The lost mass is obtained in the form of energy. For example, by the fusion of two nuclei of heavy hydrogen or deuterium (${}_1H^2$), the following is possible

$${}_1H^2 + {}_1H^2 \longrightarrow {}_1H^3 + {}_1H^1 + 4.0 \text{ MeV}$$

The nucleus of tritium (${}_1H^3$), so formed can again fuse with a deuterium nucleus.

$${}_1H^3 + {}_1H^2 \longrightarrow {}_2He^4 + {}_0n^1 + 17.6 \text{ MeV}$$

The net result of these two reactions is that three deuterium nuclei fuse together to form a helium nucleus and liberates 21.6 MeV energy.

Q. 7. Explain the processes of nuclear fission and nuclear fusion by using the plot of binding energy per nucleon (BE/A) versus the mass number A.★

Ans. Plot of binding energy per nucleon mass number :

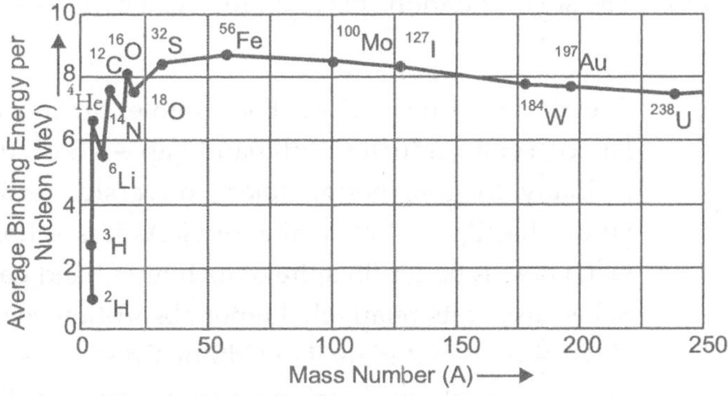

(i) When we move from the heavy nuclei region to the middle region of the plot, we find that there will be a gain in the overall binding energy and hence release of energy. This indicates that energy can be released when a heavy nucleus (A ≈ 240) breaks into two roughly equal fragments. This process is called nuclear fission.

(ii) Similarly, when we move from lighter nuclei to heavier nuclei, we again find that there will be gain in the overall binding energy and hence release of energy. This indicates that energy can

★ are board based questions from previous year

be released when two or more lighter nuclei fuse together to form a heavy nucleus. This process is called nuclear fusion.

Q. 8. (i) Write symbolically the process expressing the β + decay of ${}^{22}_{11}$Na. Also write the basic nuclear process underlying this decay.

(ii) Is the nucleus formed in the decay of the nucleus ${}^{22}_{11}$Na, an isotope or isobar ?★

Ans. (i) The β^+ decay of ${}_{11}N^{22}$ is expressed as below :

$${}_{11}N^{22} \rightarrow {}_{10}Ne^{22} + 1e^{0} + \nu$$

The neutron-proton ratio is also an important factor that determines the stability of a nucleus. A proton-rich nucleus undergoes positron emission (β^+ decay), so as to improve the neutron-proton ratio. The positron emission decreases proton number relative to neutron number and the product nucleus becomes relatively more stable. In β^+ decay, neutrino (ν) is emitted so as to conserve spin.

(ii) The nucleus so formed is an isobar of ${}^{22}_{11}$Na because the mass number is same, but the atomic numbers are different.

Q. 9. (a) Define one Becquerel.★

(b) A radioactive substance disintegrates into two types of daughter nuclei, one type with disintegration contant λ_1 and the other type with disintegration constant λ_2. Determine the half-life of the radioactive substance.★

Ans. (a) One Becquerel (Bq) is defined as the activity of a quantity of radioactive sample in which one nucleus decays per second.

(b) $X \xrightarrow{\lambda_1} D_1$

$X \xrightarrow{\lambda_2} D_2$

(Where, X is a radioactive nuclei and D_1, D_2 are the daughter nuclei)

$$\lambda_{effective} = \lambda_1 + \lambda_2$$

$$t_{1/2} = \frac{\log 2}{\lambda_{effective}} = \frac{0.693}{\lambda_1 + \lambda_2}$$

Chapter 14. Semiconductor Electronics : Materials, Devices and Simple Circuits

Q. 1. Why are Si and GaAs are preferrerd materials for solar cells ? **[NCERT]**

Ans. The solar radiation spectrum received by us is shown in Fig. —

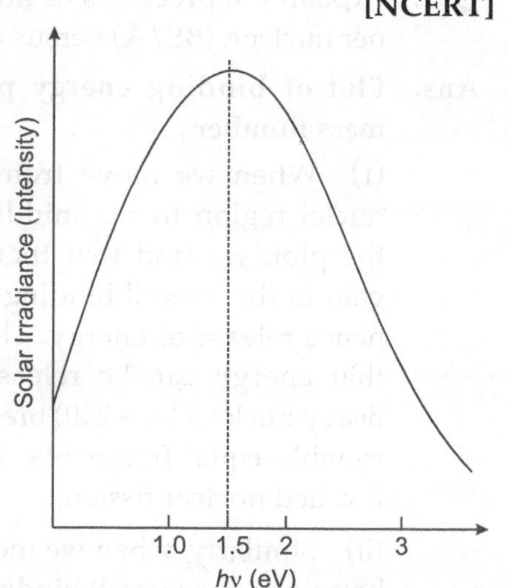

The maxima is near 1.5 eV. For photo-excitation, $h\nu > E_g$. Hence, semiconductor with band gap ~ 1.5 eV or lower is likely to give better solar conversion efficiency. Silicon has E_g ~ 1.1 eV while for GaAs it is ~ 1.53 eV. In fact. GaAs is better (in spite of its higher band gap) than Si because of its relatively higher absorption coefficient. If we choose materials like CdS or CdSe (E_g ~ 2.4 eV), we can use only the high energy component of the solar energy for photo-conversion and a significant part of energy will be of no use.

★ are board based questions from previous year

Q. 2. Distinguish between intrinsic and extrinsic semiconductors.★

Ans.

	Intrinsic Semiconductors	Extrinsic Semiconductors
(i)	It is a pure, natural semiconductor, such as pure Ge and pure Si.	It is prepared by adding a small quantity of impurity to a pure semiconductor, such as *n*-type and *p*-type semiconductors.
(ii)	Intrinsic charge carriers are electrons and holes with equal concentration.	In *p*-type semiconductor majority charge carriers are holes and minority charge carriers are electrons, while vice-versa in *n*-type semiconductor.
(iii)	Its electrical conductivity is very low.	Its electrical conductivity is significantly high.
(iv)	Its conductivity cannot be controlled.	Its conductivity can be controlled by adjusting the quantity of the impurity added.
(v)	Its conductivity increases exponentially with temperature.	Its conductivity also increases with temperature but not exponentially.

Q. 3. Distinguish between *n*-type and *p*-type semiconductors.★

Ans.

	n-type Semiconductors	*p*-type Semiconductors
(i)	It is an extrinsic semiconductor obtained by adding a pentavalent impurity to a pure intrinsic semiconductor.	It is also an extrinsic semiconductor obtained by adding a trivalent impurty to a pure intrinsic semiconductor.
(ii)	The impurity atoms added provide extra free electrons to the crystal lattice and are called donor atoms.	The impurity atoms added create holes in the crystal lattice and are called acceptor atoms because they created holes accept electrons.
(iii)	The electrons are majority carriers and the holes are minority carriers.	The holes are majority carriers and the electrons are minority carriers.
(iv)	The electrons concentration is much more than the hole concentration ($n_e >> n_h$).	The hole concentration is much more than the electron concentration ($n_h >> n_e$).

Q. 4. Draw circuit diagrams to illustrate forward biasing and reverse biasing of a diode ?

Ans. **Forward biasing of a diode :** In this arrangement, the positive terminal of battery is connected to *p*-end and negative terminal to *n*-end of the crystal, so that an external electric field E is established directed from *p* to *n*-end to oppose the internal field E_i. Thus, the junction is said to be conducting.

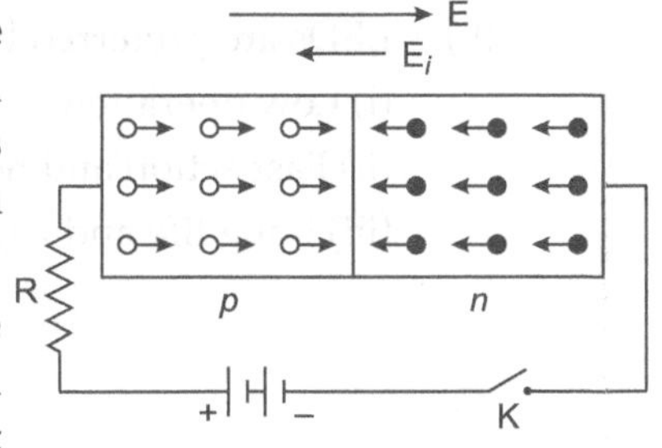

Reverse biasing of a diode : In this arrangement the positive terminal of battery is connected to *n*-end and negative to *p*-end of the crystal, so that the external field is established to support the internal field E_i.

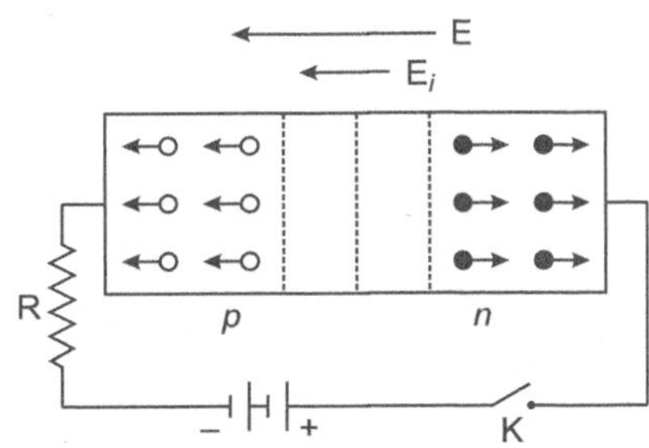

★ are board based questions from previous year

Q. 5. (a) Why is a photodiode operated under reverse bias condition?★

(b) Draw V-I characteristics curves of photo diode for incident light of intensities I_1 and I_2 ($I_1 > I_2$).★

Ans. (a) Photodiodes convert incident light to electric current more effectively in reverse bias condition than in forward bias because the width of the depletion region increases as we increase the applied reverse bias voltage across the diode (directly proportional) in a reverse biased *p-n* junction. Hence, by applying a larger voltage, more of the incident photons converted to electric current. So, photodiodes work better in reverse biased condition.

(b)

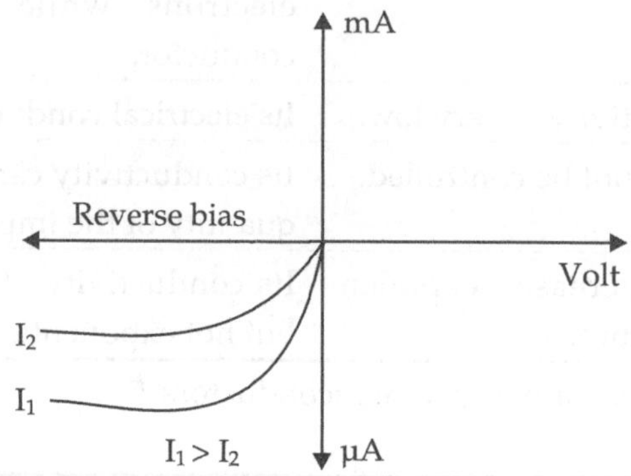

Q. 6. (a) State the level of doping and biasing condition used in light emitting diode (LED).★

(b) Write any two advantages of LED over the conventional low power lamps.★

Ans. (a) LED is a heavily doped *p-n* junction which under forward bias emits spontaneous radiations. When the diode is forward biased, electrons are sent from $n \longrightarrow p$ and holes are sent from $p \longrightarrow n$. At the junction boundary, the concentration of minority carriers increases as compared to the equilibrium conditions. Thus, at the junction boundary on either side of the junction, excess minority carriers are there, which recombine with majority carriers near the junction and on recombination, the energy is released in the form of photons. When the forward current of diode is small, the intensity of light emiting is small and increases when light reaches maximum.

Further, increase in forward current results in decrease of light intensity. LEDs are biased such that the light emiting efficiency is maximum.

(b) LEDs are preferred because :

(i) Low operational voltage and less power.

(ii) Fast action and no warmup time required.

(iii) Long life and ruggedness.

□□

★ are board based questions from previous year

Question Set 4

Assertion and Reason Based Questions

Directions: In the following questions, a statement of assertion is followed by a statement of reason. Mark the correct choice as :

(a) If both assertion and reason are true and reason is the correct explanation of assertion.

(b) If both assertion and reason are true but reason is not the correct explanation of assertion.

(c) If assertion is true but reason is false.

(d) If both assertion and reason are false.

Chapter 1. Electric Charges and Fields

1. **Assertion :** The charge on any body can be increased or decreased in terms of e.

   Reason : Quantization of charge means that the charge on a body is the integral multiple of e.

2. **Assertion :** The properties that the force with which two charges attract or repel each other are not affected by the presence of a third charge.

   Reason : Force on any charge due to a number of other charges is the vector sum of all the forces on that charge due to other charges, taken one at a time.

3. **Assertion :** When we rub a glass rod with silk, the rod gets negatively charged and the silk gets positively charged.

   Reason : On rubbing, electrons from silk cloth move to the glass rod.

4. **Assertion :** Consider two identical charges placed distance $2d$ apart, along x-axis. The equilibrium of a positive test charge placed at the point O midway between them is stable for displacements along the x-axis.

   Reason : Force on test charge is zero.

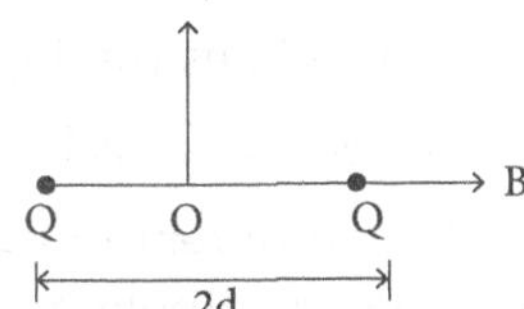

5. **Assertion :** Coulomb force is the dominating force in the universe.

   Reason : Coulomb force is weaker than the gravitational force.

6. **Assertion :** When charges are shared between any two bodies, no charge is really lost, but some loss of energy does occur.

   Reason : Some of the energy is dissipated in the form of heat, sparkling, etc.

7. **Assertion :** If there exists coulomb attraction between two bodies, both of them must be charged.

   Reason : In coulomb attraction two bodies are positively charged.

8. **Assertion :** The positive charge particle is placed in front of a spherical uncharged conductor. The number of lines of forces terminating on the sphere will be more than those emerging from it.

   Reason : The surface charge density at a point on the sphere nearest to the point charge will be negative and maximum in magnitude compared to other points on the sphere.

9. **Assertion :** A point charge is brought in an electric field, the field at a nearby point will increase or decrease, depending on the nature of charge.

   Reason : The electric field is independent of the nature of charge.

10. **Assertion :** On going away from a point charge or a small electric dipole, electric field decreases at the same rate in both the cases.

    Reason : Electric field is inversely proportional to cube of distance from the charge or an electric dipole.

11. **Assertion :** Electrostatic field lines start at positive charges and end at negative charges.
Reason : Field lines are continuous curves without any breaks and they form closed loop.

12. **Assertion :** Total flux through a closed surface is zero if no charge is enclosed by the surface.
Reason : Gauss law is true for any closed surface, no matter what its shape or size is.

13. **Assertion :** Gauss's law shows diversion when inverse square law is not obeyed.
Reason : Gauss's law is a consequence of conservation of charges.

Answers

1. **(a)** Protons and electrons are the only basic charges in the universe. All the observable charges have to be integral multiple of *e*. Thus, if a body contains n_1 electrons and n_2 protons, the total amount of charge on the body is $n_2e + n_1(-e) = (n_2 - n_1)e$. Since n_1 and n_2 are integers, their difference is also an integer. Thus, the charge on any body is always an integral multiple of *e* and can be increased or decreased in terms of *e*.
2. **(b)** Force on any charge due to a number of other charges is the vector sum of all the forces on that charge due to the other charges, taken one at a time. The individual force is unaffected due to the presence of other charges. This is the principle of superposition of charges.
3. **(d)** When we rub a glass rod with silk cloth, electrons from the glass rod are transferred to the silk cloth. Thus the rod gets positively charged and the silk gets negatively charged.
4. **(b)** If positive charge is displaced along *x*-axis, then net force will always act in a direction opposite to that of displacement and the test charge will always come back to its original position.
5. **(d)** Gravitational force is the dominating force in nature and not coulomb's force. Gravitational force is the weakest force. Also, Coulomb's force >> gravitational force.
6. **(a)** Charges are conserved by the law of conservation of charges. Energy is also conserved, if we take in account of the loss of energy by heat, sparking etc.
7. **(d)** Coulomb attraction exists even when one body is charged, and the other is uncharged.
8. **(d)** No. of lines entering the surface = No. of lines leaving the surface.
9. **(c)** The electric field will increase if positive charge is brought in an electric field.
10. **(d)** The rate of decrease of electric field is different in the two cases. In case of a point charge, it decreases as $1/r^2$ but in the case of electric dipole it decreases more rapidly, as $E \propto 1/r^3$.
11. **(c)** Electrostatic field lines are continuous curves without any breaks. They start at positive charges and end at negative charges. They cannot form closed loops.
12. **(b)** Gauss's law implies that the total electric flux through a closed surface is zero if no charge is enclosed by the surface and it is true for any closed surface, independent of its shape and size.
13. **(b)** Gauss's theorem is based on inverse square dependence of electric field on distance.

Chapter 2. Electrostatic Potential and Capacitance

1. **Assertion :** Work done in moving a charge between any two points in an electric field is dependent of the path followed by the charge, between these points.
Reason : Electrostatic force is a non-conservative force.

2. **Assertion :** The electric potential at any point on the equatorial plane of a dipole is non-zero.
Reason : The work done in bringing a unit positive charge from infinity to a point in equatorial plane is not equal for the two charges of the dipole.

3. **Assertion :** For a non-uniformly charged thin circular ring with net charge is zero, the electric field at any point on axis of the ring is zero.
Reason : For a non-uniformly charged thin circular ring with net charge zero, the electric potential at each point on axis of the ring is non-zero.

4. **Assertion :** For a point charge, concentric spheres centered at a location of the charge are equipotential surfaces.

   Reason : An equipotential surface is a surface over which potential has constant value.

5. **Assertion :** Electric potential and electric potential energy are different quantities.

   Reason : For a system of positive test charge and point charge electric potential energy = electric potential.

6. **Assertion :** Surface of a symmetrical conductor can be treated as equipotential surface.

   Reason : Charges can easily flow in a conductor.

7. **Assertion :** Two adjacent conductors of unequal dimensions, carrying the same positive charge have a potential difference between them.

   Reason : The potential of a conductor depends upon the charge given to it.

8. **Assertion :** When a dielectric slab is gradually inserted between the plates of an isolated parallel plate capacitor, the energy of the system decreases.

   Reason : The force between the plates decreases.

9. **Assertion :** Polar molecules have temporary dipole moment.

   Reason : In polar molecule, the centres of positive and negative charges coincide even when there is no external field.

10. **Assertion :** If three capacitors of capacitance $C_1 < C_2 < C_3$ are connected in parallel then their equivalent capacitance $C_p > C_s$.

    Reason : $\frac{1}{C_p} = \frac{1}{C_1} + \frac{1}{C_2} + \frac{1}{C_3}$

11. **Assertion :** If the distance between parallel plates of a capacitor is halved and dielectric constant is made three times, then the capacitor becomes six times.

    Reason : Capacity of the capacitor does not depend upon the nature of the material.

12. **Assertion :** The capacity of a given conductor remains same even if charge is varied on it.

    Reason : Capacitance depends upon nearby medium as well as size and shape of conductor.

13. **Assertion :** In Van de Graaff generator, the process of spraying the charge is called electron discharge.

    Reason : Van de Graaff generator produces high voltage and high current.

Answers

1. **(d)** Electrostatic force is conservative force.
2. **(c)** Two equipotential surfaces are not necessarily parallel to each other.
3. **(d)** For a non-uniformly charged thin circular ring with net zero charge, electric potential at each point on its axis is zero. Hence electric field at each point on its axis must be perpendicular to the axis.
4. **(b)** An equipotential surface is a surface over which potential is constant.
5. **(c)** Potential and potential energy are different quantities and cannot be equated.
6. **(a)** Potential is constant on the surface of a sphere so it behaves as an equipotential surface.
7. **(b)** Let us consider two spherical shells of radii r_1 and r_2 which are possessing the same positive charge Q. Therefore the potential on the surface of each conductor will be,

   $V_1 = \frac{kQ}{r_1}$ and $V_2 = \frac{kQ}{r_1}$ since $r_1 \neq r_2$

   $V_1 = V_2$ and consequently there will be a potential difference between the two conductors.

8. **(c)** $C' = KC$, and $U' = \frac{q^2}{2C'} = \frac{q^2}{2KC}$
9. **(d)** The molecules of a substance may be polar or non-polar. In a non-polar molecule, the centres of positive and negative charges coincide. This molecule has no permanent dipole moment.

On the other hand, a polar molecule is one is which the centres of positive and negative charges are separated, even when there is no external field. Such molecules have a permanent dipole moment.

10. (c) Equivalent capacitance of parallel combination is $C_p = C_1 + C_2 + C_3$.
11. (b) By the formula capacitance of a capacitor,

$$C_1 = \varepsilon_0 \times \frac{KA}{d}$$

$$C_1 \propto \frac{K}{d}$$

$$\therefore \quad \frac{C_1}{C_2} = \frac{K_1}{d_1} \times \frac{d_2}{K_2} = \frac{K_1}{K_2} \times \frac{d/2}{3\,K} = \frac{1}{6}$$

or $$C_2 = 6\,C_1$$

Again for capacity of a capacitor $C = Q/V$. Therefore, capacity of a capacitor does not depend upon the nature of the material of the capacitor.

12. (a) Capacitance is basically a geometrical quantity.
13. (d) In Van de Graaff generator, the process of spraying the charge is called corona discharge. Van de Graaff generator produces high voltage and low current.

Chapter 3. Current Electricity

1. **Assertion :** The current density $\vec{j}$ at any point in ohmic resistor is in direction of electric field $\vec{E}$ at that point.
   Reason : A point charge when released from rest in a region having only electrostatic field always moves along electric lines of force.
2. **Assertion :** The 200 W bulbs glows with more brightness then 100 W bulbs.
   Reason : A 100 W bulb has more resistance than a 200 W bulb.
3. **Assertion :** Bending a wire does not effect electrical resistance.
   Reason : Resistance of wire is proportional to resistivity of material.
4. **Assertion :** Fuse wire must have high resistance and low melting point.
   Reason : Fuse is used for small current flow only.
5. **Assertion :** Drift speed v_d is the average speed between two successive collisions.
   Reason : If Δl is the average distance moved between two collision and Δt is the corresponding time, then

$$v_d = \lim_{\Delta t \to 0} \frac{\Delta l}{\Delta t}.$$

6. **Assertion :** Two electric bulbs of 50 W and 100 W are given. When connected in series 50 W bulb glows more but when connected parallel 100 W bulb glows more.
   Reason : In series combination, power is directly proportional to the resistance of circuit. But in parallel combination, power is inversely proportional to the resistance of the circuit.
7. **Assertion :** When current through a bulb decreases by 0.5 %, the glow of bulb decreases by 1 %.
   Reason : Glow (Power) which is directly proportional to square of current.
8. **Assertion :** For a conductor resistivity increases with increase in temperature.
   Reason : Since $\rho = \frac{m}{ne^2\tau}$ when temperature increases the random motion of free electrons increases and vibration of ions increases which decreases τ.
9. **Assertion :** Two bulbs of same wattage, one having a carbon filament and the other having a metallic filament are connected in series. Metallic bulbs will glow more brightly than carbon filament bulb.
   Reason : Carbon is a semiconductor.

10. **Assertion** : A current flows in a conductor only when there is an electric field within the conductor.

    Reason : The drift velocity of electron in presence of electric field decreases.

11. **Assertion:** In practical application, power rating of resistance is not important.

    Reason : Property of resistance remain same even at high temperature.

12. **Assertion :** Kirchhoff's junction rule can be applied to a junction of several lines or a point in a line.

    Reason : When steady current is flowing, there is no accumulation of charges at any junction or at any point in a line.

13. **Assertion :** In meter bridge experiment, a high resistance is always connected in series with a galvanometer.

    Reason : As resistance increase current more accurately then ammeter.

14. **Assertion :** The emf of driver cell in potentiometer experiment should be greater than emf of cell to be determined.

    Reason : The fall of potential across the potentiometer wire should not be less than emf of cell to be determined.

Answers

1. **(c)** From relation $\vec{j} = \sigma \vec{E}$ the current density $\vec{j}$ at any point in ohmic resistor is in direction of electric field $\vec{E}$ at that point. In space having non-uniform electric field, charges released from rest may not move along electric line of force. Hence, assertion is correct while reason is incorrect.

2. **(a)** As we know that,

$$P = \frac{V^2}{R}$$

$$\Rightarrow \qquad R = \frac{V^2}{P}$$

$$\Rightarrow \qquad R \propto \frac{1}{P}$$

That means higher is the wattage of a bulb, lesser is the resistance and so it will glow bright.

3. **(a)** Resistance of wire $R = \rho \frac{l}{A}$

where ρ is resistivity of material which does not depend on the geometry of wire. Since when wire is bent resistivity, length and area of cross-section do not change, therefore resistance of wire also remains same.

4. **(c)** Fuse wire must have high resistance because in series current remains same, therefore according to Joule's law,

$$H = \frac{I^2Rt}{4.2} \text{ cal}$$

Heat produced is high if R is high. The melting point must be low so that wire may melt with increase in temperature. As the current equal to maximum safe value, flows through the fuse wire, it heats up, melts and break the circuit.

5. **(c)** Drift speed is the average speed between two successive collision.

6. **(a)** Resistance of 50 W bulb is two times the resistance of 100 W bulb. When bulbs are connected in series, 50 W bulb will glow more as $P = i^2R$ (current remains same in series). In parallel, the 100 W bulb will glow more as $P = \frac{V^2}{R}$ (potential difference remains same in parallel).

7. **(a)** As we know that,

Glow = Power (P) = I^2R

$$\therefore \quad \frac{dP}{P} = 2\left(\frac{dI}{I}\right) = 2 \times 0.5 = 1\%.$$

8. **(a)** When temperature increases the random motion of electrons and vibration of ions increases which results in more frequent collisions of electrons with the ions. Due to this the average time between the successive collisions, denoted by τ, decreases which increases ρ.

9. **(d)** When two bulbs are connected in series, the resistance of the circuit increases and so the voltage in each decreases, hence the brightness and the temperature also decreases. Due to decrease in temperature, the resistance of the carbon filament will slightly increase while that of metal filament will decrease. Hence, carbon filament bulb will glow more brightly $P = i^2R$. Also carbon is not a semiconductor.

10. **(c)** Before the presence of electric field, the free electrons move randomly in the conductor, so their drift velocity is zero and therefore there is no current in the conductor. In the presence of electric field, each electron in the conductor experience a force in a direction opposite to the electric field. Now the free electrons are accelerated from negative and to the positive end of the conductor and hence a current starts to flow from the conductor.

11. **(d)** Because of heat production every resistance has a maximum power rating, the maximum power that can be dissipated without overheating the device. When this rating is exceeded, heat is produced, due to which resistance may change unpredictably.

12. **(a)** **Junction rule or Kirchhoff's first law or Kirchhoff's current law (KCL)** states that the algebraic sum of the currents meeting at a junction (point) in an electrical circuit is always zero. Or, the sum of currents flowing towards the junction is equal to sum of currents leaving the junction.

$$\Sigma I = 0$$

$$I_1 + I_3 = I_2 + I_4$$

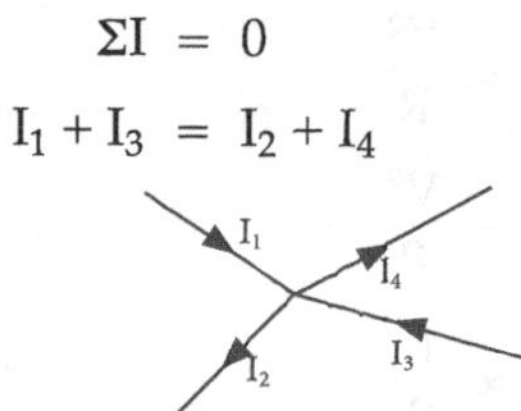

13. **(c)** The resistance of the galvanometer is fixed. In meter bridge experiments, to protect the galvanometer from a high current, high resistance is connected to the galvanometer in order to protect it from damage.

14. **(a)** If either emf of the driver cell or potential difference across the whole potentiometer wire is lesser than the emf of then experimental cell, then balance point will not obtained.

Chapter 4. Moving Charges and Magnetism

1. **Assertion :** When a test charge moves through the magnetic field, its momentum changes but kinetic energy remains constant.

Reason : The magnetic force acts as a centripetal force, which is perpendicular to the instantaneous velocity and so does no work.

2. **Assertion :** Cyclotron does not accelerate electron.

Reason : Mass of the electron is very small.

3. **Assertion :** The frequency of circular motion of a charged particle in cyclotron is independent of the mass of the particle.

Reason : Smaller the mass of the particle less will be the frequency of the particle.

4. **Assertion :** Magnetic field interacts with a moving charge and not with a stationary charge.

Reason : A moving charge produces a magnetic field.

5. **Assertion :** Free electron always keeps on moving in a conductor even then no magnetic force act on them in magnetic field unless a current is passed through it.

   Reason : The average velocity of free electron is zero.

6. **Assertion:** Ampere's law used for the closed loop shown in figure is written as $\oint \vec{B} . \vec{dl} = \mu_0 (i_1 - i_2)$. Right side of it does not include i_3 because it produces no magnetic field at the loop.

   Reason : The line integral of magnetic field produced by i_3 over the closed loop is non zero.

7. **Assertion :** Two beams of electrons travelling in the same direction repel each other.

   Reason : The electrostatic interaction is less than the magnetic interaction.

8. **Assertion :** If the current in a solenoid is reversed in direction while keeping the same magnitude, the magnetic field energy stored in the solenoid decreases.

   Reason : Magnetic field energy density is proportional to square of current.

9. **Assertion :** The magnetic field produced by a current carrying solenoid is independent of its length and cross-sectional area.

   Reason : The magnetic field inside the solenoid is uniform.

10. **Assertion :** If an electron, while coming vertically from outer space, enters the earth's magnetic field, it is deflected towards west.

    Reason : Electron has negative charge.

11. **Assertion :** If two long wires, hanging freely are connected to a battery in series, they come closer to each other.

    Reason : Force of attraction acts between the two wires carrying current.

12. **Assertion :** In a shunted galvanometer only 10% current passes through the galvanometer. The resistance of the galvanometer is G. Then resistance of the shunt is G/9.

    Reason : If S is the resistance of the shunt, then voltage across S and G is same.

13. **Assertion :** To convert a galvanometer into an ammeter a small resistance is connected in parallel with it.

    Reason : The small resistance increases the combined resistance of the combination.

14. **Assertion :** An ammeter is always connected in series whereas a voltmeter is connected in parallel.

    Reason : An ammeter is a low resistance galvanometer while a voltmeter is high resistance galvanometer.

15. **Assertion :** If an electron and proton enter a perpendicular magnetic field with equal momentum, then radius of curve for electron is; more than that of proton.

    Reason : Electron has less mass than proton.

Answers

1. **(a)** Kinetic energy of the charged particle remains same in the circular path while velocity and momentum of the particle changes because of continuous change in the direction of motion.
2. **(a)** Cyclotron is suitable for accelerating heavy particles like protons, α-particles etc, and not for electrons because of low mass. Because electrons acquire very high velocities very near to velocity of light and appreciable variation in their mass, occurs.
3. **(d)** The frequency of revolution is given by,

$$\nu = \frac{qB}{2\pi m}$$

4. **(a)** A moving charge experiences a force in magnetic field. It is because of interaction of two magnetic fields, one which is produced due to the motion of charge and other in which charge is moving.

5. **(a)** In the absence of the electric current, the free electrons in a conductor are in a state of random motion, like molecules in a gas. Their average velocity is zero *i.e.*, they do not have any net velocity in a direction. As a result, there is no net magnetic force on the free electrons in the magnetic field. On passing the current, the free electrons acquire drift velocity in a definite direction, hence magnetic force acts on them, unless the field has no perpendicular component.
6. **(d)** The magnetic field at any point on the closed loop is due to all the three currents, but line integral of i_3 over the closed loop will be zero.
7. **(c)** Two beams of electrons travelling in the same direction repel each other because the electrostatic interaction is more than the magnetic interaction.
8. **(d)** Reversing the direction of the current reverses the direction of the magnetic field. However, it has no effect on the magnetic-field energy density, which is proportional to the square of the magnitude of the magnetic field.
9. **(b)** We know that,

$$B = \mu_0 n I$$

which is independent of length and cross-sectional area.

10. **(b)** We know that the direction of the earth's magnetic field is toward north and the velocity of electron is vertically downward. Applying Fleming's left hand rule, the direction of force is towards west. Therefore, an electron coming from outer space will be deflected toward west.
11. **(d)** When two long parallel wires, are connected to a battery in series. They carry currents in opposite directions, hence they repel each other.
12. **(b)** Potential drop across galvanometer = Potential drop across the shunt.

i.e., $$I_g G = (I - I_g) S$$

$\Rightarrow$ $$S = \frac{I_g}{I - I_g} G$$

For $$I_g = \frac{I}{10}$$

$$S = \frac{\frac{I}{10}}{\left(I - \frac{I}{10}\right)} G = \frac{G}{9}$$

13. **(c)** An ammeter should have a low resistance which we get when we connect low resistance in parallel with galvanometer.
14. **(a)** An ammeter is a low resistance device and is connected in series so as the whole circuit current flows through it for an accurate measurement. A voltmeter is a device of having high resistance. So, if we connect it in series, it would hinder the current flow in the circuit hence open circuit results.
15. **(b)** When a charged particle enters in perpendicular magnetic field than radius of curved path is given by,

$$r = \frac{mv}{qB} = \frac{p}{qB}$$

As momentum p as constant,

$$r \propto \frac{1}{q}$$

As e^- and proton have same charge,

$$\frac{\bar{r}_e}{\overset{+}{r}_p} = \frac{\bar{q}_0}{\overset{+}{q}_e} = 1$$

Assertion is false, but reason is true.

Chapter 5. Magnetism and Matter

1. **Assertion :** Magnetic field on the axial line at a certain distance is twice as compared to that on the equatorial line at the same distance.
Reason : Electric field due to a dipole varies inversely as the cube of the distance.
2. **Assertion :** Two short magnets are placed on a cork which floats on water. The magnets are placed such that the axis of one produced bisects the axis of other at right angles. Then the cork has neither translational nor rotational motion.

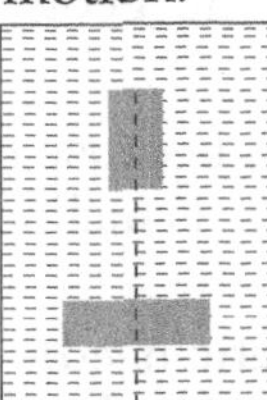

Reason : Net force on the cork is zero.
3. **Assertion :** When a magnetic dipole is placed in a non uniform magnetic field, only a torque acts on the dipole.
Reason : Force would not act on dipole if magnetic field were uniform.
4. **Assertion :** Magnetic moment of helium atom is zero.
Reason : All the electron are paired in helium atom orbitals.
5. **Assertion :** Gauss theorem is not applicable in magnetism.
Reason : Mono magnetic pole does not exist.
6. **Assertion :** The true geographic north direction is found by using a compass needle.
Reason : The magnetic meridian of the earth is along the axis of rotation of the earth.
7. **Assertion :** In the northern hemisphere the north pole of the dip needle dips downwards.
Reason : The north pole of earth as a bar magnet lies in the northern hemisphere.
8. **Assertion :** If a compass needle be kept at magnetic north pole of the earth, the compass needle may stay in any direction.
Reason : Dip needle will stay vertical at the north pole of earth.
9. **Assertion :** The susceptibility of diamagnetic materials does not depend upon temperature.
Reason : Every atom of a diamagnetic material is not a complete magnet in itself.
10. **Assertion :** To protect any instrument from external magnetic field, it is put inside an iron box.
Reason : Iron is a magnetic substance.
11. **Assertion :** A disc-shaped magnet is levitated above a superconducting material that has been cooled by liquid nitrogen.
Reason : Superconductors repel a magnet.
12. **Assertion :** For making permanent magnets, steel is preferred over soft iron.
Reason : As retentivity of steel is smaller.

Answers

1. **(b)** In a short bar magnet the value of magnetic field at a point on the axial line of the magnet is,
$$B_1 = \frac{\mu_0}{4\pi} \cdot \frac{2M}{r^3}$$
and on equatorial line at the same distance is,
$$B_2 = \frac{\mu_0}{4\pi} \cdot \frac{M}{r^3}$$
Therefore, $B_1 = 2\,B_2$.
2. **(a)** Since both the magnets exert equal and opposite force/torque on each other. Hence, net force/torque on cork will be zero.

3. **(d)** In a non uniform magnetic field, both a torque and a net force acts on the dipole. If magnetic field were uniform, net force on dipole would be zero.
4. **(a)** Helium atom has paired electrons so their electron spin are opposite to each other and hence its net magnetic moment is zero.
5. **(a)** The magnetic flux though any closed surface is zero.
6. **(d)** The compass needle enables us to locate magnetic north pole. If magnetic declination at that particular place is known, then true geographic north-south direction can be located. Therefore, the assertion is false.

   Similarly, reason is also false. This is because magnetic meridian is the vertical plane passing through magnetic axis. Magnetic axis is inclined at a certain angle θ to geographical axis and earth rotates about geographic axis.
7. **(c)** In the northern hemisphere, magnetic needle comes to rest along north-south direction. So that a greater dip angle is expected in northern hemisphere.
8. **(b)** At magnetic poles of the earth, the only vertical component of the earth's field acts, horizontal component is zero. A compass needle is free to rotate in horizontal plane and is affected by horizontal component only. Thus there will be no effect on the magnetic field on the compass needle. So the needle may stop in any direction. The angle of dip at the magnetic north pole is 90° and hence the dip needle will become vertical.
9. **(a)** As every atom of a diamagnetic material is not a complete magnet in itself, its susceptibility is not affected by the temperature.
10. **(b)** Iron is ferromagnetic in nature. Lines of force due to external magnetic field prefer to pass through iron.
11. **(b)** Superconductors occur only below a particular temperature and above that temperature they behave as normal material. If a magnet is placed above superconductor and cooled by using nitrogen, the magnetic field inside the superconductor is exactly equal and opposite in direction to applied magnetic field. They are cancelled within the superconductor and these poles repel each other. Due to this force of repulsion, magnet is repelled.
12. **(b)** Steel is preferred over soft iron for making permanent magnets, because coercivity of steel is larger.

Chapter 6. Electromagnetic Induction

1. **Assertion :** Faraday's laws are consequence of conservation of energy.

   Reason : In a purely resistive AC circuit, the current lags behind the emf in phase.
2. **Assertion :** Lenz's law is based on the principle of conservation of energy.

   Reason : Induced emf always opposes the change in magnetic flux responsible for its production.
3. **Assertion :** Two identical loops, one of copper and another of aluminium are rotated with the same speed in the same magnetic field. The emf induced in both the loop will be same.

   Reason : The magnitude of induced emf is directly proportional to the rate of change of magnetic flux linked with the circuit.
4. **Assertion :** An aircraft flies along the meridian, the potential at the ends of its wings will be the same.

   Reason : Whenever there is change in the magnetic flux emf induces.
5. **Assertion :** An induced current will be developed in a conductor, if it is moved in a direction parallel to magnetic field.

   Reason : Whenever there is relative motion between loop and magnet an induced current is produced in the loop.
6. **Assertion :** Eddy currents are produced in any metallic conductor when magnetic flux is changed around it.

   Reason : Electric potential determines the flow of charge.

7. **Assertion :** An artificial satellite with a metal surface is moving above the earth in a circular orbit. A current will be induced in satellite if the plane of the orbit is inclined to the plane of the equator.

   Reason : The current will be induced only when the speed of satellite is more than 8 km/sec.

8. **Assertion :** An induced emf is generated when magnet is withdrawn from the solenoid.

   Reason : The relative motion between magnet and solenoid induces emf.

9. **Assertion :** Only a change in magnetic flux will maintain an induced current in the coil.

   Reason : The presence of large magnetic flux through a coil maintains a current in the coil if the circuit is continuous.

10. **Assertion :** Mutual inductance of a pair of coils depends on their separation as well as their relative orientation.

    Reason : Mutual inductance depends upon the length of the coil only.

11. **Assertion :** In the phenomenon of mutual induction, self induction of each of the coil persists.

    Reason : Self induction arises due to change in current in the coil itself. In mutual induction current changes in both the individual coil.

12. **Assertion :** The self-inductance of a long solenoid is proportional to the area of cross-section and length of the solenoid.

    Reason : Self-inductance of a solenoid is independent of the number of turns per unit length.

13. **Assertion :** When two coils are wound on each other, the mutual induction between the coils is maximum.

    Reason : Mutual induction does not depend on the orientation of the coils.

14. **Assertion :** Inductance coil are made of copper.

    Reason : Induced current is more in wire having less resistance.

Answers

1. **(b)** According to Faraday's laws, the conversion of mechanical energy into electrical energy. This is in accordance with the law of conservation of energy. It is also clearly known that in pure resistance, the emf is in phase with the current.
2. **(a)** Lenz's Law is based on conservation of energy and induced emf opposes the cause of it *i.e.,* change in magnetic flux.
3. **(a)** Since both the loops are identical (same area and number of turns) and moving with a same speed in same magnetic field. Therefore, same emf is induced in both the coils. But the induced current will be more in die copper loop as its resistance will be lesser as compared to that of the aluminium loop.
4. **(c)** As the aircraft flies, magnetic flux changes through its wings due to the vertical component of the earth's magnetic field. Due to this, induced emf is produced across the wings of the aircraft. Therefore, the wings of the aircraft will not be at the same potential.
5. **(d)** Induced current will not be developed in a conductor, if it is moved in direction parallel to magnetic field. It is because, in this case, the Lorentz force on free electrons in the conductor is zero. The induced emf is produced only when the magnetic flux linked with the loop changes.
6. **(b)** When a metallic conductor is moved in a magnetic field; magnetic flux is varied. It disturbs the free electrons of the metal and set up an induced emf in it. As there are no free ends of the metal *i.e.,* it will be closed in itself so there will be induced current.
7. **(c)** When the satellite moves in inclined plane with equatorial plane (including orbit around the poles), the value of magnetic field will change both in magnitude and direction. Due to this, the magnetic flux through the satellite will change and hence induced currents will be produced in the metal of the satellite. But no current will induced if satellite orbits in the

equatorial plane because the magnetic flux does not change through the metal of the satellite in this plane.

8. (a) Emf is induced when there is change in magnetic flux.
9. (c) Presence of magnetic flux cannot produce current.
10. (c) The mutual inductance in case of a medium of relative permeability present is,

$$M = \frac{\mu_0 \mu_1 n_1 n_2 r_1^2}{l}$$

11. (c) Mutual inductance is the phenomenon according to which an opposing emf produces flux in a coil as a result of change in current on magnetic flux linked with a neighbouring coil. But When two coils are inductively coupled in additions to induced emf produced due to mutual induction, also induced emf is produced in each of two coils due to self induction.
12. (c) The self-inductance of a solenoid is given by, $L = \mu_r \mu_0 n^2 AI$.
13. (c) The manner in which the two coils are oriented, determines the coefficient of coupling between them.

$$M = K^2 \cdot L_1 L_2.$$

When the two coils are wound on each other, the coefficient of coupling is maximum and hence mutual inductance between the coil is maximum.

14. (a) The inductance coils made of copper will have very small ohmic resistance. Due to change in magnetic flux a large induced current will be produced in such an inductance, which will offer appreciable opposition to the flow of current.

Chapter 7. Alternating Current

1. **Assertion :** AC is more dangerous in use than DC.

   Reason : It is because the peak value of DC is greater than indicated value.
2. **Assertion :** Average value of AC over a complete cycle is always zero.

   Reason : Average value of AC is always defined over half cycle.
3. **Assertion:** The alternating current lags behind the emf by a phase angle of $\frac{\pi}{2}$, when AC flows through an inductor.

   Reason : The inductive reactance increases as the frequency of AC source decreases.
4. **Assertion :** Capacitor serves as a block for DC and offers an easy path to AC.

   Reason : Capacitive reactance is inversely proportional to frequency.
5. **Assertion :** In series LCR resonance circuit, the impedance is equal to the ohmic resistance.

   Reason : At resonance, the inductive reactance exceeds the capacitive reactance.
6. **Assertion :** An alternating current shows magnetic effect.

   Reason : Alternating current varies with time.
7. **Assertion :** In series LCR circuit resonance can take place.

   Reason : Resonance takes place if inductance and capacitive reactance are equal and opposite.
8. **Assertion :** AC is more dangerous than DC.

   Reason : Frequency of AC is dangerous for human body.
9. **Assertion :** Power factor correction is must in heavy machinery.

   Reason : A low power factor implies larger power loss in transmission.
10. **Assertion :** Choke coil is preferred over a resistor to adjust current in an AC circuit.

    Reason : Power factor for inductance is zero.
11. **Assertion :** When AC circuit contain resistor only, its power is minimum.

    Reason : Power of a circuit is independent of phase angle.

12. **Assertion :** The power is produced when a transformer steps up the voltage.
Reason : In an ideal transformer VI = constant.

13. **Assertion :** A transformer cannot work on DC supply.
Reason : DC changes neither in magnitude nor in direction.

14. **Assertion :** A laminated core is used in transformers to increase eddy currents.
Reason : The efficiency of a transformer increases with increase in eddy currents.

15. **Assertion :** Soft iron is used as a core of transformer.
Reason : Area of hysteresis loop for soft iron is small.

16. **Assertion :** A given transformer can be used to step up or step-down the voltage.
Reason : The output voltage depends upon the ratio of the number of turns of the two coils of the transformer.

17. **Assertion :** An ac generator is based on the phenomenon of electromagnetic induction.
Reason : In single coil, we consider self-induction only.

Answers

1. **(a)** AC is more dangerous in use than DC. It is because the peak value of AC is greater than indicated value.

2. **(b)** The mean or average value of alternating current or emf during a half cycle is given by,

$$I_m = 0.636\, I_0$$

or $$E_m = 0.636\, E_0$$

During the next half cycle, the mean value of AC will be equal in magnitude but opposite in direction. For this reason the average value of AC over a complete cycle is always zero. So the average value is always defined over a half cycle of AC.

3. **(c)** When AC flows through an inductor current lags behind the emf, by phase of $\frac{\pi}{2}$ inductive reactance,

$$X_L = \omega L$$
$$= 2\pi f L$$

So, when frequency increases correspondingly inductive reactance also increases.

4. **(a)** The capacitive reactance of capacitor is given by,

$$X_C = \frac{1}{\omega C}$$
$$= \frac{1}{2\pi f C}$$

So this is infinite for DC ($f = 0$) and has a very small value for AC. Therefore, a capacitor blocks DC.

5. **(c)** In series resonance circuit, inductive reactance is equal to capacitive reactance.

i.e., $$\omega L = \frac{1}{\omega C}$$

$\therefore$ $$Z = \sqrt{R^2 + \left(\omega L - \frac{1}{\omega C}\right)^2}$$
$$= R$$

6. **(b)** Like direct current, an alternating current also produces magnetic field. But the magnitude and direction of the field goes on changing continuously with time.

7. **(a)** At resonant frequency,

$$X_L = \frac{X_C}{Z}$$

$$= R \text{ (minimum)}$$

Therefore, current in the circuit is maximum.

8. **(a)** The effect of AC on the body depends largely on the frequency. Low frequency currents of 50 to 60 Hz (cycles per second or cps), which are commonly used, are usually more dangerous than high frequency currents and are 3 to 5 times more dangerous than DC of same voltage and ampere (current). The usual frequency of 50 cps (or 60 cps) is extremely dangerous as it corresponds to the fibrillation frequency of the myocardium. This results in ventricular fibrillation and instant death.

9. **(b)** A heavy machinery requires a large power.

   The average power is given by,

$$P_{av} = E_{rms}\, I_{rms} \cos\phi$$

   The required power can be supplied to the heavy machinery either by supplying larger current or by improving power factor. The first method is costly therefore second one is used.

10. **(a)** We can use a capacitor of suitable capacitance as a choke coil, because average power consumed per cycle in an ideal capacitor is zero. Therefore, like a choke coil, a condenser can reduce AC without power dissipation.

11. **(d)** The power of an AC circuit is given by,

$$P = EI \cos\phi$$

   where, $\cos\phi$ is power factor and ϕ is phase angle. In case of circuit containing resistance only, phase angle is zero and power factor is equal to one. Therefore, power is maximum in case of circuit containing resistor only.

12. **(a)** Transformers cannot produce power, but it transfers from primary to secondary.

13. **(a)** Transformer works on AC only, AC changes in magnitude as well as in direction and induces emf.

14. **(d)** Large eddy currents are produced in non-laminated iron core of the transformer by the induced emf, as the resistance of bulk iron core is very small. By using thin iron sheets as core the resistance is increased. Laminating the core substantially reduces the eddy currents. Eddy current heats up the core of the transformer. More the eddy currents greater is the loss of energy and the efficiency goes down.

15. **(a)** Hysteresis loss in the core of transformer is directly proportional to the hysteresis loop area of the core material. Since, soft iron has narrow hysteresis loop area, that is why soft iron core is used in the transformer.

16. **(a)** The ratio of the number of turns on each coil, called the turn ratio determines the ratio of the voltages.

17. **(b)** According to electromagnetic induction, whenever the magnetic flux changes an emf will be induced in the coil.

Chapter 8. Electromagnetic Waves

1. **Assertion :** The pairs E_x, B_y and E_y, B_x, components of space and time varying electric field ($\vec{E}$) magnetic field ($\vec{B}$) can generate a plane electromagnetic wave travelling along the Z-direction.

   Reason : Electromagnetic waves propagate in the direction of vector $\vec{E} \times \vec{B}$.

2. **Assertion :** A wave is motion of matter as a whole in a medium.

   Reason : Wind is same as sound wave in air.

3. **Assertion :** The displacement current is always equal to the conduction current.

Reason : The displacement current is given by,

$$I_D = \varepsilon_0 \frac{d\phi_E}{dt}$$

It leads to,

$$I_D = \frac{dq}{dt} \ldots \left(\because \phi_E = E \times A = \frac{q}{\varepsilon_0 A} \times A = \frac{q}{\varepsilon_0} \right)$$

$$I_D = I_C$$

4. **Assertion :** No work is done if the displacement is zero.

   Reason : Work is done by the force is defined to be the product of component of the force in the direction of the displacement and the magnitude of displacement.

5. **Assertion :** The velocity of electromagnetic waves depends on electric and magnetic properties of the medium.

   Reason : Velocity of electromagnetic waves in free space is constant.

6. **Assertion :** A portable AM radio set must be kept horizontal to receive the signals properly.

   Reason : Radio waves are polarised electromagnetic waves.

7. **Assertion :** Sound waves cannot propagate through vacuum but light waves can.

   Reason : Sound waves cannot be polarised but light waves can be polarised.

8. **Assertion :** Environmental damage has increased the amount of ozone in the atmosphere.

   Reason : Increase of ozone increases the amount of ultraviolet radiation on earth.

9. **Assertion :** Microwaves are better carrier of signals than optical waves.

   Reason : Microwaves move faster than optical waves.

Answers

1. **(a)** E is the electric field vector, and B is the magnetic field vector of the EM wave. For electromagnetic waves E and B are always perpendicular to each other and perpendicular to the direction of propagation. The direction of propagation is in the direction of $\vec{E} \times \vec{B}$.

2. **(d)** A wave is not motion of matter as a whole in a medium. Wind is different from the sound wave in air. The former involves motion of air from one place to the other. The latter involves compressions and rarefactions of layers of air.

3. **(a)** Whenever a conduction current is oscillating in time, the displacement current is equal to the conduction current in case of parallel plate capacitor.

4. **(c)** If the displacement is perpendicular to the direction of force or if it is zero then work done is equal to zero. If your cos θ is zero *i.e.*, your force applied and displacement thus produced are perpendicular then work done is zero.

5. **(b)** As it passes through different media, it continues to cause vibrations at the same frequency. The propagation of an electromagnetic wave depends on the interaction of electric and magnetic fields. Given the frequency of a wave, and the speed in a given medium, the wavelength is the speed divided by the frequency.

6. **(b)** The radio waves are polarised electromagnetic waves. The antenna of portable AM radio is sensitive to only magnetic components of electromagnetic waves. Hence the set should be put horizontal and in proper situation so that the signals are received properly from radio station.

7. **(b)** Sound waves cannot propagate through vacuum because sound waves are mechanical waves. Light waves can propagate through vacuum because light waves are electromagnetic waves. Since, sound waves are longitudinal waves, the particles move in the direction of propagation, therefore these waves cannot be polarized.

8. **(d)** The ozone layer absorbs 97 to 99 percent of the Sun's medium-frequency ultraviolet light (from about 200 nm to 315 nm wavelength), which otherwise would potentially damage exposed life forms near the surface.
9. **(b)** Microwaves are better carrier signals than radio waves for long range signals. Microwaves have higher frequency than radio waves to pass through Earth's Ionosphere to cover up-to long distances. Microwaves have shorter wavelength and can easily pass through earth's atmosphere.

Chapter 9. Ray Optics and Optical Instruments

1. **Assertion :** A convex mirror is preferred over a plane mirror in vehicles to observe traffic coming from behind.
   Reason : Images formed by convex mirrors are erect and diminished in size.
2. **Assertion :** Plane mirror may form real image.
   Reason : Plane mirror forms virtual image, if object is real.
3. **Assertion :** The size of the mirror does not affect the nature of the image.
   Reason : Small mirrors always form a virtual image.
4. **Assertion :** The focal length of the convex mirror will increase, if the mirror is placed in water.
   Reason : The focal length of a convex mirror of radius
   R is equal to, $f = \frac{R}{2}$.
5. **Assertion :** The radius of curvature of a mirror is half of the focal length.
   Reason : A concave mirror of focal length f in air is used in a medium of refractive index 2. Then the focal length of mirror in medium becomes $2f$
6. **Assertion :** Different colours travel with different speed in vacuum.
   Reason : Wavelength of light depends on refractive index of medium.
7. **Assertion :** Critical angle is minimum for red colour.
   Reason : Because critical angle
   $$\theta_c = \sin^{-1}\left(\frac{1}{2\mu}\right) \text{ and } \mu \propto \frac{1}{\lambda}.$$
8. **Assertion :** Optical fibres make use of total internal reflection.
   Reason : Light undergoes successive total internal reflections as it moves through an optical fibre.
9. **Assertion :** Critical angle of light passing from glass to air is minimum for violet colour.
   Reason : The wavelength of blue light is greater than the light of other colours.
10. **Assertion :** The focal length of lens does not change when red light is replaced by blue light.
    Reason : The focal length of lens does not depends on colour of light used.
11. **Assertion :** Combination of lenses helps to obtain diverging or converging lenses of desired magnification.
    Reason : It enhances sharpness of the image.
12. **Assertion :** If the angles of the base of the prism are equal, then in the position of minimum deviation, the refracted ray will pass parallel to the base of prism.
    Reason : In the case of minimum deviation, the angle of incidence is equal to the angle of emergence.

Answers

1. **(a)** A convex mirror produces smaller images which increases the field of view as compared to a plane mirror of the same size. Also in case of convex mirror the image is erect like in plane mirror.
2. **(b)** Plane mirror may form real image, if object is virtual.

3. (c) The size of the mirror does not affect the nature of the image except that a bigger mirror forms a brighter image as it gathers more light rays due to wider aperture.

4. **(d)** Focal length of the spherical mirror does not depend on the medium in which it placed.

5. **(d)** Focal length of a mirror does not depend on the medium in which mirror is held. It depends on the radius of the sphere from which it is made. However, focal length of a lens does depend on medium.

6. **(d)** The velocity of light of different colours (all wavelengths) is same in vacuum and $\mu \propto \frac{1}{\lambda}$.

7. **(d)** Critical angle is minimum for violet colour, critical angle = $\theta_c = \sin^{-1}\left(\frac{1}{\mu}\right)$.

8. **(b)** Optical fibres are fabricated with high quality composite glass or quartz fibres. Each fibre consists of a core and cladding. The refractive index of the material of the core is higher than that of the cladding. When a signal in the form of light is directed at one end of the fibre at a suitable angle, it undergoes repeated total internal reflections along the length of the fibre and finally comes out at the other end.

9. **(c)** $\mu \propto \frac{1}{\lambda} \propto \frac{1}{\theta_c}$, λ_v is least so θ_c for violet is also least. Also the greatest wavelength is for red colour.

10. **(d)** Focal length of the lens depends upon it's refractive index as $\frac{1}{f} \propto (\mu - 1)$. Since, $\mu_b > \mu_r$ so $f_b < f_r$, therefore, the focal length of a lens decreases when red light is replaced by blue light.

11. **(b)** Combination of lenses helps to obtain desired magnification. It also enhances sharpness of the image. Since, the image formed by the first lens becomes the object for the second, the total magnification of the combination is a product of magnification of individual lenses.

12. **(a)** In case of minimum deviation of a prism $\angle i = \angle e$ so $\angle r_1 = \angle r_2$.

Chapter 10. Wave Optics

1. **Assertion :** The frequencies of incident, reflected and refracted beam of monochromatic light at the interface of two media are same.
   Reason : The incident, reflected and refracted rays are not coplanar.
2. **Assertion :** Wavefronts obtained from light emitted by a point source in an isotropic medium are always spherical.
   Reason : Speed of light in isotropic medium is not constant.
3. **Assertion :** When a plane wave passes through a thin prism, the emerging wavefront gets tilted.
   Reason : Speed of light is less in glass than in air.
4. **Assertion :** Light shows the phenomena of reflection and refraction.
   Reason : Light behaves as corpuscles.
5. **Assertion :** Phase difference between any two points on a wavefront is 180°.
   Reason : It depends on distance of the two points from the source.
6. **Assertion :** Speed of light in a medium depends upon the nature of the source.
   Reason : Speed of light does not depend upon nature of the medium.
7. **Assertion :** When light from a source travels from denser to rarer medium, its velocity increases.
   Reason : It happens because energy carried by the wave increases when wave travels from denser to rarer medium.
8. **Assertion :** No interference pattern is detected when two coherent sources are infinitely close to each other.
   Reason : The fringe width is inversely proportional to the distance between the two sources.

9. **Assertion :** Interference is not observed if the two coherent slit sources are broad.
Reason : A broad source is equivalent to many narrow slit sources.
10. **Assertion :** In Young's double slit experiment interference pattern disappear when one of the slits is closed.
Reason : Interference occurs due to superimposition of light waves from two coherent sources.
11. **Assertion :** An excessively thin film appear black in reflected light.
Reason : The film absorb all the radiations falling on it.
12. **Assertion :** In interference, intensity of light at a point changes with time periodically.
Reason : Light energy is propagated exactly in straight line.
13. **Assertion :** No interference is seen, when light reflects from the two surfaces of a window - pane.
Reason : The surface of most window glass is very different than that of optically flat surface required to produce interference fringes.
14. **Assertion :** Intensity pattern of interference and diffraction are not same.
Reason : When there are few sources of light, then the result is usually called diffraction but if there is a large number of them, the word interference is more often used.
15. **Assertion :** If θ_c is the angle between the planes of transmission of two analysers then the intensity of emergent light is proportional to $\cos^2 \theta_c$.
Reason : In the propagation of electromagnetic waves, the angle between the direction of propagation and plane of polarisation is $\frac{\pi}{2}$

Answers

1. **(d)** By using law of reflection. The incident, reflected and refracted rays are coplanar.
2. **(c)** When the speed of light is independent of direction, the secondary waves are spherical.
3. **(a)** As we know that,

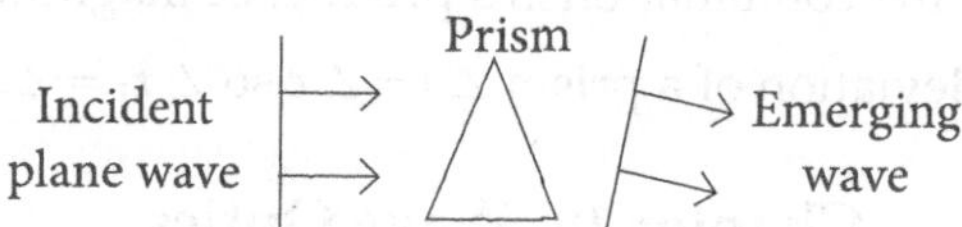

Since, the speed of light is less in glass, the lower portion of the incoming wavefront which travel through the greatest thickness of glass gets delayed. Therefore, the emerging wavefront gets tilted.

4. **(a)** Light behaves mainly like a wave but it can also be considered to consist of tiny packages of energy called photons. Light shows the phenomena of reflection and refraction.
5. **(d)** Phase difference between any two points on a wavefront is zero. Also two points on a given wavefront are equidistant from source hence independent of distance.
6. **(d)** Speed of light depends upon nature of the medium.
7. **(c)** When light from a source travels from denser to rarer medium, the energy carried by wave remains constant. It is because energy depends upon amplitude of wave and amplitude of wave remains constant when light travels from the denser to rarer medium.
8. **(a)** As we know that,

$\beta = \frac{D\lambda}{d}$, When $d \to 0$, $\beta \to \infty$, and so fringes will not be seen over the screen.

9. **(a)** Coherent light from two narrow parallel slits will produce an interference pattern on a screen. Abroad source is equivalent to a large number of parallel narrow slits, lying side by side. As there are many slits, there will be many overlapping interference patterns may be formed. The overlapping will be to such an extent that all traces of a fringe system will be lost. Thus, almost even illumination will be observed. There will be no interference pattern seen. Thus, for obtaining a distinct interference fringe pattern, a narrow slit is required.

10. **(a)** When two light waves of exactly equal frequency having phase difference which is constant with respect to time travel in same direction and superimpose, then interference occurs. When one of the slits is closed then illumination from a single source is obtained. Thus, interference does not occur.

11. **(c)** In case of an excessively thin film, the path difference is $\frac{\lambda}{2}$. As the path difference between two rays is $\frac{\lambda}{2}$, the film appears dark.

12. **(d)** In interference, intensity of light at any point does not change with time. The straight line propagation of light energy is just approximation.

13. **(b)** From window pane, many wavelengths evenly distributed over the entire spectrum will be reflected and the net result would be a white reflected light.

14. **(c)** When there are few sources of light, then the result is usually called interference but if there is a large number of them, the word diffraction is more often used.

15. **(c)** In the propagation of electromagnetic waves, plane of polarisation contains the direction of propagation.

Chapter 11. Dual Nature of Radiation and Matter

1. **Assertion** : For frequency greater than threshold frequency, photoelectric current increases with the increase in intensity of incident light.

   Reason : The number of photoelectrons emitted from a surface is independent of frequency of incident light.

2. **Assertion** : The kinetic energy of photoelectrons emitted from metal surface does not depend on the intensity of incident photon.

   Reason : The ejection of electrons from metallic surface is not possible with frequency of incident photons below the threshold frequency.

3. **Assertion** : The charge carriers in a discharge tube at very low pressure are electrons.

   Reason : The kinetic energy of photoelectrons emitted by a photosensitive surface depends on the intensity of incident light.

4. **Assertion** : In process of photoelectric emission, all emitted electrons do not have same kinetic energy.

   Reason : If radiation falling on photosensitive surface of a metal consists of different wavelength then energy acquired by electrons absorbing photons of different wavelengths shall be different.

5. **Assertion** : Photoelectric effect supports the quantum nature of light.

   Reason : It is because of the fact that electric charge of photoelectrons is quantised.

6. **Assertion** : Photoelectric effect demonstrates the wave nature of light.

   Reason : The number of photoelectrons is proportional to the frequency of light.

7. **Assertion** : An electron is not deflected on passing through certain region of space. This observation confirms that there is no magnetic field in that region.

   Reason : The deflection of electron depends on angle between velocity of electron and direction of magnetic field.

8. **Assertion** : Light is produced in gases in the process of electric discharge through them at high pressure.

   Reason : At high pressure electrons of gaseous atoms collide and reach to excited state.

9. **Assertion** : In photoelectric effect, on increasing the intensity of light, both the number of electrons emitted and kinetic energy of each of them get increased but photoelectric current remains unchanged.

   Reason : The photoelectric current depends only on wavelength of light.

10. **Assertion** : The De-Broglie wavelength equation has significance for any microscopic and sub microscopic particles.

    Reason : De-Broglie wavelength is inversely proportional to the mass of the object its velocity is constant.

11. **Assertion** : The De-Broglie wavelength of a molecule varies inversely as the square root of temperature.

    Reason : The root mean square velocity of the molecule depends on the temperature.

12. **Assertion** : When certain wavelength of light falls on metal surface it ejects electron.

    Reason : Light was wave nature.

Answers

1. **(b)** For frequency greater than threshold frequency, photoelectric current increases with increase in intensity of incident light. Number of photoelectrons emitted from surface is independent of frequency of incident light.
2. **(b)** Kinetic energy of photoelectrons emitted from metal surface does not depend on intensity of incident photon. Ejection of electrons from metallic surface is not possible with frequency of incident photons below threshold frequency.
3. **(d)** Positive ions are charge carriers at very low pressure inside discharge tube. Kinetic energy of photoelectrons emitted depends upon frequency of incident photon.
4. **(a)** In process of photoelectric emission, all emitted electrons do not have same kinetic energy. If radiation falling on photosensitive surface of a metal consists of different wavelength then energy acquired by electrons absorbing photons of different wavelengths shall be different.
5. **(c)** Photoelectric effect supports quantum nature of light because when metal surface is faintly illuminated, photoelectrons leave surface immediately.
6. **(d)** Photoelectric effect demonstrates particle nature of light. Number of emitted photoelectrons depends upon intensity of light.
7. **(d)** If electron is moving parallel to the magnetic field, then electron is not deflected i.e., if electron is not deflected we cannot be sure that there is no magnetic field in that region.
8. **(d)** Light is produced in gases by electric discharge at low pressure. When accelerated electrons collide with atoms of the gas, atoms get excited. The excited atoms return to their normal state and in this process light radiations are emitted.
9. **(d)** On increasing the intensity of incident light, the current in photoelectric cell will increase. The energy of the photons ($h\nu$) will, however not increase with increase in intensity, and hence the kinetic energy of the emitted electrons will not increase.
10. **(a)** As De-Broglie wavelength,

    $$\lambda = \frac{h}{mv}$$

    so, for constant velocity,

    $$\lambda \propto \frac{1}{m}$$

    Lesser the mass greater will be its De-Broglie wavelength.
11. **(a)** De-Broglie wavelength associated with gas molecules varies as,

    $$\lambda \propto \frac{1}{\sqrt{T}}$$
12. **(b)** Light is a transverse, electromagnetic wave that can be seen by the typical human. The wave nature of light was first illustrated through experiments on diffraction and interference. Like all electromagnetic waves, light can travel through a vacuum.

Chapter 12. Atoms

1. **Assertion :** According to electromagnetic theory an accelerated particle continuously emits radiation.

   Reason : According to classical theory, the proposed path of an electron in Rutherford atom model will be parabolic.

2. **Assertion :** In alpha particle scattering number of alpha particles undergoing head on collision is small.

   Reason : Small fraction of the number of incident particles rebound back.

3. **Assertion :** Atoms of each element are stable and emit characteristic spectrum.

   Reason : The spectrum provides useful information about the atomic structure.

4. **Assertion :** Amongst alpha, beta and gamma rays, α-particle has maximum penetrating power.

   Reason : The alpha particle is heavier than beta and gamma rays.

5. **Assertion :** Most of the mass of the atom is concentrated in its nucleus.

   Reason : All alpha particles striking a gold sheet are scattered in different directions.

6. **Assertion :** Hydrogen atom consists of only one electron but its emission spectrum has many lines.

   Reason : Only Lyman series is found in the absorption spectrum of hydrogen atom whereas in the emission spectrum, all the series are found.

7. **Assertion :** Balmer series lies in the visible region of electromagnetic spectrum.

   Reason : Wavelength emission observed in Balmer region lies between 500 Å to 1000 Å.

8. **Assertion :** It is essential that all the lines available in the emission spectrum will also be available in the absorption spectrum.

   Reason : The spectrum of hydrogen atom is only absorption spectrum.

9. **Assertion :** The electron in the hydrogen atom passes from energy level $n = 4$ to the $n = 1$ level. The maximum and minimum number of photons that can be emitted are six and one respectively.

   Reason : The photons are emitted when electron make a transition from the higher energy state to the lower energy state.

10. **Assertion :** Bohr's postulates states that the electrons in stationary orbits around the nucleus do not radiate.

    Reason : According to classical physics, all moving electrons radiate.

Answers

1. **(c)** According to classical electromagnetic theory, an accelerated charge continuously emits radiation. As electrons revolving in circular paths are constantly experiencing centripetal acceleration, hence they will be losing their energy continuously and the orbital radius will go on decreasing and form spiral and finally the electron will fall on the nucleus.
2. **(d)** The penetrating power is maximum in case of gamma rays because gamma rays are an electromagnetic radiation of very small wavelength.
3. **(b)** Atoms of each element are stable and emit characteristics spectrum. The spectrum consists of a set of isolated parallel lines is termed as line spectrum. It provides useful information about the atomic structure.
4. **(d)** The penetrating power is maximum in case of gamma rays because gamma rays are an electromagnetic radiation of very small wavelength.
5. **(b)** At minimum impact parameter α-particles rebound back ($\theta \approx \pi$) and suffers large scattering.
6. **(b)** Every atom has certain definite energy level. In the normal state, the electron in the hydrogen atom stays in lowest energy level When the atom gets appropriate energy from outside, then this electron rises to some higher energy level *i.e.,* atom is excited. Within nearly 10^{-8}s, the electron leaves the higher energy level. Now, it can return either directly to the lowest energy level (or the ground state) or come to the ground state after passing through other lower

energy levels. Since there are a large number of atoms in a light source (hydrogen lamp), all possible transitions take place in the source and many lines are seen in the spectrum. The slit gives the shape of the spectrum and large number of lines are obtained because a large number of atoms are getting excited and de-excited to different energy levels.

7. **(c)** For Balmer series,

$$\frac{1}{\lambda} = R\left[\frac{1}{2^2} - \frac{1}{n^2}\right] \quad \text{...(where, } n = 3, 4, 5\text{.....)}$$

$$\lambda_{min} \approx 3646 \text{ Å}$$

and $$\lambda_{max} \approx 6563 \text{ Å}$$

8. **(d)** Emission transitions can take place between any higher energy level and any energy level below it while absorption transitions start from the lowest energy level only and may end at any higher energy level. Hence number of absorptions transitions between two given energy levels is always less than the number of emission transitions between same two levels.

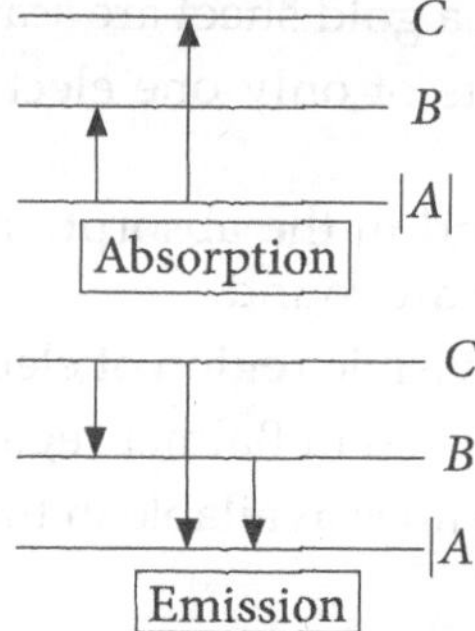

9. **(b)** Maximum number of photons is given by all the transitions possible ${}^4C_2 = 6$ Minimum number of transitions = 1, that is directly jump from 4 to 1.

Chapter 13. Nuclei

1. **Assertion :** Isobars are the element having same mass number but different atomic number.
   Reason : Neutrons and protons are present inside nucleus.
2. **Assertion :** The density of the nuclei of all the atoms is same.
   Reason : It is because density of nuclei is independent of mass number.
3. **Assertion :** For the scattering of α-particles at a large angle, only the nucleus of the atom is responsible.
   Reason : Nucleus is very heavy in comparison to electrons.
4. **Assertion :** Isotopes of an element can be separated by using a mass spectrometer.
   Reason : Separation of isotopes is possible because of difference in electron number of isotopes.
5. **Assertion :** The whole mass of the atom is concentrated in the nucleus.
   Reason : The mass of a nucleus can be either less than or more than the sum of the masses of nucleons present in it.
6. **Assertion :** Neutrons penetrate matter more readily as compared to protons.
   Reason : Neutrons are slightly more massive than protons.
7. **Assertion :** Binding energy per nucleon is nearly constant for element in the range $A = 30$ to $A = 170$.
   Reason : The nuclear force between two nucleons falls rapidly to zero as their distance is more than a few femtometres.
8. **Assertion :** All the radioactive elements are ultimately converted in lead.
   Reason : All the elements above lead are unstable.
9. **Assertion :** Electron capture occurs more often than positron emission in heavy elements.
   Reason : Heavy elements exhibit radioactivity.

10. **Assertion :** Radioactivity of 10^8 undecayed radioactive nuclei of half-life of 50 days is equal that of 1.2×10^8 number of undecayed nuclei of some other material with half-life of 60 days.

    Reason : Radioactivity is proportional to half-life.

11. **Assertion :** Fragments produced in the fission of U^{235} are radioactive.

    Reason : The fragments have abnormally high proton to neutron ratio.

12. **Assertion :** The phenomenon of nuclear fission can be carried out both in a controlled and in an uncontrolled way.

    Reason : The energy released per fission as well as the energy released per unit mass of the fuel in nuclear fission are both greater than the corresponding quantities for nuclear fusion.

13. **Assertion :** Fusion of hydrogen nuclei into helium nuclei is the source of energy of all stars.

    Reason : In fusion heavier nuclei split to form lighter nuclei.

14. **Assertion :** It is not possible to use ^{35}Cl as the fuel for fusion energy.

    Reason : The binding energy of ^{35}Cl is too small.

Answers

1. **(b)** Atoms of chemical elements having same atomic mass but a different atomic number are called Isobars. The sum of the number of protons and neutrons together forms the atomic mass. Therefore, we can also say the number of nucleons present in the nucleus is equal to the atomic mass of an atom.

2. **(a)** As we know that,

$$\text{Density of nucleus} = \frac{\text{Mass of nucleus}}{\text{Volume of nucleus}}$$

$$= \frac{A \times 1.66 \times 10^{-27}}{\frac{4}{3}\pi(1.1 \times 10^{-15})^3 \times A}$$

$$= 2.97 \times 10^{17} \text{ kg m}^{-3}$$

Since, density of nucleus is independent of mass number, hence density of all nuclei is same.

3. **(a)** We knows that an electron is very light particle as compared to an α-particle. Hence, electron cannot scatter the α-particle at large angles. According to law of conservation of momentum. α-particle on the other hand, mass of nucleus is comparable with the mass of α-particle, hence only the nucleus of atom is responsible for scattering of α-particles.

4. **(c)** Number of electrons in isotopes of an element is same. If two atoms have the same number of protons but different number of neutrons, their atomic (proton) number are equal but not their mass (nucleon) numbers. In mass spectrograph, the deflecting electric and magnetic fields were arranged so that all particles of the same mass, irrespective of their speed, were brought to a line focus. When ions of different mass were present, a series of lines, *i.e.*, a mass spectrum, obtained on a photographic film. The relative intensities of the lines enabled an estimate to be made of the relative amounts of isotopes.

5. **(c)** The mass of a nucleus is always less than the sum of the masses of the nucleons present in it. When nucleons combine to form a nucleus, some energy is liberated, and this is the binding energy of the nucleus. The mass of the nucleus cannot be more than the total mass of the nucleons because then stable nucleus cannot be formed.

6. **(a)** The proton carries one unit of positive electric charge while the neutron has no electric charge. The term nucleon is used for either a proton or a neutron. Neutrons penetrate matter more readily as compared to protons. Neutrons are slightly more massive than protons. Protons and neutrons are bound together to form a nucleus by the nuclear force.

7. **(a)** The nuclear force between two nucleons falls rapidly to zero as their distance is more than a few femtometres. This leads to saturation of forces in a medium or a large-sized nucleus, which is the reason for the constancy of the binding energy per nucleon.

8. **(c)** All those elements which are heavier than lead is radioactive. This is because in the nuclei of heavy atoms, besides the nuclear attractive forces, repulsive forces between the protons are also effective and these forces reduce the stability of the nucleus. Hence, the nuclei of heavier elements are being converted into lighter and lighter elements by emission of radioactive radiation. When they are converted into lead, the emission is stopped because the nucleus of lead is stable (or lead is most stable elements in radioactive series).

9. **(b)** Electron capture occurs more often than positron emission in heavy elements. This is because if positron emission is energetically allowed, electron capture is necessarily allowed, but the reverse is not true *i.e.*, when electron capture is energetically allowed, positron emission is not necessarily allowed.

10. **(c)** We know,

$$\text{Radioactivity} = \frac{dN}{dt} = \lambda N$$

$$= \frac{0.693\ N}{T}$$

$$= \frac{0.693 \times 10^8}{50} = \frac{0.693 \times 1.2 \times 10^8}{60}$$

$$= 0.693 \times 2 \times 10^6$$

Radioactivity is proportional to $\frac{1}{T_{1/2}}$, and not to $T_{1/2}$.

11. **(c)** Fragments produced in the fission of U^{235} are radioactive. When uranium undergoes fission, barium and krypton are not the only products. Over 100 different isotopes of more than 20 different elements have been detected among fission products. All of these atoms are, however, in the middle of the periodic table, with atomic numbers ranging from 34 to 58. Because the neutron- proton ratio needed for stability in this range is much smaller than that of the original uranium nucleus, the residual nuclei called fission fragments, always have too many neutrons for stability. A few free neutrons are liberated during fission and the fission fragments undergo a series of beta decays (each of which increases Z by one and decreases N by one) until a stable nucleus is reached. During decay of the fission fragments, an average of 15 MeV of additional energy is liberated.

12. **(b)** The reason is incorrect because the amount of energy released per unit mass of the fuel is much more for fusion than for fission.

13. **(c)** In fusion lighter nuclei combine to form a larger nucleus. Fusion of hydrogen nuclei into helium nuclei is the source of energy of all stars including our sun.

14. **(c)** In fusion, lighter nuclei are used so, fusion is not possible with ^{35}Cl. Also binding energy of ^{35}Cl is not too small.

Chapter 14. Semiconductor Electronics : Materials, Devices and Simple Circuits

1. **Assertion :** If there is some gap between the conduction band and the valence band, electrons in the valence band all remain bound and no free electrons are available in the conduction band. Then the material is an insulator.

   Reason : Resistance of insulators is very low.

2. **Assertion :** Diode lasers are used as optical sources in optical communication.

   Reason : Diode lasers consume less energy.

3. **Assertion :** A pure semiconductor has negative temperature coefficient of resistance.

   Reason : In a semiconductor on raising the temperature, more charge carriers are released, conductance increases and resistance decreases.

4. **Assertion :** If the temperature of a semiconductor is increased then it's resistance decreases.

   Reason : The energy gap between conduction band and valence band is very small.

5. **Assertion :** At a fixed temperature, silicon will have a minimum conductivity when it has a smaller acceptor doping.

Reason : The conductivity of an intrinsic semiconductor is slightly higher than that of a lightly doped *p*-type semiconductor.

6. **Assertion :** An *n*-type semiconductor has a large number of electrons but still it is electrically neutral.

   Reason : An *n*-type semiconductor is obtained by doping an intrinsic semiconductor with a pentavalent impurity.

7. **Assertion :** Semiconductors do not Obey's Ohm's law.

   Reason : Current can not be determined by the rate of flow of charge carriers.

8. **Assertion :** The energy gap between the valence band and conduction band is greater in silicon than in germanium.

   Reason : Thermal energy produces fewer minority carriers in silicon than in germanium.

9. **Assertion :** Zener diode works on principle of breakdown voltage.

   Reason : Current increases suddenly after breakdown voltage.

10. **Assertion :** The value of current through p-n junction in the given figure will be 10 mA.

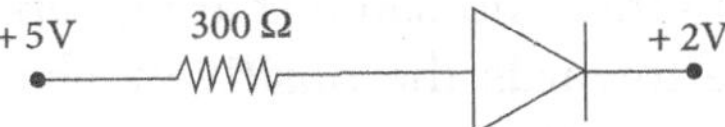

   Reason : In the above figure, *p*-side is at higher potential than *n* side.

Answers

1. **(c)** If there is some gap between the conduction band and the valence band, electrons in the valence band all remain bound and no free electrons in conduction band then it makes the material an insulator. Resistance of insulators is very high.
2. **(c)** High-power laser diodes are used in industrial applications such as heat treating, cladding, seam welding and for pumping other lasers, such as diode-pumped solid-state lasers. Diode lasers consume high energy.
3. **(a)** In semiconductors, by increasing temperature, covalent bond breaks and conduction hole and electrons increase.
4. **(a)** In semiconductors, the energy gap between conduction band and valence band is small ($\approx$ 1 eV). Due to temperature rise, electron in the valence band gain thermal energy and may jump across the small energy gap, (to the conduction band). Thus, conductivity increases and hence resistance decreases.
5. **(c)** Conductivity of an intrinsic semiconductor is less than that of a lightly doped p-type semiconductor.
6. **(b)** As we know that,
   Intrinsic (Semiconductor) + Pentalvalent(Impurity) $\Rightarrow$ N-type
7. **(d)** The assertion is not true. In fact, semiconductor Obeys Ohm's law for low values of electric field (~ 106 V/m). Above this, the current becomes almost independent of electric field.
8. **(b)** The energy gap is greater in silicon than in germanium because the minimum energy required to break a covalent bond is higher in Si (1.1 eV) . than Ge (0.72 eV).
9. **(b)** Zener diodes are specially designed junction diodes, which can operate in the reverse breakdown voltage region continuously without being damaged. This breakdown voltage occur when reverse bias is increased beyond a particular value, the reverse current increases suddenly on increasing the reverse bias even slightly. In normal diode after this breakdown voltage, low power rating will get destroyed but zener diode is specially designed to work in this region of biasing.
10. **(b)** As we know that,

$$V = 5 - 2 = 3\text{ V}$$

$$I = \frac{V}{R} = \frac{3}{300} = \frac{1}{100}\text{ A}$$

$$= 10\text{ mA}$$

❑❑

Question Set 5

Long Answers-I

Chapter 1. Electric Charges and Fields

Q. 1. Figure shows tracks of three charged particles in a uniform electrostatic field. Give the signs of the three charges. Which particle has the highest charge to mass ratio ? **[NCERT]**

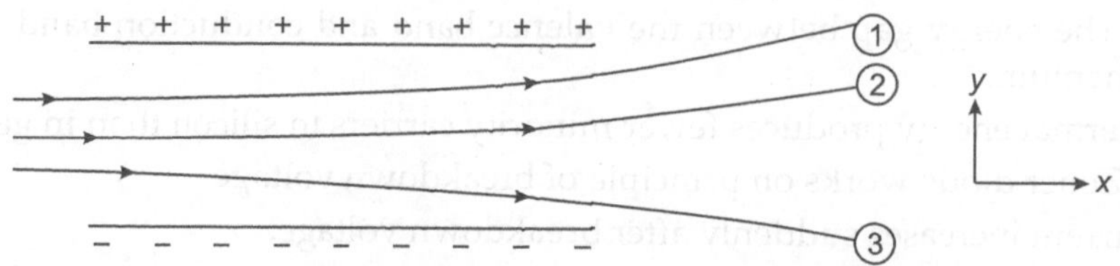

Ans. Opposite charges attract each other and same charges repel each other. It can be observed that particles 1 and 2 both move towards the positively charged plate and repel away from the negatively charged plate. Hence, these two particles are negatively charged. It can also be observed that particle 3 moves towards the negatively charged plate and repels away from the positively charged plate. Hence, particle 3 is positively charged.

The charge to mass ratio (q/m) is directly proportional to the displacement or amount of deflection for a given velocity. Since the deflection of particle 3 is the maximum, it has the highest charge to mass ratio.

Q. 2. (i) Obtain the expression for the torque $\vec{\tau}$ experienced by an electric dipole of dipole moment $\vec{p}$ in a uniform electric field, $\vec{E}$.★

(ii) What will happen if the field were not uniform ?★

Ans. (i)

$$\tau = \text{Force} \times \text{Perpendicular distance}$$
$$\tau = F \times BC$$
$$\tau = qE \times 2l \sin\theta$$
$$\tau = pE \sin\theta \qquad (\because p = q \times 2l)$$
$$\vec{\tau} = \vec{p} \times \vec{E}$$

(ii) If the field is non-uniform then there will be a net force acting on the dipole. Also, a net torque acting on the dipole which depends on the location of the dipole.

Q. 3. Use Gauss's law to find the electric field due to a uniformly charged infinite plane sheet. What is the direction of field for positive and negative charge densities ?★

Ans. Consider a thin infinite uniformly charged plane sheet having the surface charge density of σ. The electric field is normally outward to the plane sheet and is same in magnitude but opposite in direction. Now, draw a Gaussian surface in the form of cylinder around an axis. Let its cross-sectional area be A. The cylinder is made from three surfaces A, S_2, and A′ and the electric flux linked with S_2 is 0. So, the total electric flux linked through the Gaussian surface is

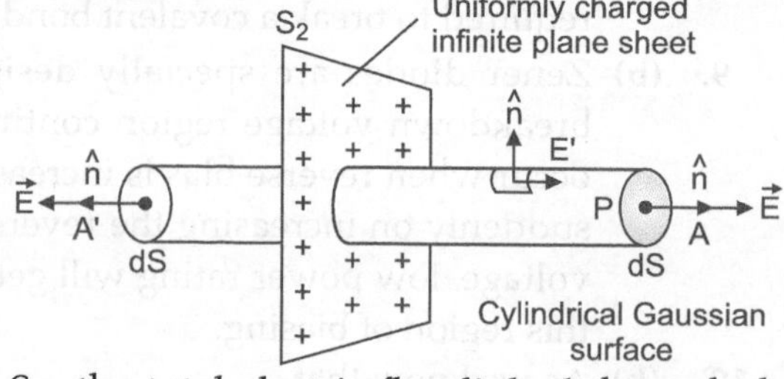

ϕ_E = electric flux through A + electric flux through S_2 + electric flux through A′

★ are board based questions from previous year

$\phi_E = EA \cos 0° + 0 + EA \cos 0°.$

$$\phi = 2EA \quad ...(i)$$

According to Gauss theorem,

$$\phi = \frac{q}{\varepsilon_0}$$

$$\phi = \frac{\sigma A}{\varepsilon_0} \quad (\because q = \sigma A) \;...(ii)$$

From equations (i) and (ii)

$$2EA = \frac{\sigma A}{\varepsilon_0}$$

$$E = \frac{\sigma}{2\varepsilon_0}$$

The direction of field for positive charge density is in outward direction and perpendicular to the plane infinite sheet whereas for the negative charge density the direction becomes inward and perpendicular to the sheet.

Q. 4. A charge is distributed uniformly over a ring of radius '*a*'. Obtain an expression for the electric intensity E at a point on the axis of the ring. Hence show that for points at large distances from the ring, it behaves like a point charge.★

Ans. Suppose we have a ring of radius *a* that carries a uniformly distributed positive charge *q*.

As the total charge *q* is uniformly distri-buted, the charge *dq* on the element *dl* is

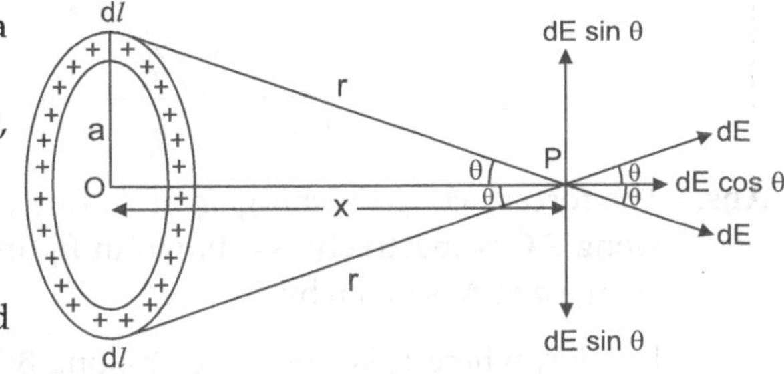

$$dq = \frac{q}{2\pi a} \cdot dl$$

∴ The magnitude of the electric field produced by the element *dl* at the axial point P is

$$dE = k \cdot \frac{dq}{r^2} = \frac{kq}{2\pi a} \cdot \frac{dl}{r^2}$$

The electric field *d*E has two components.

(i) The axial components *d*E cos θ and

(ii) The perpendicular component *d*E sin θ.

Since the perpendicular component of any two diametrically opposite elements are equal and opposite, they cancel out in pairs. Only the axial components will add up to produce the resultant field.

E at point P is given by

$$E = \int_0^{2\pi a} dE \cos\theta$$

[∵ Only the axial components contribute towards E]

$$E = \int_0^{2\pi a} \frac{kq}{2\pi a} \cdot \frac{dl}{r^2} \cdot \frac{x}{r} \quad \left[\because \cos\theta = \frac{x}{r}\right]$$

$$= \frac{kqx}{2\pi a} \cdot \frac{1}{r^3} \int_0^{2\pi a} dl$$

$$= \frac{kqx}{2\pi a} \cdot \frac{1}{r^3} (l)_0^{2\pi a}$$

$$= \frac{kqx}{2\pi a} \cdot \frac{1}{(x^2 + a^2)^{3/2}} \cdot 2\pi a \quad [\because r^2 = x^2 + a^2]$$

$$E = \frac{kqx}{(x^2 + a^2)^{3/2}}$$

★ are board based questions from previous year

$$= \frac{1}{4\pi\varepsilon_0} \cdot \frac{qx}{(x^2 + a^2)^{3/2}}$$

If $x >> a$, then $x^2 + a^2 \approx x^2$

$$E = \frac{1}{4\pi\varepsilon_0} \frac{qx}{(x^2)^{3/2}}$$

$$E = \frac{1}{4\pi\varepsilon_0} \frac{q}{x^2}$$

This expression is similar to electric field due to a point charge.

Q. 5. Consider the charges q, q and $-q$ placed at the vertices of an equilateral triangle, as shown in given figure. What is the force on each charge ? **[NCERT]**

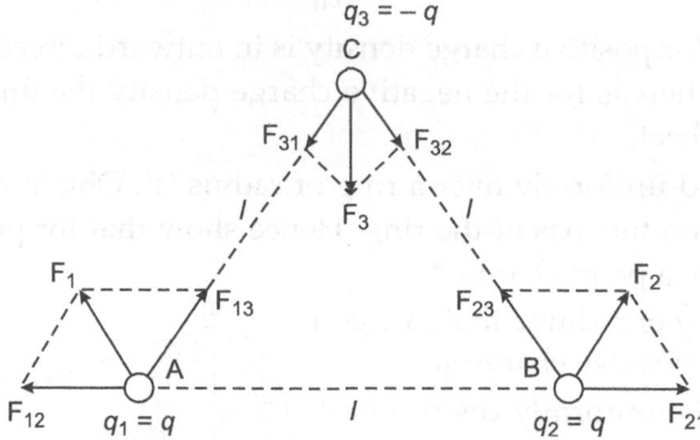

Ans. The forces acting on charge q at A, due to charges q at B and $-q$ at C are F_{12} along BA and F_{13} along AC respectively, as shown in figure from the parallelogram law, the total force F_1 on the charge q at A is given by

$F_1 = F\,\hat{r}_1$ where $\hat{r}_1$ is a unit vector along BC.

The force of attraction or repulsion for each pair of charges has the same magnitude

$$F = \frac{q^2}{4\pi\varepsilon_0\, l^2}$$

The total force F_2 on charge q at B is thus $F_2 = F\hat{r}_2$, where $\hat{r}_2$ is a unit vector along AC.

Similarly the total force on charge $-q$ at C is $F_3 = \sqrt{3}\, F\, \hat{n}$, where $\hat{n}$ is the unit vector along the direction bisecting the $\angle$ BCA.

It is interesting to see that the sum of the forces on the three charges is zero, *i.e.*,

$$F_1 + F_2 + F_3 = 0$$

Q. 6. Consider three charges q_1, q_2, q_3 each equal to q at the vertices of an equilateral triangle of side l. What is the force on a charge Q (with the same sign as q) placed at the centroid of the triangle, as shown in given figure. **[NCERT]**

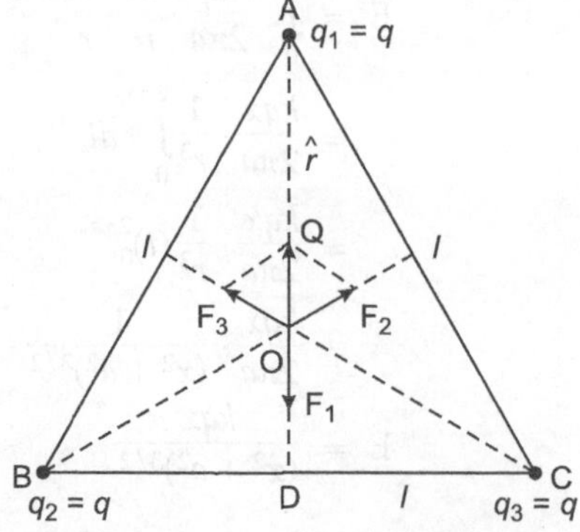

Ans. In the given equilateral triangle ABC of sides of length l, if we draw a perpendicular AD to the side BC.

AD = AC cos 30° = $(\sqrt{3}/2)\, l$ and the distance AO of the centroid O from A is (2/3)

AD = $(1/\sqrt{3})\, l$. By symmetry AO = BO = CO.

Thus,

Force F_1 on Q due to charge q at A = $\frac{3}{4\pi\varepsilon_0}\frac{Qq}{l^2}$ along AO

Force F_2 on Q due to charge q at B = $\frac{3}{4\pi\varepsilon_0}\frac{Qq}{l^2}$ along BO

Force F_3 on Q due to charge q at C = $\frac{3}{4\pi\varepsilon_0}\frac{Qq}{l^2}$ along CO

The resultant of forces F_2 and F_3 is $\frac{3}{4\pi\varepsilon_0}\frac{Qq}{l^2}$ along OA by the parallelogram law. Therefore, the total force on Q = $\frac{3}{4\pi\varepsilon_0}\frac{Qq}{l^2}(\hat{r} - r) = 0$. where $\hat{r}$ is the unit vector along OA.

It is clear also by symmetry that the sum of three forces will be zero.

Q. 7. What are different properties of charges ?

Ans. Different properties of charges are as follows :

(i) **Quantisation of charge :** Electric charge can have discrete values only, *i.e.*, it is quantised. The smallest discrete value it can have is one electronic charge

$(e = 1.6 \times 10^{-19}$ C).

Thus, charges can have value of integer multiple of e, *i.e.*, $q = \pm ne$.

It can never be $\frac{1}{2}e$ or $\frac{2}{3}e$ etc.

Note : Quark (recent invention) contains $\pm\frac{e}{3}, \pm\frac{2e}{3}$.

(ii) Charges are always exist with mass only. No charge can exist without mass but mass can exist without charge like photon, neutrino etc.

(iii) **Conservation of charges :** Charge can neither be created nor be destroyed. It can be transferred from one body to another. Equal amount of positive or negative charges can appear or disappear. This is what happens in pair production and pair annihilation. It holds good in all types of reactions.

(iv) **Invariance of charge :** Numerical value of a charge is independent of frame of reference. It means the value of charge on a body remains same whether it is moving with constant velocity or accelerating or stationary. On the other hand, mass body depends on its speed and it decreases with speed.

(v) Charge is a scalar quantity.

(vi) Like charges repel and unlike charges attract each other.

Q. 8. Four point charges $q_A = 2$ μC, $q_B = -5$μC, $q_C = 2$ μC and $q_D = -5$ μC are located at the corners of a square ABCD of side 10 cm. What is the force on a charge of 1 μC placed at the centre of the square ? **[NCERT]**

Ans. The given figure shows a square of side 10 cm with four charges placed at its corners. O is the centre of the square.

Where, (sides) AB = BC = CD = AD = 10 cm

(Diagonals) AC = BD = $10\sqrt{2}$ cm

AO = OC = DO = OB = $5\sqrt{2}$ cm

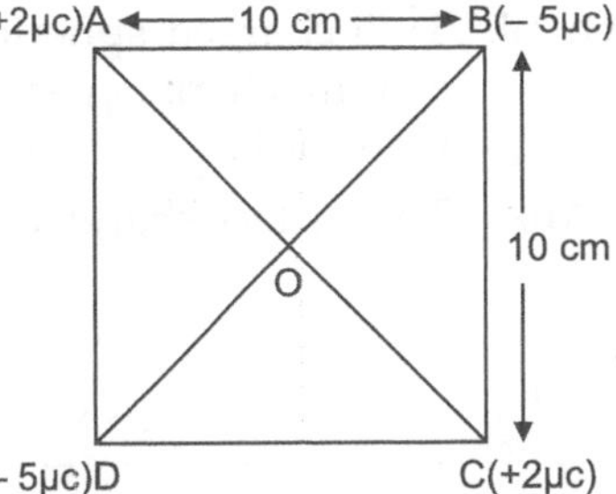

A charge of amount 1 μC is placed at point O.

Force of repulsion between charges placed at corner A and centre O is equal in magnitude but opposite in direction. Hence, they will cancel each other. Similarly, force of attraction between charges placed at corner B and centre O is equal in magnitude but opposite in direction relative to the force of attraction between the charges placed at corner D and centre O. Hence, they will also cancel each other. Therefore, net force caused by the four charges placed at the corner of the square on 1 μC charge at centre O is zero.

Q. 9. (i) Three point charges q, $-4q$ and $2q$ are placed at the vertices of an equilateral triangle ABC of side 'l' as shown in the figure. Obtain the expression for the magnitude of the resultant electric force acting on the charge q.★

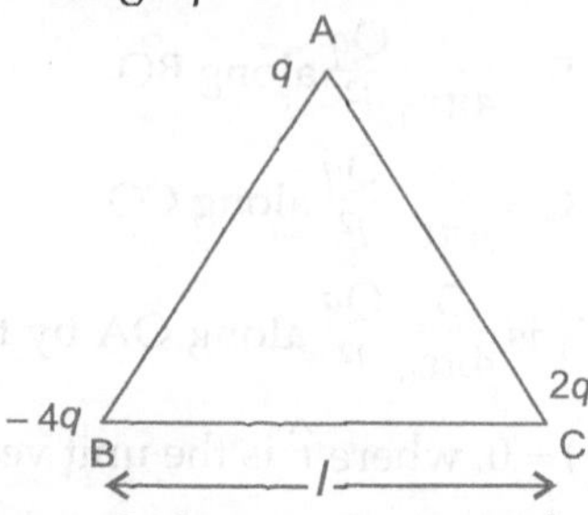

(ii) Find out the amount of the work done to separate the charges at infinite distance.★

Ans. (i) Electric force at A due to charge $2q$

$$F_C = \frac{1}{4\pi\varepsilon_0}\frac{q\times 2q}{l^2} \text{ along } \overrightarrow{CA}$$

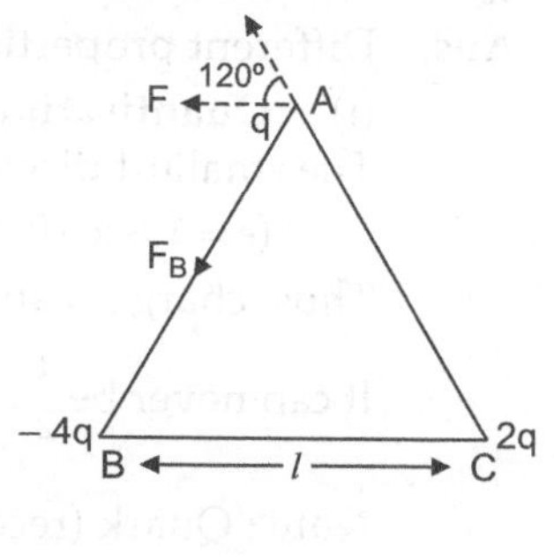

Electric force at A due to charge $(-4q)$,

$$F_B = \frac{1}{4\pi\varepsilon_0}\frac{q\times(-4q)}{l^2} \text{ along } \overrightarrow{AB}$$

$$\text{Resultant force, } F = \sqrt{F_B^2 + F_C^2 + 2F_BF_C\cos 120^\circ}$$

$$= \frac{1}{4\pi\varepsilon_0}\frac{q^2}{l^2}\sqrt{(-4)^2+(2)^2+2(-4)(2)\left(-\frac{1}{2}\right)}$$

$$= \frac{1}{4\pi\varepsilon_0}\frac{2\sqrt{7}q^2}{l^2}$$

(ii) Work done to separate the charge to infinity :

Initial potential energy,

$$U_i = \frac{1}{4\pi\varepsilon_0}\left[\frac{(-4q)q}{l}+\frac{(-4q)(2q)}{l}+\frac{(q)(2q)}{l}\right]$$

$$= \frac{1}{4\pi\varepsilon_0}\frac{q^2}{l}[-4-8+2]$$

$$= \frac{1}{4\pi\varepsilon_0}\left(\frac{-10q^2}{l}\right)$$

Final potential energy, $U_f = 0$

Thus, work done

$$U_f - U_i = 0 - \left(\frac{-10q^2}{4\pi\varepsilon_0 l}\right) = \frac{10q^2}{4\pi\varepsilon_0 l}$$

Q. 10. Two charged spherical conductors of radii R_1 and R_2 when connected by a conducting wire acquire charge q_1 and q_2 respectively. Find the ratio of their surface charge densities in terms of their radii.

Ans. The surface charge density for a spherical conductor of radius R_1 is given by :

$$\sigma_1 = \frac{q_1}{4\pi R_1^2}$$

★ are board based questions from previous year

Similarly, for spherical conductor R_2, the surface charge density is given by :

$$\sigma_2 = \frac{q_2}{4\pi R_2^2}$$

$$\therefore \qquad \frac{\sigma_1}{\sigma_2} = \frac{q_1 R_2^2}{q_2 R_1^2} \qquad \text{...(i)}$$

As the spheres are connected so the charges will flow between the spherical conductors till their potential become equal.

i.e.,

$$\frac{k_1 q_1}{R_1} = \frac{k q_2}{R_2}$$

$$\Rightarrow \qquad \frac{q_1}{q_2} = \frac{R_1}{R_2} \qquad \text{...(ii)}$$

Using (ii) in (i)

We have,

$$\frac{\sigma_1}{\sigma_2} = \frac{R_1}{R_2} \cdot \frac{R_2^2}{R_1^2} = \frac{R_2}{R_1}$$

$$\frac{\sigma_1}{\sigma_2} = \frac{R_2}{R_1}$$

Q. 11. Check that the ratio $ke^2/Gm_e m_p$ is dimensionless. Look up a table of physical constant and determine the value of this ratio. What does the ratio signify ? **[NCERT]**

Ans. The given ratio is $\frac{ke^2}{Gm_e m_p}$.

Where, G = Gravitational constant. Its unit is N m^2 kg^{-2}.

m_e and m_p = Masses of electron and proton and their unit is kg.

e = Electric charge. Its unit is C.

$k = \frac{1}{4\pi\varepsilon_0}$ and its unit is N m^2 C^{-2}.

Therefore unit of the given ratio

$$\frac{ke^2}{Gm_e m_p} = \frac{[Nm^2C^{-2}]\,[C^2]}{[Nm^2kg^{-2}]\,[kg]\,[kg]} = M^oL^oT^o$$

Hence, the given ratio is dimensionless.

$e = 1.6 \times 10^{-19}$ C

$G = 6.67 \times 10^{-11}$ N m^2/kg^{-2}

$m_e = 9.1 \times 10^{-31}$ kg

$m_p = 1.66 \times 10^{-27}$ kg

Hence, the numerical value of the given ratio is

$$\frac{ke^2}{Gm_e m_p} = \frac{9 \times 10^9 \times (1.6 \times 10^{-19})^2}{6.67 \times 10^{-11} \times 9.1 \times 10^{-31} \times 1.67 \times 10^{-27}}$$

$$\approx 2.3 \times 10^{39}$$

This is the ratio of electric force to the gravitational force between a proton and an electron, keeping distance between them constant.

Chapter 2. Electrostatic Potential and Capacitance

Q. 1. Show that energy stored in a capacitor is given by $E = \frac{1}{2}CV^2$.

Ans. Let the capacitor (C) is connected across a battery of voltage V (volts). The process of charging capacitor is equivalent to transferring charges from one plate to another plate of capacitor.

As soon as charging of capacitor starts there is a potential difference between plates. So to transfer charges against the potential difference some work is to be done and this work gets stored as electrostatic potential energy in capacitor.

Let dq be the charge transferred against the potential difference V, then work done is

$$dV = dW = Vdq$$
$$= \frac{q}{C} dq$$

Total work done to transfer charge q, is

$$U = \int_0^q dW = \int_0^q Vdq = \int_0^q \frac{q}{C} dq$$
$$U = \frac{q^2}{2C} = \frac{1}{2} CV^2$$

Hence Proved.

Q. 2. Define an equipotential surface. Draw equipotential surfaces :★

(i) in the case of a single point charge and

(ii) in a constant electric field in Z-direction.

Why the equipotential surfaces about a single charge are not equidistant ?

(iii) Can electric field exist tangential to an equipotential surface ? Give reason.

Ans. An equipotential surface is the surface which has same potential at its every point.

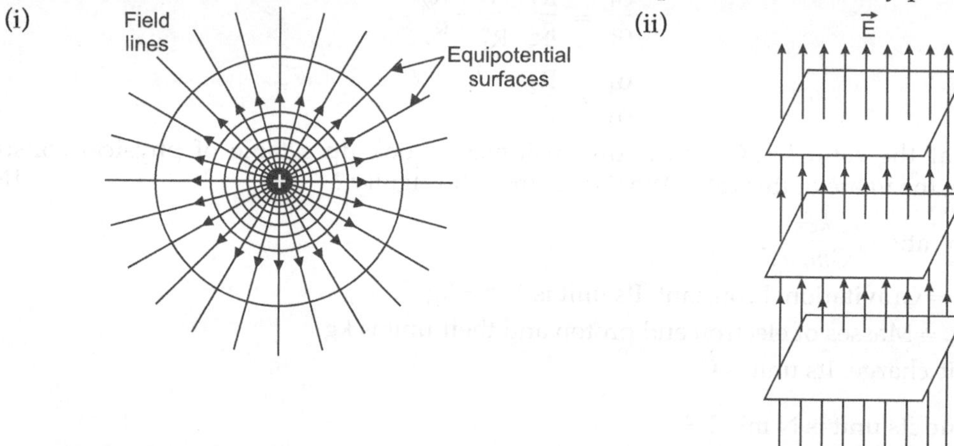

The electric field due to single charge is not constant, this is the reason why the equipotential surfaces about a single charge are not equidistant.

(iii) No, electric field cannot exist tangential to an equipotential surface. If it happen then a charged particle will experience a force along the tangential line and can move along it. As a charged particle can move only due to the potential difference *i.e.,* along the direction of charge of potential, this contradicts the concept of an equipotential surface.

Q. 3. If two similar large plates each of area A having surface charge densities + σ and – σ are separated by a distance in air. Find the expressions for★

(i) Field at points between the two plates and on outer sides of plates. Specify the direction of field in each case.

(ii) The potential difference between the plates.

(iii) The capacitance of capacitor so formed.

Ans. (i) Consider a parallel plate capacitor with two identical plates X and Y, each having an area of A, and separated by a distance d. Let the space beween the plates be filled by a dielectric medium with its dielectric constant as K and σ be the surface charge density on each of the plates.

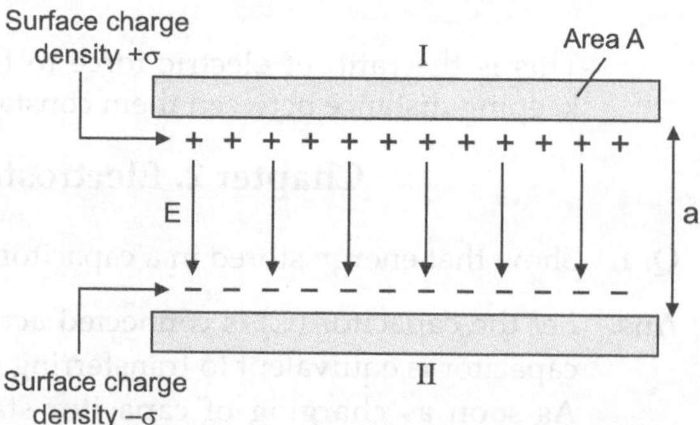

Surface charge density of plate 1

$$\sigma = \frac{Q}{A}$$

and that of plate 2 is σ.

★ are board based questions from previous year

Electric field in outer region I,

$$E = \frac{\sigma}{2\varepsilon_0} - \frac{\sigma}{2\varepsilon_0} = 0$$

Electric field in outer region II,

$$E = \frac{\sigma}{2\varepsilon_0} - \frac{\sigma}{2\varepsilon_0} = 0$$

The inner region between plates 1 and 2, In the electric fields due to the two charged plates add up. So

$$E = \frac{\sigma}{2\varepsilon_0} + \frac{\sigma}{2\varepsilon_0} = \frac{\sigma}{\varepsilon_0}$$

(ii) For uniform electric field potential difference is simply the electric field multiplied by the distance between the plates, *i.e.*,

$$V = Ed = \frac{1}{\varepsilon_0}\frac{Qd}{A}$$

(iii) Capacitance of parallel plates capacitor,

$$C = \frac{Q}{V} = \frac{Q\,\varepsilon_0 A}{Qd} = \frac{\varepsilon_0 A}{d}$$

Q. 4. Figure shows two identical capacitors, C_1 and C_2 each of 1 μF capacitance connected to a battery of 6 V. Initially switch 'S' is closed. After sometime 'S' is left open and dielectric slabs of dielectric constant $k = 3$ are inserted to fill completely the space between the plates of the two capacitors.

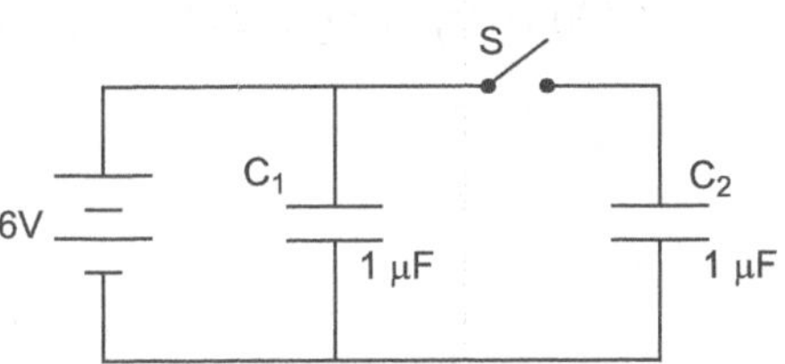

How will the (i) charge and (ii) potential difference between the plates of the capacitors be affected after the slabs are inserted.

Ans. When, switch 'S' is closed,

$$\text{Potential difference across } C_1 = \text{Potential difference across } C_2 = 6\text{ V}$$

$$V_1 = V_2 = 6V$$

$$q_1 = q_2 = 1\ \mu F \times 6V = 6\ \mu C$$

When switch 'S' is opened, the potential difference across C_1 remains 6 V, while the charge on capacitor C_2 remains 6 μC.

After insertion of dielectric slab ($k = 3$) between the plates of each capacitor, the new capacitance of each capacitor becomes

$$C'_1 = 3 \times 1\ \mu F = 3\ \mu F$$

$$C'_2 = 3\ \mu F$$

(i) Now, charge on capacitor C_1, $q'_1 = C'_1V_1 = 3 \times 6 = 18\ \mu C$

(ii) When S is open, potential difference across C_2

$$V'_2 = \frac{q_2}{C'_2} = \frac{6\ \mu C}{3\ \mu F} = 2\text{ V}$$

Chapter 3. Current Electricity

Q. 1. Write the principle of working of a potentiometer. Describe briefly, with the help of a circuit diagram, how a potentiometer is used to determine the internal resistance of a given cell.

Ans. **Principle of working :** The voltage drop along the wire is directly proportional to the length of the wire. The potentiometer works without drawing any current from the voltage source

We can also use a potentiometer to measure internal resistance of a cell. For this the cell (emf E) whose internal resistance (r) is to be determined is connected across a resistance box through a key K_2, as shown in the figure.

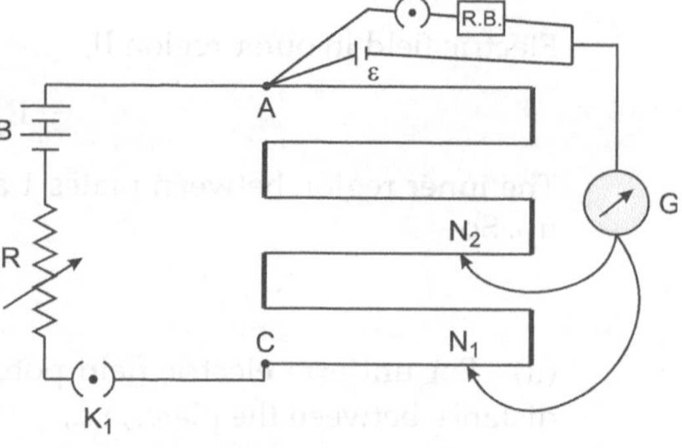

With key K_2 open, balance is obtained at length L_1 (AN_1).

Then,
$$E = \phi L_1$$

When key K_2 is closed, the cell sends a current (I) through the resistance box (R). If V is the terminal potential difference of the cell and balance is obtained at length L_2 (AN_2), then

$$V = \phi L_2$$

So, we have
$$\frac{E}{V} = \frac{L_1}{L_2}$$

But, $E = I(r + R)$ and $V = IR$. This gives

$$\frac{E}{V} = \frac{(r+R)}{R}$$

On comparing the above equations, we get

$$\frac{(R+r)}{R} = \frac{L_1}{L_2}$$

$$r = R\left(\frac{L_1}{L_2} - 1\right)$$

Using the above equation, we can find the internal resistance of a given cell.

Q. 2. Answer the following questions :

(i) A steady current flows in a metallic conductor of non-uniform cross-section. Which of these quantities is constant along the conductor : current, current density, electric field, drift speed ?

(ii) Is ohm's law universally applicable for all conducting elements ? If not, give examples of elements which do not obey ohm's law.

(iii) A low voltage supply from which one needs high currents must have very low internal resistance. Why ?

(iv) A high tension (HT) supply of, say, 6 kV must have a very large internal resistance. Why ?

Ans. (i) When a steady current flows in a metallic conductor of non-uniform cross-section, the current flowing through the conductor is constant. Current density, electric field, and drift speed are inversely proportional to the area of cross-section. Therefore, they will not be constant.

(ii) No, Ohm's law is not universally applicable for all conducting elements. Vacuum diodes, semi-conductors, a non-ohmic conductor, don't follow ohm's law.

(iii) According to Ohm's law, the relation for the potential is V = IR Voltage (V) is directly proportional to current (I).

R is the internal resistance of the source.

$$I = \frac{V}{R}$$

If V is low, then R must be very low, so that high current can be drawn from the source.

(iv) In order to prohibit the current from exceeding the safety limit, a high tension supply must have a very large internal resistance. If the internal resistance is not large, then the current drawn can exceed the safety limits in case of a short circuit.

Q. 3. (i) Write the principle of working of a Meter bridge.★

(ii) In a meter bridge, the balance point is found at a distance l_1 with resistances R and S as shown in the figure.

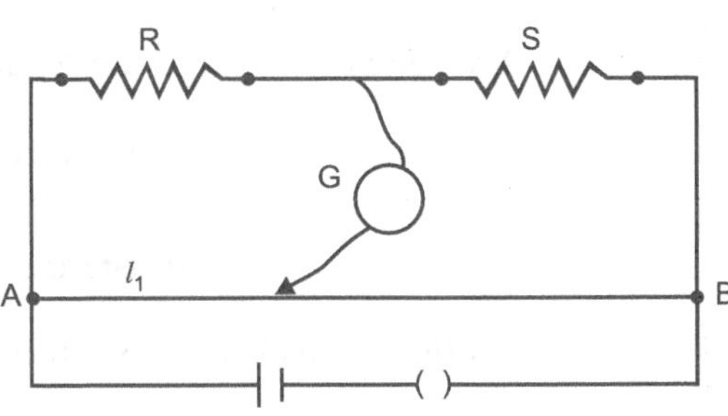

An unknown resistance X is now connected in parallel to the resistance S and the balance point is found at a distance l_2. Obtain a formula for X in terms of l_1, l_2 and S.★

Ans. (i) Meter bridge works on the principle of balanced Wheatstone bridge *i.e.*, when the bridge is balanced,

$$\frac{R_1}{R_2} = \frac{R_3}{R_4}$$

Where R_1, R_2, R_3 and R_4 are resistances connected in four arms of Wheatstone's bridge.

(ii) In first case, $$\frac{R}{S} = \frac{l_1}{100 - l_1} \quad \text{...(i)}$$

In second case,

$$\frac{R}{XS/(X+S)} = \frac{l_2}{100 - l_2} \quad \text{...(ii)}$$

Dividing (ii) by (i),

$$\frac{X+S}{X} = \frac{l_2}{l_1}\left(\frac{100 - l_1}{100 - l_2}\right)$$

$$1 + \frac{S}{X} = \frac{l_2}{l_1}\left(\frac{100 - l_1}{100 - l_2}\right)$$

$$X = \frac{S}{\frac{l_2}{l_1}\left(\frac{100 - l_1}{100 - l_2}\right) - 1}$$

$$X = \frac{l_1\,(100 - l_2)}{100\,(l_2 - l_1)}S$$

Q. 4. (a) Differentiate between electrical resistance and resistivity of a conductor.★

(b) Two metallic rods, each of length L, area of cross A_1 and A_2, having resistivities ρ_1 and ρ_2 are connected in parallel across a d.c. battery. Obtain the expression for the effective resistivity of this combination.★

Ans. (a)

Electrical Resistance	Resistivity
1. Resistance opposes the flow of free electrons charges in the conductor.	1. Resistivity is the property of the material which defines the resistance of the material having specific dimension.
2. It is measured by the ratio of the p.d.(V) across it to the current (I) flowing through it. $R = \frac{V}{I}$	2. It is numerically equal to the resistance of a sample of unit length and unit area of cross-section at certain temperature.
3. Its S.I. unit is Ω.	3. Its S.I. unit is Ω m.

(b) Here, length of each rod is L.

★ are board based questions from previous year

Area of cross sections are A_1 and A_2 and resistivities are ρ_1 and ρ_2 respectively.

$$\text{Resistance of 1st metallic rod, } R_1 = \frac{\rho_1 L}{A_1}$$

$$\text{Resistance of 2nd metallic rod, } R_2 = \frac{\rho_2 L}{A_2}$$

∴ Effective resistance when connected in parallel.

$$R_{eq} = \frac{R_1 R_2}{R_1 + R_2}$$

$$\frac{\rho_{eq} L}{A_1 + A_2} = \frac{\rho_1 \rho_1 L^2 A_1 A_2}{A_1 A_2 L(\rho_1 A_2 + \rho_2 A_1)}$$

$$\rho_{eq} = \frac{\rho_1 \rho_1 (A_1 + A_2))}{\rho_1 A_2 + \rho_2 A_1}$$

Chapter 4. Moving Charges and Magnetism

Q. 1. Write ampere's circuital law. State its one application.

Ans. The line integral of the magnetic field along any closed path is equal to μ_0 times the current enclosed by the path.

$$\oint \vec{B} \cdot \vec{dl} = \mu_0 i$$

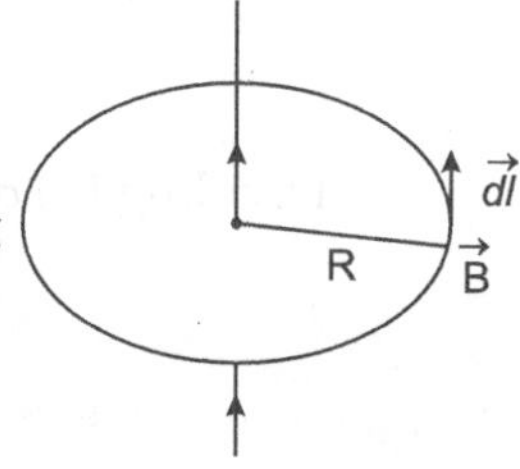

Application of ampere's circuital law :

Magnetic field at a point due to a long straight wire—Assume a long straight wire carrying a current *i*, P is point at a distance 'R' from the wire. B-magnetic field at P.

As per ampere circuital law we have

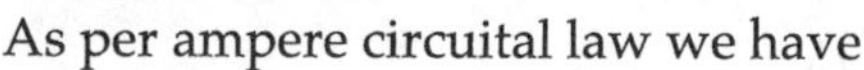

$$\oint B \,.\, dl = \mu_0 i$$

Amperian loop at P–Here it in a circle of radius R

Since B is constant at any point on the loop, $B\oint dl = \mu_0 i$

$$\Rightarrow \quad B \times 2\pi R = \mu_0 i$$

$$B = \frac{\mu_0 i}{2\pi R}$$

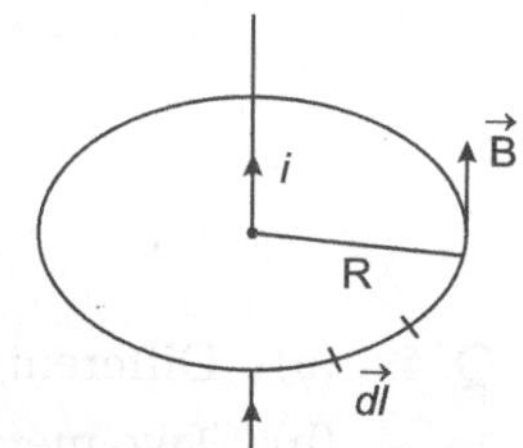

The direction of the magnetic field is found by right hand thumb rule.

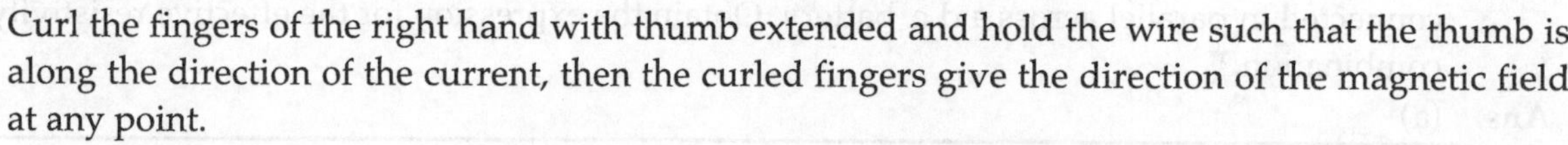

Curl the fingers of the right hand with thumb extended and hold the wire such that the thumb is along the direction of the current, then the curled fingers give the direction of the magnetic field at any point.

Q. 2. (i) A current-carrying circular loop lies on a smooth horizontal plane. Can a uniform magnetic field be set up in such a manner that the loop turns around itself (*i.e.,* turns about the vertical axis).

(ii) A current-carrying circular loop is located in a uniform external magnetic field. If the loop is free to turn, what is its orientation of stable equilibrium ? Show that in this orientation, the flux of the total field (external field + field produced by the loop) is maximum.

(iii) A loop of irregular shape carrying current is located in an external magnetic field. If the wire is flexible, why does it change to a circular shape ? **[NCERT]**

Ans. (i) No, because that would require τ to be in the vertical direction. But $\tau = IA \times B$, and since A of the horizontal loop is in the vertical direction, τ would be in the plane of the loop for any B.

(ii) Orientation of stable equilibrium is one where the area vector A of the loop is in the direction of external magnetic field. In this orientation, the magnetic field produced by the loop is in the same direction as external field, both normal to the plane of the loop, thus giving rise to maximum flux of the total field.

(iii) It assumes circular shape with its plane normal to the field to maximize flux, since for a given perimeter, a circle encloses greater area than any other shape.

Q. 3. (i) State Biot–Savart law and express this law in the vector form.★

(ii) Two identical circular coils, P and Q each of radius R, carrying currents 1 A and $\sqrt{3}$ A respectively, are placed concentrically and perpendicular to each other lying in the XY and YZ planes. Find the magnitude and direction of the net magnetic field at the centre of the coils.

Ans. (i) **Biot-Savart Law :** It states that the magnetic field strength (dB) produced due to a current element I and length dl at a point having position vector $\vec{r}$ relative to current element is

(a) directly proportional to the current I *i.e.,* $dB \propto I$.

(b) directly proportional to the length dl of the element *i.e.,* $dB \propto dl$.

(c) directly proportional to sin θ, where θ is the angle between dl and r, *i.e.,* $dB \propto \sin\theta$.

(d) inversely proportional to the square of the distance r from the current element

i.e., $dB \propto \frac{1}{r^2}$.

$$\therefore \quad dB \propto \frac{Idl \sin\theta}{\pi^2}$$

$$dB = \frac{\mu_0}{4\pi}\frac{Idl \sin\theta}{r^2}$$

In vector form,

$$\vec{dB} = \frac{\mu_0}{4\pi}\cdot\frac{I\vec{dl}\times\vec{r}}{r^3}$$

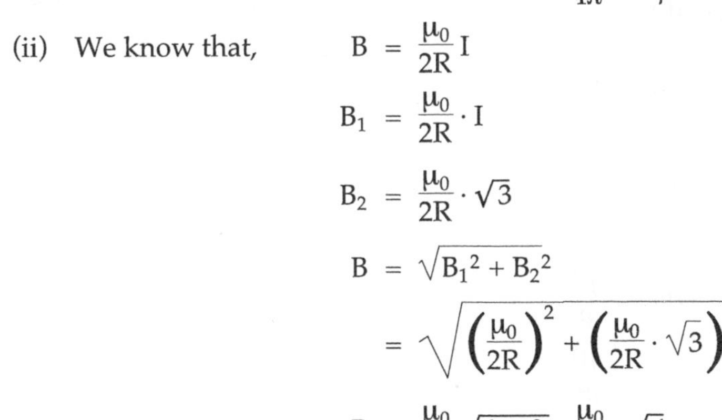

(ii) We know that,

$$B = \frac{\mu_0}{2R}I$$

$$B_1 = \frac{\mu_0}{2R}\cdot I$$

$$B_2 = \frac{\mu_0}{2R}\cdot\sqrt{3}$$

$$B = \sqrt{B_1^2 + B_2^2}$$

$$= \sqrt{\left(\frac{\mu_0}{2R}\right)^2 + \left(\frac{\mu_0}{2R}\cdot\sqrt{3}\right)^2}$$

$$B = \frac{\mu_0}{2R}\sqrt{1+3} = \frac{\mu_0}{2R}\cdot\sqrt{4}$$

$$B = \frac{\mu_0}{R}$$

$$\tan\theta = \frac{B_1}{B_2} = \frac{1}{\sqrt{3}}$$

$$\Rightarrow \quad \theta = 30°$$

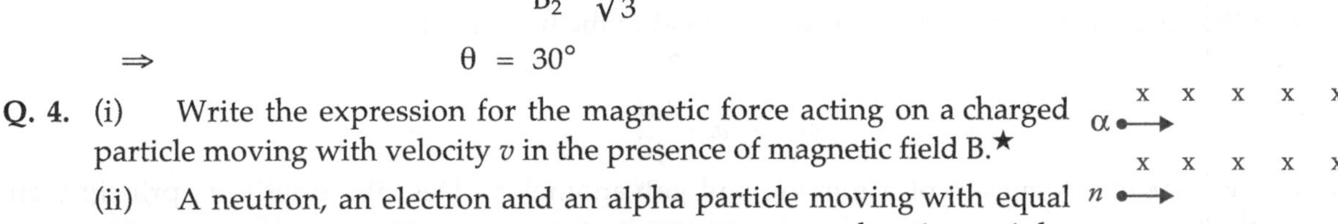

Q. 4. (i) Write the expression for the magnetic force acting on a charged particle moving with velocity v in the presence of magnetic field B.★

(ii) A neutron, an electron and an alpha particle moving with equal velocities, enter a uniform magnetic field going into the plane of the paper as shown. Trace their paths in the field and justify your answer.★

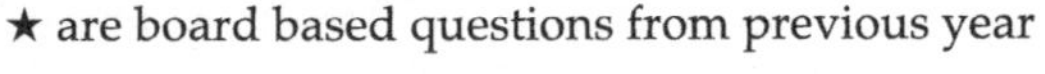

★ are board based questions from previous year

Ans. (i) A charge particle having charge q is moving with velocity 'v' in a magnetic field of field strength 'B' then the force acting on it is given by the formula

$F = q(\vec{v} \times \vec{B})$ and $F = qvB \sin\theta$ (where θ is the angle between velocity vector of magnetic field). Direction of force is given by the cross product of velocity and magnetic field.

(ii) α-particle will trace circular path in anticlockwise direction as it's deviation will be in the direction of $(\vec{v} \times \vec{B})$

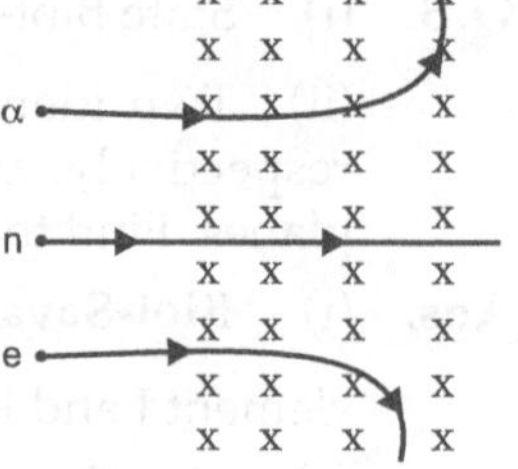

Neutron will pass without any deviation as magnetic field does not exert force on neutral particle.

Electron will trace circular path in clockwise direction as its deviation will be in the direction opposite to $(\vec{v} \times \vec{B})$ with a smaller radius due to large charge/mass ratio as $r = \frac{mv}{qB}$.

Q. 5. A uniform magnetic field B is set up along the positive X-axis. A particle of charge 'q' and mass 'm' moving with a velocity v enters the field at the origin in X-Y plane such that it has velocity components both along and perpendicular to the magnetic field B. Trace, giving reason, the trajectory followed by the particle. Find out the expression for the distance moved by the particle along the magnetic field in one rotation.★

Ans. If component v_x of the velocity vector is along the magnetic field, and remain constant, the charge particle will follow a helical trajectory; as shown in fig.

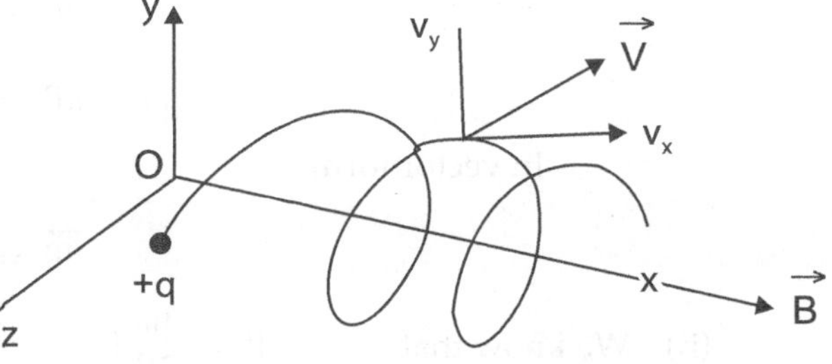

If the velocity component v_y is perpendicular to the magnetic field B, the magnetic force acts like a centripetal force qv_y B.

So,
$$qv_y B = \frac{mv_y^2}{r}$$
$$v_y = \frac{qBr}{m}$$

Since tangent velocity $v_y = r\omega$

$$\Rightarrow \quad r\omega = \frac{qBr}{m}$$
$$\Rightarrow \quad \omega = \frac{qB}{m}$$

Time taken for one revolution,
$$T = \frac{2\pi}{\omega} = \frac{2\pi m}{qB}$$

and the distance moved along the magnetic field in the helical path is
$$x = v_x \cdot T = v_x \cdot \frac{2\pi m}{qB}$$

Q. 6. Draw a labelled diagram of a moving coil galvanometer. Describe briefly its principle and working.

Ans. **Moving coil galvanometer :** It is a device used for detecting and measuring small electric current.

★ are board based questions from previous year

Principle : The working is based upon the principle, when a current carrying coil suspended in a magnetic field experiences a torque.

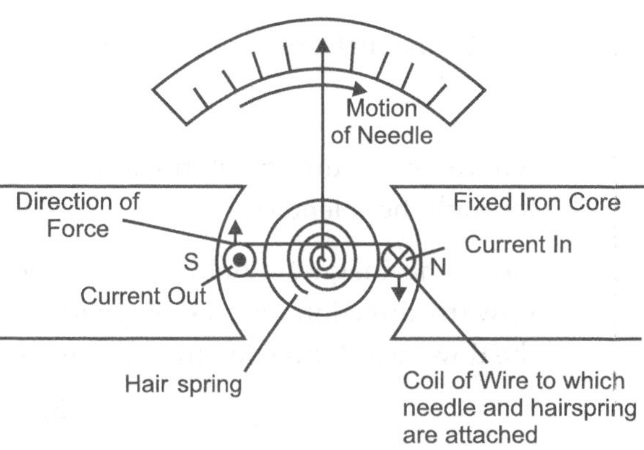

Construction : It consists of a coil with a large number of turns of insulated copper wire wounded on a metallic frame.

The coil is suspended by means of a phosphor bronze strip and a horse shoe magnet's NS surrounds it. The lower end of the coil is attached with a hair spring. The scale of the pointer is attached to the other end of the spring.

Working : When current is passed, the couple acts on it. Since the plane remains parallel to the magnetic field in all position of the coil, the force on the vertical arms always remains perpendicular to the place of the coil.

Let I = the current flowing through the coil,

B = magnetic field supposed to be uniformed and always parallel to the coil, and

A = area of the coil.

Deflection acting on the coil is

$$\tau = NIBA \sin 90° \qquad [\because \sin 90° = 1]$$

Due to the deflection torque, the coil rotates and suspended wire gets twisted. The suspension fiber experiences a restoring torque. If ϕ is angle through the coil rotates and k is the restoring torque per unit angular twist, then

Restoring torque, $\tau = k\phi$

In equilibrium, Deflection torque = Restoring torque

$$NIBA = k\phi$$

$$\phi = \left(\frac{NBA}{k}\right) \cdot I$$

$$\Rightarrow \qquad \phi = \left(\frac{1}{G}\right) \cdot I$$

$$I = G\phi$$

where $$G = \frac{k}{NBA}$$ known as galvanometer constant.

$$\therefore \qquad \phi \propto I$$

This provides a linear scale for the galvanometer.

Q. 7. Two long coaxial insulated solenoids, S_1 and S_2 of equal lengths are wound one over the other as shown in the figure. A steady current 'I' flow through the inner solenoid S_1 to the other end B, which is connected to the outer solenoid S_2 through which the same current "I" flows in the opposite direction so as to come out at end A.

If n_1 and n_2 are the number of turns per unit length, find the magnitude and direction of the net magnetic field at a point (a) inside on the axis and (b) outside the combined system.

Ans. (a) The magnetic field due to a current carrying solenoid :

$$B = \mu_0 n i$$

where, n = number of turns per unit length, i = current through the solenoid.

Now, the magnetic field due to solenoid S_1 will be in the upward direction and the magnetic field due to S_2 will be in the downward direction (by right-hand screw rule).

$$B_{net} = B_1 - B_2$$

$$B_{net} = \mu_0 n_1 I - \mu_0 n_2 I$$

$$= \mu_0 I (n_1 - n_2)$$

(b) The magnetic field is zero outside a solenoid.

Q. 8. State the underlying principle of a cyclotron. Write briefly how this machine is used to accelerate charged particles to high energies.★

Ans. The underlying principle of a cyclotron is that an oscillating electric field can be used to accelerate a charged particle to high energy.

A cyclotron involves the use of an electric field to accelerate charge particles across the gap between the two D-shaped magnetic field regions. The magnetic field is perpendicular to the paths of the charged particles that makes them follow in circular paths within the two Dee's. An alternating voltage accelerates the charged particles each time they cross the Dee's. The radius of each particle's path increases with its speed. So, the accelerated particles spiral toward the outer wall of the cyclotron.

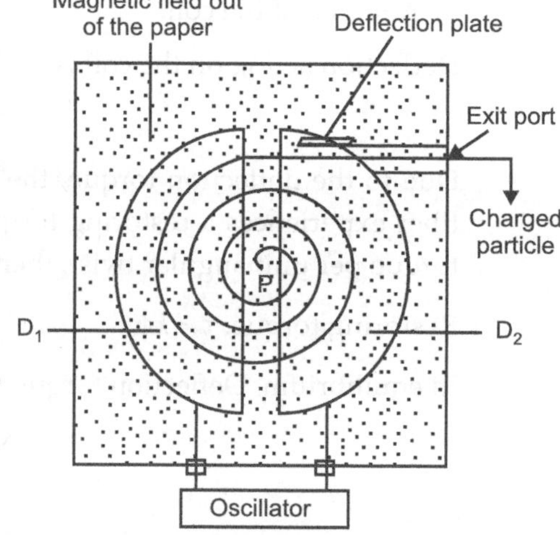

[The Dee's are the semi-circular structures (D_1 and D_2) between which the charges move. The accelerating voltage is maintained across the opposite halves of the Dee's Square wave electric fields are used to accelerate the charged particles in a cyclotron.

The accelerating electric field reverses just at the time the change particle finishes its half circle so that it gets accelerated across the gap between the Dee's.

The particle gets accelerated again and again, and its velocity increases. Therefore, it attains high kinetic energy.

The positively charged ion adopts a circular path with a constant speed v, under the action of magnetic field B, which is perpendicular to the planes of Dee's of radius r.

$$r = \frac{mv}{qB}$$

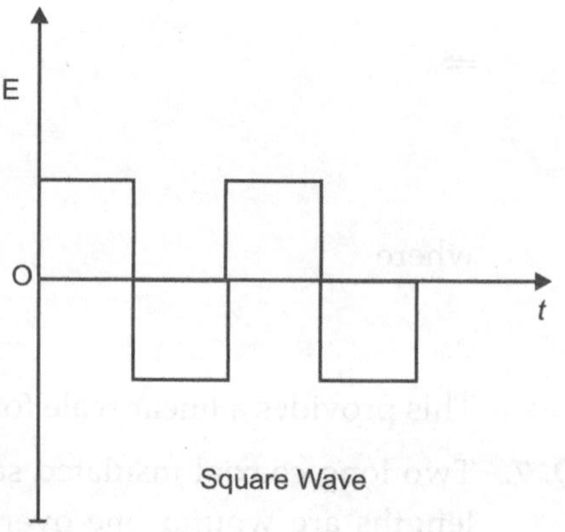

Q. 9. (i) Obtain the expression for the cyclotron frequency.★

(ii) A deuteron and a proton are accelerated by the cyclotron. Can both be accelerated with the same oscillator frequency ? Give reason to justify your answer.★

Ans. (i) Suppose the positive ion with charge q moves in a dee with a velocity v, then

★ are board based questions from previous year

$$\frac{mv^2}{r} = qvB$$

$$r = \frac{mv}{qB}$$

Where m is the mass and r is the radius of the path of ion in the dee and B is the strength of the magnetic field.

The time taken by the ion,

$$T = \frac{2\pi r}{v}$$

$$\frac{T}{2\pi} = \frac{r}{v}$$

$$\omega = \frac{v}{r} = \frac{v \times qB}{mv}$$

$$\therefore \quad \omega = \frac{qB}{m}$$

$$\therefore \quad f = \frac{qB}{2\pi m}$$

This frequency is called the cyclotron frequency.

(ii) We know that, $\omega \propto \frac{q}{m}$

Deuteron and proton have different masses hence cannot be accelerated with the same oscillator frequency.

Q. 10. State the principle of working of a galvanometer.

A galvanometer of resistance G is converted into a voltmeter to measure upto V volts by connecting a resistance R_1 in series with the coil. If a resistance R_2 is connected in series with it, then it can measure upto V/2 volts. Find the resistance, in terms of R_1 and R_2, required to be connected to convert it into a voltmeter that can read upto 2 V. Also find the resistance G of the galvanometer in terms of R_1 and R_2.

Ans. **Principle :** When a current-carrying coil is placed in a magnetic field, it experiences a torque. From the measurement of the deflection of the coil, the strength of the current can be computed. A high resistance is connected in series with the galvanometer to convert it into voltmeter. The value of the resistance is given by

$$R = \frac{V}{I_g} - G$$

Here,

V = Potential difference across the terminals of the voltmeter

I_g = Current through the galvanometer

G = Resistance of the galvanometer

When the resistance R_1 is connected in series with the galvanometer.

$$V = I_g (G + R_1)$$

When the resistance R_2 is connected in series with the galvanometer

$$\frac{V}{2} = I_g (G + R_2)$$

$$\Rightarrow \quad 2 = \frac{G + R_1}{G + R_2}$$

$$G = R_1 - 2R_2$$

Let R_3 be the resistance required for conversion of galvanometer into voltmeter of range 2V

$$\therefore \quad 2V = I_g (G + R_3)$$

Also

$$V = I_g (G + R_1)$$

$$\therefore \quad 2 = \frac{G+R_3}{G+R_1}$$

$\therefore R_3 = G + 2R_1 = R_1 - 2R_2 + 2R_1 = 3R_1 - 2R_2$

Chapter 5. Magnetism and Matter

Q. 1. Explain following terms of terrestrial magnetism.

Ans. **(i) Geographic axis :** It is a straight line passing through the geographical poles of the earth. It is the axis of rotation of the earth. It is also known as polar axis.

(ii) Geographic meridian : At any place is a vertical plane passing through the geographic north and south poles of the earth.

(iii) Geographic equator : It is a great circle on the surface of the earth, in a plane perpendicular to the geographic axis. All the points on the geographic equator are at equal distances from the geographic poles.

(iv) Magnetic axis : It is a straight line passing through the magnetic poles of the earth. It is inclined to geographic axis nearly at an angle of 17°.

(v) Magnetic meridian : At any place is a vertical plane passing through the magnetic north and south poles of the earth.

(vi) Magnetic equator : It is a great circle on the surface of the earth, in a plane perpendicular to the magnetic axis. All the points on the magnetic equator are at equal distances from the magnetic poles.

(vii) Dip or Inclination (δ) : The angle between the horizontal component of earth's magnetic field and the earth's resultant magnetic field at a place is dip or inclination at that place.

It is zero at the equator and 90° at the poles. Lines drawn up on a map through places that have the same dip are called isoclinic lines.

The line drawn through places that have zero dip is known as an aclinic line.

It is the magnetic equator.

Q. 2. Figure shows a small magnetised needle P placed at a point O. The arrow shows the direction of its magnetic moment. The other arrows show different positions (and orientations of the magnetic moment) of another identical magnetised needle Q.

(i) In which configuration the system is not in equilibrium ?

(ii) In which configuration is the system in (a) stable and (b) unstable equilibrium ?

(iii) Which configuration corresponds to the lowest potential energy among all the configuration shown ?

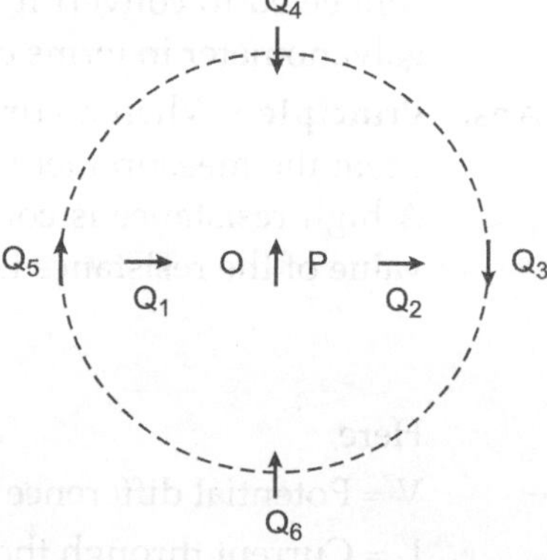

[NCERT]

Ans. Potential energy of the configuration arises due to the potential energy of one dipole (say, Q) in the magnetic field due to (P). The field due to P is given by the expression

$$B_P = -\frac{\mu_0}{4\pi}\frac{m_P}{r^3} \quad \text{(on the normal bisector)}$$

$$B_P = \frac{\mu_0}{4\pi}\frac{2\,m_P}{r^3} \quad \text{(on the axis)}$$

where m_P is the magnetic moment of the dipole P.

Equilibrium is stable when m_Q is parallel to B_P, and unstable when it is anti-parallel to B_P.

For the configuration Q_3 for which Q is along the perpendicular bisector of the dipole P, the magnetic moment of Q is parallel to the magnetic field at the position 3. Hence, Q_3 is stable. Thus,

(i) PQ_1 and PQ_2 (ii) (a) PQ_3, PQ_6 (stable); (b) PQ_5, PQ_4 (unstable) (iii) PQ_6

Q. 3. Draw magnetic field lines when a (i) diamagnetic, (ii) paramagnetic substance is placed in an external magnetic field. Which magnetic property distinguishes this behaviour of the field lines due to the two substances ? **[NCERT]**

Ans. Figure 1 shows magnetic field lines when diamagnetic materials are placed in external magnetic field.

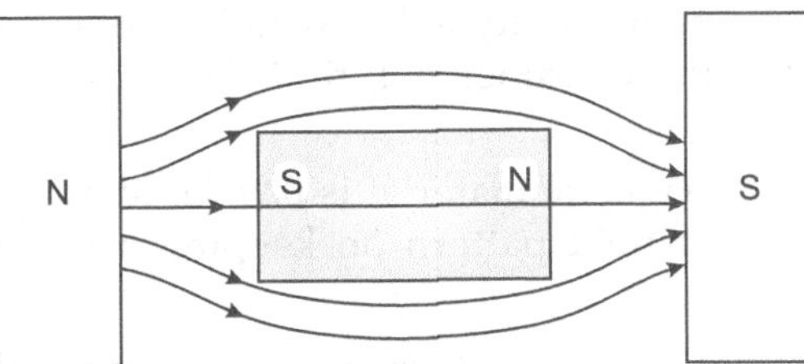

Figure 1

Figure 2 shows magnetic field lines, when paramagnetic materials are placed in external magnetic field.

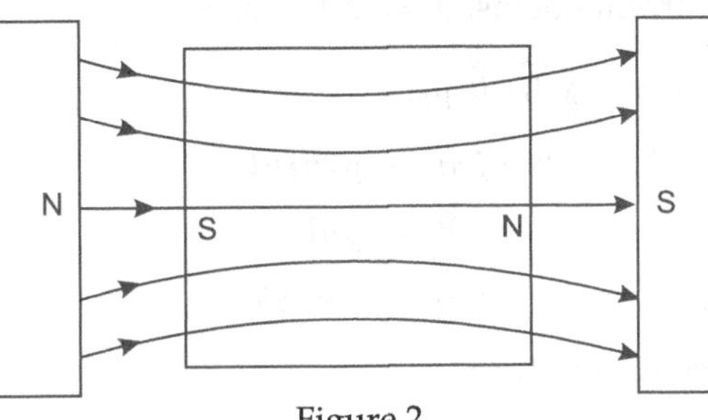

Figure 2

Magnetic dipole moment of materials are responsible for this diamagnetic material has zero magnetic moment.

Paramagnetic material has non-zero magnetic moment.

Q. 4. Answer the following :★

(i) Why is it necessary to introduce a cylindrical soft iron core inside the coil of a galvanometer.

(ii) Increasing the current sensitivity of a galvanometer may not necessarily increase its voltage sensitivity. Explain, giving reason.

Ans. (i) When a soft iron core is used the magnetic field lines tend to crowd through the core. It is because, soft iron core is ferromagnetic in nature. As a result, the strength of the magnetic field due to the field magnet increases, which in turn increases the sensitivity of the galvanometer.

(ii) Current sensitivity of a galvanometer is defined as the deflection produced in the galvanometer when a unit current flows through it.

$$I_s = \frac{\theta}{I} = \frac{nBA}{K} \qquad ...(i)$$

Voltage sensitivity of a galvanometer is defined as the deflection produced in the galvanometer when a unit voltage is applied across two terminals.

$$V_s = \frac{\theta}{V} = \frac{\theta}{IR} = \frac{nBA}{KR} = \frac{I_s}{R} \qquad ...(ii)$$

Where n = number of turns in the coil of galvanometer

B = Magnetic field around coil

A = Area of coil

K = restoring torque per unit twist

From equation (i) we can say that current sensitivity increases by increasing n, B, A and decreases by K.

From equation (ii) we can say that voltage sensitivity increases by increasing n, B, A and decreases by K, R.

★ are board based questions from previous year

In case of current sensitivity if we increases the no. of turns n its current sensitivity increases. Since resistance of galvanometer R also increases, voltage sensitivity remains same or unchanged. Therefore increase in current sensitivity of galvanometer may not necessarily increase the voltage.

Q. 5. (i) An iron ring of relative permeability μ_r has windings of insulated copper wire of n turns per metre. When the current in the windings is I, find the expression for the magnetic field in the ring.★

(ii) The susceptibility of a magnetic material is 0.9853. Identify the type of magnetic material. Draw the modification of the field pattern on keeping a piece of this material in a uniform magnetic field.★

Ans. (i) Consider a ring of radius r having n turns per metre. If n is the number of turns per metre, then total number of turns in the ring $= 2\pi rn = N$

$$\text{Current enclosed} = NI = 2\pi rn\,I.$$

By Ampere's circuital law, $\oint \vec{B}\cdot\vec{dl}\ \mu_0 I$

$$B \times 2\pi r = \mu 2\pi rnI$$

or

$$B = \mu nI$$

(ii) Given, susceptibility, $\chi = 0.9853.$

The material is paramagnetic in nature.

If a piece of this material is kept in uniform magnetic field, then field pattern gets modified as follows :

The lines of force tend to pass through the material rather than the surrounding air.

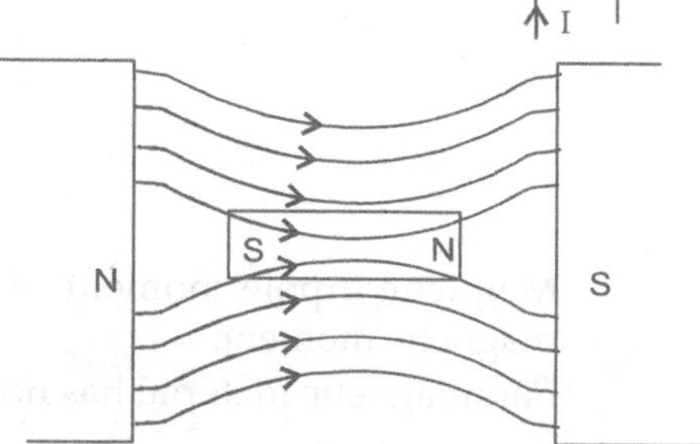

Chapter 6. Electromagnetic Induction

Q. 1. (i) A closed loop is held stationary in the magnetic field between the north and south poles of two permanent magnets held fixed. Can we hope to generate current in the loop by using very strong magnets ?

(ii) A closed loop moves normal to the constant electric field between the plates of a large capacitor. Is a current induced in the loop (a) when it is wholly inside the region between the capacitor plates, (b) when it is partially outside the plates of the capacitor ? The electric field is normal to the plane of the loop.

(iii) A rectangular loop and a circular loop are moving out of a uniform magnetic field region (Fig. 1) to a field-free region with a constant velocity v. In which loop do you expect the induced emf to be constant during the passage out of the field region ? The field is normal to the loops.

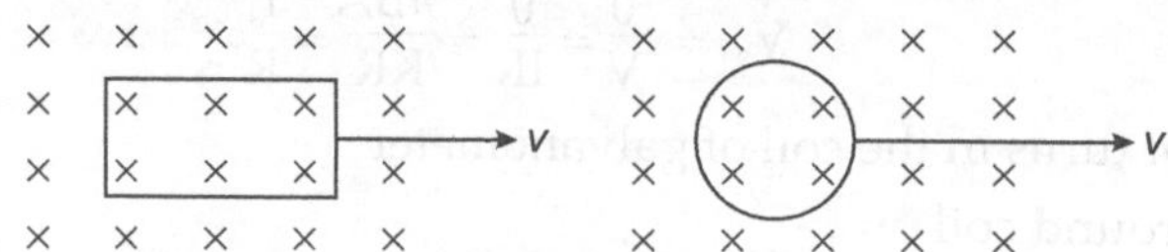

Figure 1

Ans. (i) No, however strong the magnet may be, current can be induced only by changing the magnetic flux through the loop.

(ii) No current is induced in either case. Current can not be induced by changing the electric flux.

★ are board based questions from previous year

(iii) The induced emf is expected to be constant only in the case of the rectangular loop. In the case of circular loop, the rate of change of area of the loop during its passage out of the field region is not constant, hence induced emf will vary accordingly.

Q. 2. Two concentric circular coils, one of small radius r_1 and the other of large radius r_2, such that $r_1 << r_2$, are placed co-axially with centres coinciding. Obtain the mutual inductance of the arrangement. **[NCERT]**

Ans. Let a current I_2 flow through the outer circular coil. The field at the centre of the coil is $B_2 = \mu_0 I_2 / 2r_2$. Since the other co-axially placed coil has a very small radius. B_2 may be considered constant over its cross-sectional area. Hence,

$$\phi_1 = \pi r_1^2 B_2$$

$$= \frac{\mu_0 \pi r_1^2}{2r_2} I_2 = M_{12} I_2$$

Thus,
$$M_{12} = \frac{\mu_0 \pi r_1^2}{2r_2}$$

From equation,
$$M_{12} = M_{21} = \frac{\mu_0 \pi r_1^2}{2r_2}$$

Note that we calculated M_{12} from an approximate value of ϕ_1, assuming the magnetic field B_2 to be uniform over the area πr_1^2. However, we can accept this value because $r_1 << r_2$.

Q. 3. Predict the direction of induced current in the situations described by the following figures.

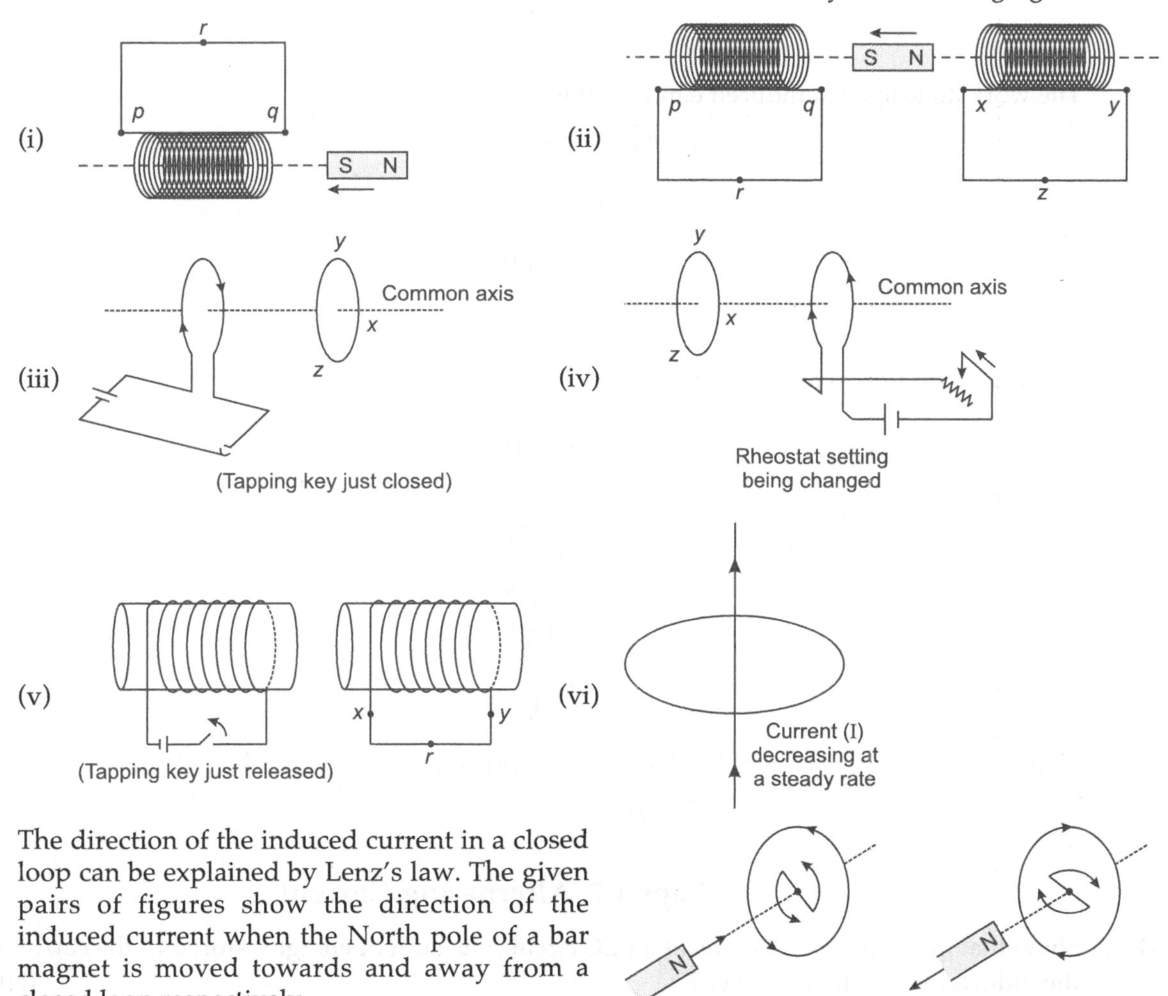

Ans. The direction of the induced current in a closed loop can be explained by Lenz's law. The given pairs of figures show the direction of the induced current when the North pole of a bar magnet is moved towards and away from a closed loop respectively.

Using Lenz's law the direction of the induced current in the given conditions can be described as follows :

(i) The direction of the induced current is in qrpq. direction.

(ii) The direction of the induced current is in prqp. direction.

(iii) The direction of the induced current is in yzxy. direction.

(iv) The direction of the induced current is in zyxz. direction.

(v) The direction of the induced current is in xryx. direction.

(vi) No current is induced since the field lines are lying in the plane of the closed loop.

Q. 4. Define the term self-inductance of a solenoid. Obtain the expression for the magnetic energy stored in an inductor of self-inductance L to build up a current I through it.★

Ans. The ratio of magnetic flux through the solenoid to the current passing through it is called self-inductance of a solenoid. It is given by

$$L = \frac{\Phi}{I}$$

Induced current

Energy stored in an inductor : When a current grows through an inductor, a back e.m.f. is set up which opposes the growth of current. So work needs to be done against back e.m.f. (e) in building up the current. This work done is stored as magnetic potential energy.

Let I be the current through the inductor L at any instant t.

The current rises at the rate $\frac{dI}{dt}$.

So the induced e.m.f. is

$$e = \frac{-LdI}{dt}$$

The work done against induced e.m.f. in dt is

$$dW = Pdt$$
$$= -eIdt$$
$$= \frac{LdI}{dt} Idt$$
$$= LIdI$$

For total work from 0 to I_0 current

$$W = \int_o^{I_o} dW$$
$$= \int_o^{I_o} LIdI$$
$$= L\left[\frac{I^2}{2}\right]_o^{I_o}$$
$$= \frac{1}{2}LI_0^2$$

Hence this work done is stored as the magnetic potential energy U in the inductor

$$U = \frac{1}{2}LI^2$$

Chapter 7. Alternating Current

Q. 1. Show that in the free oscillations of an LC circuit, the sum of energies stored in the capacitor and the inductor is constant in time. **[NCERT]**

★ are board based questions from previous year

Ans. Let q_0 be the initial charge on a capacitor. Let the charged capacitor be connected to an inductor of inductance L. LC circuit will sustain an oscillation with frequency

$$\omega\left(=2\pi\nu=\frac{1}{\sqrt{LC}}\right)$$

At an instant t, charge q on the capacitor and the current i are given by :

$$q(t) = q_0 \cos \omega t$$
$$i(t) = -q_0 \omega \sin \omega t$$

Energy stored in the capacitor at time t is

$$U_E = \frac{1}{2}CV^2 = \frac{1}{2}\frac{q^2}{C} = \frac{q_0^2}{2C}\cos^2(\omega t)$$

Energy stored in the inductor at time t is

$$U_M = \frac{1}{2}L i^2 = \frac{1}{2}L q_0^2 \omega^2 \sin^2(\omega t) = \frac{q_0^2}{2C}\sin^2(\omega t) \qquad (\because \omega^2 = 1/\sqrt{LC})$$

Sum of energies

$$U_E + U_M = \frac{q_0^2}{2C}[\cos^2 \omega t + \sin^2 \omega t] = \frac{q_0^2}{2C}$$

This sum is constant in time as q_0 and C, both are time-independent.

Q. 2. (i) For circuits used for transporting electric power, a low power factor implies large loss in transmission. Explain.

(ii) Power factor can often be improved by the use of a capacitor of appropriate capacitance in the circuit. Explain. **[NCERT]**

Ans. (i) As P = IVcos ϕ where, cos ϕ is the power factor. To supply a given power at a given voltage, if cos ϕ is small we have to increase current accordingly. But this will lead to large power loss (I^2R) in transmission.

(ii) Suppose in a circuit, current I lags the voltage by an angle ϕ. Then power factor cos ϕ = R/Z.

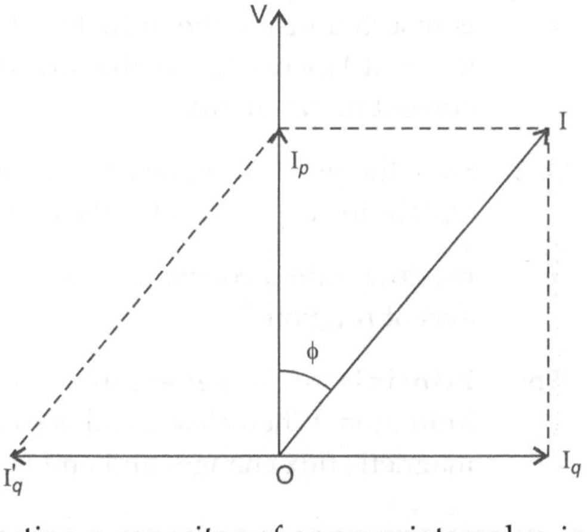

To improve the power factor (tending to 1) by making Z tend to R. In phasor diagram, resolve I into two components. I_p along the applied voltage V and I_q perpendicular to the applied voltage. I_q is called the wattless component since corresponding to this component of current, there is no power loss. I_p is known as the power component because it is in phase with the voltage and corresponding to power loss in the circuit.

To improve power factor, we must completely neutralize the lagging wattless current I_q by an equal leading wattless current I'_q. This can be done by connecting a capacitor of appropriate value in parallel so that I_q and I'_q cancel each other and P is effectively I_p V.

Q. 3. Draw a labelled diagram of a step-up transformer. Obtain the ratio of secondary to primary voltage in terms of number of turns and currents in the two coils.

Ans. We have,

$$e_s = -N_s \frac{d\phi}{dt} \quad \ldots(i)$$

$$e_p = -N_p \frac{d\phi}{dt} \quad \ldots(ii)$$

Dividing (i) by (ii), we get

$$\frac{e_s}{e_p} = \frac{N_s}{N_p} \quad \ldots(iii)$$

Step up transformer

According to the law of conservation of energy,

$$e_s I_s = e_p I_p$$

$$\frac{e_s}{e_p} = \frac{I_p}{I_s} \quad \ldots(iv)$$

From equation (iii) and (iv),

$$\frac{e_s}{e_p} = \frac{N_s}{N_p} = \frac{I_p}{I_s}$$

Q. 4. (i) When an AC source is connected to an ideal inductor show that the average power supplied by the source over a complete cycle is zero.★

(ii) A lamp is connected in series with an inductor and an AC source. What happens to the brightness of the lamp when the key is plugged in and an iron rod is inserted inside the inductor ? Explain.★

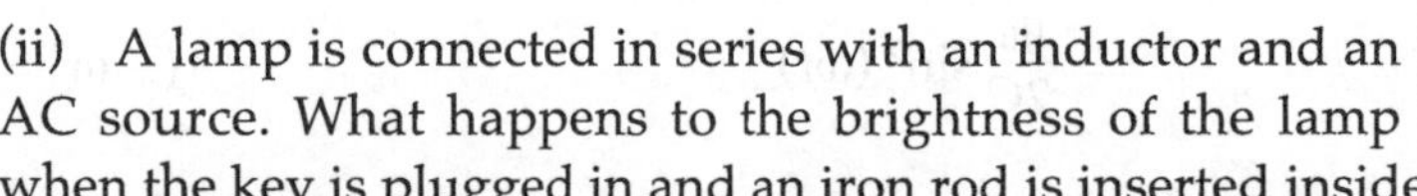

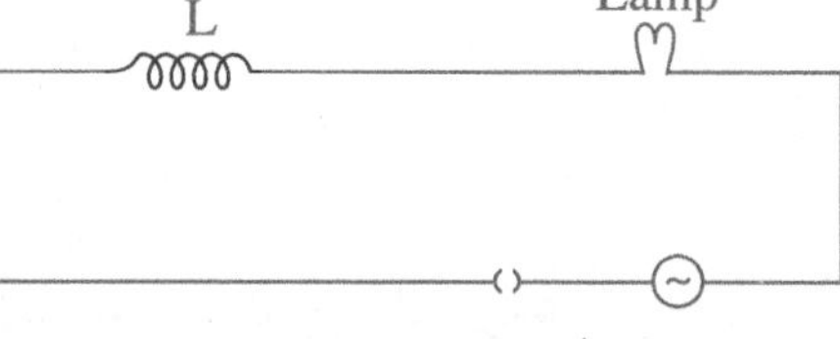

Ans. (i) The average power supplied by the source over a complete cycle is

$$P_{av} = E_{rms}.I_{rms}.\cos\phi$$

When the circuit contains an ideal inductor, then the phase difference between the current and voltage is $\pi/2$.

So, $$\phi = \frac{\pi}{2}. \text{ So, } \cos\phi = \cos\frac{\pi}{2} = 0.$$

Hence $$P_{av} = 0.$$

So, when an ac source is connected to an ideal inductor, the average power supplied by the source over a complete cycle is zero.

(ii) The brightness of the lamp will decrease. When the key is plugged in and the iron rod is inserted inside the inductor, it increases the inductance. Hence, the reactance of the inductor ($X_L = \omega L$) increases. So, the impedance of the circuit ($Z = R + j\omega L$) increases, which decreases the current in the circuit.

Q. 5. State the principle of an ac generator and explain its working with the help of a labelled diagram. Obtain the expression for the emf induced in a coil having N turns each of cross-sectional area A, rotating with a constant angular speed 'ω' in a magnetic field $\vec{B}$, directed perpendicular to the axis of rotation.★

Ans. **Principle of ac generator :** The ac generator is based on the principle of electromagnetic induction. When closed coil is rotated in a uniform field with its axis perpendicular to field, then magnetic flux changes and emf is induced.

★ are board based questions from previous year

Working : When the armature coil rotates, the magnetic flux linked with it changes and produces induced current. If initially, coil PQRS is in vertical position and rotated clockwise, then PQ moves down and SR moves up. By Fleming's right hand rule, induced current flows from Q to P and S to R which is the first half rotation of coil. Brush B_1 is positive terminal and B_2 is negative. In second half rotation, PQ moves up and SR moves down. So induced current reverses and the alternating current is produced in this manner by the generator.

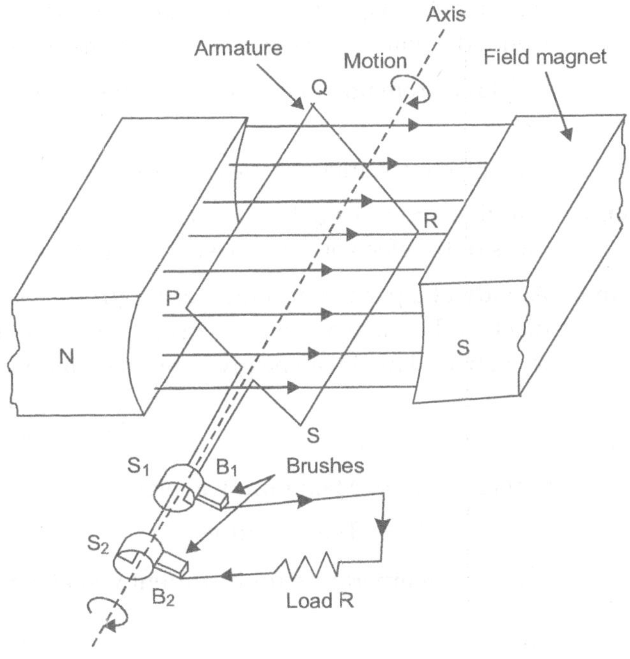

Expression for emf induced : If number of turns in coil = N, cross-section = A, angular speed of rotation = ω, magnetic field = $\vec{B}$, then to find emf induced,

Flux through the coil when its normal makes angle θ with the field,

$$\phi = BA \cos \theta$$

When coil rotates with angular velocity ω and turns through θ in time t then $\theta = \omega t$

$$\Rightarrow \quad \phi = BA \cos \omega t$$

When coil rotates, ϕ changes to set an induced emf.

$$\varepsilon = \frac{-d\phi}{dt}$$

$$= -\frac{d}{dt}(BA \cos \omega t)$$

$$= BA\,\omega \sin \omega t$$

For N turns, total induced emf,

$$\varepsilon = NBA\,\omega \sin \omega t$$

Chapter 8. Electromagnetic Waves

Q. 1. Write down properties of E.M. wave.

Ans. (i) Changes in electric and magnetic fields occur simultaneously. E.M. wave attain their maxima and minima at the same place and at the same time.

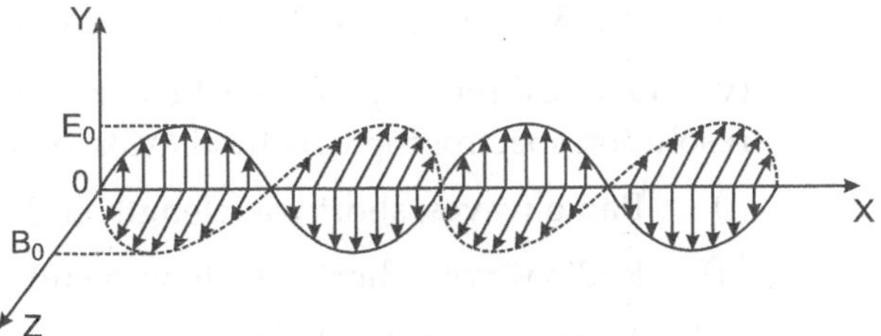

Direction of propagation

(ii) The electric and magnetic fields directions are mutually perpendicular to each other and as well as to the direction of propagation of wave.

(iii) The electric field vector (E) and magnetic field vector (B) are related by $c = E_0/B_0$, here E_0 and B_0 are the amplitudes of the respective fields and c is speed of light.

(iv) The velocity of electromagnetic waves in free space, $c = 1/\sqrt{\mu_0 \varepsilon_0}$.

(v) The velocity of electromagnetic waves in a material medium = $1/\sqrt{\mu\varepsilon}$. where μ and ε are absolute permeability and absolute permittivity of the material medium respectively.

(vi) Electromagnetic waves follow the principle of superposition.

(vii) Electromagnetic waves transfer energy as they propagate through space. This energy is divided equally between electric and magnetic fields.

(viii) Electromagnetic waves can transfer energy as well as momentum to objects placed on their paths.

(ix) Electromagnetic waves do not require material medium to travel.

Q. 2. Use the formula $\lambda_m T = 0.29$ cm K to obtain the characteristic temperature ranges for different parts of the electromagnetic spectrum. What do the numbers that you obtain tell you ? **[NCERT]**

Ans. A body at a particular temperature produces a continuous spectrum of wavelengths. In case of a black body, the wavelength corresponding to maximum intensity of radiation is given according to Planck's law. It can be given by the relation,

$$\lambda_m = \frac{0.29}{T} \text{ cm K}$$

where, λ_m = Maximum wavelength

T = Temperature

Thus, the temperature for diffrerent wavelength can be obtained as :

For $\lambda_m = 10^{-4}$ cm; $T = \frac{0.29}{10^{-4}} = 2900$ K

For $\lambda_m = 5 \times 10^{-5}$ cm; $T = \frac{0.29}{5 \times 10^{-5}} = 5800$ K

For $\lambda_m = 10^{-6}$ cm; $T = \frac{0.29}{10^{-6}} = 290000$ K and so on.

The numbers obtained tell us that temperature ranges are required for obtaining radiations in different parts of an electromagnetic spectrum. As the wavelength decreases, the corresponding temperature increases.

Q. 3. Given below are some famous numbers associated with electromagnetic radiations in different contents in physics. State the part of the electromagnetic spectrum to which each belongs.

(i) 21 cm (wavelength emitted by atomic hydrogen in interstellar spaces).

(ii) 1057 MHz (frequency of radiation arising from two close energy levels in hydrogen; known as Lamb shift).

(iii) 2.7 K [temperature associated with the isotropic radiation filling all space thought to be a relic of the 'big-bang' origin of the universe].

(iv) 5890 Å – 5896 Å [double lines of sodium].

(v) 14.4 keV [energy of a particular transition in ^{57}Fe nucleus associated with a famous high resolution spectroscopic method (Mössbauer spectroscopy)]. **[NCERT]**

Ans. (i) Radio waves; short wavelength end of the electromagnetic spectrum.

(ii) Radio waves; short wavelength end

(iii) Temperature, T = 2.7°K λ_m is given by Planck's law by

$$\lambda_m = \frac{0.29}{2.7} = 0.11 \text{ cm}$$

This wavelength corresponds to microwaves.

(iv) This is the yellow light of the visible spectrum.

(v) Transition energy is given by the relation,

$$E = h\nu$$

Here, h = Planck's constant

$= 6.6 \times 10^{-34}$ Js

ν = Frequency of radiation

$$\text{Energy, } E = 14.4 \text{ keV}$$

$$\therefore \quad \nu = \frac{E}{h}$$

$$= \frac{14.4 \times 10^3 \times 1.6 \times 10^{-19}}{6.6 \times 10^{-34}}$$

$$= 3.4 \times 10^{18} \text{ Hz.}$$

This corresponds to X-rays.

Q. 4. Write Maxwell's generalisation of Ampere's circuital law. Show that in the process of charging a capacitor, the current produced within the plates of the capacitor is $i = \varepsilon_0 \frac{d\phi_E}{dt}$ where ϕ_E is the electric flux produced during charging of the capacitor plates.★

Ans. **Correction in Ampere's circuital law (Modified Ampere's law) :** Maxwell removed the problem of current continuity and inconsistency observed in Ampere's circuital law by introducing the concept of displacement current. Displacement current arises due to change in electric flux with time and is given by $i_d = \varepsilon_0 \frac{\phi_E}{dt}$

Electric flux through the loop

$$\phi_E = EA$$

$$= \frac{\sigma}{\varepsilon_0} A = \frac{Q}{A}\frac{A}{\varepsilon_0} = \frac{Q}{\varepsilon_0} \qquad \text{(Q = charge on either plates)}$$

$$\phi_E = \frac{Q}{\varepsilon_0}$$

$$\frac{d\phi_E}{dt} = \frac{1}{\varepsilon_0}\frac{dQ}{dt}$$

$$\varepsilon_0 \frac{d\phi_E}{dt} = \frac{dQ}{dt}$$

$\frac{dQ}{dt}$ is called conduction current which is equal to $\varepsilon_0 \frac{d\phi_E}{dt}$ which is displacement current.

Hence, $i_c = i_d$

Generalization of Ampere's circuital law is :

$$\oint \vec{B}.\vec{dt} = \mu_0 (i_c + i_d)$$

Conduction current is due to the flow of charges but displacement current is not because of the flow of charges but it is due to the change in electric flux.

Q. 5. Name the parts of the electromagnetic spectrum which is★

(i) suitable for radar systems used in aircraft navigation

(ii) used to treat muscular strain

(iii) used as a diagnostic tool in medicine

Write in brief, how these waves can be produced.

Ans. (i) Microwaves are suitable for radar systems that are used in aircraft navigation.

★ are board based questions from previous year

These rays are produced by special vacuum tubes, namely klystrons, magnetrons and Gunn diodes.

(ii) Infrared waves are used to treat muscular strain. These rays are produced by hot bodies and molecules.

(iii) X-rays are used as a diagnostic tool in medicine. These rays are produced when high energy electrons are stopped suddenly on a metal of high atomic number.

Q. 6. Answer the following :★

(i) Name the e.m. waves which are suitable for radar systems used in aircraft navigation. Write the range of frequency of these waves.

(ii) If the earth did not have atmosphere, would its average surface temperature be higher or lower than what it is now ? Explain.

(iii) An e.m. wave exerts pressure on the surface on which it is incident. Justify.

Ans. (i) Microwaves are suitable for radar systems used in aircrafts navigation. The range of frequency for these waves is 10^9 Hz to 10^{12} Hz.

(ii) In the absence of atmosphere, there would be no greenhouse effect on the surface of the Earth. As a result, the temperature of the Earth would decrease rapidly, making it difficult for human survival.

(iii) The momentum transported by electromagnetic waves is given by

$$p = \frac{U}{c}$$

where U is the energy transported by electromagnetic waves in a given time and c is speed of electromagnetic waves in free space. As a result, when these waves strike a surface, pressure and hence force is exerted by them on the surface.

Q. 7. Answer the following :★

(i) Name the e.m. waves which are used for the treatment of certain forms of cancer. Write their frequency range.

(ii) Why is the amount of the momentum transferred by the e.m. waves incident on the surface so small ?

Ans. (i) Gamma rays are used for the treatment of certain forms of cancer. Their frequency range is 3×10^{19} to 5×10^{20} Hz.

(ii) The momentum transported by electromagnetic waves is given by

$$p = \frac{U}{c} = \frac{h\nu}{c}$$

where U is the energy transported by electromagnetic waves in a given time and c is speed of electromagnetic waves in free space.

Now, $h = 6.62 \times 10^{-34}$ Js, $c = 3 \times 10^8$ ms^{-1}

Therefore, even for γ-rays ($\nu \approx 10^{20}$ Hz),

$$p = \frac{6.62 \times 10^{-34} \times 10^{20}}{3 \times 10^8}$$

$$= 2.2 \times 10^{-22} \text{ kg ms}^{-1}$$

Thus, the amount of the momentum transferred by the e.m. waves incident on a surface is very small.

Q. 8. (a) We feel the warmth of the sun light but not the pressure on our hands. Explain.★

(b) Which out of wavelength, frequency and speed of an electromagnetic wave does not change on passing from one medium to another?★

(c) A thin ozone layer in the upper atmosphere is crucial for human survival on earth, why?★

★ are board based questions from previous year

Ans. (a) Because the photons of light have only energy and do not exhibit mass.

(b) When electromagnetic waves passes from one medium to another, frequency remains constant. Wavelength and speed both changes when medium of the wave changes.

(c) The thin ozone layer around the earth's atmosphere is very crucial for human survival. It protects the Earth from harmful ultraviolet (UV) rays from the sun. Without the ozone layer in the atmosphere, life on Earth would be very difficult. Plants cannot live and grow in heavy ultraviolet radiations. Ultraviolet radiations can cause skin cancer, cataracts and impaired immune systems in humans.

Chapter 9. Ray Optics and Optical Instruments

Q. 1. (i) Draw a ray diagram depicting the formation of the image by an astronomical telescope in normal adjustment.★

(ii) You are given the following three lenses. Which two lenses will you use as an eyepiece and as an objective to construct an astronomical telescope ? Give reason.★

Lenses	Power (D)	Aperture (cm)
L_1	3	8
L_2	6	1
L_3	10	1

Ans. (i)

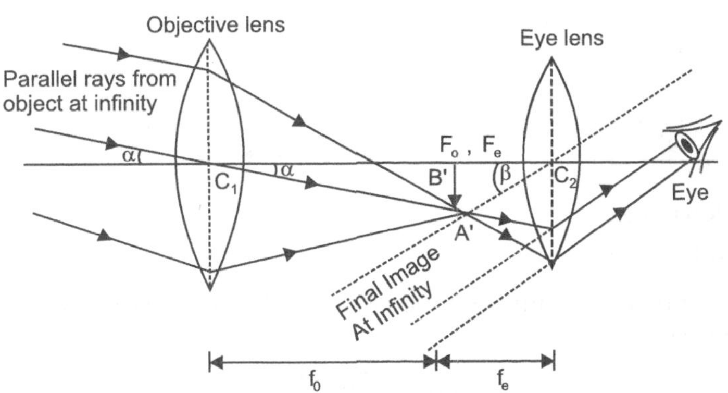

(ii) An astronomical telescope should have an objective of large aperture and longer focal length while an eyepiece of small aperture and small focal length.
Therefore, we will use L_1 as an objective and L_3 as an eyepiece.

Q. 2. (i) Draw a ray diagram showing the formation of image by a reflecting telescope.★

(ii) Write two advantages of a reflecting telescope over a refracting telescope.★

Ans. (i)

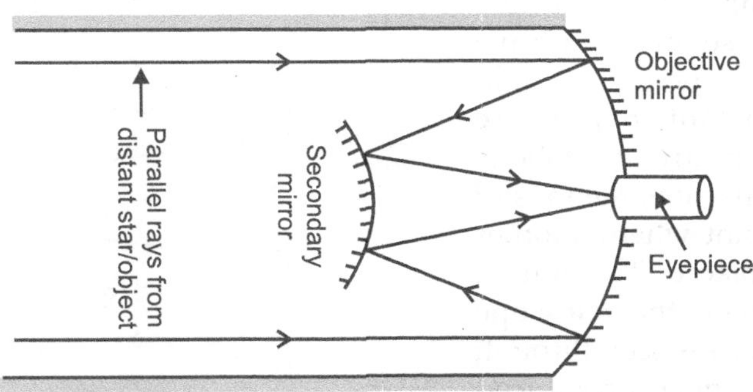

(ii) (a) The image is free from chromatic aberration.

(b) Spherical aberration can be eliminated by using parabolic mirror.

★ are board based questions from previous year

Q. 3. (i) Draw a ray diagram for the formation of image by a compound microscope.★

(ii) You are given the following three lenses. Which two lenses will you use as an eyepiece and as an objective to construct a compound microscope ?★

Lenses	Power (D)	Aperture (cm)
L_1	3	8
L_2	6	1
L_3	10	1

(iii) Define resolving power of a microscope and write one factor on which it depends.

Ans. (i)

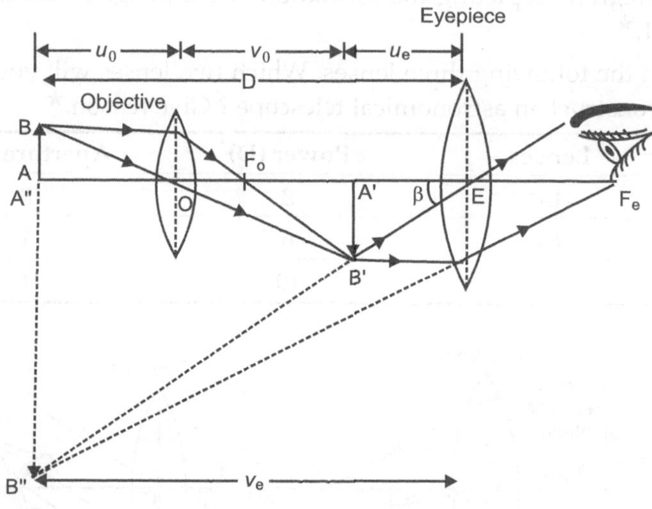

(ii) Objective lens → L_3

Eye lens → L_2

(iii) The resolving power of a microscope is its ability to form distinctly separate images of two neighbouring objects.

$$\text{R.P.} = \frac{2\mu \sin \theta}{\lambda}$$

It depends on the wavelength (λ).

Q. 4. Draw a schematic ray diagram of reflecting telescope showing how rays coming from a distant object are received at the eyepiece. Write its two important advantages over a refracting telescope.★

Ans. **Reflecting Telescope** : The reflecting telescope make use of a concave mirror as objective. The rays of light coming from distant object are incident on the objective (parabolic reflector). After reflection, the rays of light meet at a point where another convex mirror is placed. This mirror focusses light inside the telescope tube. The final image is seen through the eye-piece. The images produced by the reflecting telescope is very bright and its resolving power is high.

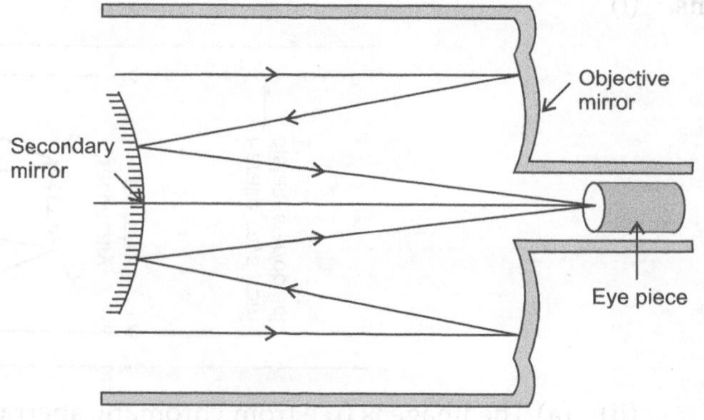

★ are board based questions from previous year

Advantages :

(i) The resolving power (the ability to observe two object distinctly) is high, due to the large diameter of the objective.

(ii) There is no chromatic aberration as the objective is a mirror.

Q. 5. Draw a ray diagram to show the image formation by a combination of two thin convex lenses in contact. Obtain the expression for the power of this combination in terms of the focal lengths of the lenses.★

Ans. (i) For first lens,

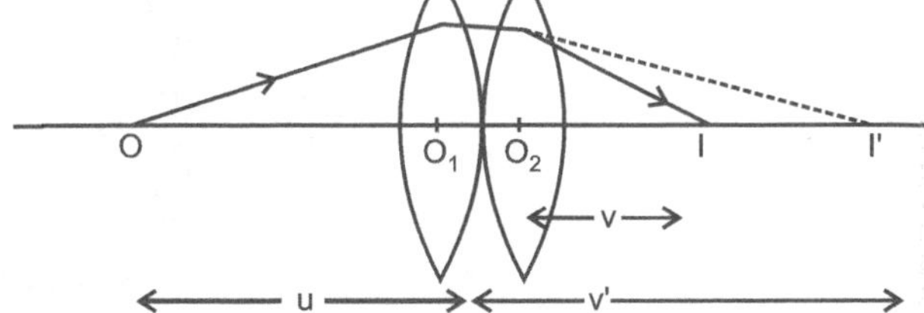

$$\frac{1}{f_1} = \frac{1}{v'} - \frac{1}{u} \quad \text{...(i)}$$

For second lens,

$$\frac{1}{f_2} = \frac{1}{v} - \frac{1}{v'} \quad \text{...(ii)}$$

Adding (i) and (ii),

$$\frac{1}{f_1} + \frac{1}{f_2} = \frac{1}{v} - \frac{1}{u} \quad \text{...(iii)}$$

If f is the combined focal length then

$$\frac{1}{f} = \frac{1}{v} - \frac{1}{u}$$

From equation (iii),

$$\frac{1}{f} = \frac{1}{f_1} + \frac{1}{f_2}$$

$$P = \frac{1}{f_1} + \frac{1}{f_2} \quad \left[\because \frac{1}{f} = P\right]$$

$$P = \frac{f_1 + f_2}{f_1 f_2}$$

Q. 6. (i) Draw a labelled ray diagram to obtain the real image formed by an astronomical telescope in normal adjustment position. Define its magnifying power.

(ii) You are given three lenses of power 0.5 D, 4 D and 10 D to design a telescope.

(a) Which lenses should be used as objective and eyepiece ? Justify your answer.

(b) Why is the aperture of the objective preferred to be large ?

Ans. (i) **Magnifying power :** The magnifying power of a refracting type astronomical telescope is defined as the ratio of angle subtended by the final image at eye to the angle subtended by the object at eye.

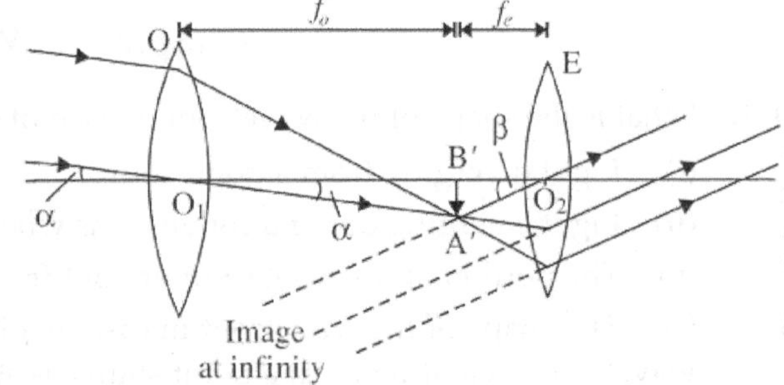

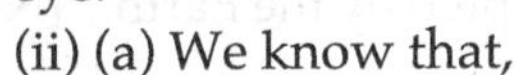

(ii) (a) We know that,

Magnification,

$$m = \frac{f_o}{f_e} = \frac{P_e}{P_o}$$

Therefore, the lens of 0.5 D should be used as objective and the lens of 10 D should be used as eye-piece in order to achieve higher magnification.

(b) The aperture of the objective lens is made larger so, that it receives as much light as coming from the distant object and the resolving power of the telescope increases.

Q. 7. Explain human eye and its parts with the help of neat diagram.

Ans. The human eye is one of the most valuable and sensitive sense organs in the human body. It enables us to see the wonderful world and colour around us. It consists of a lens, which is made up of living tissues.

★ are board based questions from previous year

Structure : A human eye has the following main parts :

- **Cornea :** It is transparent spherical membrane covering the front of the eye. Light enters the eye through this membrane. Most of the refraction of light rays entering the eye occurs at the outer surface of the cornea.

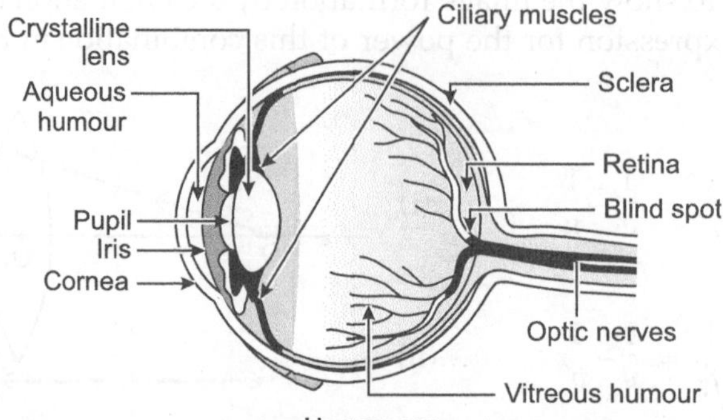

Human eye

- **Lens :** It is a central part of the eye that facilitates the image formation.
- **Iris :** It is a dark muscular diaphragm between the cornea and the lens. It controls the size of the pupil. It is the colour of the iris that we call colour of the eye.
- **Pupil :** It is a small hole in the iris through which light enters the eye. In dim light, it opens up completely due to contraction of iris muscles, but in bright light it becomes very small due to relaxation of iris muscles.
- **Ciliary muscles :** They hold the lens in position and help in modifying the curvature of the lens.
- **Retina :** It is light-sensitive surface of eye on which the image is formed. it contains light-sensitive cells, rods and cones. Rod cells respond to the intensity of light. Cones respond differently to different wavelength and responsible for colour vision. These cells generate signals which are transmitted to the brain through optical nerves.
- **Optic nerves :** It transmits visual information from the retina to the brain.
- **Sclera :** It is an opaque, fibrous, protective, outer layer of an eye containing collagen and elastic fibre. It is also known as white of the eye.
- **Blind spot :** It is the point at which the optic nerves leave the eye. It contains no rods and cones, so an image formed at this point is not sent to the brain.

Chapter 10. Wave Optics

Q. 1. What is the shape of the wavefront in each of the following cases :

(i) Light diverging from a point source.

(ii) Light emerging out of a convex lens when a point source is placed at its focus.

(iii) The portion of the wavefront of light from a distance star intercepted by the Earth. **[NCERT]**

Ans. (i) The shape of the wavefront in case of a light diverging from a point source is spherical. The wavefront emanating from a point source is shown in the given figure.

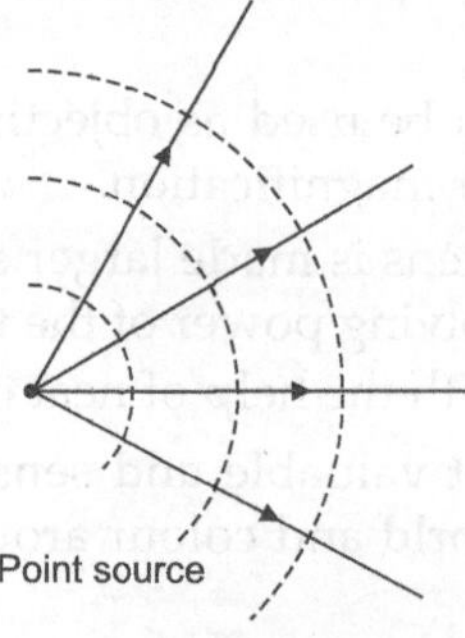

Point Source

(ii) The shape of the wavefront in case of a light emerging out of a convex lens when a point source is placed at its focus is a parallel grid. This is shown in the given figure.

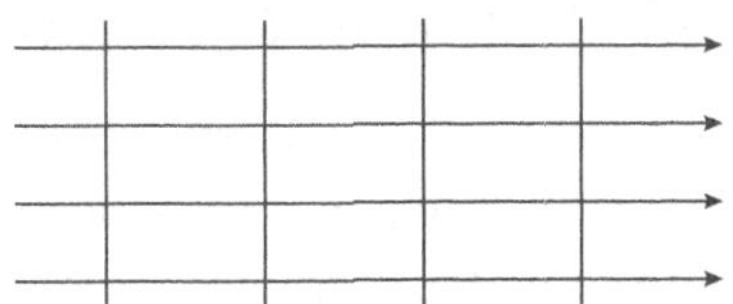

Wavefront

(iii) The portion of the wavefront of light from a distant star intercepted by the Earth is a plane.

Q. 2. Define the term wave front. State Huygen's principle. Consider a plane wave front incident on a thin convex lens. Draw a proper diagram to show how the incident wave front traverses through the lens and after refraction focusses on the focal point of the lens, giving the shape of the emergent wave front. ★

Ans. **Wave front :** A wave front is the locus of all the points in space that reach a particular distance by a propagating wave in same phase at any instant.

Huygen's principle : It is based on two assumptions :

(i) Each point of the wavefront behaves like a source of secondary wavelets. These secondary wavelets spread out in all directions with the same speed as that of the original wave.

(ii) When we draw an envelope in the forward direction of the secondary wavelets at any instant, then this envelope tells the new position of the wavefront at that instant.

Q. 3. Use Huygen's principle to show how a plane wavelength propagates from a denser to rarer medium. Hence verify Snell's law of refraction.★

Ans. In Δ ABC,

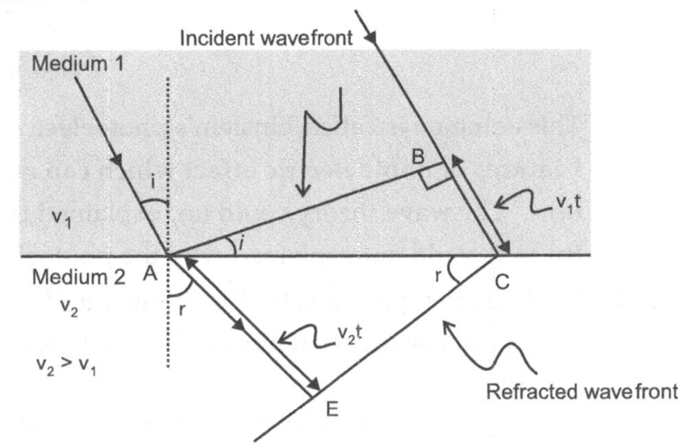

$$\sin i = \frac{BC}{AC} = \frac{v_1 t}{AC}$$

$$\sin r = \frac{AE}{AC} = \frac{v_2 t}{AC}$$

$$\therefore \quad \frac{\sin i}{\sin r} = \frac{BC}{AE} = \frac{v_1 t}{v_2 t} = \frac{v_1}{v_2}$$

$$\therefore \quad \mu_1 = \frac{c}{v_1}$$

$$\mu_2 = \frac{c}{v_2}$$

$$\therefore \quad \frac{\mu_2}{\mu_1} = \frac{v_1}{v_2}$$

$$\therefore \quad \frac{\sin i}{\sin r} = \frac{\mu_2}{\mu_1}$$

or $$\mu_2 \sin r = \mu_1 \sin i$$

It is Snell's law.

Q. 4. 'Two independent mono chromatic sources of light cannot produce a sustained interference pattern.' Give reasons.

Ans. The condition for the sustained interference is that both the sources must be coherent (*i.e.*, they must have the same wavelength and the same frequency and they must have the same phase or constant phase difference).

Two sources are monochromatic if they have the same frequency and wavelength. Since they are independent, *i.e.*, they have different phases with irregular difference, they are not coherent sources.

★ are board based questions from previous year

Chapter 11. Dual Nature of Radiation and Matter

Q. 1. Using photon picture of light, show how Einstein's photoelectric equation can be established. Write two features of photoelectric effect which cannot be explained by wave theory.★

Ans. **Einstein's photoelectric equation :** Einstein explained the various laws of photoelectric emission on the basis of Planck's quantum theory. According to Planck's quantum theory, light radiations consist of tiny packets of energy called 'quanta'. One quantum of light radiation is called a photon which travels with the speed of light.

The energy of a photon is given by,

$$E = h\nu$$

where h is Planck's constant and υ is the frequency of light radiation.

Einstein assumed that one photoelectron is ejected from a metal surface if one photon of suitable light radiation falls on it.

Consider a photon of light of frequency υ, incident on a photosensitive metal surface. The energy of the photon (= $h\nu$) is spent in two ways :

(a) A part of the energy of the photon is used in liberating the electron from the metal surface which is equal to the work function ϕ_0 of the metal.

(b) The rest of the energy of the photon is used in imparting the maximum kinetic energy K_{max} to the emitted photo electron.

If v_{max} is the maximum velocity of the emitted photoelectron and m is its mass, then

Max. kinetic energy of the photoelectron,

$$K_{max} = \frac{1}{2} m v^2_{max}$$

$$\therefore \quad h\nu = \phi_0 + \frac{1}{2} m v^2_{max}$$

This equation is called Einstein's photoelectric equation.

Features of photoelectric effect which can not be explained by wave theory :

(a) The wave theory could not explained the instantaneous process of photoelectric effect.

(b) It could not explained why the photoelectric emission is independent of intensity.

Q. 2. Explain giving reasons for the following :★

(i) Photoelectric current in a photocell increases with the increase in the intensity of the incident radiation.

(ii) The stopping potential (V_0) varies linearly with the frequency (ν) of the incident radiation for a given photosensitive surface with the slope remaining the same for different surfaces.

(iii) Maximum kinetic energy of the photoelectrons is independent of the intensity of incident radiation.

Ans. (i) Since number of photoelectrons emitted is directly proportional to the intensity of incident radiation therefore, as intensity increases the electron-hole pairs also increases.

(ii)

$$h\nu = h\nu_0 + eV_0$$

$$eV_0 = h(\nu - \nu_0)$$

$$\Rightarrow \quad V_0 = \frac{h}{e}(\nu - \nu_0)$$

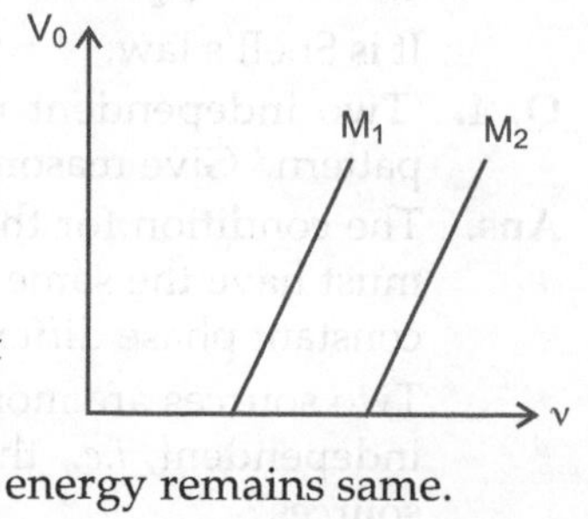

Further from graph, slope is h/e which is a constant and it does not depend on υ.

(iii) As intensity increases, the number of photons increases but the energy remains same.

★ are board based questions from previous year

Q. 3. The following graph shows the variation of photocurrent for a photosensitive metal :★

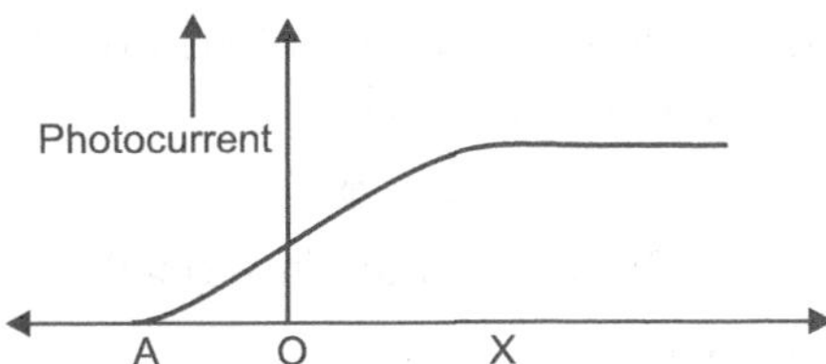

(i) Identify the variable X on the horizontal axis.

(ii) What does the point A on the horizontal axis represent ?

(iii) Draw this graph for three different values of frequencies of incident radiation v_1, v_2, and v_3 ($v_1 > v_2 > v_3$) for same intensity.

(iv) Draw this graph for three different values of intensities of incident radiation I_1, I_2 and I_3 ($I_1 > I_2 > I_3$) having same frequency.

Ans. (i) X is anode potential. (ii) Stopping potential.

(iii)

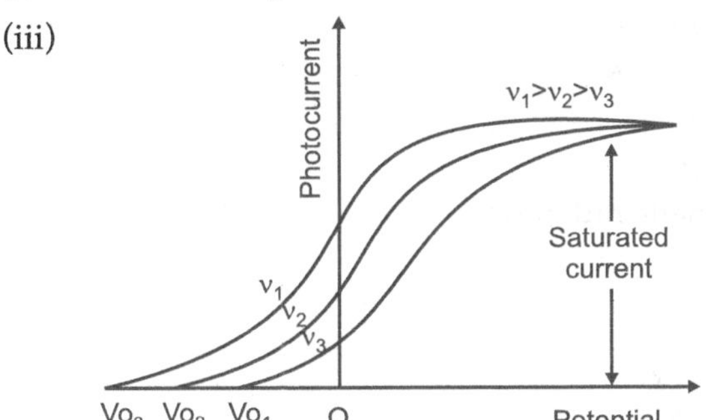

(iv)

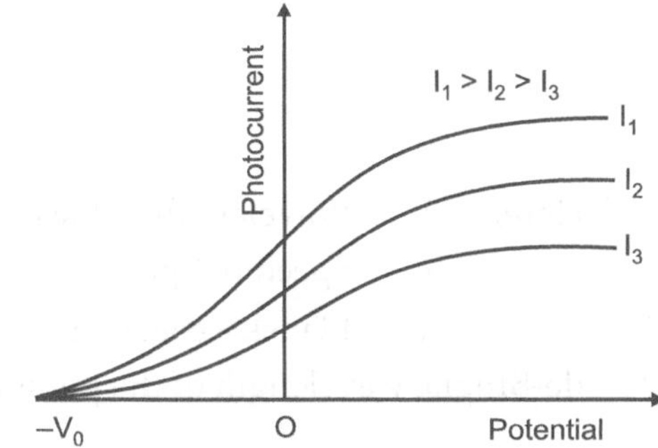

Q. 4. (i) Describe briefly three experimentally observed features in the phenomenon of photoelectric effect.★

(ii) Discuss briefly how wave theory of light cannot explain these features.★

Ans. (i) (a) The photoelectric effect will not occur when the frequency of the incident light is less than the threshold frequency. Different materials have different threshold frequencies and most elements have threshold frequencies in the ultraviolet region of the electromagnetic spectrum.

(b) The maximum kinetic energy of a stream of photoelectrons increases linearly with the frequency of the incident light above the threshold frequency.

(c) The rate at which photoelectrons are emitted from a photosensitive surface is directly proportional to the intensity of the incident light when the frequency is constant.

(ii) **Classical wave theory cannot explain :**

(a) The existence of threshold frequency because it predicts that electrons would absorb enough energy to escape and there would not be any threshold frequency.

(b) The almost immediate emission of photoelectrons as, according to this theory, electrons require a period of time before sufficient energy is absorbed by it to escape from the metal; however, such a thing does not happen practically.

(c) The independence of KE of photoelectrons on intensity and dependence on frequency because it cannot explain why maximum KE is dependent on the frequency and independent of intensity.

Q. 5. (i) Write the important properties of photons which are used to establish Einstein's photo-electric equation.★

(ii) Use this equation to explain the concept of (a) threshold frequency and (b) stopping potential.★

★ are board based questions from previous year

Ans. (i) The important property of photons that is useful in establishing Einstein's photoelectric equation is their ability to hold on to the electrons of an atom by their forces of attraction.

(ii) Einstein's photoelectric equation states that :

$$E_{max} = hv - \phi$$

If $E_{max} > 0$ then $hv - \phi > 0$

Here, v_0 is the threshold frequency.

Again, loss in KE = gain in electromagnetic PE.

or $E_{max} = eV_s$

where V_s represents the stopping potential.

Q. 6. Show that the wavelength of electromagnetic radiation is equal to the de-Broglie wavelength of its quantum (photon). **[NCERT]**

Ans. The momentum of a photon having energy (hv) is given by

$$p = \frac{hv}{c} = \frac{h}{\lambda}$$

$$\lambda = \frac{h}{p} \qquad ...(i)$$

Here, λ = Wavelength of the electromagnetic radiation

c = Speed of light

h = Planck's constant

de-Broglie wavelength of the photon is given by

$$\lambda = \frac{h}{mv}$$

But $p = mv$

$$\therefore \quad \lambda = \frac{h}{p} \qquad ...(ii)$$

Here, m = Mass of the photon

v = Velocity of the photon

Therefore it can be inferred from equations (i) and (ii) that the wavelength of the electromagnetic radiation is equal to the de-Broglie wavelength of the photon.

Chapter 12. Atoms

Q. 1. Answer the following questions, which help you to understand the difference between Thomson's model and Rutherford's model better. **[NCERT]**

(i) Is the average angle of deflection of α-particles by a thin gold foil predicted by Thomson's model much less, about the same, or much greater than that predicted by Rutherford's model ?

(ii) Is the probability of backward scattering (*i.e.*, scattering of α-particles at angles greater than 90°) predicted by Thomson's model much less, about the same, or much greater than that predicted by Rutherford's model ?

(iii) Keeping other factors fixed, it is found experimentally that for small thickness *t*, the number of α-particles scattered at moderate angles is proportional to *t*. What clue does this linear dependence on *t*-provide ?

(iv) In which model is it completely wrong to ignore multiple scattering for the calculation of average angle of scattering to α-particles by a thin foil ?

Ans. (i) The average angle of deflection of α-particles by a thin gold foil predicted by Thomson's model is of same size as predicted by Rutherford's model. This is because the average angle was taken in both models.

(ii) The probability of scattering for α-particles at angles greater than 90° predicted by Thomson's model is much less than that predicted by Rutherford's model.

(iii) Scattering is mainly due to single collisions. The chances of a single collision increases linearly with the number of target atoms. As the number of target atoms increase with an increase in thickness, the collision probability depends linearly on the thickness of the target.

(iv) It is wrong to ignore multiple scattering in Thomson's model for the calculation of average angle of scattering of α-particles by a thin foil. This is because a single collision causes very little deflection in this model. Hence, the observed average scattering angle can be explained only by considering multiple scattering.

Q. 2. On an energy level diagram of hydrogen, show by a downward or an upward arrow transition which results in :

(i) emission line of Balmer series.

(ii) emission line of Lyman series.

(iii) absorption line of Lyman series.

Ans. (i) and (ii)

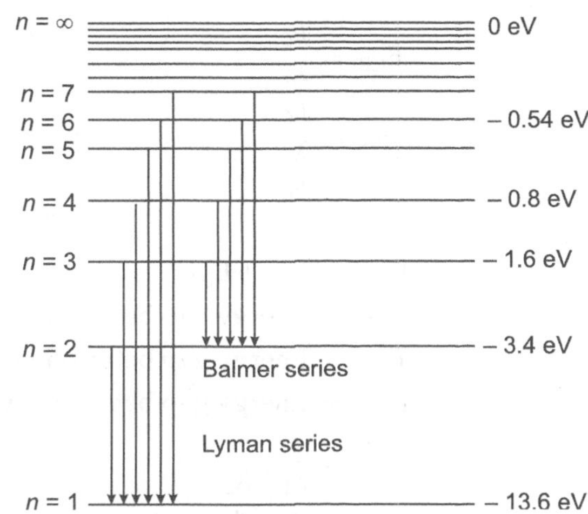

Emission lines of Lyman and Balmer series

(iii)

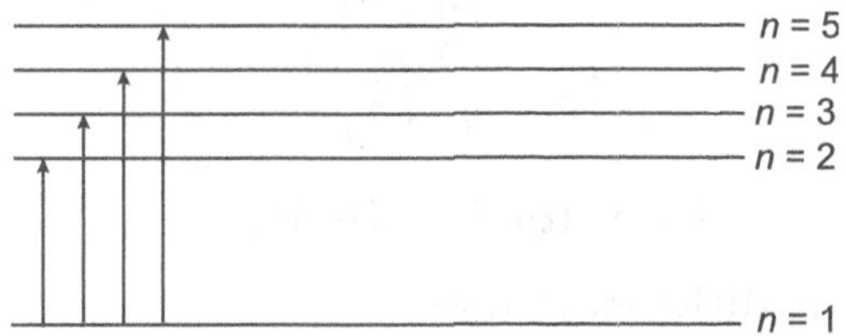

Absorption lines of Lyman series

Q. 3. (i) State Bohr's quantisation condition for defining stationary orbits. How does de-Broglie hypothesis explain the stationary orbits ?★

(ii) Find the relation between the three wavelengths λ_1, λ_2 and λ_3 from the energy level diagram shown below.★

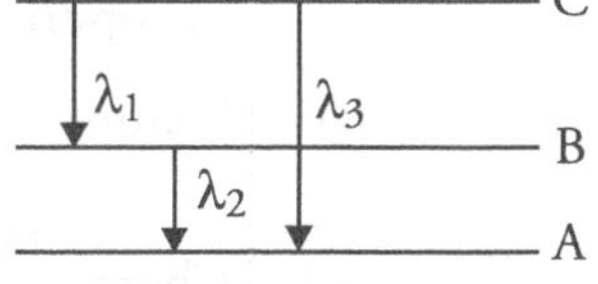

Ans. (i) **Quantisation condition** : Of all possible circular orbits allowed by the classical theory, the electrons are permitted to circulate only in those orbits in which the angular momentum of an electron is an integral multiple of $\frac{h}{2\pi}$; h being Planck's constant.

Therefore, for any permitted orbit,

$$L = mvr = \frac{nh}{2\pi};\ n = 1, 2, 3, ...$$

Where L, m, and v are the angular momentum, mass and speed of the electron, r is the radius of the permitted orbit and n is positive integer called principle quantum number.

★ are board based questions from previous year

The above equation is Bohr's famous quantum condition. When an electron of mass m is confined to move on a line of length l with velocity v, the de-Broglie wavelength λ associated with electron is :

$$\lambda = \frac{h}{mv} = \frac{h}{p}$$

or p = Linear momentum

$$\Rightarrow \quad p = \frac{h}{\lambda} = \frac{h}{2l/n} = \frac{nh}{2l}$$

When electron revolves in a circular orbit of radius 'r' then $2l = 2\pi r$.

$$\therefore \quad p = \frac{nh}{2\pi r} \text{ or } p \times r = \frac{nh}{2\pi}$$

or angular momentum $|\vec{L}| = p \times r$ is integral multiple of $h/2\pi$ which is Bohr's quantisation of angular momentum.

(ii)

$$E_{CB} = \frac{hc}{\lambda_1}$$

$$E_{BA} = \frac{hc}{\lambda_2}$$

$$E_{CA} = \frac{hc}{\lambda_3}$$

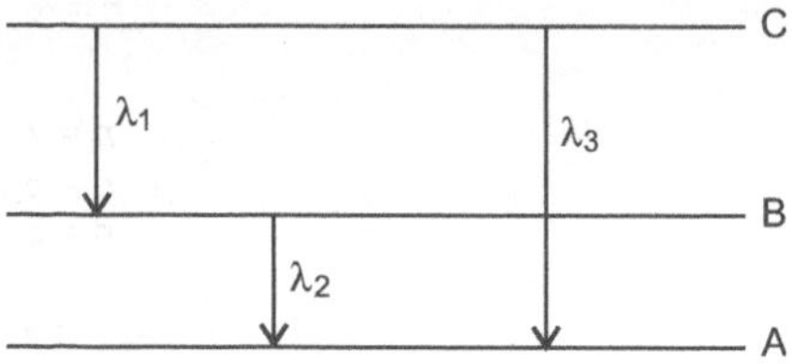

Now, $E_{CA} = E_{CB} + E_{BA}$

where, E_{CB} = Energy gap between level B and C,

E_{BA} = Energy gap between level A and B,

E_{CA} = Energy gap between level A and C.

$$\frac{hc}{\lambda_3} = \frac{hc}{\lambda_1} + \frac{hc}{\lambda_2}$$

$$\frac{1}{\lambda_3} = \frac{1}{\lambda_1} + \frac{1}{\lambda_2}$$

$$\lambda_3 = \frac{\lambda_1 \lambda_2}{\lambda_2 + \lambda_1}$$

Chapter 13. Nuclei

Q. 1. Compare properties of α, β particles and γ-rays.

Ans. Comparison of the properties of α-particles, β-particles and γ-rays :

	Property	α-particle	β-particle	γ-rays
(i)	Nature	Helium nucleus ($_2He^4$)	Fast moving electrons	Electromagnetic waves
(ii)	Charge	$+3.2 \times 10^{-19}$ C	-1.6×10^{-19} C	No charge
(iii)	Rest mass	6.6×10^{-27} kg	9.1×10^{-31} kg	Zero
(iv)	Velocity	About $\left(\frac{1}{10}\right)^{th}$ of the velocity of light	99% of the velocity of light	Equal to the velocity of light (3×10^8 ms^{-1})
(v)	Ionising power	Higher than that of β-particle	$\left(\frac{1}{100}\right)^{th}$ of α-particles	$\left(\frac{1}{100}\right)^{th}$ of that of β-particles
(vi)	Penetrating power	Smaller than that of β-particles.	100 times that of α-particles	100 times that of β-particles

Q. 2. Explain fission and state what is meant by fusion. What is mass defect ? What light does it throw on the binding energy of a nucleus ?

Ans. **Fission :** The process in which a heavy nucleus is broken into two nearly equal fragments is called 'nuclear fission'.

When a slow neutron strikes $_{92}U^{235}$ nucleus, it is absorbed by the nucleus and an isotope $_{92}U^{236}$ is formed. But $_{92}U^{236}$, being highly unstable, is immediately broken into two fragments and emits neutrons and energy. This fission can be represented by the following equation :

$$_{92}U^{235} + {_0n^1} \rightarrow {_{92}U^{236}} \rightarrow {_{56}Ba^{144}} + {_{36}Kr^{89}} + 3\,{_0n^1} + \text{energy}$$

Fusion : When two lighter nuclei moving at very high speeds fuse together to form a single heavier nucleus, then this phenomenon is called 'nuclear fusion'.

Mass defect : The mass of the atomic nucleus is slightly less than the sum of the masses of the nucleons (proton and neutrons) present in the nucleus, this mass difference is called 'mass defect'. Let mass defect be Δm.

Then, Δm = (mass of protons + mass of neutrons) – mass of nucleus.

When protons and neutrons combine to form a nucleus, an amount of mass Δm (mass defect) disappears and an equivalent amount of energy Δmc^2 is liberated, due to this energy proton and neutron remain bound in the nucleus. This is called 'Binding energy'.

Let the atomic number of the element be Z and mass number be A. Let m_p, m_n and M be the mass of the proton, neutron and the nucleus respectively. Mass of Z protons is Zm_p and the mass of (A–Z) neutron in (A – Z) m_n

$$\Delta m = [Zm_p + (A - Z)\, m_n] - M$$

$$\text{Binding energy in MeV} = \Delta m \times 931.5$$

Q. 3. What is the nature of α, β and γ-rays particles ? How will you determine experimentally the charge that these radiations carry ?

Ans. (i) **Nature of α-particles :**

(a) These particles are deflected in electric and magnetic field.

(b) The velocity of an α-particle is less than $\frac{1}{10^{th}}$ part of the velocity of light.

(c) α-particle can penetrate through matter but their penetration power is very low.

(d) The particles affect photographic plates slightly.

(ii) Nature of β-particles :

(a) β-particles are negatively charged particles.

(b) These particles are deflected in electric and magnetic field.

(c) Their velocity is from 1% to 99% of the velocity of light.

(d) They affect the photographic plates more in comparison of α-particles.

(iii) Nature of γ-Rays :

(a) γ-rays are electromagnetic waves of very short wavelength.

(b) These rays are not deflected by electric and magnetic fields.

(c) They travel with the velocity of light.

Let a radioactive substance is put in a thick walled lead box and the radioactive rays emerging from a narrow opening in the box are passed through an electrostatic field between two plates.

It was observed that radioactive rays are of three types :

(a) One which are deflected towards the negative plate, are called 'α-particles'.

(b) Second which are deflected towards the positive plate, are called 'beta-particles'.

(c) Third which go undeflected by the electric and magnetic fields are called 'gamma-rays'.

Q. 4. (i) A radioactive nucleus 'A' undergoes a series of decays as given below :

$$A \xrightarrow{\alpha} A_1 \xrightarrow{\beta} A_2 \xrightarrow{\alpha} A_3 \xrightarrow{\gamma} A_4$$

The mass number and atomic number of A_2 are 176 and 71 respectively. Determine the mass and atomic numbers of A_4 and A. ★

(ii) Write the basic nuclear process underlying β^+ and β^- decays.★

Ans. (i)

$$A \xrightarrow{\alpha} A_1 \xrightarrow{\beta} {}^{176}_{71}A_2 \xrightarrow{\alpha} A_3 \xrightarrow{\gamma} A_4$$

$${}^{176}_{71}A_2 \xrightarrow{\alpha} {}^{172}_{69}A_3 \xrightarrow{\gamma} {}^{172}_{69}A_4$$

$$A_4 \longrightarrow \begin{cases} 172 = \text{Mass number} \\ 69 = \text{Atomic number} \end{cases}$$

$${}^{180}_{72}A \xrightarrow{\alpha} {}^{176}_{70}A_1 \xrightarrow{\beta} {}^{176}_{71}A_2$$

$$A \longrightarrow \begin{cases} 180 = \text{Mass number} \\ 72 = \text{Atomic number} \end{cases}$$

(ii) The process of spontaneous emission of a positron (e^+) from nucleus is called β^+ decay.

$$p \rightarrow n + {}^{0}_{1}e + \bar{\nu}$$

The process of spontaneous emission of an electron (e^-) from nucleus is called β^- decay.

$$p \rightarrow n + {}^{0}_{-1}e + \bar{\nu}$$

Q. 5. (i) Write the basic nuclear process involved in the emission of β^+ in a symbolic form, by a radioactive nucleus.★

(ii) In the reactions given below :★

(a) $${}^{11}_{6}C \longrightarrow {}^{z}_{y}B + x + v$$

(b) $${}^{12}_{6}C + {}^{12}_{6}C \longrightarrow {}^{20}_{a}Ne + {}^{c}_{b}He$$

Find the value of *x*, *y*, *z* and *a*, *b*, *c*.

Ans. (i) In β^+-decay, the atomic number of the radioactive nucleus decreases by one and its mass number remains same. In this process, a positron (e^+) and a new particle neutrino (ν) are emitted from the nucleus.

Generally, $${}^{A}_{Z}X \longrightarrow {}_{Z-1}^{\ A}Y + e{+}{+}$$

(ii) (a) $${}^{11}_{6}C \longrightarrow {}^{z}_{y}B + x + \nu$$

$${}^{11}_{6}C \longrightarrow {}^{11}_{5}B + {}^{0}_{1}\beta + \nu$$

The corresponding *y* and *z* are 5 and 11, respectively. The *x* is the positron.

(b) $${}^{12}_{6}C + {}^{12}_{6}C \longrightarrow {}^{20}_{a}Ne + {}^{c}_{b}He$$

$$\Rightarrow \quad {}^{12}_{6}C + {}^{12}_{6}C \longrightarrow {}^{20}_{10}Ne + {}^{4}_{2}He$$

The corresponding values of *a*, *b*, *c* are 10, 2 and 4, respectively.

Q. 6. (i) Write three characteristic properties of nuclear force.

(ii) Draw a plot of potential energy of a pair of nucleons as a function of their separation. Write two important conclusions that can be drawn from the graph.

★ are board based questions from previous year

Ans. (i) (a) Nuclear forces are short range forces ;

(b) Nuclear forces are primarily attractive and extremly strong;

(c) Nuclear forces are charge independent.

(ii)

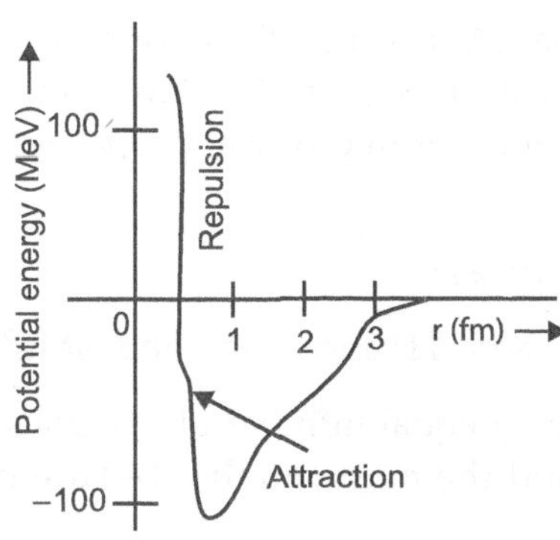

Q. 7. Write nuclear reaction equations for :

(i) α-decay of $^{226}_{88}Ra$ (ii) α-decay of $^{242}_{94}Pu$

(iii) β^--decay of $^{32}_{15}P$ (iv) β^--decay of $^{210}_{83}Bi$

(v) β^+-decay of $^{11}_{6}C$ (vi) β^+-decay of $^{97}_{43}Tc$

(vii) Electron capture of $^{120}_{54}Xe$. **[NCERT]**

Ans. α is a nucleus of helium ($_2He^4$) and β is an electron (e^- for β^- and e^- for β^+). In every α-decay, there is a loss of 2 protons and 4 nuetrons. In every β^+-decay, there is a loss of 1 proton and a neutrino is emitted from the nucleus. In ever β^--decay, there is a gain of 1 proton and an antineutrino is emitted from the nucleus.

For the given cases, the various nuclear reactions can be written by :

(i) $^{226}_{88}Ra \longrightarrow ^{222}_{86}Rn + ^{4}_{2}He$

(ii) $^{242}_{94}Pu \longrightarrow ^{238}_{92}U + ^{4}_{2}He$

(iii) $^{32}_{15}P \longrightarrow ^{32}_{16}S + ^{0}_{-1}e^- + \bar{v}$

(iv) $^{210}_{83}Bi \longrightarrow ^{210}_{84}Po + ^{0}_{-1}e^- + \bar{v}$

(v) $^{11}_{6}C \longrightarrow ^{11}_{7}B + ^{0}_{-1}e^- + \bar{v}$

(vi) $^{97}_{43}Tc \longrightarrow ^{97}_{42}Mo + ^{0}_{+1}e^+ + \bar{v}$

(vii) $^{120}_{54}Xe + ^{0}_{-1}e \longrightarrow ^{120}_{53}I + v$

Q. 8. Explain classification of nuclei in brief.

Ans. **Classification of Nuclei : Isotopes, Isobars and Isotones**

The nuclei have been classified on the basis of the number of protons (atomic number) or the total number of nucleons (mass number) in them.

(i) **Isotopes or Isoprotons :** The atoms of element whose nuclei have the same number of protons but different number of neutrons are called the 'isotopes' of that element.

The isotopes of some elements are as :

Hydrogen : $_1H^1$, $_1H^2$, $_1H^3$

Oxygen : $_8O^{16}$, $_8O^{17}$, $_8O^{17}$

Neon : $_{10}Ne^{20}$, $_{10}Ne^{21}$, $_{10}Ne^{22}$

Chlorine : $_{17}Cl^{35}$, $_{17}Cl^{31}$

Uranium : $_{92}U^{235}$, $_{92}U^{238}$

(ii) **Isobars :** The nuclei which have the same number of nucleons, but different number of protons and different number of neutrons are called 'isobars'. Their atomic numbers (Z) are different, but mass number (A) is same. Therefore, they have different places in the periodic table and differ in chemical properties also. Since, in isobars the numbers of fundamental particles are different, they differ in physical properties also. The nuclei of isobars belong to the different elements.

Some example of isobars are as :

$$_1H^3 \text{ and } _2H^3 \text{ and } _{11}He^{23} \text{ and } _{12}Mg^{23}$$

(iii) **Isotones :** The nuclei having equal number of neutrons are called 'isotones'. For them both the atomic number (Z) and the mass number (A) are different, but the value of (A – Z) is same.

Some examples of isotones are as :

$$_3Li^7 \text{ and } _4Be^8, \ _{17}Cl^{37} \text{ and } _{19}K^{39} \text{ and } _{11}Ne^{23} \text{ and } _{12}Mg^{24}$$

Chapter 14. Semiconductor Electronics : Materials, Devices and Simple Circuits

Q. 1. Draw a labelled diagram of a full wave rectifier circuit. Explain briefly the functions of its components. Sketch the input and output wave forms.

Ans. Diagram of a full wave rectifier circuit :

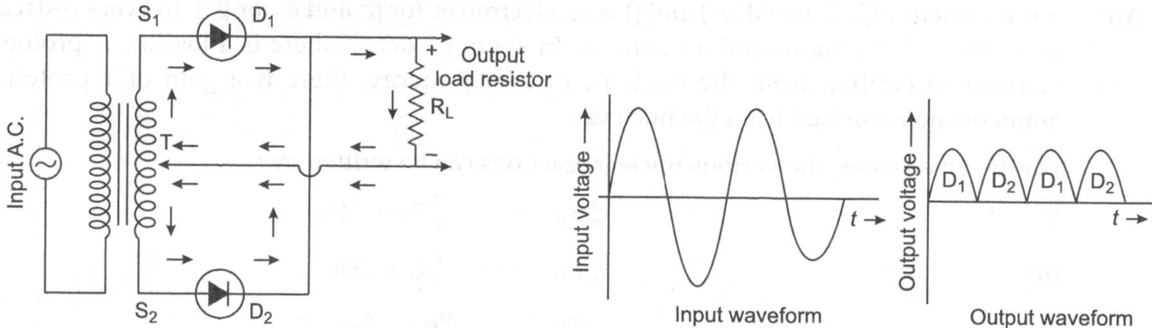

Explanation : In a full wave rectifier, unidirectional pulsating output current is obtained for both halves of the a.c. input voltage. Essentially, it requires two junction diodes so connected that one diode rectifies one half and second diode rectifies the second half of the input.

During the first half cycle of the a.c. input voltage, the terminal S_1 is positive relative to T and S_2 is negative. In this situation, the junction diode D_1 is forward biased and D_2 is reverse biased. Therefore D_1 conducts and D_2 does not. The conventional current flows through diode D_1, load R_L and the upper half of the secondary windings, as shown by the solid arrows. During the second half cycle of the input voltage, S_1 is negative relative to T and S_2 is positive. Now D_1 is reverse biased and does not conduct while D_2 is forward-biased and conducts. The current now flows through D_2, load R_L and the lower half of the secondary windings, as shown by the solid arrows. It may be seen that the current in the load R_L flows in the same direction for both the half-cycles of the a.c. input voltage. Thus, the output current is a continuous series of unidirectional pulses. However, it can be made fairly steady by means of smoothing filters.

Q. 2. What is a photodiode ? Explain in brief. Mention some uses of photodiode.

Ans. Photodiode is a *p-n* junction which is an opto-electronic device in which current carriers are generated by photons through photo-excitation, *i.e.*, photo-conduction by light.

A photodiode is a special *p-n* junction diode made of photosensitive semiconducting material. It is operated under reverse bias below the breakdown voltage. The conductivity of *p-n* junction photodiode is modulated by the absorption of incident light in or near the depletion layer which exists at the *p-n* junction. The conductivity of the *p-n* junction photodiode increases with the increase in intensity of light falling on it.

Symbolically, a photodiode is shown in the figure (i). Figure (ii) shows an experimental arrangement for the study of V-I characteristics of a photodiode in which the photodiode is reverse biased.

When the photodiode is reverse biased with a voltage less than its breakdown voltage and no light is incident on its junction, the reverse current is extremely small (almost negligible). This current is called dark current.

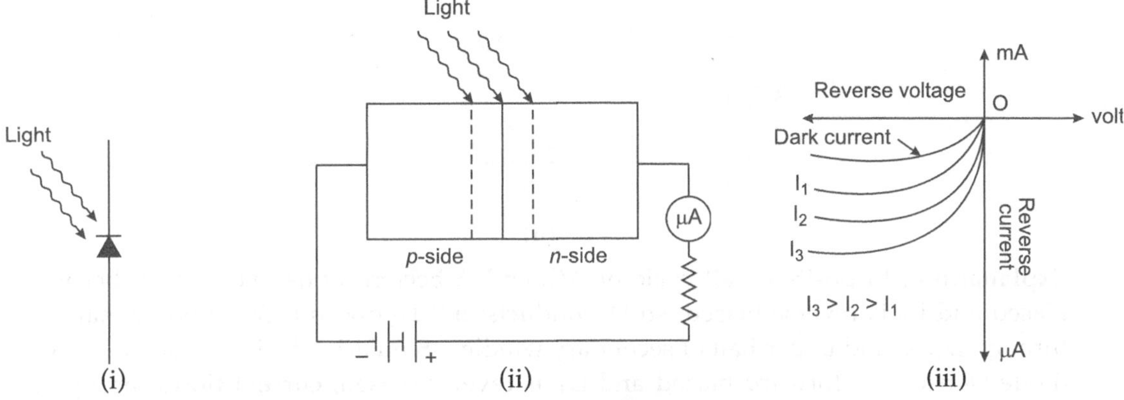

Q. 3. Show by drawing labelled diagrams, the nature of output voltages in case of an Amplifier : (In each case, input is an a.c. voltage) Circuit diagrams of these devices are not required.

Ans. **Amplifier :**

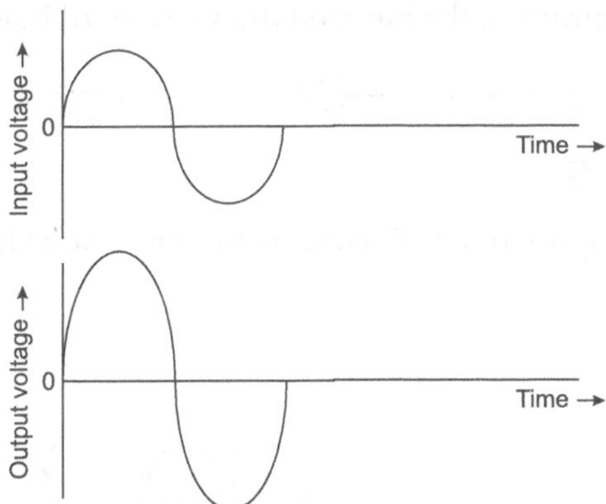

Q. 4. A student wants to use two *p-n* junction diodes to convert alternating current into direct current. Draw the labelled circuit diagram she would use and explain how it works.★

Ans. **Full wave rectifier :**

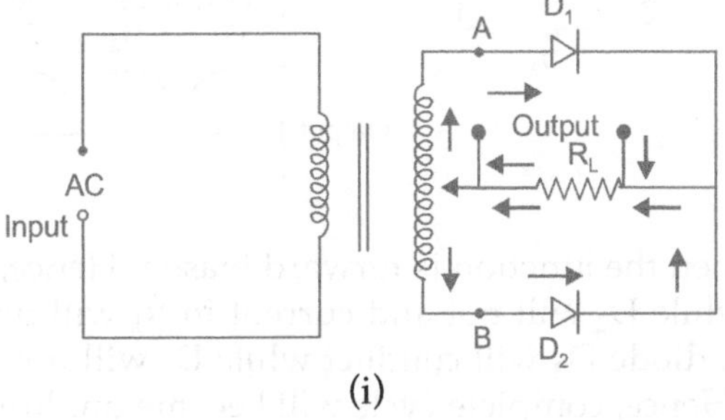

(i)

★ are board based questions from previous year

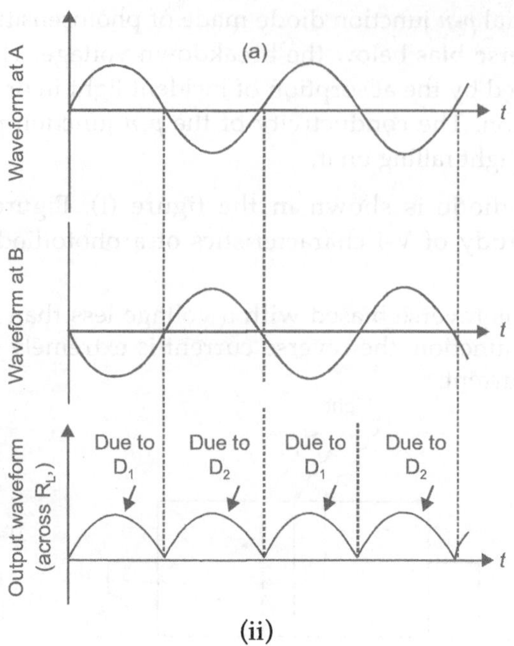

(ii)

Explanation : In positive half cycle of AC, end A becomes positive and D_1 becomes forward biased and D_2 is reverse biased, so D_1 conducts and D_2 doesn't. So conventional current flows through D_1, R_L and upper half of secondary winding. Similarly, during negative half cycle of AC, diode D_2 becomes forward biased and D_1 is reverse biased, current flows through D_2, R_L and lower half of secondary winding. Thus, current flows in same direction in both half cycles of input AC voltage.

Q. 5. (i) In the following diagram, is the junction diode forward biased or reverse biased ?★

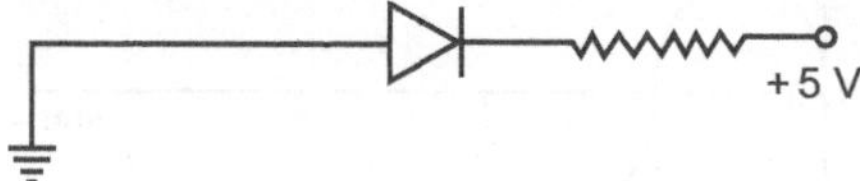

(ii) Draw the circuit diagram of a full wave rectifier and state how it works.★

Ans. (i) Reverse biased.

(ii) Full wave Rectifier :

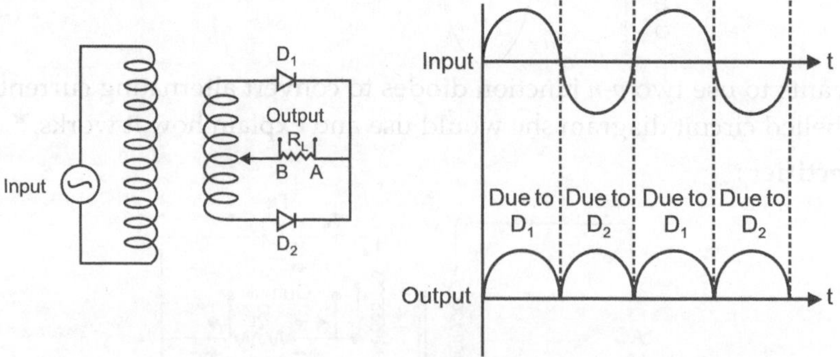

Diode conducts only when the junction is forward biased. Hence, during first half cycle of input A.C., D_1 will conduct while D_2 will not and current in R_L will flow from A to B. During second half cycle of input A.C., diode D_2 will conduct while D_1 will not conduct and current will again flow from A to B in R_L. Hence, complete cycle will become unidirectional.

★ are board based questions from previous year

Q. 6. (i) In the diagram given ahead, which bulb out of B_1 and B_2 will glow and why ?★

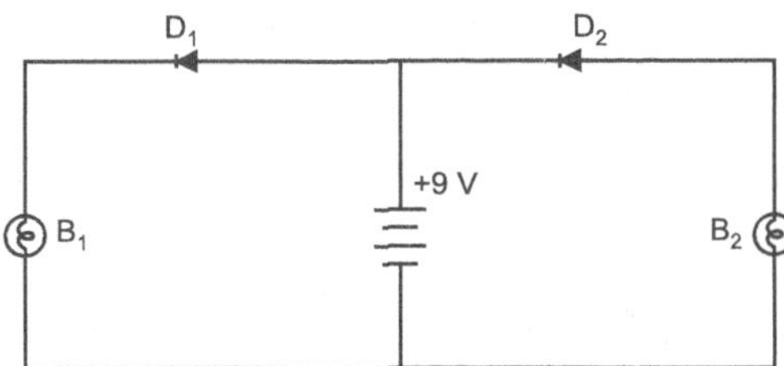

(ii) Draw a diagram of an illuminated *p-n* junction solar cell.★

(iii) Explain briefly the three processes due to which generation of emf takes place in a solar cell.★

Ans. (i) B_1 will glow because only diode D_1 is forward biased.

(ii)

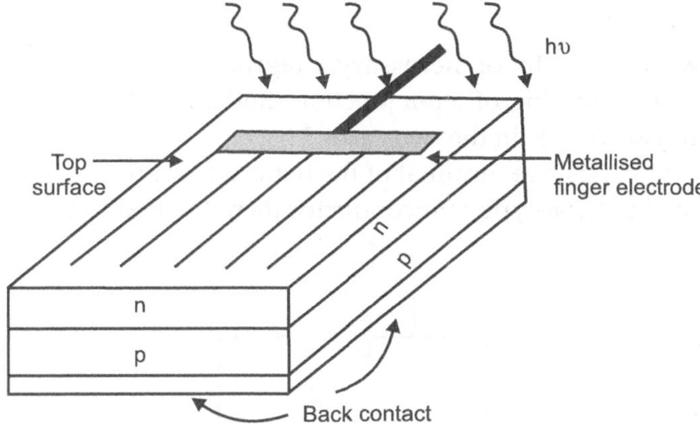

(iii) Three process due to which generation of emf takes place in solar cell are :

(a) Generation, (b) Separation, (c) Collection.

(a) **Generation :** Generation of electron-hole pairs take place due to light.

(b) **Separation :** Separation of electron-hole pairs are due to electric field of depletion region.

(c) **Collection :** Electrons reach the *n*-side and are collected in front contact and holes are collected in the back contact.

Q. 7. (i) In the following diagram 'S' is a semiconductor. Would you increase or decrease the value of R to keep the reading of the ammeter A constant when S is heated ? Give reason for your answer.

(ii) Draw the circuit diagram of a photodiode and explain its working. Draw its I – V characteristics.

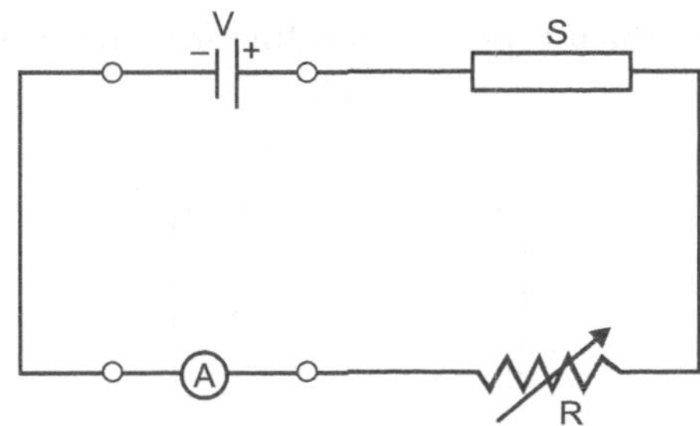

Ans. (i) The value of R has to be increase because on heating, the conductivity of a semiconductor increases.

★ are board based questions from previous year

(ii)

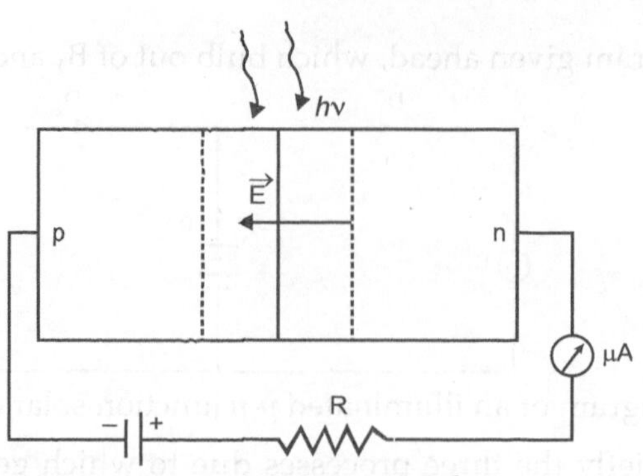

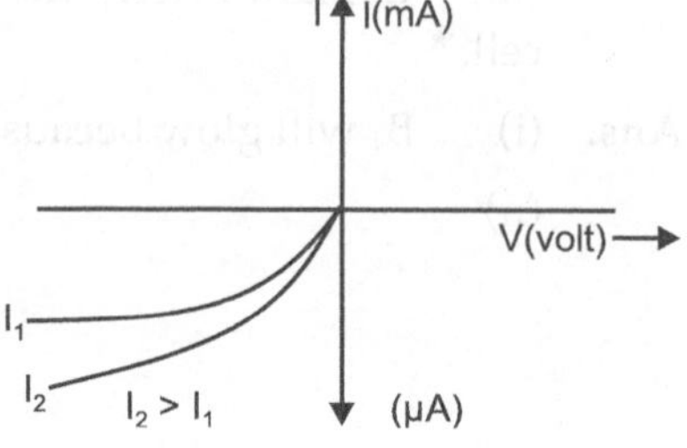

Working : In photodiode, an electric field exists across the junction from n-side to p-side. When visible light with energy $h\nu$ greater than energy gap E_g illuminates the junction, then electron-hole pairs are generated in the depletion layer. Due to electric field electron moves towards n side and holes towards p-side give rise to an emf.

Q. 8. Explain briefly with the help of necessary diagrams, the forward and the reverse biasing of a p-n junction diode. Also draw their characteristic curves in the two cases.★

Ans. Forward biasing : If the positive terminal of the battery is connected to the p-type semiconductor and negative terminal to the n-type semiconductor then it is said to be forward biased.

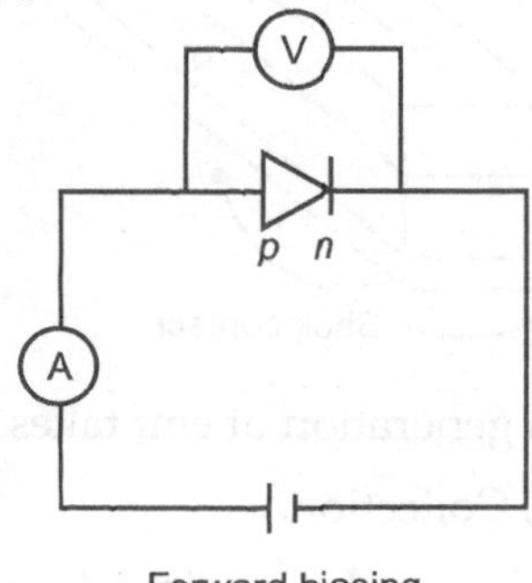

Forward biasing

Characteristics :

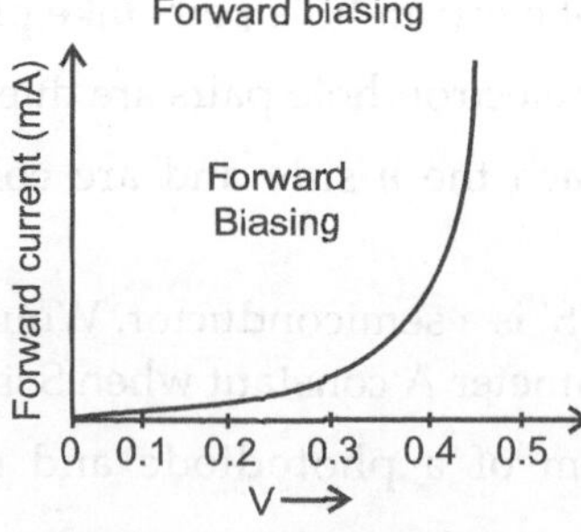

Reverse Biasing : If the positive terminal of the battery is connected to the n-type semiconductor and the negative terminal to the p-type semiconductor then it is said to be reverse biased.

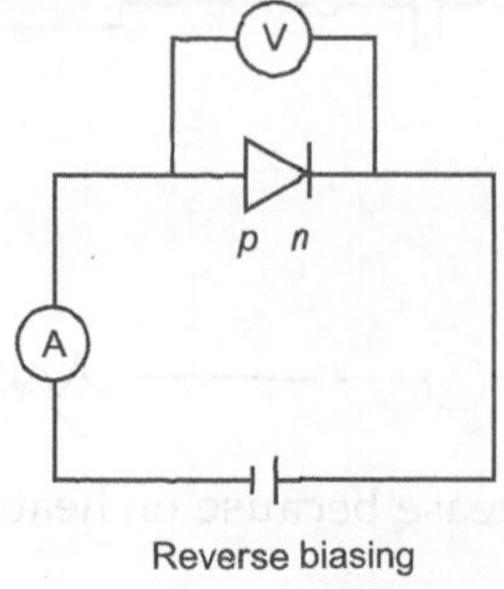

Reverse biasing

★ are board based questions from previous year

Characteristics :

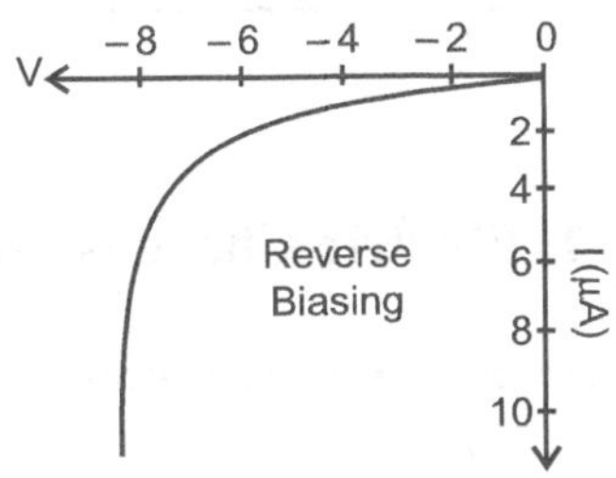

Q. 9. (i) Explain with the help of a diagram the formation of depletion region and barrier potential in a *p-n* junction.★

(ii) Draw the circuit diagram of a half wave rectifier and explain its working.★

Ans. (i) During the formation of *p-n* junction, the holes diffuse from *p*-type semiconductor to the *n*-type, and electrons diffuse from *n*-type to *p*-type. This is because of the concentration gradient across *p*-side and *n*-side.

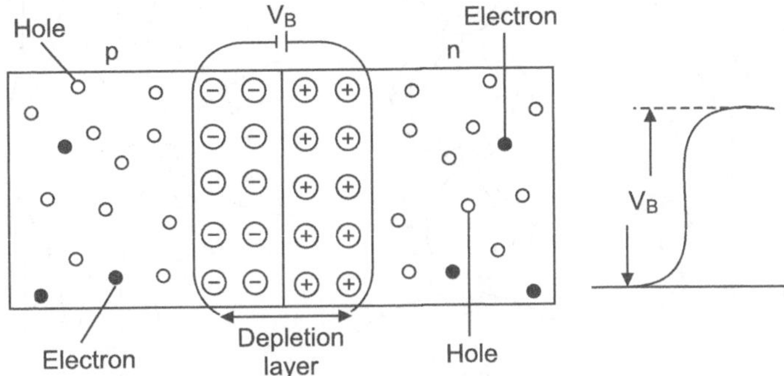

When a hole diffuse from *p* to *n*-type, it leaves an unmovable negative charge. Similarly, when an electron diffuses from *n* to *p*-type, it leaves an unmovable positive charge. When the diffusion of holes and electrons takes place continuously across the junction, a layer of unmovable positive and negative charges are developed on either side of the junction. This layer is called the depletion layer or the depletion region and the potential difference across the region is called barrier potential.

(ii) **Half wave rectifier :**

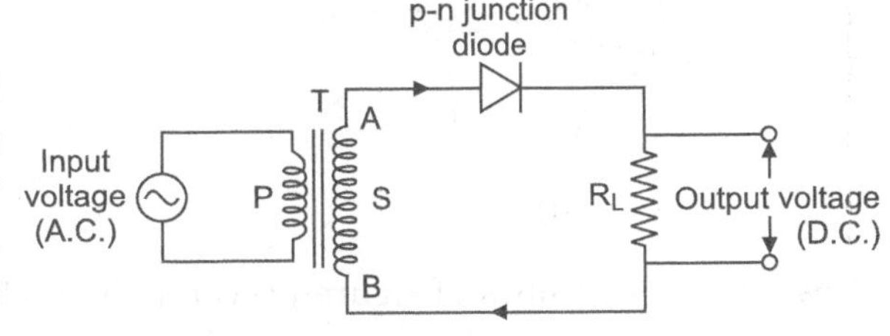

Working : When an input ac voltage is applied across the primary coil, a potential difference is developed across the ends of the secondary coil. Consider that in half cycle of input a.c. signal, the end A acts as the +ve end and B acts as the –ve end of the battery. So, the diode is in forward bias and we get output across the ends of the load resistance R_L.

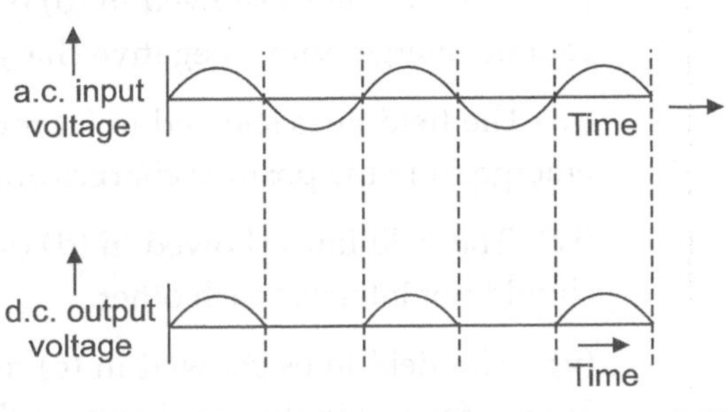

In the second half cycle, ends A and B reverse in polarity. Now, A acts as the –ve end and B acts as the +ve end. So, the diode D is in reverse bias and no output is obtained due to the high resistance offered by the diode.

So, in this process, we get output alternately, and hence the diode is called the half wave rectifier.

★ are board based questions from previous year

Question Set 6

Long Answers-II

Chapter 1. Electric Charges and Fields

Q. 1. Which among the curves shown in given figures cannot possibly represent electrostatic field lines ? **[NCERT]**

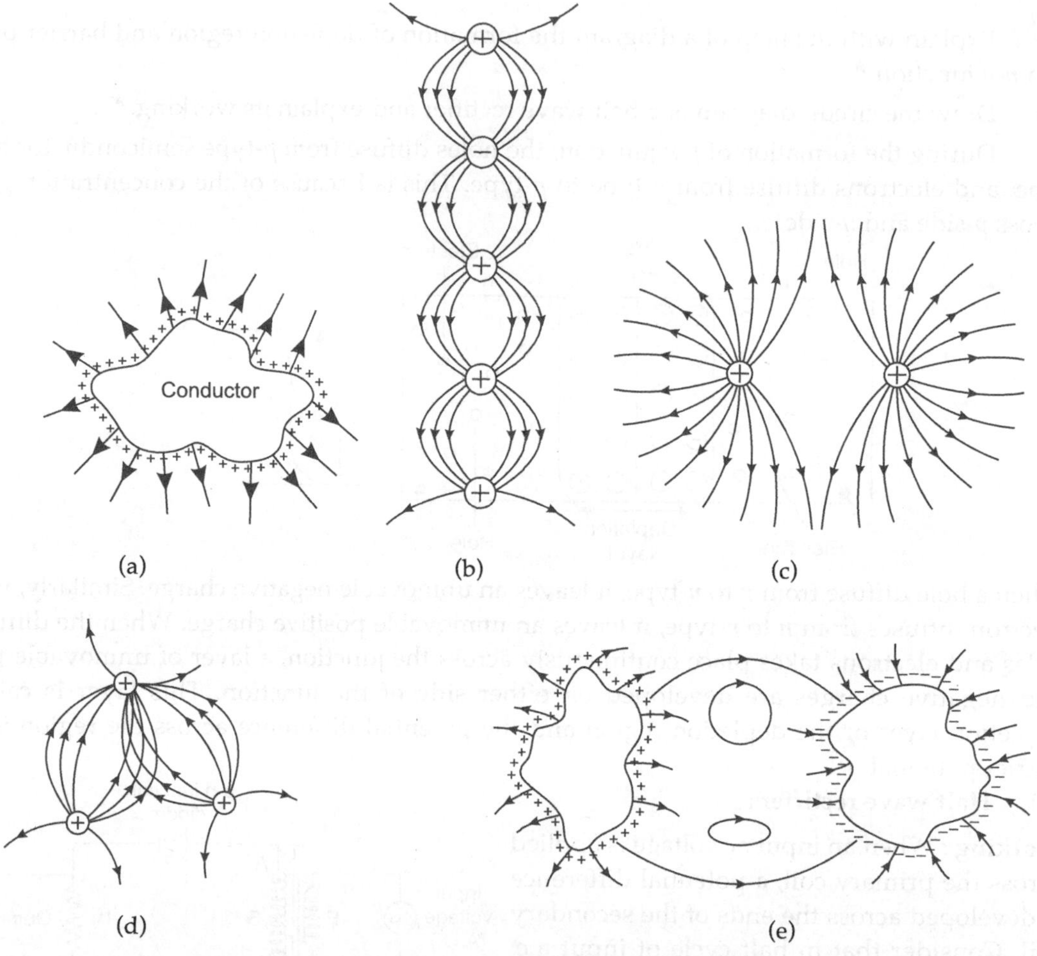

Ans. (i) Field lines of figure (a) can't show electrostatic field lines as these lines are not normal to surface.

(ii) The field lines showed in (b) do not represent electrostatic field lines because the field lines cannot emerge from a negative charge and cannot terminate at a positive charge.

(iii) The field lines showed in (c) represent electrostatic field lines. This is because the field lines emerge from the positive charges and repel each other.

(iv) The field lines showed in (d) do not represent electrostatic field lines because the field lines should not intersect each other.

(v) The field lines showed in (e) do not represent electrostatic field lines because closed loops are not formed in the area between the field lines.

Q. 2. Obtain the formula for the electric field due to a long thin wire of uniform linear charge density λ without using Gauss's law. [**Hint.** Use Coulomb's law directly and evaluate the necessary integral.] **[NCERT]**

Ans. Consider a long thin wire XY of uniform linear charge density λ. [Fig. (a)]

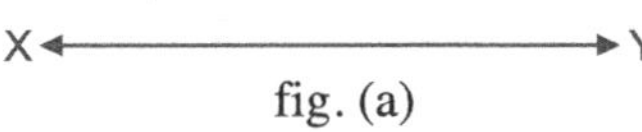

fig. (a)

Take a point P at perpendicular distance l from mid point O of wire as shown in [Fig. (b)]

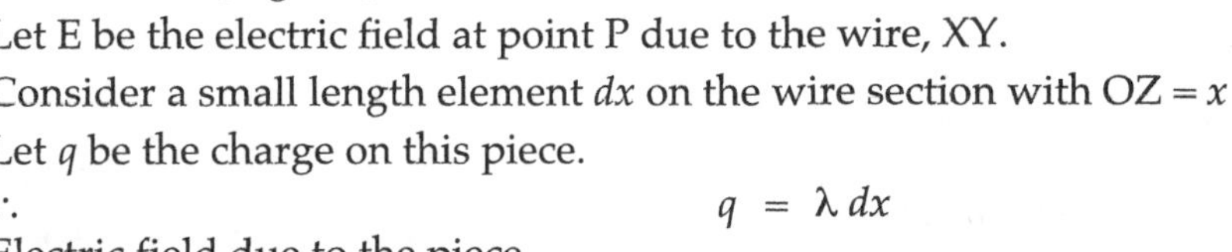

fig. (b)

Let E be the electric field at point P due to the wire, XY.

Consider a small length element dx on the wire section with $OZ = x$

Let q be the charge on this piece.

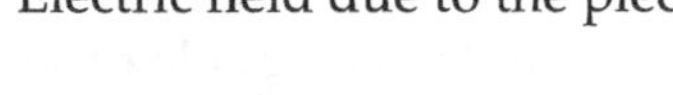

$$\therefore \quad q = \lambda\, dx$$

Electric field due to the piece,

$$dE = \frac{1}{4\pi\varepsilon_0} \cdot \frac{\lambda\, dx}{(PZ)^2}$$

However, $$PZ = \sqrt{l^2 + x^2}$$

$$\therefore \quad dE = \frac{1}{4\pi\varepsilon_0} \cdot \frac{\lambda dx}{(l^2 + x^2)}$$

The electric field is resolved into two rectangular components. $dE \cos\theta$ is the pependicular component and $dE \sin\theta$ is the parallel component. When the whole wire is considered, the component $dE \sin\theta$ is cancelled. Only the perpendicular component $dE \cos\theta$ affects point P. Hence, effective electric field at point P due to the element dx is dE_1.

$$dE_1 = \frac{1}{4\pi\varepsilon_0} \frac{\lambda\, dx \cos\theta}{(l^2 + x^2)} \qquad \text{...(i)}$$

In Δ PZO, $$\tan\theta = \frac{x}{l} \Rightarrow x = l \tan\theta \qquad \text{...(ii)}$$

$$\frac{dx}{d\theta} = l \sec^2\theta \text{ [from differentiating (ii)]}$$

$$x^2 + l^2 = l^2 \tan^2\theta + l^2 = l^2(\tan^2\theta + 1) = l^2 \sec^2\theta \qquad \text{...(iii)}$$

Now, $$dE_1 = \frac{1}{4\pi\varepsilon_0} \cdot \frac{\lambda l \sec^2\theta\, d\theta \cdot \cos\theta}{l^2 \sec^2\theta}$$

$$= \frac{1}{4\pi\varepsilon_0} \cdot \frac{\lambda \cos\theta\, d\theta}{l} \qquad \text{...(iv)}$$

The wire is so long that θ tends from $-\frac{\pi}{2}$ to $\frac{\pi}{2}$.

By integrating equation (iv), we obtain the value of field E_1 as,

$$\int_{-\pi/2}^{\pi/2} dE_1 = \int_{-\pi/2}^{\pi/2} \frac{1}{4\pi\varepsilon_0} \frac{\lambda}{l} \cos\theta\, d\theta$$

$$\Rightarrow \quad E_1 = \frac{1}{4\pi\varepsilon_0} \frac{\lambda}{l} [\sin\theta]_{-\pi/2}^{\pi/2}$$

$$\Rightarrow \quad E_1 = \frac{1}{4\pi\varepsilon_0} \frac{\lambda}{l} \times 2$$

$$\Rightarrow \quad E_1 = \frac{\lambda}{2\pi\varepsilon_0 l}$$

Q. 3. (i) Define electric flux. Is it a scalar or a vector quantity ? A point charge q is at a distance of $d/2$ directly above the centre of a square of side d, as shown in the figure. Use Gauss' law to obtain the expression for the electric flux through the square.★

(ii) If the point charge is now moved to a distance 'd' from the centre of the square is doubled, explain now the electric flux will be affected ?★

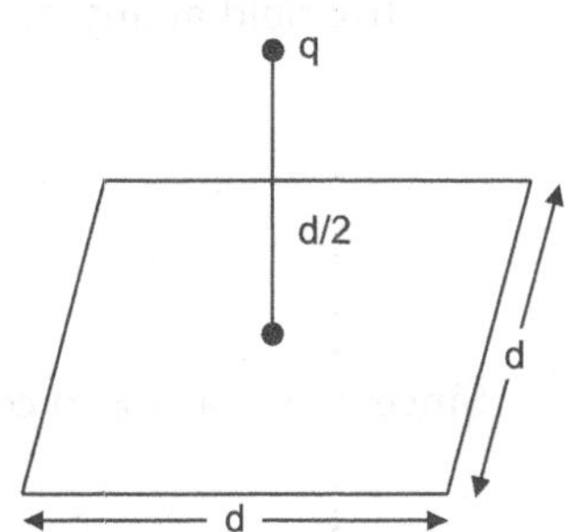

★ are board based questions from previous year

Ans. (i) **Electric flux :** Electric flux through an area is defined as the product of electric field strength E and area d S perpendicular to the field. It represents the field lines crossing the area. It is a scalar quantity.

Imagine a cube of edge d, enclosing the charge. The square surface is one of the six faces of this cube. According to Gauss' theorem in electrostatics,

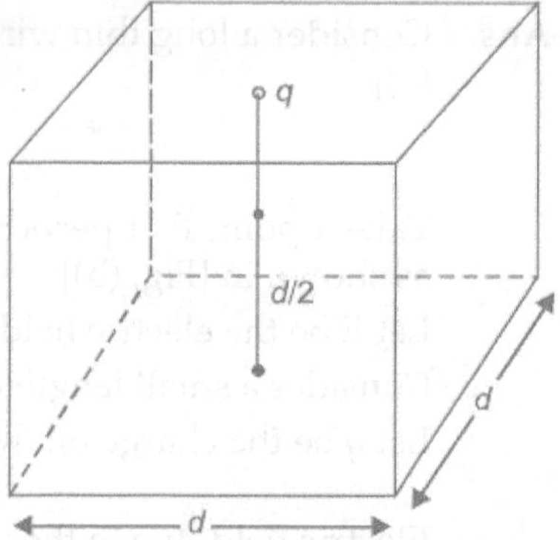

Total electric flux through the cube $= \dfrac{q}{\varepsilon_0}$

$\therefore$ Electric flux through the square surface $= \dfrac{q}{\sigma\varepsilon_0}$

(ii) On moving the charge to distance d from the centre of square and making side of square $2d$, does not change the flux at all because flux is independent of side of square or distance of charge in this case.

Chapter 2. Electrostatic Potential and Capacitance

Q. 1. (i) Show that the normal component of electrostatic field has a discontinuity from one side of a charged surface to another given by

$$(E_2 - E_1) \cdot \hat{n} = \frac{\sigma}{\varepsilon_0}$$

where $\hat{n}$ is a unit vector normal to the surface at a point and σ is the surface charge density at that point. (The direction of $\hat{n}$ is from side 1 to side 2). Hence show that just outside a conductor, the electric field is $\sigma\, \hat{n} / \varepsilon_0$.

(ii) Show that the tangential component of electrostatic field is continuous from one side a charged surface to another.

[**Hint :** For (i). Use Gauss's law. For (ii) use the fact that work done by electrostatic field on a closed loop is zero.] **[NCERT]**

Ans. (i) Electric field on one side of a charged body is E_1 and electric field on the other side of the same body is E_2. If infinite plane charged body has a uniform thickness, then electric field due to one surface of the charged body is given by,

$$\vec{E_1} = -\frac{\sigma}{2\varepsilon_0}\hat{n} \qquad \text{...(i)}$$

where,

$\hat{n}$ = Unit vector normal to the surface at a point

σ = Surface charge density at that point

Electric field due to the other surface of the charged body,

$$\vec{E_2} = \frac{\sigma}{2\varepsilon_0}\hat{n} \qquad \text{...(ii)}$$

Electric field at any point due to the two surfaces,

$$\vec{E_2} - \vec{E_1} = \frac{\sigma}{2\varepsilon_0}\hat{n} - \left(\frac{-\sigma}{2\varepsilon_0}\hat{n}\right) = \frac{\sigma}{\varepsilon_0}\hat{n}$$

$$(\vec{E_2} - \vec{E_1})\,\hat{n} = \frac{\sigma}{\varepsilon_0} \qquad \text{...(iii)}$$

Since inside a closed conductor, $\vec{E_1} = 0$

$$\therefore \qquad \vec{E} = \vec{E_2} = -\frac{\sigma}{2\varepsilon_0}\hat{n}$$

Therefore, the electric field just outside the conductor is $\dfrac{\sigma}{\varepsilon_0}\hat{n}$.

(ii) When a charged particle is moved from one point to the other on a closed loop, the work done by the electrostatic field is zero. Hence, the tangential component of electrostatic field is continuous from one side of charged surface to the other.

Q. 2. Fig. (a) and (b) show the field lines of a positive and negative point charge respectively

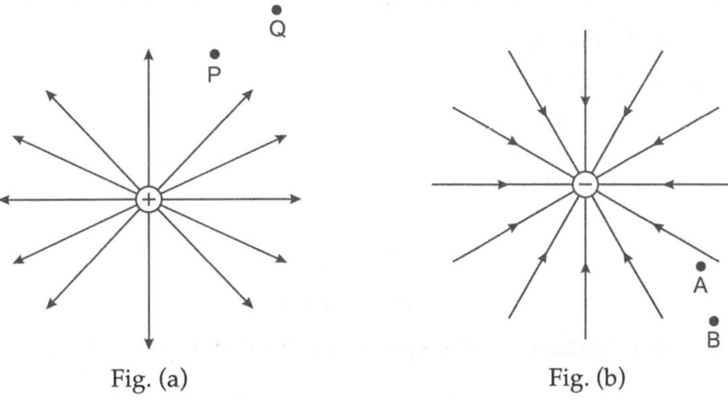

Fig. (a) Fig. (b)

(i) Give the signs of the potential difference $V_P - V_Q$, $V_B - V_A$.

(ii) Give the sign of the potential energy difference of a small negative charge between the points Q and P; A and B.

(iii) Give the sign of the work done by the field in moving a small positive charge from Q to P.

(iv) Give the sign of the work done by the external agency in moving a small negative charge from B to A.

(v) Does the kinetic energy of a small negative charge increase or decrease in going from B to A ?

[NCERT]

Ans. (i) As we know $V \propto \frac{1}{r}$

$V_P > V_Q$ [As P is closer w.r.t. Q to positive charge]

$V_A < V_B$ *i.e.*, V_A is more negative w.r.t. V_B as A is more closer to negative charge w.r.t. B.

Hence, $V_P - V_Q$ = positive

and $V_B - V_A$ = positive

(ii) A small negative charge will be attracted towards positive charge. The negative charge moves from higher potential energy to lower potential energy. Thus, sign of potential energy difference of a small negative charge between Q and P is positive.

Similar to this potential energy at A > potential energy at B, thus difference is positive.

(iii) In moving a small positive charge from Q to P. Work has to be done by an external agency against the electric field. Thus, work done by electric field is negative.

(iv) In moving a small negative charge from B to A work has to be done by the external agency. Hence, it is positive.

(v) Due to force of repulsion on negative charge, velocity decreases. Hence, the kinetic energy decreases in going from B to A.

Q. 3. Figure shows a charge array known as an electric quadrupole. For a point on the axis of the quadrupole, obtain the dependence of potential on r for $\frac{r}{a} >> 1$, and contrast your results with that due to an electric dipole and an electric monopole (*i.e.*, single charge).

[NCERT]

Ans. Four charges of same magnitude are placed at points , X, Y, Y, and Z respectively, as shown in the following figure.

A point is located at P, which is r distance away from point Y.

X a Y a Z P
q $-q-q$ q
r

The system of charges form an electric quadrupole.
It can be considered that the system of the electric quadrupole has three charges.

Charge $+q$ placed at point X

Charge $-2q$ placed at point Y

Charge $+q$ placed at point Z

XY = YZ = a

YP = r

$$PX = r + a$$

$$PZ = r - a$$

Electrostatic potential caused by the system of three charges at point P is given by,

$$V = \frac{1}{4\pi\varepsilon_0}\left[\frac{q}{XP} - \frac{2q}{YP} + \frac{q}{ZP}\right]$$

$$= \frac{1}{4\pi\varepsilon_0}\left[\frac{q}{r+a} - \frac{2q}{r} + \frac{q}{r-a}\right]$$

$$= \frac{q}{4\pi\varepsilon_0}\left[\frac{r(r-a) - 2(r+a)(r-a) + r(r+a)}{r(r+a)(r-a)}\right]$$

$$= \frac{q}{4\pi\varepsilon_0}\left[\frac{r^2 - ra - 2r^2 + 2a^2 + r^2 + ra}{r(r^2 - a)}\right]$$

$$= \frac{q}{4\pi\varepsilon_0}\left[\frac{2a^2}{r(r^2 - a^2)}\right]$$

$$= \frac{2qa^2}{4\pi\varepsilon_0 r^3\left(1 - \frac{a^2}{r^2}\right)}$$

Since $\frac{r}{a} >> 1$,

$$\therefore \quad \frac{a}{r} << 1$$

$\frac{a^2}{r^2}$ is taken as negligible.

$$\therefore \quad V = \frac{2qa^2}{4\pi\varepsilon_0 r^3}$$

It can be inferred that potential, $V \propto \frac{1}{r^3}$

However, it is known that for a dipole,

$$V \propto \frac{1}{r^2}$$

And, for a monopole, $V \propto \frac{1}{r}$

Q. 4. Show that the force on each plate of a parallel plate capacitor has a magnitude equal to $\left(\frac{1}{2}\right)$ QE, where Q is the charge on the capacitor. and E is the magnitude of electric field between the plates. Explain the origin of the factor $\frac{1}{2}$. **[NCERT]**

Ans. Consider F be the force applied to separate the plates of a parallel plate capacitor by a distance of Δx. Hence, work done by the force to do so

$$= F\,\Delta x$$

$$(W) = \text{Force} \times \text{Displacement}$$

As a result, the potential energy of the capacitor increases by an amount given as $uA\Delta x$

Where, u = Energy density

A = Area of each plate

d = Distance between the plates

V = Potential difference across the plates

The work done will be equal to the increase in the potential energy *i.e.,*

$$F\Delta x = uA\Delta x$$

$$F = uA = \left(\frac{1}{2}\varepsilon_0 E^2\right) A$$

Electric field intensity is given by,

$$E = \frac{V}{d}$$

$$\therefore \quad F = \frac{1}{2}\varepsilon_0 \left(\frac{V}{d}\right) EA$$

$$= \frac{1}{2}\left(\varepsilon_0 A \frac{V}{d}\right) E$$

However, capacitance,

$$C = \frac{\varepsilon_0 A}{d}$$

$$\therefore \quad F = \frac{1}{2}(CV)\,E$$

Charge on the capacitor is given by,

$$Q = CV$$

$$\therefore \quad F = \frac{1}{2}QE$$

The physical origin of the factor, $\frac{1}{2}$ in the force formula lies in the fact that just outside the conductor, field is E and inside it is zero. Thus, it is the average value, E/2, of the field that contributes to the force.

Q. 5. (i) If two similar large plates, each of area A having surface charge densities $+\sigma$ and $-\sigma$ are separated by a distance in air, find the expressions for★

(a) field at points between the two plates and on outer side of the plates. Specify the direction of the field in each case.

(b) the potential difference between the plates.

(c) the capacitance of the capacitor so formed.

(ii) Two metallic spheres of radii R and 2R are charged so that both of these have same surface charge density σ. If they are connected to each other with a conducting wire, in which direction will the charge flow and why ?★

★ are board based questions from previous year

Ans. (i) (a) Consider a parallel plate capacitor with two identical plates X and Y, each having an area of A, and separated by a distance *d*. Let the space between the plates be filled by a dielectric medium with its dielectric constant as K and σ be the surface charge density on each of the plates.

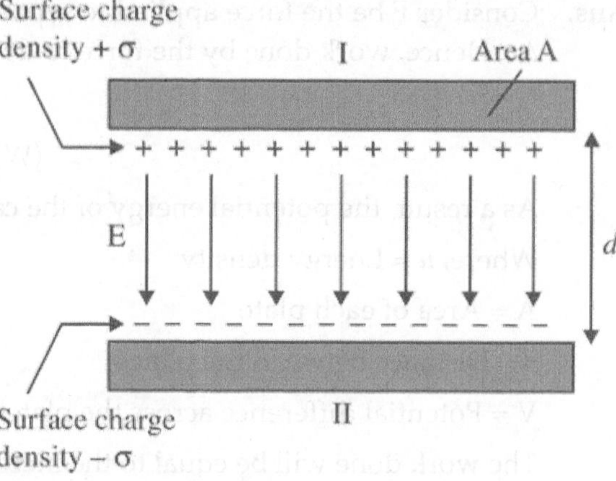

Surface charge density of plate 1

$$\sigma = Q/A$$

and that of plate 2 is – σ.

Electric field in outer region I,

$$E = \frac{\sigma}{2\varepsilon_0} - \frac{\sigma}{2\varepsilon_0} = 0$$

In the inner region between plates 1 and 2, the electric fields due to the two charged plates add up. So

$$E = \frac{\sigma}{2\varepsilon_0} + \frac{\sigma}{2\varepsilon_0} = \frac{\sigma}{\varepsilon_0}$$

(b) For uniform electric field, potential difference is simply the electric field multiplied by the distance between the plates, *i.e.*,

$$V = Ed = \frac{1}{\varepsilon_0}\frac{Qd}{A}$$

(c) Now, the capacitance of the parallel plate capacitor,

$$C = \frac{Q}{V} = \frac{Q.\varepsilon_0 A}{Qd} = \frac{\varepsilon_0 A}{d}$$

(ii) We know that the potential difference of the metallic sphere is given by,

$$V = \frac{Q}{4\pi\varepsilon_0 r}$$

where *r* is the radius of the sphere.

Now, the potential of the metallic sphere of radius R is given by,

$$V_R = \frac{Q}{4\pi\varepsilon_0 R}$$

$$V_R = \frac{\sigma(4\pi R^2)}{4\pi\varepsilon_0 R}$$

$$V_R = \frac{\sigma R}{\varepsilon_0} \quad \text{...(i)}$$

Similarly, the potential of the metallic sphere of radius 2R is given by

$$V_{2R} = \frac{Q}{4\pi\varepsilon_0 2R}$$

$$V_{2R} = \frac{\sigma(4\pi(2R)^2)}{4\pi\varepsilon_0 2R}$$

$$V_{2R} = \frac{\sigma \times 2R}{\varepsilon_0} \quad \text{...(ii)}$$

From the relation (i) and (ii) we know that $V_{2R} > V_R$.

The charge will flow from the sphere of radius of 2R to the sphere of radius R, if the spheres are connected.

Q. 6. Draw a labelled diagram of Van de Graaff generator. State its working principle to show how by introducing a small charged sphere into a larger sphere, a large amount of charge can be transferred to the outer sphere. State the use of this machine and also point out its limitations.★

Ans. Van de Graaff generator is a device used for building up high potential differences of the order of a few million volts.

★ are board based questions from previous year

Principle : The working of Van de Graaff generator is based on the following two electrostatic phenomena.

(i) Discharging action at sharp points (corona discharge) *i.e.*, electric discharge takes place in air or gases readily at the pointed ends of conductors.

(ii) If a charged conductor is brought into internal contact with a hollow conductor, all of its charge transfers to the hollow conductor, howsoever high the potential of the latter may be.

Construction : It has a big spherical conducting shell (S) kept over insulating pillars. A long narrow insulating belt is wound around two pulleys P_1 and P_2. B_1 and B_2 are two metal combs with sharp points. B_1 is known as spray comb and B_2 collecting comb.

Working : The spray comb provides positive potential by high tension source. The positive charge is sprayed on belt. As belt moves and touches the sphere, a negative charge is induced on the sharp ends of collecting comb B_2 and similar positive charge is induced on the further end of B_2. This positive charge moves immediately to the outer surface of S, because of discharging action of sharp points of B_2, the positive charge on the belt is neutralised. The uncharged belt moves downwards and collects the positive charge from B_1, which is then collected by B_2. This process is repeated and the positive charge on S goes on accumulating. In this way, voltage differences of as much as 6 or 8 million volts (with respect to the ground) can be created.

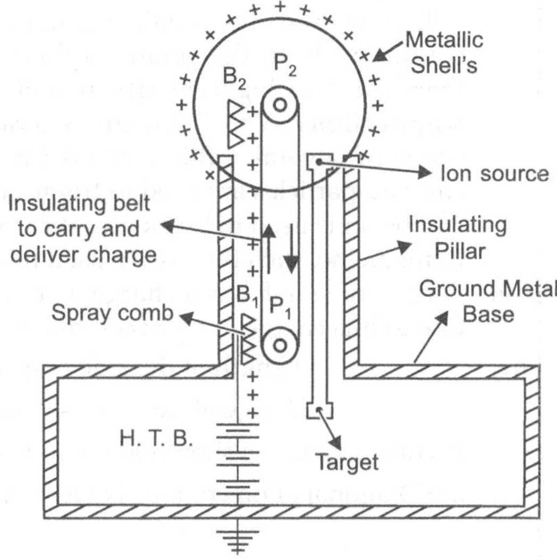

Use : Van de Graaff generator is used to create high potential differences that are used to accelerate charged particles such as electrons, protons, ions, etc., used for nuclear breakdown.

Limitations :

(1) It is a series of combination that allows only one way for moving charge.

(2) It can accelerate only the charged particles and not the uncharged particles.OSW

Q. 7. State principle of Van de Graaff generator. What are its applications ?

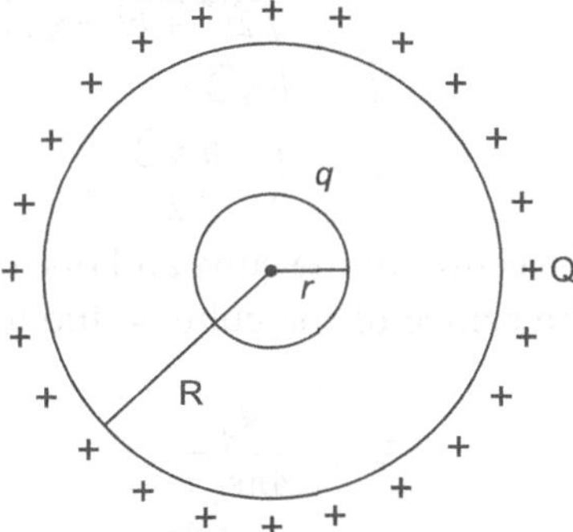

Ans. Consider two charged conducting spherical shells such that one is smaller and the other is larger. When the smaller one is kept inside the larger one and connected together, charge from the smaller one is transferred to larger shell irrespective of the higher potential of the larger shell. *i.e.*, the charge resides on the outer surface of the outer shell and the potential of the outer shell increases considerably.

Sharp pointed surfaces of a conductor have large surface charge densities and hence the electric field created by them is very high compared to the dielectric strength of the dielectric (air).

Therefore, air surrounding these conductors gets ionised and the like charges are repelled by the charged pointed conductors causing discharging action known as corona discharge or action of points. The sprayed charges moving with high speed cause electric wind.

Opposite charges are induced on the teeth of collecting comb (conductor) and again opposite charges are induced on the outer surface of the collecting sphere (Dome).

Hence, the potential difference between the surfaces of the two spheres is

$$V_r - V_R = \frac{kQ}{R} + \frac{kq}{r} - \frac{kQ}{R} - \frac{kq}{R}$$

$$= kq\left[\frac{1}{r} - \frac{1}{R}\right]$$

The above equation shows that the smaller sphere is at a higher potential as compared to the large spherical shell. So, if the two are connected through a conductor, then the electric charge will flow from the smaller sphere to the larger spherical shell. So, if charge is transferred continuously to the smaller sphere in some way, then it will accumulate on the larger shell thereby increasing its electric potential to a very large value.

Applications : Van de Graaff generator is used to produce very high potential difference (of the order of several million volts) for accelerating charged particles. The beams of accelerated charged particles are used to trigger nuclear reactions.

The beam is used to break atoms for various experiments in Physics.

In medicine, such beams are used to treat cancer. It is used for research purposes.

Q. 8. A cube of side b has a charge q at each of its vertices. Determine the potential and electric field due to this charge array at the centre of the cube. **[NCERT]**

Ans. Length of the side of a cube $= b$

Charge at each of its vertices $= q$

A cube of side b is shown in the following figure.

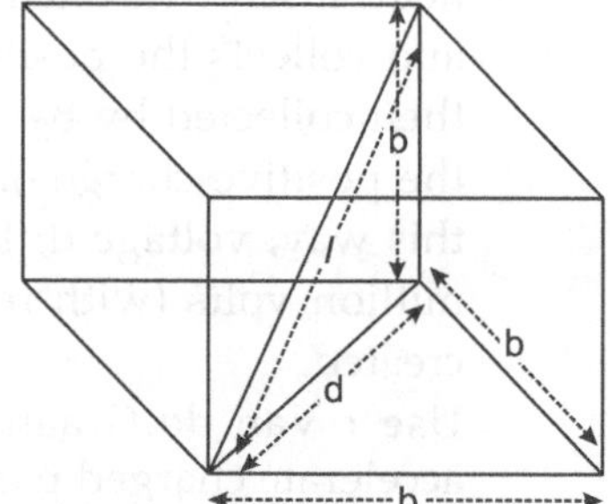

d = Diagonal of one of the six faces of the cube

$$d^2 = \sqrt{b^2 + b^2}$$

$$= \sqrt{2b^2}$$

$$d = b\sqrt{2}$$

l = Length of the diagonal of the cube

$$l = \sqrt{d^2 + b^2}$$

$$= \sqrt{(\sqrt{2}\,b)^2 + b^2}$$

$$= \sqrt{2b^2 + b^2} = \sqrt{3b^2}$$

$$l = b\sqrt{3}$$

$$r = \frac{l}{2} = \frac{b\sqrt{3}}{2}$$

This is the distance between the centre of the cube and one of the eight vertices.

The electric potential (V) at the centre of the cube is due to the presence of eight charges at the vertices.

$$V = \frac{8q}{4\pi\varepsilon_0 r}$$

$$= \frac{8q}{4\pi\varepsilon_0 \times \frac{b\sqrt{3}}{2}}$$

$$= \frac{4q}{\sqrt{3}\pi\varepsilon_0 b}$$

Therefore, the potential at the centre of the cube is $\frac{4q}{\sqrt{3}\pi\varepsilon_0 b}$.

The electric field at the centre of the cube, due to the eight charges, gets cancelled. This is because the charge are distributed symmetrically with respect to the centre of the cube. Hence, the electric field is zero at the centre.

Chapter 3. Current Electricity

Q. 1. (i) State the working principle of a potentiometer. With the help of the circuit diagram, explain how a potentiometer is used to compare the emf's of two primary cells. Obtain the required expression used for comparing the emf's.★

(ii) Write two possible causes for one sided deflection in a potentiometer experiment.★

Ans. (i) **Working principle of potentiometer :** When a constant current is passed through a wire of uniform area of cross-section, the potential drop across any portion of the wire is directly proportional to the length of that portion.

Applications of potentiometer for comparing emf's of two cells : The following figure shows an application of the potentiometer to compare the emf of two cells of emf E_1 and E_2.

E_1, E_2 are the emf of the two cells.

1, 2, 3 form a two way key.

When 1 and 3 are connected, E_1 is connected to the galvanometer (G).

Jockey is moved to N_1, which is at a distance L_1 from A, to find the balancing length.

Applying loop rule to AN_1G31A,

$$\phi l_1 + 0 - E_1 = 0 \quad \text{...(i)}$$

Where, ϕ is the potential drop per unit length.

Similarly, for E_2 is balanced against l_2 (AN_2),

$$\phi l_2 + 0 - E_2 = 0 \quad \text{...(ii)}$$

$$\frac{E_1}{E_2} = \frac{l_1}{l_2} \quad \text{...(iii)}$$

Thus, we can compare the emf's of any two sources. Generally, one of the cells is chosen as a standard cell whose emf is known to a high degree of accuracy. The emf of the other cell is then calculated from equation (iii).

(ii) (a) The emf of the cell connected in main circuit may not be more than the emf of the primary cells whose emf's are to be compared.

(b) The positive ends of all cells are not connected to the same end of the wire.

Q. 2. The current is drawn from a cell of emf E and internal resistance r connected to the network of resistors each of resistance r as shown in the figure. Obtain the expression for★

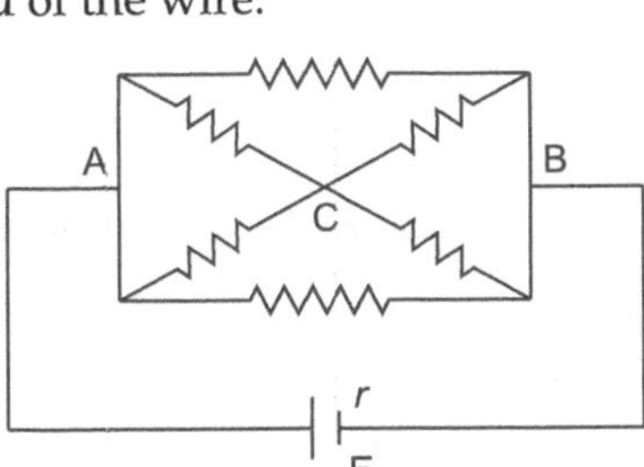

(a) the current drawn from the cell and

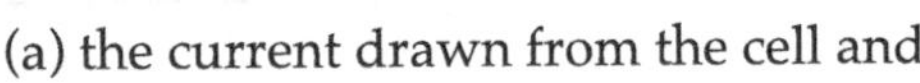

(b) the power consumed in the network.

Ans. (a) The equivalent resistance between AC and CB is given by

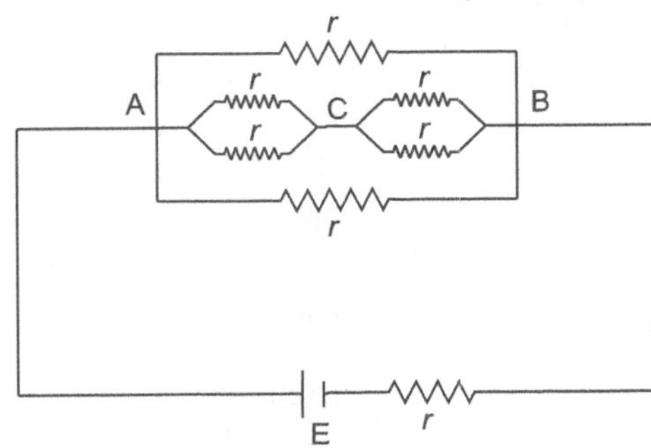

★ are board based questions from previous year

$$\frac{1}{r'} = \frac{1}{r} + \frac{1}{r}$$

$$\frac{1}{r'} = \frac{2}{r}$$

$$r' = \frac{r}{2}$$

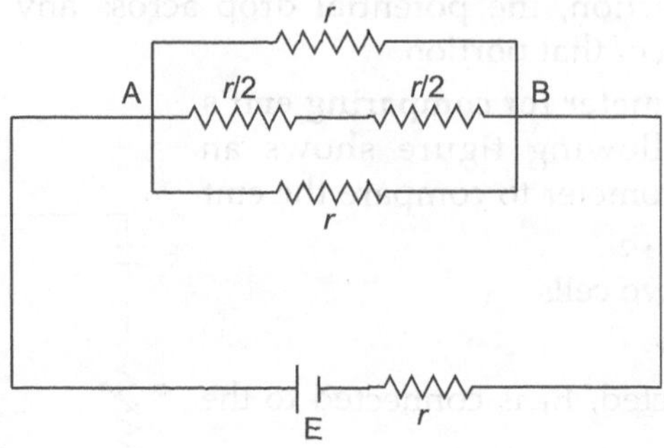

The equivalent resistance of middle branch is given by,

$$r'' = \frac{r}{2} + \frac{r}{2} = r$$

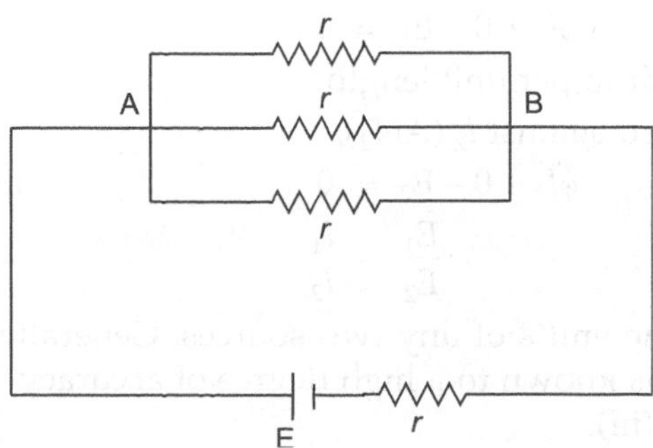

The equivalent resistance between A and B is given by

$$\frac{1}{r'''} = \frac{1}{r} + \frac{1}{r} + \frac{1}{r}$$

$$r''' = \frac{r}{3}$$

Total resistance of the circuit is given by,

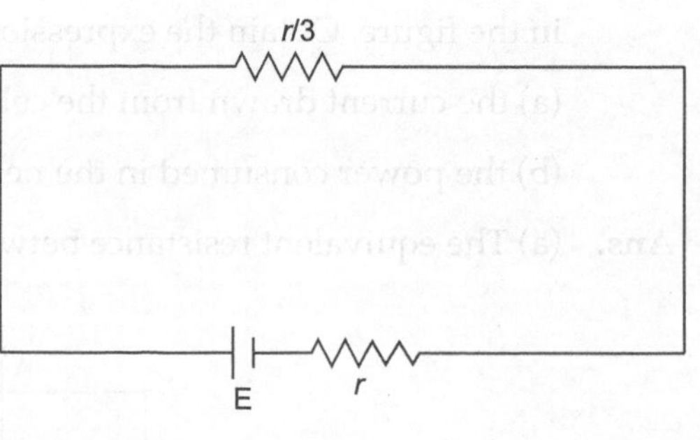

$$R = \frac{r}{3} + r$$

$$R = \frac{4r}{3}$$

Now current through the circuit is given by,

$$I = \frac{E}{\frac{4r}{3}}$$

$$I = \frac{3E}{4r}$$

(b) Power consumption $= I^2R$

$$= \left(\frac{3E}{4r}\right)^2 \frac{4r}{3}$$

$$P = \frac{3E^2}{4r}$$

Q. 3. Given figure shows a potentiometer with a cell of 2.0 V and internal resistance 0.40 Ω maintaining a potential drop across the resistor wire AB. A standard cell which maintains a constant emf of 1.02 V (for very moderate currents up to a few mA) gives a balance point at 67.3 cm length of the wire. To ensure very low currents drawn from the standard cell, a very high resistance of 600 kΩ is put in series with it, which is shorted close to the balance point. The standard cell is then replaced by a cell of unknown emf ε and the balance point found similarly, turns out to be at 82.3 cm length of the wire.

(i) What is the value ε ?

(ii) What purpose does the high resistance of 600 kΩ have ?

(iii) Is the balance point affected by this high resistance ?

(iv) Is the balance point affected by the internal resistance of the driver cell ?

(v) Would the method work in the above situation if the driver cell of the potentiometer had an emf of 1.0 V instead of 2.0 V ?

(vi) Would the circuit work well for determining an extremely small emf, say of the order of a few mV (such as the typical emf of a thermo-couple) ? If not, how will you modify the circuit ? **[NCERT]**

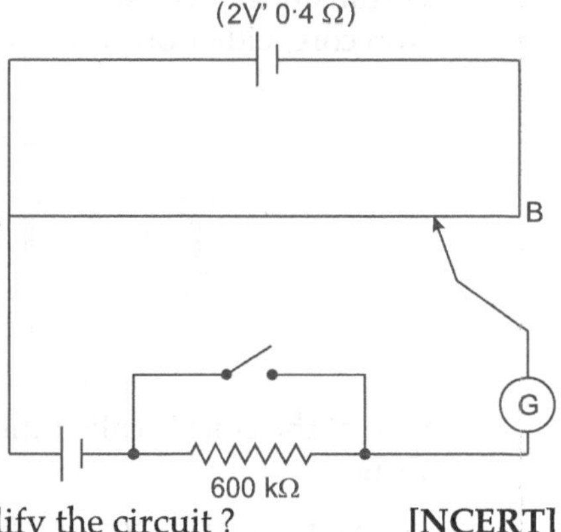

Ans. (i) Given,

constant emf of the given standard cell, $E_1 = 1.02$ V

Balance point on the wire, $l_1 = 67.3$ cm

A cell of unknown emf, ε, replaced the standard cell. Therefore, new balance point on the wire, $l =$ 82.3 cm

The relation connecting emf and balance point is,

$$\frac{E_1}{l_1} = \frac{\varepsilon}{l}$$

$$\varepsilon = \frac{l}{l_1} \times E_1$$

$$= \frac{82.3}{67.3} \times 1.02 = 1.247 \text{ V}$$

The value of unknown emf is 1.247 V.

(ii) The purposes of using the high resistance of 600 kΩ is to reduce the current through the galvanometer when the movable contact is far from the balance point.

(iii) The balance point is not affected by the presence of high resistance.

(iv) The point is not affected by the internal resistance of the driver cell.

(v) The method would not work if the driver cell of the potentiometer had an emf of 1.0 V instead of 2.0 V. This is because if the emf of the driver cell of the potentiometer is less than the emf of the other cell, then there would be no balance point on the wire.

(vi) The circuit would not work well for determining an extremely small emf. As the circuit would be unstable, the balance point would be close to end A. Hence, there would be a large percentage of error.

The given circuit can be modified if a series resistance is connected with the wire AB. The potential drop across AB is slightly greater than the emf measured. The percentage error would be small.

Chapter 4. Moving Charges and Magnetism

Q. 1. (i) With the help of labelled diagram, describe briefly the underlying principle and working of a step up transformer.★

★ are board based questions from previous year

(ii) Write any two sources of energy loss in a transformer.★

(iii) A step up transformer converts a low input voltage into a high output voltage. Does it violate law of conservation of energy ? Explain.★

Ans. (i) **Principle underlying and working of transformer :** The principle is mutual inductance. When a alternating source of voltage is introduced across a coil (which is physically coupled to another coil), the changing current through it induces an EMF across the second coil.

A transformer consists of two sets of coils, insulated from each other. They are wound on a soft-iron core, either one on top of the other, or on separate sides of the core.

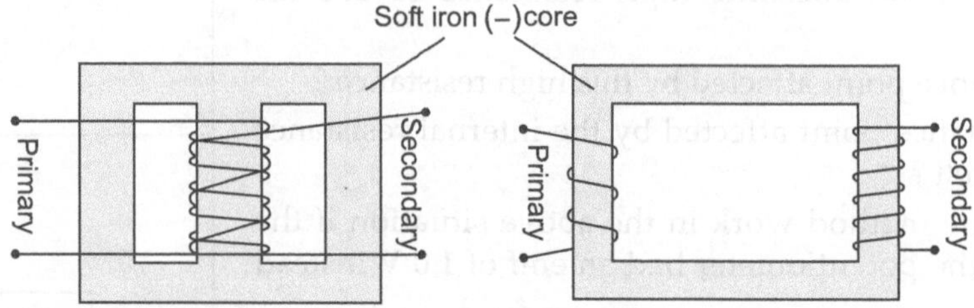

One of the coils is called the primary coil, and has N_p turns and other coil. (secondary coil) has N_s turns.

For step up transformer $N_s > N_p$ from definition, the voltage which is to be transformed is introduced across the primary coil.

When alternating voltage is applied across the primary, the resulting alternating current through it produces a varying magnetic field, whose flux through the secondary coil changes.

From Faraday's law, this varying flux induces an EMF across the secondary, whose magnitude depends on the amount of coupling of the two coils, numerically measured as mutual inductance. The more this coupling of the two coils, the more is mutual inductance, and therefore the induced EMF.

If ϕ is the flux through each turn of the core, then through N turns around the core, the total flux is $N\phi$.

Thus, the EMF induced in the secondary coil is

$$E_s = -N_s \frac{d\phi}{dt} \qquad ...(i)$$

Similarly, there will be an EMF induced in the primary coil itself, due to self inductance, given by

$$E_p = -N_p \frac{d\phi}{dt} \qquad ...(ii)$$

If the voltage applied across the primary is V_p, then if its resistance is R, the current through it will be

$$I_p = \frac{V_p - E_p}{R}$$

For negligible resistance,

$$E_p \simeq V_p \text{ and } V_p = -N_p \frac{d\phi}{dt} \qquad ...(iii)$$

If the secondary is an open circuit, no current will be drawn from it then, voltage across it will be

$$V_s = E_s = -N_s \frac{d\phi}{dt} \qquad ...(iv)$$

$$\frac{V_s}{V_p} = \frac{N_s}{N_p} \qquad \text{[From equation (iii) and (iv)]}$$

If the transformer is 100% efficient, that is, all the input power is transferred to the secondary without any leakage or losses, (conservation of energy).

★ are board based questions from previous year

$$I_p V_p = I_s V_s$$

This implies that

$$\frac{I_p}{I_s} = \frac{V_s}{V_p} = \frac{N_s}{N_p}$$

Thus from this if voltage needs to be up converted

$$N_s > N_p$$

However, in a step down transformer, there will be a greater current in the secondary as compared to the primary and vice-versa.

(ii) The possible sources of power losses in practical transformers are :

(a) **Flux Leakage :** Small amount of flux coupled from primary to secondary. To avoid this, the coils are wound over each other again and again. This leakage loss is common loss in transformer.

(b) **Resistance of windings :** The transformer coil wires cannot have absolutely zero resistance, thus small amount of loss due to heating arises in coils.

(c) **Core eddy currents :** Since the core is a very good conductor itself, currents are induced in it due to changing magnetic fields, called eddy currents. These also result in losses.

(d) **Hysteresis :** Some part of energy is frozen into the core permanently in the form of a residual magnetic field due to its ferromagnetic character.

(iii) No, it does not violate the energy conservation. When low voltage is converted to high voltage, the current is lowered, thereby conserving the total energy dissipated across the primary and secondary coil.

Q. 2. Explain, using a labelled diagram, the principle and working of a moving coil galvanometer. What is the function of (i) uniform radial magnetic field, (ii) soft iron core ?

Define the terms current sensitivity and voltage sensitivity of a galvanometer. Why does increasing the current sensitivity not necessarily increase voltage sensitivity ?★

Ans. The basic principle of a moving coil galvanometer is that when a current carrying coil is placed in a magnetic field, it experiences a torque.

When the current I is passed through the coil, the torque experienced is given by

$$\tau = NIAB \sin \theta$$

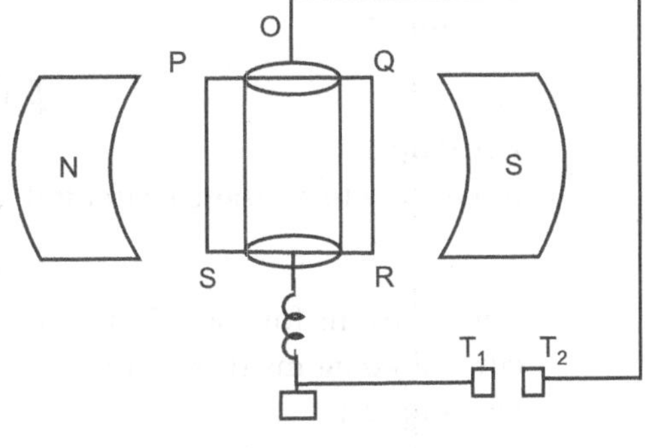

Where N = No. of turns of the coil,

A = Area of the coil

B = Magnetic field and

θ = Angle between normal of coil and magnetic field

(i) The uniform radial magnetic field is used to make the scale linear.

(ii) The soft iron core increases the strength of the magnetic field.

The current sensitivity is defined as the deflection produced in the galvanometer, while passing a current of 1 ampere. (1 amp).

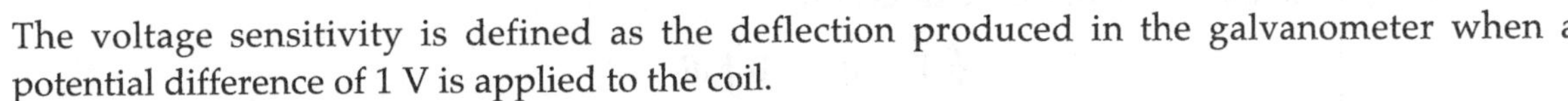

Thus, current sensitivity $\left(\frac{\alpha}{I}\right) = \frac{NBA}{K}$

The voltage sensitivity is defined as the deflection produced in the galvanometer when a potential difference of 1 V is applied to the coil.

Thus, voltage sensitivity $\left(\frac{\alpha}{V}\right) = \frac{NBA}{K}$

★ are board based questions from previous year

Where R is the resistance.

Increasing the current sensitivity does not necessarily increase the voltage sensitivity as there is an increase in the resistance as well.

Q. 3. (i) State Ampere's circuital law. Use this law to obtain the expression for the magnetic field inside an air cored toroid of average radius, having '*n*' turns per unit length and carrying a steady current I.★

(ii) An observer to the left of a solenoid of N turns each of cross section area A observes that a steady current I in it flows in the clockwise direction. Depict the magnetic field lines due to the solenoid specifying its polarity and show that it acts as a bar magnet of magnetic momentum M = NIA.★

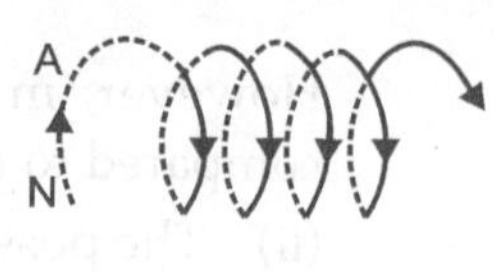

Ans. (i) Ampere's circuital law in electromagnetism is analogous to Gauss's law in electrostatics. This law states that "The line integral of resultant magnetic field along a closed plane curve is equal to μ_0 time the total current crossing the area bounded by the closed curve provided the electric field inside the loop remains constant. Thus $\oint B.dl = \mu_0 I_{enc}$ where μ_0 is permeability of free space and $I_{enc.}$ is the net current enclosed by the loop.

A toroid is a hollow circular ring on which a large number of turns of a wire are closely wound. Consider an air-cored toroid (as shown below) with centre O.

Given :

r = Average radius of the toroid

I = Current through the solenoid

n = Number of turns per unit length

To determine the magnetic field inside the toroid, we consider three amperian loops (loop 1, loop 2 and loop 3) as shown in the figure below.

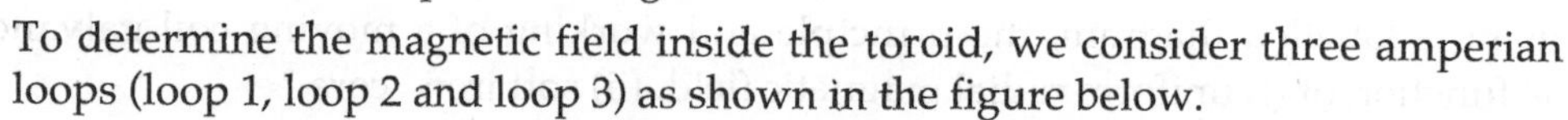

According to Ampere's circuital law, we have

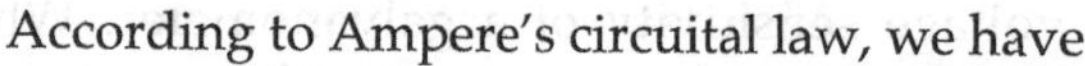

$$\oint \vec{B} \cdot \vec{dl} = \mu_0 I_{enc.} \text{ (Total current)}$$

Total current for loop 1 is zero because no current is passing through this loop

Loop 1
Loop 2
Loop 3

So,

For loop 1 :

$$\oint \vec{B} \cdot \vec{dl} = \mu_0 \text{ (Total current)}$$

For loop 3 :

According to Ampere's circuital law, we have

$$\oint \vec{B} \cdot \vec{dl} = \mu_0 \text{ (Total current)}$$

Total current for loop 3 is zero because net current coming out of this loop is equal to the net current going inside the loop.

For loop 2 :

The total current flowing through the toroid is NI, where N is the total number of turns

$$\oint \vec{B} \cdot \vec{dl} = \mu_0 \text{ (NI)} \quad \text{...(i)}$$

Now, $\vec{B}$ and $\vec{dl}$ are in the same direction

$$\oint \vec{B} \cdot \vec{dl} = B \oint dl$$

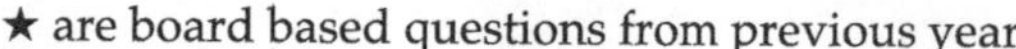

$$\Rightarrow \quad \oint \vec{B} \cdot \vec{dl} = B\,(2\pi r) \quad \text{...(ii)}$$

★ are board based questions from previous year

Comparing (i) and (ii), we get

$$B\,(2\pi r) = \mu_0 NI$$

$$\Rightarrow \qquad B = \frac{\mu_0 NI}{2\pi r}$$

Number of turns per unit length is given by

$$n = \frac{N}{2\pi r}$$

$$\therefore \qquad B = \mu_0 nI$$

This is the expression for magnetic field inside air-cored toroid.

(ii) Given that the current flows in the clockwise direction for an observer on the left side of the solenoid. This means that left face of the solenoid acts as south pole and right face acts as north pole. Inside a bar magnet the magnetic field lines are directed from south to north. Therefore, the magnetic field lines are directed from left to right in the solenoid.

Magnetic moment of single current carrying loop is given by $m' = IA$

where,

I = Current flowing through the loop

A = Area of the loop

So, Magnetic moment of the whole solenoid is given by

$$M = Nm' = N\,(IA)$$

Chapter 5. Magnetism and Matter

Q. 1. Answer the following questions :

(i) Why does a paramagnetic sample display greater magnetisation (for the same magnetising field, when cooled ?

(ii) Why is diamagnetism, in contrast almost independent of temperature ?

(iii) If a toroid uses bismuth for its core, will the field in the core be (slightly) greater or (slightly) less than when the core is empty ?

(iv) Is the permeability of a ferromagnetic material independent of the magnetic field ? If not, is it more for lower or higher fields ?

(v) Magnetic field lines are always nearly normal to the surface of a ferromagnet at every point. (This fact is analogous to the static electric field lines being normal to the surface of a conductor at every point) Why ?

(vi) Would the maximum possible magnetisation of a paramagnetic sample be of the same order of magnitude as the magnetisation of a ferromagnet ? **[NCERT]**

Ans. (i) Due to the random thermal motion of molecules, the alignments of dipoles get disrupted at high temperatures. On cooling, this disruption is reduced. Hence, a paramagnetic sample displays greater magnetisation when cooled.

(ii) The induced dipole moment in a diamagnetic substance is opposite to the magnetising field always. Hence, the internal motion of the atoms does not affect the diamagnetism of a material.

(iii) Bismuth is a diamagnetic substance. Hence, a toroid with a bismuth core has a magnetic field slightly lesser than a toroid whose core is empty.

(iv) The permeability of ferromagnetic materials is dependent on the applied magnetic field. It is greater for a lower field w.r.t. higher magnetic field.

(v) With the refraction of magnetic field lines as μ_r increases the normal component of magnetic field at surface of magnetised material increases.

For ferromagnetic materials $\mu_r >> 1$. Therefore, the tangential magnetisation near to surface is very small.

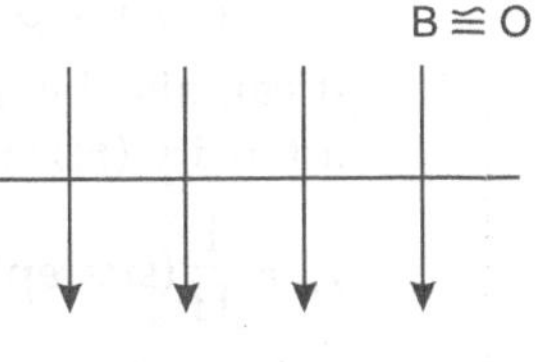

Thus, magnetic lines on the surface are nearly normal.

(vi) The maximum possible magnetisation of a paramagnetic sample can be of the same order of magnitude as the magnetisation of a ferromagnet. This requires high magnetising fields for saturation.

Q. 2. Define following terms :

(i) Magnetic intensity or magnetising force (H).

(ii) Magnetic field strength or magnetic field or magnetic induction or magnetic flux density (B).

(iii) Magnetic flux (ϕ).

(iv) Relative magnetic permeability (μ_r).

(v) What is intensity of magnetisation (I) ?

(vi) What is magnetic susceptibility (χ_m) ?

Ans. (i) Magnetic intensity at a point is the force experienced by a north pole of unit pole strength placed at that point due to pole strength of the given magnet. $H = \frac{B}{\mu}$.

It can also be defined as the force experienced by a unit positive charge flowing with unit velocity in a direction normal to the magnetic field.

Its S.I. unit is Ampere-turns per linear metre and CGS unit is Oersted.

(ii) Magnetic flux density is the number of magnetic lines of force passing normally through a unit area of a substance.

$$B = \mu H$$

Its S.I. unit is weber-m^2 or Tesla (T) and C.G.S. unit is gauss.

$$1 \text{ gauss} = 10^{-4} \text{ Tesla}$$

(iii) It is defined as the number of magnetic lines of force passing normally through a surface. Its S.I. unit is weber.

(iv) It is the ratio of magnetic flux density in a material to that in vacuum. It can also be defined as the ratio of absolute permeability of the material to that in vacuum.

$$\mu_r = \frac{B}{B_0}$$

$$\mu_r = \frac{\mu}{\mu_0}$$

Relation between B and H :

$B = \mu H$ (where μ is the permeability of the medium)

Magnetic permeability (μ) : It is the degree or extent to which magnetic lines of force can pass/enter a substance.

Its SI unit is T mA^{-1} or WbA^{-1}m^{-1} or H m^{-1}.

(v) **Intensity of magnetisation (I) :** It is the degree to which a substance is magnetised when placed in a magnetic field.

It can also be defined as the magnetic dipole moment (M) acquired per unit volume of the substance (V). It can also be defined as the pole strength (m) per unit cross-sectional area (A) of the substance.

$$I = \frac{M}{V}$$

$$I = \frac{m\,(2l)}{A\,(2l)} = \frac{m}{A}$$

S.I. unit of intensity of magnetisation is $\frac{A}{m}$, or Am^{-1}.

(vi) It is the property of the substance which shows how easily a substance can be magnetised. It can also be defined as the ratio of intensity of magnetisation (I) in a substance to the magnetic intensity (H) applied to the substance.

$\chi_m = \frac{I}{H}$ (susceptibility has no unit).

Relation between magnetic permeability (μ_r) and susceptibility (χ_m), $\mu_r = 1 + \chi_m$.

Chapter 6. Electromagnetic Induction

Q. 1. Refer to figure the arm PQ of the rectangular conductor is moved from $x = 0$, outwards. The uniform magnetic field is perpendicular to the plane and extends from $x = 0$ to $x = b$ and is zero for $x > b$. Only the arm PQ possesses substantial resistance r. Consider the situation when the arm PQ is pulled outwards from $x = 0$ to $x = 2b$, and is then moved back to $x = 0$ with constant speed v. Obtain expressions for the flux, the induced emf, the force necessary to pull the arm and the power dissipated as Joule heat. Sketch the variation of these quantities with distance.

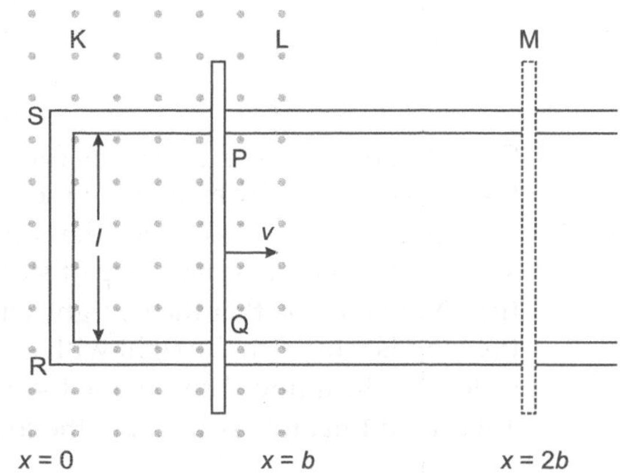

[NCERT]

Ans. Let us first consider the forward motion from $x = 0$ to $x = 2b$. The flux ϕ_B linked with the circuit SPQR is

$$\phi_B = Blx \qquad 0 \leq x < b$$
$$= Blb \qquad b \leq x < 2b$$

The induced emf is

$$\varepsilon = -\frac{d\phi_B}{dt}$$
$$= -Blv \qquad 0 \leq x < b$$
$$= 0 \qquad b \leq x < 2b$$

When the induced emf is non-zero, the current I is (in magnitude)

$$I = \frac{Blv}{r}$$

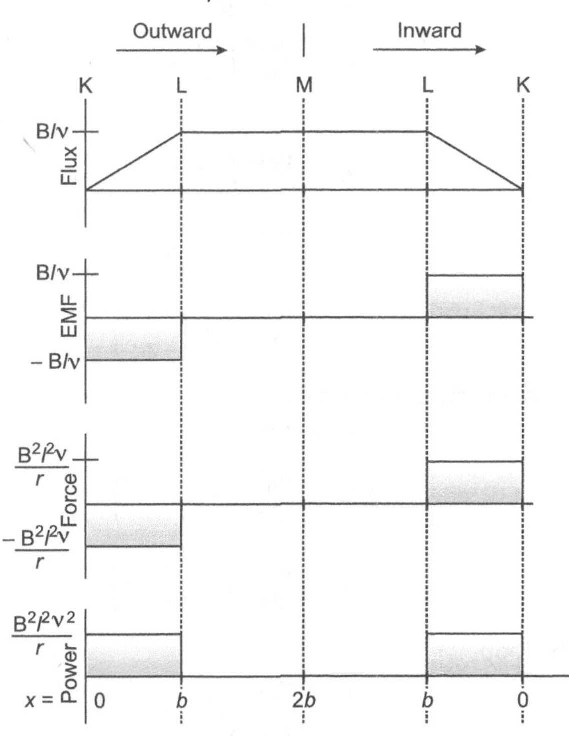

The force required to keep the arm PQ in constant motion is given by IlB. Its direction is to the left. In magnitude

$$F = \frac{B^2l^2v}{r} \qquad 0 \leq x < b$$
$$= 0 \qquad b \leq x < 2b$$

The joule heating loss is given by

$$P_J = I^2 r$$

$$= \frac{B^2 l^2 v^2}{r} \qquad 0 \le x < b$$

$$= 0 \qquad b \le x < 2b$$

One obtains similar expressions for the inward motion from $x = 2b$ to $x = 0$. One can appreciate the whole process by examining the sketch of various quantities displayed in figure.

Q. 2. (i) Obtain an expression for the mutual inductance between a long straight wire and a square loop of side a as shown in figure.

(ii) Now assume that the straight wire carries a current of 50 A and the loop is moved to the right with a constant velocity, $v = 10$ m/s. Calculate the induced emf in the loop at the instant when $x = 0.2$ m. Take $a = 0.1$ m and assume that the loop has a large resistance.

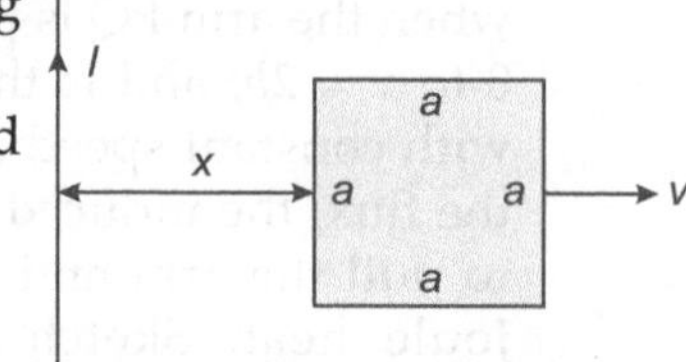

[NCERT]

Ans. (i) Consider a small element dy in the loop at a distance (v) from the long straight wire. Same has been shown in figure.

Magnetic flux associated with element dy, $d\phi = BdA$

Where,

dA = Area of element $dy = a\,dy$

B = Magnetic field at distance

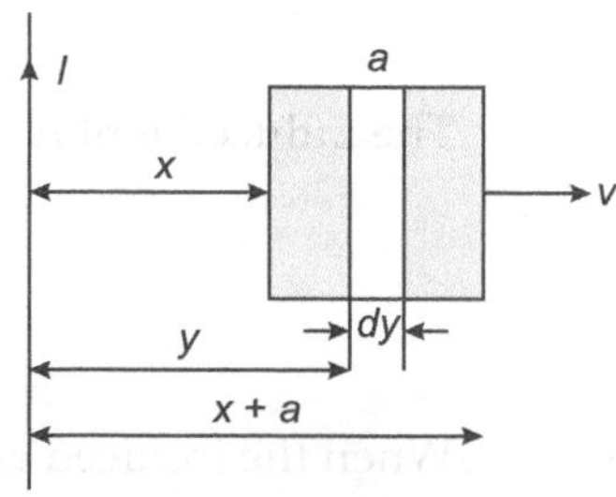

$$y = \frac{\mu_0 I}{2\pi y}$$

I = Current through wire

μ_0 = Permeability of free space = $4\pi \times 10^{-7}$ TmA^{-1}

$$\therefore \quad d\phi = \frac{\mu_0 I a}{2\pi}\frac{dy}{y}$$

$$\phi = \frac{\mu_0 I a}{2\pi}\int \frac{dy}{y}$$

y tends from x to $a + x$

$$\therefore \quad \phi = \frac{\mu_0 I a}{2\pi}\int_x^{a+x}\frac{dy}{y}$$

$$= \frac{\mu_0 I a}{2\pi}[\log_e y]_x^{a+x}$$

$$= \frac{\mu_0 I a}{2\pi}\log_e\left(\frac{a+x}{x}\right)$$

For mutual inductance M, the flux is given by

$$\phi = MI$$

$$\therefore \quad MI = \frac{\mu_0 I a}{2\pi}\log_e\left(\frac{a+x}{x}\right)$$

$$M = \frac{\mu_0 I a}{2\pi}\log_e\left(\frac{a}{x}+x\right)$$

(ii) Emf induced in the loop,

$$e = B'av$$

$$= \left(\frac{\mu_0 I}{2\pi x}\right)av$$

Given, I = 50 A

Velocity of loop (v) = 10 m/s

$$e = \frac{4\pi \times 10^{-7} \times 50 \times 0.1 \times 10}{2\pi \times 0.2}$$

$$e = 5 \times 10^{-5}\,V$$

Q. 3. (i) Explain the meaning of the term mutual inductance. Consider two concentric circular coils, one of radius r_1 and the other of radius r_2 $(r_1 < r_2)$ placed coaxially with centres coinciding with each other. Obtain the expression for the mutual inductance of the arrangement.★

(ii) A rectangular coil of area A, having number of turns N is rotated at 'f' revolutions per second in a uniform magnetic field B, the field being perpendicular to the coil. Prove that the maximum emf induced in the coil is $2\pi f$ NBA.★

Ans. (i) **Mutual inductance :** It is the property of a pair of coils due to which an e.m.f. induced in one of the coils due to the change in the current in the other coil. Let a current I_2 flow through the outer circular coil. The magnetic field at the centre of the coil is

$$B_2 = \frac{\mu_0 I_2}{2r_2} \quad \ldots(i)$$

As the inner coil placed co-axially has very small radius, therefore, B_2 may be take as constant over its cross-sectional area. Hence, flux associated with inner coil is

$$\phi_1 = \pi r_1^2 B_2$$

$$= \pi r_1^2 \frac{\mu_0 I_2}{2r_2} \quad \text{[From (i)]}$$

$$= \left(\frac{\mu_0 \pi r_1^2}{2r_2}\right) I_2$$

$$= M_{12} I_2$$

$$\therefore \quad M_{12} = \frac{\mu_0 \pi r_1^2}{2r_2}$$

Now, $$M_{21} = M_{12} = \frac{\mu_0 \pi r_1^2}{2r_2}$$

(ii) Let N be the number of turns of the rectangular coil and A be its cross-sectional area placed in a magnetic field B, then, the magnetic flux linked with the coil,

$$\phi = NBA \cos\theta$$

The induced emf,

$$e = -d\phi/dt$$

$$e = -\frac{d\phi}{dt} = \left(-NBA\,(-\sin\theta)\frac{d\theta}{dt}\right)$$

$$= NBA\,.\,\sin\theta\,(2\pi f) \quad \left[\because \theta = \omega t,\ \frac{d\theta}{dt} = \omega = 2\pi f\right]$$

For maximum induced e.m.f.

$$\sin\theta = 1$$

$$\therefore \quad e = NBA\,(2\pi f)$$

Q. 4. (i) Describe a simple experiment (or activity) to show that the polarity of emf induced in a coil is always such that it tends to produce a current which opposes the change of magnetic flux that produces it.★

(ii) The current flowing through an inductor of self inductance L is continuously increasing. Plot a graph showing the variation.★

(a) Magnetic flux versus the current

(b) Induced emf versus dI / dt

(c) Magnetic potential energy stored versus the current.

★ are board based questions from previous year

Ans. (i) **Lenz's law :** According to Lenz's law, the polarity of the induced emf is such that it opposes a change in magnetic flux responsible for its production.

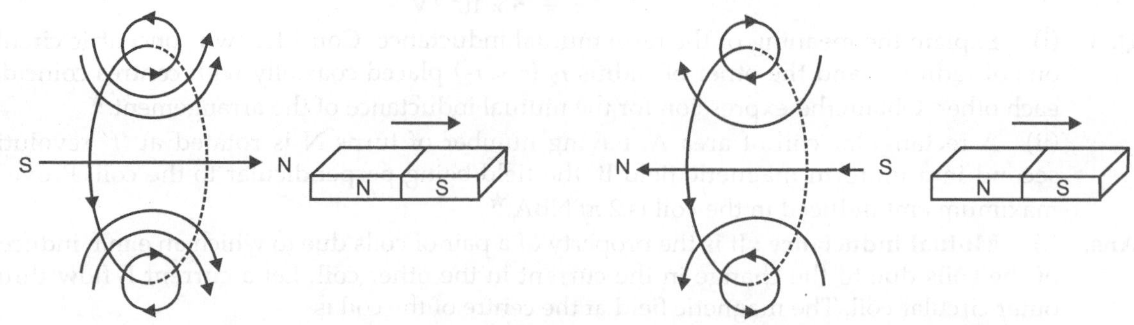

When the north pole of a bar magnet is pushed towards the coil, the amount of magnetic flux linked with the coil increase. Current is reduced in the coil from a direction such that it opposes the increase in magnetic flux. This is possible only when the current induced in the coil is in anti-clockwise direction, with respect to an observer. The magnetic moment M associated with this induced emf has north polarity, towards the north pole of the approaching bar magnet. Similarly, when the north pole of the bar magnet is moved away from the coil, the magnetic flux linked with the coil decreases. To counter this decrease in magnetic flux, current is induced in the coil in clockwise direction so that its south pole faces the receding north pole of the bar magnet. This would result in an attractive force which opposes the motion of the magnet and the corresponding decrease in magnetic flux.

(ii) (a) Since $\phi = \text{LI}$

where,

I = strength of current through the coil at any time

ϕ = Amount of magnetic flux linked with all turns of the coil at that time and,

L = Constant of proportionality called coefficient of self induction.

(b) Induced emf,

$$e = \frac{-d\phi}{dt} = -\frac{d\,(\text{LI})}{dt}$$

i.e.,

$$e = -\text{L}\left(\frac{dI}{dt}\right)$$

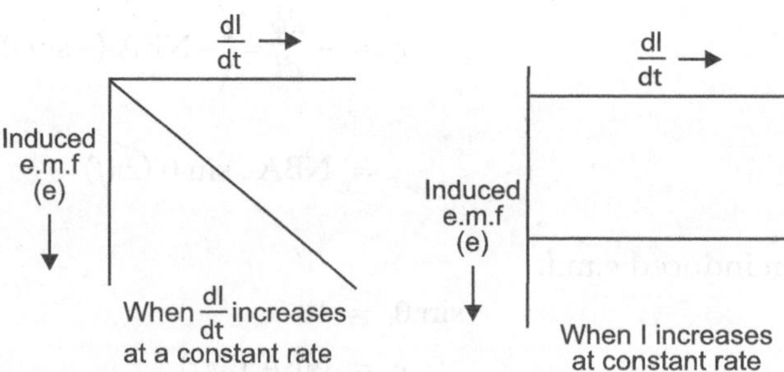

(c) Since magnetic potential energy is given by $\text{U} = \frac{1}{2}\text{LI}^2$

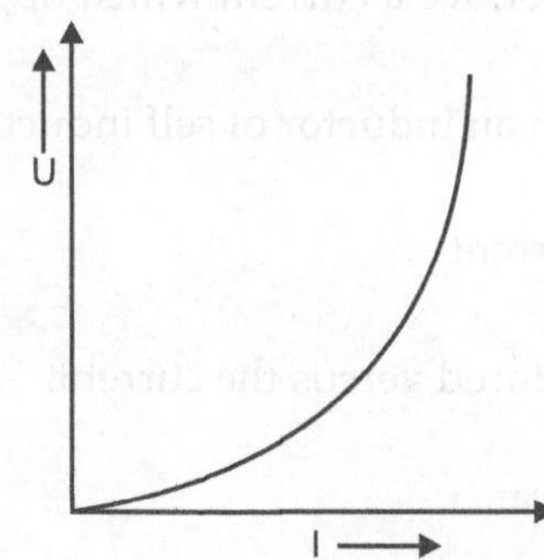

Chapter 7. Alternating Current

Q. 1. A device X is connected across an ac source of voltage $V = V_0 \sin \omega t$. The current through X is given as $I = I_0 \sin\left(\omega t + \frac{\pi}{2}\right)$.★

(i) Identify the device X and write the expression for its reactance.

(ii) Draw graphs showing variation of voltage and current with time over one cycle of ac, for X.

(iii) How does the reactance of the device X vary with freqency of the ac ? Show this variation graphically.

(iv) Draw the phasor diagram for the device X.

Ans.

$$V = V_0 \sin \omega t; I = I_0 \sin\left(\omega t + \frac{\pi}{2}\right).$$

(i) Since current leads the voltage by $\frac{\pi}{2}$ radians, X is a capacitor.

$$\text{Capacitive reactance, } X_C = \frac{1}{\omega C}$$

where ω = angular velocity, C = capacitance

(ii)

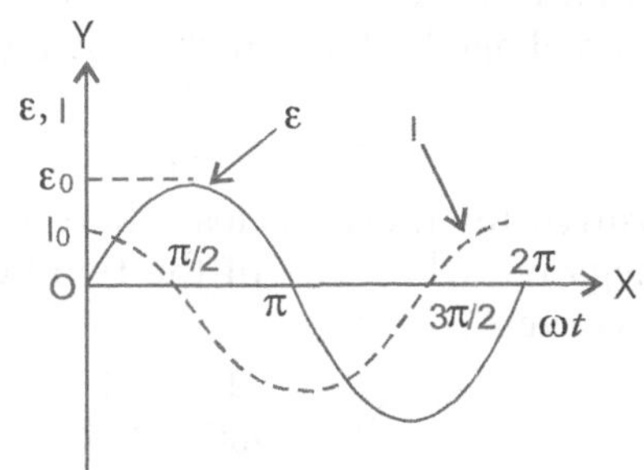

(iii) Capacitive reactance varies inversely with frequency as

$$X_C = \frac{1}{\omega C} = \frac{1}{2\pi f C}$$

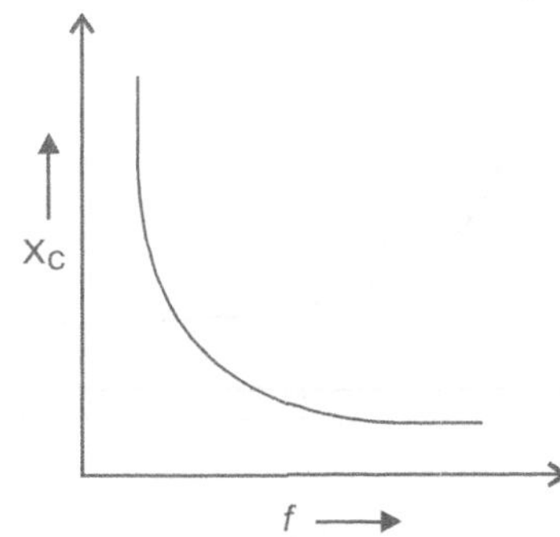

(iv)

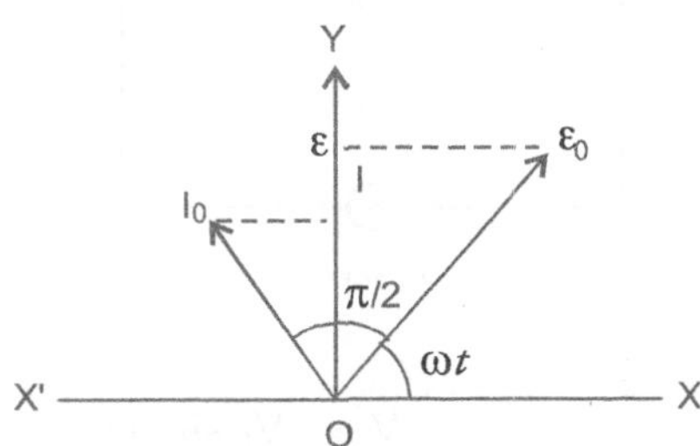

★ are board based questions from previous year

Q. 2. A device 'X' is connected to an ac source $V = V_0 \sin \omega t$. The variation of voltage, current and power in one cycle is shown in the following graph :★

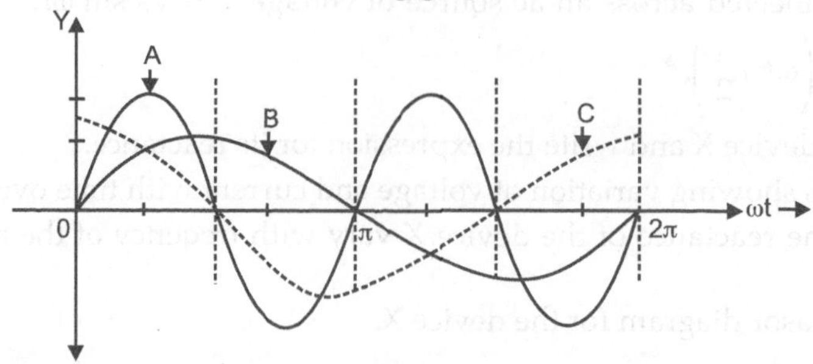

(i) Identify the device 'X'.

(ii) Which of the curves A, B and C represent the voltage, current and the power consumed in the circuit ? Justify your answer.

(iii) How does its impedance vary with frequency of the ac source ? Show graphically.

(iv) Obtain an expression for the current in the circuit and its phase relation with ac voltage.

Ans. (i) The device X is a capacitor.

(ii) The curves A,B and C represents power consumption, voltage and current respectively.

Since, it is given $V = V_0 \sin \omega t$; this sinusoidal variation is represented by the curve B. In case of capacitor, current (I) leads the voltage by 90° which is being represented by curve C. Now, we know power is given by

$$P = V \times I$$

So, the power would be positive for those cycles where both V and I are either positive or negative. Power would be negative when one of the two voltage or current, is negative. This illustration is followed by the curve A.

(iii) Capacitive reactance, $X_C = \dfrac{1}{\omega C} = \dfrac{1}{2\pi f C}$

This shows that capacitive reactance vary inversely with the frequency.

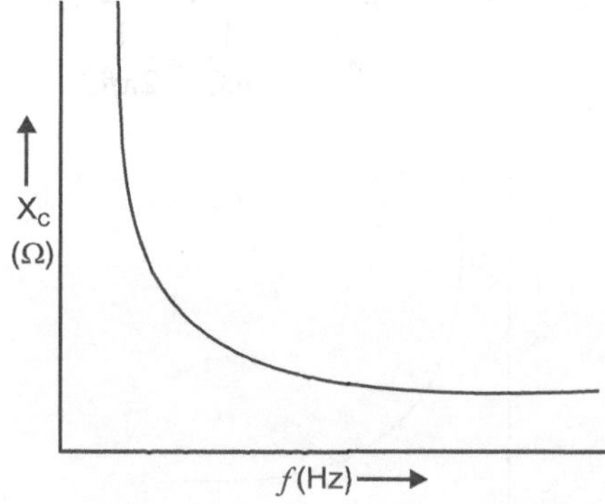

(iv)

C

$V_0 \sin \omega t$

In the given figure,

$$V = V_0 \sin \omega t$$

★ are board based questions from previous year

$$Q = CV = CV_0 \sin \omega t$$

$$I = \frac{dq}{dt} = \omega\, CV_0 \cos \omega t$$

$$= I_0 \sin (\omega t + \pi/2)$$

where $I_0 = \omega C V_0$

Current leads the voltage by a phase angle of 90°.

Q. 3. (i) Draw a labelled diagram of a step-down transformer. State the principle of its working.★

(ii) Express the turns ratio in terms of voltages.★

(iii) Find the ratio of primary and secondary currents in terms of turns ratio in an ideal transformer.★

(iv) Mention various energy losses in the device.

Ans. (i) **Principle :** A transformer works on the principle of mutual induction. Whenever the amount of magnetic flux, linked with a coil changes, an emf is induced in the neighbouring coil.

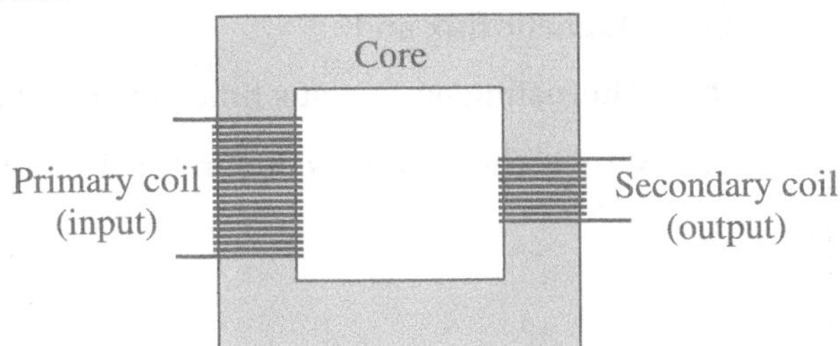

Working : When an alternating current source is connected to the ends of primary coil, the current changes continuously in the primary coil, due to which magnetic flux linked with the secondary coil changes continuously. Therefore, the alternating emf of same frequency is developed across the secondary terminals. According to Faraday's laws the e.m.f. induced in the primary coil,

$$E_P = -N_P \frac{\Delta\phi}{\Delta t} \quad \text{...(i)}$$

and emf induced in the secondary coil

$$E_S = -N_S \frac{\Delta\phi}{\Delta t} \quad \text{...(ii)}$$

From (i) and (ii)

$$\frac{E_S}{E_P} = \frac{N_S}{N_P} = K \quad \text{...(iii)}$$

For step-down transformer, $K < 1$.

$\therefore$ $E_S < E_P$

(ii) The induced emf in primary coil,

$$E_P = -N_P (d\phi/dt)$$

The induced emf in secondary coil,

$$E_S = -N_S (d\phi/dt)$$

$$E_S/E_P = N_S/N_P = K$$

Where K is the turns ratio or the transformation ratio.

(iii) If the transformer is ideal, then

Input electrical power = Output electrical power

$$E_P I_P = E_S I_S$$

$$E_S/E_P = I_P/I_S$$

$$\frac{I_P}{I_S} = \frac{E_S}{E_P} = \frac{N_S}{N_P} = K$$

(iv) **Energy losses in transformer :**

(a) **Copper loss** : Some energy is lost due to the heating of copper wires used in the primary and secondary windings. This power loss ($P = I^2R$) can be minimised by using thick copper wires of low resistance.

★ are board based questions from previous year

(b) **Eddy current loss** : The alternating magnetic flux induces eddy current in the iron core which leads to some energy loss in the form of heat. This loss can be reduced by using laminated iron core.

(c) **Hysteresis loss** : The alternating current carries the iron core through cycles of magnetisation and demagnetisation. Work done in each of these cycles and is lost as heat. This is called hysteresis loss and can be reduced by using core material having narrow hysteresis loop.

(c) **Flux leakage** : The magnetic flux produced by the primary may not fully pass through the secondary. Some of the flux may leak into air. This loss can be minimised by winding the primary and secondary coils over one another.

Q. 4. (i) Draw a schematic sketch of an ac generator describing its basic elements. State briefly its working principle. Show a plot of variation of★

(a) Magnetic flux and

(b) Alternating emf versus time generated by a loop of wire rotating in a magnetic field.

(ii) Why is choke coil needed in the use of fluorescent tubes with ac mains ?★

Ans. (i)

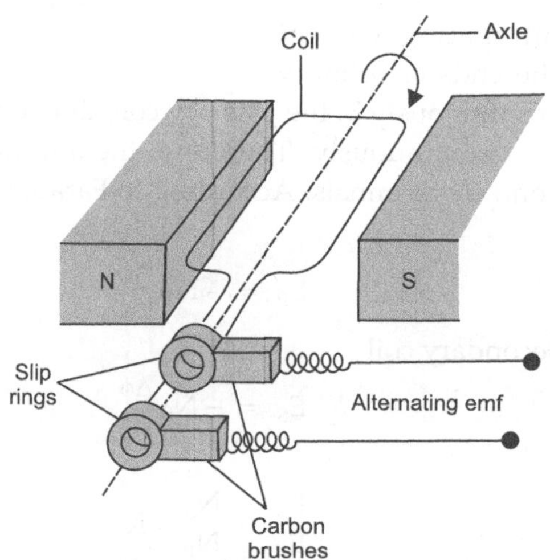

It works on the process of electromagnetic induction, *i.e.*, when a coil rotates continuously in a magnetic field, the effective area of the coil, linked (normally) with the magnetic field lines, changes continuously with time. This variation of magnetic flux with time results in the production of a (alternating) emf in the coil.

(a) **Magnetic flux versus time**

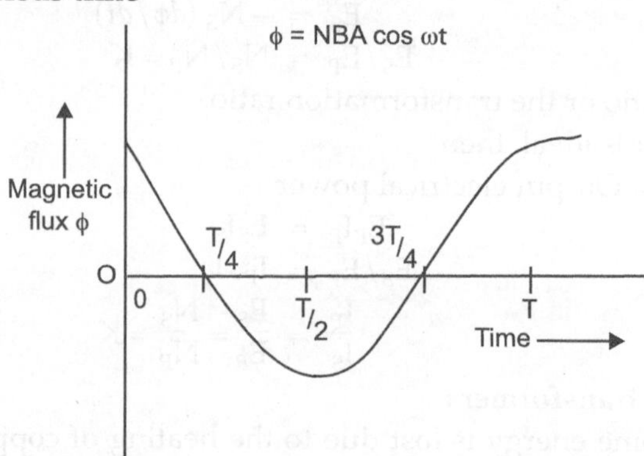

(b) **Alternating emf versus time**

$e = NAB\omega \sin \omega t = e_0 \sin \omega t$

★ are board based questions from previous year

The graph between alternating emf versus time is shown below :

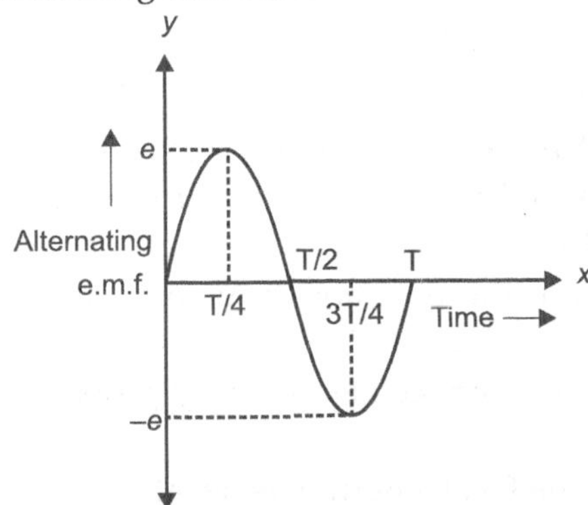

(ii) A choke coil is an electrical appliance used for controlling current in an a.c. circuit. Therefore, if we use a resistance R for the same purpose, a lot of energy would be wasted in the form of heat etc.

Chapter 8. Electromagnetic Waves

Q. 1. Explain the properties of different waves in electromagnetic spectrum.

Ans. **Electromagnetic spectrum :** Since, electromagnetic waves carry energy and momentum from source to a receiver. In early time, the only electromagnetic waves recognized were radio waves and visible light. It is now known that other forms of electromagnetic waves exist which are distinguished by their frequency and wavelength as given below :

(i) **Gamma Rays :** These are the waves emitted by radioactive nuclei and during certain nuclear reactions. Their wavelength varies as 10^{-14} m to 10^{-10} m. These waves are highly penetrating and produce serious change when absorbed by living tissues. Consequently those working near such dangerous radiations must be protected with heavily absorbing material such as thick layers of lead.

(ii) **X-rays :** These waves are commonly produced by the deceleration of high energy electron bombarding a metal target. Their wavelength varies as 10^{-13} m to 10^{-8} m. X-rays are used as a diagnostic tool in medicine and as a treatment for certain forms of cancer. X-rays are used in surgery, radiotherapy, engineering, detective departments, industries and in scientific research.

(iii) **Ultra-violet Rays :** These are the waves constituting 10% of sun light. Their wavelength varies as 3.8×10^{-7} m to 6×10^{-8} m. Ultra violet rays are used for checking the minerals samples by making use of its property of causing fluorescence. These rays are also used in the study of molecular structures. These are used in sterlizing surgical instruments.

(iv) **Visible Light :** These are the electro-magnetic waves constituting 50% of sun light. Their wavelength varies as 4×10^{-7} m to 7×10^{-7} m. These waves make the objects visible to use. These are used in many ways in our day-to-day life such as drying, decaying, making food for plants, keeping atmosphere warm etc.

(v) **Infra-red Rays :** These waves are also called heat waves. Their wavelength varies 1×10^{-3} m to 7×10^{-7} m. These waves produced by hot bodies and molecules, are readily absorbed by most materials. The infra-red energy absorbed by a substance appears as heat since the energy agitates the atoms of the body, increasing their vibrational or translational motion, which results in temperature rise. Infra-red radiation has many practical and scientific applications, including physical therapy, infra-red photography and vibrational spectroscopy.

(vi) **Micro Waves :** These are also called short wavelength radio waves. Their wavelength varies from 1 mm to 30 cm and are generated by electronic devices. Because of the short wavelength, they are well suited for the radar systems used in aircraft navigation and for studying the atomic and molecular properties of matter. Microwave ovens represent an interesting domestic application of these waves. In a recent proposal, it was suggested that solar energy could be harnessed by beaming microwaves down to earth from a solar collector in space.

(vii) **Long Waves :** These waves are produced by oscillating electronic circuits. Their wavelength is of the order of 1 m. These waves are used as carrier waves in radio broadcasting and T.V. transmission.

Chapter 9. Ray Optics and Optical Instruments

Q. 1. Use the mirror equations to deduce that :

(i) an object placed between f and $2f$ of a concave mirror produces a real image beyond $2f$.

(ii) a convex mirror always produces a virtual image independent of the location of the object.

(iii) the virtual image produced by a convex mirror is always diminished in size and is located between the focus and the pole.

(iv) an object placed between the pole and focus of a concave mirror produces a virtual and enlarged image.

[**Note :** This exercise helps you deduce algebraically properties of images that one obtains from explicit ray diagrams.] **[NCERT]**

Ans. (i) For a concave mirror, the focal length (f) is negative.

$\therefore$ $$f < 0$$

When the object is placed on the left side of the mirror, the object distance (u) is negative.

$\therefore$ $$u < 0$$

For image distance v, we can write the lens formula as :

$$\frac{1}{v}+\frac{1}{u}=\frac{1}{f}$$

$$\frac{1}{v}=\frac{1}{f}-\frac{1}{u} \quad \text{...(i)}$$

The object lies between f and $2f$.

$\therefore$ $$2f < u < f \quad (\because u \text{ and } f \text{ are negative})$$

$$\frac{1}{2f} > \frac{1}{u} > \frac{1}{f}$$

$$-\frac{1}{2f} < -\frac{1}{u} < -\frac{1}{f}$$

$$\frac{1}{f}-\frac{1}{2f} < \frac{1}{f}-\frac{1}{u} < 0 \quad \text{...(ii)}$$

From equation (i), we get

$$\frac{1}{2f} < \frac{1}{v} < 0$$

$\therefore \frac{1}{v}$ is negative, *i.e.*, v is negative.

$$\frac{1}{2f} < \frac{1}{v}$$

$$2f > v$$

$$-v > -2f$$

Therefore, the image lies beyond $2f$.

(ii) For a convex mirror, the focal length (f) is positive.

$\therefore$ $$f > 0$$

When the object is placed on the left side of the mirror, the object distance (u) is negative.

$\therefore$ $$u < 0$$

For image distance v, we have the mirror formula :

$$\frac{1}{v}+\frac{1}{u}=\frac{1}{f}$$

$$\frac{1}{v}=\frac{1}{f}-\frac{1}{u}$$

From equation (ii), we can conclude that :

$$\frac{1}{v} < 0$$

$$v > 0$$

Thus, the images is formed on the back side of the mirror.

Hence, a convex mirror always produces a virtual image, regardless of the object distance.

(iii) For a convex mirror, the focal length (f) is positive.

$\therefore$ $$f > 0$$

When the object is placed on the left side of mirror, the object distance (u) is negative,

$\therefore$ $$u < 0$$

For image distance v, we have the mirror formula :

$$\frac{1}{v}+\frac{1}{u} = \frac{1}{f}$$

$$\frac{1}{v} = \frac{1}{f}-\frac{1}{u}$$

But we have $u < 0$

$\therefore$ $$\frac{1}{v} > \frac{1}{f}$$

$$v < f$$

Hence, the image formed is diminished and is located between the focus (f) and the pole.

(iv) For a concave mirror, the focal length (f) is negative.

$\therefore$ $$f < 0$$

When the object is placed on the left side of the mirror, the object distance (u) is negative.

$\therefore$ $$u < 0$$

It is placed between the focus (f) and the pole.

$\therefore$ $$f > u > 0$$

$$\frac{1}{f} < \frac{1}{u} < 0$$

$$\frac{1}{f}-\frac{1}{u} < 0$$

For image distance v, we have the mirror formula :

$$\frac{1}{v}+\frac{1}{u} = \frac{1}{f}$$

$$\frac{1}{v} = \frac{1}{f}-\frac{1}{u}$$

$\therefore$ $$\frac{1}{v} < 0$$

$$v > 0$$

The image is formed on the right side of the mirror. Thus, it is a virtual image.

For $u < 0$ and $v > 0$, we can write.

$$\frac{1}{u} > \frac{1}{v}$$

$$v > u$$

Magnification, $$m = \frac{v}{u} > 1$$

Therefore, the formed image is enlarged.

Q. 2. Answer the following questions :

(i) You learnt that plane and convex mirrors produce virtual images of objects.

Can they produce real images under some circumstances ? Explain.

(ii) A virtual image, we always say, cannot be caught on a screen.

Yet when we 'see' a virtual image, we are obviously bringing it on to the 'screen'.

(*i.e.*, the retina) of our eye. Is there a contradiction ?

(iii) A diver under water, looks obliquely at a fisherman standing on the bank of a lake. Would the fisherman look taller or shorter to the diver than what he actually is ?

(iv) Does the apparent depth of a tank of water change if viewed obliquely ? If so, does the apparent depth increase or decrease ?

(v) The refractive index of diamond is much greater than that of ordinary glass. Is this fact of some use to a diamond cutter ? **[NCERT]**

Ans. (i) Yes

Plane and convex mirrors can produce real images also. If the object is virtual, *i.e.,* if the light rays converging at a point behind a plane mirror (or a convex mirror) are reflected to a point on a screen placed in front of the mirror, then a real image will be formed.

(ii) No

A virtual image is formed when light rays diverge. The convex lens of the eye causes these divergent rays to converge at the retina. In this case, the virtual image serves as an object for the lens to produce a real image.

(iii) The diver is in the water and the fisherman is on surface (*i.e.,* in air). Water is a denser medium than air. As diver is viewing the fisherman, this indicates that the light rays are travelling from a denser medium to a rarer medium. Hence, the refracted rays will move away from the normal resulting the fisherman will appear to be taller.

(iv) Yes; Decrease

The apparent depth of a tank of water changes when viewed obliquely because light bends on travelling from one medium to another. The apparent depth of the tank when viewed obliquely is less than the near-normal viewing.

(v) Yes

The refractive index of diamond (2.42) is higher than that of ordinary glass (1.5). The critical angle for diamond is less than that for glass. A diamond cutter uses a large angle of incidence to ensure that the light entering the diamond is totally reflected from its faces. This is the reason for the sparking effect of a diamond.

Chapter 10. Wave Optics

Q. 1. What is the effect on the interference fringes in a Young's double-slit experiment due to each of the following operations :

(i) the screen is moved away from the plane of the slits;

(ii) the (monochromatic) source is replaced by another (monochromatic) source of shorter wavelength;

(iii) the separation between the two slits is increased;

(iv) the source slit is moved closer to the double-slit plane;

(v) the width of the source slit is increased;

(vi) the monochromatic source is replaced by a source of white light ?

(In each operation, take all parameters, other than the one specified, to remain unchanged.)

[NCERT]

Ans. (i) Angular separation of the fringes remains constant ($= \lambda/d$). The actual separation of the fringes increases in proportion to the distance of the screen from the plane of the two slits.

(ii) The separation of the fringes (and also angular separation) decreases. See, however, the condition mentioned in (iv) below.

(iii) The separation of the fringes (and also angular separation) increases. See, however, the condition mentioned in (iv) below.

(iv) Let s be the size of the source and S its distance from the plane of the two slits. For interference fringes to be seen, the condition $s/S < \lambda/d$ should be satisfied : otherwise, interference patterns produced by different parts of the source overlap and no fringes are seen. Thus, as S decreases (*i.e.,* the source slit is brought closer), the interference pattern gets less and less sharp, and when the source is brought too close for this condition to be valid, the fringes disappear. Till this happens, the fringe separation remains fixed.

(v) Same as in (iv), As the source slit width increases, fringe pattern gets less and less sharp. When the source slit is so wide that the condition $s/S \leq \lambda/d$ is not satisfied, the interference pattern disappears.

(vi) The interference patterns due to different component colours of white light overlap (incoherently). The central bright fringes for different colours are at the same position. Therefore, the central fringe is white. For a point P for which $S_2P - S_1P = \lambda_b/2$, where λ_b ($\approx$ 4000 Å) represents the wavelength for the blue colour, the blue component will be absent and the fringe will appear red in colour. Slightly farther away where $S_2Q - S_1Q = \lambda_b = \lambda_r/2$ where λ_r ($\approx$ 8000 Å) is the wavelength for the red colour, the fringe will be predominantly blue.

Thus, the fringe closest on either side of the central white fringe is red and the farthest will appear blue. After a few fringes, no clear fringe pattern is seen.

Q. 2. (i) Define a wavefront. Using Huygen's principle, verify the laws of reflection at a plane surface.★

(ii) In a single slit diffraction experiment, the width of the slit is made double the original width. How does this affect the size and intensity of the central diffraction band ? Explain.★

(iii) When a tiny circular obstacle is placed in the path of light from a distant source, a bright spot is seen at the centre of the obstacle. Explain why ?★

Ans. (i) Wavefront is defined as the continuous locus of all the particles of a medium which are vibrating in the same phase.

Verification of laws of reflection using Huygen's principle : Let XY be a reflecting surface at which a wavefront is being incident obliquely. Let v be the speed of the wavefront and at time t = 0, the wavefront touches the surface XY at A. After time t, point B of wavefront reaches the point B′ of the surface.

According to Huygen's principle each point of wavefront acts as a source of secondary waves. When the point A of wavefront strikes the reflecting surface, then due to presence of reflecting surface, it cannot advance further; but the secondary wavelet originating from point A begins to spread in all directions in the first medium with speed v. As the wavefront AB advances further, its points A_1, A_2, A_3 .... etc. strike the reflecting surface successively and send spherical secondary wavelets in the first medium.

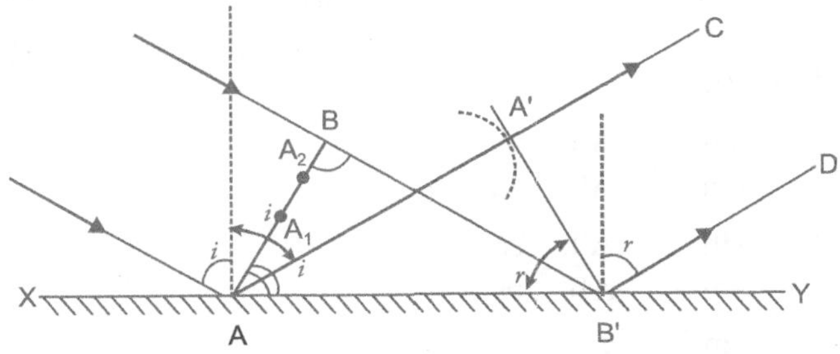

First of all the secondary wavelet starts from point A and traverses distance AA′ (= vt) in first medium in time t. In the same time t, the point B of wavefront, after travelling a distance BB′, reaches point B′ (of the surface), from where the secondary wavelet now starts. Now taking A as centre we draw a spherical arc of radius AA′(= vt) and draw tangent A′B′ on this arc from point B′. As the incident wavefront AB advances, the secondary wavelets starting from points between A and B′, one after the other and will touch A′B′ simultaneously. According to Huygen's principle wavefront A′B′ represents the new position of AB, *i.e.*, A′B′ is the reflected wavefront corresponding to incident wavefront AB.

Now in right-angled triangles ABB′ and AA′B′

$\angle ABB' = \angle AA'B'$ (both are equal to 90°)

side BB′ = side AA′ (both are equal to vt)

and side AB′ is common

i.e., both triangles are congruent.

★ are board based questions from previous year

$\therefore \qquad \angle BAB' = \angle AB'A'$

i.e, incident wavefront AB and reflected wavefront A'B' make equal angles with the reflecting surface XY. As the rays are always normal to the wavefront, therefore the incident and the reflected rays make equal angles with the normal drawn on the surface XY, *i.e.,*

$$\text{angle of incidence } i = \text{angle of reflection } r$$

This is the second law of reflection.

Since AB, A'B' and XY are all in the plane of paper, therefore the perpendiculars dropped on them will also be in the same plane. Therefore we conclude that the incident ray, reflected ray and the normal at the point of incidence, all lie in the same plane. This is the first law of reflection. Thus, Huygen's principle explains both the laws of reflection.

(ii) Width of central diffraction band $= 2D.\dfrac{\lambda}{a}$

where a is the width of the slit.

So, on doubling the width of the slit, the size of the central diffraction band reduces to half value. But, the light amplitude becomes double, which increases the intensity four fold.

(iii) If a tiny circular obstacle is kept in the path of light, a bright spot is seen at the centre of the obstacle. This is because the waves which get diffracted from the edge of the circular obstacle interfere constuctively at the centre of the shadow producing a bright spot.

Q. 3. (i) Define wavefront. Use Huygens' principle to verify the laws of refraction.★

(ii) How is linearly polarised light obtained by the process of scattering of light ? Find the Brewster angle for air-glass interface, when the refractive index of glass = 1.5.★

Ans. (i) **Wavefront :** A wavefront is a continuous locus of all the particles of a medium which are vibrating in the same phase.

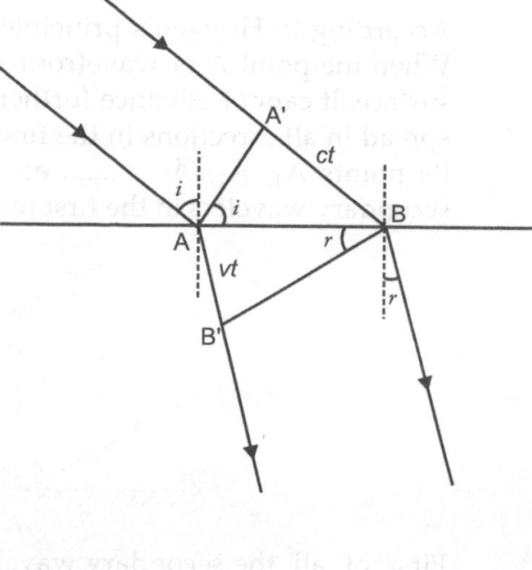

In ΔAA'B, $\qquad \sin i = \dfrac{A'B}{AB} \qquad$ ...(i)

In ΔAB'B, $\qquad \sin r = \dfrac{AB'}{AB} \qquad$ ...(ii)

Dividing equation (i) by (ii),

$$\frac{\sin i}{\sin r} = \frac{A'B/AB}{AB'/AB}$$

$$\frac{\sin i}{\sin r} = \frac{A'B}{AB'} = \frac{ct}{vt}$$

$$\frac{\sin i}{\sin r} = \frac{c}{v} = \mu$$

This is the law of refraction.

(ii) When a beam of white light is passed through a medium containing particles whose size is of the order of wavelength of light, then the beam gets scattered. When the scattered light is viewed through an analyser in a direction at the right angle to the direction of incidence, it is found to be plane polarised. This is called polarisation by scattering.

We know that, $\qquad \tan i_p = \mu$

$\Rightarrow \qquad \tan i_p = 1.5$

$\Rightarrow \qquad i_p = \tan^{-1}(1.5) = 56.31°$

Q. 4. (i) Consider two coherent sources S_1 and S_2 producing monochromatic waves to produce interference pattern. Let the displacement of the wave produced by S_1 be given by $Y_1 = a \cos \omega t$ and the displacement by S_2 be $Y_2 = a \cos (\omega t + \phi)$. Find out the expression for the amplitude of the

★ are board based questions from previous year

resultant displacement at a point and show that the intensity at that point will be $I = 4a^2 \cos^2 \phi/2$. Hence establish the conditions for constructive and destructive interference.★

(ii) What is the effect on the interference fringes in Young's double slit experiment when ★

(a) the width of the source slit is increased;

(b) the monochromatic source is replaced by a source of white light.

Ans. (i) Let S be a narrow slit illuminated by a monochromatic source of light and S_1 and S_2 two similar parallel slits very close together and equidistant from S.

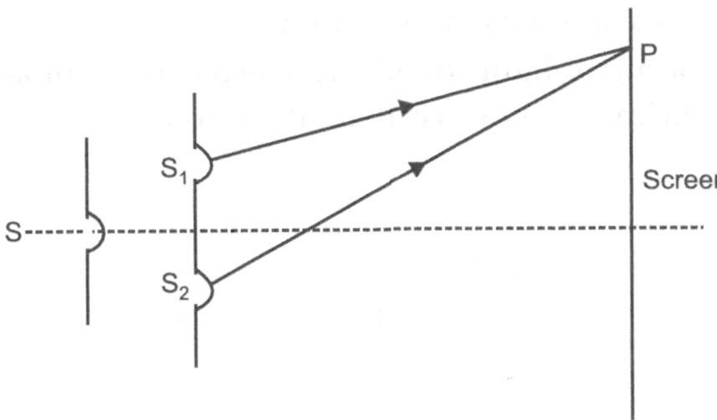

Displacement of the wave produced by S_1 is given by

$$y_1 = a \cos \omega t$$

and the displacement of the wave produced by S_2 is given by

$$y_2 = a \cos (\omega t + \phi)$$

The resultant displacement is given by

$$y = y_1 + y_2$$

$$y = a \cos \omega t + a \cos (\omega t + \phi)$$

$$y = 2a \cos \frac{\phi}{2} \cos (\omega t + \frac{\phi}{2})$$

The amplitude of the resultant displacement is $2a \cos \frac{\phi}{2}$.

The intensity of light is directly proportional to the square of amplitude of the wave. The resultant intensity is given by

$$I = 4a^2 \cos^2 \frac{\phi}{2}$$

For constructive interference the intensity of light at point P is maximum, if

$$\cos \phi = \max = +1 : \phi = 0, 2\pi, 4\pi......$$

i.e.,

$$\phi = 2n\pi \text{ where } n = 0, 1, 2, ...$$

$$\text{Path difference} = \frac{\lambda}{2\pi} \times 2n\pi$$

$$= n\lambda$$

For destructive interference the intensity of light at point P is minimum, if

$$\cos \phi = -1$$

$\therefore$

$$\phi = \pi, 3\pi, 5\pi, ...$$

or

$$\phi = (2n - 1)\pi \text{ where } n = 1, 2, 3, ...$$

The corresponding path difference between the two waves

$$x = \frac{\lambda}{2\pi}\phi = \frac{\lambda}{2\pi}(2n-1)\pi = (2n-1)\frac{\lambda}{2}$$

(ii) (a) As the width of the slits is increased, the fringe width decreases. It is because,

$$\beta \propto \frac{1}{d}$$

★ are board based questions from previous year

(b) The different colours of white light will produce different interference patterns but the central bright fringes due to all colours are at the same positions. Therefore, the central bright fringe is white in colour. Since the wavelength of the blue light is smallest, the fringe closed on the either side of the central white fringe is blue and farthest is red. Beyond a few fringes, no clear fringe pattern is visible.

Q. 5. Describe briefly how a diffraction pattern is obtained on a screen due to a single narrow slit illuminated by a monochromatic source of light. Hence obtain the conditions for the angular width of secondary maxima and secondary minima.

Ans. The phenomenon of bending of light round the sharp corners of an obstacle and spreading into the regions of the geometrical shadow is called diffraction.

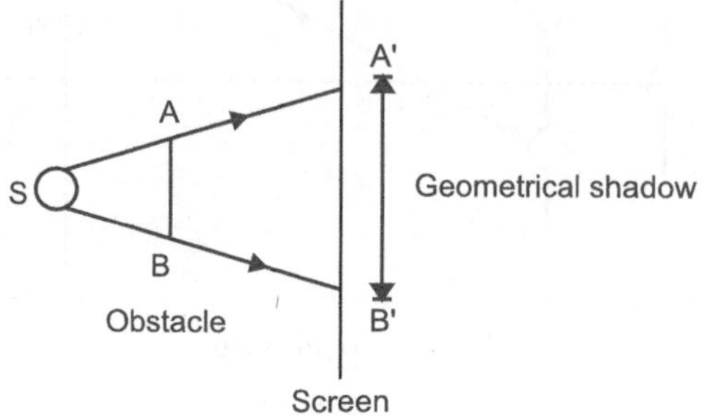

Expression for Fringes width :

Consider a parallel light beam from a lens is incident on slit AB. As diffraction happens, the pattern is focussed on screen XY with the help of lens L_2. We will get a diffraction pattern that is a central maximum at the centre O flanked by a number of dark and bright fringes known as secondary maxima and minima.

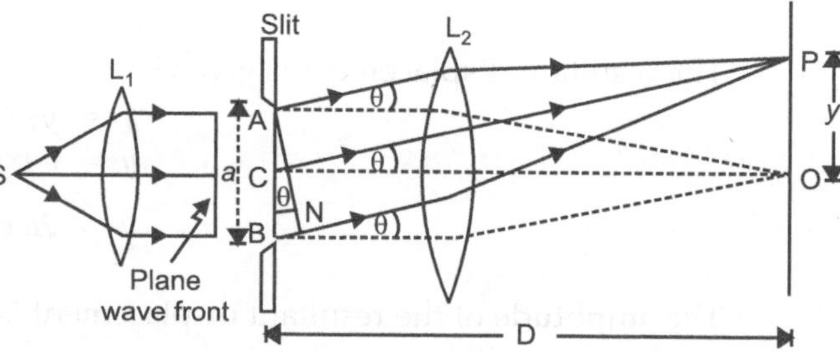

Central Maximum : Each point on the plane wave front AB sends secondary wavelets in all directions. The waves from points equidistant from the centre C kept on the upper and lower half reach point O with zero path difference and so, reinforce each other, making maximum intensity at point O.

Positions and Widths of Secondary Maxima and Minima

Consider a point P on screen at which wavelets moving in a direction making angle θ with CO are brought to focus by the lens. The wavelets from points A and B will have a path difference similar to BN :

From the right-angled ΔANB, we have :

$$BN = AB \sin\theta$$

$$BN = a \sin\theta \qquad \ldots(i)$$

Suppose $BN = \lambda$ and $\theta = \theta_1$

Then, the above equation becomes

$$\lambda = a \sin\theta_1$$

$$\sin\theta_1 = \frac{\lambda}{a} \qquad \ldots(ii)$$

Such a point on the screen will be the position of first secondary minimum,
If $BN = 2\lambda$ and $\theta = \theta_2$, then

$$\sin\theta_2 = \frac{2\lambda}{a} \qquad \ldots(iii)$$

Such a point on the screen will be the position of second secondary minimum.
In general, for n^{th} minimum at point P,

$$\sin \theta_n = \frac{n\lambda}{a} \quad \text{...(iv)}$$

If y_n is the distance of the nth minimum from the centre of the screen, fom right-angled ΔCOP, we have :

$$\tan \theta_n = \frac{OP}{CO}$$

$$\tan \theta_n = \frac{y_n}{D} \quad \text{...(v)}$$

In case θ_n is small, $\tan \theta_n \approx \sin \theta_n$

∴ Equations (iv) and (v) give

$$\frac{y_n}{D} = \frac{n\lambda}{a}$$

$$y_n = \frac{nD\lambda}{a}$$

Width of the secondary maximum,

$$\beta = y_n - y_{n-1} = \frac{nD\lambda}{a} - \frac{(n-1)D\lambda}{a}$$

$$\beta = \frac{D\lambda}{a} \quad \text{...(vi)}$$

∴ β is independent of n, all the secondary maxima are of the same width β.

If BN $= \frac{3\lambda}{2}$ and $\theta = \theta'$, from equation (i), we have :

$$\frac{3\lambda}{2} = a \sin \theta_1'$$

$$\sin \theta_1' = \frac{3\lambda}{2}$$

In general, for the nth maximum at point P,

$$\sin \theta_n' = \frac{(2n+1)\lambda}{2a} \quad \text{...(vii)}$$

If y_n' is the distance of nth maximum from the centre of the screen, then the angular position of the maximum is given by

$$\tan \theta_n' = \frac{y_n'}{D} \quad \text{...(viii)}$$

In case θ_n' is small,

$\sin \theta_n' \approx \tan \theta_n'$

$$\therefore \quad y_n' = \frac{(2n+1)\, D\lambda}{2a} \quad \text{...(ix)}$$

Width of the secondary minimum,

$$\beta' = y_n' - y'_{n-1} = \frac{(2n+1)\, D\lambda}{2a} - \frac{(2n-1)D\lambda}{2a}$$

$$\beta' = \frac{D\lambda}{a} \quad \text{...(x)}$$

Since β' is independent of n, all the secondary minima are of the same width β'.

Q. 6. (i) How does one demonstrate, using a suitable diagram, that unpolarised light when passed through a polaroid gets polarised ?★

(ii) A beam of unpolarised light is incident on a glass-air interface. Show, using a suitable ray diagram, that light reflected from the interface is totally polarised, when $\mu = \tan i_B$, where μ is the refractive index of glass with respect to air and i_B is the Brewster's angle.★

★ are board based questions from previous year

Ans. (i) The phenomenon of restricting the vibration of light (electric vector) in a particular direction perpendicular to the direction of the wave propagation is called polarisation of light.

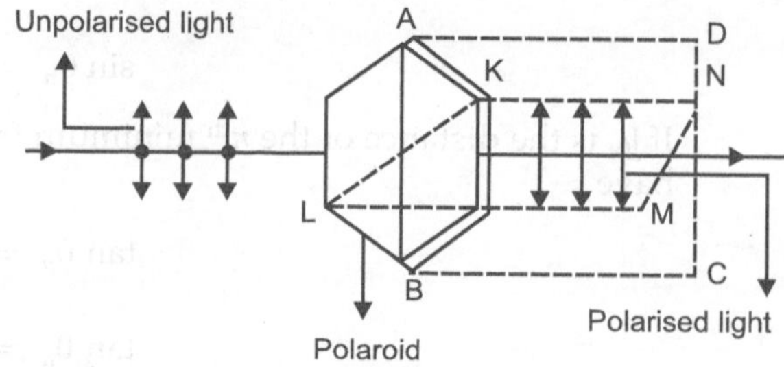

When unpolarised light is passed through a polaroid, only those vibrations of light pass through the crystal, which are parallel to the axis of the crystal (AB). All other vibrations are absorbed and that is why intensity of the emerging light is reduced.

The plane ABCD in which the vibrations of the polarised light are confined is called the plane of vibration. The plane KLMN that is perpendicular to the plane of vibration is defined as the plane of polarisation.

(ii) When unpolarised light is incident on the glass-air interference at Brewster angle i_B, then reflected light is totally polarised. This is called Brewster's Law.

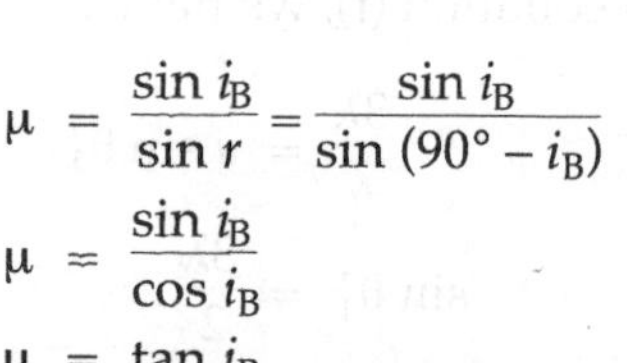

When light is incident at Brewster angle, the reflected component OB and the refracted component OC are mutually perpendicular to each other.

From the figure,

We have :

$$\angle BOY + \angle YOC = 90°$$
$$(90° - i_B) + (90° - r) = 90°$$

Where, r is angle of refraction

According to the Snell' law :

$$\mu = \frac{\sin i_B}{\sin r} = \frac{\sin i_B}{\sin (90° - i_B)}$$
$$\mu = \frac{\sin i_B}{\cos i_B}$$
$$\mu = \tan i_B$$

Q. 7. (i) Distinguish between unpolarised light and linearly polarised light. How does one get linearly polarised light with the help of a polaroid ?★

(ii) A narrow beam of unpolarised light of intensity I_0 is incident on a polaroid P_1. The light transmitted by it is then incident on a second polaroid P_2 with its pass axis making angle of 60° relative to the pass axis of P_1. Find the intenstity of the light transmitted by P_2.★

Ans. (i) **Unpolarised light :** If the electric component of light vibrates in all the direction in the plane perpendicular to the direction of propagation, the light is known as unpolarised light.

Linearly polarised light : If the vibration of electric components of light are restricted in one direction only in the plane perpendicular to the propagation of light then light is known as linearly polarised light.

A polaroid consists of long chain molecules aligned in a particular direction. The electric vectors along the direction of the aligned molecules get absorbed. So, when an unpolarised light falls on a polaroid, it lets only those of its electric vectors that are oscillating along a direction perpendicular to its aligned molecules to pass through it. The incident light thus gets linearly polarised.

(ii)

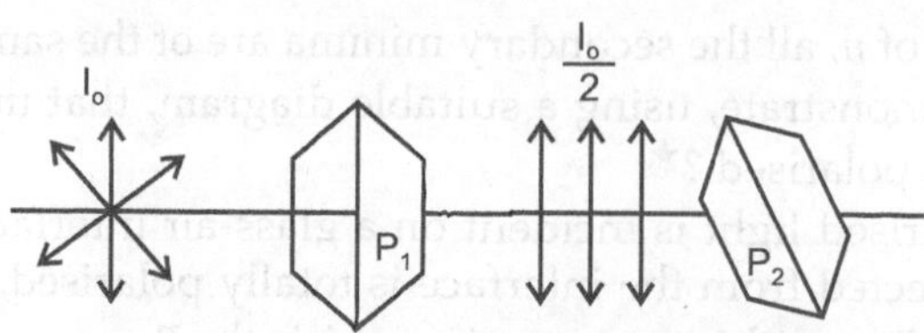

★ are board based questions from previous year

Intensity of light after passing through the polaroid P_1 $(I') = \frac{I_0}{2}$

Intensity of light after passing through the polaroid P_2,

$$I = I' \cos^2 60°$$

$$I = \frac{I_0}{2}\left(\frac{1}{2}\right)^2$$

$$I = \frac{I_0}{2} \times \frac{1}{4}$$

$$I = \frac{I_0}{8}$$

Q. 8. (i) In Young's double slit experiment, deduce the condition for (a) constructive, and (b) destructive interference at a point on the screen. Draw a graph showing variation of intensity in the interference pattern against position 'x' on the screen.★

(ii) Compare the interference pattern observed in Young's double slit experiment with single slit diffraction pattern, pointing out three distinguishing features.★

Ans. (i) Let the two waves arising from the slits A and B have the amplitudes a and b and the phase difference ϕ. Such that $y_1 = a \sin \omega t$ and $y_2 = b \sin (\omega t + \phi)$.

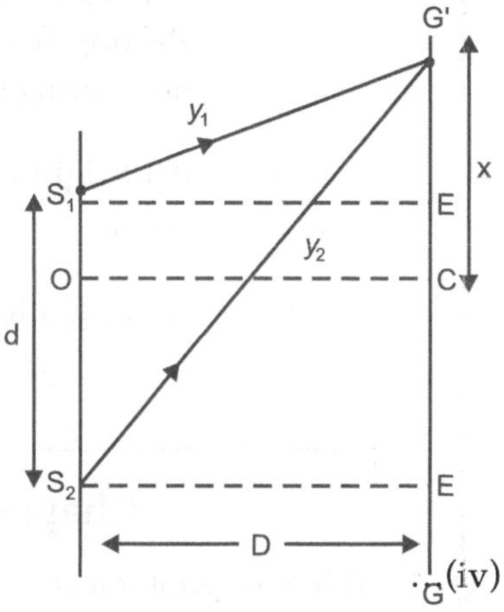

The resultant displacement is given as :

$$y = y_1 + y_2$$

$$y = a \sin \omega t + b \sin (\omega t + \phi)$$

$$y = a \sin \omega t + b \sin \omega t \cos \phi + b \cos \omega t \sin \phi$$

$$y = (a + b \cos \phi) \sin \omega t + b \sin \phi \cos \omega t \quad ...(i)$$

Let $a + b \cos \phi = A \cos \delta \quad ...(ii)$

and $b \sin \phi = A \sin \delta \quad ...(iii)$

Hence, $y = A \sin \omega t \cos \delta + A \cos \omega t \sin \delta$

$$y = A \sin (\omega t + \delta) \quad ...(iv)$$

Where the amplitude A of the resultant were can be given as :

$$A = \sqrt{a^2 + b^2 + 2ab \cos \phi} \quad ...(v)$$

and $\tan \delta = \frac{b \sin \phi}{a + b \cos \phi} \quad ...(vi)$

(a) **Constructive interference :** Intensity $I \propto A^2$ and for A to be maximum

$$\cos \phi = 1$$

or $\cos \phi = \cos 2n\pi, n = 0, 1, 2, 3, ...$

$$\phi = 2n\pi \quad ...(i)$$

and path difference

$$\Delta x = n\lambda \quad ...(ii)$$

$$A_{max} = a + b$$

$$I \rightarrow I_{max} = k(a+b)^2$$

(b) **Destructive Interference :** For I to be minimum

$$\cos \phi = -1$$

Phase difference :

$$\Delta\phi = (2n+1)\pi$$

★ are board based questions from previous year

and path difference :

$$\Delta x = (2n+1)\frac{\lambda}{2}$$

$$A_{min} = a - b$$

$$I \rightarrow I_{min} = k(a-b)^2$$

Graph showing interference pattern against position 'x' on the screen.

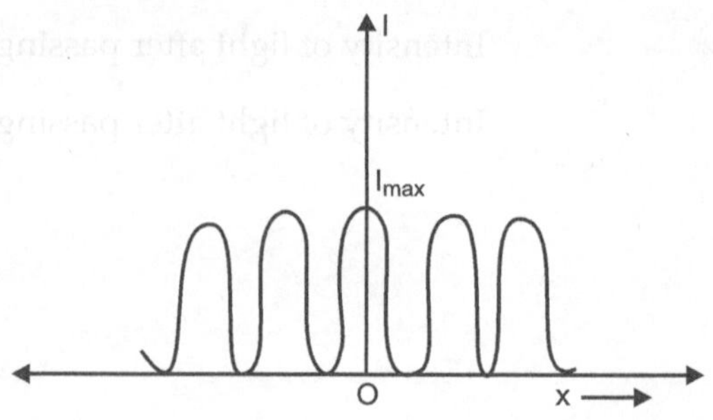

(ii) Comparison of interference pattern observed in Young's double slit and the single slit difraction :

	Interference	Diffraction
(i)	Interference is the result of superposition of secondary waves starting from two different wave fronts originating from two coherent sources.	Diffraction is the result of superposition of secondary waves starting from different part of same wavefront.
(ii)	All bright and dark fringe are of equal width.	The width of central bright fringe is twice the width of any secondary maximum.
(iii)	All bright fringes are of same intensity	Intensity of bright fringes decreases as we move away from central bright fringes on either side.

Chapter 11. Dual Nature of Radiation and Matter

Q. 1. What is photoelectric effect ? Describe an experimental set up to demonstrate the effect of intensity and frequeny of incident radiation.

Ans. The phenonomenon of ejection of eletrons from a metal surface, when light of sufficiently high frequency falls upon it, is known as 'photoelectric effect'.

The experimental set up to study the phenomenon is shown in the figure. A and B are two electrodes placed in an evacuated quartz bulb. They are connected to a voltmeter, with zero centre so that we can known by how much volt, B is positive or negative with respect to A. To provide the necessary voltage, we have a high tension battery B connected across a potential divider arrangement MN, with a key K. The centre terminal of the potential divider is connected to the plate A while sliding contact L causes varying potentials to be applied to the plate B, the current is registered by the micro-ammeter μA connected in series as shown. We make the following investigations with this experimental setup.

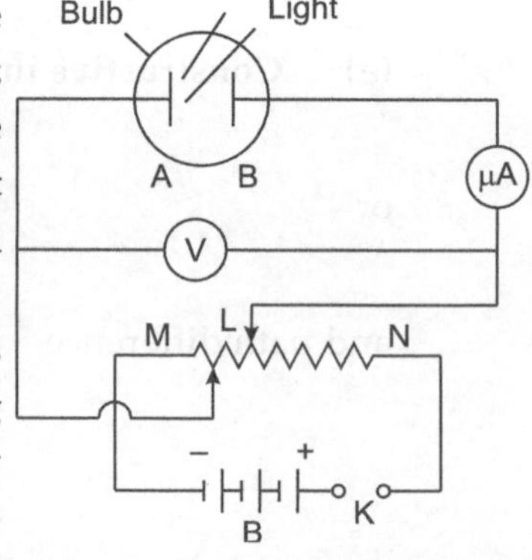

(i) **Effect of intensity of the incident radiation :** B is kept at a suitable high positive potential and A is illuminated with monochromatic radiation. The radiation causes photoelectrons to be emitted. The photoelectrons rush to the collector plate B and the photoelectric current is noted from the micro-ammeter. The intensity of the radiation is varied and the photoelectric current is carefully noted. A graph plotted between intensity I and the photoelectric current shows a straight line. This proves that the photoelectric current varies directly as the intensity of the incident radiation.

(ii) **Effect of frequency** : A particular emitter metal A was chosen (Li, Na, K etc.) and it was exposed to lights of various frequencies. The photoelectric current was observed and the negative stopping potential needed to reduce this current to zero was noted. A graph was plotted, and it came out to be a straight line. A different graph is found for a different emitting metal, but the graphs are all parallel. In the figure (ii), only one graph is shown. At ν_0 the stopping potential is zero. This means that at this frequency, the incident radiation is not able to eject any electron. So, no voltage is needed to stop it. This frequency is called the threshold frequency. It depends upon only the nature of the emitting material. If the frequency of the incident radiation is even slightly less than v_0, no emission of electrons is possible, even though, we may illuminate the surface with as much intense radiation and for as long a time as we choose. Thus, the threshold frequency for a given surface may be defined as the lowest.

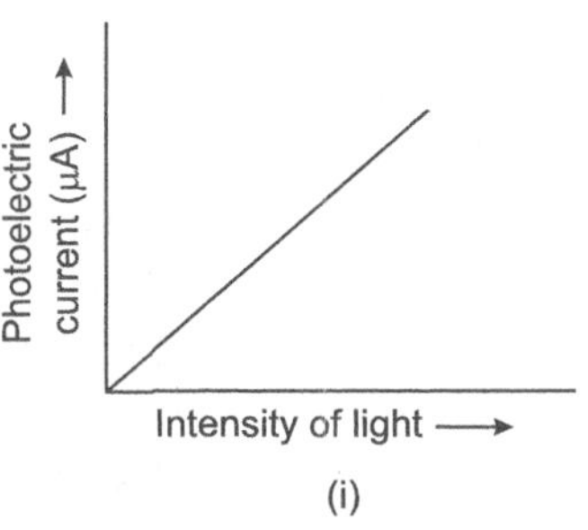

(i)

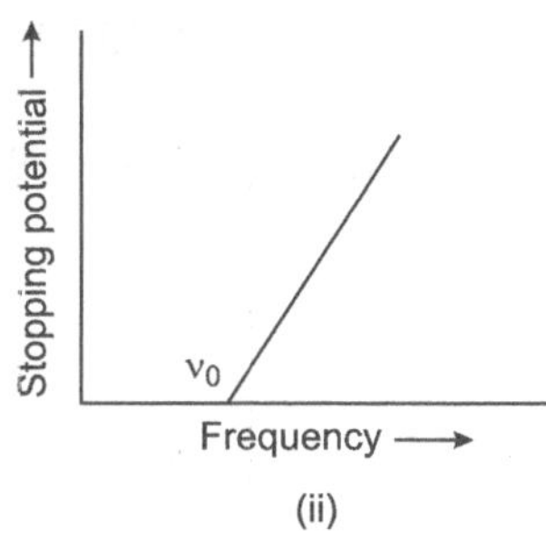

(ii)

Q. 2. Describe Davisson and Germer's experiment to demonstrate the wave nature of electrons. Draw a labelled diagram of apparatus used.

Ans. **Davisson and Germer experiment** : In 1927, Davisson and Germer performed a diffraction experiment with electron beam in analogy with X-ray diffraction to observe the wave nature of matter.

Apparatus : It consists of three parts :

(i) **Electron gun** : It gives a fine beam of electrons. de-Broglie used electron beam of energy 54 eV. de-Broglie wavelength associated with this beam

$$\lambda = \frac{h}{\sqrt{2mE_K}}$$

Here, m = Mass of electron = 9.1×10^{-31} kg

E_K = Kinetic energy of electron = 54 eV

$= 54 \times 1.6 \times 10^{-19}$ joule

$= 86.4 \times 10^{-19}$ joule

$$\therefore \quad \lambda = \frac{6.6 \times 10^{-34}}{\sqrt{2 \times 9.1 \times 10^{-31} \times 86.4 \times 10^{-19}}}$$

$$= 1.66 \times 10^{-10} \text{ m}$$

$$= 1.66 \text{ Å}$$

(ii) **Nickel crystal** : The electron beam was directed on nickel crystal against its (iii) face. The smallest separation between nickel atoms is 0.914 Å. Nickel crystal behaves as diffraction grating.

(iii) **Electron Detector** : It measures the intensity of electron beam diffracted from nickel crystal. It may be an ionisation chamber fitted with a sensitive galvanometer. The energy of electron beam, the angle of incidence of beam on nickel crystal and the position of detector can all be varied.

Method : The crystal is rotated in small steps to change the angle (α say) between incidence and scattered directions and the corresponding intensity (I) of scattered beam is measured. The variation of the intensity (I) of the scattered electrons with the angle of scattering α is obtained for different accelerating voltages.

The experiment was performed by varying the accelerating voltage from 44 V to 68 V. It was noticed that a strong peak appeared in the intensity (I) of the scattered electron for an accelerating voltage of 54 V at a scattering angle $\alpha = 50°$.

∴ From Bragg's law

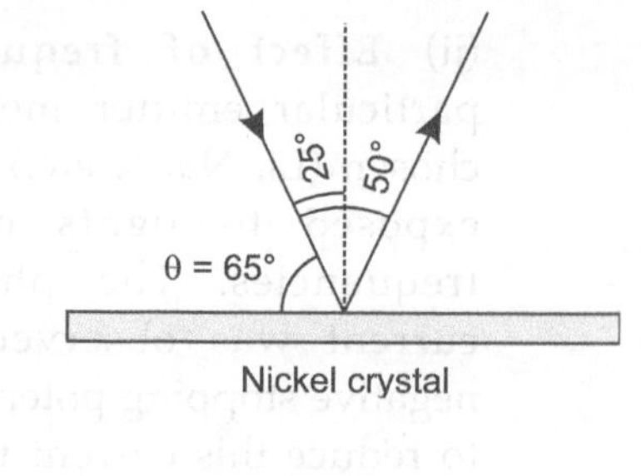

$$2d \sin \theta = n\lambda$$

Here, $n = 1$, $d = 0.914$ Å, $\theta = 65°$

$$\lambda = \frac{2d \sin \theta}{n}$$

$$= \frac{2 \times (0.914 \text{ Å}) \sin 65°}{1}$$

$$= 2 \times 0.914 \times 0.9063 \text{ Å}$$

$$= 1.65 \text{ Å}$$

The measured wavelength is in close agreement with the estimated de-Broglie wavelength. Thus, the wave nature of electron is verified. Later on G.P. Thomson demonstrated the wave nature of fast electrons. Due to their work Davisson and G.P. Thomson were awarded Nobel prize in 1937. Later on experiments showed that not only electrons but all material particles in motion (*e.g.*, neutrons, α-particles, protons etc.) show wave nature.

Chapter 12. Atoms

Q. 1. Give an account of Rutherford' experiment of α-particle scattering. What result did it yield regarding structure of an atom ?

Ans. In Rutherford's experiment, α-particle emitted from radioactive polonium were made to incident on a thin film of gold foil of thickness of 10^{-7} m which consisted of 400 layers of gold atoms. Rutherford placed a zinc-sulphide screen behind the foils as a counter of α-particles. When an α-particle strikes the screen in gives a flash of light, which is observed with a microscope. The screen can be rotated, so that α-particles can be detected in any direction.

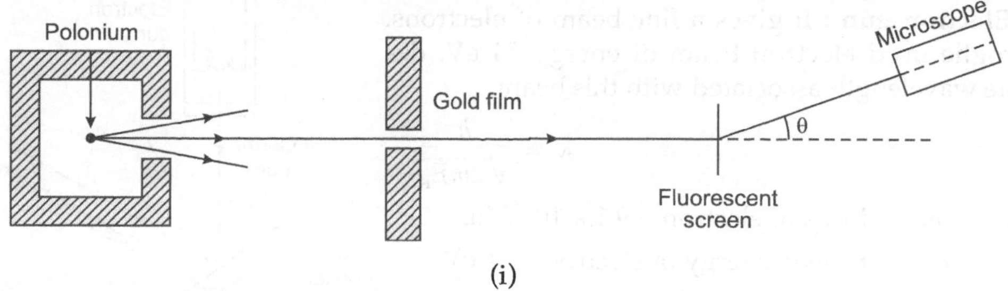

(i)

Results of Rutherford's Experiment :

(i) Most of the α-particles passed through the gold foil undeflected [fig. (ii)]. This shows that the atom is hollow.

(ii) Some α-particles were deviated through small angles but a few of them were deflected through a very large angles greater than 90°, while a still smaller number of α-particles were turned back.

These results led Rutherford to conclude that whole of the positive charge of the atom and almost whole of the mass of the atom is tightly concentrated in a very small central core, called the nucleus.

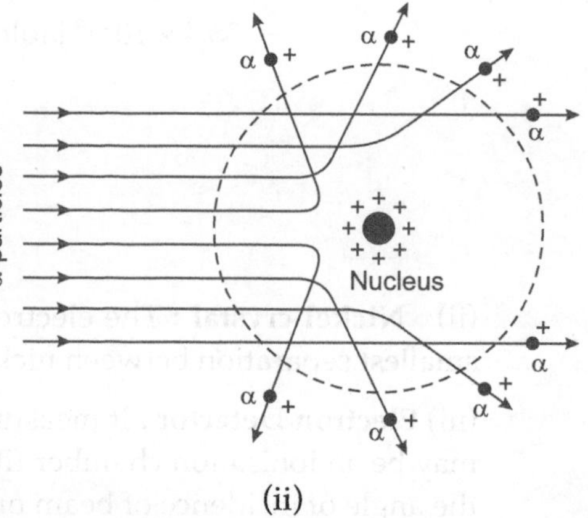

(ii)

Chapter 13. Nuclei

Q. 1. Explain the phenomenon of radioactivity stating the laws related to it.

Ans. **Radioactivity :** It is spontaneous phenomenon of disintegration of nuclei of unstable atom. Radioactivity is independent of phyiscal and chemical changes of the material.

It obeys the law of probability. It is uncertain that when particular atom will decay.

The rate of disintegration is directly proportional to the number of atoms present in the element.

$$\frac{dN}{dt} \propto -N \text{ or } \frac{dN}{dt} = -\lambda N$$

The number of atoms at any time t of the elements having initial number of atoms N_0 is given as $N = N_0 e^{-\lambda t}$ where λ is decay constant.

The S.I. unit of radioactivity is Becquerel and practical units are Curie and Rutherford.

The number of atoms decaying per second is called activity of the material.

The rate of disintegration per second is called decay constant (λ)

Decay constant is the negative slope of $N - t$ curve.

The decay constant for stable elements is zero.

The time in which a radioactive substance reduces to half of its original quantity is called half life period $T = \frac{0.6932}{\lambda}$.

Mean life $(t) = 1.44$ T

The number of half lines for a decay series (n) is given as

$$\frac{N}{N_0} = \left(\frac{1}{2}\right)^n$$

The % decreasing in activity $= \left(\frac{A_0 - A}{A_0}\right) 100$

Number of atoms left after n half lives

$$N = \frac{N_0}{2^n}$$

The % of radioactive material at time t.

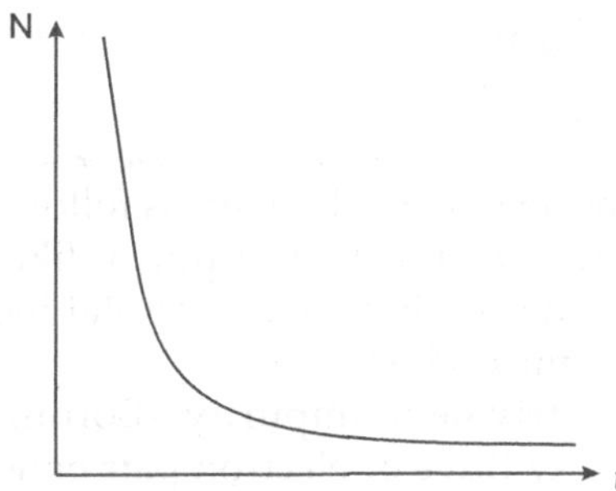

$$\left(\frac{N_0 - N}{N_0}\right)^{100} \cdot \left(\frac{1}{2}\right)^{t/1} 100$$

% of radioactive elements left after n half lives

$$= \frac{N}{N_0}(100) = \frac{1}{2^n} 100$$

The time of decay $t = n$T.

If mean life for α and β decay are T_α and T_β respectively then, the life of radioactive substance (T) is given as

$$\frac{T_\alpha - T_\beta}{T_\alpha + T_\beta}$$

Radioactive element emits α, β and γ radiations.

When one α is emitted out the atomic number and atomic mass number decrease 2 units and 4 units respectively.

When one β is emitted out the atomic number is increase by one unit and no change in atomic mass number.

$$^{A}_{Z}X \rightarrow {}^{A-4}_{Z-2}Y \rightarrow {}^{A+4}_{Z-1}Y$$

The number of α and β particles emitted can be given as

$$N(\alpha) = \frac{\Delta A}{4}, N(\beta) = \left(\frac{\Delta A - \Delta Z}{2}\right)$$ **Ans.**

Chapter 14. Semiconductor Electronics : Materials, Devices and Simple Circuits

Q. 1. Explain the formation of energy bands in solids and also explain how are *n*-type and *p*-type semiconductors obtained.

Ans. An isolated atom has well defined energy levels. However, when large number of such atoms get together to form a real solid, these individual energy levels overlap and get completely modified. Instead of discrete value of energy of electrons, the energy values lie in a certain range. The collection of these closely packed energy levels are said to form an energy band. Two types of such bands formed in solids are called valence band and conduction band. The band formed by filled energy levels is known as valence band whereas partially filled or unfilled band is known as conduction band. The two bands are generally separated by a gap called energy gap or forbidden gap. Depending upon the size of this energy gap, different materials behave as conductors, semiconductors or insulators.

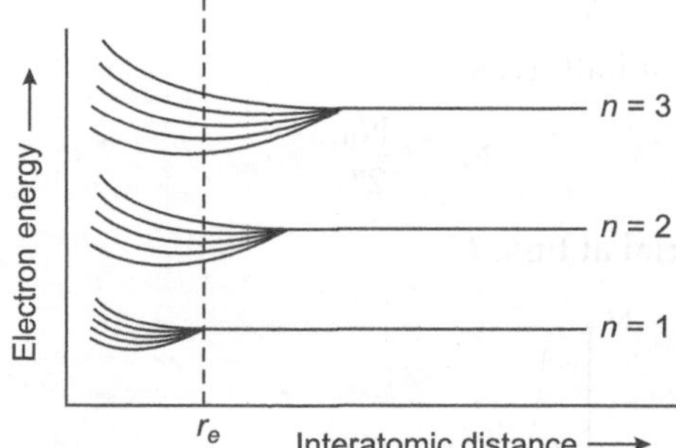

The *n* type and *p*-type semiconductors are obtained as follows :

***n*-type semiconductor** : When a pentavalent impurity like Phosphorus, Antimony, Arsenic is doped in pure Ge (or Si), then the conductivity of crystal increases due to surplus electrons and such a crystal is called *n*-type semiconductor.

***p*-type semiconductor** : If a trivalent impurity (Boron, Gallium or Indium) is added to Germanium (or Silicon) then only three of electron pair bands can be filled and the vacancy that exists in the fourth band constitutes a hole that accepts electrons. Such a crystal is called *p*-type semiconductor.

Q. 2. (i) Distinguish between conductors, insulators and semiconductors in terms of energy band diagrams. Draw the diagrams.

(ii) Why do we get a small current (in μA) through a *p-n* junction diode under reverse bias condition ? In which direction does this current flow through the junction ?

Ans. (i)

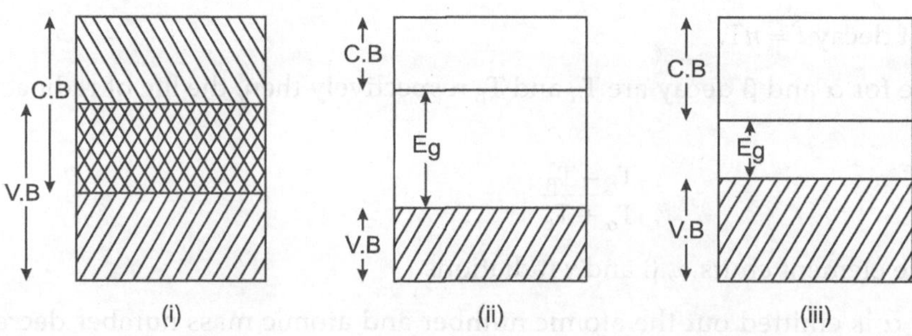

[Here, C.B. → Conduction Band, V.B. → Valence Band, E_g → Forbidden Energy gap]

Conductors : Conductors are the substances which allow the flow of electric current through them. In a conductor, the valence band and the conduction band overlap. So no energy is required to move the electrons from valence band to conduction band. Free electrons are available in conduction band and thus electric current flows through it, when a potential difference is applied across it. For example : Magnesium, Silver etc.

Insulators : Insulators are the substances which do not allow the flow of electric current through them. In insulators, the valence electrons are very rigidly held in the inter atomic bands, even at high temperatures.

In terms of band theory, insulator have a large energy gap between the conduction band and the valence band (greater then 5 eV). So electrons cannot jump from the valence band into the conduction band. (In the case of diamond the energy gap is 6 eV.) For example : Glass, Rubber etc.

Semiconductors : Semiconductors are the substances, whose conductivity lies in between that of a good conductors and an insulator (For example : Germanium, Silicon).

In terms of band theory, semiconductors have small energy gap between conduction band and valence band. At 0 K, the energy gap is 0.72 eV for germanium and 1.21 eV for silicon. This decreases with increase in temperature. At 0 K, the conduction band is empty and the semiconductor behaves as an insulator. With the increase in the temperature, the electrons from the valence band jump to conduction band and hence the conduction of electric current is possible.

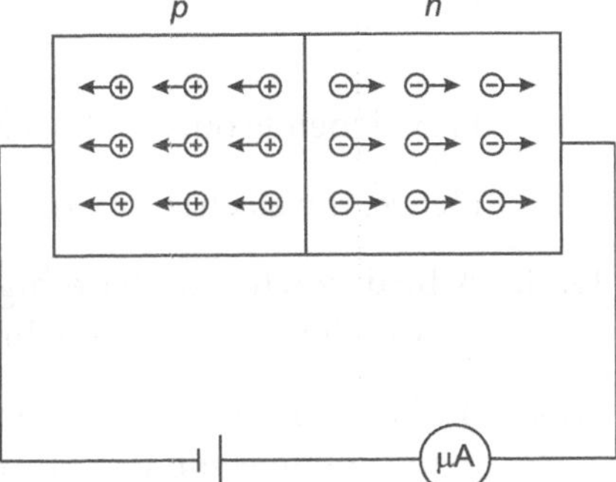

(ii) In the reverse biasing of a p-n junction diode, the depletion region, across the junction diode, increases, due to the flow of holes in p-region and electrons in n-region, away from the junction. But a very small current (order μA) flows across the junction due to the flow of minority carriers. This current flows through n-type to p-type semiconductor, across the junction.

□□

Question Set 7

Evaluation and Analysis Based

Chapter 1. Electric Charges and Fields

Q. 1. Two point charges q_1 and q_2 are placed at a distance d apart as shown in figure. The electric field intensity is zero at the point P on the line joining them as shown. Write conclusions that you can draw from this.

q_1 ←— d —→ q_2 - - - P

Ans. (i) Two point charges (q_1 and q_2) should be of opposite nature.

(ii) Magnitude of charge q_1 must be greater than magnitude of charge q_2.

Q. 2. How does the electric flux due to a point charge enclosed by a spherical Gaussian surface get affected when its radius is increased ?★

Ans. According to Gauss's law

$$\phi = \int \vec{E} \cdot \vec{ds} = \frac{q_{en}}{\varepsilon_0}$$

Flux depends only on the charge enclosed.

Hence, the electric flux remains constant.

Q. 3. A bird perches on bare high power line and nothing happens to bird. A man standing on the ground touches the same line and gets the fatal shock. Why ?

Ans. Birds whole body is at same potential. Thus no charge flows and no shock is produced in it, while man touching the ground maintains potential difference between high power line and ground and large current flows through him which creates electrolysis of blood and causes fatal shock or death.

Q. 4. If the electric field intensity is zero at a point, will the electric potential be necessarily zero at that point. Give example.

Ans. No, consider point in midway between two equal and similar charges, the electric field intensity is zero but the electric potential is not zero.

Q. 5. Why should electrostatic field be zero inside a conductor ?★

Ans. The net electric charge of a conductor resides entirely on its surface. If we consider boundaries of conductor as a Gaussian surface. Then

$$\phi = \frac{q}{\varepsilon_0} = \oint \vec{E} \cdot \vec{dS}$$

$$= \oint E \, dS \cos\theta = \frac{q}{\varepsilon_0}$$

$$E = \frac{q}{4\pi\varepsilon_0 r} = 0 \qquad [\because q = 0 \text{ inside conductor}]$$

Q. 6. Two uniformly large parallel thin plates having charge densities $+\sigma$ and $-\sigma$ are kept in X-Z plane at a distance d apart. Sketch an equipotential surface due to electric field between plates. If a

★ are board based questions from previous year

particle of mass m and charge $-q$ remains stationary between plates, what is the magnitude and direction of this field ?★

Ans. The sketch of an equipotential surface due to electric field between plates is given below :

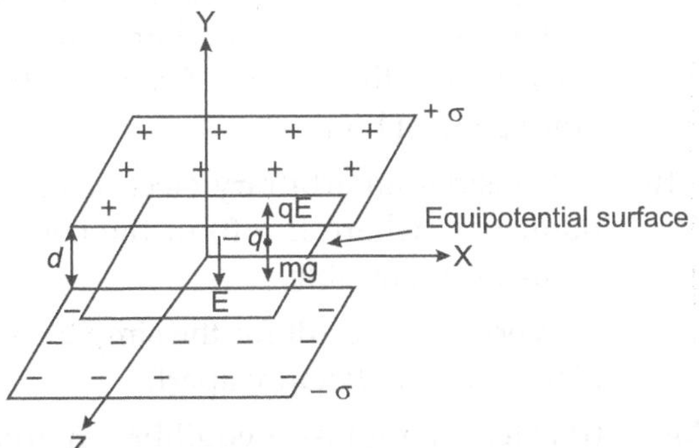

Downward force due to gravity is balanced by upward force of electric field.

$$mg = -qE$$

$$E = \frac{-mg}{q}$$

The direction of field is downwards *i.e.*, along negative Y-axis (as shown in the fig.).

Q. 7. A hollow charged conductor has a tiny hole cut into its surface. Show that the electric field in the hole is $(\sigma/2\varepsilon_0)\hat{n}$, where $\hat{n}$ is the unit vector in the outward normal direction, and σ is the surface charge density near the hole. **[NCERT]**

Ans. Let E is electric field just outside the conductor, q is the electric charge, σ charge density and ε_0 is permittivity of free space.

As per Gauss's flux, $\phi = E \cdot ds = \frac{q}{\varepsilon_0}$

$$\Rightarrow \quad E \cdot ds = \frac{\sigma \times ds}{\varepsilon_0}$$

$$\therefore \quad E = \frac{\sigma}{\varepsilon_0}\hat{n}$$

Therefore, the electric field just outside the conductor is $\frac{\sigma}{\varepsilon_0}\hat{n}$. This field is a superposition of field due to the cavity E′ and the field due to the rest of the charged conductor E′. These fields are equal and opposite inside the conductor and equal in magnitude and direction outside the conductor.

$$\therefore \quad E' + E' = E$$

$$\Rightarrow \quad E' = \frac{E}{2} = \frac{\sigma}{2\varepsilon_0}\hat{n}$$

Hence, the field due to the rest of conductor is $\frac{\sigma}{\varepsilon_0}\hat{n}$.

Q. 8. (i) Explain the meaning of the statement 'electric charge of a body is quantised'.

(ii) Why can one ignore quantisation of electric charge when dealing with macroscopic *i.e.*, large scale charges ? **[NCERT]**

Ans. (i) Electric charge of a body is quantised. This means that only integral (1, 2, ..., n) number of electrons can be transferred from one body to the other. Charges are not transferred in fraction. Hence, a body possesses total charge only in integral multiples of electric charge.

(ii) In macroscopic or large scale charges, the charges used are huge as compared to the magnitude of electric charge. Hence, quantisation of electric charge is of no use on macroscopic scale. Therefore, it is ignored and it is considered that electric charge is continuous.

Q. 9. When a glass rod is rubbed with a silk cloth, charges appear on both. A similar phenomenon is observed with many other pairs of bodies. Explain how this observation is consistent with the law of conservation of charge. **[NCERT]**

Ans. According to the law of conservation of charge, "charge can neither be created nor be destroyed but it can be transferred from one body to another body". Before rubbing both the bodies are neutral *i.e.*, the total charge of the system is zero. When the glass rod is rubbed with a silk cloth,

---

★ are board based questions from previous year

some electrons are transferred from glass rod to silk cloth. Hence, glass rod attains positive charge and silk cloth attains same amount of negative charge. Again, the total charge of the system is zero, *i.e.*, the charge before rubbing is same as the charge after rubbing. This is consistent with the law of conservation of charge. Similar phenomenon is observed with many other pairs of bodies.

Q. 10. (i) Consider an arbitrary electrostatic field configuration. A small test charge is placed at a null point (*i.e.*, where E = 0) of the configuration. Show that the equilibrium of the test charge is necessarily unstable.

(ii) Verify this result for the simple configuration of two charges of the same magnitude and sign placed certain distance apart. **[NCERT]**

Ans. (i) If test charge is in equilibrium and displaced from its position in any direction, then it will experience a restoring force towards null point (here electric field is zero). At this point, net inward flux exist. From Gauss's law, the net electric flux through a surface net enclosing any charge must be zero. Hence, the equilibrium is not stable.

(ii) Two charges of same magnitude and same sign are placed at a certain distance. The mid-point of the joining line of the charges is the null point. When a test charge is displaced along the line, it experiences a restoring force. If it is displaced normal to the joining line, then the net force takes it away from the null point. Hence, the charge is unstable because stability of equilibrium requires restoring force in all directions.

Q. 11. An electron falls through a distance of 1.5 cm in a uniform electric field of magnitude 2.0×10^4 N C^{-1} [Fig.(a)]. The direction of the field is reversed keeping its magnitude unchanged and a proton falls through the same distance [fig. (B)]. Compute the time of fall in each case. Contrast the situation with that of 'free fall under gravity'. **[NCERT]**

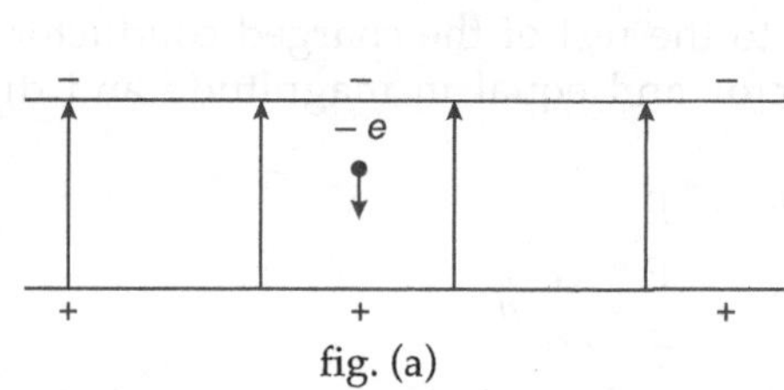

fig. (a)

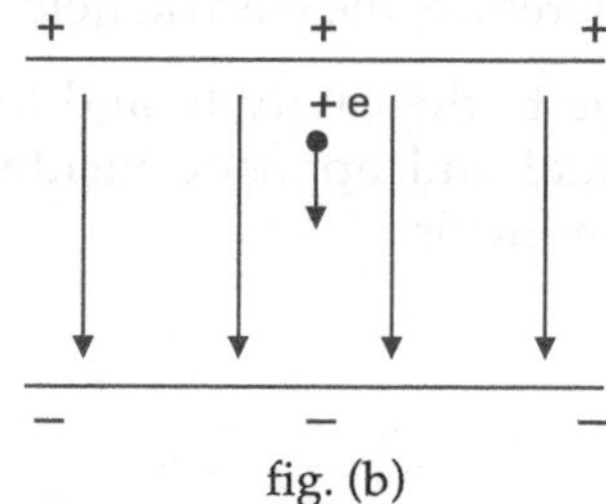

fig. (b)

Ans. In Figure (a) the field is upward, so the negatively charged electron experiences a downward force of magnitude *e*E where E is the magnitude of the electric field. The acceleration of the electron is

$$a_e = eE/m_e$$

where, m_e is the mass of the electron.

Starting from rest, the time required by the electron to fall through a distance *h* is given by

$$t_e = \sqrt{\frac{2h}{a_e}}$$

$$= \sqrt{\frac{2hm_e}{eE}}$$

For $e = 1.6 \times 10^{-19}$ C, $m_e = 9.11 \times 10^{-31}$ kg.

E = 2.0×10^4 N C^{-1}, $h = 1.5 \times 10^{-2}$ m.

$t_e = 2.9 \times 10^{-9}$ s.

In fig. (B) direction of the field is reversed than field is downward, and the positively charged proton experiences a downward force of magnitude *e*E. The acceleration of the proton is

$$a_p = eE/m_p$$

where m_p is the mass of the proton; $m_p = 1.67 \times 10^{-27}$ kg. The time of fall for the proton is

$$t_p = \sqrt{\frac{2h}{a_p}} = \sqrt{\frac{2hm_p}{eE}}$$
$$= 1.3 \times 10^{-7} \text{ s}$$

Thus, the heavier particle (proton) takes a greater time to fall through the same distance. This is in basic contrast to the situation of 'free fall under gravity' where the time of fall is independent of the mass of the body. Note that in this example we have ignored the acceleration due to gravity in calculating the time of fall. To see if this is justified. Let us calculate the acceleration of the proton in the given electric field :

$$a_p = \frac{eE}{m_p}$$
$$= \frac{(1.6 \times 10^{-19}) \times (2.0 \times 10^4)}{1.67 \times 10^{-27}}$$
$$= 1.9 \times 10^{12} \text{ ms}^{-2}$$

which is enormous compared to the value of g (9.8 m s^{-2}), the acceleration due to gravity. The acceleration of the electron is even greater. Thus, the effect of acceleration due to gravity can be ignored in this example.

Q. 12. (i) A conductor A with a cavity as shown in figure (a) is given a charge Q. Show that the entire charge must appear on the outer surface of the conductor.

(ii) Another conductor B with charge q is inserted into the cavity keeping B insulated from A. Show that the total charge on the outside surface of A is Q + q [Fig. (b)].

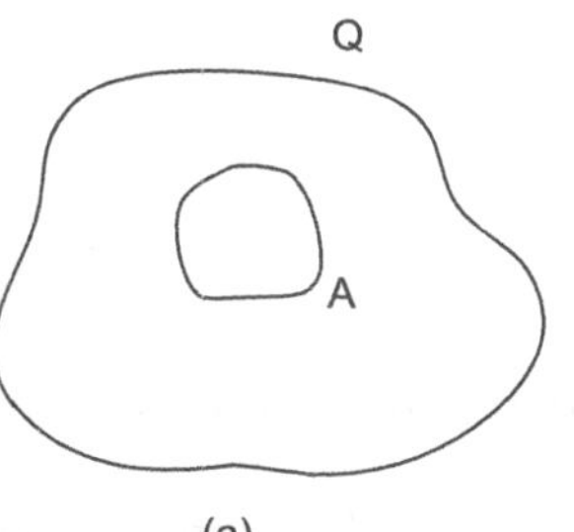

(a)

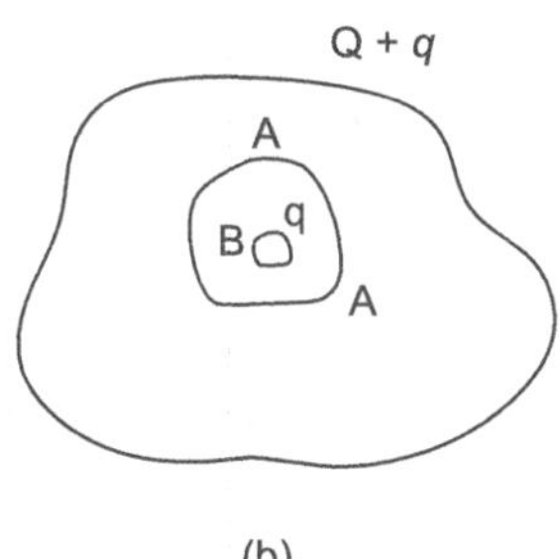

(b)

(iii) A sensitive instrument is to be shielded from the strong electrostatic fields in its environment. Suggest a possible way. **[NCERT]**

Ans. (i) Consider a Gaussian surface, completely within conductor and covered the cavity. Electric field intensity inside charged conductor is zero.

Let q is the charge inside the conductor and ε_0 is the permittivity of free space.

According to Gauss's law,

$$\text{Flux, } \phi = \vec{E} \cdot \vec{ds} = \frac{q}{\varepsilon_0}$$

Here, $$E = 0$$

$$\Rightarrow \quad \frac{q}{\varepsilon_0} = 0 \Rightarrow q = 0 \qquad [\text{as } \varepsilon_0 \neq 0]$$

Therefore, charge inside the conductor is zero.

The entire charge Q is on the outer surface of the conductor.

(ii) The outer surface of conductor A has a charge of amount Q. Another conductor B having charge + q is kept inside conductor A and it is insulated from A. Hence, a charge of amount – q will be induced in the inner surface of conductor A and + q is induced on the outer surface of conductor A. Therefore, total charge on the outer surface of conductor A is Q + q.

(iii) A sensitive instrument can be shielded from the strong electrostatic field in its environment by enclosing it fully inside a metallic surface. A closed metallic body acts as an electrostatic shield.

Q. 13. A particle of mass m and charge (– q) enters the region between the two charged plates initially moving along X-axis with speed v_x. The length of plate is L and a uniform electric field E is maintained between the plates. Show that the vertical deflection of the particle at the far edge of the plate is $qEL^2/(2mv_x^2)$. **[NCERT]**

Ans. Charge on a particle of mass, $m = -q$

Velocity of the particle = v_x

Length of plates = L

Magnitude of the uniform electric field between plates = E

$$\text{Mechanical force, F} = \text{Mass} \times \text{Acceleration}$$

$$\Rightarrow \quad a = \frac{F}{m}$$

$$\Rightarrow \quad a = \frac{qE}{m} \quad \text{...(i)}$$

Time taken by the part,
$$t = \frac{\text{Length of the plate}}{\text{Velocity of the particle}}$$

$$= \frac{L}{v_x} \quad \text{...(ii)}$$

In the vertical direction, initial velocity, $u = 0$

According to the third equation of motion, vertical deflection of the particle can be obtained as,

$$s = ut + \frac{1}{2}at^2$$

$$\Rightarrow \quad s = 0 + \frac{1}{2}\left(\frac{qE}{m}\right)\left(\frac{L}{v_x}\right)^2 \quad \text{[From (i) and (ii)]}$$

$$\Rightarrow \quad s = \frac{qEL^2}{2mv_x^2}$$

Hence, vertical deflection of the particle at the far edge of the plate is $\frac{qEL^2}{2mv_x^2}$.

This is similar to the motion of horizontal projectiles under gravity.

Q. 14. (i) Define electric flux. Write its S.I. unit. "Gauss's law in electrostatics is true for any closed surface, no matter what its shape or size is". Justify this statement with the help of a suitable example.★

(ii) Use Gauss's law to prove that electric field inside a uniformly charged spherical shell is zero.★

Ans. (i) The electric flux through an area is defined as the electric field multiplied by the area of the surface projected on a plane, perpendicular to the field. Its S.I. unit is voltmeters (Vm) or Newton metres square per coulomb ($Nm^2 C^{-1}$).

The given statement is justified because while measuring the flux, the surface area is more important than its volume or its size.

(ii) To prove that the electric field inside a uniformly charged spherical shell is zero, we place a single positive point charge 'q' at the centre of an imaginary spherical surface with radius R. The field lines of this point radiate outside equally in all directions. The magnitude E of the electric field at every point on the surface is given by $E = \frac{1}{4\pi\varepsilon_0} \times \frac{q}{R^2}$.

At each point on the surface, $\vec{E}$ is perpendicular to the surface and its magnitude is the same. Thus, the total electric flux (ϕ_E) is the product of their field magnitude E and the A.

Hence,
$$\phi_E = EA$$

$$= \frac{1}{4\pi\varepsilon_0}\frac{q}{R^2}(4\pi R^2)$$

$$= \frac{q}{\varepsilon_0}$$

★ are board based questions from previous year

If the sphere is uniformly charged, then there is zero charge inside the sphere, according to Gauss's law

When $q = 0$, $\phi_E = \frac{0}{\varepsilon_0} = 0$

Q. 15. In a certain region of space, electric field is along the Z-direction throughout. The magnitude of electric field is, however, not constant but increases uniformly along the positive Z-direction, at the rate of 10^5 NC^{-1} per metre. What are the force and torque experienced by a system having a total dipole moment equal to 10^{-7} Cm in the negative Z-direction ? **[NCERT]**

Ans. Dipole moment of the system, $p = q \times dl = -10^{-7}$ Cm

Rate of increase of electric field per unit length,

$$\frac{dE}{dl} = 10^5 \text{ N/C}$$

$$F = qE = q \times \frac{dE}{dl} \times dl$$

$$= p \times \frac{dE}{dl}$$

$$= -10^{-7} \times 10^5$$

$$= -10^{-2} \text{ N}$$

The force is -10^{-2} N in the negative Z-direction *i.e.,* opposite to the direction of electric field. Hence, the angle between electric field and dipole moment is 180°.

Torque (τ) is given by the relation, $\tau = pE \sin 180° = 0$.

Therefore, the torque experience by the system is zero.

Q. 16. It is now believed that protons and neutrons (which constitute nuclei of ordinary matter) are themselves built out of more elementary units called quarks. A proton and a neutron consist of three quarks each. Two types of quarks, the so called 'up' quark (denoted by *u*) of charge + (2/3) *e*, and the 'down' quark (denoted by *d*) of charge (– 1/3) *e*, together with electrons build up ordinary matter. (Quarks of other types have also been found which give rise to different unusual varieties of matter). Suggest a possible quark composition of a proton and neutron. **[NCERT]**

Ans. A proton has three quarks. Let there be *n* up quarks in a proton, each having a charge of $\frac{2}{3}e$.

Charge due to *n* up quarks $= \left(\frac{2}{3}e\right)n$

Number of down quarks in a proton $= 3 - n$

Each down quark has a charge of $-\frac{1}{3}e$

Charge due to $(3 - n)$ down quarks $= \left(-\frac{1}{3}e\right)(3-n)$

Total charge of a proton $= +e$

$$e = \frac{2e}{3}n + \left(-\frac{e}{3}\right)(3-n)$$

$$e = \frac{2ne}{3} - e + \frac{ne}{3}$$

$$2e = ne, n = 2$$

No. of up quark in protons $(n) = 2$, and
down quark $= 3 - n = 3 - 2 = 1$

Therefore, in the proton there are two up quarks and one down quark.

$\therefore$ Possible quark composition for proton $= uud$

A neutron has 3 quark, assume there are *n* up quarks in neutron

Charge on a neutron due to n up quarks $= \left(+\frac{2}{3}e\right)n$

Number of down quarks is $3 - n$, and each having a charge of $-\frac{1}{3}e$.

Charge on a neutron due to $(3 - n)$ down quarks $= \left(-\frac{1}{3}e\right)(3 - n)$

Total charge on a neutron = 0

$$\therefore \quad 0 = \left(\frac{2}{3}e\right)n + \left(-\frac{1}{3}e\right)(3-n)$$

$$\Rightarrow \quad 0 = \frac{2ne}{3} - e + \frac{ne}{3}$$

$$\Rightarrow \quad e = ne \Rightarrow n = 1$$

Number of up quarks in a neutron, $n = 1$

Number of down quarks in a neutron $= 3 - n = 2$

Therefore, a neutron can be represented as *'udd'*.

Q. 17. A long charged cylinder of linear charged density λ is surrounded by a hollow co-axial conducting cylinder. What is the electric field in the space between the two cylinders ? **[NCERT]**

Ans. Charge density of the long charged cylinder of length L and radius r is λ. Another cylinder of same length surrounds the previous cylinder. The radius of this cylinder is R.

Let E be the electric field produced in the space between the two cylinders. Electric flux through the Gaussian surface is given by Gauss's theorem, as,

$$\phi = E\,(2\pi d)\,L$$

Where, d = Distance of a point from the common axis of the cylinders.

Let q be the total charge on the cylinder.

It can be written as

$$\phi = E\,(2\pi dL) = \frac{q}{\varepsilon_0}$$

Where, q = Charge on the inner sphere of the outer cylinder.

ε_0 = Permittivity of free space

$$E\,(2\pi dL) = \frac{\lambda L}{\varepsilon_0}$$

$$E = \frac{\lambda}{2\pi\varepsilon_0 d}$$

Therefore, the electric field in the space between the two cylinders is $\frac{\lambda}{2\pi\varepsilon_0 d}$.

Q. 18. A spherical conducting shell of inner radius r_1 and outer radius r_2 has a charge Q.

(i) A charge q is placed at the centre of the shell. What is the surface charge density on the inner and outer surfaces of the shell ?

(ii) Is the electric field inside a cavity [with no charge] zero, even if the shell is not spherical, but has any irregular shape ? Explain. **[NCERT]**

Ans. (i) Charge placed at the centre of a shell is $+q$. Hence, a charge of magnitude $-q$ will be induced to the inner surface of the shell. Therefore, total charge on the inner surfaces of the shell is $-q$.

Surface charge density at the inner surface of the shell is given by the relation,

$$\sigma_1 = \frac{\text{Total charge}}{\text{Inner surface area}} = \frac{-q}{4\pi r_1^2} \quad \text{...(i)}$$

A charge of $+q$ is induced on the outer surface of the shell. A charge of magnitude Q is placed on the outer surface of the shell. Therefore, total charge on the outer surface of the shell is $Q + q$. Surface charge density at the outer surface of the shell.

$$\sigma_2 = \frac{\text{Total charge}}{\text{Inner surface area}} = \frac{Q+q}{4\pi r_2^2} \quad \text{...(ii)}$$

(ii) Yes, the electric field intensity inside a cavity is zero, even if the shell is not spherical and has any irregular shape. Consider a closed loop such that a part of it is inside the cavity along a field line while the rest is inside the conductor. Work done by the field in carrying a test charge over a closed loop is zero because the field inside the conductor is zero. Hence, electric field is zero, whatever is the shape.

Q. 19. A thin metallic spherical shell of radius carries a charge Q on its surface. A point charge $\frac{Q}{2}$ is placed at its centre C and another charge + 2Q is placed outside the shell at a distance x from the centre as shown in the figure. Find (i) the force on the charge at the centre of shell and at the point A, (ii) the electric flux through the shell.★

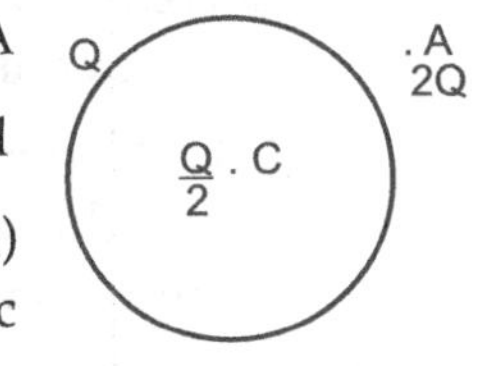

Ans. (i) Net force on the charge $\frac{Q}{2}$, placed at the centre of the shell is zero.

Force on charge 2Q kept at a point A

$$F = E \times 2Q = \frac{1 \cdot \left(\frac{3Q}{2}\right) 2Q}{4\pi\varepsilon_0 r^2} = \frac{k \times 3Q^2}{r^2}$$

where $k = \frac{1}{4\pi\varepsilon_0}$

(ii) Electric flux through the shell,

$$\phi = \frac{Q}{2\varepsilon_0}$$

Chapter 2. Electrostatic Potential and Capacitance

Q. 1. Why is electrostatic potential constant throughout the volume of the conductor and has the same value (as inside) on its surface ?

Ans. As electric field inside a conductor is zero, so E = 0. This is the reason why electrostatic potential is constant.

As $$E = \frac{-dV}{dr}$$

If $$E = 0$$

Then $$\frac{-dV}{dr} = 0$$

Hence, $$V = \text{constant}$$

Q. 2. A small sphere of radius r_1 and charge q_1 is enclosed by a spherical shell of radius r_2 and charge q_2. Show that if q_1 is positive, charge will necessarily flow from the sphere to the shell (when the two are connected by a wire) no matter what the charge q_2 on the shell is. **[NCERT]**

Ans. According to Gauss's law, the electric field between a sphere and a shell is determined by the charge q_1 on a small sphere. Hence, the potential difference, V, between the sphere and the shell is independent of charge q_2. For positive charge q_1, potential difference V is always positive.

Q. 3. The electric potential is zero at a point in an electric field. Will the electric field be necessarily zero at that point.

Ans. No, zero potential doesn't mean zero electric field. *e.g.*, At a point on the equatorial line of dipole, the potential is zero, but the electric field is not certainly zero.

★ are board based questions from previous year

Q. 4. A capacitor 'C', a variable resistor 'R' and a bulb 'B' are connected in series to the ac mains in circuit as shown. The bulb glows with some brightness. How will the glow of the bulb change if

(i) a dielectric slab is introduced between the plates of the capacitor, keeping resistance R to be the same.

(ii) the resistance R is increased keeping the same capacitance ?

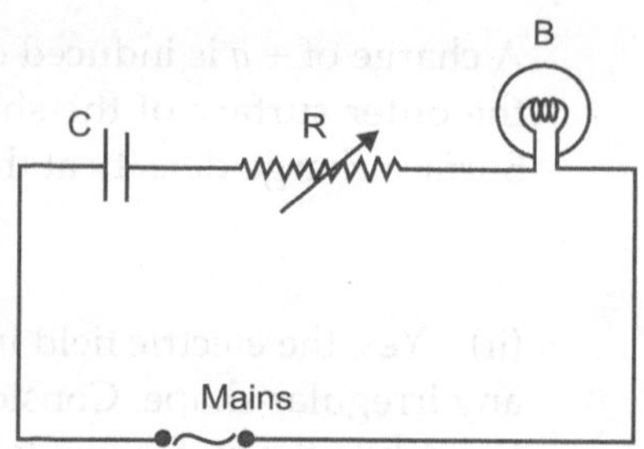

Ans. (i) As the dielectric slab is introduced between the plates of the capacitor, its capacitance will increase. Hence, the potential drop across the capacitor will decrease (V= Q/C). As a result, the potential drop across the bulb will increase (since both are connected in series). So, its brightness will increase.

(ii) As the resistance (R) is increased, the potential drop across the resistor will increase. As a result, the potential drop across the bulb will decrease (since both are connected in series). So, its brightness will decrease.

Q. 5. (i) A comb run through one's dry hair attracts small bits of paper. Why ?

What happens if the hair is wet or if it is a rainy day ? (Remember, a paper does not conduct electricity).

(ii) Ordinary rubber is an insulator. But special rubber tyres of aircraft are made slightly conducting. Why is this necessary ?

(iii) Vehicles carrying inflammable materials usually have metallic ropes touching the ground during motion. Why ? **[NCERT]**

Ans. (i) This is because the comb gets charged by friction. The molecules in the paper gets polarised by the charged comb, resulting in a net force of attraction. If the hair is wet, or if it is rainy day, friction between hair and the comb reduces. The comb does not get charged and thus it will not attract small bits of paper.

(ii) To enable them to conduct charge (produced by friction) to the ground as too much of static electricity accumulated may result in spark and result in fire.

(iiii)It is for conduction of electrostatic charge (developed by friction) to ground so that spark will not occur.

Q. 6. A spherical capacitor consists of two concentric spherical conductors, held in position by suitable insulating supports (fig.). Show that the capacitance of a spherical capacitor is given by

$$C = \frac{4\pi\varepsilon_0 r_1 r_2}{r_1 - r_2}$$

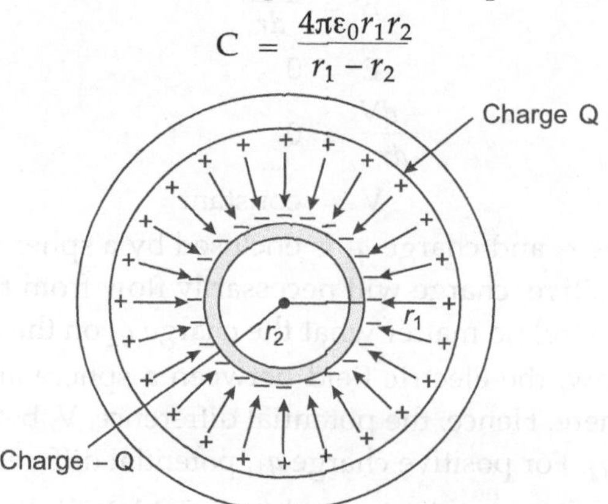

where r_1 and r_2 are the radii of the outer and inner spheres, respectively. **[NCERT]**

Ans. Radius of the outer shell $= r_1$

Radius of the inner shell $= r_2$

The inner surface of the outer shell has charge + Q.

The outer surface of the inner shell has induced charge – Q.

Potential difference between the two shells is given by,

$$V = \frac{Q}{4\pi\varepsilon_0 r_2} - \frac{Q}{4\pi\varepsilon_0 r_1}$$

where, ε_0 = Permittivity of free space

$$V = \frac{Q}{4\pi\varepsilon_0}\left[\frac{1}{r_2} - \frac{1}{r_1}\right]$$

$$= \frac{Q(r_1 - r_2)}{4\pi\varepsilon_0 r_1 r_2}$$

Capacitance of the given system is given by,

$$C = \frac{\text{Charge (Q)}}{\text{Potential difference (V)}}$$

$$= \frac{4\pi\varepsilon_0 r_1 r_2}{(r_1 - r_2)}$$

Hence Proved.

Q. 7. Answer carefully :

(i) Two large conducting spheres carrying charges Q_1 and Q_2 are brought close to each other. Is the magnitude of electrostatic force between them exactly given by $Q_1 Q_2/4\pi\varepsilon_0 r^2$, where r is the distance between their centres ?

(ii) If Coulomb's law involved $1/r^3$ dependence (instead of $1/r^2$), would Gauss's law be still true ?

(iii) A small test charge is released at rest at a point in an electrostatic field configuration. Will it travel along the field line passing through that point ?

(iv) What is the work done by the field of a nucleus in a complete circular orbit of the electron ? What if the orbit is elliptical ?

(v) We know that electric field is discontinuous across the surface of a charged conductor. Is electric potential also discontinuous there ?

(vi) What meaning would you give to the capacitance of a single conductor ?

(vii) Guess a possible reason why water has a much greater dielectric constant (= 80) than say. mica (= 6). **[NCERT]**

Ans. (i) No, because charge distributions on the spheres will not be uniform.

(ii) Gauss's law will not be true, if Coulomb's law involved $\frac{1}{r^3}$ dependence, instead of $\frac{1}{r^2}$.

(iii) Yes, If a small test charge is released at rest at a point in an electrostatic field configuration, then it will travel along the field lines passing through the point, only if the field lines are straight. This is because the field lines give the direction of acceleration and not of velocity.

(iv) Whenever the electron completes an orbit, either circular or elliptical, the work done by the field of a nucleus is zero.

(v) No

(vi) The capacitance of a single conductor is considered as a parallel plate capacitor with one of its two plates at infinity.

(vii) Water has an unsymmetrical space as compared to mica. Since it has a permanent dipole moment, it has a greater dielectric constant than mica.

Q. 8. Answer the following :

(i) The top of the atmosphere is at about 400 kV with respect to the surface of the earth, corresponding to an electric field that decreases with altitude. Near the surface of the earth, the field is about 100 Vm^{-1}. Why then do we not get an electric shock as we step out of our house into the open ? (Assume the house to be a steel cage so there is no field inside !).

(ii) A man fixes outside his house one evening a two metre high insulating slab carrying on its top a large aluminium sheet of area 1 m^2. Will he get an electric shock if he touches the metal sheet next morning ?

(iii) The discharging current in the atmosphere due to the small conductivity of air is known to be 1800 A on an average over the globe. Why then does the atmosphere not discharge itself completely in due course and become electrically neutral ? In other words, what keeps the atmosphere charged ?

(iv) What are the forms of energy into which the electrical energy of the atmosphere is dissipated during a lightning ?

[Hint : The earth has an electric field of about 100 Vm^{-1} at its surface in the downward direction, corresponding to a surface charge density = – 10^{-9} C m^2. Due to the slight conductivity of the atmosphere up to about 50 km (beyond which it is good conductor), about + 1800 C is pumped every second into the earth as a whole. The earth, however, does not get discharged since thunderstorms and lightning occuring continually all over the globe pump an equal amount of negative charge on the earth.) **[NCERT]**

Ans. (i) We do not get an electric shock as we step out of our house because the original equipotential surfaces of open air changes, keeping our body and the ground at the same potential.

(ii) Yes, the man will get an electric shock if he touches the metal slab next morning. The steady discharging current in the atmosphere charges up the aluminium sheet. As a result, its voltage rises gradually. The raise in the voltage depends on the capacitance of the capacitor formed by the aluminium slab and the ground.

(iii) The occurance of thunderstorms and lightning charges the atmosphere continuously. Hence, even with the presence of discharging current of 1800 A, the atmosphere is not discharged completely. The two opposing currents are in equilibrium and the atmosphere remains electrically neutral.

(iv) During lightning and thunderstorm, light energy, heat energy and sound energy are dissipated in the atmosphere.

Q. 9. A parallel plate capacitor of capacitance C is charged to a potential V. It is then connected to another uncharged capacitor having the same capacitance. Find out the ratio of the energy stored in the combined system to that stored initially in the single capacitor.★

Ans. Let 'q' be the charge on the charged capacitor.

Energy stored in it is,

$$U = \frac{q^2}{2C}$$

When another similar uncharged capacitor is connected, the net capacitance of the system is,

$$C' = 2C$$

The charge on the system is constant. So, the energy stored in the system now is,

$$U' = \frac{q^2}{2(C')^2}$$

$$\Rightarrow \quad U' = \frac{q^2}{2(2C)}$$

$$\Rightarrow \quad U' = \frac{q^2}{4C}$$

Thus, the required ratio is,

$$\frac{U'}{U} = \frac{\frac{q^2}{4C}}{\frac{q^2}{2C}}$$

$$\Rightarrow \quad \frac{U'}{U} = \frac{1}{2}$$

★ are board based questions from previous year

Q. 10. Two charges 5×10^{-8} C and -3×10^{-8} C are located 16 cm apart. At what point(s) on the line joining the two charges is the electric potential zero ? Take the potential at infinity to be zero.

[NCERT]

Ans. There are two charges.

$$q_1 = 5 \times 10^{-8} \text{C}$$
$$q_2 = -3 \times 10^{-8} \text{C}$$

Distance between the two charges, $d = 16 \text{ cm} = 0.16 \text{ m}$

Consider a point P on the line joining the two charges, as shown in the given figure.

Where, r = Distance of point P from charge q_1

Let the electric potential (V) at point P be zero.

Potential at point P is the sum of potentials caused by charges q_1 and q_2 respectively.

0.16 m
r
q_1 P q_2

$$\therefore \quad V = \frac{1}{4\pi\varepsilon_0} \cdot \frac{q_1}{r} + \frac{1}{4\pi\varepsilon_0} \cdot \frac{q_2}{(d-r)} \qquad \text{...(i)}$$

Where, ε_0 = Permittivity of free space

For V = 0, equation (i) reduces to

$$0 = \frac{1}{4\pi\varepsilon_0} \cdot \frac{q_1}{r} + \frac{1}{4\pi\varepsilon_0} \cdot \frac{q_2}{(d-r)}$$

$$\Rightarrow \quad \frac{1}{4\pi\varepsilon_0} \cdot \frac{q_1}{r} = -\frac{1}{4\pi\varepsilon_0} \cdot \frac{q_2}{(d-r)}$$

$$\Rightarrow \quad \frac{q_1}{r} = -\frac{q_2}{(d-r)}$$

$$\Rightarrow \quad \frac{5 \times 10^{-8}}{r} = -\frac{-3 \times 10^{-8}}{(0.16-r)}$$

$$\Rightarrow \quad 5(0.16-r) = 3r$$

$$\Rightarrow \quad 0.8 = 8r \Rightarrow r = 0.1 \text{ m} = 10 \text{ cm}$$

IInd case : Let point P is outside the system of two charges at a distance s from the negative charge, where potential is zero, as shown in the following figure.

For this arrangement, potential is given by,

$$V = \frac{1}{4\pi\varepsilon_0} \cdot \frac{q_1}{s} + \frac{1}{4\pi\varepsilon_0} \cdot \frac{q_2}{(s-d)} \qquad \text{...(ii)}$$

Where,

ε_0 = Permittivity of free space

For V = 0, equation (ii) reduces to

S
d
q_1 q_2 P

$$0 = \frac{1}{4\pi\varepsilon_0} \cdot \frac{q_1}{s} + \frac{1}{4\pi\varepsilon_0} \cdot \frac{q_2}{(s-d)}$$

$$0 = \frac{1}{4\pi\varepsilon_0} \cdot \frac{q_1}{s} + \frac{1}{4\pi\varepsilon_0} \cdot \frac{q_2}{(s-d)}$$

$$\Rightarrow \quad \frac{1}{4\pi\varepsilon_0} \cdot \frac{q_1}{s} = -\frac{1}{4\pi\varepsilon_0} \cdot \frac{q_2}{(s-d)}$$

$$\Rightarrow \quad \frac{q_1}{s} = -\frac{q_2}{(s-d)}$$

$$\Rightarrow \quad \frac{5 \times 10^{-8}}{s} = -\frac{-3 \times 10^{-8}}{(s-0.16)}$$

$$\Rightarrow \quad 5(s-0.16) = 3s$$

$$\Rightarrow \quad 0.8 = 2s \Rightarrow s = 0.4 \text{ m} = 40 \text{ cm}$$

Q. 11. Two identical parallel plate capacitors A and B are connected to battery of V volts with the switch S closed. The switch is now opened and the free space between the plates of the capacitors is filled with a dielectric of dielectric constant K. Find the ratio of the total electrostatic energy stored in both capacitors before and after the introduction of the dielectric.★

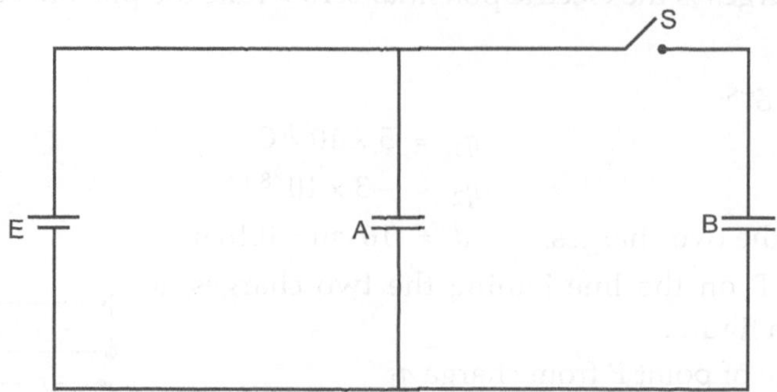

Ans. Energy stored before introduction of dielectric,

$$U_{iA} = U$$
$$U_{iB} = U$$
$$U_i = U + U = 2U$$

After introduction of dielectric,

$$U_{fA} = KU$$
$$U_{fB} = \frac{U}{K}$$
$$U_f = KU + \frac{U}{K} = U(K + \frac{1}{K})$$
$$\text{Ratio} = \frac{U_i}{U_f} = \frac{2U}{U\left(K + \frac{1}{K}\right)}$$
$$= \frac{2K}{K^2 + 1}$$

Q. 12. Two parallel plate capacitors X and Y have the same area of plates and same separation between them. X has air between the plates while Y contains a dielectric medium of $\varepsilon_r = 4$.★

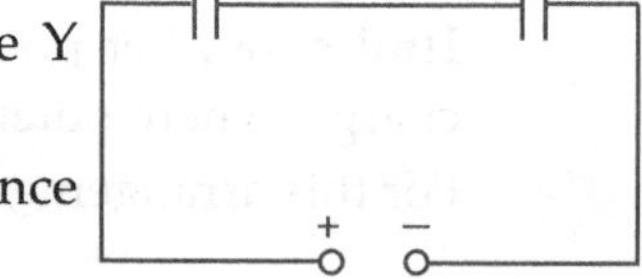

(i) Calculate capacitance of each capacitor if equivalent capacitance of the combination is 4 μF.

(ii) Calculate the potential difference between the plates of X and Y.

(iii) Estimate the ratio of electrostatic energy stored in X and Y.

Ans. (i) Let capacitance of X be C_1 and capacitance of Y be C_2.

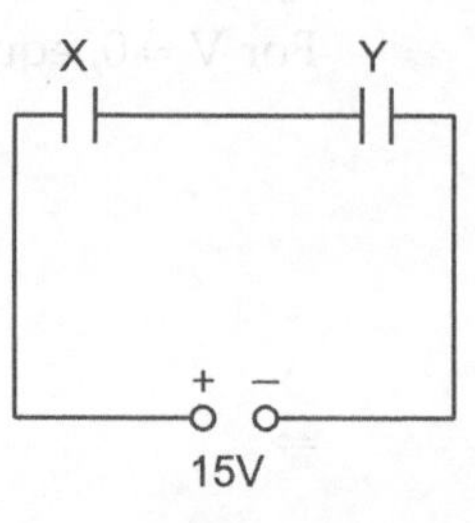

$$C_1 = \frac{\varepsilon_0 A}{d}$$
$$C_2 = \frac{\varepsilon_r \varepsilon_0 A}{d}$$
$$\frac{C_1}{C_2} = \frac{1}{\varepsilon_r}$$
$$\Rightarrow \quad C_2 = \varepsilon_r C_1$$
$$C_1 = C$$
$$C_2 = 4C \qquad [\because \varepsilon_r = 4]$$

Since two capacitance are connected in series so, equivalent capacitance will be

★ are board based questions from previous year

$$\frac{1}{C_{eq}} = \frac{1}{C_1} + \frac{1}{C_2}$$

$$C_{eq} = \frac{C_1 C_2}{C_1 + C_2}$$

$$4\mu F = \frac{C \times 4C}{C + 4C}$$

$$\Rightarrow \quad C = 5\mu F$$

So, $C_1 = 5\mu F$ and $C_2 = 20\ \mu F$

(ii) $C_{eq} V_{net} = Q_{Total}$

$$4\mu F \times 15V = Q_{Total}$$

$$Q_{Total} = 60\ \mu C$$

Since in series configuration charge on each capacitor is equal.

Hence, $Q_1 = Q_2 = Q_{Total}$

$$= 60\mu C$$

Using, $Q = CV$

$$V_1 = \frac{Q_1}{C_1} = \frac{60\ \mu C}{5\mu F} = 12V$$

$$V_2 = \frac{Q_2}{C_2} = \frac{60\mu C}{20\mu F} = 3V$$

(iii) $$U_1 = \frac{1}{2}\frac{Q_1^2}{C_1} = \frac{1}{2}\frac{(60\ \mu C)^2}{5\mu F} = 360\ \mu J$$

$$U_2 = \frac{1}{2}\frac{Q_2^2}{C_2} = \frac{1}{2}\frac{(60\ \mu C)^2}{20\mu F} = 90\ \mu J$$

$$\Rightarrow \quad \frac{U_1}{U_2} = \frac{4}{1}$$

$$\Rightarrow \quad V_1 : V_2 = 4 : 1.$$

Q. 13. A charge of 8 mC is located at the origin. Calculate the work done in taking a small charge of -2×10^{-9} C from a point P (0, 0, 3 cm) to a point Q (0, 4 cm, 0), via a point R (0, 6 cm, 9 cm). **[NCERT]**

Ans. Charge located at the origin,

$$q = 8\text{ mC} = 8 \times 10^{-3}\text{ C}$$

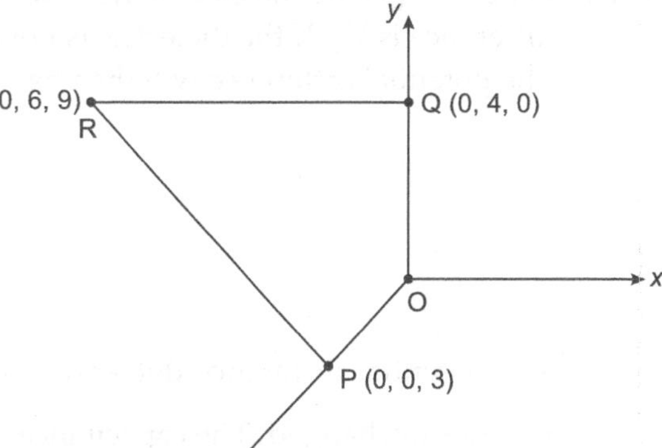

Magnitude of a small charge, which is taken from a point P to point R to point Q, $q_1 = -2 \times 10^{-9}$ C.

All the points are represented in the given figure,

Point P is at a distance, $d_1 = 3$ cm, from the origin along z-axis.

Point Q is at a distance, $d_2 = 4$ cm, from the origin along y-axis.

Potential at point P, $$V_1 = \frac{q}{4\pi\varepsilon_0 d_1}$$

Potential at point Q, $$V_2 = \frac{q}{4\pi\varepsilon_0 d_2}$$

Work done (W) by the electrostatic force is independent of the path.

$$\therefore \quad W = q_1 [V_2 - V_1]$$

$$= q_1\left[\frac{q}{4\pi\varepsilon_0 d_2} - \frac{q}{4\pi\varepsilon_0 d_1}\right]$$

$$= \frac{qq_1}{4\pi\varepsilon_0}\left[\frac{1}{d_2} - \frac{1}{d_1}\right] \quad \text{...(i)}$$

Where, $\frac{1}{4\pi\varepsilon_0} = 9 \times 10^9 \text{ N m}^2 \text{ C}^{-2}$

$\therefore$

$$W = 9 \times 10^9 \times 8 \times 10^{-3} \times (-2 \times 10^{-9})\left[\frac{1}{0.04} - \frac{1}{0.03}\right]$$

$$= -144 \times 10^{-3} \times \left(\frac{-25}{3}\right)$$

$$= 1.27 \text{ J}$$

Therefore, work done during the process is 1.27 J.

Q. 14. A molecule of a substance has a permanent electric dipole moment of magnitude 10^{-29} C m. A mole of this substance is polarised (at low temperature) by applying a strong electrostatic field of magnitude 10^6 V m^{-1}. The direction of the field is suddenly changed by an angle of 60°. Estimate the heat released by the substance in aligning its dipoles along the new direction of the field. For simplicity, assume 100% polarisation of the sample.

Ans. Dipole moment of each molecules = 10^{-29} C m

As 1 mole of the substance contains 6×10^{23} molecules.

Total dipole moment of all the molecules, $p = 6 \times 10^{23} \times 10^{-29}$ C m

$= 6 \times 10^{-6}$ C m

Initial potential energy, $U_i = -pE\cos\theta = -6 \times 10^{-6} \times 10^6 \cos 0° = -6$ J

Final potential energy (when $\theta = 60°$). $U_f = -6 \times 10^6 \times 10^6 \cos 60° = -3$J

Change in potential energy = – 3J – (– 6J) = 3J

So, there is loss in potential energy. This must be the energy released by the substance in the form of heat in aligning its dipoles.

Q. 15. A slab of material of dielectric constant k has the same area as the plates of a parallel plate capacitor but has a thickness (3/4) d. where d is the separation of the plates. How is the capacitance changed when the slab is inserted between the plates ? **[NCERT]**

Ans. Let $E_0 = V_0/d$ be the electric field between the plates when there is no dielectric and the potential difference is V_0. If the dielectric is now inserted, the electric field in the dielectric will be $E = E_0/K$. The potential difference will then be

$$V = E_0\left(\frac{1}{4}d\right) + \frac{E_0}{K}\left(\frac{3}{4}d\right)$$

$$= E_0 d\left(\frac{1}{4} + \frac{3}{4K}\right) = V_0\frac{K+3}{4K}$$

The potential difference decreases by the factor $\frac{(K+3)}{4K}$ while the free charge Q_0 on the plates remains unchanged. The capacitance thus increases

$$C = \frac{Q_0}{V} = \frac{4K}{K+3}\cdot\frac{Q_0}{V_0} = \frac{4K}{K+3}C_0$$

Q. 16. Two charged conducting spheres of radii a and b are connected to each other by a wire. What is the ratio of electric fields at the surfaces of the two spheres ? Use the result obtained to explain why charge density on the sharp and pointed ends of a conductor is higher than on its flatter portions. **[NCERT]**

Ans. Let a be the radius of a sphere A, Q_A be the charge on the sphere and C_A be the capacitance of the sphere. Let B be the radius of a sphere b, Q_B be the charge on the sphere, and C_B be the capacitance of the sphere. Since the two spheres are connected with a wire, their potential (V) will becomes equal.

Let E_A be the electric field of sphere A and E_B be the electric field of sphere B.

Their ratio, $$E_A = \frac{Q_A}{4\pi\varepsilon_0 a^2}$$

and $$E_B = \frac{Q_B}{4\pi\varepsilon_0 b^2}$$

$$\frac{E_A}{E_B} = \frac{Q_A}{4\pi\varepsilon_0 a^2} \times \frac{b^2 4\pi\varepsilon_0}{Q_B}$$

$$\frac{E_A}{E_B} = \frac{Q_A}{Q_B} \times \frac{b^2}{a^2} \quad \text{...(i)}$$

However, $$\frac{Q_A}{Q_B} = \frac{C_A V}{C_B V}$$

and $$\frac{C_A}{C_B} = \frac{a}{b}$$

$\therefore$ $$\frac{Q_A}{Q_B} = \frac{a}{b} \quad \text{...(ii)}$$

Substituting the value of of (ii) in (i), we obtain

$$\frac{E_A}{E_B} = \frac{ab^2}{ba^2} = \frac{b}{a}$$

Therefore, the ratio of electric fields at the surface is $\frac{b}{a}$.

Q. 17. Two charges $-q$ and $+q$ are located at points $(0, 0, -a)$ and $(0, 0, a)$, respectively.

(i) What is the electrostatic potential at the points $(0, 0, 2)$ and $(x, y, 0)$?

(ii) Obtain the dependence of potential on the distance r of a point from the origin when $r/a >> 1$.

(iii) How much work is done in moving a small test charge from the point $(5, 0, 0)$ to $(-7, 0, 0)$ along the x-axis ? Does the answer change if the path of the test charge between the same points is not along the x-axis. **[NCERT]**

Ans. (i) Charge $-q$ is located at $(0, 0, -a)$ and charge $+q$ is located at $(0, 0, a)$. Hence, they form a dipole which is placed along z-axis. Point $(0, 0, z)$ is on the axis of this dipole and point $(x, y, 0)$ is normal to the axis of the dipole. Hence, electrostatic potential at point $(x, y, 0)$ is zero. Fig. shows representation of condition as per question

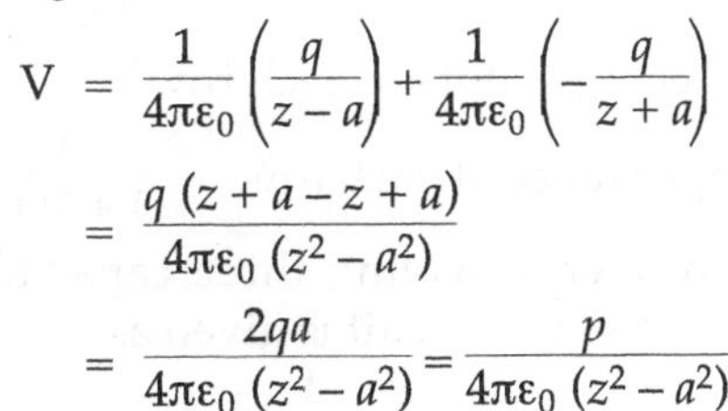

Electrostatic potential at point $(0, 0, z)$ is given by,

$$V = \frac{1}{4\pi\varepsilon_0}\left(\frac{q}{z-a}\right) + \frac{1}{4\pi\varepsilon_0}\left(-\frac{q}{z+a}\right)$$

$$= \frac{q\,(z + a - z + a)}{4\pi\varepsilon_0\,(z^2 - a^2)}$$

$$= \frac{2qa}{4\pi\varepsilon_0\,(z^2 - a^2)} = \frac{p}{4\pi\varepsilon_0\,(z^2 - a^2)}$$

where, ε_0 = Permittivity of free space,

p = dipole moment of the system of two charges = $2qa$

(ii) We know that, $$V = \frac{p\cos\theta}{4\pi\varepsilon_0\,(r^2 - a^2\cos^2\theta)}$$

If $\frac{r}{a} >> 1$ then $a << r$

$$\therefore \quad V = \frac{p \cos \theta}{4\pi\varepsilon_0 r^2}$$

$$\therefore \quad V \propto \frac{1}{r^2}$$

The potential is inversely proportional to the square of the distance.

(iii) A test charge is moved from point (5, 0, 0) to point (– 7, 0, 0) along the x-axis. Electrostatic potential (V_1) at point (5, 0, 0) is given by,

$$V_1 = \frac{-q}{4\pi\varepsilon_0} \frac{1}{\sqrt{(5-0)^2 + (-a)^2}} + \frac{q}{4\pi\varepsilon_0} \frac{1}{\sqrt{(5-0)^2 + a^2}}$$

$$= \frac{-q}{4\pi\varepsilon_0 \sqrt{25 + a^2}} + \frac{q}{4\pi\varepsilon_0 \sqrt{25 + a^2}}$$

$$V_1 = 0$$

Electrostatic potential V_2, at point (– 7, 0, 0) is given by,

$$V_2 = \frac{-q}{4\pi\varepsilon_0} \frac{1}{\sqrt{(-7)^2 + (a)^2}} + \frac{q}{4\pi\varepsilon_0} \frac{1}{\sqrt{(-7)^2 + a^2}}$$

$$= \frac{-q}{4\pi\varepsilon_0 \sqrt{49 + a^2}} + \frac{q}{4\pi\varepsilon_0 \sqrt{49 + a^2}}$$

$$= 0$$

As $V_1 = V_2$, hence no work is done in moving a small test charge from point (5, 0, 0) to point (– 7, 0, 0) along the X-axis.

The answer does not change because work done by the electrostatic field in moving a test charge between the two points is independent of the path connecting the two points.

Q. 18. An electrical technician requires a capacitance of 2 μF in a circuit across a potential difference of 1 kV. A large number of 1 μF capacitors are available to him each of which can withstand a potential difference of not more than 400 V. Suggest a possible arrangement that requires the minimum number of capacitors. **[NCERT]**

Ans. Total required capacitance, $C = 2\ \mu F$

Potential difference, $V = 1\ kV = 1000\ V$

Capacitance of each capacitor, $C_1 = 1\ \mu F$

Each capacitor can withstand a potential difference, $V_1 = 400\ V$

Suppose a number of capacitors are connected in series and these series circuits are connected in parallel (row) to each other. The potential difference across each row must be 1000 V and potential difference across each capacitor must be 400 V.

Hence, number of capacitors in each row is given as

$$n = \frac{1000}{400} = 2.5$$

Hence, there are three capacitors in each row.

$$\text{Capacitance of each row} = \frac{1}{1+1+1} = \frac{1}{3}\ \mu F$$

Let there are m rows, each having three capacitors, which are connected in parallel. Hence, equivalent capacitance of the circuit is given as

$$\frac{1}{3} + \frac{1}{3} + \frac{1}{3} + \ldots\ldots\ldots\ldots\ m \text{ terms}$$

$$= \frac{m}{3}$$

However, capacitance of the circuit is given as 2 μF,

$$\therefore \quad \frac{m}{3} = 2$$

$$m = 6$$

Hence, 6 rows of three capacitors are present in the circuit. A minimum of 6 × 3 *i.e.,* 18 capacitors are required for the given arrangement.

Q. 19. The plates of a parallel plate capacitor have an area of 90 cm^2 each and are separated by 2.5 mm. The capacitor is charged by connecting it to a 400 V supply.

(i) How much electrostatic energy is stored by the capacitor?

(ii) View this energy as stored in the electrostatic field between the plates and obtain the energy per unit volume u. Hence arrive at a relation between and the magnitude of electric field E between the plates. **[NCERT]**

Ans. (i) Capacitance of parallel plate capacitor is given by

$$C = \frac{\varepsilon_0 A}{d}$$

Electrostatic energy stored in the capacitor is given by the relation,

$$E_1 = \frac{1}{2}CV^2 \quad \ldots(i)$$

$$= \frac{1}{2}\frac{\varepsilon_0 A}{d}V^2 \quad \text{[from substituting value of C in eq. (i)]}$$

Where,

ε_0 = Permittivity of free space = $8.85 \times 10^{-12}\ C^2\ N^{-1}\ m^{-2}$

$$\therefore \quad E_1 = \frac{1 \times 8.85 \times 10^{-12} \times 90 \times 10^{-4} \times (400)^2}{2 \times 2.5 \times 10^{-3}}$$

$$= 2.55 \times 10^{-6}\ J$$

(ii) Volume of the given capacitor,

$$V' = A \times d$$

$$= 90 \times 10^{-4} \times 2.5 \times 10^{-3} = 2.25 \times 10^{-5}\ m^3$$

Energy stored in the capacitor per unit volume is given by

$$u = \frac{E_1}{V'} = \frac{2.55 \times 10^{-6}}{2.25 \times 10^{-4}} = 0.113\ J\ m^{-3}$$

Again, $$u = \frac{E_1}{V'} = \frac{\frac{1}{2}CV^2}{Ad} = \frac{\frac{1}{2}\frac{\varepsilon_0 A}{2d}V^2}{Ad} = \frac{1}{2}\varepsilon_0\left(\frac{V}{d}\right)^2$$

Where, $$\frac{V}{d} = \text{Electric field intensity} = E$$

$$\therefore \quad u = \frac{1}{2}\varepsilon_0 E^2$$

Q. 20. A parallel plate capacitor is to be designed with a voltage rating 1 kV, using a material of dielectric constant 3 and dielectric strength about 10^7 Vm^{-1}. (Dielectric strength is the maximum electric field a material can tolerate without breakdown, *i.e.,* without starting to conduct electricity through partial ionisation.) For safety, we should like the field never to exceed, say 10% of the dielectric strength. What minimum area of the plates is required to have a capacitance of 50 pF ? **[NCERT]**

Ans. Potential rating of a parallel plate capacitor, V = 1 kV = 1000 V

Dielectric constant of a material, $\varepsilon_r = 3$

Dielectric strength = 10^7 V/m

For safety, the field intensity should never exceeds 10% of the dielectric strength.

Hence, electric field intensity, E = 10% of 10^7 = 10^6 V/m

From question, capacitance of the parallel plate capacitor, $C = 50\ pF = 50 \times 10^{-12}\ F$

Distance between the plate is given by,

$$d = \frac{V}{E} = \frac{1000}{10^6} = 10^{-3}\ m$$

Capacitance is given by the relation,

$$C = \frac{\varepsilon_0 \varepsilon_r A}{d}$$

where A = Area of each plate

ε_0 = Permittivity of free space = $8.85 \times 10^{-12}\ C^2\ N^{-1}\ m^{-2}$

$$\therefore \quad A = \frac{Cd}{\varepsilon_0 \varepsilon_r} = \frac{50 \times 10^{-12} \times 10^{-3}}{8.85 \times 10^{-12} \times 3} = 19\ cm^2$$

Hence, the area of each plate is about $19\ cm^2$.

Q. 21. A parallel plate capacitor with air between the plates has a capacitance of 8 pF ($1\ pF = 10^{-12}\ F$). What will be the capacitance if the distance between the plates is reduced by half and the space between them is filled with a substance of dielectric constant 6 ? **[NCERT]**

Ans. Capacitance between the parallel plates of the capacitor, C = 8 pF

Initially, distance between the parallel plates was d and it was filled with air. Dielectric constant of air, $k = 1$

Capacitance of parallel plate capacitor is given by

$$C = \frac{k\,\varepsilon_0 A}{d} \qquad \text{...(i)}$$

Where, A = Area of each plate.

ε_0 = Permittivity of free space

If distance between the plates is reduced to half, then new distance, $d' = \frac{d}{2}$

Dielectric constant of the substance filled in between the plates, $k = 6$

Hence, capacitance of the capacitor becomes

$$C' = \frac{k\,\varepsilon_0 A}{d'} = \frac{6\varepsilon_0 A}{\frac{d}{2}} \qquad \text{...(ii)}$$

Taking ratios of equations (i) and (ii), we obtain

$$C' = 2 \times 6C = 12C = 12 \times 8 = 96\ pF$$

Therefore, the capacitance between the plates is 96 pF.

Chapter 3. Current Electricity

Q. 1. Why an ammeter is always connected in series while the voltmeter is connected in parallel with the rest of the circuit.

Ans. As ammeter has very low resistance, therefore connection of ammeter in series will not change in actual current which is required to measure. That's why it is connected in series to the circuit. While voltmeter connected in parallel or shunt so that voltage (potential difference) which is to be measured will not change during measurement as voltmeter has very high resistance.

Q. 2. A nichrome heating element across 230 V supply consumes 1 kW of power and heats up to a temperature of 750°C. A tungsten bulb across the same supply operates at much higher temperature of 1600°C in order to be able to emit light. Does it means the tungsten bulb necessarily consumes more power ?

Ans. No, as temperature of nichrome and tungsten bulbs are different for emitting light and heat.

Q. 3. What conclusion can you draw from the following observations on a resistor made of alloy manganin ? **[NCERT]**

Current (A)	Voltage (V)	Current (A)	Voltage (V)
0.2	3.94	3.0	59.2
0.4	7.87	4.0	78.8
0.6	11.8	5.0	98.6
0.8	15.7	6.0	118.5
1.0	19.7	7.0	138.2
2.0	39.4	8.0	158.0

Ans. From the given table it is observed that the ratio of voltage with current is a constant, for all values, which is equal to 19.7. It means manganin alloy obeys ohms law. Hence, manganin is an ohmic conductor.

Q. 4. A cell of emf 'E' and internal resistance 'r' is connected across a variable resistor 'R'. Plot a graph showing variation of terminal voltage 'V' of the cell versus the current 'I'. Using the plot, show how the emf of the cell and its internal resistance can be determined ?

Ans. The terminal voltage 'V' of the cell is given by

$$V = -Ir + E$$

Where E is the emf of the cell,

r is the internal resistance of the cell.

and, I is the current through the circuit.

Comparing with the equation of a straight line $y = mx + c$, we get

$$y = V; x = I; m = -r; c = E$$

E

V (Terminal voltage)

slope = r

I (Current) →

Graph showing variation of terminal voltage 'V' of the cell versus the current 'I'.

Emf of the cell = Intercept on V axis.

Internal resistance = slope of the line.

Q. 5. Answer the following :★

(i) Why are the connections between the resistors in a meter bridge made of thick copper strips ?

(ii) Why is it generally preferred to obtain the balance point in the middle of the meter bridge wire ?

(iii) Which material is used for the meter bridge wire and why ?

Ans. (i) The connection between the resistors in a meter bridge is made of thick copper strips because the resistivity of a copper wire is very low. As, the connections are thick, so the area becomes large and the resistance of the wires becomes almost negligible.

(ii) It is preferred to obtain the balance point in the middle of the meter bridge wire because it improves the sensitivity of the meter bridge.

(iii) Constantan is used for meter bridge wire because its temperature coefficient of resistance is almost negligible due to which the resistance of the wire does not get affected on increasing temperature of the wire due to flow of current.

Q. 6. Plot a graph showing the variation of current density (j) versus the electric field (E) for two conductors of different materials. What information from this plot regarding the properties of the conducting material, can be obtained which can be used to select suitable materials for use in making (i) standard resistance and (ii) connecting wires in electric circuits ?

★ are board based questions from previous year

Electron drift speed is estimated to be of the order of mm s^{-1}. Yet large current of the order of few amperes can be set-up in the wire. Explain briefly.★

Ans. The current density j is related to electric field as

$$E = \sigma E$$

Here, σ is the conductivity of the material

The above relation is equivalent to Ohm's law.

Now, for a given material, σ is a constant.

Hence, the plot of *j* versus E will be a straight line starting from the origin.

The slope of the graph gives the conductivity σ of the material.

The graph with a greater slope is a better conductor and the graph with a lesser slope is a poor conductor than the other.

(i) **To make a standard resistance :** A resistor should allow only a limited curent to flow through it. Hence, from the graph, we come to know that the material whose plot is similar to plot (2) should be used to make a resistor.

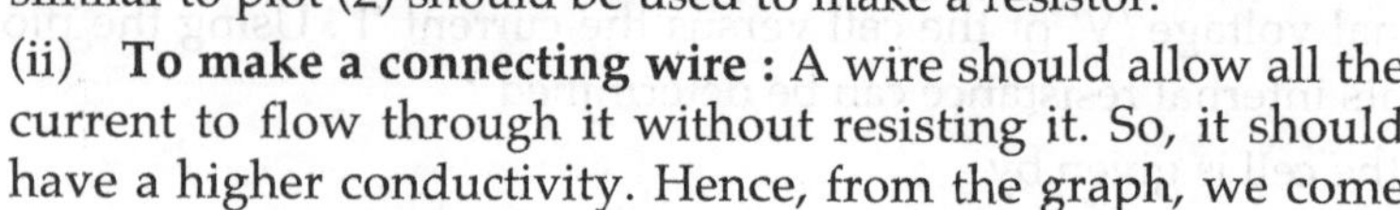

(ii) **To make a connecting wire :** A wire should allow all the current to flow through it without resisting it. So, it should have a higher conductivity. Hence, from the graph, we come to know that the material whose plot is similar to plot (1) should be used to make a wire.

Electron drift speed is estimated to be of the order of mm s^{-1}. However the current set-up in the wires is of the order of few amperes. This is because the electron density is very large in a material. It is of the order $10^{28}/m^3$ of the wire. Hence, all these electrons contribute to the total current. Therefore, despite having small drift speeds, the current set-up in wires is large.

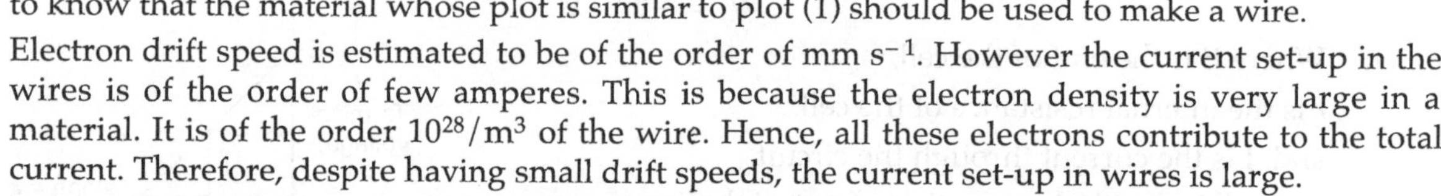

Q. 7. (i) The electron drift speed is estimated to be only a few mm s^{-1} for currents in the range of a few amperes ? How then is current established almost the instant a circuit is closed ?

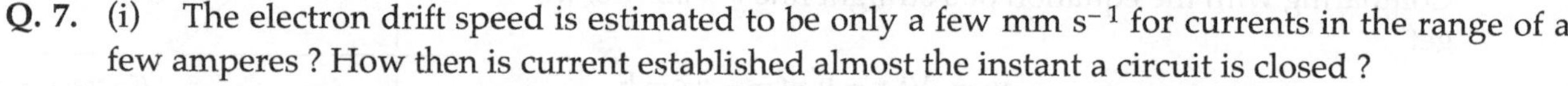

(ii) The electron drift arises due to the force experienced by electrons in the electric field inside the conductor. But force should cause acceleration. Why then do the electrons acquire a steady average drift speed ?

(iii) If the electron drift speed is so small and the electron's charge is small, how can we still obtain large amounts of current in a conductor ?

(iv) When electrons drift in a metal from lower to higher potential, does it mean that all the 'free' electrons of the metal are moving in the same direction ?

(v) Are the paths of electrons straight lines between successive collisions (with the positive ions of the metal) in the (a) absence of electric field, (b) presence of electric field ? **[NCERT]**

Ans. (i) Electric field is established throughout the circuit, almost instantly causing at every point a local electron drift. Establishment of a current does not have to wait for eletrons from one end of the conductor travelling to the other end. However, it does take a little while for the current to reach its steady value.

(ii) Each 'free' electron does accelerate, increasing its drift speed until it collides with a positive ion of the metal. It loses its drift speed after collision but starts to accelerate and increases its drift speed again only to suffer a collision again and so on. Therefore, electrons acquire only a drift speed (on average).

(iii) Because the electron number density is enormous, ~ 10^{29} m^{-3}.

(iv) No, the drift velocity is superposed over the large random velocities of electrons.

(v) (a) In the absence of electric field, the paths are straight lines, (b) in presence of electric field, the paths are, curved (generally).

Q. 8. Two wires of equal length, one of aluminium and the other of copper have the same resistance. Which of the two wires is lighter ? Hence explain why aluminium wires are preferred for

---

★ are board based questions from previous year

overhead power cables. ($\rho_{Al} = 2.63 \times 10^{-8}\ \Omega$ m, $\rho_{cu} = 1.72 \times 10^{-8}\ \Omega$ m, Relative density of Al = 2.7, of Cu = 8.9). **[NCERT]**

Ans. Given,

Resistivity of Al, $\rho_{Al} = 2.63 \times 10^{-8}\ \Omega$ m

Relative density of Al, $d_1 = 2.7$

Let I_1 be the length of Al wire and m_1 be its mass.

Resistance of the Al wire = R_1

Area of cross-section of the Al wire = A_1

Resistivity of copper, $\rho_{Cu} = 1.72 \times 10^{-8}\ \Omega$ m

Relative density of copper, $d_2 = 8.9$

Let I_2 be the length of copper wire and m_2 be its mass.

Resistance of the copper wire = R_2

Area of cross-section of the copper wire = A_2

Therefore, $$R_1 = \rho_1 \frac{l_1}{A_1} \quad \text{...(i)}$$

$$R_2 = \rho_2 \frac{l_2}{A_2} \quad \text{...(ii)}$$

It is given that, $$R_1 = R_2$$

$$\rho_l \frac{l_1}{A_1} = \rho_2 \frac{l_2}{A_2}$$

and, $$I_1 = I_2$$

$\therefore$ $$\frac{\rho_1}{A_1} = \frac{\rho_2}{A_2}$$

$$\frac{A_1}{A_2} = \frac{\rho_1}{\rho_2}$$

$$= \frac{2.63 \times 10^{-8}}{1.72 \times 10^{-8}} = \frac{2.63}{1.72}$$

Q. 9. An electric toaster uses nichrome for its heating element. When a negligibly small current passes through it, its resistance at room temperature (27.0 °C) is found to be 75.3 Ω. When the toaster is connected to a 230 V supply, the current settles, after a few seconds, to a steady value of 2.68 A. What is the steady temperature of the nichrome element ? The temperature coefficient of resistance of nichrome averaged over the temperature range involved, is $1.70 \times 10^{-4}\ °C^{-1}$. **[NCERT]**

Ans. When the current through the element is very small, heating effects can be ignored and the temperature T_1 of the element is the same as room temperature. When the toaster is connected to the supply, its initial current will be slightly higher than its steady value of 2.68 A. But due to heating effect of the current, the temperature will rise. This will cause an increase in resistance and a slight decrease in current. In a steady state resistance R_2 at the steady temperature T_2 is

$$R_2 = \frac{230 \text{ V}}{2.68 \text{ A}} = 85.8\ \Omega$$

From $$R_2 = R_1 [1 + \alpha (T_2 - T_1)]$$

(temperature dependence of resistance)

$\alpha = 1.70 \times 10^{-4}\ °C^{-1}$,

$$T_2 - T_1 = \frac{(85.8 - 75.3)}{(75.3) \times 1.70 \times 10^{-4}} = 820\ °C$$

$\Rightarrow$ $$T_2 = (820 + 27.0)\ °C = 847\ °C$$

Thus, the steady temperature of the heating element is 847 °C.

Q. 10. Given figure shows a potentiometer circuit for comparison of two resistances. The balance point with a standard resistor R = 10.0 Ω is found to be 58.3 cm, while that with the unknown resistance X is 68.5 cm. Determine the value of X. What might you do if you failed to find a balance point with the given cell of emf ε ? **[NCERT]**

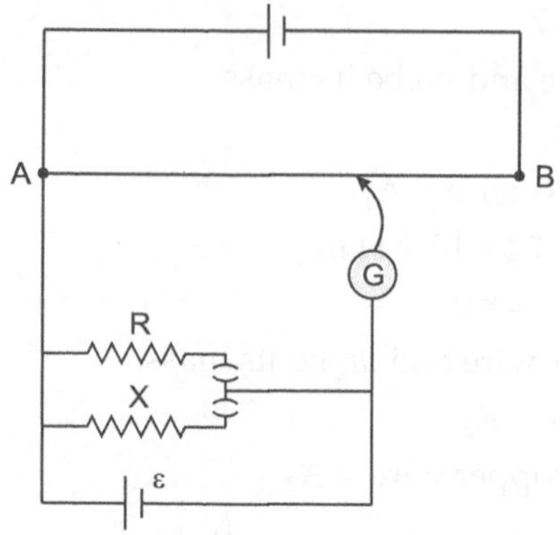

Ans. Given,

Resistance of the standard resistor, R = 10.0 Ω

Balance point for this resistance, l_1 = 58.3 cm

Current in the potentiometer wire = i

Hence, potential drop across R, $E_1 = iR$

Resistance of the unknown resistor = X

Balance point for this resistor, l_2 = 68.5 cm

Hence, potential drop across X, $E_2 = iX$

The relation connecting emf and balance point is

$$\frac{E_1}{E_2} = \frac{l_1}{l_2}$$

$$\frac{iR}{iX} = \frac{l_1}{l_2}$$

$$X = \frac{l_1}{l_2} \times R$$

$$= \frac{68.5}{58.3} \times 10 = 11.749\ \Omega$$

Therefore, the value of the unknown resistance, X is 11.75 Ω.

If we fail to find a balance point with the given cell of emf, ε, then the potential drop across R and X must be reduced by putting a resistance in series with it. Only if the potential drop across R or X is smaller than the potential drop across the potentiometer wire AB, a balance point is obtained.

Q. 11. The earth's surface has a negative surface charge density of 10^{-9} C m^{-2}. The potential difference of 400 kV between the top of the atmosphere and the surface results (due to the low conductivity of the lower atmosphere) in a current of only 1800 A over the entire globe. If there were no mechanism of sustaining atmospheric electric field, how much time (roughly) would be required to neutralise the earth's surface ? (This never happens in practice because there is a mechanism to replenish electric charges, namely the continual thunderstorms and lightning in different parts of the globe). (Radius of earth = 6.37×10^6 m). **[NCERT]**

Ans. Given,

Surface charge density of the earth, $\sigma = 10^{-9}$ C m^{-2}

Current over the entire globe, I = 1800 A

Radius of the earth, $r = 6.37 \times 10^6$ m

Surface area of the earth,

$$A = 4\pi r^2$$

$$= 4\pi \times (6.37 \times 10^6)^2$$

$$= 5.09 \times 10^{14}\ m^2$$

$$\text{Charge on the earth surface } (q) = \sigma \times A$$

$$= 10^{-9} \times 5.09 \times 10^{14}$$

$$= 5.09 \times 10^5\ C$$

Time taken to neutralise the earth's surface $= t$

$$\text{Current (I)} = \frac{q}{t}$$

$$t = \frac{q}{I}$$

$$= \frac{5.09 \times 10^5}{1800} = 282.77\ s$$

Therefore, the time taken to neutralise the earth's surface is 282.77 s.

Q. 12. Find the relation between drift velocity and relaxation time of charge carriers in a conductor. A conductor of length L is connected to a d.c. source of emf 'E'. If the length of the conductor is tripled by stretching it, keeping 'E' constant, explain how its drift velocity would be affected.★

Ans. All free electrons suffer collisions with the heavy fixed ions inside the conductor. After collisions, these electrons again emerge with the same speed, but in random directions. So, at given time, net velocity of the electrons is zero *i.e.,*

$$\vec{u}_{avg} = \frac{\vec{u}_1 + \vec{u}_2 + \ldots + \vec{u}_n}{n} = 0$$

If the electric field established inside the conductor, electrons get accelerated, so

$$a = -\frac{eE}{m}$$

Now, the average velocity of all electrons is given by

$$\vec{v}_d = \vec{u}_{avg} - \frac{eE}{m}\tau$$

$$\vec{v}_d = 0 - \frac{eE}{m}\tau$$

where τ is the average relaxation time.

But $$E = \frac{V}{l}$$

$$\vec{v}_d = -\frac{eV\tau}{ml}$$

$$v_d \propto \frac{1}{l}$$

As the drift velocity is inversely proportional to its length, so the drift velocity would be reduced by one third if the length of the conductor is tripled.

Q. 13. (i) In a meter bridge [given figure] the balance point is found to be at 39.5 cm from the end A, when the resistor Y is of 12.5 Ω. Determine the resistance of X. Why are the connections between resistors in a Wheatstone or meter bridge made of thick copper strips ?★

(ii) Determine the balance point of the bridge above if X and Y are interchanged.

(iii) What happens if the galvanometer and cell are interchanged at the balance point of the bridge ? Would the galvanometer show any current ?

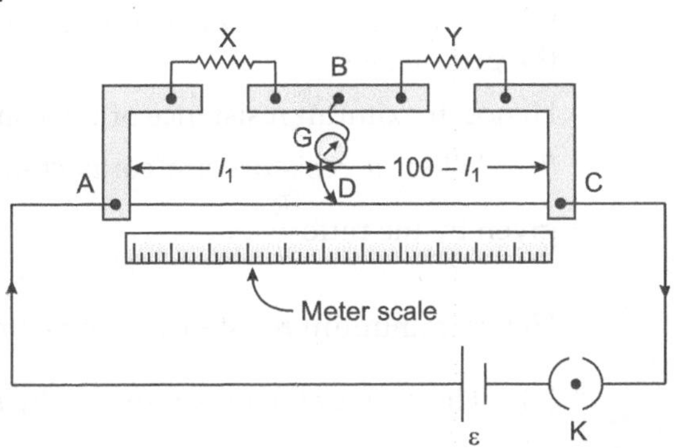

[NCERT]

★ are board based questions from previous year

Ans. A meter bridge with resistors X and Y is represented in the given figure.

(i) Balance point from end A, $l_1 = 39.5$ cm

Resistance of the resistor Y = 12.5 Ω

Condition for the balance is given as,

$$\frac{X}{Y} = \frac{100 - l_1}{l_1}$$

$$X = \frac{100 - 39.5}{39.5} \times 12.5 = 8.2\ \Omega$$

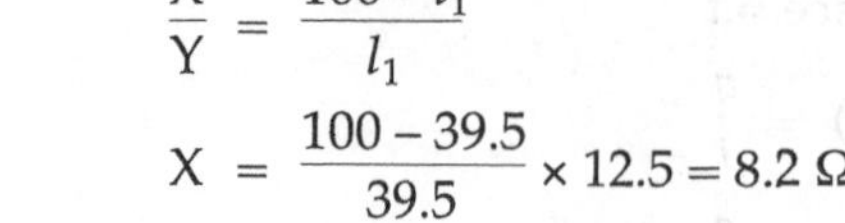

Therefore, the resistance of resistor X is 8.2 Ω.

The connection between resistors in a Wheatstone or meter bridge is made of thick copper strips to minimize the resistance, which is not taken into consideration in the bridge formula.

(ii) If X and Y are interchanged then l_1 and $100 - l_1$ get interchanged.

The balance point of the bridge will be $100 - l_1$ from A.

$$100 - l_1 = 100 - 39.5 = 60.5 \text{ cm}$$

Therefore, the balance point is 60.5 cm from A.

(iii) When the galvanometer and cell are interchanged at the balance point of the bridge, the galvanometer will show no deflection. Hence, no current would flow through the galvanometer.

Q. 14. (i) Given n resistors each of resistance R, how will you combine them to get the (a) maximum (b) minimum effective resistance ? What is the ratio of the maximum to minimum resistance ?

(ii) Given the resistances of 1 Ω, 2 Ω, 3 Ω, how will be combine them to get an equivalent resistance of (i) $\left(\frac{11}{3}\right)$ Ω (ii) $\left(\frac{11}{5}\right)$ Ω. (iii) 6 Ω. (iv) $\left(\frac{6}{11}\right)$ Ω ?

(iii) Determine the equivalent resistance of networks shown in figure. **[NCERT]**

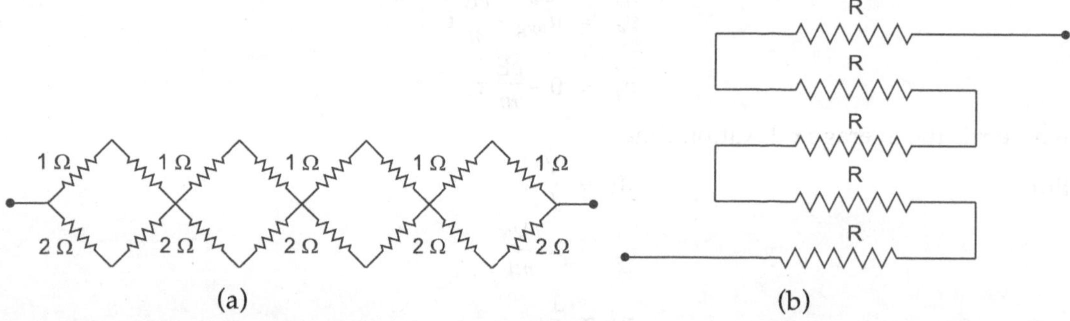

(a)   (b)

Ans. (i) Total number of resistors = n

Resistance of each resistor = R

(a) When n resistors are connected in series, equivalent resistance R_1 is the maximum, given by the product nR.

Hence, maximum resistance of the combination, $R_1 = nR$

(b) When n resistors are connected in parallel, the effective resistance (R_2) is the minimum, given by the ratio $\frac{R}{n}$.

Hence, minimum resistance of the combination, $R_2 = \frac{R}{n}$

(c) The ratio of the maximum to the minimum resistance is,

$$\frac{R_1}{R_2} = \frac{nR}{R/n} = n^2$$

(ii) The resistance of the given resistors is,

$R_1 = 1\ \Omega,\ R_2 = 2\ \Omega,\ R_3 = 3\ \Omega$

(a) Equivalent resistance, (R′) $= \frac{11}{3}\ \Omega$

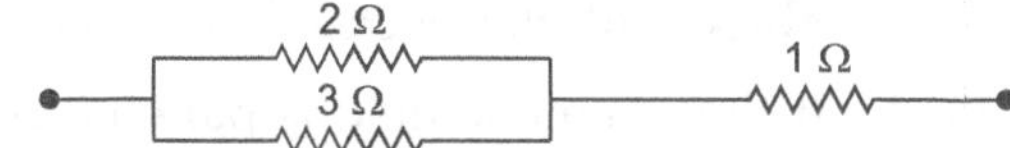

Consider the following combination of the resistors.
Equivalent resistance of the circuit is given by,

$$R' = \frac{2 \times 1}{2 + 1} + 3 = \frac{2}{3} + 3 = \frac{11}{3}\ \Omega$$

(b) Equivalent resistance, (R′) $= \frac{11}{5}\ \Omega$

Consider the following combination of the resistors.
Equivalent resistance of the circuit is given by

$$R' = \frac{2 \times 3}{2 + 3} + 1 = \frac{6}{5} + 1 = \frac{11}{5}\ \Omega$$

(c) Equivalent resistance, (R′) $= 6\ \Omega$

Consider the series combination of the resistors, as shown in the side circuit,

Equivalent resistance of the circuit is given by the sum,

$$R' = 1 + 2 + 3 = 6\ \Omega$$

(d) Equivalent resistance (R′) $= \frac{6}{11}\ \Omega$

Consider the parallel combination of the resistors, as shown in the given circuit :

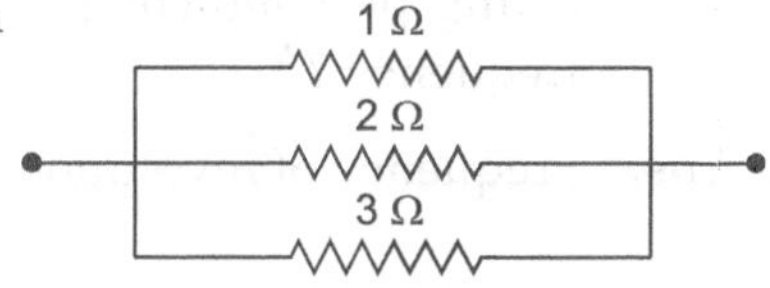

Equivalent resistance of the circuit is given by,

$$R' = \frac{1 \times 2 \times 3}{1 \times + 2 \times 3 + 3 \times 1} = \frac{6}{11}\ \Omega$$

(iii) (a) It can be observed from the given circuit that in the first small loop, two resistors of resistance 1 Ω each are connected in series.

Hence, their equivalent resistance = (1 + 1) = 2 Ω

It can also be observed that two resistors of resistance 2 Ω each are connected in series.

Hence, their equivalent resistance = (2 + 2) = 4 Ω

Therefore, the circuit can be redrawan as

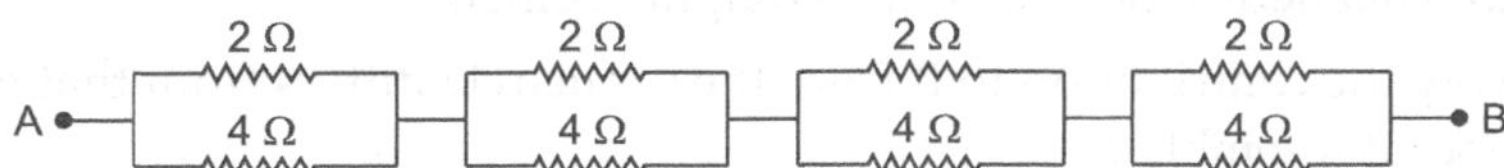

It can be observed that 2 Ω and 4 Ω resistors are connected in parallel in all the four loops. Hence, equivalent resistance (R′) of each loop is given by,

$$R' = \frac{2 \times 4}{2 + 4} = \frac{8}{6} = \frac{4}{3}\ \Omega$$

The circuit reduces to,

A $\frac{4}{3}\Omega$ $\frac{4}{3}\Omega$ $\frac{4}{3}\Omega$ $\frac{4}{3}\Omega$ B

All the four resistors are connected in series.

Hence, equivalent resistance of the given circuit is $\frac{4}{3} \times 4 = \frac{16}{3}\ \Omega$

(b) It can be observed from the given circuit that five resistors of resistance R each are connected in series.

Hence, equivalent resistance of the circuit = R + R + R + R + R = 5 R

Chapter 4. Moving Charges and Magnetism

Q. 1. Is it possible to accelerate a particle like electron using cyclotron ? Why ?

Ans. Cyclotron is not suitable for accelerating electrons. In this, the angular velocity of the particle should remain constant to keep the particle in phase with the alternating electric field. The mass of electron changes largely due to its increasing velocity. Hence, the electron soon gets out of phase with the electric field and does not reach 'dees' at the correct instant to be accelerated.

Q. 2. If the magnetic field is parallel to the positive y-axis and the charged particle is moving along the positive x-axis (Fig.) which way would the Lorentz force be for (a) an electron (negative charge). (b) a proton (positive charge). **[NCERT]**

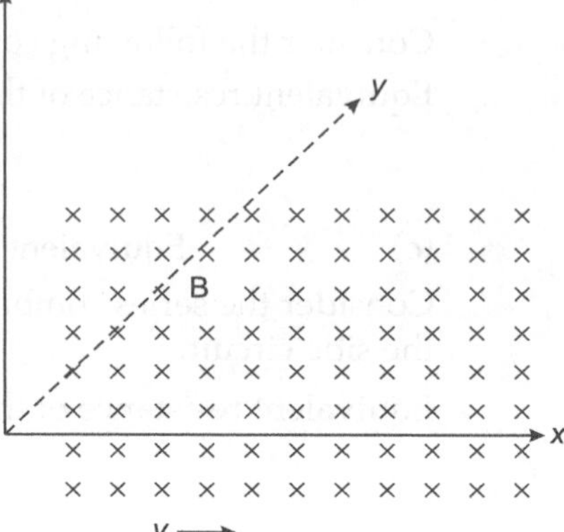

Ans. The velocity v of particle is in the direction of x-axis, while B, the magnetic field is in the direction of y-axis, so $v \times B$ is along the z-axis (screw rule or right-hand thumb rule).

(i) Thus, for electron it will be along $-z$-axis.

(ii) For a positive charge (proton) the force is along $+z$-axis.

Q. 3. A proton and an electron travelling along parallel paths enter a region of uniform magnetic field, acting perpendicular to their paths. Which of them will move in a circular path with higher frequency ?★

Ans. Frequency of revolution of a particle,

$$f = \frac{Bq}{2\pi m}$$

or

$$f \propto \frac{1}{m}$$

Since mass of electron is less than that of proton, therefore, its frequency of revolution will be higher than that of electron.

Q. 4. A straight wire carrying a current of 12 A is bent into a semi-circular arc of radius 2.0 cm as shown in fig. (a). Consider the magnetic field B at the centre of the arc.

(i) What is the magnetic field due to the straight segments ?

(ii) In what way the contribution to B from the semicircle differs from that of a circular loop and in what way does it resemble ?

(iii) Would your answer be different if the wire were bent into a semi-circular arc of the same radius but in the opposite way as shown in fig. (b) ? **[NCERT]**

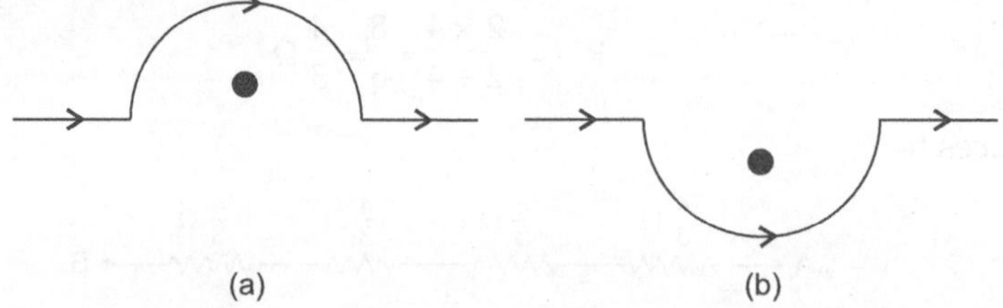

Ans. (i) dl and r for each element of the straight segments are parallel. Therefore, $dl \times r = 0$. Thus, straight segments do not contribute to $|B|$.

(ii) For all segments of the semicircular arc, $dl \times r$ are all parallel to each other (into the plane of the paper). All such contributions add up in magnitude. Hence direction of B for a semicircular

★ are board based questions from previous year

arc is given by the right-hand rule and magnitude is half that of a circular loop. Thus B is 1.9×10^{-4} T normal to the plane of the paper going into it.

(iii) From right hand thumb rule, magnitude of B will be same but direction is opposite to that in fig. (b).

Q. 5. Find the condition under which the charged particles moving with different speeds in the presence of electric and magnetic field vectors can be used to select charged particles of a particular speed.★

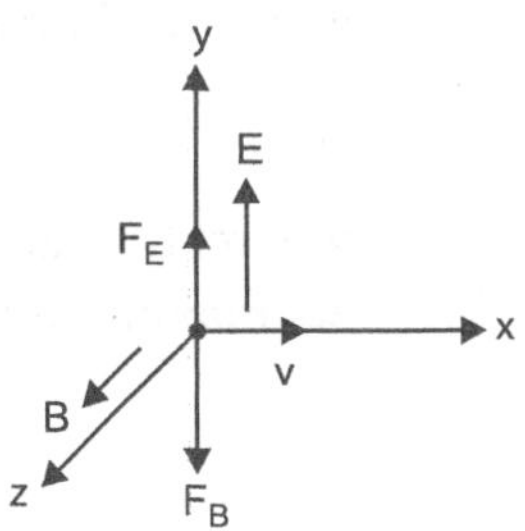

Ans. (i) The velocity $\vec{v}$, of the charged particles, and the $\vec{E}$ and $\vec{B}$ vectors, should be mutually perpendicular.

Also the forces on q, due to $\vec{E}$ and $\vec{B}$, must be oppositively directed.

(ii) If magnetic force = Electrostatic force

$$F_m = F_e$$

$$qvB = qE$$

$$v = \frac{E}{B}$$

Q. 6. Answer the following questions :

(i) A magnetic field that varies in magnitude from point to point but has a constant direction (east to west) is set up in a chamber. A charged particle enters the chamber and travels undeflected along a straight path with constant speed. What can you say about the initial velocity of the particle ?

(ii) A charged particle enters an environment of a strong and non-uniform magnetic field varying from point to point both in magnitude and direction, and comes out of it following a complicated trajectory. Would its final speed equal the initial speed if it suffered no collisions with the environment ?

(iii) An electron travelling west to east enters a chamber having a uniform electrostatic field in north to south direction. Specify the direction in which a uniform magnetic field should be set up to prevent the electron from deflecting from its straight line path. **[NCERT]**

Ans. (i) The initial velocity of the particle is in parallel or anti-parallel to the magnetic field. Than only it travels along a straight path without suffering any deflection in the field.

(ii) Yes, the final speed of the charged particle will be same to its initial speed, because magnetic force can change the direction of velocity, but its magnitude will be same.

(iii) An electron travelling from west to east enters a chamber having a uniform electrostatic field in north to south direction. Moving electron can remain undeflected if the force acting on it is equal and opposite of magnetic field. Magnetic force is directed towards the south. From Fleming's left hand rule, magnetic field should be applied in a vertically downward direction.

Q. 7. Show that cyclotron frequency is independent of energy of the particle. Is there an upper limit on the energy acquired by the particle ? Give reason.★

Ans. Centripetal force, needed by the charged particle to move in circular track, is provided by the magnetic field.

$$\frac{mv^2}{r} = qvB$$

$$\Rightarrow \quad v = \frac{qBr}{m}$$

★ are board based questions from previous year

Period of revolution, $$T = \frac{2\pi r}{v} = \frac{2\pi rm}{qBr}$$

$$\Rightarrow \quad T = \frac{2\pi m}{qB}$$

$$\Rightarrow \quad \nu = \frac{1}{T} = \frac{qB}{2\pi m}$$

Thus, frequency of revolution is independent of the energy of the particle.

Yes, there is an upper limit on the energy acquired by the charged particle. The charged particle gains maximum speed when it moves in a path of radius equal to the radius of the dees.

i.e., $$v_{max} = \frac{qBR}{m}$$

where, R = radius of the dees.

So, maximum kinetic energy

$$K.E. = \frac{1}{2} mv_{max}^2 = \frac{1}{2} m \left(\frac{qBR}{m}\right)^2 = \frac{q^2B^2R^2}{2m}$$

Q. 8. An observer to the left of a solenoid of N turns each of cross section area A observes that a steady current I in it flows in the clockwise direction. Depict the magnetic field lines due to the solenoid specifying its polarity and show that it acts as a bar magnet of magnetic momentum M = NIA.★

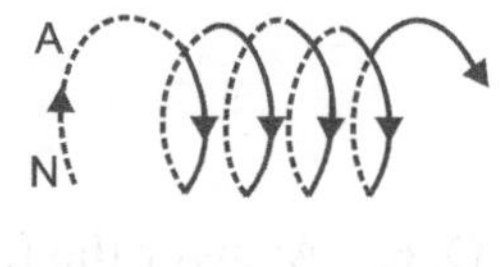

Ans. Given that the current flows in the clockwise direction for an observer on the left side of the solenoid. This means that left face of the solenoid acts as south pole and right face acts as north pole. Inside a bar magnet the magnetic field lines are directed from south to north. Therefore, the magnetic field lines are directed from left to right in the solenoid.

Magnetic moment of single current carrying loop is given by $m' = IA$

where,

I = Current flowing through the loop

A = Area of the loop

So, Magnetic moment of the whole solenoid is given by

$$M = Nm' = N(IA)$$

Q. 9. A solenoid 60 cm long and of radius 4.0 cm has 3 layers of windings of 300 turns each A 2.0 cm long wire of mass 2.5 g lies inside the solenoid (near its centre) normal to its axis; both the wire and the axis of the solenoid are in the horizontal plane. The wire is connected through two leads parallel to the axis of the solenoid to an external battery which supplies a current of 6.0 A in the wire. What value of current (with appropriate sense of circulation) in the windings of the solenoid can support the weight of the wire ? $g = 9.8 \text{ m/s}^{-2}$. **[NCERT]**

Ans. Given,

Length of the solenoid (L) = 60 cm = 0.6 m

Radius of the solenoid (r) = 4.0 cm = 0.04 m

There are 3 layers of windings of 300 turns each.

Therefore total number of turns (n) = 3 × 300 = 900

Length of the wire (l) = 2 cm = 0.02 m

Mass of the wire (m) = 2.5 g = 2.5×10^{-3} kg

Current flowing in the wire (i) = 6 A

Acceleration due to gravity, $g = 9.8 \text{ m/s}^2$

★ are board based questions from previous year

Magnetic field inside the solenoid, is given by $B = \frac{\mu_0 n I}{L}$

Where,

μ_0 = Permeability of free space = $4\pi \times 10^{-7}$ T m A^{-1}

I = Current flowing through the windings of the solenoid

Magnetic force is given by

$$F = Bil = \frac{\mu_0 n I}{L} il$$

The force on the wire is equal to the weight of the wire.

$$\therefore \quad mg = \frac{\mu_0 n I\, il}{L}$$

$$I = \frac{mgL}{\mu_0 n i l}$$

$$= \frac{2.5 \times 10^{-3} \times 9.8 \times 0.6}{4\pi \times 10^{-7} \times 900 \times 0.02 \times 6} = 108 \text{ A}$$

Hence, the current flowing through the solenoid is 108 A.

Q. 10. Figure shows a long straight wire of a circular cross-section (radius a) carrying steady current I. The current I is uniformly distributed across this cross-section. Calculate the magnetic field in the region $r > a$ and $r < a$. **[NCERT]**

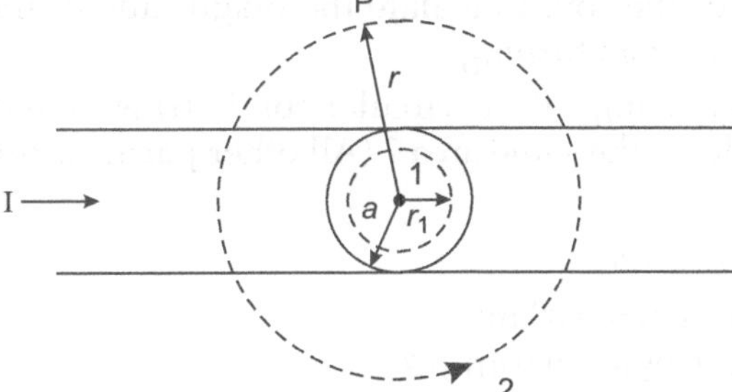

Ans. (i) Case $r > a$. The Amperian loop, labelled 2, is a circle concentric with the cross-section. For this loop,

$$L = 2\pi r$$

I_e = Current enclosed by the loop = I

The result is the familiar expression for a long straight wire

$$B(2\pi r) = \mu_0 I$$

$$B = \frac{\mu_0 I}{2\pi r}$$

$$B \propto \frac{1}{r} \ (r > a)$$

(ii) Case $r < a$. The Amperian loop is a circle labelled 1. For this loop, taking the radius of the circle to be r,

$$L = 2\pi r$$

the current enclosed I_e I, but $I_e < I$, but

Since the current distribution is uniform, the current enclosed is given by

$$I_e = I\left(\frac{\pi r^2}{\pi a^2}\right) = \frac{I r^2}{a^2}$$

Using ampere's law, $$B(2\pi r) = \mu_0 \frac{I r^2}{a^2}$$

$$B = \left(\frac{\mu_0 I}{2\pi a^2}\right) r$$

$B \propto r \ (r < a)$, Hence it can be shown in graph in side.

It shows a plot of the magnitude of B with distance r from the centre of the wire.

Q. 11. The horizontal component of the earth's magnetic field at a certain place is 3.0×10^{-5} T and the direction of the field is from the geographic south to the geographic north. A very long straight conductor is carrying a steady current of 1A. What is the force per unit length on it when it is placed on a horizontal table and the direction of the current is (i) east to west; (ii) south to north ?

Ans.

$$F = Il \times B$$
$$F = IlB \sin \theta$$

The force per unit length is $f = \frac{F}{l} = IB \sin \theta$

(i) When the current is flowing from east to west.

$$\theta = 90°$$

Hence, $f = IB = 1 \times 3 \times 10^{-5}$

$$= 3 \times 10^{-5} \text{ N m}^{-1}$$

This is more than the value 2×10^{-7} Nm^{-1} quoted in the definition of the ampere.
The direction of the force is downwards.

(ii) When the current is flowing from south to north,

$$\theta = 0°$$
$$f = 0$$

Hence, there is no force on the conductor.

Q. 12. (i) A circular coil of 30 turns and radius 8.0 cm carrying a current of 6.0 A is suspended vertically in a uniform horizontal magnetic field of magnitude 1.0 T. The field lines make an angle of 60° with the normal of the coil. Calculate the magnitude of the counter torque that must be applied to prevent the coil from turning.

(ii) Would your answer change, if the circular coil in (i) were replaced by a planar coil of some irregular shape that encloses the same area ? (All other particulars are also unaltered). **[NCERT]**

Ans. (i) Given,

Number of turns of coil (n) = 30

Radius of the coil (r) = 8.0 cm = 0.08 m

Area of the coil (πr^2) = $\pi (0.08)^2 = 0.0201$ m^2

Current flowing in the coil (I) = 6.0 A

Magnetic field strength (B) = 1 T

Angle between the field lines and normal to the coil surface (θ) = 60°

The coil experiences a torque in the magnetic field. Hence, it turns, the counter torque to prevent the coil from turning will be

$$\tau = nIBA \sin \theta$$
$$= 30 \times 6 \times 1 \times 0.0201 \times \sin 60°$$
$$= 3.133 \text{ N m}$$

(ii) From relation $\tau = nIBA \sin \theta$,

the applied torque is not dependent on the shape of the coil. It depends on the area of the coil. Therefore, the answer would not change if the circular coil in the above case is replaced by a planar coil of some irregular shape that encloses the same area.

Q. 13. For a circular coil of radius R and N turns carrying current I, the magnitude of the magnetic field at a point on its axis at a distance x from its centre is given by,

$$B = \frac{\mu_0 I R^2 N}{2 (x^2 + R^2)^{3/2}}$$

(i) Show that this reduces to the familiar result for field at the centre of the coil.

(ii) Consider two parallel co-axial circular coils of equal radius R, and number of turns N, carrying equal currents in the same direction and separated by a distance R. Show that the field on the axis around the mid-point between the coils is uniform over a distance that is small as compared to R and is given by, $B = 0.72 \frac{\mu_0 BNI}{R}$, approximately.

[Such an arrangement to produce a nearly uniform magnetic field over a small region is known as Helmholtz coils]. **[NCERT]**

Ans. Given,

Radius of circular coil = R

Number of turns on the coil = N

Current in the coil = I

Magnetic field at a point on its axis at distance x due to current in coil will be

$$B = \frac{\mu_0 IR^2 N}{2(x^2 + R^2)^{\frac{3}{2}}}$$

Here, μ_0 = Permeability of free space = $4\pi \times 10^{-7}$ T m A^{-1}.

(i) For the magnetic field at the centre of the coil $x = 0$.

$$\therefore \quad B = \frac{\mu_0 IR^2 N}{2R^3} = \frac{\mu_0 IN}{2R}$$

This is familiar result for field at the centre of the coil.

(ii) Radii of two parallel co-axial circular coils = R

Number of turns on coils = N

Current in both coils = I

Distance between both the coils = R

Take a point Q at distance d from the centre.

One coil is at a distance of $\frac{R}{2} + d$ from point Q.

Magnetic field at a point Q will be

$$\therefore \quad B_1 = \frac{\mu_0 NIR^2}{2\left[\left(\frac{R}{2} + d\right)^2 + R^2\right]^{\frac{3}{2}}}$$

the other coil is at distance of $\frac{R}{2} - d$ from point Q.

Magnetic field due to this coil will be

$$B_2 = \frac{\mu_0 NIR^2}{2\left[\left(\frac{R}{2} - d\right)^2 + R^2\right]^{\frac{3}{2}}}$$

Thus total magnetic field,

$$B = B_2 + B_1$$

$$= \frac{\mu_0 IR^2}{2}\left[\left\{\left(\frac{R}{2} - d\right)^2 + R^2\right\}^{\frac{3}{2}} + \left\{\left(\frac{R}{2} + d\right)^2 + R^2\right\}^{\frac{3}{2}}\right]$$

$$= \frac{\mu_0 IR^2}{2}\left[\left(\frac{5R^2}{4} + d^2 - Rd\right)\right]^{3/2} + \left[\left(\frac{5R^2}{4} + d^2 + Rd\right)^{3/2}\right]$$

$$= \frac{\mu_0 IR^2}{2} \times \left(\frac{5R^2}{4}\right)^{3/2}\left[\left(1 + \frac{4d^2}{5R^2} - \frac{4d}{5R}\right)^{3/2} + \left(1 + \frac{4d^2}{5R^2} + \frac{4d}{5R}\right)^{3/2}\right]$$

for $d << R$, we can neglect all $\left(\frac{d}{R}\right)^2 \left(\frac{d}{R}\right)^3$ etc., terms, therefore

$$B = \frac{\mu_0 IR^2}{2} \times \left(\frac{5R^2}{4}\right)^{3/2} \times \left[\left(1 - \frac{4d}{5R}\right)^{3/2} + \left(1 + \frac{4d}{5R}\right)^{3/2}\right]$$

$$\approx \frac{\mu_0 IR^2 N}{2R^3} \times \left(\frac{4}{5}\right)^{3/2}\left[1 - \frac{6d}{5R} + 1 + \frac{6d}{5R}\right]$$

$$B = \left(\frac{4}{5}\right)^{3/2} \times \frac{\mu_0 IN}{R} = 0.72\left(\frac{\mu_0 IN}{R}\right)$$

Thus, it is proved that the field along the axis around the mid-point between the coils is uniform.

Q. 14. A magnetic field set up using Helmholtz coils is uniform in a small region and has a magnitude of 0.75 T. In the same region, a uniform electrostatic field is maintained in a direction normal to the common axis of the coils. A narrow beam of (single species) charged particles all accelerated through 15 kV enters this region in a direction perpendicular to both the axis of the coils and the electrostatic field. If the beam remains undeflected when the electrostatic field is 9.0×10^{-5} V m^{-1}, make a simple guess as to what the beam contains. Why is the answer not unique ? **[NCERT]**

Ans. Given,

Magnetic field, B = 0.75 T

Accelerating voltage, V = 15 kV = 15×10^3 V

Electrostatic field, E = 9×10^5 V/m

Consider mass of the electron = m

Charge of the electron = e

Velocity of the electron = v

Kinetic energy of the electron = eV

$$\Rightarrow \quad \frac{1}{2}mv^2 = eV$$

$$\therefore \quad \frac{e}{m} = \frac{v^2}{2V} \quad \text{...(i)}$$

As the particle remains undeflected by electric and magnetic fields, we can conclude that the electric field has balanced the magnetic field.

$$\therefore \quad eE = evB$$

$$v = \frac{E}{B} \quad \text{...(ii)}$$

From equation (ii) and equation (i),

$$\frac{e}{m} = \frac{1}{2}\frac{\left(\frac{E}{B}\right)^2}{V} = \frac{E^2}{2VB^2}$$

$$= \frac{(9.0 \times 10^5)^2}{2 \times 15000 \times (0.75)^2}$$

$$= 4.8 \times 10^7 \text{ C/kg}$$

This value of specific charge e/m is equal to the value of deuteron or deuterium ions.

This is not a unique answer. Possible answers may be He^{++}, Li^{++}, etc.

Q. 15. A straight horizontal conducting rod of length 0.45 m and mass 60 g is suspended by two vertical wires at its ends. A current of 5.0 A is set up in the rod through the wires.

(i) What magnetic field should be set up normal to the conductor in order that the tension in the wires is zero ?

(ii) What will be the total tension in the wires if the direction of current is reversed keeping the magnetic field same as before ? (Ignore the mass of the wires.) $g = 9.8$ m s^{-2}. **[NCERT]**

Ans. Given,

Length of the rod, (l) = 0.45 m

Mass suspended by the wires, (m) = 60 g = 60×10^{-3} kg

Acceleration due to gravity (g) = 9.8 m/s^2

Current in the rod flowing in wire (I) = 5 A

(i) Force due to magnetic field (B) is equal and opposite to the weight of the wire. Thus

$$BIl = mg$$

$$\therefore \quad B = \frac{mg}{Il} = \frac{60 \times 10^{-3} \times 9.8}{5 \times 0.45} = 0.26 \text{ T}$$

A horizontal magnetic field of 0.26 T normal to the length of the conductor should be set up in order to get zero tension in the wire and from Flemings left hand rule it will be in upward magnetic force.

(ii) If the direction of the current is reversed then the force due to magnetic field and the weight of the wire acts in a vertically downward direction.

$$\therefore \quad \text{Total tension in the wire} = BIl + mg$$

$$= 0.26 \times 5 \times 0.45 + (60 \times 10^{-3}) \times 9.8$$

$$= 1.176 \text{ N}$$

Q. 16. A 100 turn closely wound circular coil of radius 10 cm carries a current of 3.2 A.

(i) What is the field at the centre of the coil ?

(ii) What is the magnetic moment of this coil ?

The coil is placed in a vertical plane and is free to rotate about a horizontal axis which coincides with its diameter. A uniform magnetic field of 2T in the horizontal direction exists such that initially the axis of the coil is in the direction of the field. The coil rotates through an angle of 90° under the influence of the magnetic field.

(iii) What are the magnitudes of the torques on the coil in the initial and final position ?

(iv) What is the angular speed acquired by the coil when it has rotated by 90° ? The moment of inertia of the coil is 0.1 kg m^2. **[NCERT]**

Ans. (i) As we know

$$B = \frac{\mu_0 NI}{2R}$$

Here, N = 100 , I = 3.2 A, and R = 0.1 m. Hence

$$B = \frac{4\pi \times 10^{-7} \times 10^2 \times 3.2}{2 \times 10^{-1}}$$

$$= \frac{4 \times 10^{-5} \times 10}{2 \times 10^{-1}} \qquad [\text{using } \pi \times 3.2 = 10]$$

$$= 2 \times 10^{-3} \text{T}$$

The direction is given by the right-hand thumb rule.

(ii) The magnetic moment is given by

$$m = NIA = NI\pi r^2$$

$$= 100 \times 3.2 \times 3.14 \times 10^{-2} = 10 \text{ A m}^2$$

The direction is given by the right hand thumb rule.

(iii)
$$\tau = |m \times B|$$

$$= mB \sin \theta$$

Initially, $\theta = 0$. Thus, initial torque $\tau_1 = 0$. Finally, $\theta = \frac{\pi}{2}$ (or 90°).

Thus, final torque $\tau_f = mB = 10 \times 2 = 20$ N m.

(iv) From Newton's second law,

$$I\frac{d\omega}{dt} = m \text{ B} \sin \theta$$

Where I is the moment of inertia of the coil.

$$\frac{d\omega}{dt} = \frac{d\omega}{d\theta}\frac{d\theta}{dt} = \frac{d\omega}{d\theta}\omega$$

Using this.

$$I\omega\, d\omega = m\,B \sin\theta\, d\theta$$

Integrating from $\theta = 0$ to $\theta = \frac{\pi}{2}$,

$$I\int_0^{\omega_f} \omega\, d\omega = mB\int_0^{\pi/2} \sin\theta\, d\theta$$

$$I\frac{\omega^2{}_f}{2} = -\,m\,B\cos\theta \Big|_0^{\pi/2} = mB$$

$$\omega_f = \left(\frac{2\,mB}{I}\right)^{\frac{1}{2}} = \left(\frac{2\times 20}{10^{-1}}\right)^{\frac{1}{2}} = 20\ s^{-1}$$

Q. 17. Two long straight parallel conductors carry steady current I_1 and I_2 separated by a distance d. If the currents are flowing in the same direction, show how the magnetic field set up if one produces an attractive force on the other. Obtain the expression for this force. Hence define one ampere.★

Ans. Magnetic field produced on the wire (carrying current I_2) due to I_1 will be

$$B = \frac{\mu_0 I_1}{2\pi d}$$

Force acting at l length is

$$F = I_2 l B$$

$$F = \frac{\mu_0 I_1 I_2 2l}{2\pi d} \text{ towards } I_1$$

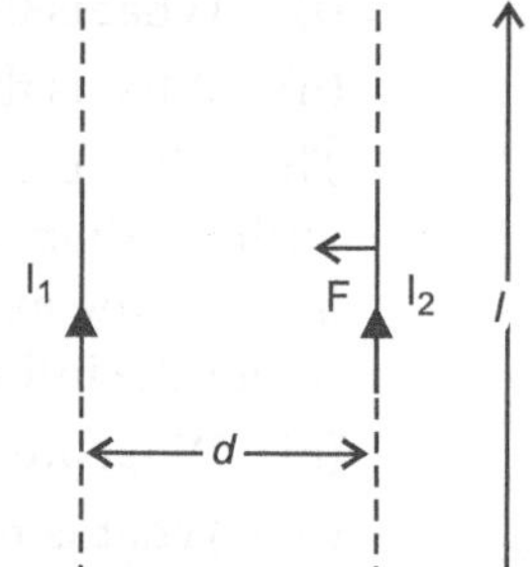

If $l = 1$m, $d = 1$m, $I_1 = I_2 = I$ and $F = 2\times 10^{-7}$N

$\Rightarrow$ $I = 1A$

So one ampere is defined as the current, which when maintained in two parallel infinite length conductors, held at a separation of one metre will produce a force of 2×10^{-7} N per metre on each conductor.

Chapter 5. Magnetism and Matter

Q. 1. Why is soft iron considered more suitable than steel for making the core of a transformer ?

Ans. The area of hysteresis loop for soft iron is very small as compared to steel. The energy dissipation during a complete cycle of magnetisation and demagnetisation is proportional to the hysteresis loop area so, smaller energy will be lost in soft iron core. Hence, it is considered more suitable as compared to steel.

Q. 2. Two spherical bobs, one metallic and the other of glass, of the same size are allowed to fall freely from the same height above the ground. Which of the two would reach earlier and why ?★

Ans. Glass bob will reach the ground earlier than the metallic bob. As the metallic bob falls, it intercepts earth's magnetic field and induced currents are set up in it which oppose its downward motion. But no such currents are induced in the glass bob.

Q. 3. Out of the two magnetic materials, 'A' has relative permeability slightly greater than unity while 'B' has less than unity. Identify the nature of the materials 'A' and 'B'. Will their susceptibilities be positive or negative ?★

Ans. For a paramagnetic material, the relative permeability lies between $1 < \mu_r < 1 + \varepsilon$ and its susceptibility lies between $0 < \chi < \varepsilon$.

Hence, 'A' is a paramagnetic material and its susceptibility is positive. This is because its relative permeability is slightly greater than unity. For a diamagnetic material, the relative permeability lies between $0 \le \mu_r < 1$ and its susceptibility lies between $-1 < \chi < 0$. Hence, 'B' is a diamagnetic material and its susceptibility is negative. This is because its relative permeability is less than unity.

Here μ_r and χ refer to the relative permeability and susceptibility.

★ are board based questions from previous year

Q. 4. (i) What happens if a bar magnet is cut into two pieces : (a) transverse to its length, (b) along its length ?

(ii) A magnetised needle in a uniform magnetic field experiences a torque but no net force. An iron nail near a bar magnet, however, experiences a force of attraction in addition to a torque. Why ?

(iii) Must every magnetic configuration have a north pole and a south pole ? What about the field due to a toroid ?

(iv) Two identical looking iron bars A and B are given, one of which is definitely known to be magnetised. (We do not know which one.) How would one ascertain whether or not both are magnetised ? If only one is magnetised, how does one ascertain which one ? [Use nothing else but the bars A and B.] **[NCERT]**

Ans. (i) In either case, one gets two magnets, each with a north and south pole.

(ii) No force if the field is uniform. The iron nail experiences a non-uniform field due to the bar magnet. There is induced magnetic moment in the nail, therefore, it experiences both force and torque. The net force is attractive because the induced south pole (say) in the nail is closer to the north pole of magnet than induced north pole.

(iii) It is not necessary, it is true only if the source of the field has a net non-zero magnetic moment. This is not true for a toroid or even for a straight infinite conductor.

(iv) Try to bring different ends of the bars closer. A repulsive force in some situation establishes that both are magnetised. If it is always attractive, then one of them is not magnetised. In a bar magnet the intensity of the magnetic field is the strongest at the two ends (poles) and weakest at the central region. This fact may be used to determine whether A or B is the magnet. In this case, to see which one of the two bars is a magnet pick up one. (Assume A) and lower one of its ends; first on one of the ends of the other (Assume B), and then on the middle of B. If you notice that in the middle of B. A experiences no force, then B is magnetised. If you do not notice any change from the end to the middle of B, then A is magnetised.

Q. 5. Many of the diagrams given below show magnetic field lines (thick lines in the figure) wrongly. Point out what is wrong with them. Some of them may describe electrostatic field lines correctly. Point out which ones.

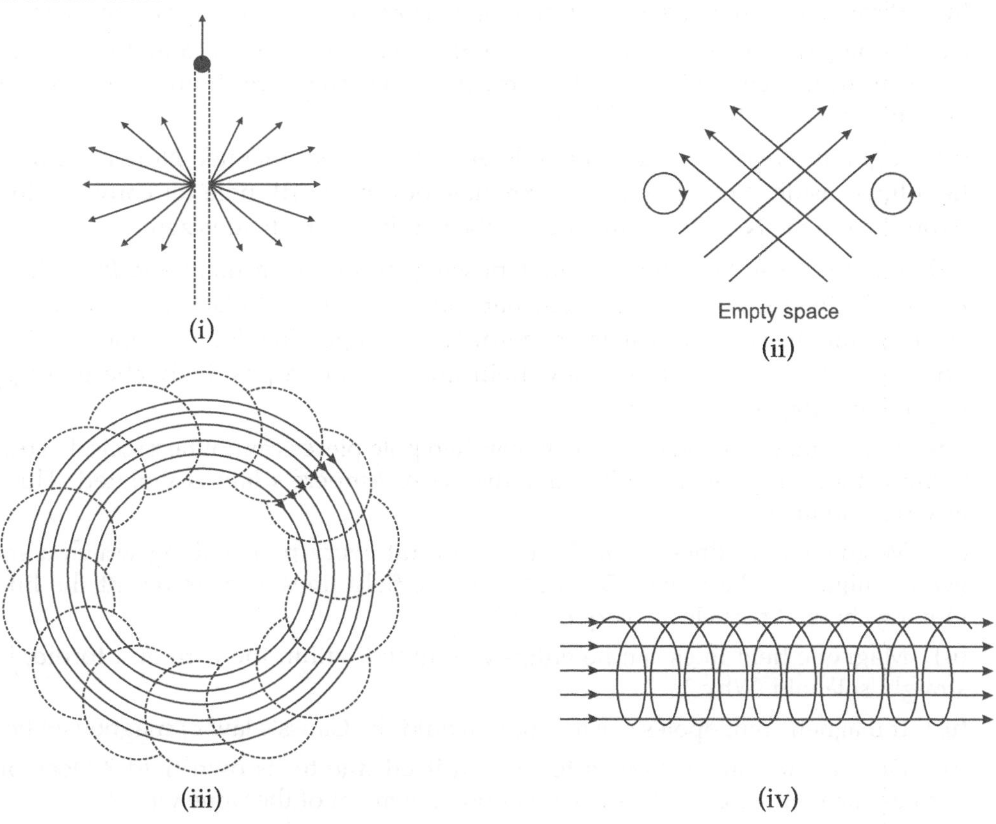

(i) (ii) (iii) (iv)

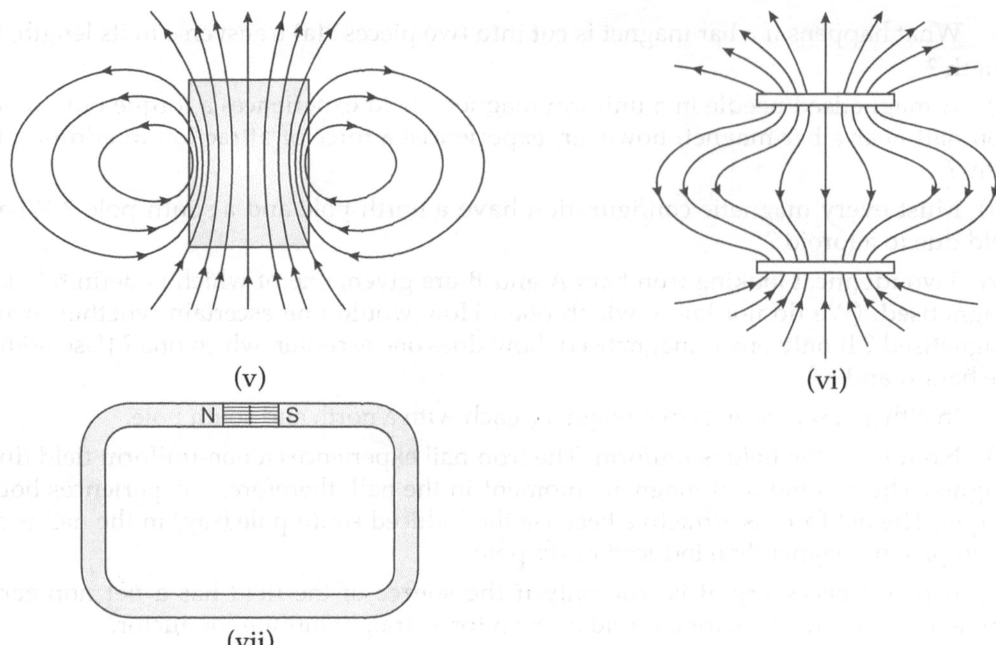

(v) (vi) (vii)

Ans. (i) Wrong, magnetic field lines can never emanate from a point, as shown in figure. Over any closed surface, the net flux of B must always be zero. The field lines shown represent electric field of a long positively charged wire. The correct magnetic field lines are circling the straight conductor.

(ii) Wrong, magnetic field lines can never cross each other, because if it do the direction of field at the point of intersection is ambiguous. There is another error in the figure. Magnetostatic field lines can never form closed loops around empty space. A closed loop of static magnetic field line must enclose a region across which a current is passing. Electrostatic field lines never form closed loops in empty space nor when loop encloses charges.

(iii) Right, magnetic lines are completely confined within a toroid. Nothing wrong here in field lines forming closed loops, since each loop encloses a region across which a current passes.

(iv) Wrong, field lines due to a solenoid at its end and outside cannot be so completely straight and confined; such a thing violates Ampere's law. The lines should curve out at both ends and meet eventually to form closed loops.

(v) Right, these are field lines outside and inside a bar magnet. Note carefully the direction of field lines inside. Not all field lines emanate out of a north pole (or converge into a south pole). Around both the N-pole and the S-pole, the net flux of the field is zero.

(vi) Wrong, these field lines cannot possibly represent a magnetic field. Look at the upper region. All the field lines seem to emanate out of the shaded plate. The net flux through a surface surrounding the shaded plate is not zero. This is impossible for a magnetic field. The given field lines, in fact, show the electrostatic field lines around a positively charged upper plate and a negatively charged lower plate.

(vii) Wrong, magnetic field lines between two pole pieces cannot be precisely straight at the ends. Some fringing of lines is invitable. Otherwise, Ampere's law is violated. This is also true for electric field lines.

Q. 6. (i) Magnetic field lines show the direction (at every point) along which a small magnetised needle aligns (at the point). Do the magnetic field lines also represent the lines of force on a moving charged particle at every point ?

(ii) Magnetic field lines can be entirely confined within the core of a toroid, but not within a straight solenoid. Why ?

(iii) If magnetic monopoles existed, how would the Gauss's law of magnetism be modified ?

(iv) Does a bar magnet exert a torque on itself due to its own field ? Does one element of a current-carrying wire exert a force on another element of the same wire ?

(v) Magnetic field arises due to charges in motion. Can a system have magnetic moments even though its net charge is zero ? **[NCERT]**

Ans. (i) No, the magnetic force is always normal to B (remember magnetic force = $qv \times B$). It is misleading to call magnetic field lines as lines of force.

(ii) If field lines were entirely confined between two ends of a straight solenoid, the flux through the cross-section at each end would be non-zero. But the flux of field B through any closed surface must always be zero. For a toroid, this difficulty is absent because it has no 'ends'.

(iii) Gauss's law of magnetism states that the flux of B through any closed surface is always zero $\oint_s B \cdot ds = 0$.

If monopoles existed, the right hand side would be equal to the monopole (magnetic charge) q_m enclosed by S. [Analogous to Gauss's law of electrostatics, $\int_s B \cdot ds = \mu_0 q_m$ where q_m is the (monopole) magnetic charge enclosed by S.]

(iv) No, there is no force or torque on an element due to the field produced by that element itself. But there is a force (or torque) on an element of the same wire. (For the special case of a straight wire, this force is zero.)

(v) Yes, the average of the charge in the system may be zero. Yet, the mean of the magnetic moments due to various current loops may not be zero. We will come across such examples in connection with paramagnetic material where atoms have net dipole moment through their net charge is zero.

Q. 7. Answer the following questions regarding earth's magnetism :

(i) A vector needs three quantities for its specification. Name the three independent quantities conventionally used to specify the earth's magnetic field.

(ii) The angle of dip at a location in southern India is about 18°. Would you expect a greater or smaller dip angle in Britain ?

(iii) If you made a map of magnetic field lines at Melbourne in Australia, would the lines seem to go into the ground or come out of the ground ?

(iv) In which direction would a compass free to move in the vertical plane point to, if located right on the geomagnetic North or South pole ?

(v) The earth's field, it is claimed, roughly approximates the field due to a dipole of magnetic moment 8×10^{22} JT^{-1} located at its centre. Check the order of magnitude of this number in some way.

(vi) Geologists claim that besides the main magnetic N-S poles, there are several local poles on the earth's surface oriented in different directions. How is such a thing possible at all ? **[NCERT]**

Ans. (i) The three independent quantities conventionally used to specify earth's magnetic field are :

(a) Magnetic declination

(b) Angle of dip, and

(c) Horizontal component of earth's magnetic field

(ii) The angle of dip at a point depends on how close or far the point is located with respect to the North Pole or the South Pole. The angle of dip would be greater in Britain as it is about 70° than in southern India because the location of Britain on the globe is closer to the magnetic North Pole.

(iii) In a map depicting earth's magnetic field lines, the field lines at Melbourne, Australia would seem to come out of the ground, because it is hypothetical that huge bar magnet is dipped inside earth with its north pole near the geographic south pole and south pole near geographic north pole.

(iv) If a compass is located on the geomagnetic North Pole or South Pole, then the compass will be free to move in the horizontal plane while earth's field is exactly vertical to the magnetic poles. In such a case, the compass can point in any direction.

(v) Magnetic moment, $M = 8 \times 10^{22}$ $J\,T^{-1}$

Radius of earth, $r = 6.4 \times 10^6$ m

As we know that

$$\text{Magnetic field strength, } B = \frac{\mu_0 M}{4\pi r^3}$$

Where,

μ_0 = Permeability of free space = $4\pi \times 10^{-7}$ T mA^{-1}

$$\therefore \quad B = \frac{4\pi \times 10^{-7} \times 8 \times 10^{22}}{4\pi (6.4 \times 10^6)^3}$$

$$= 0.3 \text{ G}$$

This is of the order of magnitude of the observed field on earth.

(vi) Yes, there are several local poles on earth's surface oriented in different direction. This is possible due to multiple magnetised mineral deposit is at different places.

Q. 8. Answer the following questions :

(i) The earth's magnetic field varies from point to point in space. Does it also change with time ? If so, on what time scale does it change appreciably ?

(ii) The earth's core is known to contain iron. Yet geologists do not regard this as a source of the earth's magnetism. Why ?

(iii) The charged currents in the outer conducting regions of the earth's core are thought to be responsible for earth's magnetism. What might be the 'battery' (*i.e.*, the source of energy) to sustain these currents ?

(iv) The earth may have even reversed the direction of its field several times during its history of 4 to 5 billion years. How can geologists know about the earth's field in such distant past ?

(v) The earth's field departs from its dipole shape substantially at large distances (greater than about 30,000 km). What agencies may be responsible for this distortion ?

(vi) Interstellar space has an extremely weak magnetic field of the order of 10^{-12} T. Can such a weak field be of any significant consequence ? Explain. **[NCERT]**

Ans. (i) Earth's magnetic field changes with time. It takes a few hundred years to change by an considerable amount.

(ii) Earth's core contains molten iron as this form of iron is not ferromagnetic. Therefore, this is not considered as a source of earth's magnetism.

(iii) The radioactivity of interior of earth is the source of energy that sustains the currents in the outer conducting regions of earth's core. These are considered to be responsible for earth's magnetism.

(iv) This has been recorded from the study of rocks as. These magnetic field got weakly recorded in rocks during their solidification. One can get clues about the geomagnetic history from the analysis of this rock magnetism.

(v) This is because of the presence of the ionosphere. In ionosphere, earth's field gets modified because of the field of ions. While in motion, these ions produce the magnetic field associated with them.

(vi) An extremely weak magnetic field can bend charged particles moving in a circle. This may not be noticeable for a large radius path. With reference to the very huge interstellar space, the deflection can affect the passage of charged particles.

Q. 9. Answer the following questions :

(i) Explain qualitatively on the basis of domain picture the irreversibility in the magnetisation curve of a ferromagnet.

(ii) The hysteresis loop of a soft iron piece has a much smaller area than that of a carbon steel piece. If the material is to go through repeated cycles of magnetisation, which piece will dissipate greater heat energy ?

(iii) A system displaying a hysteresis loop such as a ferromagnet, is a device for storing memory ? Explain the meaning of this statement.

(iv) What kind of ferromagnetic material is used for coating magnetic tapes in a cassette player or for building 'memory stores' in a modern computer ?

(v) A certain region of space is to be shielded from magnetic fields. Suggest a method. **[NCERT]**

Ans. The hysteresis curve (B-H curve) of a ferromagnetic material is shown in figure.

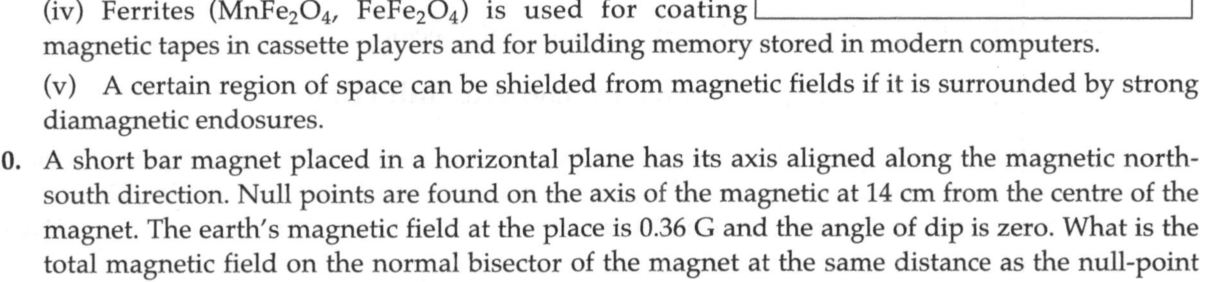

(i) It can be observed from the curve that magnetisation remains, even when the external field is removed. This reflects the irreversibility of a ferromagnet.

(ii) The dissipated heat energy is directly proportional to the area of a hysteresis (B-H) loop. A carbon steel piece has a greater hysteresis curve area. Therefore, it dissipates greater heat energy.

(iii) A ferromagnet has very high saturation, because of the high value of μ_r. Therefore, when it is magnetised it remains magnetised for long time even if external field is removed. Thus, it is said to have strong magnetic storage.

(iv) Ferrites ($MnFe_2O_4$, $FeFe_2O_4$) is used for coating magnetic tapes in cassette players and for building memory stored in modern computers.

(v) A certain region of space can be shielded from magnetic fields if it is surrounded by strong diamagnetic endosures.

Q. 10. A short bar magnet placed in a horizontal plane has its axis aligned along the magnetic north-south direction. Null points are found on the axis of the magnetic at 14 cm from the centre of the magnet. The earth's magnetic field at the place is 0.36 G and the angle of dip is zero. What is the total magnetic field on the normal bisector of the magnet at the same distance as the null-point (*i.e.*, 14 cm) from the centre of the magnet ? (At null points, field due to a magnet is equal and opposite to the horizontal component of earth's magnetic field.) **[NCERT]**

Ans. Earth's magnetic field at the given place, H = 0.36 G

The magnetic field at a distance *d*, on the axis of the magnet is given as :

$$B_1 = \frac{\mu_0}{4\pi}\frac{2M}{d^3} = H \qquad \text{...(i)}$$

Where, μ_0 = Permeability of free space

M = Magnetic moment

The magnetic field at the same distance *d*, on the equatorial line of the magnet is given as :

$$B_2 = \frac{\mu_0 M}{4\pi d^3} = \frac{H}{2} \qquad \text{[Using equation (i)]}$$

Total magnetic field,

$$B = B_1 + B_2 = H + \frac{H}{2}$$

$$= 0.36 + 0.18 = 0.54 \text{ G}$$

Hence, the magnetic field is 0.54 G in the direction of earth's magnetic field.

Q. 11. If the bar magnet in previous question is turned around by 180°, where will the new null points be located ? **[NCERT]**

Ans. Given,

The magnetic field on the axis of the magnet at a distance (d_1) = 14 cm

$$B_1 = \frac{\mu_0}{4\pi}\frac{2M}{(d_1)^3} = H \qquad \text{...(i)}$$

Where,

M = Magnetic moment

μ_0 = Permeability of free space

H = Horizontal component of the magnetic field at d_1

If the bar magnet is turned through 180°, then the neutral point will lie along equatorial line.

Hence, the magnetic field at a distance d_2, along equatorial line of the magnet can be written as :

$$B_2 = \frac{\mu_0}{4\pi}\frac{2M}{(d_2)^3} = H \qquad \ldots\text{(ii)}$$

From equations (i) and (ii), we get

$$\frac{2}{(d_1)^3} = \frac{1}{(d_2)^3}$$

$$\left(\frac{d_2}{d_1}\right)^3 = \frac{1}{2}$$

$$d_2 = d_1 \times \left(\frac{1}{2}\right)^{\frac{1}{3}}$$

$$= 14 \times 0.794 = 11.1 \text{ cm}$$

The null points will be located 11.1 cm on the normal bisector.

Chapter 6. Electromagnetic Induction

Q. 1. Given figure shows planar loops of different shapes moving out of or into a region of a magnetic field which is directed normal to the plane of the loop away from the reader. Determine the direction of induced current in each loop using Lenz's law. **[NCERT]**

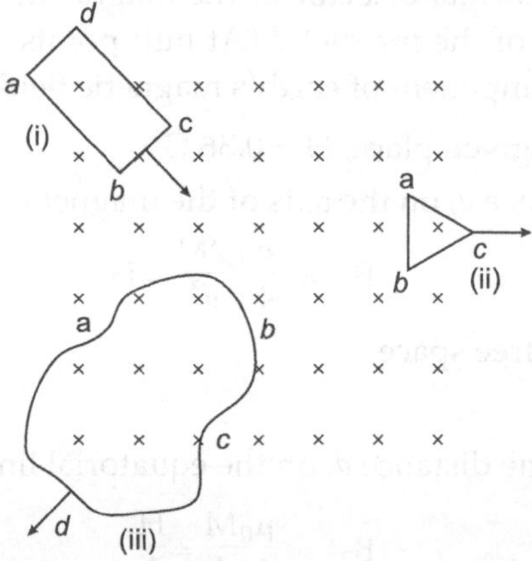

Ans. (i) The magnetic flux through the rectangular loop abcd increases, due to the motion of the loop into the region of magnetic field. The induced current must flow along the path bcdab so that it opposes the increasing flux.

(ii) Due to the outward motion, magnetic flux through the triangular loop *abc* decreases due to which the induced current flows along *bacb,* so as to oppose the change in flux.

(iii) As the magnetic flux decreases due to motion of the irregular shaped loop *abcd* out of the region of magnetic field, the induced current flows along *cdabc,* so as to oppose change in flux.

Note that there are no induced current as long as the loops are completely inside or outside the region of the magnetic field.

Q. 2. A 1.0 m long metallic rod is rotated with an angular frequency of 400 rad s^{-1} about an axis normal to the rod passing through its one end. The other end of the rod is in contact with a circular metallic ring. A constant and uniform magnetic field of 0.5 T parallel to the axis exists everywhere. Calculate the emf developed between the centre and the ring. **[NCERT]**

Ans. Given,

Length of the rod (l) = 1 m

Angular frequency (ω) = 400 rad/s

Magnetic field strength (B) = 0.5 T

As one end of the rod has zero linear velocity, but the other end has a linear velocity of Iω.

Average linear velocity of the rod, $(v) = \frac{l\omega + 0}{2} = \frac{l\omega}{2}$

Emf developed between the centre and the ring is given by

$$e = Blv = Bl\left(\frac{l\omega}{2}\right) = \frac{Bl^2\omega}{2}$$

$$= \frac{0.5 \times (1)^2 \times 400}{2} = 100 \text{ V}$$

Hence, the emf developed between the centre and the ring is 100 V.

Q. 3. A metallic rod of 1 m length is rotated with a frequency of 50 rev/s, with one end hinged at the centre and the other end at the circumference of a circular metallic ring of radius 1 m, about an axis passing through the centre and perpendicular to the plane of the ring (Fig.). A constant and uniform magnetic field of 1 T parallel to the axis is present everywhere. What is the emf between the centre and the metallic ring ? **[NCERT]**

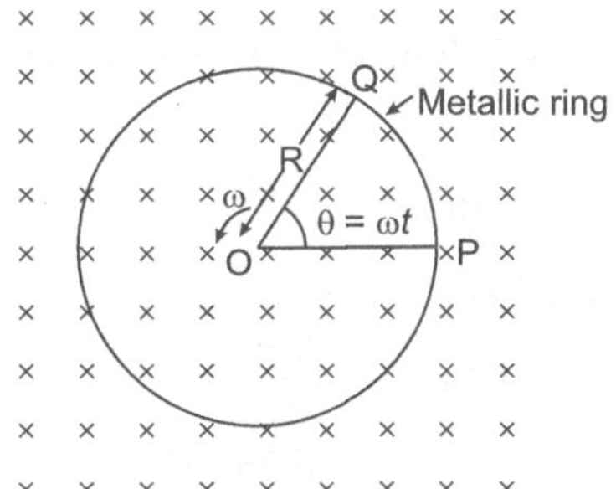

Ans. As the rod is rotated, free electrons in the rod move towards the outer end due to Lorentz force and get distributed over the ring. Thus, the resulting separation of charges produces an emf across the ends of the rod. At a certain value of emf, there is no more flow of electrons and a steady state is reached. The magnitude of the emf generated across a length dr of the rod as it moves at right angles to the magnetic field is given by

$$d\varepsilon = Bv\,dr$$

Hence,

$$\varepsilon = \int d\varepsilon = \int_0^R Bv\,dr$$

$$= \int_0^R B\omega r\,dr = \frac{B\omega R^2}{2}$$

Note that we have used $v = \omega r$. This gives

$$\varepsilon = \frac{1}{2} \times 1.0 \times 2\pi \times 50 \times (1^2)$$

$$= 157 \text{ V}$$

Q. 4. A metallic rod CD rests on thick metallic wire PQRS with arms PQ and RS parallel to each other, at a distance l = 40 cm as shown in following figure. A uniform magnetic field B = 0.1 T acts perpendicular to the plane of this paper, pointing inwards (*i.e.,* away from the reader). The rod is now made to slide towards right, with a constant velocity of v = 5 m/s. **[NCERT]**

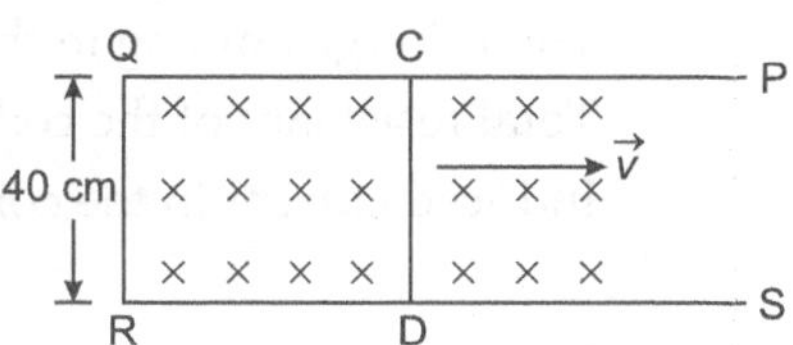

(i) How much emf is induced between the two ends of the rod CD ?

(ii) What is the direction in which the induced current flows ?

Ans. (i) $e = Bvl$

$$= 0.1 \times 5 \times 0.40 = 0.2 \text{ volt}$$

(ii) Direction of induced current = D → C → Q → R (anticlockwise)

Q. 5. A line charge λ per unit length is lodged uniformly onto the rim of a wheel of mass M and radius R. The wheel has light non-conducting spokes and is free to rotate without friction about its axis (Fig). A uniform magnetic field extends over a circular region within the rim. It is given by,

$$B = -B_0 \hat{k} \ (r \le a; a < R)$$

$$= 0 \text{ (otherwise)}$$

What is the angular velocity of the wheel after the field is suddenly switched off ?

Ans. As we know,

$$\text{line charge per unit length } (\lambda) = \frac{\text{Total charge}}{\text{Length}} = \frac{Q}{2\pi r}$$

Where, r = Distance of the point within the wheel

Mass of the wheel = M

Radius of the wheel = R

Magnetic field $\vec{B} = -B_0 \hat{k}$

At distance r, the magnetic force is balanced by the centripetal force.

Therefore,

$$BQv = \frac{Mv^2}{r}$$

Where, v = linear velocity of the wheel

$$\therefore \quad B\, 2\pi r\lambda = \frac{Mv}{r}$$

$$v = \frac{B\, 2\pi\, \lambda r^2}{M}$$

$$\therefore \quad \text{Angular velocity } (\omega) = \frac{v}{R} = \frac{B\, 2\pi\, \lambda r^2}{MR} \qquad [\because v = \omega R]$$

For $r \le$ and $a < R$, we get

$$\omega = \frac{2\, B_0\, a^2\, \lambda}{MR} \hat{k}$$

Q. 6. It is desired to measure the magnitude of field between the poles of a powerful loud speaker magnet. A small flat search coil of area 2 cm² with 25 closely wound turns, is positioned normal to the field direction, and then quickly snatched out of the field region. Equivalently, one can give it a quick 90° turn to bring its plane parallel to the field direction. The total charge flown in the coil (measured by a ballistic galvanometer connected to coil) is 7.5 mC. The combined resistance of the coil and the galvanometer is 0.50 Ω. Estimate the field strength of magnet. **[NCERT]**

Ans. Given,

Area of the small flat search coil, (A) = 2 cm² $= 2 \times 10^{-4}$ m²

Number of turns on the coil, (N) = 25

Total charge flowing in the coil, (Q) = 7.5 mC = 7.5×10^{-3} C

Total resistance of the coil and galvanometer, (R) = 0.50 Ω

Induced current in the coil, is given by

$$I = \frac{\text{Induced emf } (e)}{R} \qquad \text{...(i)}$$

Induced emf is given by

$$e = -N\frac{d\phi}{dt} \qquad \text{...(ii)}$$

Where, $d\phi$ = Change in flux

From equations (i) and (ii), we get

$$I = -\frac{N\frac{d\phi}{dt}}{R}$$

$$Idt = -\frac{N}{R}\,d\phi \qquad \text{...(iii)}$$

Initial flux through the coil, $\phi_i = BA$

Where, B = Magnetic field strength

Final flux through the coil, $\phi_f = 0$

Integrating equation (iii) on both sides, we have

$$\int I\,dt = \frac{-N}{R}\int_{\phi_i}^{\phi_f} d\phi$$

But total charge, $Q = \int I\,dt$

$$Q = \frac{-N}{R}(\phi_f - \phi_i) = \frac{-N}{R}(-\phi_i) = \frac{N\phi_i}{R}$$

$$Q = \frac{NBA}{R}$$

$$\therefore \quad B = \frac{QR}{NA}$$

$$= \frac{7.5 \times 10^{-3} \times 0.5}{25 \times 2 \times 10^{-4}}$$

$$= 0.75\ \text{T}$$

Therefore, the field strength of the magnet is 0.75 T.

Q. 7. A long straight current carrying wire passes normally through the centre of circular loop. If the current through the wire increases, will there be an induced emf in the loop ? Justify.

Ans. No, the emf will not be induced as the magnetic field lines are parallel to the plane of the circular loop.

$$\phi = BA\cos\theta = BA\cos 90° = 0$$

Hence, induced emf, $\varepsilon = -\frac{d\phi}{dt} = 0$

Chapter 7. Alternating Current

Q. 1. Why power factor correction is must in heavy machinery ?

Ans. $P_{av} = E_V I_V \cos\phi$

Heavy machinery requires large power. If I_V is increased thick wires will be required. This will increases the cost of installation, for large value of P_{av}, cos ϕ should be increased. In other words, ϕ should be reduced. This can be done by using proper capacitor. Note that X_L and X_C are out of phase. So, the use of capacitor would decrease ϕ and increase cos ϕ.

Q. 2. Why is it not possible to have electrolysis by a.c. ?

Ans. An a.c. reverses its direction after each half cycle. Therefore, on passing a.c. through a solution, the motion of the positive and negative ions become vibratory. So, ions will not be separated and electrolysis does not takes place.

Q. 3. A lamp is connected in series with a capacitor. Predict your observations for d.c. and a.c. connections. What happens in each case if the capacitance of the capacitor is reduced ? **[NCERT]**

Ans. When a d.c. source is connected to a capacitor, the capacitor gets charged and after charging no current flows in the circuit and the lamp will not glow. There will be no change even if C is reduced. With a.c. source, the capacitor offers capacitative reactance ($1/\omega C$) and the current flows in the circuit. Consequently, the lamp will shine. Reducing C will increase reactance and the lamp will shine less brightly than before.

Q. 4. A light bulb and an open coil inductor are connected to an a.c. source through a key as shown in following fig.

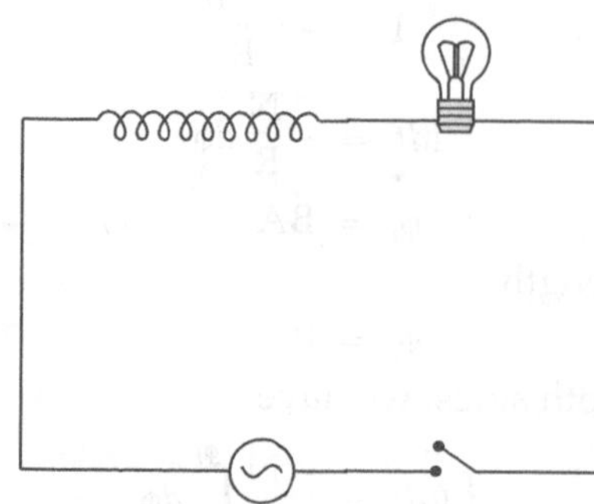

The switch is closed and after sometime, an iron rod is inserted into the interior of the inductor. The glow of the light bulb : (i) increases, (ii) decreases, (iii) is unchanged, as the iron rod is inserted. Give your answer with reasons. **[NCERT]**

Ans. As soon as iron rod is inserted, the magnetic field inside the coil magnetizes the iron which increases the magnetic field inside it. Therefore, the inductance of the coil increases. Consequently, the inductive reactance of the coil increases. Hence, a larger fraction of the applied a.c. voltage appears across the inductor, leaving less voltage across the bulb. Therefore, the glow of the light bulb decreases.

Q. 5. Answer the following questions :

(i) In any a.c. circuit, is the applied instantaneous voltage equal to the algebraic sum of the instantaneous voltages across the series elements of the circuit ? Is the same true for rms voltage ?

(ii) A capacitor is used in the primary circuit of an induction coil.

(iii) An applied voltage signal consists of a superposition of a d.c. voltage and an a.c. voltage of high frequency. The circuit consists of an inductor and a capacitor in series. Show that the d.c. signal will appear across C and the a.c. signal across L.

(iv) A choke coil in series with a lamp is connected to a d.c. line. The lamp is seen to shine brightly. Insertion of an iron core in the choke causes no change in the lamp's brightness. Predict the corresponding observations if the connection is to an a.c. line.

(v) Why is choke coil needed in the use of fluorescent tubes with a.c. mains ? Why can we not use an ordinary resistor instead of the choke coil ? **[NCERT]**

Ans. (i) Yes; the statement is not true for r.m.s. voltage.

It is true that in any a.c. circuit, the applied voltage is equal to the average sum of the instananeous voltages across the series elements of the circuit. However, this is not true for r.m.s. voltage because voltages across different elements may not be in phase.

(ii) High induced voltage is used to charge the capacitor.

A capacitor is used in the primary circuit of an induction coil. This is because when the circuit is broken, a high induced voltage is used to charge the capacitor to avoid sparks.

(iii) The d.c. signal will appear across capacitor C because for d.c. signals, the impedance of an inductor (L) is negligible while the impedance of a capacitor (C) is very high almost infinite. Hence, a d.c. signal appears across C.

For an a.c. signal of high frequency, the impedance of L is high and that of C is very low. Hence, an a.c. signal of high frequency appears across L.

(iv) If an iron core is inserted in the choke coil (which is in series with a lamp connected to the a.c. line), then the lamp will glow dim. This is because the choke coil and the iron core increase the impedance of the circuit.

(v) A choke coil is needed in the use of fluorescent tubes with a.c. mains because it reduces the voltage across the tube without wasting much power. An ordinary resistor cannot be used instead of a choke coil for this purpose because it wastes power in the form of heat.

Q. 6. A resistor of 200 Ω and a capacitor of 15.0 μF are connected in series to a 220 V, 50 Hz a.c. source.

(i) Calculate the current in the circuit

(ii) Calculate the voltage (r.m.s.) across the resistor and the capacitor. Is the algebraic sum of these voltages more than the source voltage ? If yes, resolve the paradox. **[NCERT]**

Ans. Given,

$R = 200\ \Omega$, $C = 15.0\ \mu F = 15.0 \times 10^{-6}$ F, $V = 220$ V, $\nu = 50$ Hz

(i) To find out current, the impedance of the circuit is required, that is

$$Z = \sqrt{R^2 + X_C^2}$$
$$= \sqrt{R^2 + (2\pi\nu C)^{-2}}$$
$$= \sqrt{(200\ \Omega)^2 + (2 \times 3.14 \times 50 \times 10^{-6}\ F)^{-2}}$$
$$= \sqrt{(200\ \Omega)^2 + (212\ \Omega)^2}$$
$$= 291.5\ \Omega$$

Therefore, the current in the circuit is

$$I = \frac{V}{Z} = \frac{220\ V}{291.5\ \Omega} = 0.755\ A$$

(ii) As the current is the same throughout the circuit, we have

$$V_R = IR = (0.755\ A)\ (200\ \Omega)\ 151\ V$$
$$V_C = IX_C = (0.755\ A)\ (212.3\ \Omega) = 160.3\ V$$

The algebraic sum of the two voltages, V_R and V_C is 311.3 V which is more than the source voltage of 220 V. How to resolve this paradox ? As you have learnt in the text, the two voltages are not in the same phase. Therefore, they cannot be added like ordinary numbers. The two voltages are out of phase by ninety degrees. Therefore, the total of these voltages must be obtained using the Pythagorean theorem :

$$V_{R+C} = \sqrt{V_R^2 + V_C^2}$$
$$= 220\ V$$

Thus, if the phase difference between two voltages is properly taken into account, the total voltage across the resistor and the capacitor is equal to the voltage of the source.

Q. 7. An LC circuit contains a 20 mH inductor and a 50 μF capacitor with an initial charge of 10 mC. The resistance of the circuit is negligible. Let the instant the circuit is closed be $t = 0$.

(i) What is the total energy stored initially ? Is it conserved during LC oscillations ?

(ii) What is the natural frequency of the circuit ?

(iii) At what time is the energy stored (a) completely electrical (*i.e.*, stored in the capacitor) ? (b) completely magnetic (*i.e.*, stored in the inductor) ?

(iv) At what times is the total energy shared equally between the inductor and the capacitor ?

(v) If a resistor is inserted in the circuit, how much energy is eventually dissipated as heat ?

[NCERT]

Ans. Given,

Inductance of inductor (L) = 20 mH = 20×10^{-3} H

Capacitance of capacitor (C) = 50 μF = 50×10^{-6} F

Inititial charge on the capacitor (Q) = 10 mC = 10×10^{-3} C

(i) Total energy stored initially in the circuit is given by

$$E = \frac{1}{2}\frac{Q^2}{C} = \frac{(10 \times 10^{-3})^2}{2 \times 50 \times 10^{-6}}$$
$$= 1\ J$$

Thus, the total energy stored in the LC circuit will be conserved because there is no resistor connected in the circuit.

(ii) Natural frequency of the circuit is given by

$$\nu = \frac{1}{2\pi\sqrt{LC}}$$

$$= \frac{1}{2\pi\sqrt{20\times10^{-3}\times50\times10^{-6}}} = \frac{10^3}{2\pi} = 159.24 \text{ Hz}$$

Natural angular frequency,

$$\omega_r = \frac{1}{\sqrt{LC}}$$

$$= \frac{1}{\sqrt{20\times10^{-3}\times50\times10^{-6}}} = \frac{1}{\sqrt{10^{-6}}} = 10^3 \text{ rad/s}$$

Thus, the natural frequency of the circuit is 10^3 rad/s.

(iii) (a) For time period $\left(T = \frac{1}{\nu} = \frac{1}{159.24} = 6.28 \text{ ms}\right)$, total charge on the capacitor at any time (t).

$$Q' = Q\cos\frac{2\pi}{T}t$$

For energy stored is electrical, $(Q') = Q$

Thus, it can be concluded that the energy stored in the capacitor is completely electrical at time, $t = 0, \frac{T}{2}, \frac{3T}{2}, \ldots\ldots$

(b) Magnetic energy is maximum when electrical energy (Q') is equal to 0.

Hence, it can be inferred that energy stored in the capacitor is completely magnetic at time, $t = \frac{T}{4}, \frac{3T}{4}, \frac{5T}{4}, \ldots\ldots$

(iv) Q' = Charge on the capacitor when total energy is equally shared between the capacitor and the inductor at time t.

When total energy is equally shared between the inductor and capacitor, the energy stored in the capacitor $= \frac{1}{2}$ (maximum energy)

$$\Rightarrow \quad \frac{1}{2}\frac{(Q')^2}{C} = \frac{1}{2}\left(\frac{1}{2}\frac{Q^2}{C}\right) = \frac{1}{4}\frac{Q^2}{C}$$

$$Q' = \frac{Q}{\sqrt{2}}$$

But

$$Q' = Q\cos\frac{2\pi}{T}t$$

$$\frac{Q}{\sqrt{2}} = Q\cos\frac{2\pi}{T}t$$

$$\cos\frac{2\pi}{T}t = \frac{1}{\sqrt{2}} = \cos(2n+1)\frac{\pi}{4}; \qquad \text{where } n = 0, 1, 2, \ldots\ldots$$

$$t = (2n+1)\frac{T}{8}$$

Thus, total energy is equally shared between the inductor and the capacitor at time,

$$t = \frac{T}{8}, \frac{3T}{8}, \frac{5T}{8} \ldots\ldots$$

(v) If a resistor is inserted in the circuit, then total initial energy is dissipated as heat energy in the circuit. The resistance damps out the LC oscillation.

Chapter 8. Electromagnetic Waves

Q. 1. Bees can see objects in the ultraviolet light while human beings can't do so. Why ?

Ans. Ultraviolet (UV) light has wavelength shorter than that of visible light. Bees have some retinal cones that are sensitive to UV light, so they can see objects in UV light. Human eyes do not posses retinal cones sensitive to UV light, so human beings can't see objects in UV light. In other words, human beings are UV blind.

Q. 2. Why does current in a steady state not flow in a capacitor connected across a battery ? However momentary current does flow during charging or discharging of the capacitor. Explain.

Ans. When there is change in the electric flux there will be a displacement current and when flux is fixed the displacement as well as conduction current will be zero.

$$I_D = \varepsilon_0 \frac{d\phi_E}{dt}$$

Q. 3. Answer the following :

(i) Welders wear special glass goggles while working. Why ? Explain.

(ii) Why are infrared waves often called as heat waves ? Give their one application.

Ans. (i) Welders wear special glass goggles while working so that they can protect their eyes from harmful electromagnetic radiation.

(ii) Infrared waves are often called as heat waves because they induce resonance in molecules and increase internal energy in a substance.

Infrared waves are used in burglar alarms, security lights and remote controls for television and DVD players.

Q. 4. Answer the following questions :

(i) Long distance radio broadcasts use short-wave bands. Why ?

(ii) It is necessary to use satellites for long distance TV transmission. Why ?

(iii) Optical and radio telescopes are built on the ground but X-ray astronomy is possible only from satellites orbiting the earth. Why ?

(iv) If the earth did not have an atmosphere, would it's average surface temperature be higher or lower than what it is now ?

(v) Some scientists have predicted that a global nuclear war on the earth would be followed by a severe 'nuclear winter' with a devastating effect on life on earth. What might be the basis of this prediction ? **[NCERT]**

Ans. (i) Long distance radio broadcasts use shortwave bands because only shortwaves are refracted by the ionosphere.

(ii) It is necessary to use satellites for long distance TV transmissions because. These signals are not reflected by the ionosphere. Hence, satellites are helpful in reflecting TV signals. Also, they help in long distance TV transmissions.

(iii) With reference to X-ray astronomy, X-rays are absorbed by the atmosphere. However, visible and radiowaves can penetrate it. Hence, optical and radio telescopes are built on the ground, while X-rays astronomy is possible only with the help of satellite orbiting the Earth.

(iv) In absence of an atmosphere, there would be no greenhouse effect on the surface of the Earth. As a result, the temperature of the Earth would decrease rapidly, making it chilly and difficult for human survival.

(v) A global nuclear war on the surface of the Earth would have disastrous consequences. Post-nuclear war, the Earth will experience severe winter as the war will produce clouds of smoke that would cover maximum parts of the sky, thereby preventing solar light form reaching the atmosphere. Also, it will lead to the depletion of the ozone layer.

Chapter 9. Ray Optics and Optical Instruments

Q. 1. What should be the position of an object relative to biconvex lens so that it behaves like a magnifying lens ? **[NCERT]**

Ans. For biconvex lens to behave as a magnifying lens, the object must be placed between the focal point and optical centre of lens.

Q. 2. How does the angle of minimum deviation of a glass prism vary, if the incident violet light is replaced by red light ? Give reason.★

Ans. We have,

$$\mu = \frac{\sin\left(\frac{A + \delta_m}{2}\right)}{\sin (A/2)}$$

$\therefore$ $$\mu \propto \delta_m$$

And $$\mu_{red} < \mu_{violet}$$

$\therefore$ δ_m will decrease.

Q. 3. Why does bluish colour predominate in a clear sky ?★

Ans. While light from the sun reaches the atmosphere that is comprised of the tiny particles of the atmosphere. These act as a prism and cause the different components to scatter. As blue light travels is shorter and smaller waves in comparison to the other colours of the spectrum, it is scattered the most, causing the sky to appear bluish.

Q. 4. Why are mirrors used in search lights are parabolic and not concave spherical ?

Ans. In search lights, we need an intense parallel beam of light. If a source is placed at the focus of a concave spherical mirror, only paraxial rays are rendered parallel. Due to large aperture of mirror, marginal rays give a divergent beam.

But in case of parabolic mirror, when source is at the focus, beam of light produced over the entire cross-section of the mirror is a parallel beam.

Q. 5. Explain why white light is dispersed while passing through a prism.

Ans. The refractive index of prism-material depends on the wavelength λ of light. It is inversely proportional to square of wavelength *i.e.*, $n \propto \frac{1}{\lambda^2}$. Accordingly refractive index is maximum for violet (λ = 4000 Å) and minimum for red (λ = 7500 Å). So, the deviation caused by prism $f = (n - 1)$ A is maximum for violet and minimum for red; hence on passage through the prism; the different colours are separated; thus causing dispersion of white light.

Q. 6. Define the magnifying power of a compound microscope when the final image is formed at infinity. Why must both the objective and the eyepiece of a compound microscope has short focal lengths ? Explain.★

Ans. **Magnifying Power :** The magnifying power of a compound microscope is defined as the ratio of the angle subtended at the eye by the final image to the angle subtended at the eye by the object, when both the final image and the object are situated at the least distance of distinct vision from the eye.

Magnifying power of the compound microscope when the final image is at infinity,

$$m = -\frac{L}{f_o} \times \frac{D}{f_e}$$

From the above equation, we can see that to achieve a large magnification, the objective and eyepiece should have small focal lengths.

Q. 7. A convex lens is placed in contact with a plane mirror. A point object at a distance of 20 cm on the axis of this combination has its image coinciding with itself. What is the focal length of the lens ?★

★ are board based questions from previous year

Ans. The figure shows a convex lens L placed in contact with a plane mirror M. P is the point object, kept in front of this combination at a distance of 20 cm, from it. As the image coincides with itself, the rays from the object, after refraction from lens, should fall normally on the mirror M, so that they retrace their path. For this, the rays from P, after refraction from the lens must form a parallel beam perpendicular to M. For clarity, M has been shown at a small distance from L (in diagram). As the rays from P, form a parallel beam after refraction, P must be at the focus of the lens. Hence, the focal length of the lens is 20 cm.

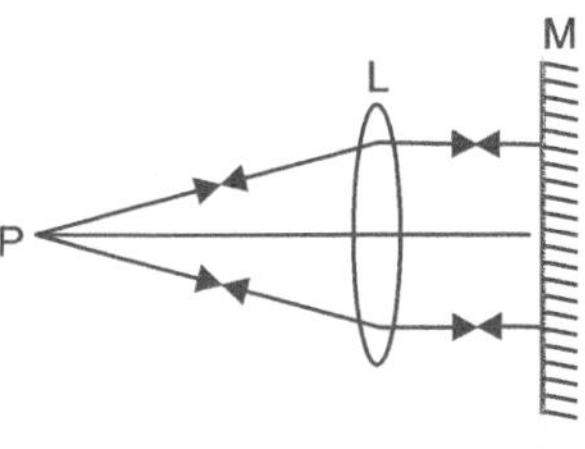

Q. 8. (i) The near point of a hypermetropic person is 75 cm from the eye. What is the power of the lens required to enable the person to read clearly a book held at 25 cm from the eye ?

(ii) In what way does the corrective lens help the above person ? Does the lens magnify objects held near the eye ?

(iii) The above person prefers to remove the spectacles while looking at the sky. Explain why ?

[NCERT]

Ans. (i) **Given :**

$$u = -25 \text{ cm}, v = -75 \text{ cm},$$

$$\frac{1}{f} = \frac{1}{25} - \frac{1}{75}, \; i.e., f = 37.5 \text{ cm}.$$

The corrective lens needs to have a converging power of + 2.67 dioptres.

(ii) The corrective lens produces a virtual image (at 75 cm) of an object at 25 cm. The angular size of this image is the same as that of the object. In this sense, the lens does not magnify the object but merely brings the object to the near point of the hypermetric eye, which then gets focussed on the retina. However, the angular size is greater than that of the same object at the near point (75 cm) viewed without the spectacles.

(iii) A hypermetropic eye may have normal far point *i.e.,* it may have enough converging power to focus parallel rays from infinity on the retina of the shortened eyeball. Wearing spectacles of converging lenses (used for near vision) will amount to more converging power than needed for parallel rays. Hence, the person prefers not to use the spectacles for far objects.

Q. 9. (i) A mobile phone lies along the principal axis of a concave mirror. Show, with the help of a suitable diagram, the formation of its image. Explain why magnification is not uniform.★

(ii) Suppose the lower half of the concave mirror's reflecting surface is covered with an opaque material. What effect this will have on the image of the object ? Explain.★

Ans. (i) The image of the mobile phone formed by the convave mirror will be as shown in fig. given below :

The magnification produced by a concave mirror,

$$m = \frac{I}{O} = \frac{f}{f-u}$$

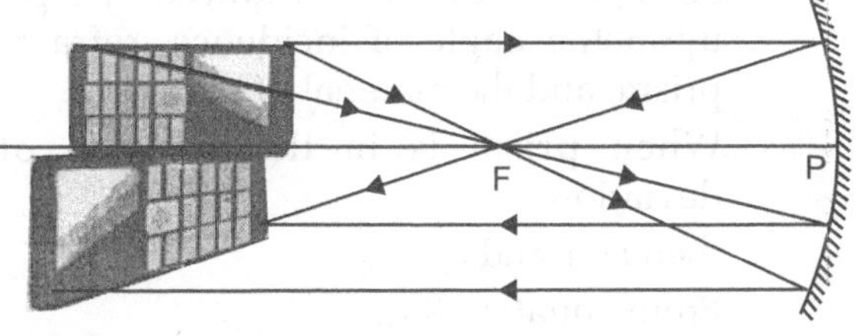

Since, the object distance for the different parts of the mobile phone along its length are different, the different parts will be magnified differently. Hence, the magnification is not uniform.

(ii) As the laws of reflection are true for all points of the mirror, the height of the whole image will be produced. However, as the area of the reflecting surface has been reduced, the image intensity will be reduced. In other words, the image produced will be less bright.

Q. 10. A ray 'PQ' of light is incident on the face AB of a glass prism ABC (as shown in the figure) and emerges out of the face AC. Trace the path of the ray. Show that $\angle i + \angle e = \angle A + \angle \delta$. When δ and e denote the angle of deviation and angle of emergence respectively.

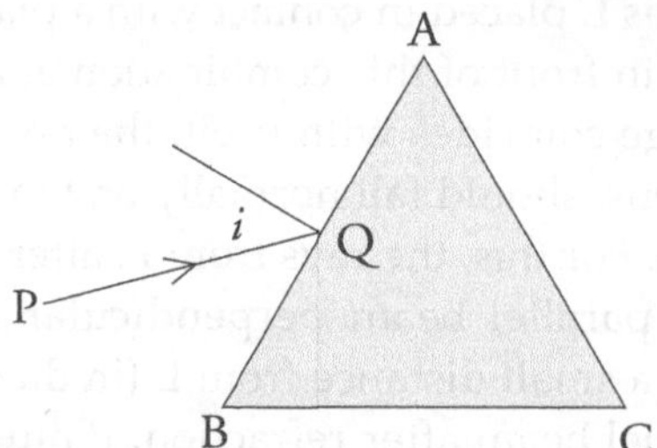

Plot a graph showing the variation of the angle of deviation as a function of angle of incidence. State the condition under which ∠δ is minimum.★

Ans. Let the incident ray meet refracting face AB of the prism at point P. Ray PQ is the refracted ray inside the prism and δ_2 and r_1 are the angle of the deviation and refraction at interface AB. At interface AC the ray goes out of the prism. Let e be the angle of emergence. The angle of deviation at point Q is δ_2 as shown in figure.

Using geometry, we see that at point P,

$$i = \delta_1 + r_1$$

$$\therefore \delta_1 = i - r_1$$

and at point Q,

$$e = \delta_2 + r_2$$

$$\delta_2 = e - r_2$$

The total deviation δ, suffered by the incident ray is equal to $\delta_1 + \delta_2$.

or

$$\delta = \delta_1 + \delta_2$$

$$= (i - r_1) + (e - r_2)$$

$$= (i + e) - (r_1 + r_2) \quad \text{...(i)}$$

In quadrilateral POQA, the sum of all four angle is 360°

$$P + O + Q + A = 360°$$

as P and Q both are right angles

$$P + Q = 180°$$

$$O + A = 180° \quad \text{...(ii)}$$

In triangle POQ

$$O + r_1 + r_2 = 180° \quad \text{...(iii)}$$

Comparing equations (ii) and (iii), we have

$$A = r_1 + r_2$$

Substituting this value in equation (i)

$$\delta = i + e - A$$

$$\delta + A = i + e \quad \text{...(iv)}$$

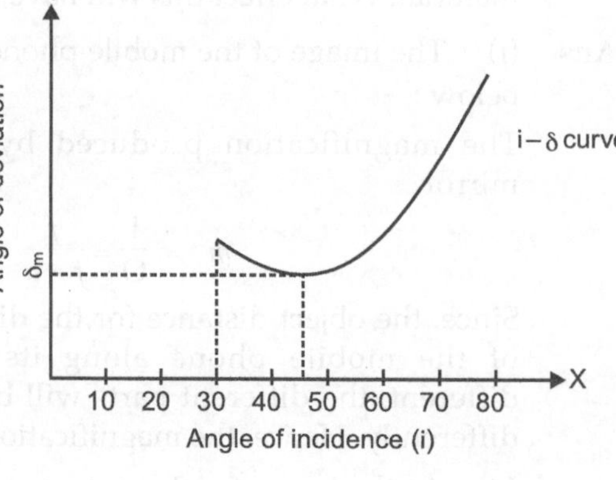

So angle of deviation produced by a prism depends upon the angle of incidence, refracting angle of prism, and the material of the prism.

When prism is in the position of minimum deviation,

then $i = e$ and $r_1 = r_2$

From equation (iv),

$$\delta_m + A = i + i$$

$$i = \frac{(\delta_m + A)}{2}$$

Q. 11. (i) The far point of a myopic person is 80 cm in front of the eye. What is the power of the lens required to enable him to see very distant object clearly ?

★ are board based questions from previous year

(ii) In what way does the corrective lens help the above person ? Does the lens magnify very distant object ? Explain carefully.

(iii) The above person prefers to remove his spectacles while reading a book. Explain why ?

[NCERT]

Ans. (i) From lens formula $$\frac{1}{f} = \frac{1}{v} - \frac{1}{u}$$

$u = -80$ cm, $v = 2u$ in this case

$v = -160$ cm

Thus, $$\frac{1}{f} = \frac{-1}{160} + \frac{1}{80} = \frac{-1}{80} \text{ cm}$$

$$f = -80 \text{ cm}$$

We find that the person should use a concave lens of focal length = – 80 cm, *i.e.*, of power = – 1.25 dioptres.

$$P(\text{dip}) = \frac{1}{0.8} = -1.25 \text{ dioptres.}$$

(ii) No. The concave lens, in fact, reduces the size of the object, but the angle subtended by the distant object at the eye is the same as the angle subtended by the image (at the far point) at the eye. The eye is able to see distant objects not because the corrective lens magnifies the object, but because it brings the object (*i.e.*, it produces virtual image of the object) at the far point of the eye which then can be focussed by the eye-lens on the retina.

(iii) The myopic person may have a normal near point, *i.e.*, about 25 cm (or even less). In order to read a book with the spectacles, such a person must keep the book at a distance greater than 25 cm so that the image of the book by the concave lens is produced not closer than 25 cm. The angular size of the book (or its image) at the greater distance is evidently less than the angular size when the book is placed at 25 cm and no spectacles are needed. Hence, the person prefers to remove the spectacles while reading.

Q. 12. Answer the following questions :

(i) The angle subtended at the eye by an object is equal to the angle subtended at the eye by the virtual image produced by a magnifying glass. In what sense then does a magnifying glass provide angular magnification ?

(ii) In viewing through a magnifying glass, one usually positions one's eyes very close to the lens. Does angular magnification change if the eye is moved back ?

(iii) Magnifying power of a simple microscope is inversely proportional to the focal length of the lens. What then stops us from using a convex lens of smaller and smaller focal length and achieving greater and greater magnifying power ?

(iv) Why must both the objective and the eyepiece of a compound microscope have short focal lengths ?

(v) When viewing through a compound microscope, our eyes should be positioned not on the eyepiece but a short distance away from it for best viewing. Why ? How much should be that short distance between the eye and eyepiece ? **[NCERT]**

Ans. (i) Though the image size is bigger than the object, the angular size of the image is equal to the angular size of the object. A magnifying glass helps one see the objects placed closer than the least distance of distinct vision (*i.e.*, 25 cm). A closer object causes a larger angular size. A magnifying glass provides angular magnification. Without magnification, the object cannot be placed closer to the eye. With magnification, the object can be placed much closer to the eye.

(ii) Yes, the angular magnification changes. When the distance between the eye and a magnifying glass is increased, the angular magnification decreases a little. This is because the angle subtended at the eye is slightly less than the angle subtended at the lens. Image distance does not have any effect on angular magnification.

(iii) The focal length of a convex lens cannot be decreased by a greater amount. This is because making lenses having very small focal lengths is not easy. Spherical and chromatic aberrations are produced by a convex lens having a very small focal length.

(iv) The angular magnification produced by the eyepiece of a compound microscope is $\left[\left(\frac{25}{f_e}\right)+1\right]$.

Where, f_e = Focal length of the eyepiece

It can be inferred that if f_e is small, then angular magnification of the eyepiece will be large.

The angular magnification of the objective lens of a compound microscope is given by

$$\frac{1}{(u_o/f_o)}$$

Where, u_o = Object distance for the objective lens

f_o = Focal length of the objective

The magnification is large when $u_o > f_o$. In the case of a microscope, the object is kept close to the objective lens. Hence, the object distance is very little. Since, u_o is small, f_o will be even smaller. Therefore, f_e and f_o both small in the given condition.

(v) When we place our eyes too close to the eyepiece of a compound microscope, we are unable to collect much refracted light. As a result, the field of view decreases substantially. Hence, the clarity of the image gets blurred.

The best position of the eye for viewing through a compound microscope is at the eyering attached to the eyepiece. The precise location of the eye depends on the separation between the objective lens and the eyepiece.

Q. 13. The figure shows a ray of light falling normally on the face AB of an equilateral glass prism having refractive index $\frac{3}{2}$, placed in water of refractive index $\frac{4}{3}$. Will this ray suffer total internal reflection on striking the face AC ? Justify your answer.★

Ans. Given, ${}^a\mu_g = \frac{3}{2}$ and ${}^a\mu_w = \frac{4}{3}$

$${}^w\mu_g = \frac{{}^a\mu_g}{{}^a\mu_w} = \frac{3/2}{4/3} = \frac{9}{8} = 1.125$$

We know at critical angle C,

$${}^w\mu_g = \frac{1}{\sin C}$$

or

$$\sin C = \frac{1}{{}^w\mu_g} = \frac{1}{1.125} = 0.88$$

i.e.,

$$C = 62.73°$$

But from the figure, light will incident on face AC making an angle 60° with the normal *i.e.*, $i = 60°$

So, $i < C$

∴ Light will not suffer total internal reflection.

Q. 14. A tank is filled with water to a height of 12.5 cm. The apparent depth of a needle lying at the bottom of the tank is measured by a microscope to be 9.4 cm. What is the refractive index of water ? If water is replaced by a liquid of refractive index 1.63 upto the same height, by what distance would the microscope have to be moved to focus on the needle again ? **[NCERT]**

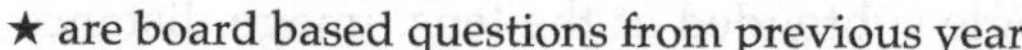

★ are board based questions from previous year

Ans. Given,

Actual depth of the needle in water (h_1) = 12.5 cm

Apparent depth of the needle in water (h_2) = 9.4 cm

Refractive index of water = μ

The value of μ can be obtained as follows :

$$\mu = \frac{h_1}{h_2} = \frac{12.5}{9.4} = 1.33$$

Thus, the refractive index of water is about 1.33.

Water is replaced by a liquid of refractive index, $\mu = 1.63$

The actual depth of the needle remains the same, but its apparent depth changes.

Let y be the new apparent depth of the needle. Hence, the relation will be as follows.

$$\mu' = \frac{h_1}{y}$$

$$\therefore \quad y = \frac{h_1}{\mu'} = \frac{12.5}{1.63} = 7.67 \text{ cm}$$

Therefore, the new apparent depth of the needle is 7.67 cm. It is less than h_2.

Hence, to focus the needle again, the microscope should be moved up.

$\therefore$ Distance by which the microscope should be moved up = 9.4 –7.67 = 1.73 cm

Q. 15. A symmetric biconvex lens of radius of curvature R and made of glass of refractive index 1.5, is placed on a layer of liquid placed on top of a plane mirror as shown in the figure. An optical needle with its tip on the principal axis of the lens is moved along the axis until its real, inverted image coincides with the needle itself. The distance of the needle from the lens is measured to be x. On removing the liquid layer and repeating the experiment, the distance is found to be y. Obtain the expression for the refractive index of the liquid in terms of x and y.★

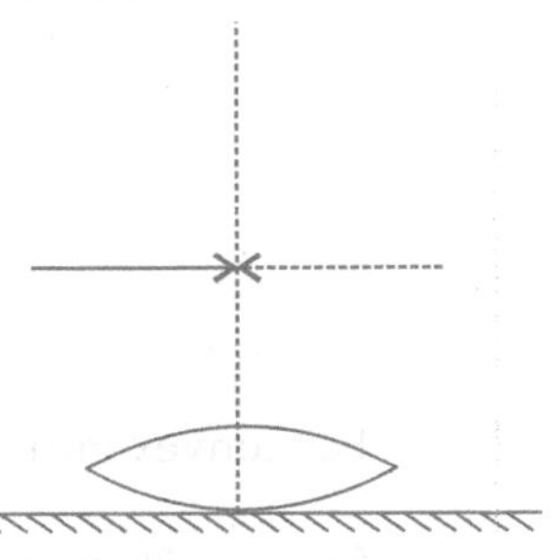

Ans. Given,

refractive index of lens, $\mu_g = 1.5$

The distance of the needle from the lens in the first case = The focal length of the combination of convex lens and plano-concave lens formed by the liquid, $f = x$.

And, the distance measured in the second case = Focal length of the convex lens, $f_1 = y$

If the focal length of plano-concave lens formed by the liquid be f_2, then

$$\frac{1}{f} = \frac{1}{f_1} + \frac{1}{f_2}$$

$$\Rightarrow \quad \frac{1}{f_2} = \frac{1}{f} - \frac{1}{f_1} = \frac{1}{x} - \frac{1}{y} = \frac{y-x}{xy}$$

$$\text{or} \quad f_2 = \frac{xy}{y-x}$$

Now, refractive index of lens, $\mu = 1.5$, radius of curvature of one surface = R and radius of curvature of the other surface = – R

$$\therefore \quad \frac{1}{f_1} = (\mu_g - 1)\left(\frac{1}{R} - \frac{1}{-R}\right)$$

$$\frac{1}{y} = (1.5 - 1)\left(\frac{2}{R}\right) \Rightarrow R = y$$

And, if refractive index of liquid is μ_l,

Radius of curvature on the side of plane mirror = ∞

★ are board based questions from previous year

Radius of curvature on the side of lens = – R

$$\therefore \quad \frac{y-x}{xy} = (\mu_l - 1)\left(\frac{1}{-R} - \frac{1}{\infty}\right) = (\mu_l - 1)\left(-\frac{1}{y}\right)$$

$$\Rightarrow \quad (\mu_l - 1) = -\left(\frac{y-x}{x}\right)$$

$$\text{or} \quad \mu_l = \frac{x-y}{x} + 1 = \frac{2x-y}{x}$$

Q. 16. Suppose while sitting in a parked car, you notice a jogger approaching towards you in the side view mirror of R = 2 m. If the jogger is running at a speed of 5 m s^{-1}, how fast the image of the jogger appear to move when the jogger is (i) 39 m, (ii) 29 m, (iii) 19 m and (iv) 9 m away.

[NCERT]

Ans. From the mirror equation,

$$\frac{1}{f} = \frac{1}{v} + \frac{1}{u}$$

$$\frac{1}{v} = \frac{1}{f} - \frac{1}{u}$$

$$\frac{1}{v} = \frac{u-f}{uf}$$

$$v = \frac{fu}{u-f}$$

For convex mirror, since R = 2 m, f = 1 m. Then

for $u = -39$ m, $v = \frac{(-39)\times 1}{-39-1} = \frac{39}{40}$ m

As the jogger moves at a constant speed of 5 m s^{-1}, after 1 s the position of the image v (for $u = -39 + 5 = -34$) is (34/35) m.

The shift in the position of image in 1 s is

$$\frac{39}{40} - \frac{34}{35} = \frac{1365-1360}{1400} = \frac{5}{1400} = \frac{1}{280} \text{ m}$$

Therefore, the average speed of the image when the jogger is between 39 m and 34 m from the mirror, is (1/280) m s^{-1}.

Similarly, it can be seen that for $u = -29$ m, – 19 m and – 9 m, the speed with which the image appears to move is

$$\frac{29}{40} - \frac{24}{35} = \frac{29\times 35 - 24\times 40}{1400} = 140 \text{ m}$$

$\frac{1}{150}$ m s^{-1}, $\frac{1}{60}$ ms^{-1} and $\frac{1}{10}$ ms^{-1} respectively.

Although the jogger has been moving with a constant speed, the speed of his/her image appears to increase substantially as he/she moves closer to the mirror. This phenomenon can be noticed by any person sitting in a stationary car or a bus. In case of moving vehicles, a similar phenomenon could be observed if the vehicle in the rear is moving closer with a constant speed.

Q. 17. How does resolving power of a telescope change on decreasing the aperture of its object lens ? Justify your answer.

Ans. Resolving power of telescope $= \frac{D}{1.22\,\lambda}$ · Therefore, on decreasing aperture (D) of the objective lens, the resolving power of the telescope decreases.

Chapter 10. Wave Optics

Q. 1. What is the effect on the interference pattern in Young's double slit experiment due to each of the following operations ?

(i) The widths of the slits are increased equally.

(ii) The whole apparatus is kept in a denser medium.

Ans. (i) As the widths of the slits are increased equally, the fringe width decreases. It is because,

$$\beta \propto \frac{1}{d}$$

(ii) When the apparatus is immersed in denser medium of refractive index μ, the wavelength of the light used decreases. Since $\beta \propto \lambda$, the fringe width decreases.

Q. 2. How does an unpolarised light get polarised, when it is passed through a polaroid ?

Ans. When unpolarised light is passed through a polaroid, only those electric field vectors, which are parallel to the crystallographic axis of the polaroid, emerge out of it. As such, the emerging light possesses vibrations in one plane only and is said to be plane polarised.

Q. 3. Show using a proper diagram how unpolarised light can be linearly polarised by reflection from a transparent glass surface.

Ans. Polarisation of light by reflection from a transparent glass surface :

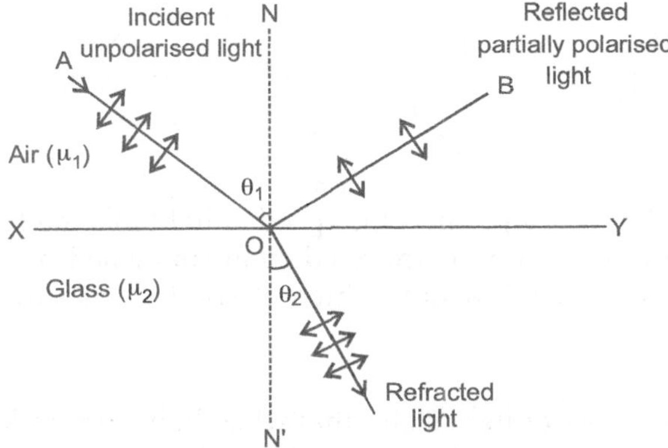

Q. 4. For sound waves, the Doppler formula for frequency shift differs slightly between the two situations : (i) source at rest; observer moving, and (ii) source moving; observer at rest.

The exact Doppler formulas for the case of light waves in vacuum are, however, strictly identical for these situations. Explain why this should be so. Would you expect the formulas to be strictly identical for the two situations in case of light travelling in a medium ? **[NCERT]**

Ans. No, sound waves can propagate only through a medium. The two given situations are not scientifically identical because the motion of an observer relative to a medium is different in the two situations. Hence, the Doppler formulas for the two situations cannot be the same.

In case of light waves, sound can travel in a vacuum. In a vacuum, the above two cases are identical because the speed of light is independent of the motion of the observer and the motion of the source. When light travels in a medium, the above two cases are not identical because the speed of light depends on the wavelength of the medium.

Q. 5. Answer the following questions :

(i) When a low flying aircraft passes overhead, we sometimes notice a slight shaking of the picture on our TV screen. Suggest a possible explanation.

(ii) As you have learnt in the text, the principle of linear superposition of wave displacement is basic to understanding intensity distributions in diffraction and interference patterns. What is the justification of this principle ? **[NCERT]**

Ans. (i) Weak radar signals sent by a low flying aircraft can interfere with the TV signals received by the antenna. This may distort TV signals. Thus, when a low flying aircraft passes overhead, we sometimes notice a slight shaking of the picture on our TV screen.

(ii) The principle of linear superposition of wave displacement is essential to our understanding of intensity distributions and interference patterns. This is because superposition follows from the linear character of a differential equation that governs wave motion. If y_1 and y_2 are the solutions of the second order wave equation, then any linear combination of y_1 and y_2 will also be the solution of the wave equation.

Q. 6. Using the phenomenon of polarisation, show how transverse nature of light can be demonstrated.

Ans. Suppose that an ordinary light is incident normally on a pair of crystals C_1 and C_2. When the incident ray of light passes through crystal C_1, it gets plane polarised in the direction perpendicular to the length of crystal. Now, we see that when the axis of two crystals are parallel, the intensity of the emerging light will be maximum. When the second crystal is placed perpendicular with respect to the first crystal, the intensity of light observed is zero. This is due to the electric field of the plane polarised light obtained from C_1 can vibrate only in one direction. Hence, when the axis of the crystal is perpendicular to its direction of vibration of electric field, it gets blocked. This shows the transverse nature of light.

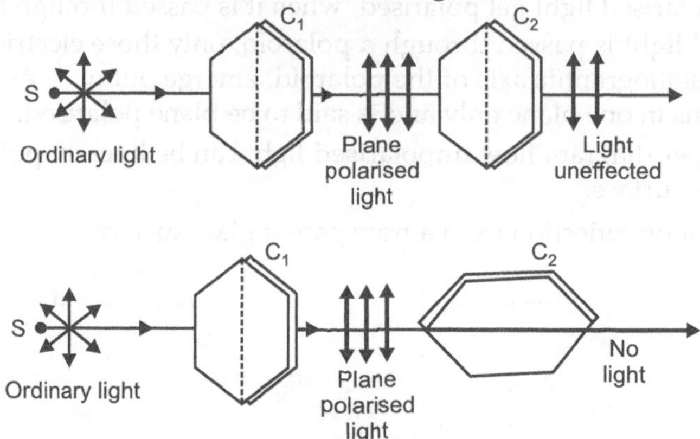

Q. 7. Explain how corpuscular theory predicts the speed of light in a medium, say, water, to be greater than the speed of light in vacuum. Is the prediction confirmed by experimental determination of the speed of light in water ? If not, which alternative picture of light is consistent with experiment ? **[NCERT]**

Ans. No, Wave theory

Newton's corpuscular theory of light states that when light corpuscles strike the interface of two media from a rarer (air) to a denser (water) medium, the particles experience forces of attraction normal to the surface. Hence, the normal component of velocity increases while the component along the surface remains unchanged. Hence, we can write the expression

$$c \sin i = v \sin r \quad \text{...(i)}$$

where, i = Angle of incidence

r = Angle of reflection

c = Velocity of light in air

v = velocity of light in water

We have the relation for relative refractive index of water with respect to air as—

$$\mu = \frac{v}{c}$$

Hence, equation (i) reduces to

$$\frac{v}{c} = \frac{\sin i}{\sin r} = \mu \quad \text{...(ii)}$$

But, $\mu > 1$

Hence, it can be inferred from equation (ii) that $v > c$. This is not possible since this prediction is opposite to the experimental results of $c > v$. The wave picture of light is consistent with the experimental results.

Q. 8. You have learnt in the text how Huygens' principle leads to the laws of reflection and refraction. Use the same principle to deduce directly that a point object placed in front of a plane mirror produces a virtual image whose distance from the mirror is equal to the object distance from the mirror. **[NCERT]**

Ans. Let an object at O be placed in front of a plane mirror of MO′ at a distance r (as shown in the given figure).

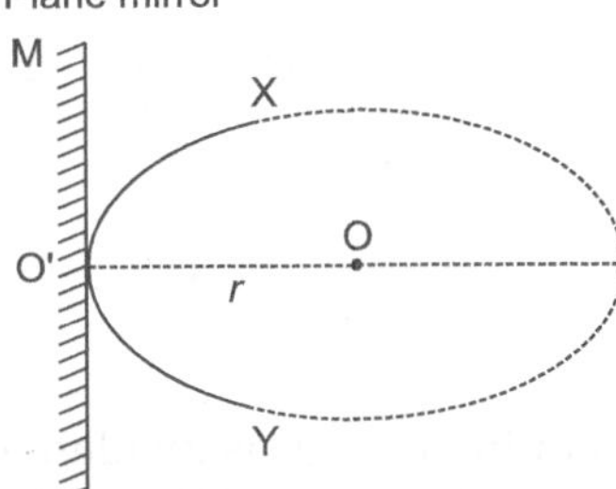

A circle is drawn from the centre (O) such that it just touches the planes mirror at point O′. According to Huygens' principle, XY is the wavefront of incident light.

If the mirror is absent, then a similar wavefront X′Y′ (as XY) would form behind O′ at distance r (as shown in the given figure).

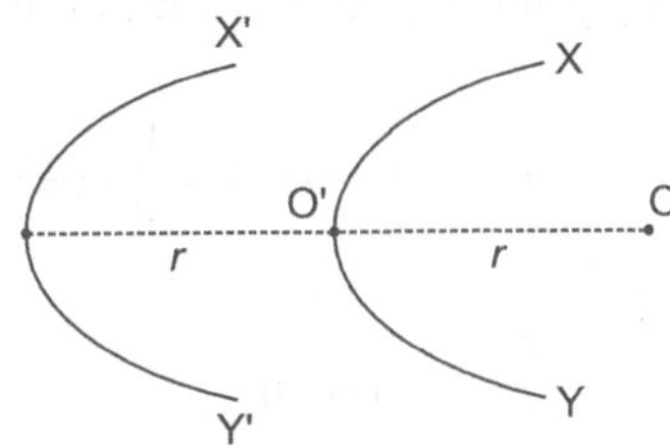

X′Y′ can be considered as a virtual reflected ray for the plane mirror.

Hence, a point object placed in front of the plane mirror produces a virtual image whose distance from the mirror is equal to the object distance (r).

Q. 9. (i) When monochromatic light is incident on a surface separating two media, the reflected and refracted light both have the same frequency as the incident frequency. Explain why ?

(ii) When light travels from a rarer to a denser medium, the speed decreases. Does the reduction in speed imply a reduction in the energy carried by the light wave ?

(iii) In the wave picture of light, intensity of light is determined by the square of the amplitude of the wave. What determines the intensity of light in the photon picture of light. **[NCERT]**

Ans. (i) Reflection and refraction arise through interaction of incident light with the atomic constituents of matter. Atoms may be viewed as oscillators, which take up the frequency of the external light causing forced oscillations. The frequency of light emitted by a charged oscillator equals its frequency of oscillation. Thus, the frequency of scattered light equals the frequency of incident light.

(ii) No. Energy carried by a wave depends on the amplitude of the wave, not on the speed of wave propagation.

(iii) For a given frequency, intensity of light in the photon picture is determined by the number of photons crossing an unit area per unit time.

Q. 10. State clearly how an unpolarised light gets linearly polarised when passed through a polaroid.★

(i) Unpolarised light of intensity I_0 is incident on a polaroid P_1 which is kept near another polaroid P_2 whose pass axis is parallel to that of P_1. How will the intensities of light, I_1 and I_2, transmitted by the polaroids P_1 and P_2 respectively, change on rotating P_1 without disturbing P_2 ?

(ii) Write the relation between the intensities I_1 and I_2.

Ans. Polaroid filters are made of a special material that is capable of blocking of the two planes of vibration of an electromagnetic wave. In this sense, a polaroid serves as a device that filter out one-half of the vibrations upon transmission of the light through the filter. When unpolarised light is transmitted through a polaroid filter, it emerge with one-half the intensity and with vibrations in a single plane; it emerges as polarised light.

★ are board based questions from previous year

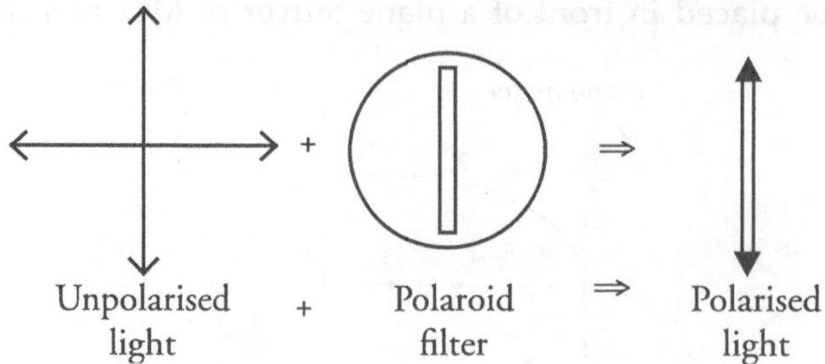

(i) When an unpolarised light of intensity I_0 is incident on a polaroid P_1 which is kept near. To another P_2 whose pass axis is parallel to that of P_1 the light intensity I_1 will become half of the incident intensity.

$$I_1 = \frac{I_0}{2}$$

and when it will pass through P_2 polaroid there is no change in the intensity but the angle will change by $\cos^2 \theta$

$$I_2 = I_1 \cos^2 \theta = \frac{I_0}{2} \cos^2 \theta$$

(ii) The relation between I_1 and I_2 is given by

$$\frac{I_2}{I_1} = \cos^2 \theta$$

Q. 11. For a single slit of width a, the first minimum of the interference pattern of a monochromatic light of wavelength λ occurs at an angle of $\frac{\lambda}{a}$. At the same angle of $\frac{\lambda}{a}$, we get a maximum for two narrow slits separated by a distance a. Explain.★

Ans. When a single slit is used, the interference pattern is due to the diffraction phenomenon. In case of diffraction from a single slit of width a using monochromatic light of wavelength λ, the first minimum of the interference pattern occurs at an angle θ_1, which is given by

$$\sin \theta_1 = \frac{\lambda}{a}$$

Since λ is very small, θ_1 will also be very small.

$$\therefore \quad \theta_1 \approx \sin \theta_1 = \frac{\lambda}{a} \quad \text{...(i)}$$

In case of interference from two narrow slits separated by a distance a using monochromatic light of wavelength λ, the first maximum occurs at a distance y_1 from the centre of the screen, which is given by.

$$y_1 = \frac{D\lambda}{a}$$

Here, D is the distance of the screen from the centre of the slits.

If the first maximum occurs at an angle θ_1', then

$$\tan \theta_1' = \frac{y_1}{D} = \frac{\lambda}{a}$$

Again as λ is very small, θ_1' will also be very small.

$$\therefore \quad \tan \theta_1' \approx \theta_1' = \frac{\lambda}{a} \quad \text{...(ii)}$$

From the equations (i) and (ii), we have

$$\theta_1 = \theta_1' = \frac{\lambda}{a}$$

Hence, it proves the result.

★ are board based questions from previous year

Q. 12. Answer the following questions :

(i) In a single slit diffraction experiment, the width of the slit is made double the original width. How does this affect the size and intensity of the central diffraction band ?

(ii) In what way is diffraction from each slit related to the interference pattern in a double slit experiment ?

(iii) When a tiny circular obstacle is placed in the path of light from a distance source, a bright spot is seen at the centre of the shadow of the obstacle. Explain why ?

(iv) Two students are separated by a 7 m partition wall in a room 10 m high. If both light and sound waves can bend around obstacles, how is it that the students are unable to see each other even though they can converse easily.

(v) Ray optics is based on the assumption that light travels in a straight line. Diffraction effects (observed when light propagates through small apertures/slits or around small obstacles) disprove this assumption. Yet the ray optics assumption is so commonly used in understanding location and several other properties of images in optical instruments. What is the justification ?

[NCERT]

Ans. (i) In a single slit diffraction experiment, if the width of the slit is made double the original width, then the size of the central diffraction band reduces to half and the intensity of the central diffraction band increases up to four times.

(ii) The interference pattern in a double-slit experiment is modulated by diffraction from each slit. The pattern is the result of the interference of the diffracted wave from each slit.

(iii) When a tiny circular obstacle is placed in the path of light from a distant source, a bright spot is seen at the centre of the shadow of the obstacle. This is because light wave are diffracted from the edge of the circular obstacle, which interferes constructively at the centre of the shadow. This constructive interference produces a bright spot.

(iv) Bending of waves by obstacles by a large angle is possible when the size of the obstacle is comparable to the wavelength of the waves.

On the one hand, the wavelength of the light waves is too small in comparison to the size of the obstacle. Thus, the diffraction angle will be very small. Hence, the students are unable to see each other. On the other hand, the size of the wall is comparable to the wavelength of the sound waves.Thus, the bending of the waves takes place at a large angle. Hence, the students are able to hear each other.

(v) The justification is that in ordinary optical instruments, the size of the aperture involved is much larger than the wavelength of the light used.

Q. 13. Discuss the intensity of transmitted light when a polaroid sheet is rotated between two crossed polaroids. In which position of the polaroid sheet will the transmitted intensity be maximum ?

[NCERT]

Ans. Consider I_0 be the intensity of polarised light after passing through the first polariser P_1. Then the intensity of light after passing through second polariser P_2 will be

$$I = I_0 \cos^2 \theta.$$

where θ is the angle between pass axes of P_1 and P_2. Since P_1 and P_3 are crossed the angle between the pass axes of P_2 and P_3 will be $\left(\frac{\pi}{2} - \theta\right)$. Hence, the intensity of light emerging from P_3 will be

$$I = I_0 \cos^2 \theta \cos^2\left(\frac{\pi}{2} - \theta\right)$$

$$= I_0 \cos^2 \theta \sin^2 \theta = (I_0/4) \sin^2 2\theta$$

Therefore, the transmitted intensity will be maximum when $\theta = \pi/4$.

Q. 14. (i) The refractive index of glass is 1.5. What is the speed of light in glass ? Speed of light in vacuum is 3.0×10^8 m/s.

(ii) Is the speed of light in glass independent of the colour of light ? If not, which of the two colours red and violet travels slower in a glass prism ? **[NCERT]**

Ans. (i) Refractive index of glass, $\mu = 1.5$.

Speed of light, $c = 3 \times 10^8$ m/s

Speed of light in glass is given by

$$v = \frac{c}{\mu} = \frac{3 \times 10^8}{1.5} = 2 \times 10^8 \text{ m/s}$$

Thus, the speed of light in glass is 2×10^8 m/s.

(ii) The speed of light in glass is dependent of the colour of light.

The refractive index of a violet component of white light is greater than the refractive index of a red component. Therefore, the speed of violet light is less than the speed of red light in glass. Thus, violet light travels slower than red light in a glass prism.

Q. 15. In Young's double-slit experiment using monochromatic light of wavelength (λ) the intensity of light at a point on the screen where path difference is λ, is K units. What is the intensity of light at a point where path difference is $\lambda/3$? **[NCERT]**

Ans. Consider I_1 and I_2 be the intensity of the two light waves. Their resultant intensities can be obtained by

$$I' = I_1 + I_2 + 2\sqrt{I_1 I_2} \cos \phi$$

where, ϕ = Phase difference between the two waves

For monochromatic light waves,

$$I_1 = I_2$$

$$\therefore \quad I' = I_1 + I_1 + 2\sqrt{I_1 I_1} \cos \phi$$

$$= 2I_1 + 2I_1 \cos \phi$$

$$\text{Phase difference} = \frac{2\pi}{\lambda} \times \text{Path difference}$$

As path difference = λ,

Phase difference, $\phi = 2\pi$

$\therefore \quad I' = 2I_1 + 2I_1 = 4I_1$

From question $I' = K$

$$\therefore \quad I_1 = \frac{K}{4} \quad \text{...(i)}$$

When path difference = $\frac{\lambda}{3}$

Phase difference, $\phi = \frac{2\pi}{3}$

Hence, resultant intensity,

$$(I'_R) = I_1 + I_1 + 2\sqrt{I_1 I_1} \cos \frac{2\pi}{3}$$

$$= 2I_1 + 2I_1\left(-\frac{1}{2}\right) = I_1$$

Using equation (i), $I_R = I_1 = \frac{K}{4}$

Thus, the intensity of light of a point where the path difference is $\frac{\lambda}{3}$ is $\frac{K}{4}$ units.

Q. 16. A monochromatic light of wavelength 500 nm is incident normally on a single slit of width 0.2 mm to produce a diffraction pattern. Find the angular width of the central maximum obtained on the screen. Estimate the number of fringes obtained in Young's double slit experiment with fringe width 0.5 mm, which can be accommodated with the region of total angular spread of the central maximum due to single slit.★

Ans. Angular width of central maxima $= \dfrac{2\lambda}{a} = \dfrac{2 \times 500 \times 10^{-9}}{0.2 \times 10^{-3}}$

$$= 5 \times 10^{-3} \text{ rad.}$$

$$\beta = \frac{\lambda D}{d}$$

Linear width of central maxima in the diffraction pattern

$$w' = \frac{2\lambda D}{a}$$

Let 'n' be the number of interference fringes which can be accommodated in the central maxima.

$$n \times \beta = w'$$

$$n = \frac{2\lambda D}{a} \times \frac{d}{\lambda D}$$

$$n = \frac{2d}{a}$$

Chapter 11. Dual Nature of Radiation and Matter

Q. 1. The work function of aluminium is 4.2 eV. If two photons each of energy 2.5 eV strike an electron of aluminium, will the emission of electron be possible ?

Ans. The emission of electron is not possible. This is only possible when the energy of each photon is more than the work function of the surface.

Q. 2. When monochromatic radiation of wavelength 2000 Å falls upon a nickel plate, the later acquires a positive charge. When the wavelength is increased, at 3400 Å the effect is found to cease, however intense the incident radiation may be. Explain it.

Ans. 3400 Å is the threshold wavelength for nickel. Therefore, for the incident radiation of wavelength greater than 3400 Å, photoelectrons are not emitted from the plate and so it no more acquires positive charge.

Q. 3. Draw graphs showing variation of photoelectric current with applied voltage for two incident radiations of equal frequency and different intensities. Mark the graph for the radiation of higher intensity.★

Ans. Graph for photoelectric current (I) versus applied potential for radiations of same frequency and varying intensity.

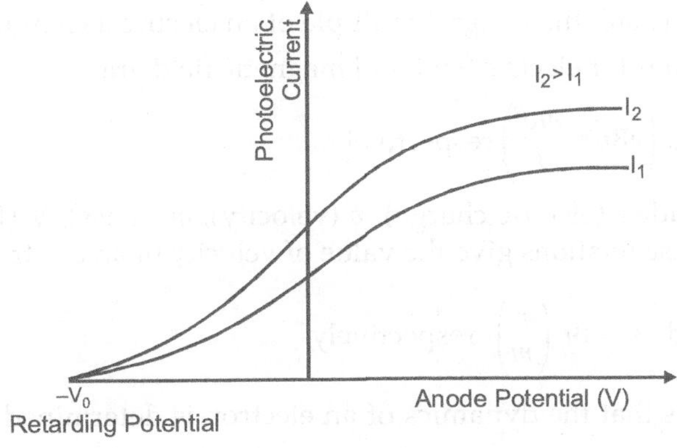

★ are board based questions from previous year

Q. 4. Sketch the graphs showing variation of stopping potential with frequency of incident radiations for two photosensitive materials A and B having threshold frequencies $\nu_A > \nu_B$

(i) In which case is the stopping potential more and why ?★

(ii) Does the slope of the graph depend on the nature of the material used ? Explain.★

Ans. (i) We know that,

$$V_0 = \frac{h}{e}(\nu - \nu_0)$$

For the same value of ν, stopping potential is more for which threshold frequency (ν_0) is less.

Given, $\nu_A > \nu_B$

Therefore, threshold frequency for metal B is less than the threshold frequency for metal A. Hence, stopping potential is more for metal B.

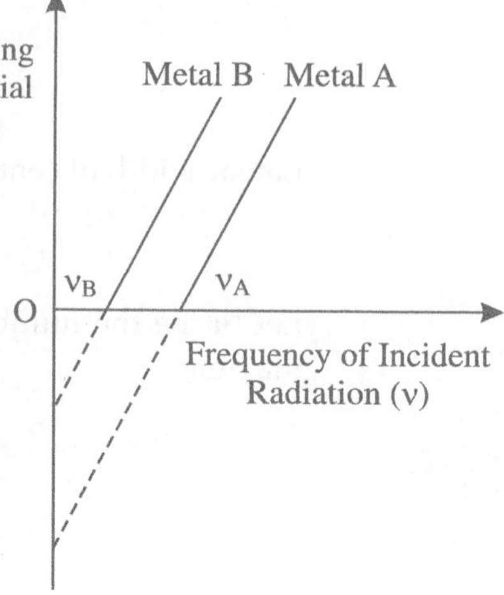

(ii) No, the slope of the graph tells us the value of $\frac{h}{e}$ which is same for both the materials. So, it does not depend on the nature of the materials.

Q. 5. Answer the following questions :

(i) Quarks inside protons and neutrons are thought to carry fractional charges $\left[\left(+\frac{2}{3}\right)e, \left(-\frac{1}{3}\right)e\right]$. Why do they not show up in Millikan's oil-drop experiment ?

(ii) What is so special about the combination e/m^2 ? Why do we not simply talk of e and m separately ?

(iii) Why should gases be insulators at ordinary pressures and start conducting at very low presssures ?

(iv) Every metal has a definite work function. Why do all photoelectrons not come out with the same energy if incident radiation is monochromatic ? Why is there an energy distribution of photoelectrons ?

(v) The energy and momentum of an electron are related to the frequency and wavelength of the associated matter wave by the relations :

$$E = h\nu, p = \frac{h}{\lambda}$$

But while the value of λ is physically significant, the value of ν (and therefore, the value of the phase speed νλ) has no physical significance. Why ? **[NCERT]**

Ans. (i) Quarks inside protons and neutrons carry fractional charges, its due to nuclear force increases extremely if they are pulled apart. Hence, fractional charges may exist in nature; observable charges are still the integral multiple of an electrical charge.

(ii) The basic relations for electric field and magnetic field are

$$\left(eV = \frac{1}{2}mv^2\right) \text{ and } \left(eBv = \frac{mv^2}{r}\right) \text{ respectively.}$$

These relations include e (electric charge), v (velocity), m (mass), V (Potential), r (radius), and B (Magnetic field). These relations give the value of velocity of an electron as

$$v = \sqrt{2V\left(\frac{e}{m}\right)} \text{ and } v = Br\left(\frac{e}{m}\right) \text{ respectively.}$$

From above relations that the dynamics of an electron is determined not by e and m separately, but by the ratio e/m.

★ are board based questions from previous year

(iii) At atmospheric pressure, the ions of gases have no chance of reaching their respective electrons because of collision and recombination with other gas molecules. Hence, gases are insulators at atmospheric pressure. At low pressures, ions have a chance of reaching their respective eletrodes and constitute a current. Hence, they conduct electricity at these pressures.

(iv) The work function of a metal is the minimum energy required for a conduction electron to get out of the metal surface. All the electrons in an atom do not have the same energy level. When a ray having some photon energy is incident on a metal surface, the electrons come out from different levels with different energies. Hence, these emitted electrons show different energy distributions.

(v) The absolute value of energy of a particle is arbitrary within the additive constant. Hence, wavelength (λ) is significant, but the frequency (ν) associated with an electron has no direct physical significance.

Therefore, the product νλ (phase speed) has no physical significance.

Group speed is given as

$$v_r = \frac{dv}{dk}$$

$$= \frac{dv}{d\left(\frac{1}{\lambda}\right)} = \frac{dE}{dp} = \frac{d\left(\frac{p^2}{2m}\right)}{dp} = \frac{p}{m}$$

The quantity has a physical meaning.

Q. 6. A mercury lamp is a convenient source for studying frequency dependence of photoelectric emission, since it gives a number of spectral lines ranging from the UV to the red of the visible spectrum. In our experiment with rubidium photo-cell, the following lines from a mercury source were used : $\lambda_1 = 3650$ Å, $\lambda_2 = 4047$ Å, $\lambda_3 = 4358$ Å, $\lambda_4 = 5461$ Å, $\lambda_5 = 6907$ Å. The stopping voltages, respectively, were measured to be : $V_{01} = 1.28$ V, $V_{02} = 0.95$ V, $V_{03} = 0.74$ V, $V_{04} = 0.16$ V, $V_{05} = 0$ V.

Determine the value of Planck's constant *h*, the threshold frequency and work function for the material.

[**Note :** You will notice that to get *h* from the data, you will need to know *e* (which you can take to be 1.6×10^{-19} C). Experiments of this kind on Na, Li, K, etc. were performed by Millikan, who, using his own value of *e* (from the oil-drop experiment) confirmed Einstein's photoeletric equation and at the same time gave an independent estimate of the value of *h*.] **[NCERT]**

Ans. Einstein's photoelectric equation is given by

$$eV_0 = h\nu - \phi_0$$

$$V_0 = \frac{h}{e}\nu - \frac{\phi_0}{e} \qquad \text{...(i)}$$

Here, V_0 = Stopping potential

h = Planck's constant

e = Charge on an electron

ν = Frequency of radiation

ϕ_0 = Work function of a material

From equation (i) potential V_0 is directly proportional to frequency ν.

$$\text{Frequency } (\nu) = \frac{\text{Speed of light } (c)}{\text{Wavelength } (\lambda)}$$

This relation can be used to obtain the frequencies of the various lines of the given wavelengths.

$$\nu_1 = \frac{c}{\lambda_1} = \frac{3 \times 10^8}{3650 \times 10^{-10}} = 8.219 \times 10^{14} \text{ Hz}$$

$$\nu_2 = \frac{c}{\lambda_2} = \frac{3 \times 10^8}{4047 \times 10^{-10}} = 7.412 \times 10^{14} \text{ Hz}$$

$$\nu_3 = \frac{c}{\lambda_3} = \frac{3 \times 10^8}{4358 \times 10^{-10}} = 6.884 \times 10^{14} \text{ Hz}$$

$$\nu_4 = \frac{c}{\lambda_4} = \frac{3 \times 10^8}{5461 \times 10^{-10}} = 5.493 \times 10^{14} \text{ Hz}$$

$$\nu_5 = \frac{c}{\lambda_5} = \frac{3 \times 10^8}{6907 \times 10^{-10}} = 4.343 \times 10^{14} \text{ Hz}$$

The given quantities can be listed in tabular form as :

Frequency × 10^{14} Hz	8.219	7.412	6.884	5.493	4.343
Stopping potential V_0	1.28	0.95	0.74	0.16	0

The following figures shows a graph between ν and V_0.

It can be observed that the obtained curve is a straight line. It intersects the ν-axis at 5×10^{14} Hz, which is the threshold frequency (ν_0) of the material. Point D corresponds to a frequency less than the threshold frequency. Hence, there is no photoelectric emission for the λ_5 line, and therefore, no stopping voltage is required to stop the current.

$$\text{Slope of the straight line} = \frac{AB}{CB} = \frac{1.28 - 0.16}{(8.214 - 5.493) \times 10^{14}}$$

From equation (i), the slope $\frac{h}{e}$ can be written as :

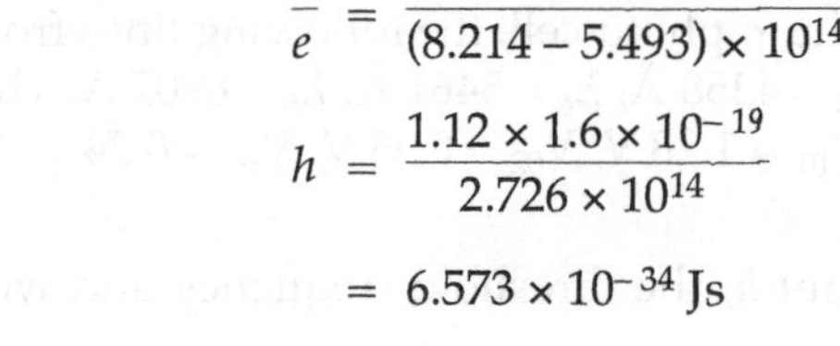

$$\frac{h}{e} = \frac{1.28 - 0.16}{(8.214 - 5.493) \times 10^{14}}$$

$$\therefore \quad h = \frac{1.12 \times 1.6 \times 10^{-19}}{2.726 \times 10^{14}}$$

$$= 6.573 \times 10^{-34} \text{ Js}$$

The work function of the metal is given as :

$$\phi_0 = h\nu_0$$
$$= 6.573 \times 10^{-34} \times 5 \times 10^{14}$$
$$= 3.286 \times 10^{-19} \text{ J}$$
$$= \frac{3.286 \times 10^{-19}}{1.6 \times 10^{-18}} = 2.054 \text{ eV}$$

Q. 7. When an electron in hydrogen atom jumps from the third excited state to the ground state, how would the de-Broglie wavelength associated with the electron change ? Justify your answer.★

Ans. de-Broglie wavelength associated with a moving charge particle having a kinetic energy 'K' can be given as

$$\lambda = \frac{h}{p} = \frac{h}{\sqrt{2mK}} \quad \text{...(i)}$$

$$\left[K = \frac{1}{2} mv^2 = \frac{p^2}{2m}\right]$$

The kinetic energy of the electron in any orbit of hydrogen atom can be given as

$$K = -E = -\left(\frac{13.6}{n^2} \text{ eV}\right) = \frac{13.6}{n^2} \text{ eV} \quad \text{...(ii)}$$

Let K_1 and K_4 be the kinetic energy of the electron in ground state and third excited state, where $n_1 = 1$ shows ground state and $n_2 = 4$ shows third excited state.

★ are board based questions from previous year

Using the concept of equations (i) and (ii), we have

$$\frac{\lambda_1}{\lambda_4} = \sqrt{\frac{K_4}{K_1}} = \sqrt{\frac{\frac{1}{n_2}}{n_2^2}}$$

$$\frac{\lambda_1}{\lambda_4} = \sqrt{\frac{1^2}{4^2}} = \frac{1}{4}$$

$$\Rightarrow \qquad \lambda_1 = \frac{\lambda_4}{4}$$

i.e., the wavelength in the ground state will decrease.

Chapter 12. Atoms

Q. 1. In the Rutherford's nuclear model of the atom, the nucleus (radius about 10^{-15} m) is analogous to the sun about which the electron move in orbit (radius $\approx 10^{-10}$ m) like the earth orbits around the sun. If the dimensions of the solar system had the same proportions as those of the atom, would the earth be closer to or farther away from the sun than actually it is ? The radius of earth's orbit is about 1.5×10^{11} m. The radius of sun is taken as 7×10^{8} m. **[NCERT]**

Ans. The ratio of the radius of electron's orbit to the radius of nucleus is $\frac{(10^{-10}\text{ m})}{(10^{-15}\text{ m})} = 10^5$, that is, the radius of the electron's orbit is 10^5 times larger than the radius of nucleus. If the radius of the earth's orbit around the sun were 10^5 times larger than the radius of the sun, the radius of the earth's orbit would be $10^5 \times 7 \times 10^8$ m $= 7 \times 10^{13}$ m. This is more than 100 times greater than the actual orbital radius of earth. Thus, the earth would be much farther away from the sun.

It implies that an atom contains a much greater fraction of empty space than our solar system does.

Q. 2. If Bohr's quantisation postulate (angular momentum $= nh/2\pi$) is a basic law of nature, it should be equally valid for the case of planetary motion also. Why then do we never speak of quantisation of orbits of planets around the sun ? **[NCERT]**

Ans. Quantisation of orbits of planets around the Sun never comes in discussion because the angular momentum associated with planetary motion is largely relative to the value of Planck's constant (h). The angular momentum of the Earth in its orbit is of the order of $10^{70}\, h$. This leads to a very high value of quantum levels n of the order of 10^{70}. For large values of n, successive energies and angular momenta are relatively small. Hence, the quantum levels for planetary motion are considered continuous.

Q. 3. Suppose you are given a chance to repeat the alpha-particle scattering experiment using a thin sheet of solid hydrogen in place of the gold foil. (Hydrogen is a solid at temperatures below 14 K). What results do you expect ? **[NCERT]**

Ans. In the alpha-particle scattering experiment, if a thin sheet of solid hydrogen is used in place of a gold foil, then the scattering angle would not be large enough, because the mass of hydrogen (1.67×10^{-27} kg) is less than the mass of incident α-particles. Thus, the mass of the scattering particle is more than the target nucleus (hydrogen).

As a result, the α-particles would not bounce back if solid hydrogen is used in the α-particle scattering experiment.

Q. 4. In the study of Geiger-Marsdon experiment on scattering of α-particles by a thin foil of gold, draw the trajectory of α-particles in the Coulomb field of target nucleus. Explain briefly how one gets the information on the size of the nucleus from this study.

From the relation $R = R_0 A^{1/3}$, where R_0 is constant and A is the mass number of the nucleus, show that nuclear matter density is independent of A.★

Ans. From this experiment, the following is observed :

(i) Most of the alpha particles pass straight through the gold foil. It means that they do not suffer any collision with gold atoms.

(ii) About one alpha particle in every 8000 alpha particles deflects by more than 90°.

As most of the alpha particles go undeflected and only a few get deflected, this shows that most of the space in an atom is empty and at the centre of the atom, there exists a nucleus. By the number of the alpha particles deflected, the information regarding size of the nucleus can be known.

Trajectory of a Particles in Coulomb Field of Target Nucleus

If m is the average mass of a nucleon and R is the nuclear radius, then mass of nucleus = mA, where A is the mass number of the element.

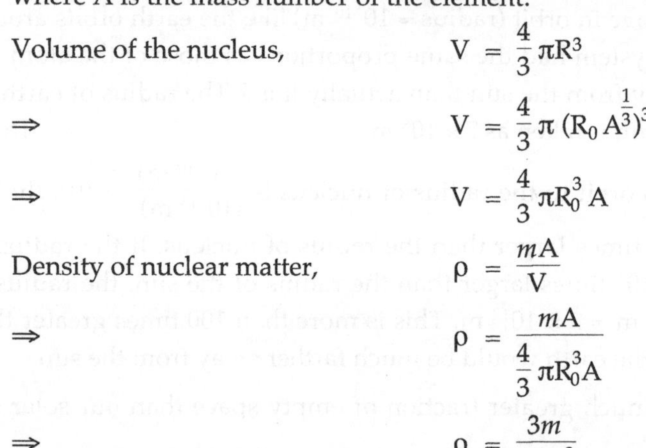

Volume of the nucleus, $V = \frac{4}{3}\pi R^3$

$\Rightarrow \quad V = \frac{4}{3}\pi (R_0 A^{\frac{1}{3}})^3$

$\Rightarrow \quad V = \frac{4}{3}\pi R_0^3 A$

Density of nuclear matter, $\rho = \frac{mA}{V}$

$\Rightarrow \quad \rho = \frac{mA}{\frac{4}{3}\pi R_0^3 A}$

$\Rightarrow \quad \rho = \frac{3m}{4\pi R_0^3}$

This shows that the nuclear density is independent of A.

Q. 5. Define ionisation energy. How would the ionisation energy change when electron in hydrogen atom replaced by a particle of mass 200 times that of the electron but having the same charge ?★

Ans. Ionization energy is defined as the amount of energy needed to remove the valence electron of an isolated gaseous atom.

The ionisation energy of hydrogen atom is

$$E_H \propto m_e \quad \text{..(i)}$$

When mass of electron is replaced by a particle having mass = 200 m_e, then

$$E_H' \propto 200\, m_e \quad \text{...(ii)}$$

From (i) and (ii), we have

$$E_H' = 200\, E_H$$

$$= 200 \times (-13.6)$$

$$= -2720 \text{ eV} \qquad (\because E_H = -13.6 \text{ eV})$$

Q. 6. The gravitational attraction between electron and proton in a hydrogen atom is weaker than the coulomb attraction by a factor of about 10^{-40}. An alternative way of looking at this fact is to estimate the radius of the first Bohr orbit of a hydrogen atom if the electron and proton were bound by gravitational attraction. You will find the answer interesting. **[NCERT]**

★ are board based questions from previous year

Ans. Radius of the first Bohr orbit is given by

$$r_1 = \frac{4\pi\varepsilon_0\left(\frac{h}{2\pi}\right)^2}{m_e e^2} \quad \text{...(i)}$$

Here, ε_0 = Permittivity of free space

h = Planck constant $= 6.63 \times 10^{-34}$ Js

m_e = Mass of an electron $= 9.1 \times 10^{-31}$ kg

e = Charge of an electron $= 1.9 \times 10^{-19}$ C

m_p = Mass of a proton $= 1.67 \times 10^{-27}$ kg

r = Distance between the electron and the proton

Coulomb attraction between an electron and a proton is given by

$$F_C = \frac{e^2}{4\pi\varepsilon_0 r^2} \quad \text{...(ii)}$$

Gravitational force of attraction between an electron and a proton is given by

$$F_G = \frac{Gm_p m_e}{r^2} \quad \text{...(iii)}$$

Here, G = Gravitational constant $= 6.67 \times 10^{-11}$ N m^2/kg^2

It the electrostatic (Coulomb) force and the gravitational force between an electron and a proton are equal, then we can write :

$$\therefore \quad F_G = F_C$$

$$\frac{Gm_p m_e}{r^2} = \frac{e^2}{4\pi\varepsilon_0 r^2}$$

$$\therefore \quad \frac{e^2}{4\pi\varepsilon_0} = Gm_p m_e \quad \text{...(iv)}$$

From equation (iv) and equation (i),

$$r_1 = \frac{\left(\frac{h}{2\pi}\right)^2}{Gm_p m_e^2}$$

$$= \frac{\left(\frac{6.63 \times 10^{-34}}{2 \times 3.14}\right)^2}{6.67 \times 10^{-11} \times 1.67 \times 10^{-27} \times (9.1 \times 10^{-31})^2}$$

$$\approx 1.21 \times 10^{29} \text{ m}$$

It is known that the universe is 156 billion light years wide or 1.5×10^{27} m wide. Hence, we can conclude that the radius of the first Bohr orbit is much greater than the estimated size of the whole universe.

Chapter 13. Nuclei

Q. 1. The mass of β-particles is higher than the mass of electrons obtained by other means, whereas β-particles are also electrons. Why ?

Ans. β-particles are emitted with very high velocity (upto 0.99 c). So, according to Einstein's theory of relativity, the mass of β-particle is much higher as compared to its rest mass (m_0). The velocity of electrons obtained by other means is very small as compared to velocity of light (c). So, the mass of electron remains nearly m_0.

Q. 2. Tritium has a half-life of 12.5 years undergoing beta decay. What fraction of a sample of pure tritium will remain undecayed after 25 years. **[NCERT]**

Ans. By definition of half-life, half of the initial sample will remain undecayed after 12.5 yr. In the next 12.5 yr., one-half of these nuclei would have decayed. Hence, one fourth of the sample of the initial pure tritium will remain undecayed.

Q. 3. Four nuclei of an element undergo fusion to form a heavier nucleus, with release of energy. Which of the two— the parent or the daughter nucleus—would have higher binding energy per nucleon ? ★

Ans. When lighter nuclei combine to form a heavier nucleus, binding energy per nucleon increases and energy is released. Thus, the daughter nucleus would have higher binding energy per nucleon.

Q. 4. Why is it found experimentally difficult to detect neutrinos in nuclear β-decay ?★

Ans. Neutrinos are difficult to detect experimentally in β-decay because these are uncharged particles with almost zero mass and also they interact weakly with matter.

Q. 5. For the β^+ (positron) emission from a nucleus, there is another competing process known as electron capture (electron from an inner orbit, say, the K-shell, is captured by the nucleus and a neutrino is emitted).

$$_{-1}e^0 + {}^A_Z X \longrightarrow {}_{Z-1}^{\;\;A}Y + V$$

Show that if β^+ emission is energetically allowed, electron capture is necessarily allowed but not vice-versa. **[NCERT]**

Ans. For the nucleus $_ZX^A$, the β^+-emission may be represented as

$$_ZX^A \longrightarrow {}_{Z-1}Y^A + {}_{+1}e^0 + v + E_1,$$

where E_1 is the energy released.

$$\therefore \quad E_1 = [m_N(_ZX^A) - m_N(_{Z-1}Y^A) - m_e] \times 931.5 \text{ MeV}$$

Now, $$m_N(_ZX^A) = m(_ZX^A) - Zm_e$$

and $$m_N(_{Z-1}Y^A) = m(_{Z-1}Y^A) - (Z-1)m_e$$

$$\therefore \quad E_1 = [\{m(_ZX^A) - Zm_e\} - \{m(_{Z-1}Y^A) - (Z-1)m_e\} - m_e] \times 931.5 \text{ MeV}$$

or $$E_1 = [m(_ZX^A) - m(_{Z-1}Y^A) - 2m_e] \times 931.5 \text{ MeV} \quad \text{...(i)}$$

For the nucleus $_ZX^A$, the electron capture is represented as

$$_{-1}e^0 + {}_ZX^A \longrightarrow {}_{Z-1}Y^A + v + E_2,$$

where E_2 is the energy released.

$$\therefore \quad E_2 = [m_e + m_N(_ZX^A) - m_N(_{Z-1}Y^A)] \times 931.5 \text{ MeV}$$

Now, $$m_N(_ZX^A) = m(_ZX^A) - Z\,m_e$$

and $$m_N(_{Z-1}Y^A) = m(_{Z-1}Y^A) - (Z-1)\,m_e$$

$$\therefore \quad E_2 = [m_e + \{m(_ZX^A) - Z\,m_e\} - \{m(_{Z-1}Y^A) - (Z-1)\,m_e\}] \times 931.5 \text{ MeV}$$

or $$E_2 = \{m(_ZX^A) - m(_{Z-1}Y^A)\} \times 931.5 \text{ MeV} \quad \text{...(ii)}$$

From the equations (i) and (ii), it follows that

if $$E_1 > 0, \text{ then } E_2 > 0$$

i.e., if β^+-emission is energetically allowed, then the electron capture is necessarily allowed. On the other hand, if $E_2 > 0$, then it does not necessarily mean that $E_1 > 0$. Thus, if the electron capture is energetically allowed, then the β^+-emission is not necessarily allowed.

Q. 6. Answer the following questions :

(i) Are the equations of nuclear reactions 'balanced' in the sense a chemical equation (*e.g.*, $2H_2 + O_2 \rightarrow 2H_2O$) is ? If not, in what sense are they balanced on both sides ?

★ are board based questions from previous year

(ii) If both the number of protons and the number of neutrons are conserved in each nuclear reaction. In what way is mass converted into energy (or vice-versa) in a nuclear reaction ?

(iii) A general impression exists that mass-energy interconversion takes place only in nuclear reaction and never in chemical reaction. This is strictly speaking, incorrect. Explain. **[NCERT]**

Ans. (i) A chemical equation is balanced in the sense that the number of atoms of each element is the same on both sides of the equation. A chemical reaction merely alters the original combinations of atoms. In a nuclear reaction, elements may be transmuted. Thus, the number of atoms of each element is not necessarily conserved in a nuclear reaction. However, the number of protons and the number of neutrons are both separately conserved in a nuclear reaction. (Actually, even this is not strictly true in the realm of very high energies–what is strictly conserved is the total charge and total 'baryon number'. In nuclear reactions, the number of protons and the number of neutrons are the same on the two sides of the equations.

(ii) We know that the binding energy of a nucleus gives a negative contribution to the mass of the nucleus (mass defect). Now, since proton number and neutron number are conserved in a nuclear reaction, the total rest mass of neutrons and protons is the same on either side of a reaction. But the total binding energy of nuclei on the left side need not be the same as that on the right hand side. The difference in these binding energies appears as energy released on absorbed in a nuclear reaction. Since binding energy contributes to mass. We say that the difference in the total mass of nuclei on the two sides get converted into energy or vice-versa. It is in these sense that a nuclear reaction is an example of mass-energy interconversion.

(iii) From the point of view of mass-energy interconversion, a chemical reaction is similar to a nuclear reaction in principle. The energy released or absorbed in a chemical reaction can be traced to the difference in chemical (not nuclear) binding energies of atoms and molecules on the two sides of a reaction. Since, strictly speaking, chemical binding energy also gives a negative contribution (mass defect) to the total mass of an atom or molecule, we can equally well say that the difference in the total mass of atoms or molecules, on the two sides of the chemical reaction gets converted into energy or vice-versa. However, the mass defects involved in a chemical reaction are almost a million times smaller than those in a nuclear reaction. This is the reason for the general impression, (which is incorrect) that mass-energy interconversion does not take place in a chemical reaction.

Q. 7. From the relation $R = R_o A^{1/3}$, where R_o is a constant and A is the mass number of a nucleus, show that the nuclear matter density is nearly constant (*i.e.*, independent of A). **[NCERT]**

Ans. Nuclear radius, $R = R_o A^{\frac{1}{3}}$

Here, R_o = Constant.

A = Mass number of the nucleus

$$\text{Nuclear matter density, } (\rho) = \frac{\text{Mass of the nucleus}}{\text{Volume of the nucleus}}$$

Let m be the average mass of nucleus.

Therefore, mass of the nucleus = mA

$$\therefore \quad \rho = \frac{mA}{\frac{4}{3}\pi R^3} = \frac{3mA}{4\pi (R_0 A^{1/3})^3}$$

$$= \frac{3mA}{4\pi R_o^3 A} = \frac{3\,m}{4\pi R_o^3}$$

Therefore, the nuclear matter density is independent of A. It is nearly constant.

Chapter 14. Semiconductor Electronics : Materials, Devices and Simple Circuits

Q. 1. C, Si and Ge have same lattice structure. Why is C insulator while Si and Ge intrinsic semiconductors ? **[NCERT]**

Ans. The 4 bonding electrons of C, Si or Ge lie, respectively, in the second, third and fourth orbit. Hence, energy required to take out an electron from these atoms (*i.e.*, ionisation energy E_g) will be least for Ge, followed by Si and highest for C. Hence, number of free electrons for conduction in Ge and Si are significant but negligibly small for C.

Q. 2. Carbon, silicon and germanium have four valence electrons each. These are characterised by valence and conduction bands separated by energy band gap respectively equal to $(E_g)_C$, $(E_g)_{Si}$ and $(E_g)_{Ge}$. Which of the following statements is true ? **[NCERT]**

(i) $(E_g)_{Si} < (E_g)_{Ge} < (E_g)_C$ (ii) $(E_g)_C < (E_g)_{Ge} > (E_g)_{Si}$

(iii) $(E_g)_C > (E_g)_{Si} > (E_g)_{Ge}$ (iv) $(E_g)_C = (E_g)_{Si} = (E_g)_{Ge}$

Ans. (iii) The energy band gap of carbon is the maximum among given elements and that of germanium is the least.

Thus energy band gap of these elements are related as : $(E_g)_C > (E_g)_{Si} > (E_g)_{Ge}$

Q. 3. What happens to the conductivity and resistivity of a semiconductor on increasing temperature ?

Ans. In a semiconductor, conductivity is due to the movement of electron-hole pairs, temperature is increased due to which covalent bonds breaks and more electron-hole pairs are available for conduction. Thus, conductivity increases and resistivity, the reciprocal of conductivity decreases.

Q. 4. Why does a semiconductor gets damaged when a heavy current flows through it ?

Ans. When a heavy current flows, the semiconductor gets heated up. Many covalent bonds breaks liberating a large number of free electrons. Therefore, the semiconductor loses its property of controlled conduction and becomes an ordinary conductor.

Q. 5. The current in the forward bias is known to be more (~mA) than the current in the reverse bias (~μA). What is the reason then to operate the photodiodes in reverse bias ? **[NCERT]**

Ans. Consider the case of an n-type semiconductor. Obviously, the majority carrier density (n) is considerbly larger than the minority hole density p (*i.e.*, $n >> p$). On illumination, let the excess electrons and holes generated be Δn and Δp, respectively.

$$n' = n + \Delta n$$

$$p' = p + \Delta p$$

Here n' and p' are the electron and hole concentrations at any particular illumination and n and p are carriers concentration when there is no illumination. Remember $\Delta n = \Delta p$ and $n >> p$. Hence, the fractional change in the majority carriers (*i.e.*, $\Delta n / n$) would be much less than that in the minority carriers (*i.e.*, $\Delta p / p$). In general, we can state that the fractional change due to the photo-effects on the minority carrier dominated reverse bias current is more easily measurable than the fractional change in the forward bias current. Hence, photodiodes are preferably used in the reverse bias condition for measuring light intensity.

Q. 6. A zener diode is fabricated by heavily doping both p and n- sides of the junction. Explain, why ? Briefly explain the use of zener diode as a dc voltage regulator with the help of a circuit diagram.

Ans. By heavily doping both p and n sides of the junction, depletion region formed is very thin, *i.e.*, $< 10^{-6}$ m. Hence, electric field across the junction is very high ($- 5 \times 10^6$ V/m) even for a small reverse bias voltage. This can lead to a break down during reverse biasing.

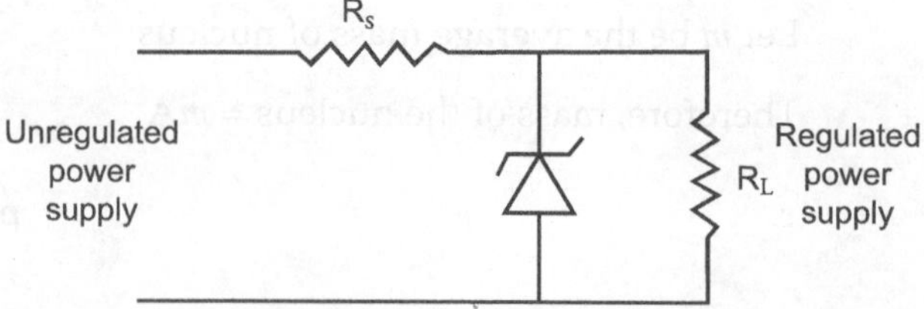

If the input voltage increases/decreases then current through resistor R_s and Zener diode also increases/decreases. This increases/decreases the voltage drop across R_s without any change in voltage across the zener diode.

This is because in the breakdown region, zener voltage remains constant even though the current through the zener diode changes.

Q. 7. In an intrinsic semiconductor the energy gap E_g is 1.2 eV. Its hole mobility is much smaller than electron mobility and independent of temperature. What is the ratio between conductivity at 600 K and that 300 K ? Assume that the temperature dependence of intrinsic carrier concentration n_i is given by

$$n_i = n_0 \exp\left[-\frac{E_g}{2k_B T}\right]$$

Where n_0 is a constant. **[NCERT]**

Ans. Given,

Energy gap of the intrinsic semiconductor, $(E_g) = 1.2$ eV.

The temperature dependence of the intrinsic carrier concentration is given by

$$n_i = n_0 \exp\left[-\frac{E_g}{2k_B T}\right]$$

Here, k_B = Boltzman constant = 8.62×10^{-5} eV/K (T = Temperature, n_0 = Constant)

Initial temperature $(T_1) = 300$ K.

The intrinsic carrier-concentration at this temperature can be given by

$$n_{i_1} = n_0 \exp\left[-\frac{E_g}{2k_B \times 300}\right] \quad \text{...(i)}$$

Final temperature $(T_2) = 600$ K.

The intrinsic carrier-concentration at this temperature can be given as

$$n_{i_2} = n_0 \exp\left[-\frac{E_g}{2k_B \times 600}\right] \quad \text{...(ii)}$$

The ratio between the conductivity at 600 K and at 300 K is equal to the ratio between the respective intrinsic carrier-concentrations at these temperature.

□□

Question Set 8

Case Based Questions

Chapter 1. Electric Charges and Fields

1. Faraday Cage:

A Faraday cage or Faraday shield is an enclosure made of a conducting material. The fields within a conductor cancel out with any external fields so the electric field within the enclosure is zero. These Faraday cages act as big hollow conductors you can put things in to shield them from electrical fields. Any electrical shocks the cage receives, pass harmlessly around the outside of the cage.

(i) Which of the following material can be used to make a Faraday cage?

(a) Plastic (b) Glass

(c) Copper (d) Wood

(ii) Example of a real-world Faraday cage is:

(a) Car (b) Plastic box

(c) Lighting rod (d) Metal rod

(iii) What is the electrical force inside a Faraday cage when it is struck by lightning?

(a) The same as the lightning (b) Half that of the lightning

(c) Zero (d) A quarter of the lightning

(iv) An isolated point charge $+q$ is placed inside the Faraday cage. Its surface must have change equal to:

(a) Zero (b) $+q$

(c) $-q$ (d) $+2q$

(v) A point charge of 2C is placed at centre of Faraday cage in the shape of cube with surface of 9 cm edge. The number of electric field lines passing through the cube normally will be:

(a) 1.9105 Nm^2/C entering the surface (b) 1.9105 Nm^2/C leaving the surface

(c) 2.0105 Nm^2/C leaving the surface (d) 2.0105 Nm^2/c entering the surface

Ans. (i) (c) Copper, as Faraday cage is made from conducting material.

(ii) (a) Car, as it can save us from lightning.

(iii) (c) Zero, as electrical field inside then is zero.

(iv) (b) $-q$, charge is $+q$ inside it. so, $-q$ will be induced on outer surface

(v) Correct option is not given.

$$F = \frac{q}{\varepsilon_0} = \frac{2}{8.85 \times 10^{-12}}$$

$$= 2.25 \times 10^{11} \ Nm^2/C$$

leaving the surface

2. Gauss theorem: Gauss theorem is mainly used to find out the electric flux linked to a closed surface. It does not depend upon the shape or size of surface. According to this theorem the electric flux linked to a closed surface is equal to $\left(\frac{1}{\varepsilon_0}\right)$ times the charge enclosed by the surface. Let we have a charge q, now if we want to find out the net flux linked to a closed surface around it them,

$$\text{Electric flux } \phi = \oint_s \vec{E} \cdot \vec{ds} = \frac{q}{\varepsilon_0}$$

(i) Gauss theorem is used to find out:

(a) Electric force (b) Electric flux

(c) Electric potential (d) None of these.

(ii) This theorem is applied over a........surface:

(a) Closed surface (b) Open surface

(c) Both (a) and (b) (d) None of these

(iii) Gauss theorem does not depends upon the.........of surface:

(a) Shape (b) Size

(c) Area (d) Both (a) and (b)

(iv) If we increase the charge enclosed by the surface then electric flux will:

(a) Increases (b) Decrease

(c) Remain same (d) Both (a) and (b)

(v) Net flux linked to a closed surface around a charge particle is.............times the charge.

(a) ε_0 (b) $\frac{1}{\varepsilon_0}$

(c) ε_0^2 (d) None of these.

Ans. (i) (b) Electric flux (ii) (c) Closed surface

(iii) (a) both (a) and (b) (iv) (a) increase

(v) (b) $\frac{1}{\varepsilon_0}$

Chapter 2. Electrostatic Potential and Capacitance

1. The electric field at a point is the force experienced by unit positive charge at that point. It is a vector quantity. The electric field due to a point charge q at a distance r from it is given by:

$$E = \frac{1}{4\pi\varepsilon_0}\frac{q}{r^2}$$

Its direction is towards the charge when it is negative and away from the charge when the charge is positive. If there are a system of charges, then field due to each charge will get added vectorially. Electric potential at a point is the work done in bringing unit positive charge from infinity to that point. The potential due to a point charge at a distance r from it is given by:

$$V = \frac{1}{4\pi\varepsilon_0}\frac{q}{r}$$

If there are a number of charge in the vicinity of the point, then the potential due to each charge gets add up algebraically (one need to specify the sign of the charge in calculation).

Now consider the following situation.

Four charges $+q$, $+q$, $-q$ and $-q$ are placed at the four corners A, B, C and D respectively of a square of side a arranged in the same order. Midpoint of BC is E and that of CD is F. O is the centre of the square.

(i) The direction of the net electric field at O is towards:

(a) AB (b) BC (c) CD (d) AD

(ii) The magnitude of the electric field at O is:

(a) $\frac{q}{\sqrt{2\pi\varepsilon_0 a^2}}$ (b) $\frac{q}{\sqrt{3\pi\varepsilon_0 a^2}}$

(c) $\frac{\sqrt{3}q}{\pi\varepsilon_0 a^2}$ (d) $\frac{\sqrt{2}q}{\pi\varepsilon_0 a^2}$

(iii) The electric potential at O is:

(a) $\frac{\sqrt{2}q}{\pi\varepsilon_0 a}$ (b) $\frac{\sqrt{3}q}{\pi\varepsilon_0 a}$

(c) $\frac{q}{\pi\varepsilon_0 a}$ (d) 0

(iv) The work done in carrying a charge from O to E is:

(a) $\frac{\sqrt{2}qe}{\pi\varepsilon_0 a}$ (b) $\frac{qe}{\pi\varepsilon_0 a}\left(\frac{1}{\sqrt{5}}-1\right)$

(c) $\frac{qe}{\pi\varepsilon_0 a}\left(\frac{1}{\sqrt{5}}+1\right)$ (d) 0

(v) The work done in carrying a charge from O to F is:

(a) $\frac{\sqrt{2}qe}{\pi\varepsilon_0 a}$ (b) $\frac{qe}{\pi\varepsilon_0 a}\left(\frac{1}{\sqrt{5}}-1\right)$

(c) $\frac{qe}{\pi\varepsilon_0 a}\left(\frac{1}{\sqrt{5}}+1\right)$ (d) 0

Ans. (i) (c) CD

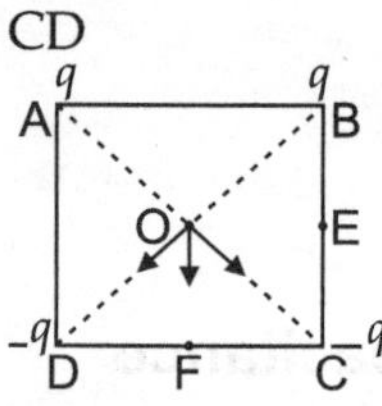

(ii) (d) $\frac{\sqrt{2}q}{\pi\varepsilon_0 a^2}$

$$E_O = 4 \times \frac{1}{4\pi\varepsilon_0}\frac{q}{(a/\sqrt{2})^2} \times \frac{1}{\sqrt{2}} = \frac{\sqrt{2}q}{\pi\varepsilon_0 a^2}$$

(iii) (d) O

Charges at A and C and those at B and B and D symmetrical with O.

(iv) (b) $\frac{q_e}{\pi\varepsilon_0 a}\left(\frac{1}{\sqrt{5}}-1\right)$

$V_E = 0$. Therefore, no potential difference between O and E. Work require to shift change from O to E is 0.

(v) (a) $\frac{\sqrt{2}qe}{\pi\varepsilon_0 a}$ This depend on the potential at F.

$$V_F = 2 \times \frac{1}{4\pi\varepsilon_0}\left[\frac{q}{(a^2 + a^2/4)^{1/2}} - \frac{q}{a/2}\right] = \frac{q}{\pi\varepsilon_0 a}\left(\frac{1}{\sqrt{5}}-1\right)$$

$$W_{OF} = e\,(V_F - V_O) = \frac{qe}{\pi\varepsilon_0 a}\left(\frac{1}{\sqrt{5}}-1\right)$$

Chapter 3.Current Electricity

1. The resistance of a conductor depends on its length L and area of cross section A. It also depends on the nature of the material and temperature. If the temperature is kept constant, the resistance of a conductor is directly proportional to its length and inversely to its area of cross section.

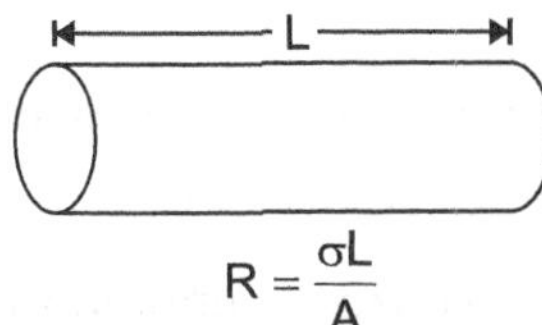

This can be given : $R = \frac{\rho L}{A}$, where ρ is a constant called the resistivity of the material of the conductor. As temperature increases, the resistivity of a conductor will also increase. But it is not a linear relationship. In SI system, resistance is expressed in ohm (Ω) and resistivity in ohm-metre (Ω – m).

(i) A cylindrical wire has a resistance of 18Ω. The resistance of another wire of the same material with the same cross section, but 1.5 times the length is:

(a) 18Ω (b) 24Ω

(c) 12Ω (d) 9Ω

(ii) If a cuboidal conductor is as follows, then its resistance is the most between which pair of points?

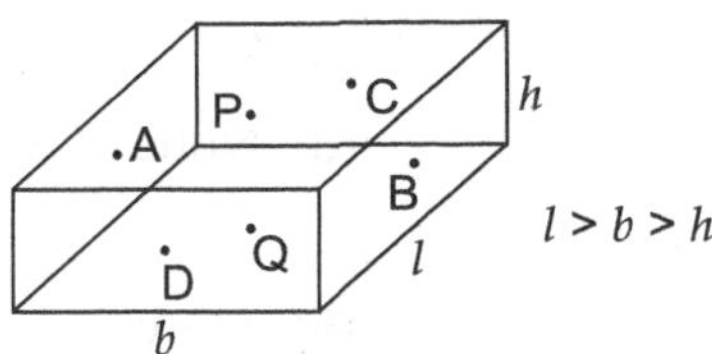

P and Q are the centres of top and bottom surfaces, A and B are those of longer sides and C and D are centres of shorter sides.

(a) AB (b) CD

(c) PQ (d) All have the same resistance

(iii) R is the resistance of a cylindrical wire. If it is stretched to twice of its length, keeping the volume constant, the new resistance will be:

(a) R (b) 2R

(c) 4R (d) R/2

(iv) In an experiment, a sample of wire has a resistance 20Ω at 15° C. If the experiment is conducted keeping the same wire sample at a temperature of 30° C, then the possible value of its resistance can be:

(a) 20Ω (b) 10Ω

(c) 15Ω (d) 22.5Ω

(v) In an experiment, a particular wire sample (A) has a resistance 10Ω. The person who conducts the experiment has three more sample of wires (B, C and D) of three different materials. All the four samples have the same length and same area of cross section. Resistance of sample B is 15Ω, that of C is 18Ω and that of D is 5Ω.

Choose the incorrect statement from the following.

(a) Resistivity of material of A is less than that of C

(b) Resistivity of material of A twice that of D

(c) Resistivity of B in thrice the resistivity of D

(d) Resistivity of material of C is greater than that of B

Ans. (i) (b) $R \propto L$

(ii) (b) $R \propto \frac{L}{A}$; L is the most and A is the least between C and D.

(iii) (c) $R \propto \frac{L}{A}$; Length gets doubled and A becomes halved, resistance becomes 4 times.

(iv) (d) As temperature increases, resistivity increases, thereby increasing the resistance.

(v) (c) Resistivity of B is thrice the resistivity of D.

2. When an inductor is connected to a source of emf and the switch is closed, initially the inductor offers infinite reactance to the current. As a result, first after closing the switch, the current in the circuit will be zero. But as time passes, the reactance reduces and after a long time the current through the inductor will be maximum.

Now consider the following circuit.

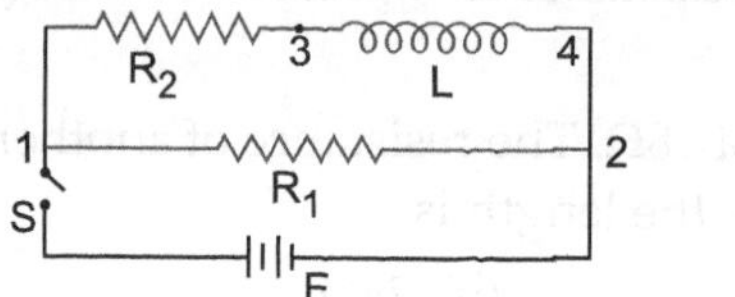

$E = 150$ V.

$R_1 = 25\ \Omega$

$R_2 = 50\ \Omega$

$L = 0.15$ H

Switch S is closed at $t = 0$. Just after switch is closed.

(i) Which point 1 or 2 is at a higher potential.

(a) 1 (b) 2

(c) Both at same potential (d) None of these.

(ii) The potential difference V_{12} across R_1 is

(a) 75 V (b) 125 V

(c) 150 V (d) 112.5 V

(iii) The potential difference V_{34} across L is:

(a) 75 V (b) 125 V

(c) 150 V (d) 112.5 V

(iv) Now the switch is opened after a long time just after opening the switch, the potential difference across R_1 is

(a) 37.5 V (b) 75 V

(c) 64 V (d) 112.5 V

(v) Which point 3 or 4 is at lights potential?

(a) 3 (b) 4

(c) 3 and 4 at the same potential (d) None of these

Ans. (i) (a) Just after the switch is closed, the inductor offers infinite resistance. Current occur across R1 only point 1 is at a higher potential.

(ii) (c) $V_{12} = E = 150$ V

(iii) (c) $V_{34} = V_{12} = 150$ V

(iv) (b) Just before opening the switch, the current through the inductor

$$i_0 = \frac{E}{R_2} = \frac{150}{50} = 3\text{A}$$

(v) (a) Point 3 will be at a higher potential.

Chapter 4. Moving Charges and Magnetism

1. A charged particle enters into a uniform magnetic field and follows a circular path as shown. This happens when the particle enters perpendicular to the magnetic field. It gets deflected by magnetic Lorentz force. Arrows indicate the direction of motion of the charged particle.

If $\vec{v}$ the velocity of the particle of charge q and B is the magnetic field, then the force acting on the particle is given by: $\vec{F} = q(\vec{v} \times \vec{B})$.

If the particle velocity is perpendicular to the magnetic field, then the magnitude of the force is F = qvB. This force provides the necessary centripetal force for the particle and the radius of the circular path is given by: $r = \frac{mv}{qB}$, where m is the mass of the particle.

The direction of deflection also depends on the nature of the charge.

If the particle enters the fields at an angle other than 90°, then, the path of the particle will be helical (not circular).

(i) Which of the following particles will not get affected if sent into a magnetic field normal to it?

(a) Alpha (b) Positron

(c) Neutron (d) Electron

(ii) When a proton and an electron enter a magnetic field normal to it, which of the following statements is correct (assume, their speeds are equal)?

(a) They will get deflected in opposite directions

(b) Radius of circular path of electron will be less

(c) Radius of circular path of proton will be less

(d) Both (a) and (b)

(iii) A deuteron (nucleus of deuterium) enters a magnetic field parallel to it. Its path inside the field will be (consider, no other force can act on it):

(a) Straight line (b) Circle

(c) Helix (d) Parabola

(iv) If a proton, a deuteron and an alpha particle are projected into a magnetic field perpendicular to it with the same speed, which of them will have the path of highest radius?

(a) Proton (b) Deuteron

(c) Alpha particle (d) Both (b) and (c)

(v) An electron enters a magnetic field making an angle of 45° with it. The path of the electron will be:

(a) Straight line (b) Circle

(c) Helix (d) Parabola

Ans. (i) (c) Neutron

Neutral particles are not affected by magnetic field.

(ii) (d) Both (a) and (b)

Opposite deflection because of opposite charge. Lighter particle gets deflected more.

(iii) (a) Straight line

$\vec{F} = q(v \times \vec{B})$; If velocity is parallel or anti-parallel to the field, the particle goes unaffected.

(iv) (d) Both (b) and (c)

$$r = \frac{mv}{qB} \Rightarrow r \propto \frac{m}{q}$$

(v) (c) Helix

Angle with the field is not 0°, 90° or 180°. Hence path will be helical.

2. **Biot-Savart's law:** Biot Savart law was given by Biot and Savart after doing many experiments. This law is related with the magnetic field induced at a point due to a small current carrying element (conductor). According to the law the magnetic field induced at a point near the current carrying element is directly proportional to the current flowing in conductor, length of the element, sin θ and inversely proportional to the square of the distance of point from the element.

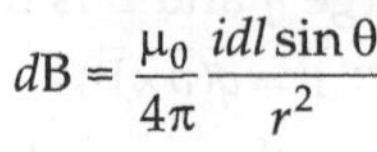

$$dB = \frac{\mu_0}{4\pi}\frac{idl\sin\theta}{r^2}$$

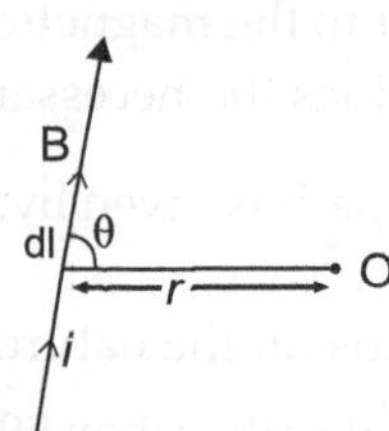

(i) Biot Savart law was given by :

(a) Oersted (b) Ampere

(c) Biot and Savart (d) Maxwell

(ii) Biot Savart law is related with the ________ induced at a point near current carrying element.

(a) Magnetic field (b) Gravitational field

(c) Electric field (d) None of these

(iii) Induced magnetic field is directly proportional to:

(a) i (b) dl

(c) sin θ (d) All of these

(iv) The magnetic field induced at a point is inversely proportional to:

(a) r^2 (b) r^3

(c) $\frac{1}{r}$ (d) $\frac{1}{r^2}$

(v) As the distance between the point and current carrying element decreases, dB:

(a) Increases (b) Decreases

(c) Remains same (d) Both (a) and (b)

Ans. (i) (c) Biot and Savart

(ii) (a) Magnetic field

(iii) (d) All of these; $\because dB = \frac{\mu_0 i\, dl\sin\theta}{4\pi\, r^2}$

(iv) (a) r^2

(v) (a) Increases; $\because dB \propto \frac{1}{r^2}$

Chapter 5. Magnetism and Matter

1. A suspended magnet gets aligned with the earth's magnetic field at that place. Along the equator, a suspended magnet will align horizontally. The orientation of the magnet depends on the dip angle at that place. Angle of dip is the angle between the net magnetic field of earth and the horizontal. At equator, angle of dip is 0 (zero) and at poles its value is 90°. This means, at equator, the earth's magnetic field is completely horizontal and at poles. The Earth's magnetic field is vertical.

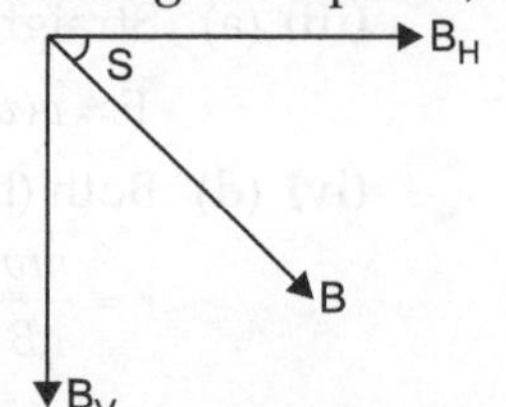

If S is the angle of dip at a place and B is the Earth's magnetic field, then the horizontal component of Earth's magnetic field $B_H = B\cos\delta$ the vertical component $B_V = B\sin\delta$

$$\Rightarrow \qquad B = \sqrt{B_H{}^2 + B_{V^2}}$$

$$\text{Also} \qquad \tan\delta = \frac{B_V}{B_H}$$

(i) A magnet stands vertical at a place when suspended that place will be:

(a) At equator (b) At poles

(c) At 30° latitude (d) At 60° latitude

(ii) The angle of dip of a face where the magnet gets aligned completely horizontal is:

(a) 0° (b) 90°

(c) 30° (d) 45°

(iii) The horizontal and vertical components of the magnetic field at a place are $\frac{\sqrt{3}}{2}$ R and $\frac{B}{2}$ respectively. When a magnet is left free at that place, the angle made by its magnetic axis with the horizontal is:

(a) 45° (b) 30°

(c) 60° (d) 15°

(iv) Earth's magnetic field always has a vertical component except at:

(a) Poles (b) Equator

(c) 30° latitude (d) 75° latitude

(v) The magnetic field due to earth has a horizontal component of 26μT at a place where the dip is 60°. Vertical component of the field at that point is:

(a) 25 μT (b) 30 μT

(c) 45μT (d) 54 μT

Ans. (i) (b) At poles.

Net magnetic field is vertical at poles.

(ii) (a) 0°

Earth's magnetic field is completely horizontal at equator. Angle of dip is 0° at equator.

(iii) (b) 30°

$$B\cos\delta = \frac{\sqrt{3}}{2}B \Rightarrow \cos\delta = \frac{\sqrt{3}}{2} \Rightarrow \delta = 30°$$

(iv) (b) Equator

At equator, the magnetic field of earth is completely vertical.

(v) (c) 45μT

$$\tan\delta = \frac{B_V}{B_H},\ \delta = 60°$$

$$B_V = B_H \tan\delta$$

$$= 26\ \mu \times \tan 60°$$

$$= 45\ \mu T.$$

Chapter 6. Electromagnetic Induction

1. Lenz's Law

This law is mostly used to find the direction current induced in a circuit. According to this law the polarity of e.m.f. induced in the circuit is such that it opposes or checks the variation in magnetic flux responsible for it. It we move north pole of a bar magnet towards coil, then magnetic flux linked with the coil changes (*i.e.,* increases). Due to this current is induced in the coil in anticlockwise direction. The magnetic moment related with the induced current has north polarity towards the north pole. If we move away the bar magnet then magnetic moment has south polarity.

Lenz's law obeys the law of conservation of energy.

(i) Lenz's law is used to find the direction of:

(a) Electric field (b) Force

(c) Induced current (d) Electrostatic force

(ii) Lenz law follows the law of:

(a) Conservation of force (b) Conservation of mass

(c) Conservation of momentum (d) Conservation of energy

(iii) What will be the polarity of magnetic moment towards the north pole if we move the north pole bar magnet towards coil ?

(a) North (b) South

(c) Both (a) and (b) (d) None of these

(iv) If we move away the bar magnet from coil then what will be the direction of induced current in coil.

(a) Clockwise (b) Anticlockwise

(c) Both (a) and (b) (d) None of these

(v) As we move the north pole of magnet towards coil, the magnetic flux linked with the coil.

(a) Increases (b) Decreases

(c) Remain same (d) None of these

Ans. (i) (c) Induced current

(ii) (d) Conservation of energy

(iii) (a) North

(iv) (a) Clockwise

(v) (a) Increases

Chapter 7. Alternating Current

1. The figure shows a series LCR circuit:

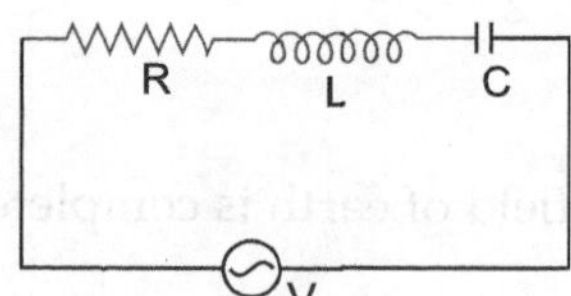

For such a circuit, the impedance Z is given by. $Z = \sqrt{R^2 + (X_L - X_C)^2}$, where X_L and X_C are inductive and capacitive resistances respectively. As the frequency of a.c. is increased, at a particular frequency. X_L become, equal to X_C. For that frequency maximum current occurs. This is because impedance becomes equal is its least value R. Current through the circuit I = V/R. The circuit behaves like a pure resistive circuit and current and voltage will be in phase. This is called resonance. Frequency of a.c. at which resonance occurs is called resonant frequency. If frequency is less than the resonant frequency, then the capacitive reactance will be more. The circuit will be capacitive in nature and current leads voltage. On the other hand, if frequency is more than the resonant frequency inductive reactance will be more. Circuit is inductive in nature and current lags the voltage.

An LCR circuit with a resistance 50 Ω has a resonant angular frequency 2×10^3 rad/s. At resonance, the voltage across the resistance and inductance are 25V and 20V respectively. Then

(i) The value of inductance is:

(a) 20 mH (b) 10 mH

(c) 40 mH (d) 25 mH

(ii) The value of capacitance is:

(a) 25 μF (b) 1 μ F

(c) 2 μ F (d) 12.5 μF

(iii) The impedance at resonance is:

(a) 50 W (b) 16 W

(c) 64 W (d) 25 W

(iv) Which of the following angular frequency of a.c. will see the circuit as inductive in nature?

(a) 1.5×10^3 red/s (b) 10^3 red/s

(c) 2×10^3 red/s (d) 5×10^3 red/s

(v) At angular frequency 10^3 red/s, the nature of circuit:

(a) Inductive (b) Capacitive

(c) Resistive (d) None of these

Ans. (i) (a)

$$X_L = \frac{V_L}{I}$$

$$I = \frac{V_R}{R} = \frac{25}{50} = \frac{1}{2}$$

$$X_L = \frac{20}{1/2} = 40\,\Omega$$

But
$$X_L = \omega_L \Rightarrow L = \frac{X_L}{\omega} = \frac{40}{2 \times 10^3}$$

$$= 20 \times 10^{-3} = 20\ \mu F$$

(ii) (d)
$$\omega^2 = \frac{1}{L_C} \Rightarrow C = \frac{1}{\omega^2 L}$$

$$= \frac{1}{(2 \times 10^3) \times 20 \times 10^{-13}} = 12.5\ \mu F$$

(iii) (a) At resonance, the impedance equal just resistance.

(iv) (d) For inductive nature $\omega > \omega_r$.

(v) (b) If $\omega < \omega_r$, the circuit will be capacitive in nature.

Chapter 8. Electromagnetic waves

1. Electromagnetic waves: These waves are produced by accelerated or ascillating charges. These are non mechanical waves so these does not require any kind of material medium for propagation. The speed of these waves in free space is equal to the speed of light. These are the sinusoidal variation in the magnetic as well as electric field vectors $(\vec{E}$ and $\vec{B})$. Electromagnetic waves are transverse in nature. The velocity of electromagnetic waves in the dielectrics is always less than the speed of light ($c = 3 \times 10^8$ m/s).

   The energy of electromagnetic waves is equally divided between electric field and magnetic field vectors.

   In these waves the electric vector is responsible for optical effects of electromagnetic waves. So electric vector is also called light vector.

(i) Which of the following e-m wave has highest wavelength?

(a) X-ray (b) UV rays

(c) infrared rays (d) microwaves

(ii) Who first demonstrated the existence of em waves:

(a) Hertz (b) Maxwell

(c) Ampere (d) Faraday

(iii) The dimensions of E/B are same as that of:

(a) charge (b) current

(c) velocity (d) acceleration

(iv) An em-wave of frequency, v= 3MHz passes from vacuum into dielectric medium with permettivity Er = 4, then:

(a) wavelength is doubled and frequency is unchanged

(b) wavelength is doubled and frequency becomes half

(c) wavelength is halved and frequency remains unchanged

(d) wavelength and frequency both remains unchanged

(v) If u_e and u_m are the electric and magnetic field densities in an electromagnetic wave, then c is the speed of light:

(a) $u_e = u_m$ (b) $u_e = Cu_m$

(c) $u_e = \frac{u_m}{C}$ (d) None of these

Ans. (i) (c) infrared rays

(ii) (a) Hertz

(iii) (c) Velocity

(iv) (c) wavelength is halved and frequency remains unchanged. As frequency remains same on passing from one medium to another.

(v) (a) $u_e = u_m$

2. Infrared waves : The discovery of these waves was done by Herschell. The frequency of these waves is in between 3×10^{11} Hz to 4×10^{14} Hz. These are also known as heat waves because due to absorption of these waves heating effect in the bodies is observed. These waves are produced by hot bodies. The infrared waves are used in physical therapy to reduce or treat muscular strain. These waves also give electrical energy to satellites with the help of solar cells.

These waves are also helpful in taking photographs during fog, smoke etc. These also keep warm the plants in green houses. Weather forecasting is also done by infrared photography. These are also helpful in the study of molecular structure.

(i) These waves give ....energy to satellites.

(a) Mechanical (b) Magnetic

(c) Electrical (d) None

(ii) Waves in decreasing order of their wavelength are:

(a) X-rays, infrared rays, visible rays, radio waves

(b) radio waves, visible rays, infrared rays, X-rays

(c) radiowaves, infrared rays, visible rays, X-rays

(d) radiowaves, U-V rays, visible rays, X-rays

(iii) The e-m waves used to treat muscular pain is:

(a) VHF radiowaves (b) X-rays

(c) UV rays (d) infrared rays

(iv) Find the ratio of the speed of the microwaves and infrared rays in the vaccum.

(a) 1:4 (b) 4:1

(c) 1:1 (d) 1:2

(v) An infrared waves is incident on a material surface. The wave delivers momentum P and energy E. Then,

(a) $P \neq 0, E \neq 0$ (b) $P = 0, E = 0$

(c) $P = 0, E \neq 0$ (d) $P \neq 0, E = 0$

Ans. (i) (c) Electrical

(ii) (c) radiowaves, infrared rays, visible rays, X-rays

(iii) (d) infrared rays are used to treat muscular pain

(iv) (c) 1:1 As speed of all e-m waves is constant in vaccum. i.e. speed of light $c = 3 \times 10^8$m/s.

(v) (a) When plane *e-m* wave strikes the material surface, it delivers some momentum and energy to the surface. Hence, $P \neq 0, E \neq 0$.

Chapter 9. Ray Optics and Optical Instruments

1. Sparking Brilliance of Diamond:

The total internal reflection of the light is used in polishing diamonds to create a sparking brilliance. By polishing the diamond with specific cuts, it is adjusted the most of the light rays approaching the surface are incident with an angle of incidence more than critical angle. Hence, they suffer multiple reflections and ultimately come out of diamond from the top. This gives the diamond a sparking brilliance.

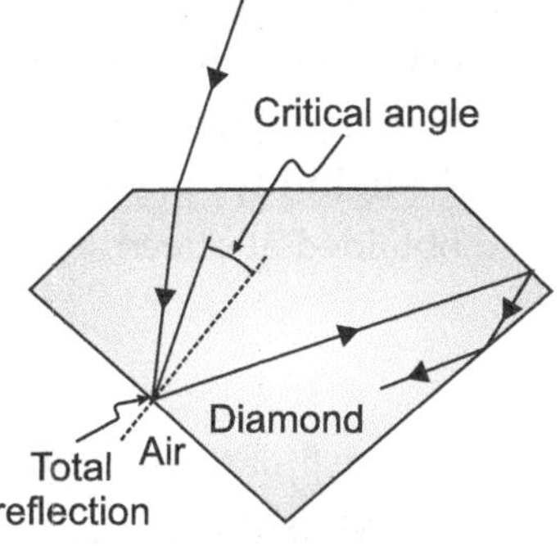

(i) Light cannot easily escape a diamond without multiple internal reflections. This is because:

(a) Its critical angle with reference to air is too large

(b) Its critical angle with reference to air is too small

(c) The diamond is transparent

(d) Rays always enter at angle greater than critical angle

(ii) The critical angle for a diamond is 24.4°. Then its refractive index is:

(a) 2.42 (b) 0.413

(c) 1 (d) 1.413

(iii) The basic reason for the extraordinary sparkle of suitably cut diamond is that:

(a) It has low refractive index (b) It has high transparency

(c) It has high refractive index (d) It is very hard

(iv) A diamond is immersed in a liquid with a refractive index greater than water. Then the critical angle for total internal reflection will :

(a) Will depend on the nature of the liquid (b) Decrease

(c) Remain the same (d) Increases

(v) The following diagram shows same diamond cut in two different shapes.

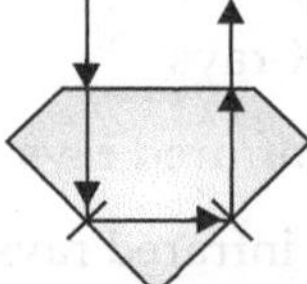
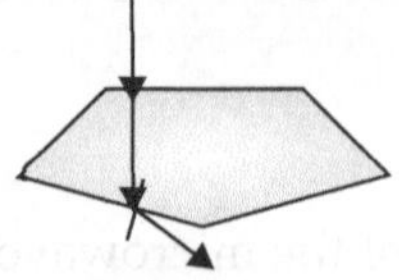

The brilliance of diamond in the second diamond will be:

(a) Less than the first (b) Greater than first

(c) Same as first (d) Will depend on the intensity of light

Ans. (i) (b) Its critical angle with respect to air is too small. So it suffers multiple internal reflections easily.

(ii) (d) Rays always enters at angle greater than critical angle.

(iii) (c) High refractive index.

It has high refractive index as compared to glass. The sparkling isn't just the high refractive index, it shape matters too, which cause multiple internal reflection.

(iv) (d) Increase, as the refractive index of liquid is greater than diamond, so critical angle increases.

(v) (a) Less than first

As multiple reflections, is observed in first case and not in second case.

So brilliance of diamond will be more in first case.

Chapter 10. Wave Optics

1. Young's Double slit experiment this experiment was done by Thomas Young to demonstrate experimentally the interference of light. In this, coherent sources of monochromatic light are used. Here two sources of monochromatic light are obtained by placing two slits before a single light source. As light rays passes from the slits and interference of light takes place. The interference fringes are obtained at screen. The fringe obtained at the centre of screen is a bright fringe

$$\text{Fringe width } (\beta) = \frac{\lambda D}{d},$$

Where β - Fringe width, D – Distance between screen and slit, *d* – Distance between slits.

(i) What kind of light sources are used in this experiment?

(a) Monochromatic (b) Dichromatic

(c) Trichromatic (d) Any

(ii) How many slits are used in this experiment?

(a) Two (b) One

(c) Three (d) Many

(iii) The central fringe in the experiment is:

(a) Dark (b) Very dark

(c) Bright (d) None of these

(iv) What kind of phenomena of light is demonstrated with this experiment?

(a) Reflection (b) Refraction

(c) Interference (d) Diffraction

(v) What will be the ratio of fringe width B_1 and B_2 obtained with red light of $\lambda_1 = 660$ nm and ultraviolet light of $\lambda_2 = 165$ nm

(a) 1 : 4 (b) 4 : 1

(c) 2 : 1 (d) 1 : 2

Ans. (i) (a) Monochromatic (ii) (a) Two

(iii) (c) bright (iv) (c) Interference

(v) (b) 4 : 1,

Given: $\lambda_1 = 660$ nm, $\lambda_2 = 165$ nm

As $\beta = \dfrac{\lambda D}{d}$; D, d are same for lights

$$\therefore \quad \frac{B_1}{B_2} = \frac{\lambda_1}{\lambda_2} = \frac{660}{165} = 4:1$$

2. Diffraction of light

The diffraction of light takes place when an object or obstacle comes in the path of light. In this phenomena bending of light takes place around the edges of the obstacle. The bending of light from its path is larger when the size of the obstacle is comparable to the wavelength of light.

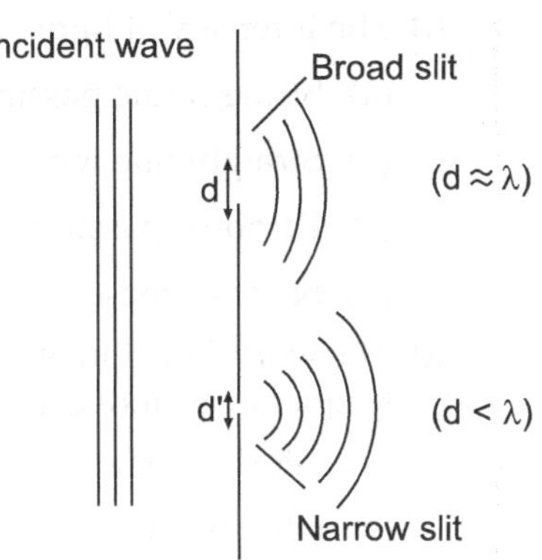

It we place a narrow slit in the path of wavefront of light, then bending of diffracted wavefront will be more. The diffraction of sound waves also takes place. For visible light, diffraction can happen when an object of $\lambda \approx 10^{-6}$ m order comes in its path.

(i) The phenomenon of bending of light around the edges is called ________.

(a) Interference (b) Refraction

(c) Diffraction (d) Polarisation

(ii) The diffraction of an incident wave will not happen when size aperture will be ________.

(a) Small (b) Very small

(c) Very large (d) None of these

(iii) The diffraction of light waves increases as the size of aperture of a hole ________.

(a) Increases (b) Decreases

(c) Both (a) and (b) (d) Remain same

(iv) The diffraction of light can take place through ________.

(a) Sharp edges (b) Very small apertures

(c) Very large apertures (d) Both (a) and (b)

(v) In which case diffraction of light will be more for an aperture (d).

(a) $d > \lambda$ (b) $d \approx \lambda$

(c) $d < \lambda$ (d) None of these

Ans. (i) (c) Diffraction (ii) (c) Very large

(iii) (b) Decreases (iv) (d) Both (a) and (b)

(v) (c) $d < \lambda$

Chapter 11. Dual Nature of Radiation and Matter

1. Photoelectric effect refers to the emission of electrons from certain metal surfaces when irradiated with high frequency light (Em radiation). (UV rays falling on cesium metal is a good example).

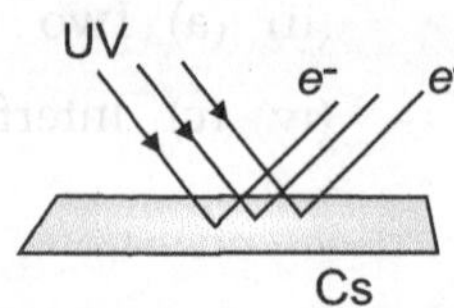

Number of electrons emitted is proportional to the intensity of light. KE of electron emitted depends on the frequency of incident light. KE does not depend on intensity of light. The minimum frequency of light required to initial photoelectric effect is called threshold frequency (ν_0). The minimum energy required to eject one electron from a metal surface is called its work function and is given by $\phi = h\nu_0$. Maximum KE of emitted electron when light of frequency ν falls on the metal surface is given by $KE_{max} = h\nu - \phi$ (ϕ is the work function) Negative potential can be applied to the electrons to stop them first outside the metal surface.

(i) The intensity of light is plotted against number of photoelectrons . The graph is a:

(a) Straight line passing through origin

(b) Straight line with $a + y$ intercept

(c) Parabola passing through the origin

(d) None of these

(ii) The work function of a metal corresponds to a frequency 2×10^{15}Hz of light. Which of the following frequency cannot cause photoelectric effect from the metal?

(a) $1{\cdot}5 \times 10^{16}$ Hz (b) $8{\cdot}2 \times 10^{14}$ Hz

(c) $6{\cdot}5 \times 10^{15}$ Hz (d) $2{\cdot}2 \times 10^{15}$ Hz

(iii) A particular metal can emit electrons when green light falls on it. In which of the following case, the maximum KE of emitted electrons cannot be more, compared to this case?

(a) Using blue light with the same metal

(b) Using a metal of less work function with the same light (green)

(c) Using red light with the first metal

(d) Using UV light with the second metal

(iv) If frequency of incident light becomes n times the initial frequency ($\nu > \nu_0$), then KE of emitted electron becomes.

(a) n times the initial KE (b) More than n times the initial KE

(c) Less than n times the initial KE (d) No change in KE

(v) Stopping potential (negative potential required to stop the electron just outside the metal surface is related to.

(a) Mean wavelength (b) Shortest frequency

(c) Maximum KE of emitted electrons (d) Intensity of light incident

Ans. (i) (a) Straight line points through origin

n µ I No. electrons are emitted without suitable light $\Rightarrow n = 0$ for I = 0

(ii) (b) 8.2×10^{14} Hz

Frequency of light should be more than threshold frequency for photoelectric emission.

(iii) (c) Using red light with first metal

$KE_{max} = (h\nu - \phi)$. Increasing ν and reducing ϕ will make KE_{max} more.

(iv) (b) More than n time the initial KE.

$KE_1 = h\nu - \phi : n\,KE_1 = nh\nu - n\phi$

$KE_2 = h\nu - \phi \Rightarrow KE_2 > n\,KE_1$

(v) (b) Shortest frequency

Stopping potential is related to the maximum KE of emitted electrons.

2. In photoelectric effect, the energy of the incident light causes the ejection of electron. This is the result of particle nature of light. Similarly all material particles shows wave nature also (while in motion). This is observed in the case of tiny particles moving at speed comparable with that of light. Electrons can be sent through YDSE set up and observations similar to that of light can be made. The figure shows a cesium sample irradiated with light of wavelength λ.The photoelectrons emitted are directly perpendicular to the plane of slits of a YDSE set up. The electrons are accelerated before passing through the slits. Wavelength of material particles in motion is called de Broglie wavelength and it is given by $\lambda = \frac{h}{p}$ where h is the Planck's constant and p is the momentum of the particle.

Charge of electron is e and its mass is m.

(i) If momentum of a particle increases its de Broglie wavelength:

(a) Increases (b) Decreases

(c) No change (d) Variables

(ii) The fringe width due to electron rays in the mentioned example is (V is the accelerating voltage):

(a) $\frac{\lambda D}{2d}$ (b) $\frac{\lambda D}{2d}$

(c) $\frac{hD}{d\sqrt{2meV}}$ (d) $\frac{\lambda D^2}{dm}$

(iii) If a light of less wavelength (then that of light falling on (s sample) is used in the experiment, then fringe width on the screen will:

(a) Increase (b) Decrease

(c) First increases, then decreases (d) No change occurs

(iv) If the experiment was conducted with protons instead of electrons, then fringe width and will be:

(a) More (b) Less

(c) Remaining same (d) Irregular and varnishing quickly

(v) If electrons were moving randomly before lifting the slit plane, then the central maximum shifts:

(a) Upwards (b) Downwards

(c) No shift in position (d) No fringes formed

Ans. (i) (b) $\lambda \alpha \frac{1}{p}$

(ii) (c) $\beta = \frac{D\lambda}{d} = \lambda \frac{h}{p} = \frac{h}{\sqrt{2meV}}$

(iii) (b) λ decreases as momentum increases and so β decreases.

(iv) (b) l decreases as momentum increases so b decreases.

(v) (c) No change in position of central maximum.

Chapter 12. Atoms

1. Rutherford's model of atom: This model of atom was given by Rutherford. According to him each atom have tiny core at its centre which is called nucleus of the atom. In the nucleus whole mass and positive charge of the nucleus is concentrated. The size of the nucleus is of the order of 10^{-15} m and atomic size is about 10^{-10} m.

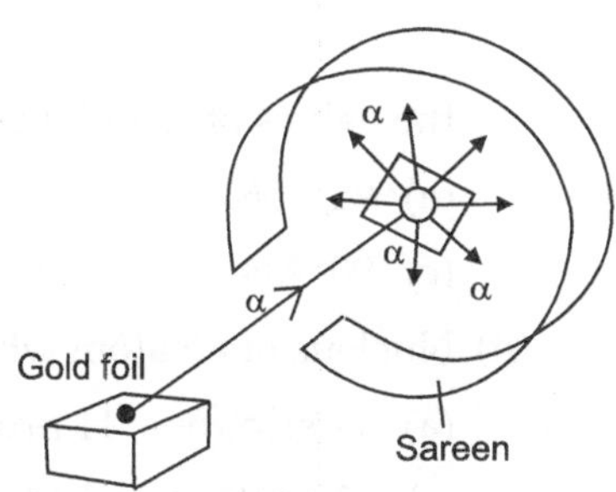

The nucleus of the atom is surrounded by a number of electrons. But whole atom is electrically neutral. These electrons revolve in certain orbits around the nucleus of an atom the electrons get centripetal force from the electrostatic force between nucleus and electrons.

(i) The tiny core at the centre of atom is:

(a) Electron (b) Atom

(c) Nucleus (d) Proton

(ii) Whole mass of the atom is concentrated at

(a) Periphery of atom (b) Nucleus

(c) Proton (d) None of these

(iii) The size of the nucleus is of the order of:

(a) 10^{15} m (b) 10^{-5} m

(c) 10^{-15}m (d) 10^{10} m

(iv) The nucleus of the atom is surrounded by a number of

(a) Proton (b) Neutron

(c) Electron (d) None of these

(v) Atom is electrically:

(a) Positive (b) Negative

(c) Earthed (d) Neutral

Ans. (i) (c) Nucleus (ii) (b) Nucleus

(iii) (c) 10^{-15}m (iv) (c) Electron

(v) (d) Neutral

Chapter 13. Nuclei

1. Nature of α-rays: α-rays contains α-particles. The charge on these particles is double the charge of proton. Their charge is also equal to the charge present on helium nucleus. The mass of α-particles is approximately four times that of hydrogen atom.

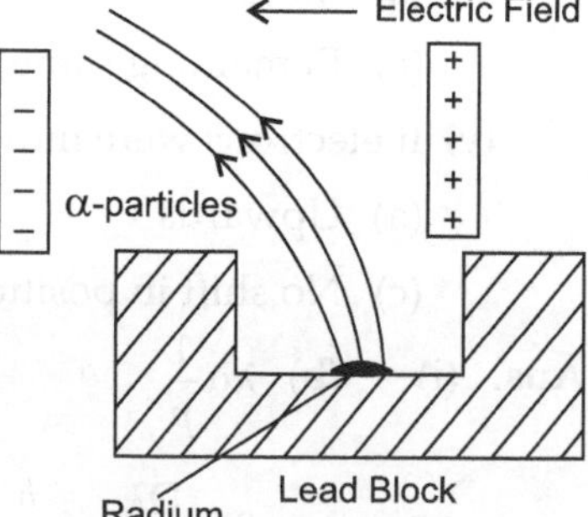

The velocity of α-particles is in between $1{\cdot}4 \times 10^7$ m/sec to $2{\cdot}1 \times 10^7$ m/sec and it also depends upon the source. Due to large mass and large velocity, α-particles have large ionising power. Each α-particle can create thousands of ions before absorption.

α-particles produce fluorescence in some substances e.g., zinc sulphide. α-particles can affect the photographic plates. These are deflected by electric and magnetic fields.

(i) In the disintegration series :

$$^{238}_{92}U \xrightarrow{-\alpha} X \xrightarrow{-\beta} {}^{A}_{Z}Y$$

the values of Z and A respectively will be:

(a) 92, 236 (b) 88, 230

(c) 90, 234 (d) 91, 234

(ii) Nucleus of an atom whose atomic mass is 24 consists of :

(a) 11 electrons, 11 protons and 13 neutrons (b) 11 electrons, 13 protons and 11 neutrons

(c) 11 protons and 13 neutrons (d) 11 protons and 13 electrons

(iii) The same radioactive nucleus may emit.

(a) all the three α, β and γ one after another

(b) all the three α, β and γ radiations simultaneously

(c) only α and β simultaneously

(d) only one α, β and γ at a time

(iv) Particles which can be added to the nucleus of an atom without changing its chemical properties are called:

(a) neutrons (b) protons

(c) electrons (d) alpha particles

(v) Consider α and β particles and γ-rays each having an energy of 0·5 MeV. In the increasing order of penetrating power, the radiation are respectively:

(a) α, β, γ (b) α, γ, β

(c) β, γ, α (d) γ, β, α

Ans. (i) (d) 91, 234

$${}^{238}_{92}U \xrightarrow{\alpha} {}^{234}_{90}X \xrightarrow{\beta} {}^{234}_{91}Y$$

(ii) (c) 11 protons and 13 neutrons since nucleus does not contain neutrons.

(iii) (d) only one α, β and γ at a time.

(iv) (a) chemical properties are unaffected with addition of neutrons to the nucleus.

(v) (a) α, β, γ

2. Nature of γ-rays

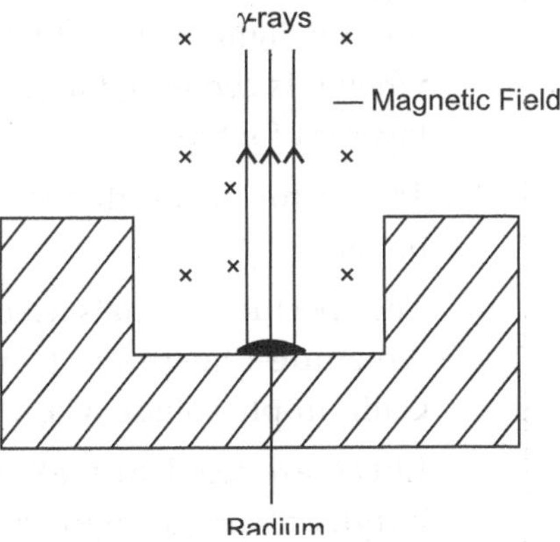

These are made up of γ-particles. In these, the γ-ray photon has zero rest mass. γ-rays travels with the speed of light. These are electromagnetic waves. The wavelength of γ-ray photon is smaller than the X-rays.

γ-rays are not deflected by electric and magnetic fields. These rays does not have any charge. These rays have very large penetrating power. These can pass through many centimetre of lead. These rays have small ionising power. These can affect the photographic plate. γ-rays can cause nuclear reactions. These can produce fluarscence in willemite. These can emit electron from the surface of a metal.

(i) In a γ-ray emission from a nucleus.

(a) only the proton number changes

(b) both the neutron number and the proton number change

(c) there is no change in the proton number and the neutron number

(d) only the neutron number change

(ii) The γ-radiation are:

(a) e-m radiation with high energy (b) e-m radiation with low energy

(c) charged particles emitted (d) electrons orbiting the nucleus

(iii) A nuclei having same number of neutron but different number of protons/atomic number are called:

(a) Isobars (b) Isomers

(c) Isotones (d) Isotopes

(iv) When a nucleus in an atom undergoes a radioative decay, the electronic energy levels of the atom:

(a) do not change for any type of radioactivity

(b) change for α, β radioactivity but not for X-radioactivity

(c) change for α-radioactivity but not for others

(d) change for β-radioactivity but not for others

(v) In the nuclear reaction

$$_{6}^{11}C \longrightarrow _{5}^{11}B + \beta^{+} + X$$

what does X stand for:

(a) Electron (b) Proton

(c) Neutron (d) Neutrino

Ans. (i) (c) As the γ-rays have no charge and no mass.

(ii) (a) e-m radiation with high energy

(iii) (c) isotones

(iv) (b) α particles-2 unit of positive charge

β particles-1 unit of positive charge

γ particles-No charge

Due to this reason, energy levels of the atom change for α and β but not for others.

(v) (d) By conservation of mass A = 0 and by conservation of charge Z = 0. Hence X is neutrino.

Chapter 14. Semiconductor Electronics : Materials, Devices and Simple Circuits

1. Light Emitting Diode (LED): LED is the photoelectronic device which converts electrical energy into the light energy. It has heavily doped *p-n* junction diode and it gives spontaneous radiation when it is connected in forward bias. In this the upper layer is of *p*-type semiconductor and lower layer is of *n*-type.

To control the brightness of light emitted by LED, a resistance is connected in the circuit with battery.

The specific materials used for making LED's are Gallium-Arsenide-Phosphide (GaAsP) for yellow or red light, Gallium-phosphide (GaP) for red or green light etc.

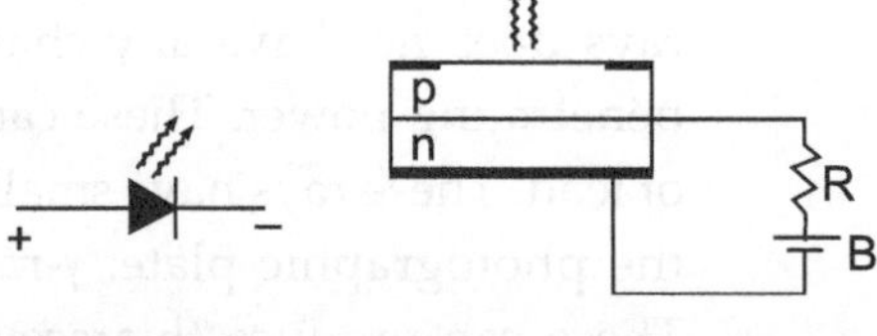

LED's are used in making calculators digital watches, burglar alarms, computers, picture phones, remote control and traffic light etc.

(i) LED is a:

(a) Photo device (b) Electrical device

(c) Mechanical device (d) Photo electronic device

(ii) *p-n* junction diode used in LED's is:

(a) Lightly doped (b) Moderately doped

(c) Heavily doped (d) None of these

(iii) The brightness of light is controlled by using __________ in the circuit.

(a) Inductor (b) Resistance

(c) Capacitor (d) None of these

(iv) The lower layer of LED's is of __________ semiconductor.

(a) p-type (b) n-type

(c) q-type (d) Both (a) and (b)

(v) What kind of material in used in making LED's.

(a) GaAs P (b) GaP

(c) GaAs (d) All of these

Ans. (i) (d) Photo electronic device (ii) (c) Heavily doped

(iii) (b) Resistance (iv) (b) *n* type

(v) (d) All of these

2. A logic gate is a digital circuit that follows certain logical relationship between the input and output voltages. There are only two voltage states 1 (high) is 0 (low). A gate can have one or more inputs. Depending on the logical operation, there are three basic gates OR, AND and NOT. OR gate has two inputs and the output is high if any of the input is high. AND gate has two inputs and the output is high only when both the inputs are high. NOT gate has only one input and the output is the opposite of the inputs (for high input, low output and for low input, a high output)

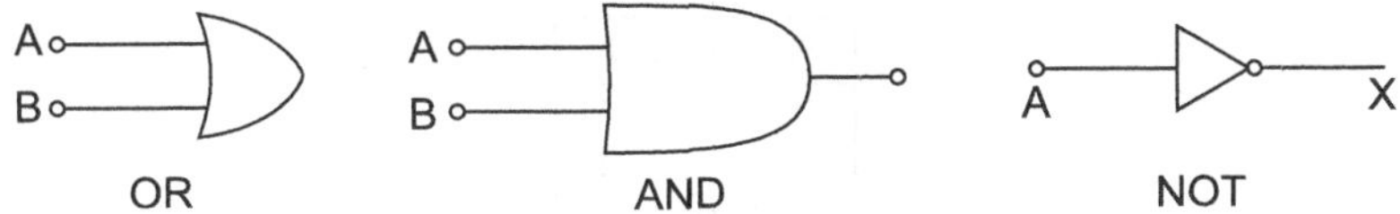

There are two more gates we commonly use: NAND and NOR. NAND gate is the combination of AND and NOT. Output of AND will be reversed a new output is obtained. Similarly NOR gate is a combination of OR and NOT gates.

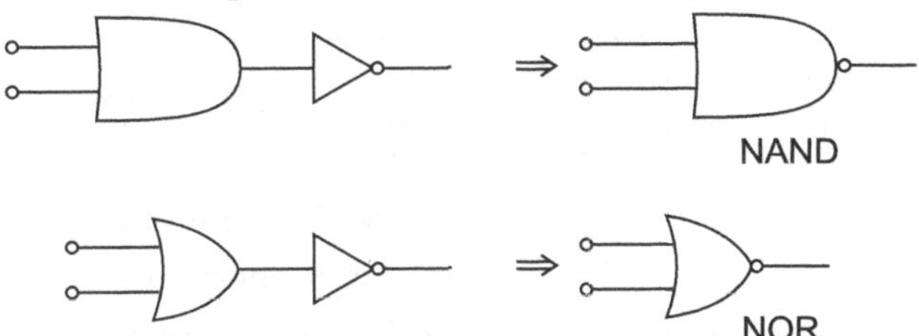

Now answer the following questions:

(i) For the following gate, which input combination gives an output 1?

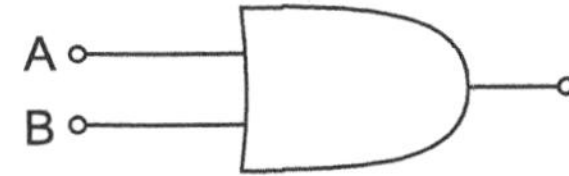

(a) A = 0, B = 0 (b) A = 1, B = 1

(c) A = 1, B = 0 (d) B = 1

(ii) The truth table of NOR gate is:

(a)

A	B	X
0	0	0
0	1	1
1	0	1
1	1	1

(b)

A	B	X
0	0	1
0	1	0
1	0	0
1	1	0

(c)

A	B	X
0	0	1
0	1	0
1	0	0
1	1	1

(d)

A	B	X
0	0	0
0	1	1
1	0	1
1	1	0

(iii) If the input shown in the wave form is fed through a NOT gate, what will be the output wave form?

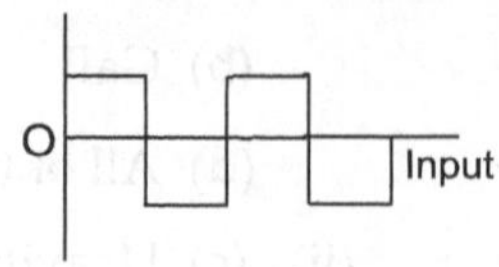

(a) 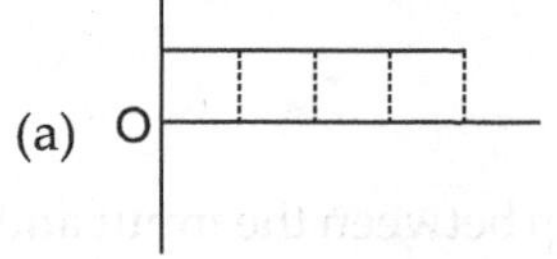

(b) 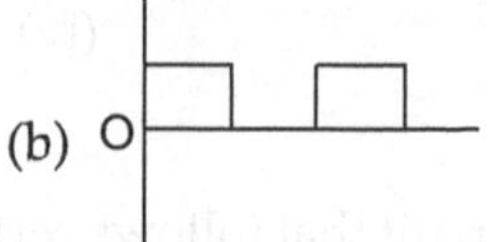

(c) 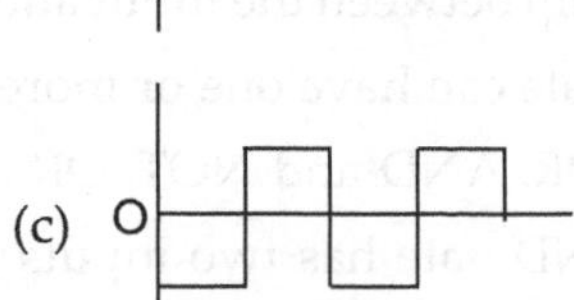

(d) 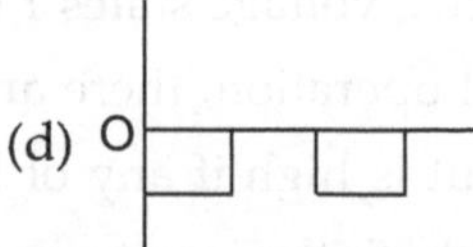

(iv) Which gate has the following truth table?

A	B	X
0	0	1
0	1	1
1	0	1
1	1	0

(a) AND gate (b) NAND gate
(c) NOR gate (d) OR gate

(v) The combination of gates shown in the figure below will result in:

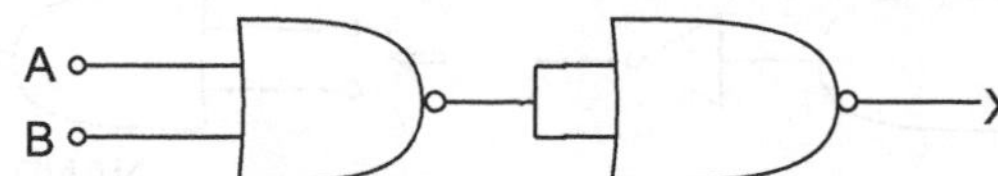

(a) AND operation (b) NAND operation
(c) NOT operation (d) NOR operation

Ans. (i) (b) AND gives high output only when both the inputs are high

(ii) (b) NOR = (OR) + (NOT)

(iii) (c) For NOT gate

A	X
0	1
1	0

(iv) (b) NAND = AND + (NOT)

(v) (b) This shows how AND gate is obtained using NAND gates.

❑❑

Question Set 9

Derivations

Chapter 1. Electric Charges and Fields

Q. 1. A thin straight infinitely long conducting wire having charge density (λ) is enclosed by a cylindrical surface of radius r and length L, its axis coinciding with the length of the wire. Find the expression for the electric flux through the surface of the cylinder.

Ans. Electric flux through the cylindrical surface,

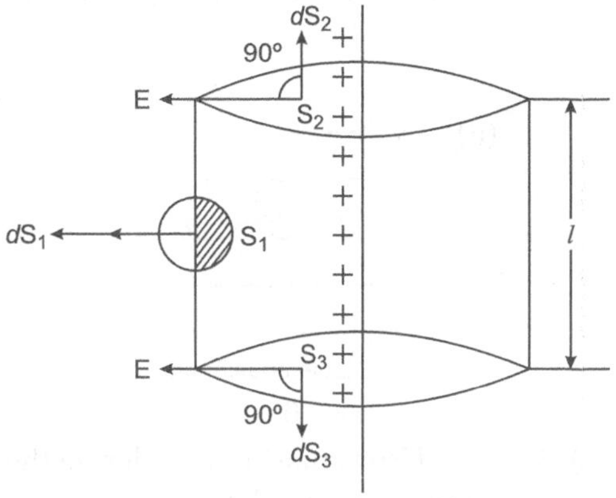

$$\phi_E = \oint \vec{E} \cdot \vec{dS}$$

$$= \int_{S_1} \vec{E} \cdot \vec{dS_1} + \int_{S_2} \vec{E} \cdot \vec{dS_2} + \int_{S_3} \vec{E} \cdot \vec{dS_3}$$

$$= \int_{S_1} E\, dS_1 \cos 0° + \int_{S_2} E\, dS_2 \cos 90° + \int_{S_3} E\, dS_1 \cos 90°$$

$$= E \int_{S_1} E\, dS_1 + 0 + 0$$

$$= E \times 2\pi r l \quad \text{[Since area of curved surface} = 2\pi r l]$$

But $$E = \frac{\lambda}{2\pi\varepsilon_0 r}$$

∴ $$\phi_E = \frac{\lambda}{2\pi\varepsilon_0 r} \cdot 2\pi r l = \frac{\lambda l}{\varepsilon_0}$$

Q. 2. (i) Derive the expression for electric field at a point on the equatorial line of an electric dipole.★

(ii) Depict the orientation of the dipole in (a) stable, (b) unstable equilibrium in a uniform electric field.★

Ans. (i) Consider a point P on broad side position of dipole formed of charges $+q$ and $-q$ at separation $2l$. The distance of point P from mid point (O) of electric dipole is r. Let $\vec{E_1}$ and $\vec{E_2}$ be the electric field strengths due to charges $+q$ and $-q$ of electric dipole.

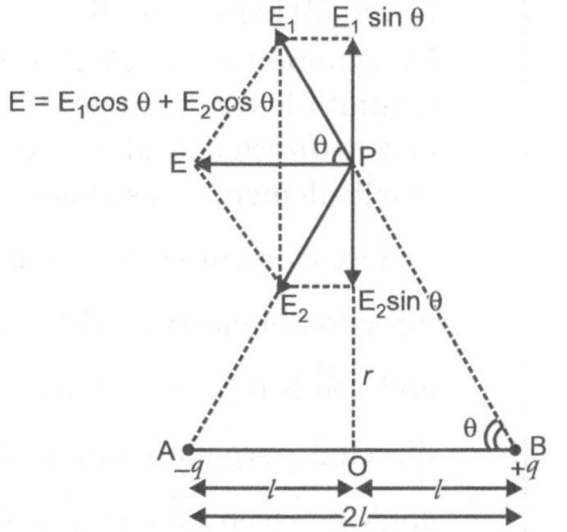

From the figure, $$AP = BP = \sqrt{r^2 + l^2}$$

∴ $$\vec{E_1} = \frac{1}{4\pi\varepsilon_0} \frac{q}{r^2 + l^2} \text{ along B to P}$$

$$\vec{E_2} = \frac{1}{4\pi\varepsilon_0} \frac{q}{r^2 + l^2} \text{ along P and A}$$

The components of $\vec{E_1}$ and $\vec{E_2}$ perpendicular to AB : $E_1 \sin\theta$ and $E_2 \sin\theta$ being equal and opposite cancel each other, while the components of $\vec{E_1}$ and $\vec{E_2}$ parallel to AB : $E_1 \cos\theta$ and $E_2 \cos\theta$ being in the same direction add up and give the resultant electric field whose direction is parallel to $\vec{BA}$

∴ Resultant electric field at P is $E = E_1 \cos\theta + E_2 \cos\theta$

But $$E_1 = E_2 = \frac{1}{4\pi\varepsilon_0} \frac{q}{(r^2 + l^2)}$$

★ are board based questions from previous year

From the figure,

$$\cos\theta = \frac{OB}{PB} = \frac{l}{\sqrt{r^2 + l^2}}$$

$$= \frac{l}{(r^2 + l^2)^{1/2}}$$

$$E = 2E_1 \cos\theta$$

$$= 2 \times \frac{1}{4\pi\varepsilon_0} \frac{q}{(r^2 + l^2)} \cdot \frac{l}{(r^2 + l^2)^{1/2}}$$

$$= \frac{1}{4\pi\varepsilon_0} \frac{2ql}{(r^2 + l^2)^{3/2}}$$

But $q.2l = p$ = electric dipole moment

$$E = \frac{1}{4\pi\varepsilon_0} \frac{p}{(r^2 + l^2)^{3/2}}$$

(ii) (a) $\vec{E}$ $\vec{p}$ $-q$ $+q$ (b) $\vec{E}$ $\vec{p}$ $+q$ $-q$

Stable equilibrium *i.e.,* $\theta = 0°$ Unstable equilibrium *i.e.,* $\theta = 180°$

Q. 3. (i) Use Gauss' law to derive the expression for the electric field (E) due to a straight uniformly charged infinite line of charge density λ C/m.★

(ii) Draw a graph to show the variation of E with perpendicular distance r from the line of charge.★

(iii) Find the work done in bringing a charge q from perpendicular distance r_1 to r_2 $(r_2 > r_1)$.★

Ans. (i) **Electric field ($\vec{E}$) due to a straight uniformly charged infinite line of charge density λ :** Consider a cylindrical Gaussian surface of radius r and length l coaxial with line charge. The cylindrical Gaussian surface may be divided into three parts : (1) curved surface S_1 (2) flat surface S_2 and (3) flat surface S_3.

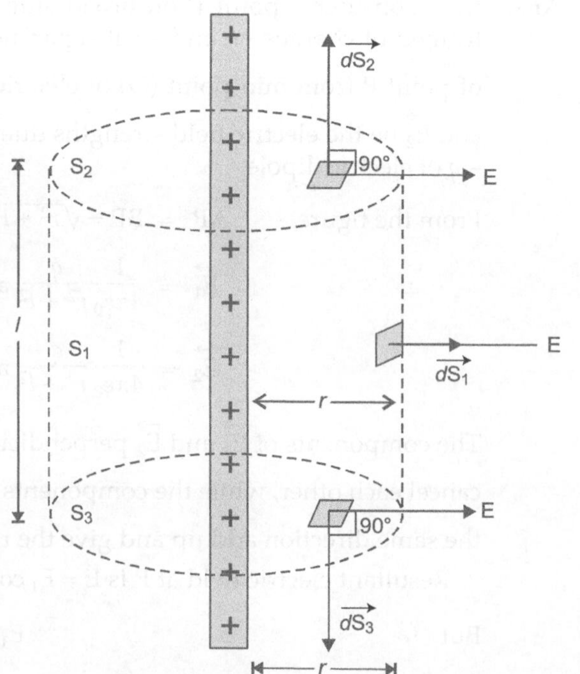

By symmetry, the electric field has the same magnitude E at each point of curved surface S_1 and is directed radially outward. We consider small elements of surfaces S_1, S_2 and S_3. The surface element vector $\vec{dS_1}$ is directed along the direction of electric field (*i.e.,* angle between $\vec{E}$ and $\vec{dS_1}$ is 0°); the elements $d\vec{S_2}$ and $d\vec{S_3}$ are directed perpendicular to field vector $\vec{E}$ (*i.e.,* angle between $\vec{dS_2}$ and $\vec{E}$ and $\vec{dS_3}$ and $\vec{E}$ is 90°).

Electric flux through the cylindrical surface,

$$\oint_S \vec{E}.\vec{dS} = \oint_{S_1} \vec{E}.\vec{dS_1} + \oint_{S_2} \vec{E}.\vec{dS_2} + \oint_{S_3} \vec{E}.\vec{dS_3}$$

$$= \oint_{S_1} E\,dS_1 \cos 0° + \oint_{S_2} E\,dS_2 \cos 90° + \oint_{S_3} E\,dS_3 \cos 90°$$

$$= E\int dS_1 + 0 + 0$$

$$= E \times \text{area of curved surface} = E \times 2\pi rl$$

★ are board based questions from previous year

$$\text{Charge enclosed, } q = \lambda l$$

By Gauss' theorem,

$$\phi_E = \frac{q}{\varepsilon_0}$$

$$E.\ 2\pi r l = \frac{\lambda l}{\varepsilon_0}$$

or

$$E = \frac{\lambda}{2\pi\varepsilon_0 r}$$

(ii) **Graph showing variation of E with perpendicular distance from line of charge :** The electric field is inversely proportional to distance 'r' from line of charge. Thus,

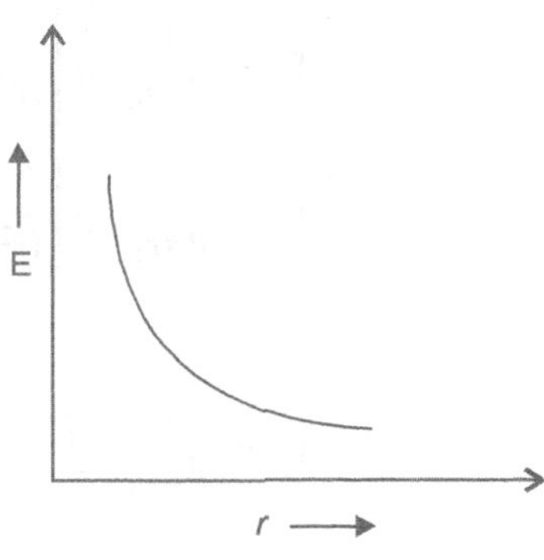

(iii) Work done in bringing a charge q from perpendicular distance r_1 to r_2 ($r_2 > r_1$), in field of a charge q_0,

$$dW = \vec{F}.\vec{dx}$$

$$\int_0 dW = \int_{r_1}^{r_2} \frac{1}{4\pi\varepsilon_0}\frac{qq_0}{x^2}\,dx$$

$$W = \frac{1}{4\pi\varepsilon_0} qq_0 \left[-\frac{1}{x}\right]_{r_1}^{r_2}$$

$$= \frac{1}{4\pi\varepsilon_0} qq_0 \left[\frac{1}{r_1} - \frac{1}{r_2}\right]$$

Q. 4. (i) Derive an expression for the electric field E due to a dipole of length '$2a$' at a point at distance r from the centre of the dipole on the axial line.★

(ii) Draw a graph of E versus r for $r >> a$.★

(iii) If this dipole were kept in a uniform external electric field E_0, diagrammatically represent the position of the dipole in stable and unstable equilibrium and write the expressions for the torque acting on the dipole in both the cases.★

Ans. (i) Consider an electric dipole AB. The charges $-q$ and $+q$ of dipole are situated at A and B respectively, as shown in the figure. The separation between the charges is $2a$. Electric dipole moment is given by

$$p = q.\,2a. \qquad ..(i)$$

Consider a point P on the axis of dipole at a distance r from mid point O of electric dipole.

The distance of point P from charge $+q$ at B is, BP $= r - a$ and distance of point P from charge $-q$ at A is, AP $= r + a$

Let E_1 and E_2 be the electric field strengths at point P due to charges $+q$ and $-q$ respectively.

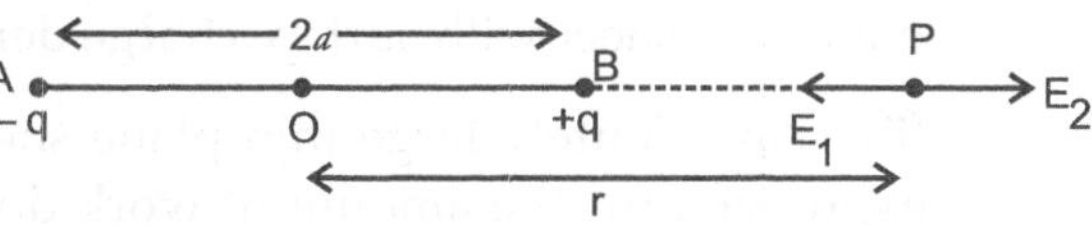

The resultant electric field due to electric dipole is given by

---

★ are board based questions from previous year

$$E = E_2 - E_1$$

$$= \frac{1}{4\pi\varepsilon_0} \cdot \frac{q}{(r-a)^2} - \frac{1}{4\pi\varepsilon_0} \cdot \frac{q}{(r+a)^2}$$

$$E = \frac{1}{4\pi\varepsilon_0} \cdot q \left[\frac{1}{(r-a)^2} - \frac{1}{(r+a)^2}\right]$$

$$E = \frac{1}{4\pi\varepsilon_0} \cdot q \left[\frac{(r+a)^2 - (r-a)^2}{(r-a)^2 (r+a)^2}\right]$$

$$E = \frac{1}{4\pi\varepsilon_0} \cdot q \left[\frac{4ra}{(r^2 - a^2)^2}\right]$$

$$E = \frac{1}{4\pi\varepsilon_0} \cdot q \left[\frac{2pr}{(r^2 - a^2)^2}\right] \quad \text{[From (i)]}$$

(ii) If $r >> a$,

$$E = \frac{1}{4\pi\varepsilon_0} \cdot \frac{2p}{r^3}$$

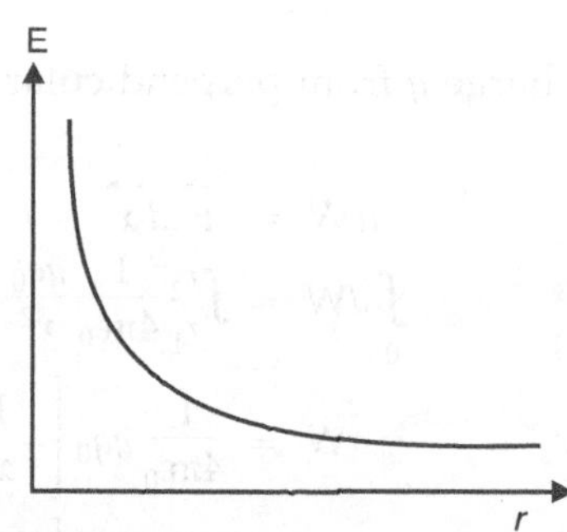

(iii) **Position of dipole in stable equilibrium :**

In stable equilibrium, $\theta = 0°$

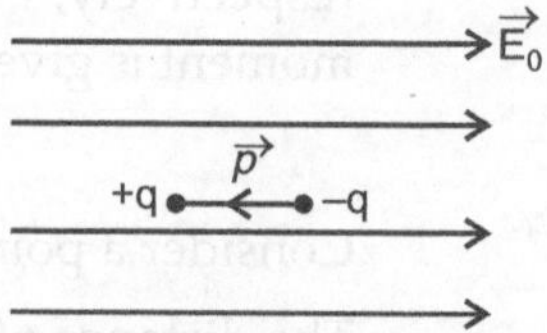

$$\vec{\tau} = \vec{p} \times \vec{E}$$

$$= pE \sin\theta$$

$$= pE \sin 0°$$

$$\vec{\tau} = 0$$

Position of dipole in unstable equilibrium :

In unstable equilibrium, $\theta = 180°$

$$\vec{\tau} = pE \sin 180°$$

$$\vec{\tau} = 0$$

Q. 5. (i) Use Gauss's theorem to find the electric field due to a uniformly charged infinitely large plane thin sheet with surface charge density σ.★

(ii) An infinitely large thin plane sheet has a uniform surface charge density +σ. Obtain the expression for the amount of work done in bringing a point charge *q* from infinity to a point, distant *r*, in front of the charged plane sheet.★

★ are board based questions from previous year

Ans. (i) Let electric charge be uniformly distributed over the surface of a thin non-conducting infinite sheet. Let the surface charge density be σ.

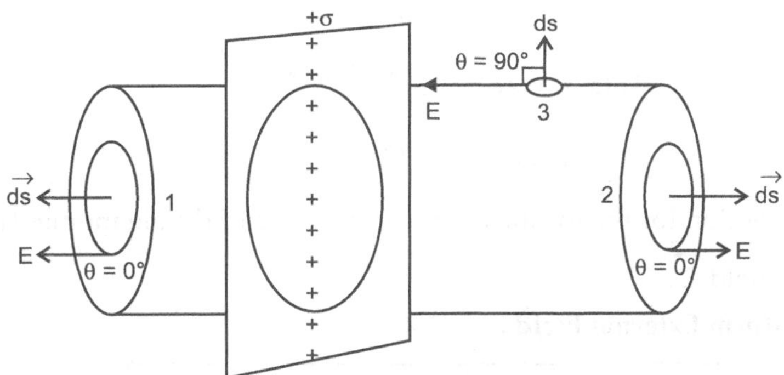

According to Gauss theorem :

$$\int_1 \vec{E}\cdot\vec{ds} + \int_2 \vec{E}\cdot\vec{ds} + \int_3 \vec{E}\cdot\vec{ds} = \frac{q}{\varepsilon_0}$$

$$\int_1 Eds \cos 0^\circ + \int_2 Eds \cos 0^\circ + \int_3 Eds \cos 90^\circ = \frac{q}{\varepsilon_0}$$

$$\int_1 Eds + \int_2 Eds = \frac{q}{\varepsilon_0}$$

$$2\int Eds = \frac{q}{\varepsilon_0}$$

$$2E.S = \frac{\sigma S}{\varepsilon_0}$$

$$E = \frac{\sigma}{2\varepsilon_0}$$

(ii) The electric field due to a uniformly charged infinitely large thin sheet with surface charge density σ is,

$$E = \frac{\sigma}{2\varepsilon_0} \qquad \text{...(i)}$$

The amount of work done in bringing a point charge q from infinity to a point, at a distance r is given by

$$W = \int_\infty^r F.dr$$

Where F is the force experienced by charge q.

Now,
$$W = q\int_\infty^r -E.dr \qquad [\because F = -Eq]$$

$$W = -q\frac{\sigma}{2\varepsilon_0}\int_\infty^r dr \qquad \text{[From (i)]}$$

$$W = q\frac{\sigma}{2\varepsilon_0}[\infty - r] = \infty.$$

Q. 6. Given the electric field in the region $\vec{E} = 2\,\hat{x}l$, find the net electric flux through the cube and the charge enclosed by it.

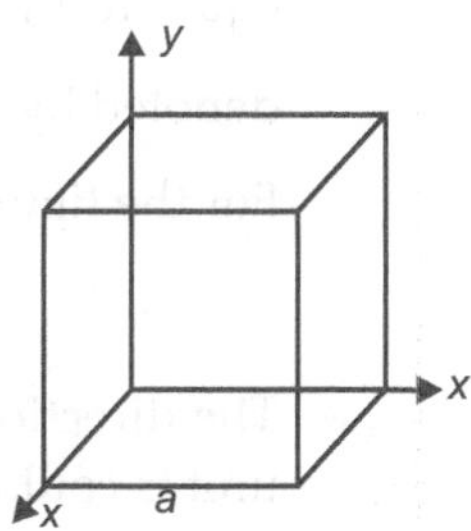

Ans. Since, the electric field is parallel to the faces parallel to xy and xz planes, the electric flux through them is zero.

Electric flux through the left face

$$\phi_L = (E_L)(a^2)\cos 180^\circ$$
$$= (0)(a^2)\cos 180^\circ = 0$$

Electric flux through the right face,

$$\phi_R = (E_R)(a^2)\cos 0°$$
$$= (2a)(a^2) \times 1$$
$$= 2a^3$$

$$\text{Total flux } (\phi) = 2a^3 = \frac{q_{enclosed}}{\varepsilon_0}$$

$$\therefore \quad q_{enclosed} = 2a^3\varepsilon_0$$

Q. 7. Deduce the expression for the torque acting on a dipole of dipole moment $\vec{p}$ in the presence of a uniform electric field $\vec{E}$.★

Ans. **Dipole in a Uniform External Field :**

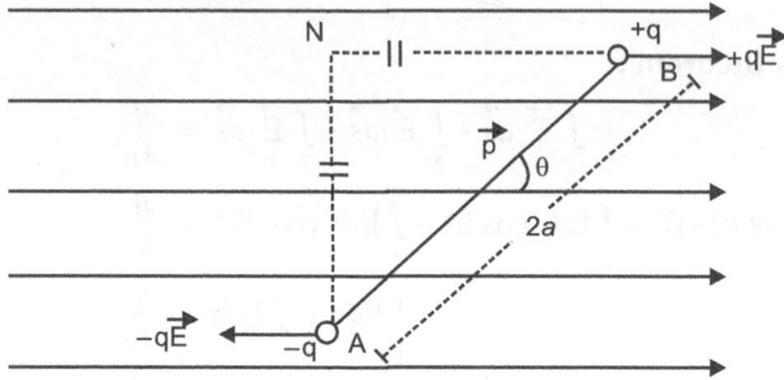

Consider an electric dipole consisting of charges $-q$ and $+q$ and of length $2a$ placed in a uniform electric field $\vec{E}$ making an angle θ with electric field.

Force on charge $-q$ at A $= -q\vec{E}$ (opposite to $\vec{E}$)

Force on charge $+q$ at B $= +q\vec{E}$ (along $\vec{E}$)

Electric dipole is under the action of two equal and unlike parallel force, which give rise to a torque on the dipole.

τ = Force × Perpendicular distance between the two forces

$$\tau = qE\,(AN) = qE \times (2a \sin\theta)$$

$$\tau = pE \sin\theta$$

$$\vec{\tau} = \vec{p} \times \vec{E}$$

Q. 8. Define electric dipole moment. Is it a scalar or vector' of a ? Derive the expression for the electric field of a dipole at a point on equatorial plane of the dipole.

Ans. **Electric dipole moment :** It is the measure of the strength of electric dipole and its magnitude is equal to the product of magnitude of either charge of dipole with distance between them. It is denoted by $\vec{p}$. It is a vector quantity

For the figure shown dipole moment is given by

(+) ←— 2a —→ (−)

$$\vec{p} = q \times 2a$$

The direction of dipole moment is from negative electric charge to positive electric charge and its unit is coulomb metre.

★ are board based questions from previous year

Electric field of dipole at a point on equatorial plane : Consider a point M on equatorial plane of electric dipole AB formed of charges $+q$ and $-q$ at separation $2a$.

The distance of point M from mid point (O) of electric dipole is r.

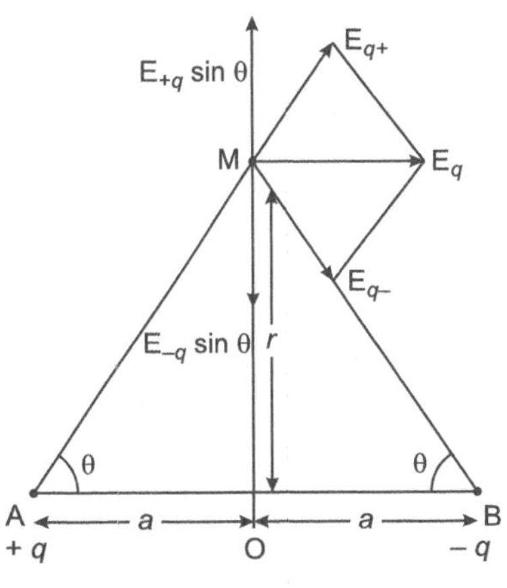

From figure,

$$AM = BM = \sqrt{r^2 + a^2}$$

Electric field at M due to $+q$ is

$$\vec{E}_{+q} = \frac{1}{4\pi\varepsilon_0}\left(\frac{q}{r^2 + a^2}\right)$$

[Direction from A to M]

Electric field at M due to $-q$ is

$$\vec{E}_{-q} = \frac{1}{4\pi\varepsilon_0}\left(\frac{q}{r^2 + a^2}\right)$$

[Direction from M to B]

Thus net electric field will be sum of horizontal components of electric fields only as vertical components cancel each other

$$\vec{E} = \vec{E}_{+q} + \vec{E}_{-q}$$

$$E = E_{+q}\cos\theta + E_{-q}\cos\theta$$

But $$E_{+q} = E_{-q} = \frac{1}{4\pi\varepsilon_0}\frac{q}{(r^2 + a^2)}$$

From the figure, $$\cos\theta = \frac{AO}{AM} = \frac{OB}{MB} = \frac{a}{\sqrt{a^2 + r^2}}$$

$$E = 2E_{+q}\cos\theta$$

$$= \frac{1}{4\pi\varepsilon_0}\frac{q}{r^2 + a^2}\cdot 2\cos\theta$$

$$= \frac{2q}{4\pi\varepsilon_0 (r^2 + a^2)}\frac{a}{\sqrt{a^2 + r^2}}$$

$$= \frac{2qa}{4\pi\varepsilon_0 (r^2 + a^2)^{3/2}}$$

In vector form,

$$\vec{E} = -\frac{\vec{p}}{4\pi\varepsilon_0 (r^2 + a^2)^{3/2}}$$

The negative sign shows that the direction is opposite to direction of electric field due to dipole.

Chapter 2. Electrostatic Potential and Capacitance

Q. 1. Derive equivalent capacitances for

(i) Combination of capacitors in Series.

(ii) Combination of capacitors in Parallel.

Ans. (i)

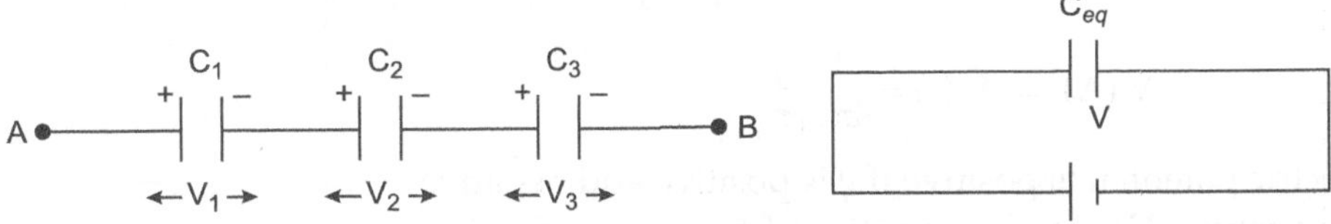

When capacitors are connected in series, the voltage drop across them will be as follows :

$$V = V_1 + V_2 + V_3$$

$$= \frac{Q}{C_1} + \frac{Q}{C_2} + \frac{Q}{C_3}$$

As same charge will flow in series connections.

$$\frac{Q}{C_{eq}} = Q\left[\frac{1}{C_1} + \frac{1}{C_2} + \frac{1}{C_3}\right]$$

$$\frac{1}{C_{eq}} = \frac{1}{C_1} + \frac{1}{C_2} + \frac{1}{C_3} \text{ Farad.}$$

(ii)

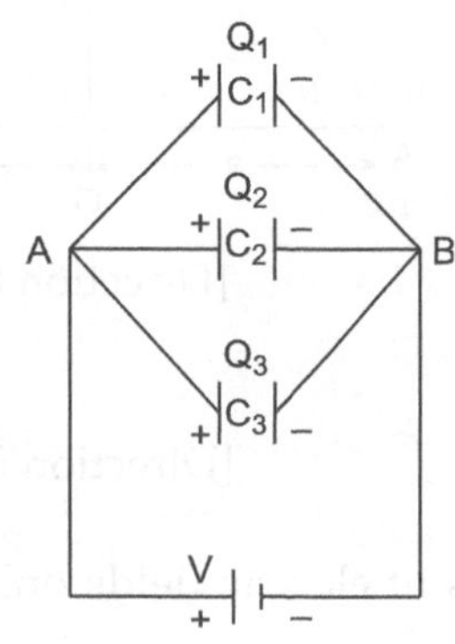

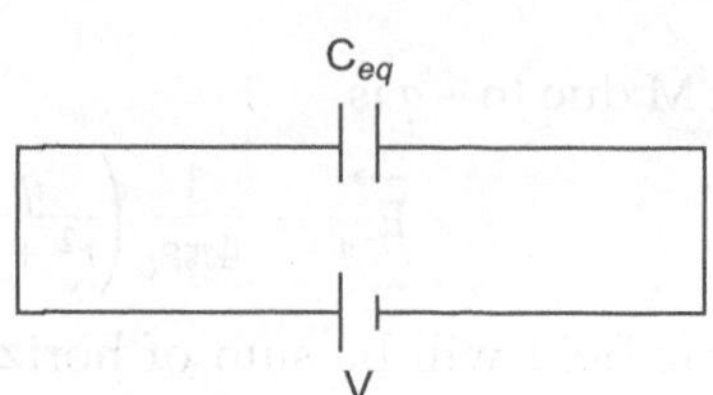

In parallel configuration, voltage across capacitors are same but total charge gets distributed among all parallel capacitors.

$$\therefore \quad Q = Q_1 + Q_2 + Q_3$$

$$= C_1V + C_2V + C_3V$$

$$C_{eq}V = V(C_1 + C_2 + C_3)$$

$$C_{eq} = (C_1 + C_2 + C_3) \text{ Farad.}$$

Q. 2. Derive expression for electrostatic potential due to electric field of
(i) Point charge (ii) Electric dipole

Ans. (i) The electric field at any point A having $\vec{r}$ as its position-vector due to a point charge q placed at the origin of the coordinate system is given by

$$\vec{E}(\vec{r}) = \frac{1}{4\pi\varepsilon_0}\frac{q}{r^2}\hat{r} = \frac{kq}{r^2}\hat{r} \qquad \left[\because k = \frac{1}{4\pi\varepsilon_0} = 9 \times 10^9 \text{ Nm}^2/\text{C}^2\right]$$

The electric potential at point A as shown in the figure is given by

$$V(A) = -\int_{\infty}^{A} \vec{E} \cdot \vec{dl}$$

= Amount of work done to bring unit positive charge from ∞ to A.

$$\therefore \quad V(A) = -\int_{\infty}^{A} \left|\frac{kq}{r^2}\hat{r}\right| \cdot (dr\,\hat{r})$$

$$= -kq\int_{\infty}^{A}\frac{dr}{r^2} = -kq\left[-\frac{1}{r}\right]_{\infty}^{r} = \frac{kq}{r}$$

$$V(A) = V(r) = \frac{1}{4\pi\varepsilon_0}\frac{q}{r}$$

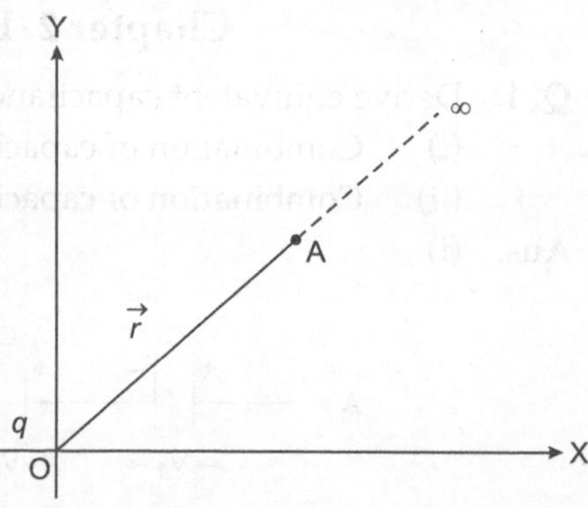

The electric potential is positive if q is positive and negative if q is negative. Hence, for negative charge negative sign should be used for q in the above equation.

(ii) **Electrostatic potential due to electric dipole.**

Let the two charges $-q$ and $+q$ be placed at A and B respectively with a distance $2a$ between them.

Let P be a unit at a distance r from the centre of this dipole and making an angle θ with its direction such that $AP = r_1$ and $BP = r_2$.

The electric potential at P is given by

$$V(r) = \frac{1}{4\pi\varepsilon_0}\frac{q}{r_1} - \frac{1}{4\pi\varepsilon_0}\frac{q}{r_2} \qquad \ldots(i)$$

Now, by geometry,

$$r_1^2 = r^2 + a^2 - 2ar\cos\theta$$

$$r_2^2 = r^2 + a^2 + 2ar\cos\theta$$

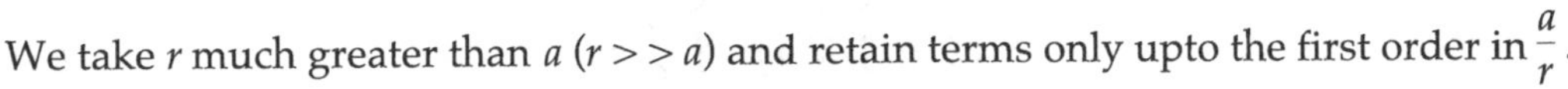

We take r much greater than a ($r >> a$) and retain terms only upto the first order in $\frac{a}{r}$.

$$r_1^2 = r^2\left(1 - \frac{2a\cos\theta}{r} + \frac{a^2}{r^2}\right)$$

$$= r^2\left(1 - \frac{2a\cos\theta}{r}\right)$$

Similarly, $$r_2^2 = r^2\left(1 + \frac{2a\cos\theta}{r}\right)$$

Using the Binomial theorem and retaining terms upto the first order in $\frac{a}{r}$, we obtain,

$$\frac{1}{r_1} \cong \frac{1}{r}\left(1 - \frac{2a\cos\theta}{r}\right)^{\frac{-1}{2}} \cong \frac{1}{r}\left(1 + \frac{a}{r}\cos\theta\right) \qquad \ldots(ii)$$

$$\frac{1}{r_2} \cong \frac{1}{r}\left(1 + \frac{2a\cos\theta}{r}\right)^{\frac{-1}{2}} \cong \frac{1}{r}\left(1 - \frac{a}{r}\cos\theta\right) \qquad \ldots(iii)$$

Puting equation (ii) and (iii) in equation (i), we get

$$\therefore \quad V(r) = \frac{q}{4\pi\varepsilon_0}\left(\frac{2a\cos\theta}{r^2}\right)$$

$$= \frac{1}{4\pi\varepsilon_0}\frac{p\cos\theta}{r^2} \qquad (\because p = 2aq)$$

As $$p\cos\theta = \vec{p}\cdot\vec{r}$$

$$V(\vec{r}) = \frac{1}{4\pi\varepsilon_0}\frac{\vec{p}\cdot\hat{r}}{r^2} \qquad [\text{for } |\vec{r}| >> 2a]$$

Q. 3. (i) Derive the expression for the energy stored in a parallel plate capacitor. Hence obtain the expression for the energy density of the electric field.★

(ii) A fully charged parallel plate capacitor is connected across an uncharged identical capacitor. Show that the energy stored in the combination is less than that stored initially in the single capacitor. ★

★ are board based questions from previous year

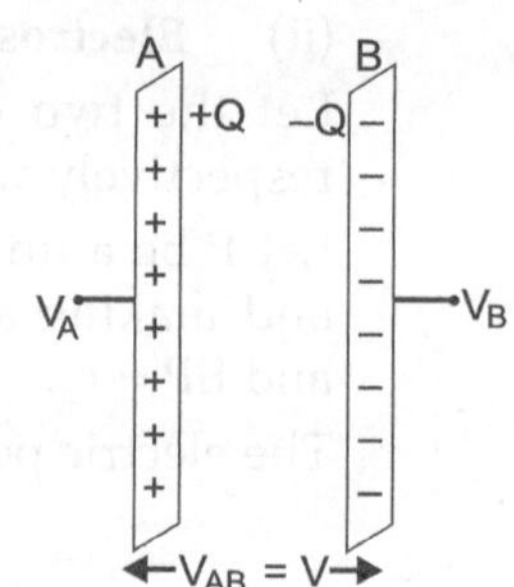

Ans. (i) Consider a capacitor of capacitance C. Initial charge on capacitor is zero. Initial potential difference between capacitor plates is zero. Let a charge Q be given to it in small steps. When charge is given to capacitor, the potential difference between its plates increases. Let at any instant when charge on capacitor be q, the potential difference between its plates $V = \frac{q}{C}$.

Now work done in giving an additional infinitesimal charge dq to capacitor

$$dW = V\,dq = \frac{q}{C}\,dq$$

The total work done in giving charge from 0 to Q will be equal to the sum of all such infinitesimal works, which may be obtained by integration. Therefore, total work

$$W = \int_0^Q V\,dq = \int_0^Q \frac{q}{C}\,dq$$

$$= \frac{1}{C}\left[\frac{q^2}{2}\right]_0^Q = \frac{1}{C}\left(\frac{Q^2}{2} - \frac{0}{2}\right) = \frac{Q^2}{2C}$$

If V is the final potential difference between capacitor plates, then Q = CV

$$\therefore \quad W = \frac{(CV)^2}{2C} = \frac{1}{2}CV^2 = \frac{1}{2}QV$$

This work is stored as electrostatic potential energy of capacitor *i.e.*, Electrostatic potential energy,

$$U = \frac{Q^2}{2C} = \frac{1}{2}CV^2 = \frac{1}{2}QV$$

Energy density : Consider a parallel plate capacitor consisting of plates, each of area A, separated by a distance d. If space between the plates is filled with a medium of dielectric constant K, then

Capacitance of capacitor, $C = \frac{K\varepsilon_0 A}{d}$

If σ is the surface charge density of plates, then electric field strength between the plates

$$E = \frac{\sigma}{K\varepsilon_0} \Rightarrow \sigma = K\,\varepsilon_0\,E$$

Charge on each plate of capacitor, $Q = \sigma A = K\,\varepsilon_0\,E\,A$

∴ Energy stored by capacitor,

$$U = \frac{Q^2}{2C} = \frac{(K\varepsilon_0 EA)^2}{2\left(K\varepsilon_0 \frac{A}{d}\right)}$$

$$= \frac{1}{2}K\,\varepsilon_0\,E^2\,A\,d$$

But Ad = Volume of space between capacitor plates

∴ Energy stored, $U = \frac{1}{2}K\,\varepsilon_0\,E^2\,Ad$

Electrostatic energy stored per unit volume,

$$u_e = \frac{U}{Ad} = \frac{1}{2}K\varepsilon_0\,E^2$$

This is expression for electrostatic energy density in medium of dielectric constant K.

(ii) Initially, if we consider a charged capacitor, then its charge would be

$$Q = CV$$

A ——V——||—— B
C

and energy stored is

$$U_1 = \frac{1}{2}CV^2 \quad \text{...(i)}$$

Then, this charged capacitor is connected to uncharged capacitor.

Let the common potential be V. The charge flows from first capacitor to the other capacitor unless both the capacitors attain common potential.

$$\Rightarrow \quad Q_1 = CV_1 \text{ and } Q_2 = CV_2$$

Applying conservation of charge,

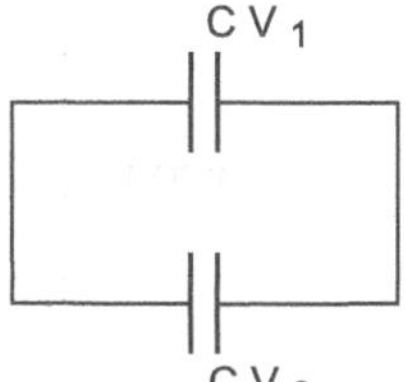

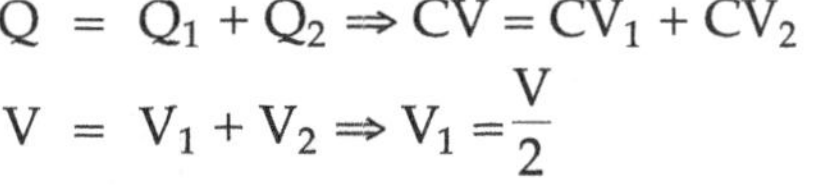

$$Q = Q_1 + Q_2 \Rightarrow CV = CV_1 + CV_2$$

$$\Rightarrow \quad V = V_1 + V_2 \Rightarrow V_1 = \frac{V}{2}$$

Total energy stored,

$$U_2 = \frac{1}{2}CV_1^2 + \frac{1}{2}CV_2^2$$

$$U_2 = \frac{1}{2}C\left(\frac{V}{2}\right)^2 + \frac{1}{2}C\left(\frac{V}{2}\right)^2$$

$$\Rightarrow \quad U_2 = \frac{1}{4}CV^2 \quad \text{...(ii)}$$

From equations (i) and (ii), we get

$$U_2 < U_1$$

$\Rightarrow$ Energy stored in the combination is less than that stored initially in single capacitor.

Q. 4. (i) Derive the expression for the electric potential due to an electric dipole at a point on its axial line.★

(ii) Depict the equipotential surfaces due to an electric dipole.★

Ans. (i) Consider an electric dipole having charges $-q$ and $+q$ at separation '$2a$'. The dipole moment of dipole is $\vec{p} = q(2\vec{a})$, directed from $-q$ to $+q$.

The electric potential due to dipole is the algebraic sum of potentials due to charges $+q$ and $-q$.

If r_1 and r_2 are distances of any point P from charge $+q$ and $-q$ respectively as shown in the figure, then the potential due to electric dipole at point P, is

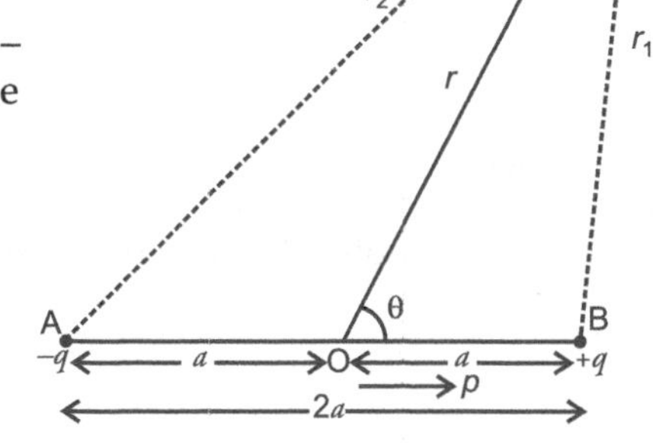

$$V = \frac{1}{4\pi\varepsilon_0}\frac{q}{r_1} - \frac{1}{4\pi\varepsilon_0}\frac{q}{r_2}$$

$$= \frac{q}{4\pi\varepsilon_0}\left[\frac{1}{r_1} - \frac{1}{r_2}\right] \quad \text{...(i)}$$

If (r, θ) are polar coordinates of point P with respect to mid-point O of dipole, then

By geometry, $$r_1^2 = r^2 + a^2 - 2ar\cos\theta \quad \text{...(ii)}$$

and $$r_2^2 = r^2 + a^2 + 2ar\cos\theta \quad \text{...(iii)}$$

From (ii), $$r_1^2 = r^2\left[1 - \frac{2a\cos\theta}{r} + \frac{a^2}{r^2}\right]$$

If $r >> a$ *i.e.*, $\frac{a}{r} << 1$, then it is sufficient to retain terms only upto first order in $\left(\frac{a}{r}\right)$.

$$\therefore \quad r_1^2 = r^2\left[1 - \frac{2a\cos\theta}{r}\right]$$

★ are board based questions from previous year

$$\Rightarrow \qquad r_1 = r\left[1-\frac{2a\cos\theta}{r}\right]^{1/2} \qquad \text{...(iv)}$$

Similarly from (iii),

$$r_2^2 = r^2\left[1+\frac{2a\cos\theta}{r}\right]$$

$$\Rightarrow \qquad r_2 = r\left[1-\frac{2a\cos\theta}{r}\right]^{1/2} \qquad \text{...(v)}$$

From (iv) and (v),

$$\frac{1}{r_1} = \frac{1}{r}\left[1-\frac{2a\cos\theta}{r}\right]^{-1/2}$$

and

$$\frac{1}{r_2} = \frac{1}{r}\left[1+\frac{2a\cos\theta}{r}\right]^{-1/2}$$

Using binomial theorem and retaining terms upto first order in $\frac{a}{r}$ only, we have

$$\frac{1}{r_1} = \frac{1}{r}\left[1-\left(-\frac{1}{2}\right)\frac{2a\cos\theta}{r}\right]$$

$$= \frac{1}{r}\left[1+\frac{a}{r}\cos\theta\right] \qquad \text{...(vi)}$$

and

$$\frac{1}{r_2} = \frac{1}{r}\left[1-\frac{a}{r}\cos\theta\right] \qquad \text{...(vii)}$$

Substituting these values in (i), we get

$$\text{V} = \frac{q}{4\pi\varepsilon_0}\left[\frac{1}{r}\left(1+\frac{a}{r}\cos\theta\right)-\frac{1}{r}\left(1-\frac{a}{r}\cos\theta\right)\right]$$

$$= \frac{1}{4\pi\varepsilon_0}\frac{q}{r}\left[1+\frac{a}{r}\cos\theta-1+\frac{a}{r}\cos\theta\right]$$

$$= \frac{1}{4\pi\varepsilon_0}\frac{q}{r}\left[\frac{2a}{r}\cos\theta\right] = \frac{1}{4\pi\varepsilon_0}\frac{(q.2a)\cos\theta}{r^2}$$

or

$$\text{V} = \frac{1}{4\pi\varepsilon_0}\frac{p\cos\theta}{r^2} \qquad \text{...(viii)}$$

When point P lies on the axis of dipole, then $\theta = 0°$

$$\therefore \qquad \cos\theta = \cos 0° = 1$$

$$\therefore \qquad \text{V} = \frac{1}{4\pi\varepsilon_0}\cdot\frac{p}{r^2}$$

(ii)

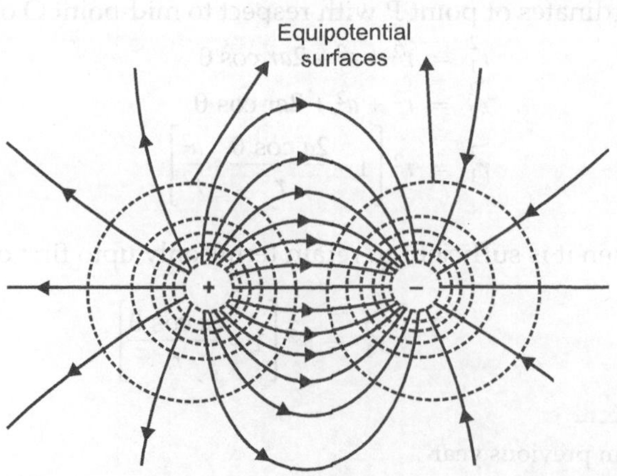

Equipotential surfaces get closer to each other near the point charges, as strong electric field is produced there.

Q. 5. A slab of material of dielectric constant k has same area as that of the plates of parallel plate capacitor but has the thickness $d/2$, where d is the separation between the plates. Derive expression for its capacitance when the slab is inserted between the plates of the capacitor.★

Ans. Initially when there is vaccum between the two plates, the capacitance of the two parallel plates is

$$C = \frac{\varepsilon_0 A}{d}$$

Here, A is area of plates and d is separation between the plates.

Let capacitor is connected with battery. Thus electric field inside the capacitor will be produced and it is given by E.

As soon as dielectric slab of thickness $t = \frac{d}{2}$ inserted the electric field inside slab (t) will be E_0 and in $(d - t)$ will be E_0, $(E_0 = kE)$.

Now potential difference between the plates will be as follows :

$$V = Et + E_0\,(d - t) \qquad \left[\text{as } t = \frac{d}{2}\right]$$

$$= \frac{Ed}{2} + \frac{E_0 d}{2} = \frac{d}{2}(E + E_0)$$

$$= \frac{d}{2}\left(\frac{E_0}{k} + E_0\right) \qquad \left[\because \frac{E_0}{E} = k\right]$$

$$= \frac{E_0 d}{2}\frac{(k+1)}{k}$$

Now,
$$E_0 = \frac{\sigma}{\varepsilon_0} = \frac{q}{\varepsilon_0 A}$$

$$V = \frac{d}{2k}\frac{q}{\varepsilon_0 A}(k+1)$$

As
$$C = \frac{q}{V}$$

Thus,
$$C = \frac{2\,k\,\varepsilon_0 A}{d\,(k+1)}$$

Q. 6. (i) Obtain the expression for the energy stored per unit volume in a charged parallel plate capacitor.★

(ii) The electric field inside a parallel plate capacitor is E. Find the amount of work done in moving a charge q over a closed rectangular loop *abcda*.★

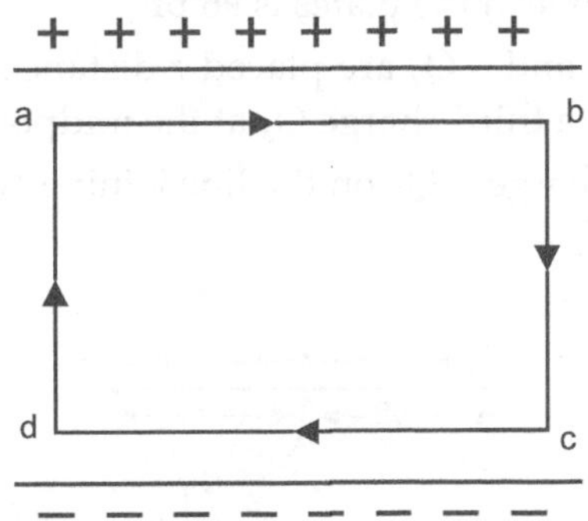

★ are board based questions from previous year

Ans. (i) Let us consider a parallel-plate capacitor of plate area A. If separation between plates is d metre, capacitance C is given by

$$C = \frac{\varepsilon_0 A}{d}$$

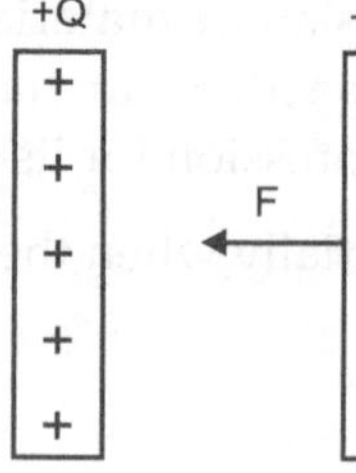

We know that, the magnitude of the electric field between the charged plates of the capacitor is

$$E = \frac{\sigma}{\varepsilon_0}$$

Where σ in surface density of either plate. Therefore, the plate charge is

$$Q = \sigma A = \varepsilon_0 EA$$

Now, the energy stored in the capacitor is

$$U = \frac{1}{2}\frac{Q^2}{C} = \frac{1}{2}\frac{(\varepsilon_0 EA)^2}{\varepsilon_0 A/d}$$

$$U = \frac{1}{2}\varepsilon_0 E^2 (Ad)\ \text{J}$$

The volume between the plates is Ad m³. Therefore, the energy per unit volume is given by

$$U' = \frac{U}{Ad} = \frac{1}{2}\varepsilon_0 E^2\ \text{J/m}^3$$

(ii) Work done, $W = F.d$

Here, F is the force exerted on the charge (q) due to electric field (E) and is given by,

$$F = qE$$

Net displacement, $d = 0$

∴ $W = 0$

Where $\varepsilon_0 \rightarrow$ permittivity of free space

$A \rightarrow$ Area of plates

$d \rightarrow$ Separation between plates

$k =$ dielectric constant

If distance between the plates is reduced to half, then new distance $d' = d/2$

Dielectric constant of the substance filled in between the plates, $k_1 = 6$

Hence, capacitance of the capacitor becomes

$$C_1 = \frac{k_1 \varepsilon_0 A}{d_1} = \frac{6\varepsilon_0 A}{d/2} = \frac{12\varepsilon_0 A}{d} \quad \text{...(ii)}$$

Taking ratios of equations (i) and (ii), we obtain

$C_1 = 2 \times 6\ C = 12\ C = 12 \times 8\ pF = 96\ pF$

Therefore, the capacitance between the plates is 96 pF.

Q. 7. (a) Two point charges + Q_1 and – Q_2 are placed r distance apart. Obtain the expression for the amount of work done toplace a third charge Q_3 at the midpoint of the line joining two charges.★

(b) At what distance from charge + Q_1 on the line joining two charges (in terms of Q_1, Q_2 and r) will this work done by zero.★

Ans. According to the question,

Q_1 ← r → Q_2

← $r/2$ → ← $r/2$ →

Q_3 (at ∞)

★ are board based questions from previous year

(a) Work done :

$$W = \frac{k\,Q_1Q_3}{r/2} - \frac{k\,Q_3Q_2}{r/2}$$

$$W = \frac{2r\,Q_1Q_3}{r} - \frac{2r\,Q_3Q_2}{r}$$

$$= \frac{2r\,Q_3(Q_1 - Q_2)}{r}$$

$$= \frac{2\,Q_3(Q_1 - Q_2)}{4\pi\varepsilon_0 r}$$

(b) **For work done to be zero :** Work done will be zero for the point where the electric potential is zero consider a point P to be lying in between the two charges, where electric potential is zero. Let the distance of the point P from the charge Q_1 be x.

$$V = \frac{k\,Q_1}{x} - \frac{k\,Q_2}{(r-x)} = 0$$

$$\Rightarrow \quad \frac{Q_1}{x} - \frac{Q_2}{r-x} = 0$$

$$\Rightarrow \quad r\,Q_1 - x\,Q_1 - x\,Q_2 = 0$$

$$\Rightarrow \quad x = \frac{r\,Q_1}{(Q_1 + Q_2)}$$

Chapter 3. Current Electricity

Q. 1. A resistance of R draws current from a potentiometer. The potentiometer wire, AB, has a total resistance of R_0. A voltage V is supplied to the potentiometer. Derive an expression for the voltage across R when the sliding contact is in the middle of potentiometer wire.

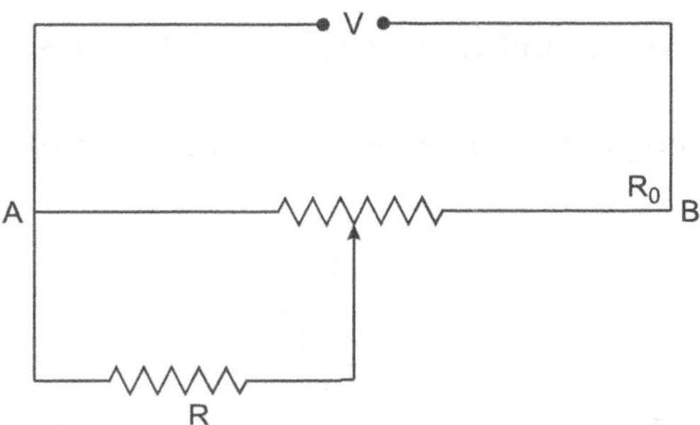

Ans. The equivalent resistance between A and C,

$$R' = \frac{R\frac{R_0}{2}}{R + \frac{R_0}{2}} = \frac{RR_0}{2R + R_0} \qquad \ldots(i)$$

Equivalent resistance between A and B,

$$R_{eq} = \frac{RR_0}{2R + R_0} + \frac{R_0}{2}$$

$$R_{eq} = \frac{R_0\,(4R + R_0)}{2\,(2R + R_0)} \qquad \ldots(ii)$$

Current in the circuit,

$$I = \frac{V}{R_{eq}} = \frac{V.2\,(2R + R_0)}{R_0\,(4R + R_0)}$$

$\therefore$ Voltage across R,

$$V_R = I.R'$$

$$V_R = \frac{V.2\,(2R + R_0)}{R_0\,(4R + R_0)} \times \frac{RR_0}{2R + R_0}$$

$$V_R = \frac{2VR}{4R + R_0}$$

Q. 2. (i) Derive an expression for drift velocity of free electrons.★

(ii) How does drift velocity of electrons in a metallic conductor vary with increase in temperature ? Explain.★

Ans. (i) Consider a conductor in which an electric field E is produced. Let a free electron experience a force (– eE) in this electric field. So, the acceleration of free electron is

$$a = \frac{F}{m} = \frac{-eE}{m} \quad \text{...(i)}$$

Here, e = Charge on electron.

m = Mass of an electron.

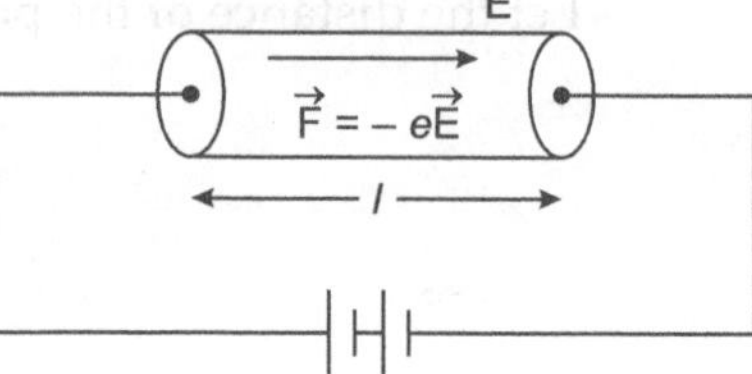

So, the final velocity of the free electron in time interval t_1 is,

$$v_1 = u_1 + at_1$$

For n free electrons, the final velocities be $v_1, v_2, v_3, ..., v_n$.

So, the average velocity of the free electrons or the drift velocity

$$v_d = \frac{(v_1 + v_2 + v_3 + ... + v_n)}{n}$$

or $$v_d = \frac{(u_1 + at_1 + u_2 + at_2 + ... + u_n + at_n)}{n}$$

or $$v_d = \frac{[(u_1 + u_2 + ... + u_n) + (at_1 + at_2 + ... + at_n)]}{n}$$

or $$v_d = \frac{[(u_1 + u_2 + ... + u_n) + a\,(t_1 + t_2 + ... + t_n)]}{n}$$

But, $\frac{(u_1 + u_2 + ... + u_n)}{n}$ = average initial velocity of free electrons = 0.

and $\frac{(t_1 + t_2 + ... + t_n)}{n}$ = average time taken between two consecutive collision = τ

where τ is relaxation time.

So, $$v_d = a\tau$$

or $$v_d = \frac{-e\vec{E}}{m}\tau \quad \text{[from (i)]}$$

This is the required relation.

(ii) The drift velocity of free electrons in a metallic conductor decreases with increase in temperature, because, if we increase the temperature of the metallic conductor the collision between the electrons and ions increases, which decreases relaxation time. Hence, drift velocity decreases.

Q. 3. (i) Derive an expression for drift velocity of electrons in a conductor. Hence deduce ohm's law.★

(ii) A wire whose cross-sectional area is increasing linearly from its one end to the other, is connected across a battery of V volts. Which of the following quantities remain constant in the wire ?★

(a) Drift speed (b) Current density

(c) Electric current (d) Electric field

Justify your answer.

★ are board based questions from previous year

Ans. (i) **Expression of drift velocity :** It is defined as the average velocity gained by the free electrons of a conductor in the opposite direction of the externally applied electric field. If an electron have initial velocity u_1 and accelerated for time τ_1 then velocity attain by it will be

$$\vec{v_1} = \vec{u_1} + \vec{a}\,\tau_1$$

Similarly
$$\vec{v_2} = \vec{u_2} + \vec{a}\,\tau_2$$
$$\vec{v_3} = \vec{u_3} + \vec{a}\,\tau_3$$

....

....

$$\vec{v_n} = \vec{u_n} + \vec{a}\,\tau_n$$

Then average velocity,

$$\vec{v_d} = \frac{\vec{v_1} + \vec{v_2} + \vec{v_3} + \ldots \vec{v_n}}{n}$$

$$\vec{v_d} = \frac{(\vec{u_1} + \vec{a}\,\tau_1) + (\vec{u_2} + \vec{a}\,\tau_2) + \ldots (\vec{u_n} + \vec{a}\,\tau_n)}{n}$$

$$\vec{v_d} = \frac{(\vec{u_1} + \vec{u_2} + \ldots \vec{u_n})}{n} + \frac{(\tau_1 + \tau_2 + \ldots \tau_n)}{n}$$

$$\vec{v_d} = 0 + \vec{a\tau}$$

$$\vec{v_d} = \vec{a} \cdot \tau$$

where $\tau = \dfrac{\tau_1 + \tau_2 + \ldots \tau_n}{n}$ = average relaxation time.

$$\vec{v_d} = \frac{-e\vec{E}\tau}{m} \qquad \left[\because a = \frac{-e\vec{E}}{m}\right]$$

$$|\vec{v_d}| = \frac{-e\vec{E}\tau}{m}$$

Now,
$$v_d = \frac{eE\tau}{m}$$

$$v_d = \frac{eV\tau}{ml} \qquad \left[\because E = \frac{V}{l}\right]$$

and
$$I = neAv_d$$

$$I = neA\frac{eV\tau}{ml}$$

Then
$$\frac{V}{I} = \frac{ml}{ne^2\tau A}$$

If physical condition remain unchanged then right hand side will be constant.

$$\therefore \qquad \frac{V}{I} = \text{Constant}$$

$$\frac{V}{I} = R \text{ or } V = IR$$

Hence, Ohm's law is derived.

(ii) The electric current will remain constant in the wire as it does not depend upon the cross-sectional area whereas the drift speed, current density, electric field depends upon the increasing area of cross-section with the following relations :

$$\text{Drift speed, } v_d = \frac{I}{neA}$$

$$\text{Current density, } J = \frac{I}{A}$$

$$\text{Electric field, } E = \frac{J}{\sigma}$$

Q. 4. (i) Define the term 'conductivity' of a metallic wire. Write its SI unit.★

(ii) Using the concept of free electrons in a conductor, derive the expression for the conductivity of a wire in terms of number density and relaxation time. Hence obtain the relation between current density and the applied electric field E.★

Ans. (i) Conductivity of a metallic wire is defined as its ability to allow electric charges or heat to pass through it.

Numerically, conductivity is reciprocal of resistivity.

SI unit : $ohm^{-1}\ m^{-1}$ or mho m^{-1} or Sm^{-1}.

(ii) Consider a potential difference V be applied across a conductor of length l and cross-section A.

Electric field inside the conductor,

$$E = \frac{V}{l}.$$

Due to the external field the free electrons inside the conductor drift with velocity v_d.

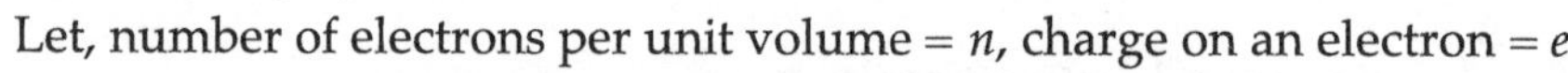

Let, number of electrons per unit volume = n, charge on an electron = e

$\therefore$ Total electrons in length, $l = nAl$

And, total charge, $q = neAl$

Time taken by electrons to enter and leave the conductor.

$$t = \frac{l}{v_d}$$

$$\text{Current, } I = \frac{q}{t} = \frac{neAl}{l/v_d}$$

$$= neAv_d$$

$$\text{Current density, } J = \frac{I}{A} = nev_d \qquad \text{...(i)}$$

We known, $$v_d = \frac{eE\tau}{m} = \frac{eV\tau}{ml}$$

$$\therefore \quad I = neAv_d = \frac{neAVe\tau}{ml}$$

$$\Rightarrow \quad \frac{V}{I} = \frac{ml}{ne^2\tau A}$$

$$\Rightarrow \quad R = \frac{ml}{ne^2\tau A} \qquad \left[\because \frac{V}{I} = R; \text{ Ohm's law}\right]$$

Resistivity; $$\rho = \frac{RA}{l} = \frac{mlA}{ne^2\tau A}$$

or $$\rho = mlne^2\tau$$

Since conductivity, $$\sigma = \frac{l}{\rho}$$

$$\therefore \quad \sigma = \frac{ne^2\tau}{m} \qquad \text{...(ii)}$$

★ are board based questions from previous year

Relation between current density and field :

For an electron, charge $q = -e$

And, current density, $J = \frac{I}{A} = -nev_d$ [from (i)]

$$J = (-ne)\left(\frac{-eE\tau}{m}\right) = \left(\frac{ne^2\tau}{m}\right)E$$

$\Rightarrow$ $J = \sigma E$ [from (ii)]

which is the required relation.

Q. 5. (i) Define the term drift velocity.★

(ii) On the basis of electron drift, derive an expression for resistivity of a conductor in terms of number density of free electrons and relaxation time. On what factors does resistivity of a conductor depend ?★

Ans. (i) Drift velocity is defined as the average velocity with which the free electrons are drifted towards the positive terminal under the effect of applied electric field. Thermal velocities are randomly distributed and average thermal velocity is zero.

$$\frac{\vec{u_1} + \vec{u_2} \ldots\ldots + \vec{u_N}}{N} = 0$$

$$v_d = -\frac{eE\tau}{m}$$

(ii) We know that the current flowing through the conductor is :

$$I = nAev_d$$

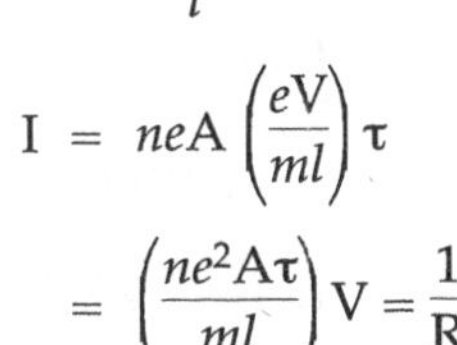

$\therefore$ $$I = neA\left(-\frac{eE\tau}{m}\right)$$

Using $$E = -\frac{V}{l}$$

$$I = neA\left(\frac{eV}{ml}\right)\tau$$

$$= \left(\frac{ne^2A\tau}{ml}\right)V = \frac{1}{R}V$$

$I \propto V \rightarrow$ Which is Ohm's law.

Where $R = \frac{ml}{nAe^2\tau}$ is a constant for a particular conductor at a particular temperature and is called the resistance of the conductor.

$$R = \left(\frac{m}{ne^2\tau}\right)\frac{l}{A} = \frac{\rho l}{A}$$

$$\rho = \left(\frac{m}{ne^2\tau}\right)$$

Where, ρ is the specific resistance or resistivity of the material of the wire. It depends on number of free electron per unit volume and temperature.

Chapter 4. Moving Charges and Magnetism

Q. 1. State Biot-Savart law, giving the mathematical expression for it.

Use this law to derive the expression for the magnetic field due to a circular coil carrying current at a point along its axis.

How does a circular loop carrying current behave as a magnet ?★

★ are board based questions from previous year

Ans. Biot-Savart law states that the magnitude of magnetic field $d\vec{B}$ due to current element is directly proportional to the current I, the element length $|dl|$ and inversely proportional to the square of the distance r of the field point. Its direction is perpendicular to the plane containing $\vec{dl}$ and $\vec{r}$.

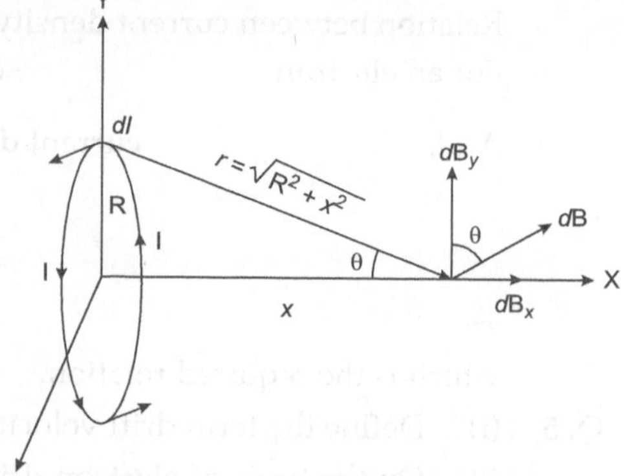

$$d\vec{B} \propto \frac{I \cdot \vec{dl} \times \vec{r}}{r^3}$$

According to Biot-Savart's law

$$dB = \frac{\mu_0}{4\pi}\frac{I \cdot \vec{dl} \times \vec{r}}{r^3}$$

$$dB = \frac{\mu_0}{4\pi}\frac{Idl \sin\theta}{r^2}$$

$$dB = \frac{\mu_0}{4\pi}\frac{Idl}{r^2} \qquad \left[\because \vec{dl} \perp \hat{r}\right]$$

dB has two components dB_x and dB_y, dB_y cancel out and only dB_x component remains.

$$\therefore \quad dB_x = dB \sin\theta$$

$$= \frac{\mu_0 Idl}{4\pi r^2} \times \frac{R}{(x^2+R^2)^{\frac{1}{2}}}$$

$$dB_x = \frac{\mu_0 Idl}{4\pi} \times \frac{R}{(x^2+R^2)^{\frac{3}{2}}} \qquad [\because r^2 = x^2 + R^2]$$

Total magnetic field at P

$$B = \int dB_x$$

$$= \frac{\mu_0 IRdl}{4\pi (x^2+R^2)^{\frac{3}{2}}}$$

$$\int dl = 2\pi R$$

$$\therefore \quad B = \frac{\mu_0 IR^2}{2(x^2+R^2)^{\frac{3}{2}}}$$

$$\vec{B} = \frac{\mu_0 IR^2}{2(x^2+R^2)^{\frac{3}{2}}}\hat{i}$$

Explanation : A magnetic needle placed at the centre or axis of a circular coil shows deflection. This implies that a circular coil behaves as a magnet.

Q. 2. (i) Write using Biot-Savart law, the expression for the magnetic field $\vec{B}$ due to an element $\vec{dl}$ carrying current I at a distance $\vec{r}$ from it in a vector form.
Hence derive the expression for the magnetic field due to a current carrying loop of radius R at a point P distant X from its centre along the axis of the loop.★
(ii) Explain how Biot-Savart law enables one to express the Ampere's circuital law in the integral form, *viz.*,

★ are board based questions from previous year

$$\oint \vec{B} \cdot \vec{dl} = \mu_0 I$$

Where I is the total current passing through the surface.★

Ans. (i) Suppose we have a conductor of length l in which current i is flowing. We need to calculate the magnetic field at a point P in vacuum. If $i\vec{dl}$ is one of the infinitely small current element, the magnetic field $d\vec{B}$ at point P is given by

$$d\vec{B} \propto \frac{i\vec{dl} \times \vec{r}}{r^3}$$

$$d\vec{B} = \frac{\mu_0}{4\pi} \frac{i\vec{dl} \times \vec{r}}{r^3}$$

Where $\frac{\mu_0}{4\pi}$ is a proportionally constant.

Suppose there is a circular coil of radius R, carrying a current i. Let P be a point at the axis of the coil at a distance x from the centre, at which the field in required.

Consider a conducting element dl of the loop. The magnetic field due to dl is given by the Biot-Savart law,

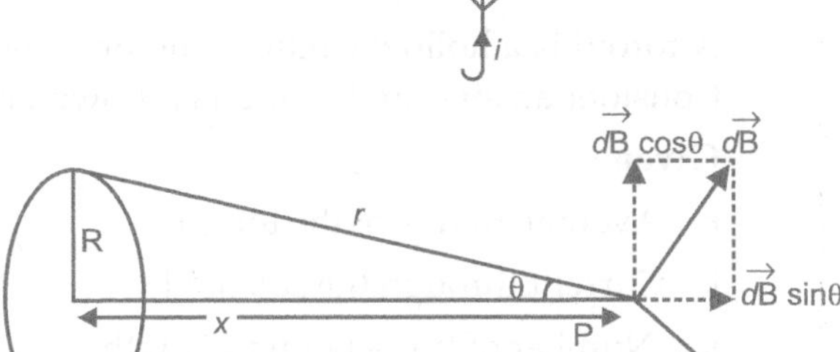

$$dB = \frac{\mu_0}{4\pi} \frac{i \, | \vec{dl} \times \vec{r} \, |}{r^3}$$

$$dB = \frac{\mu_0}{4\pi} \frac{idl}{(R^2 + x^2)}$$

The direction of dB is perpendicular to the plane formed by dl and r. It has an X-component dB_x and a component perpendicular to X-axis, $dB_{\perp}$. When the components perpendicular to the X-axis are summed over, they cancel out and we obtain null result. Thus, only the X-component survives.

So the resultant field $\vec{B}$ at P is given by

$$B = \int dB \sin\theta$$

$$B = \frac{\mu_0}{4\pi} \frac{i}{r^2} \int dl \sin\theta$$

$$B = \frac{\mu_0}{4\pi} \frac{iR}{r^3} \int dl \qquad \left[\because \sin\theta = \frac{R}{r}\right]$$

But $\int dl = 2\pi R$ and $r = (R^2 + x^2)^{\frac{1}{2}}$

$$\therefore \qquad B = \frac{\mu_0}{4\pi} \frac{2\pi i R^2}{(R^2 + x^2)^{\frac{3}{2}}}$$

If the coil has N turns, then each turn will contribute equally to B. Then,

$$B = \frac{\mu_0 N i R^2}{2\,(x^2 + R^2)^{\frac{3}{2}}}$$

(ii) According to Biot-Savart law the line integral of the magnetic field $\vec{B}$ around any 'closed' path is equal to μ_0 times the net current I threading through the area enclosed by the path.

★ are board based questions from previous year

i.e., $$\oint \vec{B} \cdot \vec{dl} = \mu_0 I$$

Where μ_0 in the permeability of free space. Ampere's circuital law in electromagnetism is analogous to Gauss' law in electrostatics.

Q. 3. State Ampere's circuital law. Use this law to obtain the expression for the magnetic field inside an air cored toroid of average radius, having '*n*' turns per unit length and carrying a steady current I.★

Ans. Ampere's circuital law in electromagnetism is analogous to Gauss's law in electrostatics. This law states that "The line integral of resultant magnetic field along a closed plane curve is equal to μ_0 time the total current crossing the area bounded by the closed curve provided the electric field inside the loop remains constant. Thus $\oint B.dl = \mu_0 I_{enc}$ where, μ_0 is permeability of free space and $I_{enc.}$ is the net current enclosed by the loop.

A toroid is a hollow circular ring on which a large number of turns of a wire are closely wound. Consider an air-cored toroid (as shown below) with centre O.

Given :

r = Average radius of the toroid

I = Current through the solenoid

n = Number of turns per unit length

To determine the magnetic field inside the toroid, we consider three amperian loops (loop 1, loop 2 and loop 3) as shown in the figure below.

According to Ampere's circuital law, we have

$$\oint \vec{B} \cdot \vec{dl} = \mu_0 I_{enc.} \text{ (Total current)}$$

Total current for loop 1 is zero because no current is passing through this loop

So,

For loop 1 :

$$\oint \vec{B} \cdot \vec{dl} = \mu_0 \text{ (Total current)}$$

For loop 3 :

According to Ampere's circuital law, we have

$$\oint \vec{B} \cdot \vec{dl} = \mu_0 \text{ (Total current)}$$

Total current for loop 3 is zero because net current coming out of this loop is equal to the net current going inside the loop.

For loop 2 :

The total current flowing through the toroid is NI, where N is the total number of turns

$$\oint \vec{B} \cdot \vec{dl} = \mu_0 \text{ (NI)} \qquad \text{...(i)}$$

Now, $\vec{B}$ and $\vec{dl}$ are in the same direction

$$\oint \vec{B} \cdot \vec{dl} = B \oint dl$$

$$\Rightarrow \quad \oint \vec{B} \cdot \vec{dl} = B\,(2\pi r) \qquad \text{...(ii)}$$

Comparing (i) and (ii), we get

$$B\,(2\pi r) = \mu_0 NI$$

★ are board based questions from previous year

$$\Rightarrow \qquad B = \frac{\mu_0 NI}{2\pi r}$$

Number of turns per unit length is given by

$$n = \frac{N}{2\pi r}$$

$$\therefore \qquad B = \mu_0 nI$$

This is the expression for magnetic field inside air-cored toroid.

Q. 4. Deduce an expression for the frequency of revolution of a charged particle in a magnetic field and show that it is independent of velocity or energy of the particle.★

Ans. When a charged particle having charge q moves inside a magnetic field $\vec{B}$ having velocity v, it experiences a force, which is given by :

$$\vec{F} = q(\vec{v} \times \vec{B})$$

Here,

$\vec{v}$ is perpendicular to $\vec{B}$,

$\vec{F}$ is the force on the charged particle which behaves as the centripetal force and make it move in a circular path.

Let m be the mass of the charged particle and r be the radius of the circular path.

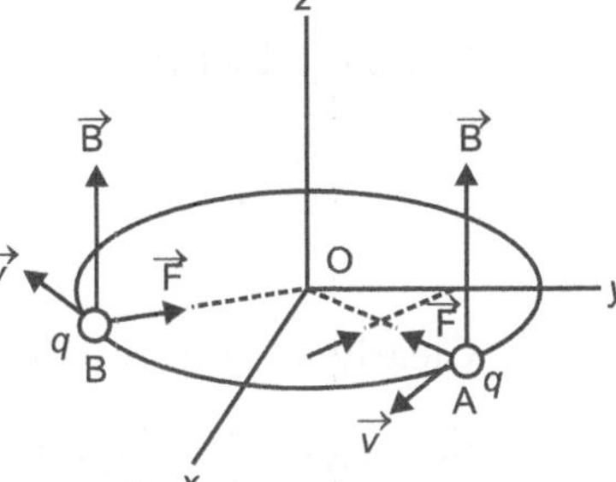

$$\therefore \qquad q(\vec{v} \times \vec{B}) = \frac{mv^2}{r}$$

v and B are at right angles :

$$\therefore \qquad qvB = \frac{mv^2}{r}$$

$$r = \frac{mv}{Bq}$$

Time period of circular motion of the charged particle can be calculated by,

$$T = \frac{2\pi r}{v}$$

$$= \frac{2\pi \, mv}{v \;\; Bq}$$

$$T = \frac{2\pi m}{Bq}$$

$\therefore$ Angular frequency is

$$\omega = \frac{2\pi}{T}$$

$$\therefore \qquad \omega = \frac{Bq}{m}$$

Therefore, the frequency of the revolution of the charged particle is independent of the velocity or the energy of the particle.

Chapter 5. Magnetism and Matter

Q. 1. The magnetic moment vectors μ_s and μ_l associated with the intrinsic spin angular momentum S and orbital angular momentum l, respectively, of an electron are predicted by quantum theory (and verified experimentally to a high accuracy) to be given by :

$$\mu_s = -(e/m)\,S,$$

★ are board based questions from previous year

$$\mu_l = -(e/2m)\, l$$

Which of these relations is in accordance with the result expected classically ? Outline the derivation of the classical result.

Ans.

$$\vec{\mu_1} = -\left(\frac{c}{2m}\right)\vec{l}$$

It follows from the definition of μ_1 and I.

$$\mu_1 = iA = \left(\frac{e}{T}\right)\pi r^2$$

$$l = mvr = m\left(\frac{2\pi r}{T}\right)r$$

Where, r is the radius of the circular orbit, which the electron of mass m and change $(-e)$ completes in time T.

Divide (i) by (ii)

$$\frac{\mu_1}{l} = \frac{-e}{T}\pi r^2 \times \frac{T}{m \times 2\pi r^2} = \frac{-e}{2m}$$

$$\vec{\mu_1} = -\left(\frac{-e}{2m}\right)\vec{l}$$

Clearly $\vec{\mu_1}$ and $\vec{l}$ will be antiparallel (both being normal to the plane of the orbit)

In contrast $\frac{\mu_s}{s} = \frac{e}{m}$. It is obtained on the basis of the quantum mechanics.

Chapter 6. Electromagnetic Induction

Q. 1. Define mutual inductance between a pair of coils. Derive an expression for the mutual inductance of two long coaxial solenoids of same length wound one over the other.★

Ans. Mutual inductance is numerically equal to the induced e.m.f in the secondary coil when the current in the primary coil changes by unity.

Suppose two long co-axial solenoids each of length l. We denote the area of the inner solenoid S_1 by A_1 and the number of turns per unit length by n_1. The corresponding quantities for the outer solenoid S_2 are A_2 and n_2 respectively. Let N_1 and N_2 be the total number of turns of coils S_1 and S_2 respectively.

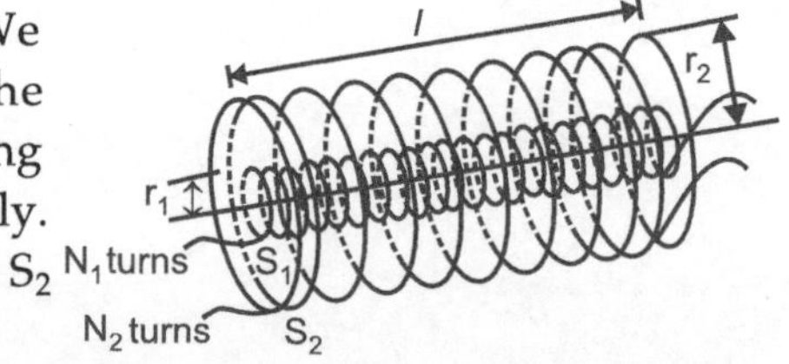

When a current I_2 is set up through S_2, it in turn sets up a magnetic flux through S_1. Let us denote it by ϕ_1.

The magnetic field due to current I_1 in S_1 is given by

$$B_1 = \mu_0 n_1 I_1$$

Flux linked with each coil of second solenoid,

$$\phi_2 = B_1 A_1$$

∴ Total flux linked with N_2 turns,

$$\phi_2 = B_1 A_1 N_2$$

$$\phi_2 = \mu_0 n_1 I_1 A_1 N_2 \quad \text{...(i)}$$

Also $$\phi_2 = M_{21} I_1 \quad \text{...(ii)}$$

Comparing, $$M_{21} = \mu_0 n_1 N_2 A_1$$

★ are board based questions from previous year

$$M_{21} = \mu_0 n_1 n_2 l A_1 \qquad [\because N_2 = n_2 l]$$

Similarly
$$M_{12} = \mu_0 n_1 n_2 l A_1$$

Q. 2. Define the term 'mutual inductance' between the two coils. Obtain the expression for mutual inductance of a pair of long coaxial solenoids each of length l and radii r_1 and r_2 ($r_2 >> r_1$). Total number of turns in the two solenoids are N_1 and N_2 respectively.★

Ans. Mutual inductance of two coils is equal to the e.m.f. induced in one coil when rate of change of current through the other coil is unity.

Mutual inductance of two co-axial solenoids :

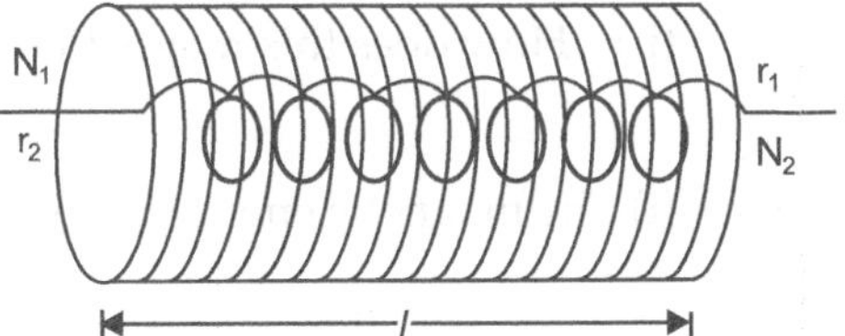

Consider two long co-axial solenoid each of length l with number of turns N_1 and N_2 wound one over the other. Number of turns per unit length in solenoid, $n = \frac{N_1}{l}$. If I_1 is the current flowing in primary solenoid, the magnetic field produced within this solenoid.

$$B_1 = \frac{\mu_0 N_1 I_1}{l}$$

The flux linked with each turn of inner solenoid coil is $\phi_2 = B_1 A_2$, where A_2 is the cross-sectional area of inner solenoid. The total flux linkage with inner coil of N_2 turns.

$$\begin{aligned} \phi_2 &= N_2 \phi_2 \\ &= N_2 B_1 A_2 \\ &= N_2 \left(\frac{\mu_0 N_1 I_1}{l} \right) A_2 \\ &= \frac{\mu_0 N_1 N_2}{l} \end{aligned}$$

Mutual Inductance,
$$M_{21} = \frac{\phi_2}{I_1} = \frac{\mu_0 N_1 N_2 A_2}{l}$$

If n_1 is number of turns per unit length of outer solenoid and r_2 is radius of inner solenoid, then

$$M = \mu_0 n_1 N_2 \pi r_2^2$$

Q. 3. A rod of length 'l' is moved horizontally with a uniform velocity 'v' in a direction perpendicular to its length through a region in which a uniform magnetic field is acting vertically downward. Derive the expression for the emf induced across the ends of the rod.★

Ans. Consider a rod PQ of length l, moving in a magnetic field $\vec{B}$ with a constant velocity $\vec{v}$. The length of the rod is perpendicular to the magnetic field and also the velocity is perpendicular to both the rod and field. The free electrons of the rod also move at this velocity $\vec{v}$ because of which it experiences a magnetic force.

$$\vec{F_b} = q\vec{v} \times \vec{B}$$

This force is towards Q to P.

Thus, the free electrons will move towards P and positive charge will appears at Q. An electrostatic field E is developed within the wire from Q to P. This field exerts a force.

$$\vec{F_e} = q\vec{E}$$

on each free electron. the charge keeps on gathering until

$$\vec{F_b} = \vec{F_e}$$

$$\Rightarrow \qquad |q\vec{v} \times \vec{B}| = |q\vec{E}|$$

$$vB = E$$

★ are board based questions from previous year

After this, resultant force on the free electrons of the wire PQ becomes zero. The potential difference between the ends Q and P is given by,

$$V = El = vBl$$

Thus, the potential difference is maintained by the magnetic force on the moving free electron and hence, produces an emf, $e = Bvl$

Q. 4. (i) Obtain the expression for the magnetic energy stored in a solenoid in terms of magnetic field B, area A and length l of the solenoid.

(ii) How does this magnetic energy compare with the electrostatic energy stored in a capacitor ?

[NCERT]

Ans. (i) The magnetic energy is given by

$$U_B = \frac{1}{2}LI^2$$

$$= \frac{1}{2}L\left(\frac{B}{\mu_0 n}\right)^2 \quad \text{(since } B = \mu_0 nI\text{, for a solenoid)}$$

$$= \frac{1}{2}(\mu_0 n^2 Al)\left(\frac{B}{\mu_0 n}\right)^2 \quad (\text{as } L = \mu_0 n^2 Al)$$

$$= \frac{1}{2\mu_0}B^2 Al$$

(ii) The magnetic energy per unit volume is,

$$u_B = \frac{U_B}{V} \quad \text{(where V is volume that contains flux)}$$

$$= \frac{U_B}{Al} \quad \text{[From part (i) of question]}$$

$$= \frac{B^2}{2\mu_0} \quad \text{...(i)}$$

We have already obtained the relation for the electrostatic energy stored per unit volume in a parallel plate capacitor.

$$u_E = \frac{1}{2}\varepsilon_0 E^2 \quad \text{...(ii)}$$

In both the cases energy is proportional to the square of the field strength. Equations (i) and (ii) have been derived for special cases : a solenoid and a parallel plate capacitor, respectively. But they are general and valid for any region of space in which a magnetic field or/and an electric field exist.

Q. 5. (i) Define mutual inductance and write its S.I. units.★

(ii) Derive an expression for the mutual inductance of two long co-axial solenoids of same length wound one over the other.★

(iii) In an experiment two coils c_1 and c_2 are placed close to each other. Find out the expression for the emf induced in the coil c_1 due to a change in the current through the coil c_2.★

Ans. (i) Mutual inductance is the property of two coils by the virtue of which each opposes any change in the value of current flowing through the other by developing an induced emf. The S.I. unit of mutual inductance is henry and its symbol is H.

(ii) Consider two long solenoids S_1 and S_2 of same length l such that solenoid S_2 surrounds solenoid S_1 completely.

★ are board based questions from previous year

Let

n_1 = Number of turns per unit length of S_1

n_2 = Number of turns per unit length of S_2

I_1 = Current passed through solenoid S_1

ϕ_{21} = Flux linked with S_2 due to current flowing through S_1

$$\phi_{21} \propto I_1$$

$$\phi_{21} = M_{21}I_1$$

where M_{21} = Coefficient of mutual induction of the two solenoids

When current is passed through solenoid S_1, an emf is induced in solenoid S_2.

Magnetic field produced inside solenoid S_1 on passing current through it is given by

$$B_1 = \mu_0 n_1 I_1$$

Magnetic flux linked with each turn of solenoid S_2 will be equal to B_1 times the area of cross-section of solenoid S_1.

Magnetic flux linked with each turn of the solenoid $\phi_{21} = B_1A$

Therefore, total magnetic flux linked with the solenoid S_2 is given by

$$\phi_{21} = B_1A = \mu_0 n_1 I_1 A$$

$$M_{21} = \frac{N_2\phi_{21}}{I_1}$$

$$M_{21} = \frac{N_2\mu_0 n_1 I_1 A}{I_1} = \mu_0 n_1 N_2 A$$

where N_2 is total number of turns wound over the secondary coil.

$$\therefore \quad M_{21} = \mu_0 n_1 N_2 A$$

Similarly the mutual inductance between the two solenoids when current is passed through solenoid S_2 and induced emf is produced in solenoid S_1 is given by

$$M_{21} = \mu_0 N_1 n_2 A$$

where N_1 total numbers of turns wound over primary coil.

(iii) It is found that,

$$\phi \propto I$$

$$\phi = MI$$

Where, I is the strength of current in coil 2, and ϕ is the total amount of magnetic flux linked with coil 1.

E.m.f. induced in neighbouring coil C_1 is,

$$e = -\frac{d\phi}{dt}$$

$$= -\frac{d\,(MI)}{dt} = -\frac{MdI}{dt}$$

Chapter 7. Alternating Current

Q. 1. A voltage $V = V_0 \sin \omega t$ is applied to a series LCR circuit. Derive the expression for the average power dissipated over a cycle. Under what condition is (i) no power dissipated even though the current flows through the circuit, (ii) maximum power dissipated in the circuit ?★

Ans. Voltage $V = V_0 \sin \omega t$ is applied to a series LCR circuit.

Current is $$I = I_0 \sin(\omega t + \phi)$$

$$I_0 = \frac{V_0}{Z}$$

★ are board based questions from previous year

$$\phi = \tan^{-1}\left(\frac{X_C - X_L}{R}\right)$$

Instantaneous power supplied by the source is

$$P = VI = (V_0 \sin \omega t) \times (I_0 \sin (\omega t + \phi))$$

$$= \frac{V_0 I_0}{2}[(\cos \phi - \cos (2\omega t + \phi)]$$

The average power over a cycle is average of the two terms on the R.H.S. of the above equation. The second term is time dependent, so, its average is zero.

So, $$P = \frac{V_0 I_0}{2} \cos \phi$$

$$= \frac{V_0 I_0}{\sqrt{2}\sqrt{2}} \cos \phi = VI \cos \phi$$

$$P = I^2 Z \cos \phi$$

$\cos \phi$ is called the power factor.

Case 1.

For pure inductive circuit or pure capacitive circuit, the phase difference between current and voltage *i.e.*, ϕ is $\frac{\pi}{2}$.

$\therefore$ $$\phi = \frac{\pi}{2}$$

So $$\cos \phi = 0$$

Therefore, no power is dissipated.

Case 2.

For power dissipated at resonance in an LCR circuit,

$$X_C - X_L = 0, \ \phi = 0$$

$\therefore$ $$\cos \phi = 1$$

So, maximum power is dissipated.

Q. 2. What do you mean by the impedance of LCR circuit, derive an expression for it.

Ans. Impedance : The opposition offered by the combination of a resistor and reactive components to flow of alternating current is called impedance.

It is ratio of rms voltage applied and rms current produced in the circuit.

$$Z = \frac{V}{I}$$

Consider that an inductor of inductance L, a capacitor of capacitance C and resistor of resitance R are connected in series to an alternating source of emf (E). It is given by

$$E = E_0 e^{j\omega t} \quad \text{...(i)}$$

The instantaneous complex current (I) is given by

$$IR + I(j\omega L) + I\left(\frac{1}{j\omega C}\right) = E$$

$$I = \frac{E}{R + j\omega L + \frac{1}{j\omega C}}$$

R L C E

The complex quantity $R + j\omega L + \frac{1}{j\omega C}$ is called complex impedance of LCR circuit. It can be expressed as

$$R + j\omega L + \frac{1}{j\omega C} = R + j\left(\omega L - \frac{1}{\omega C}\right) = ze^{j\phi} \quad \text{...(ii)}$$

where $$Z = \sqrt{R^2 + \left(\omega L - \frac{1}{\omega C}\right)^2} \quad \ldots(iii)$$

The argument of complex impedance is given by

$$\tan\phi = \frac{\omega L - \frac{1}{\omega C}}{R}$$

From equation (i) and (ii), $$I = \frac{E_0\, e^{j\omega t}}{z e^{j\phi}} = \frac{E_0}{z}\, e^{j(\omega t - \phi)} = I_0\, e^{j(\omega t - \phi)} \quad \ldots(iv)$$

Where, $$I_0 = \frac{E_0}{\sqrt{R^2 + \left(\omega L - \frac{1}{\omega C}\right)^2}}$$

is the peak value of current.

From (i) and (iv), it follows that in LCR circuit emf leads the current by phase angle ϕ.

Q. 3. Draw a labelled diagram of AC generator. Derive the expression for the instan taneous value of the emf induced in the coil.★

Ans. (i) Where B_1, B_2 are brushes.

R_1, R_2 are slip rings.

R_L is load resistance.

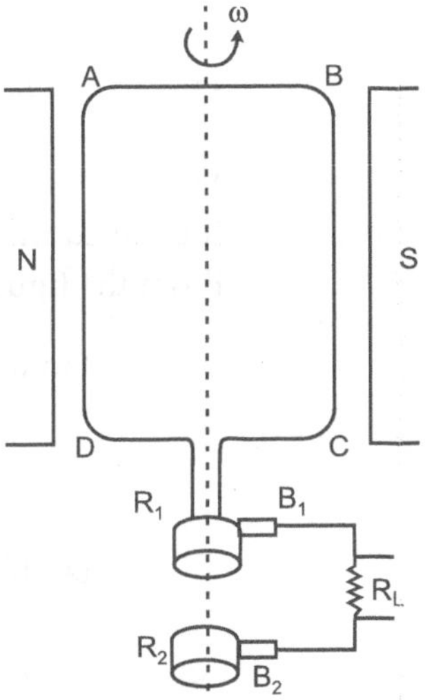

ABCD is armature.

If N is the number of turns in coil, A area of coil and B the magnetic induction then flux ϕ is given by

$$\phi = B.A$$

$$\phi = BA\cos\omega t$$

The emf induced in the coil is given by,

$$e = -N\frac{d\phi}{dt}$$

$$= -N\frac{d}{dt}(BA\cos\omega t)$$

$$= -NBA(-\omega\sin\omega t)$$

$$= NBA\,\omega\sin\omega t$$

$$e = e_0\sin\omega t \qquad (\text{where } e_0 = NBA\omega)$$

Q. 4. (i) An a.c. source of voltage $V = V_0 \sin\omega t$ is connected to a series combination of L, C and R. Use the phasor diagram to obtain expressions for impedance of the circuit and phase angle between voltage and current. Find the condition when current will be in phase with the voltage. What is the circuit in this condition called ?★

(ii) In a series LR circuit $X_L = R$ and power factor of the circuit is P_1. When capacitor with capacitance C such that $X_L = X_C$ is put in series, the power factor becomes P_2. Calculate $\frac{P_1}{P_2}$.★

★ are board based questions from previous year

Ans. (i) Let a series LCR circuit is connected to an a.c. source V (Fig). We take the voltage of the source to be $V = V_0 \sin \omega t$

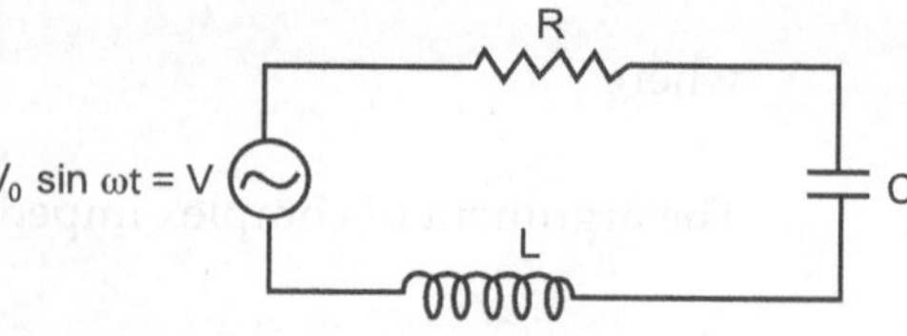

The a.c. current in each element is the same at any time, having the same amplitude and phase. It is given by,

$$I = I_0 \sin(\omega t + \phi)$$

Let V_L, V_R, V_C and V represent the voltage across the inductor, resistor, capacitor and the source respectively.

Let $$V_L > V_C$$

$\therefore$ $$V_0^2 = V_R^2 + (V_L - V_C)^2$$

$$V_0^2 = (I_0R)^2 + (I_0X_L - I_0X_C)^2$$

$$V_0^2 = I_0^2 [R^2 + (X_L - X_C)^2]$$

and $$I_0 = \frac{V_0}{\sqrt{R^2 + (X_L - X_C)^2}}$$

$\Rightarrow$ $$I_0 = \frac{V_0}{Z}$$

Where $$Z = \sqrt{R^2 + (X_L - X_C)^2}$$

It is called the impedance in an a.c. circuit.

From the figure,

$$\tan\theta = \frac{V_L - V_C}{V_R}$$

$$= \frac{I_0X_L - I_0X_C}{I_0R}$$

$$\tan\theta = \frac{X_L - X_C}{R}$$

$\Rightarrow$ $$\theta = \tan^{-1}\left(\frac{X_L - X_C}{R}\right)$$

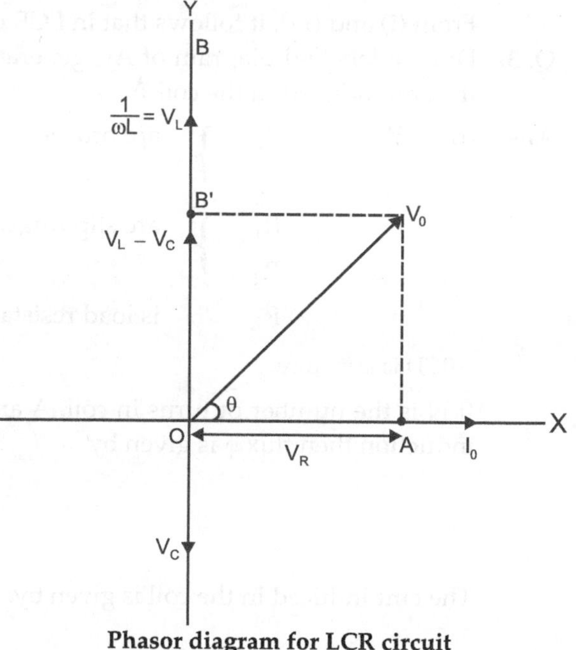

Phasor diagram for LCR circuit

Condition : The current will be in phase with the voltage at resonance condition.

At resonance condition,

$$X_L = X_C$$

$$\omega L = \frac{1}{\omega C}$$

$$\omega = \frac{1}{\sqrt{LC}}$$

$$2\pi f = \frac{1}{\sqrt{LC}}$$

$$f = \frac{1}{2\pi\sqrt{LC}}$$

(ii) As $$\cos\theta = \frac{R}{Z}$$

In LR circuit, $$P_1 = \cos\phi$$

$$P_1 = \frac{R}{\sqrt{R^2 + X_L^2}} = \frac{R}{\sqrt{2R^2}} \qquad [\because X_L = R]$$

$$P_1 = \frac{1}{\sqrt{2}}$$

In LCR circuit when, $X_L = X_C$

$$P_2 = \frac{R}{\sqrt{R^2 + (X_L - X_C)^2}}$$

$$= \frac{R}{R} = 1 \qquad [\because X_L = X_C]$$

$$\therefore \quad \frac{P_1}{P_2} = \frac{1}{\sqrt{2}}$$

Chapter 9. Ray Optics and Optical Instruments

Q. 1. (i) Draw a ray diagram to show image formation when the concave mirror produces a real, inverted and magnified image of the object.★

(ii) Obtain the mirror formula and Derive the expression for the linear magnification.★

Ans. (i) Concave mirror produces real, inverted and magnified image for object placed between F and C.

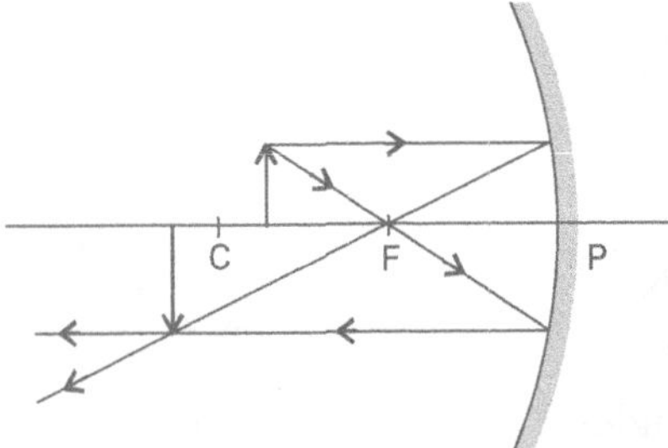

(ii) **Derivation for mirror formula and magnification :** Consider an object AB be placed in front of a concave mirror beyond centre of curvature C.

If F is the focus, focal length = f

Object distance = u

Image distance = v

As Δs ABC and A′B′C are similar,

$$\frac{AB}{A'B'} = \frac{CB}{CB'} \qquad \text{...(i)}$$

Again, as Δs ABP and A′B′P are similar

$$\frac{AB}{A'B'} = \frac{PB}{B'P} \qquad \text{...(ii)}$$

$$\Rightarrow \quad \frac{CB}{CB'} = \frac{PB}{PB'} \qquad \text{...(iii)}$$

$$\therefore \quad \text{From (iii), } \frac{PB - PC}{PC - PB'} = \frac{PB}{PB'}$$

$$\Rightarrow \quad \frac{-u + R}{-R + v} = \frac{-u}{v} \qquad [\because PB = -u, PC = -R, PB' = -v]$$

$$\Rightarrow \quad uR + vR = 2uv$$

★ are board based questions from previous year

$$\Rightarrow \quad \frac{1}{v}+\frac{1}{u} = \frac{2}{R}$$

$$\Rightarrow \quad \boxed{\frac{1}{v}+\frac{1}{u} = \frac{1}{f}} \qquad (\because R = 2f)$$

$$\text{Linear magnification, } m = \frac{\text{Image height } (h_i)}{\text{Object height } (h_o)}$$

$$\text{From equation (ii), } \frac{A'B'}{AB} = \frac{PB'}{PB}$$

Using new Cartesian sign conventions, $A'B' = -h_2$, $AB = +h_1$, $PB' = -v$, $PB = -u$

$$\therefore \quad \frac{-h_2}{h_1} = \frac{-v}{-u} = \frac{v}{u}$$

$$\text{or} \quad m = \frac{h_2}{h_1} = \frac{-v}{u}$$

Q. 2. Derive the mathematical relation between refractive indices n_1 and n_2 of two media and radius of curvature R for refraction at a convex spherical surface. Consider the object to be a point since lying on the principal axis in rarer medium of refractive index n_1 and a real image formed in the denser medium of refractive index n_2. Hence, derive Lens Maker's formula.★

Ans.

$$\tan \angle NOM = \frac{MN}{OM};$$

$$\tan \angle NCM = \frac{MN}{MC};$$

$$\tan \angle NIM = \frac{MN}{MI}$$

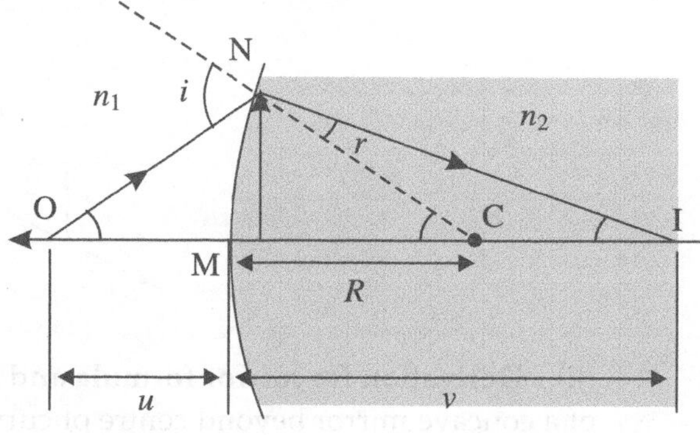

For ΔNOC, i is the exterior angle.

Assuming the incident ray is very close to the principal axis, all the angles are very small. Hence, for very small angles,

$$\tan x \approx x \text{ and } \sin x \approx x$$

$$\therefore \quad i = \angle NOM + \angle NCM$$

$$i = \frac{MN}{OM}+\frac{MN}{MC} \qquad \ldots(i)$$

Similarly,

$$r = \angle NCM - \angle NIM$$

i.e.,

$$r = \frac{MN}{MC}-\frac{MN}{MI} \qquad \ldots(ii)$$

By Snell's law, $\quad n_1 \sin i = n_2 \sin r$

For small angles, $\quad n_1 i = n_2 r$

On substituting the values of i and r in equations, we get

$$n_1\left(\frac{MN}{OM}+\frac{MN}{MC}\right) = n_2\left(\frac{MN}{MC}-\frac{MN}{MI}\right)$$

$$\frac{n_1}{OM}+\frac{n_2}{MI} = \frac{n_2-n_1}{MC} \qquad \ldots(iii)$$

On applying new Cartesian sign conventions,

$$OM = -u, MI = +v, MC = +R$$

Substituting these values in equation (iii), we get

★ are board based questions from previous year

$$\frac{n_2}{v} - \frac{n_1}{u} = \frac{n_2 - n_1}{R} \qquad ...(iv)$$

In deriving Lens maker's formula, we adopt the coordinate geometry sign convention and make the assumptions :

(a) The lens is thin so that the distances measured from the poles of its two surfaces can be taken as equal to the distances from its optical centre.

(b) The aperture of the lens is small.

(c) The object is a point-object placed on the principal axis of the lens.

(d) The incident and the refracted rays make small angles with the principal axis.

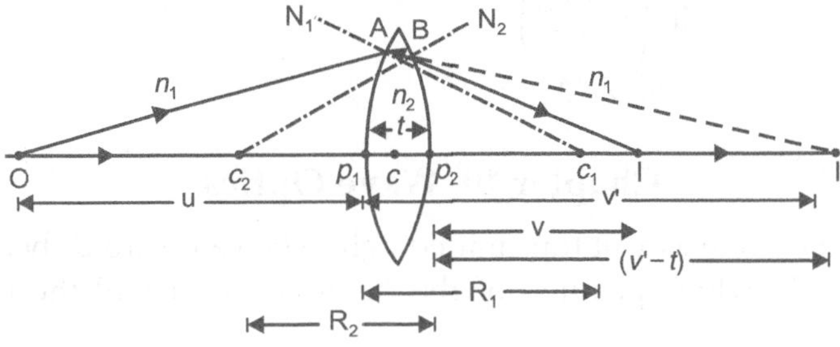

Refraction at the first surface,

$$\frac{n_2}{v'} - \frac{n_1}{u} = \frac{n_2 - n_1}{R_1} \qquad \text{[by eq}^n \text{ (iv)]}$$

Refraction at the second surface,

$$\frac{n_1}{v} - \frac{n_2}{v' - t} = \frac{n_1 - n_2}{R_2}$$

The lens is 'then', hence $t << v'$ and can be ignored. Then, we have

$$\frac{n_1}{v} - \frac{n_2}{v'} = \frac{n_1 - n_2}{R_2} \qquad ...(ii)$$

Adding equation (i) and (ii), we get

$$\frac{n_1}{v} - \frac{n_1}{u} = (n_2 - n_1)\left(\frac{1}{R_1} - \frac{1}{R_2}\right)$$

Putting $\frac{n_2}{n_1} = n$, the refractive index of the material of the lens with respect to the surrounding medium, we have

$$\frac{1}{v} - \frac{1}{u} = (n - 1)\left(\frac{1}{R_1} - \frac{1}{R_2}\right) \qquad ...(iii)$$

When the object is at infinity, the image will be formed at the principal focus of the lens, *i.e.*, when $u = \infty$,

$$\frac{1}{f} - \frac{1}{\infty} = (n - 1)\left(\frac{1}{R_1} - \frac{1}{R_2}\right)$$

$$\frac{1}{f} = (n - 1)\left(\frac{1}{R_1} - \frac{1}{R_2}\right)$$

This is the Lens Maker's formula.

Q. 3. Plot a graph to show variation of the angle of deviation as a function of angle of incidence for light passing through a prism. Derive an expression for refractive index of the prism in terms of angle of minimum deviation and angle of prism.★

★ are board based questions from previous year

Give relevant calculations.

Ans. From figure, $\delta = \delta_m$, $i = e$ which implies $r_1 = r_2$

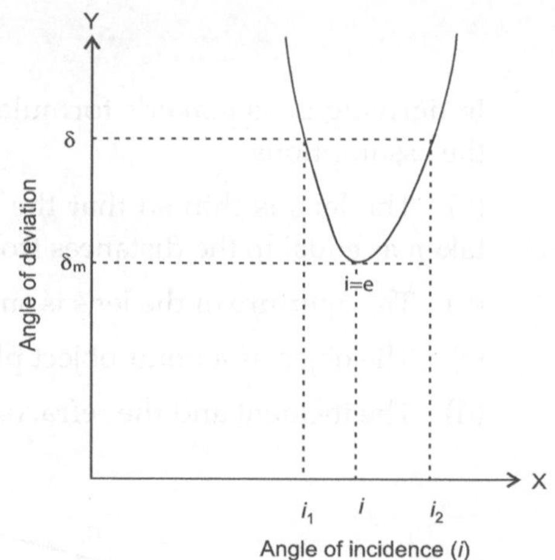

$$2r = A, \text{ or } r = A/2$$

Using
$$\delta = 2i - A$$

$$i = \frac{A + \delta_m}{2}$$

$$\mu = \frac{\sin i}{\sin r}$$

$$= \frac{\sin\left(\frac{A + \delta_m}{2}\right)}{\sin A/2}$$

Chapter 10. Wave Optics

Q. 1. Find an expression for intensity of transmitted light when a polaroid sheet is rotated between two crossed polaroids. In which position of the polaroid sheet will the transmitted intensity be maximum ?★

Ans. Let us consider two crossed polarisers P_1 and P_2 with a polaroid sheet P_3 placed between them.

Let I_0 be the intensity of polarised light after passing through the first polarizer P_1. If θ is the angle between the axes of P_1 and P_3, then the intensity of the polarised light after passing through P_3 will be $I = I_0 \cos^2 \theta$.

As P_1 and P_2 are crossed the angle between the axes of P_1 and $P_2 = 90°$.

∴ Angle between the axes of P_2 and $P_3 = (90° - \theta)$

The intensity of light emerging from P_2 will be given by

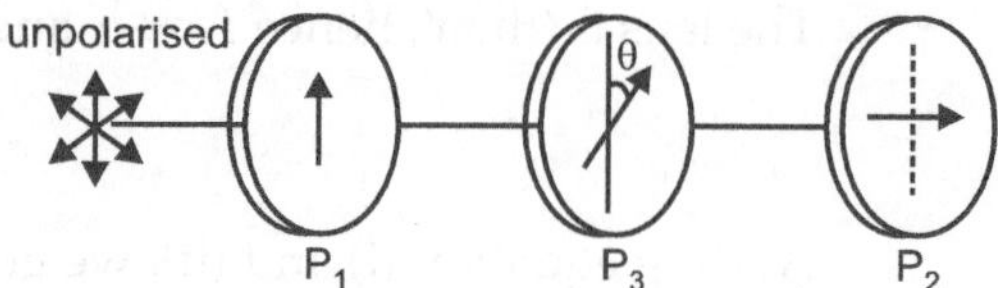

$$I = [I_0 \cos^2 \theta] \cos^2 (90° - \theta)$$

$$\Rightarrow \quad I = [I_0 \cos^2 \theta] \sin^2 \theta$$

$$\Rightarrow \quad I = \frac{I_0}{4}(4 \cos^2 \theta \sin^2 \theta)$$

$$\Rightarrow \quad I = \frac{I_0}{4}(2 \sin \theta \cos \theta)^2$$

$$\Rightarrow \quad I = \frac{I_0}{4}(\sin^2 2\theta)$$

The intensity of polarised light transmitted from P_2 will be maximum when

$$\sin 2\theta = \text{maximum} = 1$$

$$\Rightarrow \quad \sin 2\theta = \sin 90°$$

$$\Rightarrow \quad 2\theta = 90°$$

$$\Rightarrow \quad \theta = 45°$$

Also the maximum transmitted intensity will be given by

$$I = \frac{I_0}{4}$$

Q. 2. (i) State Huygen's principle. Using this principle draw a diagram to show how a plane wave front incident at the interface of the two media gets refracted when it propagates from a rarer to a denser medium. Hence verify Snell's law of refraction.★

★ are board based questions from previous year

(ii) When monochromatic light travels from a rarer to a denser medium, explain the following, giving reasons :★

(a) Is the frequency of reflected and refracted light same as the frequency of incident light ?

(b) Does the decreases in speed imply a reduction in the energy carried by light wave ?

Ans. (i) **Huygen's principle :**

(a) Each point on the primary wave front acts as a source of secondary wavelets, transferring out disturbance in all directions in the same way as the original source of light does.

(b) The new position of the wave front at any instant is the envelope of the secondary wavelets at that instant.

Refraction on the basis of wave theory :

(i) Consider any point Q on the incident wave front.

(ii) Suppose when disturbance from point P on incident wave front reaches point P′ on the refracted wave front, the disturbance from point Q reaches Q′ on the refracting surface XY.

Since P′A′ represents the refracted wave front the time taken by light to travel from a point on incident wavefront to the corresponding point on refracted wave front should always be the same. Now, time taken by light to go from Q to Q′ will be

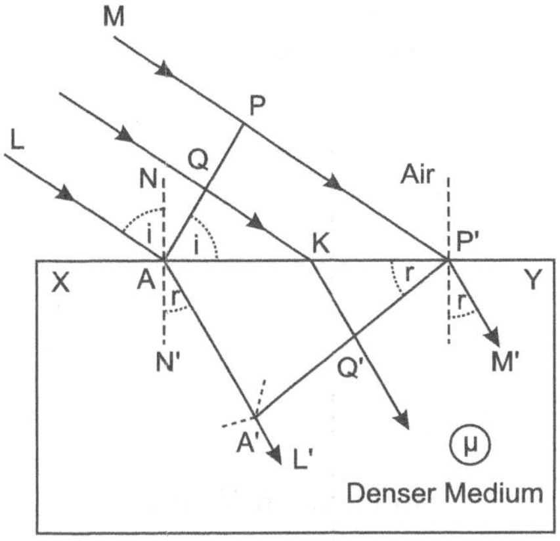

$$t = \frac{QK}{c} + \frac{Q'K}{v} \quad \ldots(i)$$

In right-angled ΔAQK, ∠QAK = i

$$\therefore \quad QK = AK \sin i \quad \ldots(ii)$$

In right-angled ΔP′Q′K, ∠Q′P′K = r and Q′K = KP′ sin r ...(iii)

Substituting (ii) and (iii) in equation (i),

$$t = \frac{AK \sin i}{c} + \frac{KP' \sin r}{v}$$

$$t = \frac{AK \sin i}{c} + \frac{(AP' - AK) \sin r}{v} \qquad [\because KP' = AP' - AK]$$

$$t = \frac{AP'}{v} \sin r + AK\left(\frac{\sin i}{c} - \frac{\sin r}{v}\right) \quad \ldots (iv)$$

The rays from different points on the incident wave front will take the same time to reach the corresponding points on the refracted wave front *i.e.,* given equation (iv) is independent of AK. It will happen so, if

$$\frac{\sin i}{c} - \frac{\sin r}{v} = 0$$

$$\Rightarrow \quad \frac{\sin i}{\sin r} = \frac{c}{v}$$

$$\Rightarrow \quad n = \frac{\sin i}{\sin r}$$

This is the Snell's law for refraction of light.

(ii) (a) The frequency of reflected and refracted light remains constant as the frequency of incident light because frequency only depends on the source of light.

(b) Since the frequency remains same, hence there is no reduction in energy.

★ are board based questions from previous year

Q. 3. In Young's double slit experiment, describe briefly how bright and dark fringes are obtained on the screen kept in front of a double slit. Hence obtain the expression for the fringe width.★

Ans. In Young's double slit experiment, the wave fronts from the two illuminated slits superpose on the screen. This results in formation of alternate bright and dark fringes because of constructive and destructive interference, respectively. The intensity of light is maximum at the centre C of the screen and it is called central maxima.

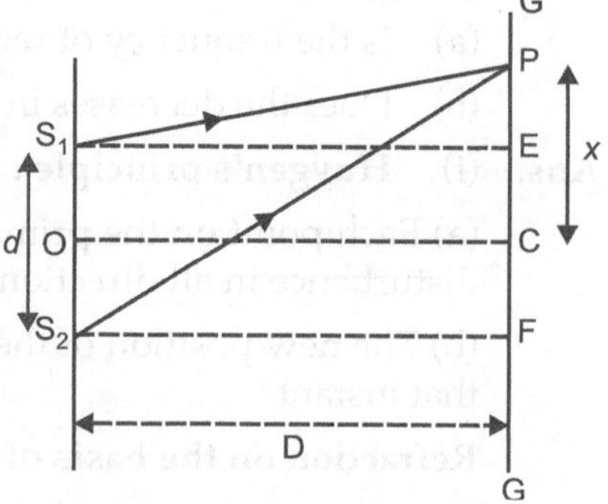

Let S_1 and S_2 be two slits separated by a distance d. GG' is the screen at a distance D from the slits S_1 and S_2. Both the slits are equidistant from point C. The intensity of light will be maximum at this point due to the path difference of the waves reaching this point will be zero.

At point P, the path difference between the rays coming from the slits S_1 and S_2 is $S_2P - S_1P$.

Now, $$S_1S_2 = d$$

$$EF = d_1 \text{ and } S_2F = D$$

∴ In ΔS_2PF,

$$S_2P = [S_2F^2 + PF^2]^{\frac{1}{2}}$$

$$S_2P = \left[D^2 + \left(x + \frac{d}{2}\right)^2\right]^{\frac{1}{2}}$$

$$= D\left[1 + \frac{\left(x + \frac{d}{2}\right)^2}{D^2}\right]^{\frac{1}{2}}$$

Similarly, in ΔS_1PE

$$S_1P = D\left[1 + \frac{\left(x - \frac{d}{2}\right)^2}{D^2}\right]^{\frac{1}{2}}$$

$$S_2P - S_1P = D\left[1 + \frac{\left(x + \frac{d}{2}\right)^2}{D^2}\right]^{\frac{1}{2}} - D\left[1 + \frac{\left(x - \frac{d}{2}\right)^2}{D^2}\right]^{\frac{1}{2}}$$

On expanding it binomially,

$$S_2P - S_1P = \frac{1}{2D}\left[4x \times \frac{d}{2}\right] = \frac{xd}{D}$$

For bright fringes (constructive interference), the path difference is an integral multiple of wavelengths, *i.e.*, path difference is $n\lambda$.

$$\therefore \quad n\lambda = \frac{xd}{D}$$

$$x = \frac{n\lambda D}{d}, \text{ where } n = 0, 1, 2, 3, 4, \ldots\ldots$$

For $\quad n = 0, \quad x_0 = 0$

$\quad n = 1, \quad x_1 = \dfrac{\lambda D}{d}$

★ are board based questions from previous year

$$n = 2, \qquad x_2 = \frac{2\lambda D}{d}$$

$$n = 3, \qquad x_3 = \frac{3\lambda D}{d}$$

$$\vdots$$

$$n = n, \qquad x_n = \frac{n\lambda D}{d}$$

The separation between the centres of two consecutive bright interference fringes is the width of a dark fringe.

$$\therefore \qquad \boxed{\beta_1 = x_n - x_{n-1} = \frac{\lambda D}{d}}$$

$$x_n = (2n-1)\frac{\lambda D}{2d}$$

For $\quad n = 1, \qquad x_1 = \dfrac{\lambda D}{2d}$

For $\quad n = 2, \qquad x_2 = \dfrac{3\lambda D}{2d}$

The separation between the centres of two consecutive dark interference fringes is the width of a bright fringe.

$$\therefore \qquad \boxed{\beta_2 = x_n - x_{n-1} = \frac{\lambda D}{d}}$$

$$\beta_1 = \beta_2$$

All the bright and dark fringes are of equal width as $\beta_1 = \beta_2$.

Chapter 11. Dual Nature of Radiation and Matter

Q. 1. Describe de-Broglie concept of matter waves and derive an expression for the momentum and wavelength of the particle. **[NCERT]**

Ans. According to the de-Broglie concept, all the material particles exhibit wave properties and the wavelength of the wave associated with any material particles is given by,

$$\lambda = \frac{h}{p} \qquad \text{...(i)}$$

On the analogy of light radiation, the expression for the wavelength λ can be easily derived. The momentum of the photon is,

$$p = \frac{h\nu}{c} = \frac{h}{\lambda} \qquad \text{...(ii)}$$

λ or $\dfrac{h}{p}$ is called the de-Broglie wavelength.

If m is the mass and v the velocity of the material particle, then

$$p = mv$$

Hence, $$\lambda = \frac{h}{mv} \qquad \text{...(iii)}$$

If E is the kinetic energy of the material particle, then

$$E = \frac{1}{2}mv^2 \Rightarrow E = \frac{p^2}{2m} \text{ [from equation (iii)]} \qquad \text{...(iv)}$$

$\therefore$ The de-Broglie wavelength is given by

$$\lambda = \frac{h}{\sqrt{2mE}} \qquad \text{[from equation (iv)]}$$

and for electron, $$\lambda = \frac{12.27}{\sqrt{V}} \text{Å}.$$

Q. 2. Derive Einstein's photoelectric equation $\frac{1}{2}mv^2 = h\nu - h\nu_0$.

Ans. **Einstein's explanation of photoelectric effect :** Einstein explained photoelectric effect on the basis of quantum theory. The main points are :

(i) Light is propagated in the form of bundles of energy. Each bundle of energy is called a quantum or photon and has energy $h\nu$, where, h = Planck's constant and ν = frequency of light.

(ii) The photoelectric effect is due to collision of a photon of incident light and a bound electron of the metallic cathode.

(iii) When a photon of incident light falls on the metallic surface, it is completely absorbed. Before being absorbed it penetrates through a distance of nearly 10^{-8} m (or 100 Å). The absorbed photon transfers its whole energy to a single electron. The energy of photon goes in two parts : a part of energy is used in releasing the electron from the metal surface (*i.e.*, in overcoming work function) and the remaining part appears in the form of kinetic energy of the same electron.

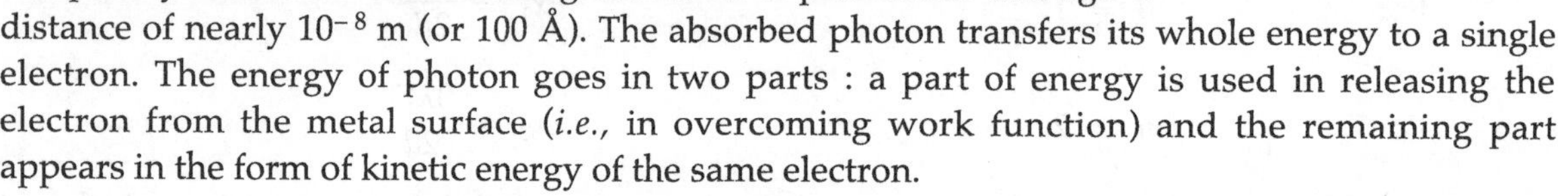

If ν be the frequency of incident light, the energy of photon = $h\nu$. If W be the work function of metal and E_K the maximum kinetic energy of photoelectron, then according to Einstein's explanation.

$$h\nu = W + E_K$$

or $$E_K = h\nu - W \qquad \text{...(i)}$$

This is called Einstein's photoelectric equation.

If ν_0 be the threshold frequency, then if frequency of incident light is less then ν_0, no electron will be emitted and if the frequency of incident light be ν_0, then $E_K = 0$; so from equation (i)

$$0 = h\nu_0 - Q \text{ or } W = h\nu_0$$

If λ_0 be the threshold wavelength, then, $\nu_0 = \frac{c}{\lambda_0}$,

where c is the speed of light in vacuum

$$\therefore \quad \text{Work function, } W = h\nu_0 = \frac{hc}{\lambda_0} \qquad \text{...(ii)}$$

Substituting this value in equation (i), we get

$$E_K = h\nu - h\nu_0 \qquad \text{...(iii)}$$

This is another form of Einstein's photoelectric equation.

Q. 3. Write Einstein's photoelectric equation and mention which important features in photoelectric effect can be explained with the help of this equation.

The maximum kinetic energy of the photoelectrons gets doubled when the wavelength of light incident on the surface changes from λ_1 to λ_2. Derive the expressions for the threshold wavelength λ_0 and work function for the metal surface.★

Ans. Einstein's photoelectric equation is given by

$$K_{max} = \frac{1}{2}mv^2_{max}$$

$$K_{max} = h\nu - \phi_0$$

★ are board based questions from previous year

or $$h\nu = h\nu_0 + \frac{1}{2} mv^2_{max}$$

where

K_{max} = Maximum kinetic energy of the photoelectron

v_{max} = Maximum velocity of the emitted photoelectron

m = Mass of the photoelectron

ν = Frequency of the light radiation

ϕ_0 = Work function

h = Planck's constant

If ν_0 is the threshold frequency, then the work function can be written as

$$\phi_0 = h\nu_0$$

$$\Rightarrow \quad K_{max} = \frac{1}{2} mv^2_{max} = h\nu - h\nu_0 = h(\nu - \nu_0)$$

The above equations explains the following results :

(i) If $v < v_0$, then the maximum kinetic energy is negative, which is impossible. Hence, photoelectric emission does not take place for the incident radiation below the threshold frequency. Thus, the photoelectric emission can take place if $\nu > \nu_0$.

(ii) The maximum kinetic energy of emitted photoelectrons is directly proportional to the frequency of the incident radiation. This means that maximum kinetic energy of photoelectron depends only on the frequency of incident light.

According to the photoelectric equation,

$$K_{max} = \frac{1}{2} mv^2_{max} = h\nu - \phi_0$$

$$K_{max} = \frac{hc}{\lambda_1} - \phi_0 \quad \text{...(i)}$$

Let the maximum kinetic energy for the wavelength λ_2 be K_2.

$$K_2 = \frac{hc}{\lambda_2} - \phi_0 \quad \text{...(ii)}$$

$$K_2 = 2K_1 \quad \text{(given)}$$

From (i) and (ii), we have

$$\frac{hc}{\lambda_2} - \phi_0 = 2\left(\frac{hc}{\lambda_1} - \phi_0\right)$$

$$\Rightarrow \quad \phi_0 = hc\left(\frac{2}{\lambda_1} - \frac{1}{\lambda_2}\right)$$

$$\Rightarrow \quad h\nu_0 = hc\left(\frac{2}{\lambda_1} - \frac{1}{\lambda_2}\right)$$

$$\Rightarrow \quad \frac{c}{\lambda_0} = c\left(\frac{2}{\lambda_1} - \frac{1}{\lambda_2}\right)$$

$$\Rightarrow \quad \frac{1}{\lambda_0} = \left(\frac{2}{\lambda_1} - \frac{1}{\lambda_2}\right)$$

$$\Rightarrow \quad \lambda_0 = \frac{\lambda_1\lambda_2}{2\lambda_2 - \lambda_1} \quad \text{...(iii)}$$

Work function is the energy required to eject a photoelectron from the metal.

$$W = \frac{hc}{\lambda_0}$$

$$W = \frac{hc\,(2\lambda_2 - \lambda_1)}{\lambda_1\lambda_2} \quad \text{(from (iii))}$$

Chapter 12. Atoms

Q. 1. Show that the radius of the orbit in hydrogen atom varies as n^2. Where, n is the principal quantum number of the atom.★

Ans. According to the Bohr's theory of hydrogen atom, the angular momentum of a revolving electron is given by

$$mvr = \frac{nh}{2\pi} \quad \text{...(i)}$$

Here, m = Mass of the electron

v = Velocity of the electron

r = Radius of the orbit

h = Planck's constant

n = Principal quantum number of the atom

If an electron of mass m and velocity v is moving in a circular orbit of radius r, then the centripetal force required is given by

$$F = \frac{mv^2}{r}$$

Also, if the charge on the nucleus is Ze, then the force of electrostatic attraction between the nucleus and the electron will provide the necessary centripetal force.

$$F = \frac{1}{4\pi\varepsilon_0}\frac{(Ze)(e)}{r^2} = \frac{KZe^2}{r^2}$$

where $$K = \frac{1}{4\pi\varepsilon_0}$$

$$\therefore \quad \frac{mv^2}{r} = \frac{KZe^2}{r^2} \quad \text{...(ii)}$$

From (i), we get

$$v = \frac{nh}{2\pi mr}$$

Putting this value in (ii), we get

$$\frac{m}{r}\frac{n^2h^2}{4\pi^2m^2r^2} = \frac{KZe^2}{r^2}$$

$$\Rightarrow \quad r = \frac{n^2h^2}{4\pi^2mKZe^2}$$

$$\Rightarrow \quad r \propto n^2$$

Q. 2. Using Rutherford model of the atom, derive the expression for the total energy of the electron in hydrogen atom. What is the significance of total negative energy possessed by the electron ?★

Ans. In a hydrogen atom, an electron having charge $-e$ revolves around the nucleus having charge $+e$ in a circular orbit of radius r.

Let,

F_c = Centripetal force required by the electron to move in circular orbit of radius r.

F_e = Electrostatic force of attraction between revolving electron and nucleus.

The electrostatic force of attraction (F_e) provides the necessary centripetal force,

$$F_c = F_e$$

$$\frac{mv^2}{r} = \frac{(e)(e)}{4\pi\varepsilon_0 r^2} \quad \text{...(i)}$$

K.E. of electron in the orbit,

★ are board based questions from previous year

$$\text{K.E.} = \frac{1}{2}mv^2$$

From equation (i),
$$\text{K.E.} = \frac{e^2}{8\pi\varepsilon_0 r}$$

Potential energy of electron in orbit,

$$\text{P.E.} = \frac{(e)(-e)}{4\pi\varepsilon_0 r} = \frac{-e^2}{4\pi\varepsilon_0 r}$$

∴ Total energy of electron in hydrogen atom

$$E = \text{K.E.} + \text{P.E.} = \frac{e^2}{8\pi\varepsilon_0 r} - \frac{e^2}{4\pi\varepsilon_0 r}$$

$$E = -\frac{e^2}{8\pi\varepsilon_0 r}$$

Here, negative sign indicates that the revolving electron is bound to the positive nucleus.

Q. 3. Using Bohr's postulates of the atomic model, derive the expression for radius of nth electron orbit. Hence obtain the expression for Bohr's radius.★

Ans. Suppose m be the mass of an electron and v_n be its speed in n^{th} orbit of radius r_n. From Rutherford model, the centripetal force for revolution is produced by electrostatic attraction between electron and nucleus.

$$\frac{mv_n^2}{r_n} = \frac{1}{4\pi\varepsilon_0}\frac{Ze^2}{r_n^2} \quad \text{...(i)}$$

From Bohr's postulate for quantisation of angular momentum of nth orbit.

$$mv_n r_n = \frac{nh}{2\pi} \Rightarrow v_n = \frac{nh}{2\pi m r_n}$$

Substituting this value in equation (i), we get

$$\frac{m}{r_n}\left[\frac{nh}{2\pi m r_n}\right]^2 = \frac{1}{4\pi\varepsilon_0}\frac{Ze^2}{r_n^2}$$

or
$$r_n = \frac{\varepsilon_0 n^2 h^2}{\pi m Ze^2}$$

For Bohr's radius $n = 1$,

$$r_1 = \frac{\varepsilon_0 h^2}{\pi m Ze^2}$$

This is the expression for Bohr's radius.

Q. 4. What do you mean by the distance of the closest approach of an α-particle in Rutherford experiment ? Derive its expression.

Ans. An α-particle directed towards the centre of the nucleus will move close upto a distance r_0 where its kinetic energy will appear as potential energy. This distance r_0 is called 'distance of closest approach'.

Kinetic energy of an α-particle is given by,

$$\text{K.E.} = \frac{1}{2}mv^2 \quad \text{...(i)}$$

The electrostatic potential energy of the α-particle of charge $2e$ at a distance r_0 is given by,

$$\text{P.E.} = \frac{1}{4\pi\varepsilon_0}\cdot\frac{(Ze)\times 2e}{r_0}$$

At distance r_0 closest approach, both P.E. and K.E. are balanced. Thus,

★ are board based questions from previous year

$$\text{P.E.} = \text{K.E.}$$

$$\frac{1}{4\pi\varepsilon_0}\frac{Ze \times 2e}{r_0} = \frac{1}{2}mv^2$$

$$r_0 = \frac{1}{4\pi\varepsilon_0}\frac{4Ze^2}{mv^2}$$

$$r_0 = \frac{Ze^2}{\pi\varepsilon_0\, mv^2}$$

Q. 5. Obtain an expression for the energy in n^{th} stationary orbit using Bohr's postulates.

Ans. **Energy of the electron in stationary orbits :** The total energy E of an electron in an orbit is the sum of kinetic energy and potential energy.

The kinetic energy of electron is :

$$\text{K.E.} = \frac{1}{2}mv^2$$

$$= \frac{Ze^2}{8\pi\varepsilon_0 r}$$

The potential energy of the electron in an orbit of radius r due to the electrostatic attraction by the nucleus is given by :

$$U = \frac{1}{4\pi\varepsilon_0}\frac{Ze(-e)}{r}$$

$$= -\frac{1}{4\pi\varepsilon_0}\cdot\frac{Ze^2}{r}$$

The total energy of the electron is :

$$E_n = \text{K.E.} + U$$

$$= \frac{Ze^2}{8\pi\varepsilon_0 r} - \frac{Ze^2}{4\pi\varepsilon_0}$$

$$= -\frac{Ze^2}{8\pi\varepsilon_0 r}$$

But $$r = n^2\frac{h^2\varepsilon_0}{\pi m Z e^2}$$

$\therefore$ $$E_n = -\frac{mZ^2e^4}{8\pi\varepsilon_0^2 h^2}\left(\frac{1}{n^2}\right)$$

Substituting the values of the constants, we get

$$E_n = \frac{-13{\cdot}6\,Z^2}{n^2}\,\text{eV}$$

This is the expression for the energy of the electron in the nth orbit.

Q. 6. Give the expression for the velocity of electron in the first orbit of hydrogen atom and for the radius in the ground state of hydrogen atom. Hence, derive the expression for the magnetic field ($\vec{B}$) at the centre of the nucleus due to the circular motion of the electron. [Give your answer in terms of fundamental constants].

Ans. Consider an element of atomic number Z. The charge on the nucleus is Ze. Let r be the radius of hydrogen atom and v be the velocity of electron in hydrogen orbit.

As we know for the circular motion of electron in n^{th} orbit, the conditions are :

$$\frac{mv^2}{r} = \frac{1}{4\pi\varepsilon_0}\frac{Ze \times e}{r^2} \quad \text{...(i)}$$

According to Bohr's first postulate, angular momentum of the electron,

$$mvr = \frac{nh}{2\pi}$$

or $$v = \frac{nh}{2\pi mr} \quad \text{...(ii)}$$

Putting equation (ii) in equation (i), we get

$$\frac{m}{r}\left(\frac{nh}{2\pi mr}\right)^2 = \frac{1}{4\pi\varepsilon_0}\cdot\frac{e \times e}{r^2} \quad \text{(For hydrogen atom Z = 1)}$$

or $$\frac{mn^2h^2}{4\pi^2m^2r^2} = \frac{1}{4\pi\varepsilon_0}\cdot\frac{e^2}{r}$$

or $$r = \frac{n^2h^2\varepsilon_0}{\pi me^2}$$

For ground state of hydrogen atom, $n = 1$

$$r = \frac{h^2\varepsilon_0}{\pi me^2} \quad \text{...(iii)}$$

As, $$v = \frac{nh}{2\pi mr} \quad \text{[From equation (ii)]}$$

$$v = \frac{nh}{2\pi m\left(\frac{n^2h^2\varepsilon_0}{\pi me^2}\right)} = \frac{e^2}{2\varepsilon_0\, nh}$$

For 1st orbit of Hydrogen atom, $n = 1$

$$v = \frac{e^2}{2\varepsilon_0 h} \quad \text{...(iv)}$$

Magnetic field at the centre of the nucleus due to circular motion of the electron,

$$B = \frac{\mu_0 ni}{2r}$$

$$B = \frac{\mu_0 n}{2r}\left(\frac{e}{T}\right) \quad \left[\because i = \frac{\text{Charge}}{\text{Time}} = \frac{e}{T}\right]$$

$$B = \frac{\mu_0 n}{2r}\,\frac{e \times v}{(2\pi r)} \quad \left[\because T = \frac{2\pi r}{v}\right]$$

$$B = \frac{\mu_0 nev}{4\pi r^2} = \frac{\mu_0 ev}{4\pi r^2} \quad [\because n = 1]$$

From equation (iii) and equation (iv),

$$B = \frac{\mu_0 e}{4\pi}\left(\frac{\frac{e^2}{2\varepsilon_0 h}}{\frac{h^4\,\varepsilon_0^2}{\pi^2m^2e^4}}\right)$$

$$= \frac{\mu_0 e}{4\pi}\times\frac{e^6\pi^2m^2}{2h^5\varepsilon_0^3}$$

$$= \frac{\mu_0\pi e^7 m^2}{8\varepsilon_0^3h^5}\,\text{Wb/m}^2.$$

Q. 7. Obtain an expression for the frequency of radiation emitted when a hydrogen atom de-excites from level n to level $(n - 1)$. For large n, show that this frequency equals the classical frequency of revolution of the electron in the orbit. **[NCERT]**

Ans. It is given that a hydrogen atom de-excites from an upper level (n) to a lower level ($n - 1$). We have the relation for energy (E_1) of radiation at level n as :

$$E_1 = h\nu_1$$

$$= \frac{hme^4}{(4\pi)^3\,\varepsilon_0^2\left(\frac{h}{2\pi}\right)^3}\times\left(\frac{1}{n^2}\right) \quad \text{...(i)}$$

Here, ν_1 = Frequency of radiation at level n

h = Planck's constant

m = Mass of hydrogen atom

e = Charge on an electron

ε_0 = Permittivity of free space

Now, the relation for energy (E_2) of radiation at level $(n-1)$ is given by

$$E_2 = h\nu_2$$

$$= \frac{hme^4}{(4\pi)^3 \varepsilon_0^2 \left(\frac{h}{2\pi}\right)^3} \times \frac{1}{(n-1)^2} \quad \text{...(ii)}$$

Here, ν_2 = Frequency of radiation at level $(n-1)$

Energy (E) released as a result of de-excitation :

$$E = E_2 - E_1$$

$$h\nu = E_2 - E_1 \quad \text{...(iii)}$$

Here, ν = Frequency of radiation emitted

Putting values from equations (i) and (ii) in equation (iii), we get :

$$\nu = \frac{me^4}{(4\pi)^3 \varepsilon_0^2 \left(\frac{h}{2\pi}\right)^3} \left[\frac{1}{(n-1)^2} - \frac{1}{n^2}\right]$$

$$= \frac{me^4 (2n-1)}{(4\pi)^3 \varepsilon_0^2 \left(\frac{h}{2\pi}\right)^3 n^2 (n-1)^2}$$

For large n, we can write $(2n-1) \simeq 2n$ and $(n-1) \approx n$

$$\therefore \quad \nu = \frac{me^4}{32\pi^3 \varepsilon_0^2 \left(\frac{h}{2\pi}\right)^3 n^3} \quad \text{...(iv)}$$

Classical relation of frequency of revolution of an electron is given by

$$\nu_e = \frac{v}{2\pi r} \quad \text{...(v)}$$

Here, velocity of the electron in the nth orbit is given by

$$v = \frac{e^2}{4\pi \varepsilon_0 \left(\frac{h}{2\pi}\right) n} \quad \text{...(vi)}$$

And, radius of the nth orbit is given by

$$r = \frac{4\pi \varepsilon_0 \left(\frac{h}{2\pi}\right)^2}{me^2} n^2 \quad \text{...(vii)}$$

From equations (vi), (vii) and equation (v), we get :

$$\nu_e = \frac{me^4}{32\pi^3 \varepsilon_0^2 \left(\frac{h}{2\pi}\right)^3 n^3} \quad \text{...(viii)}$$

Thus, the frequency of radiation emitted by the hydrogen atom is equal to its classical orbital frequency.

Chapter 13. Nuclei

Q. 1. Deduce the expression, $N = N_0 e^{-\lambda t}$, for the law of radioactive decay.★

Ans. Suppose initially the number of atoms in radioactive element is N_0 and N the number of atoms after time t. After time t, let dN be the number of atoms which disintegrate in a short interval dt, then rate of disintegration will be, $\frac{dN}{dt}$, this is also called the activity of the element.

According to Rutherford-Soddy law,

$$\frac{dN}{dt} \propto N$$

or

$$\frac{dN}{dt} = -\lambda N \qquad \text{...(i)}$$

where λ is a constant, called decay constant or disintegration constant of the element. Its unit is s^{-1}. Negative sign shows that the rate of disintegration decreases with increase of time. For a given element λ is a constant and is different for different elements. Equation (i) may be rewritten as

$$\frac{dN}{N} = -\lambda dt$$

Integrating

$$\log_e N = -\lambda t + C \qquad \text{...(ii)}$$

where C is a constant of integration.

At $t = 0, N = N_0$

$\therefore$ $\log_e N_0 = 0 + C \Rightarrow C = \log_e N_0$

$\therefore$ Equation (ii) gives $\log_e N = -\lambda t + \log_e N_0$

or $\log_e N - \log_e N_0 = -\lambda t$

or $\log_e \frac{N}{N_0} = -\lambda t$

$$\frac{N}{N_0} = e^{-\lambda t}$$

$$N = N_0 e^{-\lambda t}$$

□□

★ are board based questions from previous year

Question Set 10

Numericals

Chapter 1. Electric Charges and Fields

Q. 1. What is the force between two point charges of 1 C each separated by 1 m in air ?

Ans. By using Coulomb's law,

$$F = \frac{1}{4\pi\varepsilon_0} \cdot \frac{q_1 q_2}{r^2}$$

$$= \frac{(9 \times 10^9) \times 1 \times 1}{(1)^2}$$

$$= 9 \times 10^9 \text{ N}$$

Q. 2. A point charge of 2.0 μC is at the centre of a cubic Gaussian surface 9.0 cm on edge. What is the net electric flux through the surface ? **[NCERT]**

Ans. Net electric flux (ϕ_{Net} through the cubic surface is given by

$$\phi_{Net} = \frac{q}{\varepsilon_0}$$

Where, ε_0 = Permittivity of free space = 8.854×10^{-12} C^2 N^{-1} m^{-2}.

q = Net charge contained inside the cube = 2.0 μC = 2×10^{-6} C

$$\therefore \quad \phi_{Net} = \frac{2 \times 10^{-6}}{8.854 \times 10^{-12}}$$

$$= 2.26 \times 10^5 \text{ N m}^2/\text{C}$$

The net electric flux through the surface is 2.26×10^5 N m^2/C.

Q. 3. How much positive and negative charge is there in a cup of water. **[NCERT]**

Ans. Consider that the mass of one cup of water is 250 gm. The molecular mass of water is 18 gm.

1 mole of water = 18 gm = 6.023×10^{23} molecules

Therefore, number of molecules in one cup of water

$$= \frac{250}{18} \times 6.023 \times 10^{23}$$

Each molecule of water contains two hydrogen atoms and one oxygen atom, *i.e.*, 10 electrons and 10 protons. Hence, the total positive and total negative charge has the same magnitude which is given by

$$= \frac{250}{18} \times 6.023 \times 10^{23} \times 10 \times 1.6 \times 10^{-19} \text{ C}$$

$$= 1.34 \times 10^7 \text{ C}$$

But net charge will always be zero.

Q. 4. An infinite line charge produces a field of 9×10^4 N/C at a distance of 2 cm. Calculate the linear charge density. **[NCERT]**

Ans. Electric field produced by the infinite line charges at a distance d having linear charge density λ is given by the relation,

$$E = \frac{\lambda}{2\pi\varepsilon_0 d}$$

$$\Rightarrow \quad \lambda = 2\pi\varepsilon_0 d \times E$$

where, d = 2 cm = 0.02 m

E = 9×10^4 N/C

ε_0 = Permittivity of free space and

$\frac{1}{4\pi\varepsilon_0} = 9 \times 10^9$ C^2 N^{-1} m^{-2}

$$\therefore \quad \lambda = \frac{0.02 \times 9 \times 10^4}{2 \times 9 \times 10^9}$$

$$= 10 \ \mu\text{C/m}$$

Therefore, the linear charge density is 10 μC/m.

Q. 5. A system has two charges $q_A = 2.5 \times 10^{-7}$ C and $q_B = -2.5 \times 10^{-7}$ C located at points A : (0, 0, – 15 cm) and B : (0, 0, + 15 cm), respectively. What are the total charge and electric dipole moment of the system ? **[NCERT]**

Ans. Both the charges can be drawn in a coordinate frame of reference as shown in the figure.

At A, amount of charge, $q_A = 2.5 \times 10^{-7}$ C

At B, amount of charge $q_B = -2.5 \times 10^{-7}$ C

Total charge of the system $q = q_A + q_B$

$$= 2.5 \times 10^{-7} \text{ C} - 2.5 \times 10^{-7} \text{ C} = 0$$

Distance between two charges at point A and B

$$d = 15 + 15 = 30 \text{ cm} = 0.3 \text{ m}$$

Electric dipole moment of the system is given by,

$$p = q_A \times d = q_B \times d = 2.5 \times 10^{-7} \times 0.3 \text{ m}$$

$$= 7.5 \times 10^{-8} \text{ Cm along positive Z-axis.}$$

Therefore, the electic dipole moment of the system is 7.5×10^{-8} Cm along positive Z-axis.

Q. 6. An electric dipole of length 4 cm, when placed with its axis making an angle of 60° with a uniform electric field, experiences a torque of $4\sqrt{3}$ Nm. Calculate the potential energy of the dipole, if it has charge ± 8 nC.★

Ans. As

$$\tau = pE \sin\theta$$

$$\therefore \quad 4\sqrt{3} = pE \sin 60°$$

$$\Rightarrow \quad pE \times \frac{\sqrt{3}}{2} = 4\sqrt{3}$$

$$\Rightarrow \quad pE = 8$$

Potential energy of dipole,

$$U = -pE\cos\theta$$

$$= -pE\cos 60°$$

$$= -8 \times \frac{1}{2}$$

$$= -4\text{J}.$$

Q. 7. What is the force between two small charged spheres having charges of 2×10^{-7} C and 3×10^{-7} C placed 30 cm apart in air ? **[NCERT]**

Ans. Charge on the sphere are $q_1 = 2 \times 10^{-7}$ C and $q_2 = 3 \times 10^{-7}$C

Distance between the sphere, $r = 30$ cm $= 0.3$ m

Thus, the electrostatic force between the spheres is given by the relation

$$F = \frac{1}{4\pi\varepsilon_0}\frac{q_1 q_2}{r^2}$$

Where, ε_0 = Permittivity of free space and

$$\frac{1}{4\pi\varepsilon_0} = 9 \times 10^9 \text{ Nm}^2/\text{C}^2$$

$$\therefore \quad \text{Force, } F = \frac{9 \times 10^9 \times 2 \times 10^{-7} \times 3 \times 10^{-7}}{(0.3)^2}$$

$$= 6 \times 10^{-3}\text{N}$$

Hence, force between the two small charged spheres is 6×10^{-3} N. The charges are of same nature. Hence, force between them will be repulsive.

★ are board based questions from previous year

Q. 8. Electric dipole moment of $CuSO_4$ molecule is 3.2×10^{-32} cm. Find the separation between copper and sulphate ions.

Ans. Electric dipole moment is given by $p = ql$.

$$p = 3.2 \times 10^{-32} \text{ cm.}$$

The ions of $CuSO_4$ are Cu^{2+} and SO_4^{2-} so, q would be two times of the electron.

$$\therefore \quad q = 2e = 2 \times 1.6 \times 10^{-19} \text{ C} = 3.2 \times 10^{-19} \text{ C}$$

So, separation between Cu and SO_4 ions can be calculated by using

$$l = \frac{p}{q} = \frac{3.2 \times 10^{-32}}{3.2 \times 10^{-19}} = 10^{-13} \text{ m}$$

Q. 9. An electric dipole with dipole moment 4×10^{-9} C m is aligned at 30°C with the direction of a uniform electric field of magnitude 5×10^4 N/C. Calculate the magnitude of the torque acting on the dipole. **[NCERT]**

Ans. Electric dipole moment, $p = 4 \times 10^{-9}$ C m

Angle made by p with a uniform electric field, $\theta = 30°$

Electric field, $E = 5 \times 10^4$ N/C

Torque acting on the dipole is given by

$$\tau = pE \sin \theta = 4 \times 10^{-9} \times 5 \times 10^4 \times \sin 30° = 20 \times 10^{-5} \times 0.5 = 10^{-4} \text{ Nm}$$

The magnitude of the torque acting on the dipole is 10^{-4} N m.

Q. 10. A conducting sphere of radius 10 cm has an unknown charge. If the electric field 20 cm from the centre of the sphere is 1.5×10^3 N/C and points radially inward, what is the net charge on the sphere ? **[NCERT]**

Ans. Electric field intensity (E) at a distance (d) from the centre of a sphere containing net charge q is given by the relation,

$$E = \frac{1}{4\pi\varepsilon_0} \cdot \frac{q}{d^2}$$

Where, Net electric field (E) $= 1.5 \times 10^3$ N/C

d = Distance from the centre = 20 cm = 0.2 m

ε_0 = Permittivity of free space and $\frac{1}{4\pi\varepsilon_0} = 9 \times 10^9$ Nm^2/C^2

$$\therefore \quad q = E \cdot 4\pi\varepsilon_0 d^2 = \frac{1.5 \times 10^3}{9 \times 10^9} \times 0.2 \times 0.2 = 6.67 \times 10^{-9} \text{ C.}$$

Therefore, the net charge on the sphere is 6.67 nC.

Q. 11. A point charge + 10 μC is a distance 5 cm directly above the centre of a square of side 10 cm, as shown in Figure. What is the magnitude of the electric flux through the square ? (**Hint :** Think of the square as one face of a cube with edge 10 cm.) **[NCERT]**

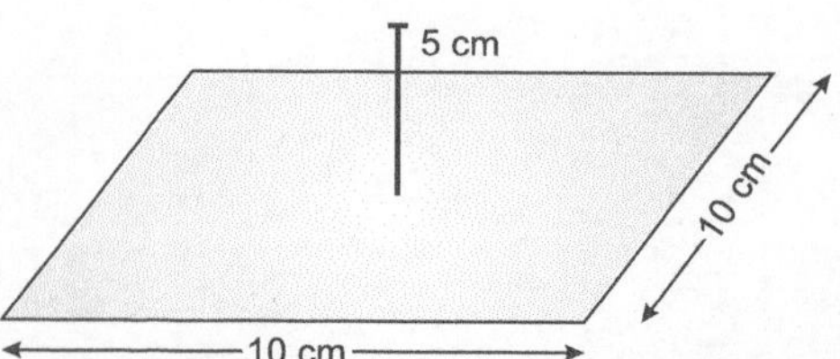

Ans. The square can be considered as one face of a cube of edge 10 cm with a centre where charge q is placed. According to Gauss's theorem for a cube, total electric flux is through all its six faces.

$$\phi_{Total} = \frac{q}{\varepsilon_0}$$

Hence, electric flux through one face of the cube *i.e.*, through the square is

$$\phi = \frac{\phi_{Total}}{6} = \frac{q}{6 \cdot \varepsilon_0}$$

where, ε_0 = Permittivity of free space $= 8.854 \times 10^{-12}\ C^2\ N^{-1}\ m^{-2}$

$$q = 10\ \mu C = 10 \times 10^{-6}\ C$$

$$\therefore \quad \phi = \frac{1}{6} \times \frac{10 \times 10^{-6}}{8.854 \times 10^{-12}}$$

$$= 1.88 \times 10^5\ N\ m^2\ C^{-1}$$

Therefore, electric flux through the square is $1.88 \times 10^5\ N\ m^2\ C^{-1}$.

Q. 12. A polythene piece rubbed with wool is found to have a negative charge of 3×10^{-7} C.

(i) Estimate the number of electrons transferred (from which to which ?)

(ii) Is there a transfer of mass from wool to polythene ? **[NCERT]**

Ans. (i) When polythene is rubbed against wool, a number of electrons get transferred from wool to polythene. Hence, wool becomes positively charged and polythene becomes negatively charged.

Amount of charge on the polythene piece, $q = -3 \times 10^{-7}$ C

Amount of charge on an electron, $e = -1.6 \times 10^{-19}$ C

Let the number of electrons transferred from wool to polythene = n

n can be calculated using the relation,

$$q = ne$$

$$n = \frac{q}{e} = \frac{-3 \times 10^{-7}}{-1.6 \times 10^{-19}}$$

$$= 1.87 \times 10^{12}$$

Therefore, the number of electrons transferred from wool to polythene is 1.87×10^{12}.

(ii) Yes. There is a transfer of mass taking place, because an electron has mass,

$$m_e = 9.1 \times 10^{-31}\ kg.$$

Total mass transferred to polythene from wool,

$$m = m_e \times n$$

$$= 9.1 \times 10^{-31} \times 1.85 \times 10^{12}$$

$$= 1.706 \times 10^{-18}\ kg$$

A negligible amount of mass is transferred from wool to polythene.

Q. 13. Given a uniform electric field $\vec{E} = 2 \times 10^3\ \hat{i}$ N/C. Find the flux of this field through a square of side 20 cm, whose plane is parallel to the Y-Z plane. What would be the flux through the same square, if the plane makes an angle of 30° with the X-axis ?★

Ans. When the plane is parallel to the Y-Z plane

$$\phi = \vec{E}.\vec{A}.$$

Here, $\vec{E} = 2 \times 10^3 \hat{i}$

$$A = (20\ cm)^2 \hat{i} = 0.04\ m^2\ \hat{i}$$

$$\therefore \quad \phi = (2 \times 10^3\ \hat{i}) \cdot (0.04\ i)$$

$$\Rightarrow \quad \phi = 80 \text{ weber or } Nm^2C^{-1}.$$

When the plane makes a 30° angle with the X-axis, the area vector makes a 60° angle with the X-axis.

$$\phi = \vec{E} \cdot \vec{A}$$

$$\Rightarrow \quad \phi = EA \cos \theta$$

$$\Rightarrow \quad \phi = 2 \times 10^3 \times 0.04 \times \cos 60°$$

★ are board based questions from previous year

$\Rightarrow \qquad \phi = 2 \times 10^3 \times 0.04 \times \frac{1}{2}$

$\Rightarrow \qquad \phi = 40$ weber or Nm^2C^{-1}.

Q. 14. Careful measurement of the electric field at the surface of a black box indicates that the net outward flux through the surfaces of the box is $8.0 \times 10^3\ Nm^2/C$.

(i) What is the net charge inside the box ?

(ii) If the net outward flux through the surface of the box were zero, could you conclude that there were no charges inside the box ? Why or Why not ? **[NCERT]**

Ans. (i) Net outward flux through the surface of the box,

$$\phi = 8.0 \times 10^3\ N\,m^2/C$$

For a body containing net charge q, flux is given by the relation,

$$\phi = \frac{q}{\varepsilon_0}$$

Where ε_0 = Permittivity of free space

$= 8.854 \times 10^{-12}\ C^2N^{-1}m^{-2}$

$$q = \varepsilon_0\phi$$
$$= 8.854 \times 10^{-12} \times 8.0 \times 10^3\ C$$
$$= 7.08 \times 10^{-8}\ C = 0.07\ \mu C$$

Therefore, the net charge inside the box is 0.07 μC.

(ii) No, Net flux piercing out through a body depends on the net charge contained in the body. If net flux is zero, then it can be inferred that net charge inside the body is zero. The body may have equal amount of positive and negative charges.

Q. 15. A uniformly charged conducting sphere of 2.4 m diameter has a surface charge density of $80.0\ \mu C/m^2$.

(i) Find the charge on the sphere.

(ii) What is the total electric flux leaving the surface of the sphere ? **[NCERT]**

Ans. (i) Radius of the sphere, $r = 1.2$ m.

Surface charge density, $\sigma = 80.0\ \mu C/m^2 = 80 \times 10^{-6}\ C/m^2$

Total charge on the surface of the sphere.

$$Q = \text{Charge density} \times \text{Surface area}$$
$$= \sigma \times 4\pi r^2$$
$$= 80 \times 10^{-6} \times 4 \times 3.14 \times (1.2)^2$$
$$= 1.447 \times 10^{-3}\ C$$

Therefore, the charge on the sphere is 1.447×10^{-3} C.

(ii) Total electric flux (ϕ_{Total}) leaving out the surface of a sphere containing net charge Q is given by the relation.

$$\phi_{Total} = \frac{Q}{\varepsilon_0}$$

Where, ε_0 = Permittivity of free space

$= 8.854 \times 10^{-12}\ C^2\ N^{-1}\ m^{-2}$

$Q = 1.447 \times 10^{-3}$ C

$$\therefore \qquad \phi_{Total} = \frac{1.447 \times 10^{-3}}{8.854 \times 10^{-12}}$$
$$= 1.63 \times 10^8\ N\,C^{-1}\,m^2$$

Therefore, the total electric flux leaving the surface of the sphere is $1.63 \times 10^8\ N\,C^{-1}\,m^2$.

Q. 16. An oil drop of 12 excess electrons is held stationary under a constant electric field of 2.55×10^4 N/C in Millikan's oil drop experiment. The density of the oil is $1.26\ g\ cm^{-3}$. Estimate the radius of the drop. ($g = 9.81\ ms^{-2}$, $e = 1.60 \times 10^{-19}$ C)

Ans. Excess electrons on oil drop (n) = 12,

Electric field intensity (E) = 2.55×10^4 N/C.

Density of oil = 1.26 g/cm^3.

In free body diagram,

F = qE

mg

$$F = qE = mg$$

Here, q = Net charge on the oil drop = ne

m = Mass of the oil drop = Volume of the drop × Density of oil = $\frac{4}{3}\pi r^3 \times \rho$

$$r = \left[\frac{3\,Ene}{4\pi\rho g}\right]^{\frac{1}{3}}$$

$$= \left[\frac{3 \times 2.55 \times 10^4 \times 12 \times 1.6 \times 10^{-19}}{4 \times 3.14 \times 1.26 \times 10^3 \times 9.81}\right]^{\frac{1}{3}}$$

$$= [946.09 \times 10^{-21}]^{\frac{1}{3}}$$

$$= 9.82 \times 10^{-4} \text{ mm}$$

Q. 17. A point charge causes an electric flux of -1.0×10^3 Nm^2/C to pass through a spherical Gaussian surface of 10.0 cm radius centred on the charge.

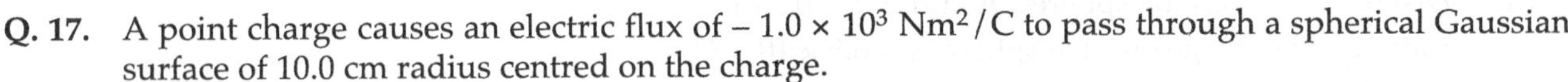

(i) If the radius of the Gaussian surface were doubled, how much flux would pass through the surface ?

(ii) What is the value of the point charge ? **[NCERT]**

Ans. (i) Electric flux, $\phi = -1.0 \times 10^3$ N m^2/C

Radius of the Gaussian surface, r = 10.0 cm.

Electric flux piercing out through a surface depends on the net charge enclosed inside a body. It does not depend on the size of the body. If the radius of the Gaussian surface is doubled, then the flux passing through the surface remains the same *i.e.*, -10^3 N m^2/C.

(ii) Electric flux is given by the relation

$$\phi = \frac{q}{\varepsilon_0}$$

Where, ε_0 = Permittivity of free space = 8.854×10^{-12} C^2 N^{-1} m^{-2}

and q = Net charge enclosed by the spherical surface = $\phi\varepsilon_0$.

$$= -1.0 \times 10^3 \times 8.854 \times 10^{-12}$$

$$= -8.854 \times 10^{-9} \text{ C}$$

$$= -8.854 \text{ nC}$$

Therefore, the value of the point charge is – 8.854 nC.

Q. 18. The electrostatic force on a small sphere of charge 0.4. μC due to another small sphere of charge – 0.8 μC in air is 0.2 N.

(i) What is the distance between the two sphere ?

(ii) What is the force on the second sphere due to the first ? **[NCERT]**

Ans. (i) Electrostatic force on the first sphere,

$$F = 0.2 \text{ N}$$

Charge on this sphere, $q_1 = 0.4\ \mu C = 0.4 \times 10^{-6}$ C

Charge on the second sphere, $q_2 = -0.8\ \mu C = -0.8 \times 10^{-6}$ C

Electrostatic force between the spheres is given by the relation,

$$F = \frac{1}{4\pi\varepsilon_0}\frac{q_1 q_2}{r^2}$$

where, ε_0 = Permittivity of free space

$$\frac{1}{4\pi\varepsilon_0} = 9 \times 10^9 \text{ N m}^2 \text{ C}^{-2}$$

$$\therefore \quad r^2 = \frac{1}{4\pi\varepsilon_0} \frac{q_1 q_2}{F}$$

$$= \frac{9 \times 10^9 \times 0.4 \times 10^{-6} \times 0.8 \times 10^{-6}}{0.2}$$

$$r^2 = 0.0144$$

$$\Rightarrow \quad r = 0.12 \text{ m}$$

The distance between the two spheres is 0.12 m.

(ii) Both the spheres attract each other with the same force. Therefore, the force on the second sphere due to the first is 0.2 N.

Q. 19. Suppose the spheres A and B in previous question have identical sizes. A third sphere of the same size but uncharged is brought in contact with the first, then brought in contact with the second, and finally removed from both. What is the new force of repulsion between A and B ? **[NCERT]**

Ans. Distance between the spheres, A and B, $r = 0.5$ m.
Initially, the charge on each sphere, $q = 6.5 \times 10^{-7}$ C.

When sphere A is touched with an uncharged sphere C, $\frac{q}{2}$ amount of charge from A will transfer to sphere C. Hence, charge on each of the spheres, A and C, is $\frac{q}{2}$.

When sphere C with charge $q/2$ is brought in contact with sphere B with charge q, total charges on the system will divide into two equal halves given as, $\dfrac{\left(\frac{q}{2}\right)+q}{2} = \dfrac{3q}{4}$.

Hence, charge on each of the spheres, C and B, is $\frac{3q}{4}$.

Force of repulsion between sphere A having charge $\frac{q}{2}$ and sphere B having charge $\frac{3q}{4}$ is

$$\frac{1}{4\pi\varepsilon_0} \frac{q_A q_B}{r^2} = \frac{1}{4\pi\varepsilon_0} \frac{q}{2} \times \frac{3q}{4r^2}$$

$$= \frac{9 \times 10^9 \times 3 \times (6.5 \times 10^{-7})^2}{8 \times (0.5)^2}$$

$$= 5.703 \times 10^{-3} \text{ N}$$

Therefore, the force of attraction between the two spheres is 5.703×10^{-3} N.

Q. 20. (i) Two insulated charged copper spheres A and B have their centres separated by a distance of 50 cm. What is the mutual force of electrostatic repulsion if the charge on each is 6.5×10^{-7} C ? The radii of A and B are negligible compared to the distance of separation.

(ii) What is the force of repulsion if each sphere is charged double the above amount, and the distance between them is halved ? **[NCERT]**

Ans. (i) Charge on sphere A $= 6.5 \times 10^{-7}$ C $= q_A$

Charge on sphere B $= 6.5 \times 10^{-7}$ C $= q_B$

Distance between the spheres, $r = 50$ cm $= 0.5$ m

Force of repulsion among the two spheres is given by

$$F = \frac{1}{4\pi\varepsilon_0} \cdot \frac{q_A q_B}{r^2}$$

Where, ε_0 = permittivity of free space and $\frac{1}{4\pi\varepsilon_0} = 9 \times 10^9 \text{ Nm}^2\text{C}^{-2}$

$$\therefore \qquad F = \frac{9 \times 10^9 \times (6.5 \times 10^{-7})^2}{(0.5)^2}$$

$$= 1.52 \times 10^{-2} \text{ N}$$

Therefore, the force between the two spheres is 1.52×10^{-2} N.

(ii) After doubling the charge,

Charge on sphere A, $q_A = 1.3 \times 10^{-6}$ C

Charge on sphere B, $q_B = 1.3 \times 10^{-6}$ C

The distance between the spheres is halved

$$\therefore \qquad r = \frac{0.5}{2} = 0.25 \text{ m}$$

Force of repulsion between the two spheres.

$$F = \frac{1}{4\pi\varepsilon_0} \cdot \frac{q_A q_B}{r^2}$$

$$= \frac{9 \times 10^9 \times 1.3 \times 10^{-6} \times 1.3 \times 10^{-6}}{(0.25)^2}$$

$$= 16 \times 1.52 \times 10^{-2}$$

$$= 0.243 \text{ N}$$

Therefore, the force between the two spheres is 0.243 N.

Q. 21. The electric field components in the given figure are $E_x = \alpha x^{1/2}$, $E_y = E_z = 0$, in which $\alpha = 800 \text{ N/Cm}^{1/2}$. Calculate :

(i) The flux through the cube

(ii) The charge within the cube. (Assume that $a = 0.1$ m.) **[NCERT]**

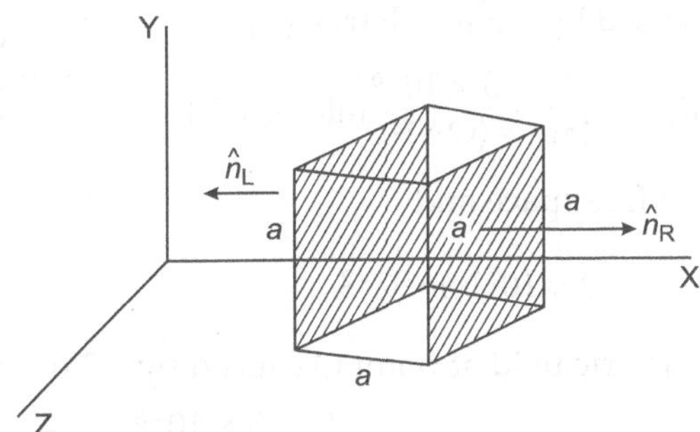

Ans. (i) Since the electric field x-component $(E_\alpha = \alpha x^{1/2})$ only, for faces perpendicular to X-direction, the angle between E and ΔS is $\pm\frac{\pi}{2}$. Therefore the flux $\phi = E.\Delta S$ is separately zero for each face of the cube except for two shaded faces. Magnitude of electric field at left face,

$$E_L = \alpha x^{1/2} = \alpha\, a^{1/2}$$

and, at right face,

$$E_R = \alpha x^{1/2} = \alpha(2a)^{1/2}$$

The flux at left face is given by,

$$\phi_L = E_L \cdot \Delta S = \Delta S\, E_L \cdot \hat{n}_L$$

$$= E_L\, \Delta S \cos\theta$$

$$= -E_L\, \Delta S \qquad \text{[Since } \theta = 180°\text{]}$$

$$\phi_L = -E_L\, a^2$$

$$\phi_R = E_R \cdot \Delta S$$

$$= E_R\, \Delta S \cos\theta$$

$$= E_R\, \Delta S \qquad \text{[Since } \theta = 0°\text{]}$$

$$= E_R a^2$$

Net flux through the cube $= \phi_R + \phi_L$

$$= E_R a^2 - E_L a^2$$

$$= a^2 (E_R - E_L) = \alpha a^2 [(2a)^{1/2} - a^{1/2}]$$

$$= \alpha a^{5/2} (\sqrt{2} - 1)$$

$$= 800 (0.1)^{5/2} (\sqrt{2} - 1)$$

$$= 1.05 \text{ Nm}^2/\text{C}$$

(ii) From Gauss's law, $\phi = \dfrac{q}{\varepsilon_0}$

where q = Charge enclosed

or $q = \phi\varepsilon_0 = 1.05 \times 8.854 \times 10^{-12}$ C

$$= 9.27 \times 10^{-12} \text{ C}$$

Q. 22. Two point charges $q_A = 3\ \mu$C and $q_B = -3\ \mu$C are located 20 cm apart in vacuum.

(i) What is the electric field at the midpoint O of the line AB joining the two charges ?

(ii) If a negative test charge of magnitude 1.5×10^{-9} C is placed at this point, what is the force experienced by the test charge ? **[NCERT]**

Ans. (i) The situation is represented in the given figure. O is the mid-point of line AB.

Distance between the two charges, AB = 20 cm

$\therefore$ AO = OB = 10 cm

Net electric field at point O = E

Electric field at point O caused by + 3 μC charge (E_1)

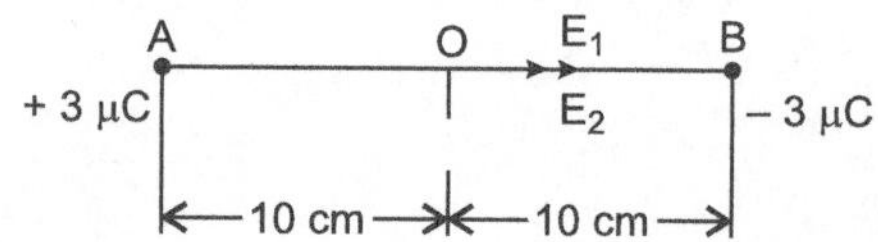

$$E_1 = \frac{1}{4\pi\varepsilon_0} \cdot \frac{3 \times 10^{-6}}{(OA)^2} \text{ (along OB)}$$

Where, ε_0 = Permittivity of free space and

$$\frac{1}{4\pi\varepsilon_0} = 9 \times 10^9 \text{ Nm}^2\text{C}^{-2}$$

Therefore, magnitude of electric field at point O caused by – 3 μC charge (E_2)

$$E_2 = \frac{1}{4\pi\varepsilon_0} \cdot \frac{3 \times 10^{-6}}{(OB)^2} \text{ (along OB)}$$

$\therefore$ $E = E_1 + E_2$ (along OB)

Total electric field

$$= \frac{2}{4\pi\varepsilon_0} \frac{3 \times 10^{-6}}{10^{-2}} \text{ N/C (along OB)}$$

[Since the magnitudes of E_1 and E_2 are equal and in the same direction]

$\therefore$ $E = 2 \times 9 \times 10^9 \times 3 \times 10^{-6} \times 10^2$ N/C

$$= 5.4 \times 10^6 \text{ NC}^{-1} \text{ (along OB)}$$

Therefore, the electric field at mid-point O is 5.4×10^6 N/C along OB.

(ii) A test charge of amount 1.5×10^{-9} C is placed at mid-point O.

$$q = 1.5 \times 10^{-9} \text{ C}$$

Force experienced by the test charge = F

$\therefore$ $F = qE$

$$= 1.5 \times 10^{-9} \times 5.4 \times 10^6$$

$$= 8.1 \times 10^{-3} \text{ N}$$

The force is directed along line OA. This is because the negative test charge is repelled by the charge placed at point B but attracted towards point A.

Therefore, the force experienced by the test charge is 8.1×10^{-3} N along OA.

Q. 23. Two charges ± 10 μC are placed 5.0 mm apart. Determine the electric field at

(i) a point P on the axis of the dipole 15 cm away from its centre O on the side of the positive charge, as shown in Fig. (a), and

(ii) a point Q, 15 cm away from O on a line passing through O and normal to the axis of the dipole, as shown in Fig. (b). **[NCERT]**

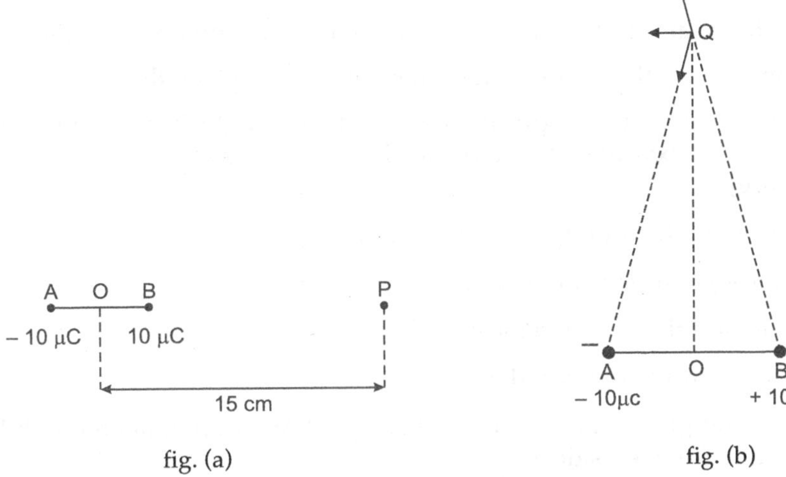

fig. (a) fig. (b)

Ans. (i) The electric field at P due to charge + 10 μC is given by

$$(\vec{E_1}) = \frac{1}{4\pi\varepsilon_0}\frac{q}{r_2}$$

$$= \frac{10^{-5}}{4\pi\,(8.854 \times 10^{-12})} \times \frac{1}{(15 - 0.25)^2 \times 10^{-4}}$$

$$= 4.13 \times 10^6 \text{ NC}^{-1} \text{ (along BP)}$$

The electric field at P due to charge – 10 μC is given by

$$\vec{E_2} = \frac{-10^{-5}}{4\pi\,(8.854 \times 10^{-12})} \times \frac{1}{(15 + 0.25)^2 \times 10^{-4}}$$

$$= -3.86 \times 10^6 \text{ N C}^{-1} \text{ (along PA)}$$

The resultant electric field at P due to the two charges at A and B is given by

$$E = \vec{E_1} + \vec{E_2}$$

$$= (4.13 - 3.86) \times 10^6$$

$$= 0.27 \times 10^6$$

$$= 2.7 \times 10^5 \text{ NC}^{-1} \text{ (along BP)}$$

(ii) Field at Q due to charge + 10 μC at B is given by

$$(E_1) = \frac{1}{4\pi\varepsilon_0}\frac{q}{r^2}$$

$$= \frac{10^{-5}}{4\pi\,(8.854 \times 10^{-12})} \times \frac{1}{[15^2 + (0.25)^2] \times 10^{-4}}$$

$$= 3.99 \times 10^6 \text{ N C}^{-1} \text{ (along BQ)}$$

Field at Q due to charge – 10 μC at A is given by

$$(E_2) = \frac{10^{-5}}{4\pi\,(8.854 \times 10^{-12})} \times \frac{1}{[15^2 + (0.25)^2] \times 10^{-4}}$$

$$= 3.99 \times 10^6 \text{ N C}^{-1} \text{ (along QA)}$$

Clearly, the components of these two forces with equal magnitudes cancel along the direction OQ (vertical) but add up along the direction parallel (horizontal) to BA. Therefore, the resultant electric field at Q due to the two charges at A and B is

$$= 2 \times \frac{0.25}{\sqrt{15^2 + (0.25)^2}} \times 3.99 \times 10^6 \text{ N C}^{-1} \text{ (along BA)}$$

$$= 1.33 \times 10^5 \text{ N C}^{-1} \text{ (along BA)}$$

Q. 24. An electric field is uniform, and in the positive x-direction for positive x, and uniform with the same magnitude but in the negative x direction for negative x. It is given that $E = 200\,\hat{i}$ N/C for $x > 0$ and $E = -200\,\hat{i}$ N/C for $x < 0$. A right circular cylinder of length 20 cm and radius 5 cm has its centre at the origin and its axis along the X-axis so that one face is at $x = \pm 10$ cm and other is at $x = -10$ cm [Given figure].

(i) What is the net outward flux through each flat face ?

(ii) What is the flux through the side of the cylinder ?

(iii) What is the net outward flux through the cylinder ?

(iv) What is the net charge inside the cylinder ? **[NCERT]**

Ans. (i) From given figure and given data it is clear that E and ΔS are parallel on both left face of cylinder. The outward flux from left side is

$$\phi_L = E \cdot \Delta S$$

$$= -200\,\hat{i} \cdot \Delta S$$

$$= +200\,\Delta S \qquad [\text{since } \hat{i} \cdot \Delta S = -\Delta S]$$

$$= 200 \times \pi\,(0.05)^2$$

$$= 1.57 \text{ N m}^2\text{C.}$$

On the right face E and ΔS are parallel thus

$$\phi_R = E \cdot \Delta S = +1.57 \text{ N m}^2/\text{C.}$$

(ii) For any point on the side of the cylinder, E is perpendicular to vector ΔS and hence $E \cdot \Delta S = 0$. Thus, flux out of this side of cylinder is zero.

(iii) Net outward flux through the cylinder is

$$\phi = \phi_L + \phi_R + \phi_{Normal}$$

$$= 1.57 + 1.57 + 0$$

$$= 3.14 \text{ N m}^2/\text{C}$$

(iv) From Gauss's law

The charge enclosed in sphere is

$$q = \varepsilon_0\,\phi$$

$$= 3.14 \times 8.854 \times 10^{-12} \text{ C}$$

$$= 2.78 \times 10^{-11} \text{ C}$$

Q. 25. Two point charges q_1 and q_2 of magnitude $+10^{-8}$ C and -10^{-8} C, respectively, are placed 0.1 m apart. Calculate the electric fields at points A, B and C as shown in Fig. below. **[NCERT]**

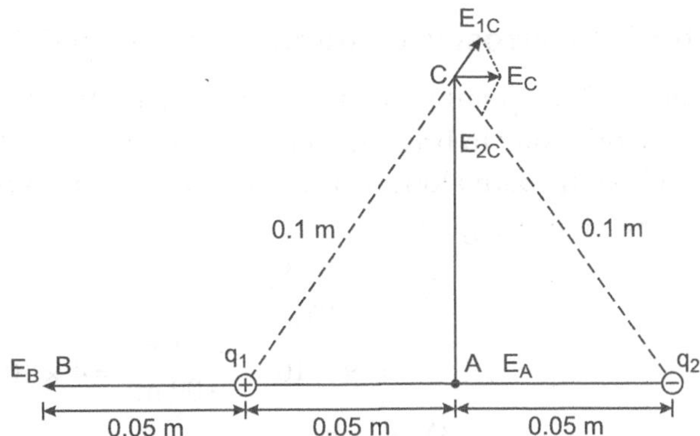

Ans. (i) The electric field vector E_{1A} at A due to the positive charge q_1 points towards the right. From expression of electric field

$$E = \frac{1}{4\pi\varepsilon_0}\frac{q}{r^2}$$

The magnitude of

$$E_{1A} = \frac{(9 \times 10^9) \times (10^{-8})}{(0.05)^2}$$

$$= 3.6 \times 10^4 \text{ N C}^{-1}$$

The electric field vector E_{2A} at A due to the negative charge q_2 points towards the right and has the same magnitude. Hence the magnitude of the total electric field E_A at A is

$$E_A = E_{1A} + E_{2A} = 7.2 \times 10^4 \text{ N C}^{-1}$$

E_A is directed towards the right.

(ii) The electric field vector E_{1B} at B due to the positive charge q_1 points towards the left and has a magnitude

$$E_{1B} = \frac{(9 \times 10^9) \times (10^{-8})}{(0.5)^2}$$

$$= 3.6 \times 10^4 \text{ N C}^{-1}$$

The electric field vector E_{2B} at B due to the negative charge q_2 points towards the right and has a magnitude

$$E_{2B} = \frac{(9 \times 10^9)(10^{-8})}{(0.15)^2}$$

$$= 4 \times 10^3 \text{ N C}^{-1}$$

The magnitude of the total electric field at B is

$$E_B = E_{1B} - E_{2B} = 3.2 \times 10^4 \text{ N C}^{-1}$$

E_B is directed towards the left.

(iii) The magnitude of each electric field vector at point C, due to charge q_1 and q_2 is

$$E_{1C} = E_{2C} = \frac{(9 \times 10^9)(10^{-8})}{(0.10)^2}$$

$$= 9 \times 10^3 \text{ N C}^{-1}$$

The directions in which these two vectors point are indicated in the given Fig. The resultant of these two vectors is

$$E_C = E_1 \cos\frac{\pi}{3} + E_2 \cos\frac{\pi}{3}$$

$$= 9 \times 10^9 \text{ N C}^{-1}$$

E_C points towards the right.

Chapter 2. Electrostatic Potential and Capacitance

Q. 1. (i) Calculate the potential at a point P due to a charge of 4×10^{-7} C located 9 cm away.

(ii) Hence obtain the work done in bringing a charge of 2×10^{-9} C from infinity to the point P. Does the answer depend on the path along which the charge is brought ? **[NCERT]**

Ans. Potential at a point P at r distance is given by

(i)
$$V = \frac{1}{4\pi\varepsilon_0}\frac{Q}{r}$$
$$= 9 \times 10^9 \times \frac{4 \times 10^{-7}}{0.09 \text{ m}} = 4 \times 10^4 \text{ V}$$

(ii)
$$W = qV$$
$$= 2 \times 10^{-9} \times 4 \times 10^4$$
$$= 8 \times 10^{-5} \text{ J}$$

No, work done will be path independent.

Q. 2. Three capacitors each of capacitance 9 pF are connected in series.

(i) What is the total capacitance of the combination ?

(ii) What is the potential difference across each capacitor if the combination is connected to a 120 V supply ? **[NCERT]**

Ans. (i) Capacitance of each of the three capacitors, C = 9 pF

Equivalent capacitance (C_{eq}) of the combination of the capacitors is given by the relation,

$$\frac{1}{C_{eq}} = \frac{1}{C} + \frac{1}{C} + \frac{1}{C} = \frac{3}{C} = \frac{3}{9} = \frac{1}{3}$$

$$\therefore \quad C_{eq} = 3 \text{ pF}$$

(ii) Supply voltage, $V = 120$ V

Potential difference (V_1) across each capacitor is equal to one-third of the supply voltage.

$$\therefore \quad V_1 = \frac{V}{3} = \frac{120}{3} = 40 \text{ V}$$

Therefore, the potential difference across each capacitor is 40 V.

Q. 3. A 12 pF capacitor is connected to a 50 V battery. How much electrostatic energy is stored in the capacitor ? **[NCERT]**

Ans. Capacitor of the capacitance, $C = 12 \text{ pF} = 12 \times 10^{-12}$ F

Potential difference, $V = 50$ V

Electrostatic energy stored in the capacitor is given by the relation.

$$E = \frac{1}{2}CV^2$$
$$= \frac{1}{2} \times 12 \times 10^{-12} \times (50)^2 \text{ J} = 1.5 \times 10^{-8} \text{ J}$$

Therefore, the electrostatic energy stored in the capacitor is 1.5×10^{-8} J.

Q. 4. (i) Calculate the potential at a point P due to charge of 4×10^{-7} C located 9 cm away.

(ii) Obtain work done in bringing a charge of 2×10^{-9} C from infinity to the point P. Does the answer depend on the path along which the charge is brought ? **[NCERT]**

Ans. (i) Since, $\frac{1}{4\pi\varepsilon_0} = 9 \times 10^9$ and charge (Q) $= 4 \times 10^{-7}$, $r = 9$ cm $= 0.09$ m

$$V = \frac{1}{4\pi\varepsilon_0}\frac{Q}{r}$$
$$= 9 \times 10^9 \times \frac{4 \times 10^{-7}}{0.09} = 4 \times 10^4 \text{ V}$$

(ii) $W = qV$
$$= 2 \times 10^{-9} \text{C} \times 4 \times 10^4 \text{ V} = 8 \times 10^5 \text{ J}$$

It doesn't depend on path.

Q. 5. A regular hexagon of side 10 cm has a charge 5 μC at each of its vertices. Calculate the potential at the centre of the hexagon.

Ans. The given figure shows six equal amount of charges, q at the vertices of a regular hexagon.

Given :

Charge on each vertex, $q = 5\ \mu C = 5 \times 10^{-6}$ C

Side of the hexagon, l = AB = BC = CD = DE = EF = FA = 10 cm

Distance of each vertex from centre O, d = 10 cm

Electric potential at point O,

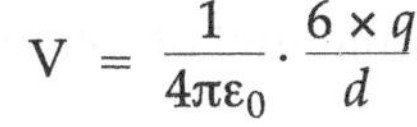

$$V = \frac{1}{4\pi\varepsilon_0} \cdot \frac{6 \times q}{d}$$

Where, ε_0 = Permittivity of free space

and $\frac{1}{4\pi\varepsilon_0} = 9 \times 10^9\ Nm^2C^{-2}$

$$\therefore \quad V = \frac{9 \times 10^9 \times 6 \times 5 \times 10^{-6}}{0.1} = 2.7 \times 10^6\ V$$

Therefore, the potential at the centre of the hexagon is 2.7×10^6 V.

Q. 6. If one of the two electrons of a H_2 molecule is removed, we get a hydrogen molecular ion H_2^+. In the ground state of an H_2^+, the two protons are separated by roughly 1.5 Å, and the electron is roughly 1 Å from each proton. Determine the potential energy of the system. Specify your choice of the zero of potential energy.

Ans. The system of two protons and one electron is represented in the given figure.

The potential energy of the system is,

$$V = \frac{q_1q_2}{4\pi\varepsilon_0 d_1} + \frac{q_2q_3}{4\pi\varepsilon_0 d_3} + \frac{q_3q_1}{4\pi\varepsilon_0 d_2}$$

Substituting $\frac{1}{4\pi\varepsilon_0} = 9 \times 10^9\ Nm^2C^{-2}$, we obtain

$$V = \frac{9 \times 10^9 \times 10^{-19} \times 10^{-19}}{10^{-10}} \left[-(1.6)^2 + \frac{(1.6)^2}{1.5} + (-1.6)^2\right]$$

$$= -30.7 \times 10^{-19}\ J$$

$$= -19.2\ eV$$

Therefore, the potential energy of the system is – 19.2 eV.

Q. 7. In a parallel plate capacitor with air between the plates, each plate has an area of $6 \times 10^{-3}\ m^2$ and the distance between the plates is 3 mm. Calculate the capacitance of the capacitor. If this capacitor is connected to a 100 V supply, what is the charge on each plate of the capacitor ?

[NCERT]

Ans. Area of each plate of the parallel plate capacitor, $A = 6 \times 10^{-3}\ m^2$

Distance between the plates, $d = 3\ mm = 3 \times 10^{-3}$ m

Supply voltage, V = 100 V

Capacitance C of a parallel plate capacitor is given by, $C = \frac{\varepsilon_0 A}{d}$

Here, ε_0 = Permittivity of free space = $8.854 \times 10^{-12}\ C^2\ N^{-1}\ m^{-2}$

A = Area of plates

d = Distance between plates

k = Dielectric constant of capacitor = 1

$$\therefore \quad C = \frac{8.854 \times 10^{-12} \times 6 \times 10^{-3}}{3 \times 10^{-3}} = 17.71 \times 10^{-12} \text{ F}$$

$$= 17.71 \text{ pF}$$

So, charge on each plate of the capacitor,

$$q = VC$$

$$= 100 \times 17.71 \times 10^{-12} \text{ C}$$

$$= 1.771 \times 10^{-9} \text{ C}$$

Therefore, capacitance of the capacitor is 17.71 pF and charge on each plate is 1.771×10^{-9} C.

Q. 8. A cylindrical capacitor has two co-axial cylinders of length 15 cm and radii 1.5 cm and 1.4 cm. The outer cylinder is earthed and the inner cylinder is given a charge of 3.5 μC. Determine the capacitance of the system and the potential of the inner cylinder. Neglect end effects (*i.e.*, bending of field lines at the ends). **[NCERT]**

Ans. Length of a co-axial cylinder, l = 15 cm = 0.15 m

Radius of outer cylinder, r_1 = 1.5 cm = 0.015 m

Radius of inner cylinder, r_2 = 1.4 cm = 0.014 m

Charge on the inner cylinder, $q = 3.5\ \mu\text{C} = 3.5 \times 10^{-6}$ C

Capacitance of a co-axial cylinder of radii r_1 and r_2 is given by the relation

$$C = \frac{2\pi\varepsilon_0 l}{\log e \frac{r_1}{r_2}}$$

where, ε_0 = Permittivity of free space = 8.85×10^{-12} C^2 N^{-1} m^{-2}

$$\therefore \quad C = \frac{2\pi \times 8.85 \times 10^{-12} \times 0.15}{2.3026 \log_{10}\left(\frac{0.15}{0.14}\right)}$$

$$= \frac{2\pi \times 8.85 \times 10^{-12} \times 0.15}{2.3026 \times 0.0299}$$

$$= 1.2 \times 10^{-10} \text{ F}$$

Potential difference of the inner cylinder is given by

$$F = \frac{q}{C} = \frac{3.5 \times 10^{-6}}{1.2 \times 10^{-10}}$$

$$= 2.92 \times 10^4 \text{ V}$$

Q. 9. (i) A 900 pF capacitor is charged by 100 V battery [Fig. (a)]. How much electrostatic energy is stored by the capacitor ?

(ii) The capacitor is disconnected from the battery and connected to another 900 pF capacitors [Fig. (b)]. What is the electrostatic energy stored by the system ? **[NCERT]**

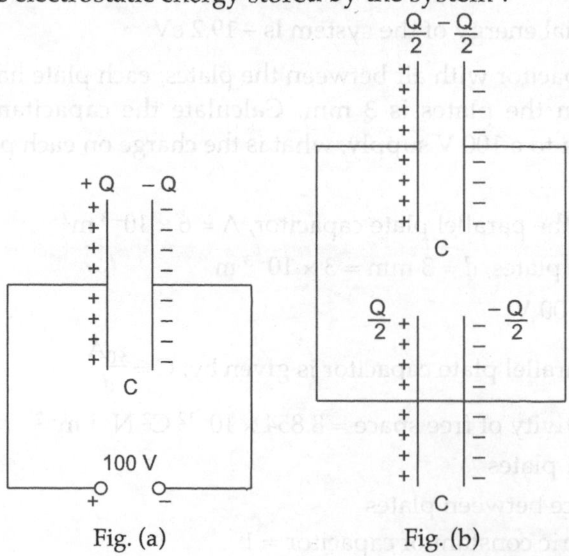

Fig. (a) Fig. (b)

Ans. (i) The charge on the capacitor is

$$Q = CV$$
$$= 900 \times 10^{-12}\ F \times 100\ V = 9 \times 10^{-8}\ C$$

Energy stored by the capacitors $= \frac{1}{2}CV^2 = \frac{1}{2}QV$

$$= \frac{1}{2} \times 9 \times 10^{-8}\ C \times 100\ V = 4.5 \times 10^{-6}\ J$$

(ii) In the steady situation, the two capacitors have their positive plates at the same potential and their negative plates at the same potential. Let the common potential difference be V′. The charge on each capacitor is then Q′ = CV′. By charge conservation, Q′ = Q/2. This implies V′ = V/2.

The total energy of the system $= 2 \times \frac{1}{2}Q'V' = \frac{1}{4}QV = 2.25 \times 10^{-6}\ J$

Thus in going from (i) to (ii), though no charge is lost : the final energy is only half the initial energy.

Q. 10. A capacitor of unknown capacitance is connected across a battery of V volts. The charge stored in it is 360 μC. When potential across the capacitor is reduced by 120 V, the charge stored in it becomes 120 μC.

Calculate :

(i) The potential V and unknown capacitance C.★

(ii) What will be the charge stored in the capacitor, if the voltage applied had increased by 120 V ?★

Ans. (i) Initial voltage, V_1 = V volts and charge stored

$$Q_1 = 360\ \mu C$$
$$Q_1 = CV_1 \quad ...(i)$$

Changed potential, $V_2 = V - 120$

$$Q_2 = 120\ \mu C$$
$$Q_2 = CV_2 \quad ...(ii)$$

By dividing (ii) from (i), we get

$$\frac{Q_1}{Q_2} = \frac{CV_1}{CV_2}$$

$\Rightarrow$ $$\frac{360}{120} = \frac{V}{V-120}$$

$\Rightarrow$ $$V = 180\ V$$

Hence, $$C = \frac{Q_1}{V_1} = \frac{360 \times 10^{-6}}{180} = 2 \times 10^{-6}\ F$$

$$= 2\ \mu F$$ **Ans.**

(ii) If the voltage applied had increased by 120 V, then

$$V_3 = 180 + 120 = 300\ V$$

Hence charge stored in capacitor, $Q_3 = CV_3$

$$= 2 \times 10^{-6} \times 300 = 600\ \mu C$$

Q. 11. A spherical conductor of radius 12 cm has a charge of 1.6×10^{-7} C distributed uniformly on its surface. What is the electric field ?

(i) inside the sphere

(ii) just outside the sphere

(iii) at a point 18 cm from the centre of the sphere ? **[NCERT]**

★ are board based questions from previous year

Ans. (i) Radius of the spherical conductor, $r = 12$ cm $= 0.12$ m

Charge is uniformly distributed over the conductor, $q = 1.6 \times 10^{-7}$ C

Electric field inside a spherical conductor is zero. This is because if there is field inside the conductor, then charge will move to neutralise it.

(ii) Electric field (E) just outside the conductor can given by

$$E = \frac{1}{4\pi\varepsilon_0}\frac{q}{r^2}$$

Just outside the conductor, $d = r$.

Therefore,

$$E = \frac{9 \times 10^{-9} \times 1.6 \times 10^{-7}}{(0.12)^2} = 10^5 \text{ NC}^{-1}$$

Therefore, the electric field just outside the sphere is 10^5 NC^{-1}.

(iii) Electric field at a point 18 cm from the centre of the sphere = E_1

Distance of the point from the centre, $d = 18$ cm $= 0.18$ m

$$E_1 = \frac{1}{4\pi\varepsilon_0} \cdot \frac{q}{d^2}$$

$$= \frac{9 \times 10^9 \times 1.6 \times 10^{-7}}{(1.8 \times 10^{-2})^2} = 4.4 \times 10^4 \text{ NC}^{-1}$$

Therefore, the electric field at a point 18 cm from the centre of the sphere is 4.4×10^4 NC^{-1}.

Q. 12. Two capacitors of unknown capacitances C_1 and C_2 are connected first in series and then in parallel across a battery of 100 V. If the energy stored in the two combinations is 0.045 J and 0.25 J respectively, determine the value of C_1 and C_2. Also calculate the charge on each capacitor in parallel combination.★

Ans. When the capacitors are connected in parallel.

Equivalent capacitance, $C_p = C_1 + C_2$

The energy stored in the combination of the capacitors,

$$E_p = \frac{1}{2}C_pV^2$$

$$\Rightarrow \quad E_p = \frac{1}{2}(C_1 + C_2)(100)^2 = 0.25 \text{ J}$$

$$\Rightarrow \quad (C_1 + C_2) = 5 \times 10^{-5} \quad \text{...(i)}$$

When the capacitors are connected in series.

Equivalent capacitance,

$$C_s = \frac{C_1C_2}{C_1 + C_2}$$

The energy stored in the combination of the capacitors,

$$E_s = \frac{1}{2}C_sV^2$$

$$\Rightarrow \quad E_s = \frac{1}{2}\frac{C_1C_2}{C_1 + C_2}(100)^2 = 0.045 \text{ J}$$

$$\Rightarrow \quad \frac{1}{2} \times \frac{C_1C_2}{5 \times 10^{-5}}(100)^2 = 0.045 \text{ J}$$

$$\Rightarrow \quad C_1C_2 = 0.045 \times 10^{-4} \times 5 \times 10^{-5} \times 2 = 4.5 \times 10^{-10}$$

$$(C_1 - C_2)^2 = (C_1 + C_2)^2 - 4C_1C_2$$

$$\Rightarrow \quad (C_1 - C_2) = \sqrt{7 \times 10^{-10}} = 2.64 \times 10^{-5}$$

$$C_1 - C_2 = 2.64 \times 10^{-5} \quad \text{...(ii)}$$

Solving (i) and (ii), we get

★ are board based questions from previous year

$$C_1 = 38.2\ \mu F \text{ and } C_2 = 11.8\ \mu F$$

When the capacitors are connected in parallel, the charge on each of them can be obtained as follows :

$$Q_1 = C_1V = 38.2 \times 10^{-6} \times 100 = 38.2 \times 10^{-4}\ C$$
$$Q_2 = C_2V = 11.8 \times 10^{-6} \times 100 = 11.8 \times 10^{-4}\ C$$

Q. 13. A 12 pF capacitor is connected to a 50 V battery. How much electrostatic energy is stored in the capacitor ? If another capacitor of 6 pF is connected in series with it with the same battery connected across the combination, find the charge stored and potential difference across each capacitor.★

Ans. Energy stored in the capacitor of capacitance 12 pF,

$$U = \frac{1}{2}CV^2$$
$$= \frac{1}{2} \times 12 \times 10^{-12} \times 50 \times 50\ J$$
$$= 1.5 \times 10^{-8}\ J$$

Let C be the equivalent capacitance of 12 pF and 6pF capacitor connected in series and it is given by

$$\frac{1}{C} = \frac{1}{12} + \frac{1}{6} = \frac{1+2}{12}$$

$\therefore$ $$C = 4pF$$

As the capacitors are connected in series, so the charge remains same

$\therefore$ Charge stored across each capacitor,

$$q = CV = 4 \times 10^{-12} \times 50$$
$$= 2 \times 10^{-10}\ C$$

Potential difference across capacitor C_1,

$$V_1 = \frac{2 \times 10^{-10}}{12 \times 10^{-12}} \text{volt} = \frac{50}{3}\ V$$

Potential difference across capacitor C_2,

$$V_2 = \frac{2 \times 10^{-10}}{6 \times 10^{-12}} \text{volt} = \frac{100}{3}\ V$$

Q. 14. (i) Find equivalent capacitance between A and B in the combination given below. Each capacitor is of 2 μF capacitance.★

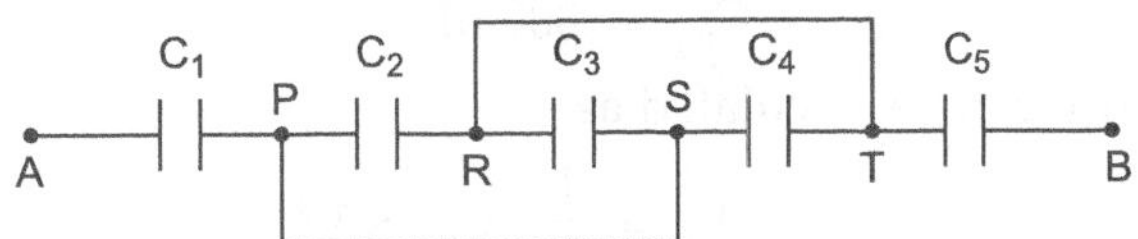

(ii) If a *dc* source of 7 V is connected across AB. How much charge is drawn from the source and what is energy stored in the network.★

Ans. (i) Redrawing the fig.

Here C_1 is in series with parallel combination of C_2, C_3, C_4 and in series with C_5.

The equivalent capacitance of the parallel combination is given by C.

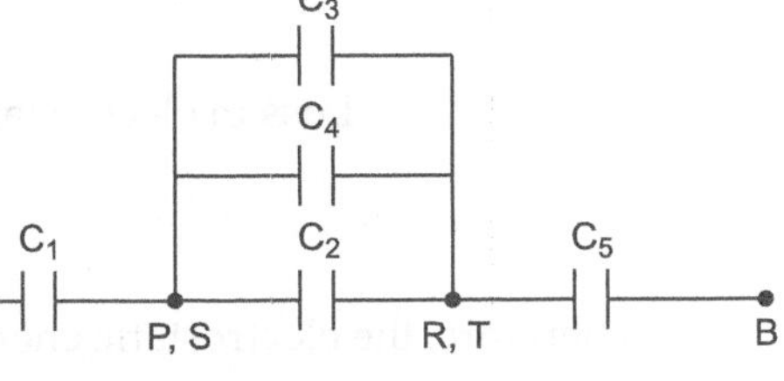

Thus, $$C' = C_2 + C_3 + C_4 = 6\ \mu F$$

$$\frac{1}{C_{eq}} = \frac{1}{C_1} + \frac{1}{C'} + \frac{1}{C_5}$$

★ are board based questions from previous year

$$= \frac{1}{2}+\frac{1}{6}+\frac{1}{2}$$

$$= \frac{7}{6}\,\mu F$$

$$C_{eq} = \frac{6}{7}\,\mu F$$

(ii) As we know charge remains same in series circuit

$$Q = C_{eq}V$$

$$= \frac{6}{7}\times 7$$

$$Q = 6\,\mu C$$

As we also know

$$\text{Potential Energy (U)} = \frac{1}{2}C_{eq}V^2$$

$$= \frac{1}{2}\times\frac{6}{7}\times 7^2 \times 10^{-6}$$

$$= 21\times 10^{-6}\,J$$

Q. 15. A 600 pF capacitor is charged by a 200 V supply. It is then disconnected from the supply and is connected to another uncharged 600 pF capacitor. How much electrostatic energy is lost in the process ? **[NCERT]**

Ans. Capacitance of the capacitor, C = 600 pF

Potential difference, V = 200 V

Electrostatic energy stored in the capacitor is given by

$$E_1 = \frac{1}{2}CV^2 = \frac{1}{2}\times(600\times 10^{-12})\times(200)^2\,J$$

$$= 1.2\times 10^{-5}\,J$$

If supply is disconnected from the capacitor and another capacitor of capacitance C = 600 pF is connected to it, then equivalent capacitance (C_{eq}) of the combination is given by,

$$\frac{1}{C_{eq}} = \frac{1}{C}+\frac{1}{C}$$

$$\Rightarrow \quad \frac{1}{C_{eq}} = \frac{1}{600}+\frac{1}{600}=\frac{2}{600}=\frac{1}{300}$$

$$\Rightarrow \quad C_{eq} = 300\,pF$$

New electrostatic energy can be calculated as

$$E_2 = \frac{1}{2}\times C_{eq.}\times V^2$$

$$= \frac{1}{2}\times 300\times 10^{-12}\times(200)^2$$

$$= 0.6\times 10^{-5}\,J$$

$$\text{Loss in electrostatic energy} = E_1 - E_2$$

$$= 1.2\times 10^{-5} - 0.6\times 10^{-5}\,J$$

$$= 0.6\times 10^{-5}\,J = 6\times 10^{-6}\,J$$

Therefore, the electrostatic energy lost in the process is 6×10^{-6} J.

Q. 16. A spherical capacitor has an inner sphere of radius 12 cm and an outer sphere of radius 13 cm. The outer sphere is earthed and the inner sphere is given a charge of 2.5 μC. The space between the concentric spheres is filled with a liquid of dielectric constant 32.

(i) Determine the capacitance of the capacitor.

(ii) What is the potential of the inner sphere ?

(iii) Compare the capacitance of this capacitor with that of an isolated sphere of radius 12 cm. Explain why the latter is much smaller. **[NCERT]**

Ans. Radius of the inner sphere, $r_2 = 12$ cm $= 0.12$ m

Radius of the outer sphere, $r_1 = 13$ cm $= 0.13$ m

Charge on the inner sphere, $q = 2.5\ \mu C = 2.5 \times 10^{-6}$ C

Dielectric constant of a liquid, $\varepsilon_r = 32$

(i) Capacitance of the capacitor is given by the relation,

$$C = \frac{4\pi\varepsilon_0\varepsilon_r r_1 r_2}{(r_1 - r_2)}$$

where, ε_0 = Permittivity of free space $= 8.85 \times 10^{-12}\ C^2N^{-1}m^{-2}$

$$\frac{1}{4\pi\varepsilon_0} = 9 \times 10^9\ N\ m^2\ C^{-2}$$

$$\therefore \quad C = \frac{32 \times 0.12 \times 0.13}{9 \times 10^9 \times (0.13 - 0.12)}$$

$$\approx 5.5 \times 10^{-9}\ F$$

Hence, the capacitance of the capacitor is approximately 5.5×10^{-9} F.

(ii) Potential of the inner sphere is given by,

$$V = \frac{q}{c} = \frac{2.5 \times 10^{-6}}{5.5 \times 10^{-3}} = 4.5 \times 10^2\ V$$

Hence, the potential of the inner sphere is 4.5×10^2 V.

(iii) Radius of an isolated sphere, $r = 12 \times 10^{-2}$ m

Capacitance of the sphere is given by the relation,

$$C = 4\pi\varepsilon_0 r$$

$$= 4\pi \times 8.85 \times 10^{-12} \times 12 \times 10^{-2}$$

$$= 1.33 \times 10^{-11}\ F$$

The capacitance of the isolated sphere is less in comparison to the concentric spheres. This is because the outer sphere of the concentric spheres is earthed. Hence, the potential difference is less and the capacitance is more than the isolated sphere.

Q. 17. In a hydrogen atom, the electron and proton are bound at a distance of about 0.53 Å :

(i) Estimate the potential energy of the system in eV, taking the zero of the potential energy at infinite separation of the electron from proton.

(ii) What is the minimum work required to free the electron, given that its kinetic energy in the orbit is half the magnitude of potential energy obtained in (i) ?

(iii) What are the answers to (i) and (ii) above if the zero of potential energy is taken at 1.06 Å separation ? **[NCERT]**

Ans. The distance between electron-proton of a hydrogen atom, $d = 0.53$ Å

Charge on an electron, $q_1 = -1.6 \times 10^{-19}$ C

Charge on a proton, $q_2 = +1.6 \times 10^{-19}$ C

(i) Potential energy of proton-electron system is

$$U = \frac{1}{4\pi\varepsilon_0}\frac{q_1q_2}{r}$$

$$= 9 \times 10^9 \times \frac{(-1.6 \times 10^{-19}) \times 1.6 \times 10^{-19}}{(0.53 \times 10^{-10})}$$

$$= -43.47 \times 10^{-19}\ J$$

$$= \frac{-43.47 \times 10^{-19}}{1.6 \times 10^{-19}} \text{ eV}$$

$$= -27.16 \text{ eV}$$

(ii) Kinetic energy is half of the magnitude of potential energy.

$$\text{Kinetic energy} = \frac{1}{2} \times (-27.2) = 13.6 \text{ eV}$$

$$\text{Total energy} = \text{Potential energy} + \text{Kinetic energy}$$

$$= 13.6 - 27.2 = -13.6 \text{ eV}$$

Therefore, the minimum work required to free the electron is 13.6 eV.

(iii) When zero of Potential Energy (P.E.) is taken at $d_1 = 1.06$ Å

$$\text{P.E. of system} = \text{P.E. at } d_1 - \text{P.E. at } d$$

$$= \frac{q_1 q_2}{4\pi\varepsilon_0 d_1} + 27.2 \text{ eV}$$

$$= \frac{9 \times 10^9 \times (1.6 \times 10^{-19} \times -1.6 \times 10^{-19})}{1.06 \times 10^{-10}} + 27.2 \text{ eV}$$

$$= -21.73 \times 10^{-19} \text{ J} + 27.2 \text{ eV}$$

$$= -13.58 \text{ eV} + 27.2 \text{ eV}$$

$$= -13.6 \text{ eV}$$

$$\text{Work required to free the electron} = 0 - (-13.6) = 13.6 \text{ eV}.$$

Q. 18. Two charges 3×10^{-8} C and -2×10^{-8} C are located 15 cm apart. At what point on the line joining the charges is the electric potential zero ? Take the potential at infinity to be zero. **[NCERT]**

Ans. Let us consider the origin O at location of positive charge. The line joining the two charges is taken to be the X-axis. The negative charge is taken to be on right side of origin.

Let Q is the point in line joining OP, where potential is zero.

Thus, the sum of potential at Q due to both charges.

$$\frac{1}{4\pi\varepsilon_0}\frac{3 \times 10^{-8}}{x \times 10^{-2}} - \frac{1}{4\pi\varepsilon_0}\frac{2 \times 10^{-8}}{(15 - x) \times 10^{-2}} = 0 \qquad \left\{\text{As } V = \frac{1}{4\pi\varepsilon_0}\frac{Q}{r}\right\}$$

$$\frac{1}{4\pi\varepsilon_0} \times \frac{10^{-8}}{10^{-2}}\left[\frac{3}{x} - \frac{2}{15 - x}\right] = 0$$

Thus,

$$\frac{3}{x} - \frac{2}{15 - x} = 0$$

$$3(15 - x) = 2x$$

$$45 - 3x = 2x$$

$$45 = 5x$$

$$x = 9 \text{ cm}$$

Let Q is at the extension of OP, the required condition is

$$\text{Voltage at Q due to + charge at O} = \frac{1}{4\pi\varepsilon_0}\frac{Q}{x}$$

$$= \frac{1}{4\pi\varepsilon_0}\frac{3 \times 10^{-8}}{x \times 10^{-2}}$$

and voltage at Q due to – charge at P $= \frac{1}{4\pi\varepsilon_0} \frac{(-2 \times 10^{-8})}{(x-15) \times 10^{-2}}$

As per question,

$$V_{QO} + V_{QP} = 0$$

$$\frac{1}{4\pi\varepsilon_0} \cdot \frac{10^{-8}}{10^{-2}} \left[\frac{3}{x} - \frac{2}{x-15}\right] = 0$$

$$\frac{3}{x} - \frac{2}{x-15} = 0$$

$$3x - 45 = 2x$$

$$x = 45 \text{ cm}$$

Q. 19. A 4 μF capacitor is charged by a 200 V supply. It is then disconnected from the supply, and is connected to another uncharged 2 μF capacitor. How much electrostatic energy of the first capacitor is lost in the form of heat and electromagnetic radiation ? **[NCERT]**

Ans. Capacitance of a charge capacitor, $C_1 = 4\mu F = 4 \times 10^{-6}$ F

Supply voltage, $V_1 = 200$ V

Electrostatic energy stored in capacitor is given by,

$$E = \frac{1}{2}CV^2$$

Thus, energy stored in C_1,

$$E_1 = \frac{1}{2} \times 4 \times 10^{-6} \times (200)^2$$

$$= 8 \times 10^{-2} \text{ J}$$

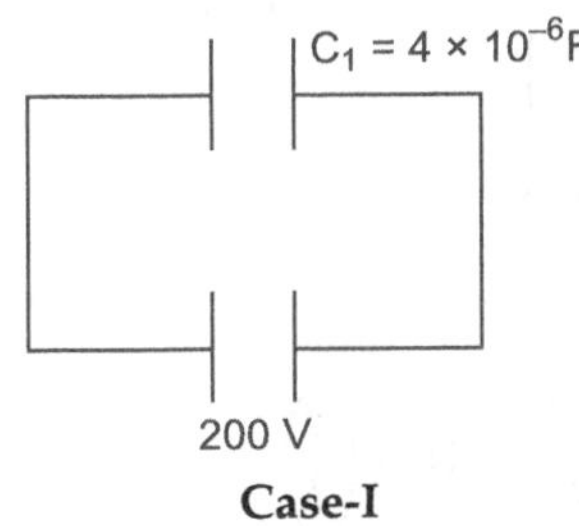

Case-I

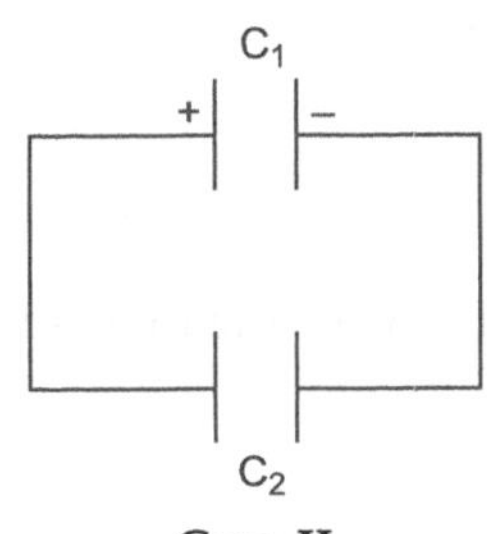

Case-II

Capacitance of an uncharged capacitor, $C_1 = 2\ \mu F = 2 \times 10^{-6}$ F

When C_2 is connected to the circuit, the potential acquired by it is V_2.

As per conservation of charge, initial charge on capacitor C_1 is equal to the final charge on capacitors, C_1 and C_2.

$$\therefore \quad V_2 (C_1 + C_2) = C_1V_1$$

$$V_2 \times (4 + 2) \times 10^{-6} = 4 \times 10^{-6} \times 200$$

$$V_2 = \frac{400}{3} \text{ V}$$

Electrostatic energy for the combination of two capacitors is given by,

$$E_2 = \frac{1}{2}(C_1 + C_2)V_2^2$$

$$= \frac{1}{2}(2+4) \times 10^{-6} \times \left(\frac{400}{3}\right)^2$$

$$\approx 5.33 \times 10^{-2} \text{ J}$$

Hence, amount of electrostatic energy lost by capacitor C_1,

$$= E_1 - E_2$$

$$= 0.08 - 0.0533 = 0.0267$$

$$= 2.67 \times 10^{-2} \text{ J}$$

Q. 20. Two tiny spheres carrying charges 1.5 μC and 2.5 μC are located 30 cm apart. Find the potential and electric field :

(i) at the mid-point of the line joining the two charges, and

(ii) at a point 10 cm from this mid-point in a plane normal to the line and passing through the mid-point. **[NCERT]**

Ans. Two charges placed at points A and B are shown in the given figure. O is the mid-point of the line joining the two charges.

1.5 μC A O B 2.5 μC ← 30 cm →

$$\text{Magnitude of charge located at A, } q_1 = 1.5\ \mu C$$

$$\text{Magnitude of charge located at B, } q_2 = 2.5\ \mu C$$

$$\text{Distance between the two charges, } d = 30\text{ cm} = 0.3\text{ m}$$

(i) Let V_1 and E_1 are the electric potential and electric field respectively at O.

V_1 = Potential due to charge at A + Potential due to charge at B

$$V_1 = \frac{q_1}{4\pi\varepsilon_0\left(\frac{d}{2}\right)} + \frac{q_2}{4\pi\varepsilon_0\left(\frac{d}{2}\right)}$$

$$= \frac{1}{4\pi\varepsilon_0\left(\frac{d}{2}\right)}[q_1 + q_2]$$

Where, ε_0 = Permittivity of free space

$$\frac{1}{4\pi\varepsilon_0} = 9 \times 10^9\ NC^2m^{-2}$$

$$V_1 = \frac{9 \times 10^9 \times 10^{-6}}{\left(\frac{0.30}{2}\right)}[2.5 + 1.5]$$

$$= 2.4 \times 10^5\ V$$

E_1 = Electric field due to q_2 – Electric field due to q_1

$$= \frac{q_2}{4\pi\varepsilon_0\left(\frac{d}{2}\right)^2} - \frac{q_1}{4\pi\varepsilon_0\left(\frac{d}{2}\right)^2}$$

$$= \frac{9 \times 10^9 \times 10^{-6}}{\left(\frac{0.30}{2}\right)^2}[2.5 - 1.5]$$

$$= 4 \times 10^5\ V\ m^{-1}$$

Therefore, the potential at mid-point is 2.4×10^5 V and the electric field at mid-point is $4 \times 10^5\ Vm^{-1}$. The field is directed from the larger charge to the smaller charge.

(ii) Consider a point Z such that normal distance OZ = 10 cm = 0.1 m, as shown in the following figure.

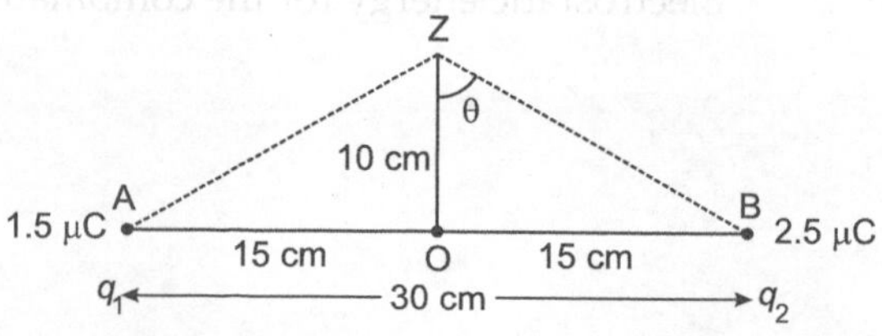

Let, V_2 and E_2 are the electric potential and electric field respectively at Z.

It can be observed from one figure that distance,

$$BZ = AZ$$

$$= \sqrt{(0.1)^2 + (0.15)^2}$$

$$= 0.18\text{ m}$$

V_2 = Electric potential due to A + Electric Potential due to B

$$= \frac{q_1}{4\pi\varepsilon_0\,(AZ)} + \frac{q_2}{4\pi\varepsilon_0\,(BZ)}$$

$$= \frac{9 \times 10^9 \times 10^{-6}}{0.18}[1.5 + 2.5]$$

$$= 2 \times 10^5 \text{ V}$$

Electric field due to q_1 at Z,

$$E_A = \frac{q_1}{4\pi\varepsilon_0 (AZ)^2}$$

$$= \frac{9 \times 10^9 \times 1.5 \times 10^{-6}}{(0.18)^2}$$

$$= 0.416 \times 10^6 \text{ V m}^{-1}$$

Electric field due to q_2 at Z,

$$E_B = \frac{q_2}{4\pi\varepsilon_0 (BZ)^2}$$

$$= \frac{9 \times 10^9 \times 2.5 \times 10^{-6}}{(0.18)^2}$$

$$= 0.69 \times 10^6 \text{ V m}^{-1}$$

The resultant field intensity at Z,

$$E = \sqrt{E_A^2 + E_B^2 + 2\, E_A E_B \cos 2\theta}$$

$$\cos\theta = \frac{0.10}{0.18} = \frac{5}{9} = 0.5556$$

$$\theta = \cos^{-1}(0.5556) = 56.25°$$

$$\therefore \quad 2\theta = 112.5°$$

$$\cos 2\theta = -0.38$$

$$E = \sqrt{(0.416 \times 10^6)^2 + (0.69 \times 10^6)^2 + 2 \times 0.416 \times 0.69 \times 10^{12} \times (-0.38)}$$

$$= 6.6 \times 10^5 \text{ V m}^{-1}$$

Therefore, the potential at a point 10 cm (perpendicular to the mid-point) is 2.0×10^5 V and electric field is 6.6×10^5 V m^{-1}.

Q. 21. Obtain the equivalent capacitance of the network in Fig. For a 300 V supply, determine the charge and voltage across each capacitor. **[NCERT]**

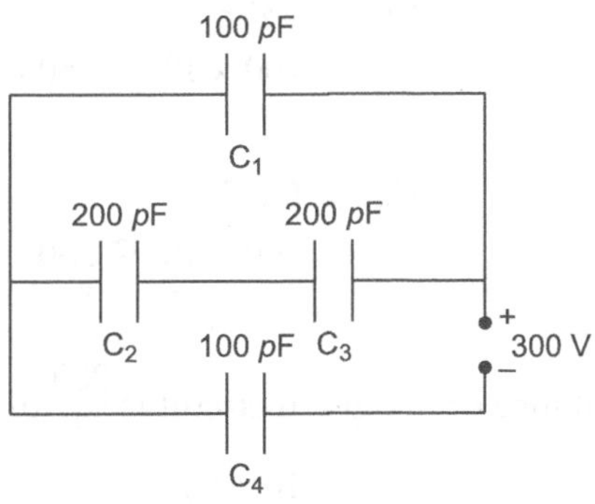

Ans. Capacitance of capacitor C_1 is 100 pF.

Capacitance of capacitor C_2 is 200 pF.

Capacitance of capacitor C_3 is 200 pF.

Capacitance of capacitor C_4 is 100 pF.

Supply potential, V = 300 V

Capacitors C_2 and C_3 are connected in series. Let their equivalent capacitance be C'

$$\therefore \quad \frac{1}{C'} = \frac{1}{200} + \frac{1}{200} = \frac{2}{200}$$

$$C' = 100 \text{ pF}$$

Capacitors C_1 and C' are in parallel. Let their equivalent capacitance be C''.

$$\therefore \quad C'' = C'_1 + C_1$$
$$= 100 + 100 = 200 \text{ pF}$$

C'' and C_4 are connected in series. Let their equivalent capacitance be C.

$$\therefore \quad \frac{1}{C'} = \frac{1}{C''} + \frac{1}{C_4} = \frac{1}{200} + \frac{1}{100} = \frac{2+1}{200}$$
$$C = \frac{200}{3} \text{ pF}$$

Hence, the equivalent capacitance of the circuit is $\frac{200}{3}$ pF.

Potential difference across $C'' = V''$

Potential difference across $C_4 = V_4$

$$\therefore \quad V'' + V_4 = V = 300 \text{ V}$$

Charge on C_4 is given by,

$$Q_4 = CV$$

(As C_4 and C'' are in series then they have same charge = Q_4)

$$= \frac{200}{3} \times 10^{-12} \times 300 = 2 \times 10^{-8} \text{ C}$$

$$\therefore \quad V_4 = \frac{Q_4}{C_4} = \frac{2 \times 10^{-8}}{100 \times 10^{-12}} = 200 \text{ V}$$

$\therefore$ Voltage across C_1 is given by,

$$V_1 = V - V_4 = 300 - 200 = 100 \text{ V}$$

Hence, potential difference, V_1 across C_1 is 100 V.

Charge on C_1 is given by,

$$Q_1 = C_1V_1$$
$$= 100 \times 10^{-12} \times 100 = 10^{-8} \text{ C}$$
$$V_2 = V_3 = 50 \text{ V}$$

Therefore, charge on C_2 is given by,

$$Q_2 = C_2V_2$$
$$= 200 \times 10^{-12} \times 50 = 10^{-8} \text{ C}$$

and charge on C_3 given by,

$$Q_3 = C_3 V_3$$
$$= 200 \times 10^{-12} \times 50$$
$$= 10^{-8} \text{ C}$$

Hence, the equivalent capacitance of the given circuit is $\frac{200}{3}$ pF with.

$$Q_1 = 10^{-8} \text{ C}, \qquad V_1 = 100 \text{ V}$$
$$Q_2 = 10^{-8} \text{ C}, \qquad V_2 = 50 \text{ V}$$
$$Q_3 = 10^{-8} \text{ C}, \qquad V_3 = 50 \text{ V}$$
$$Q_4 = 2 \times 10^{-8} \text{ C}, \qquad V_4 = 200 \text{ V}$$

Q. 22. You are given three capacitors of 2 μF, 3 μF and 4 μF, respectively.★

(a) Form a combination of all these capacitors of equivalent capacitance $\frac{13}{3}$ μF.

(b) What is the maximum and minimum value of the equivalent capacitance that can be obtained by connecting these capacitors?

★ are board based questions from previous year

Ans. Given, capacitor are 2 μF, 3 μF and 4 μF.

(a) The combination of 2 μF and 4 μF in series with 3 μF in parallel.

$$\frac{1}{C_s} = \frac{1}{2} + \frac{1}{4}$$

$$\Rightarrow \quad \frac{1}{C_s} = \frac{2+1}{4}$$

$$\Rightarrow \quad C_s = \frac{4}{3}\,\mu F$$

Now,

$$C_{equivalent} = \frac{4}{3} + 3\,\mu F$$

$$= \frac{13}{3}\,\mu F$$

So, we can get an equivalent of $\frac{13}{3}$ μF by connecting 2 μF and 4 μF in series and 3 μF in parallel.

(b) Maximum capacitance can be achieved by joining the capacitors in parallel.

i.e.,

$$C_{max} = 2 + 3 + 4 = 9\,\mu F$$

Minimum capacitance is achieved by joining the capacitors in series.

i.e.,

$$\frac{1}{C_{min}} = \frac{1}{2} + \frac{1}{3} + \frac{1}{4}$$

$$= \frac{6+4+3}{12}$$

$$C_{min} = \frac{12}{13}\,\mu F$$

Chapter 3. Current Electricity

Q. 1. Estimate the average drift speed of conduction electrons in a copper wire of cross-sectional area 2.5×10^{-7} m^2 carrying a current of 2.7 A. Assume the density of conduction electrons to be 9×10^{28} m^{-3}.

Ans. We know that drift velocity,

$$v_d = \frac{1}{nAq}$$

Where I is the current, *n* is charge density, *q* is charge of electron and A is cross-sectional area.

$$v_d = \frac{2.7}{9 \times 10^{28} \times 2.5 \times 10^{-7} \times 1.6 \times 10^{-19}}$$

$$v_d = 7.5 \times 10^{-4}\ \text{m/s}$$

or

$$v_d = 0.75\ \text{mm s}^{-1}$$

This is the required average drift velocity.

Q. 2. The storage battery of a car has an emf of 12 V. If the internal resistance of the battery is 0.4 Ω, what is the maximum current that can be drawn from the battery ?

Ans. Given,

Emf of the battery, E = 12 V

Internal resistance of the battery, $(r) = 0.4\ \Omega$

Let maximum current drawn from the battery = I

As per ohm's law,

$$E = Ir$$

$$I = \frac{E}{r}$$

$$I = \frac{12}{0.4} = 30\ A$$

The maximum current drawn from the given battery is 30 A.

Q. 3. In a potentiometer arrangement, a cell of emf 1.25 V gives a balance point at 35.0 cm length of the wire. If the cell is replaced by another cell and the balance point shifts to 63.0 cm., what is the emf of the second cell ? **[NCERT]**

Ans. Given,

Emf of the cell, $E_1 = 1.25$ V

Balance point of the potentiometer, $l_1 = 35$ cm

The cell is replaced by another cell of emf E_2.

New balance point of the potentiometer, $l_2 = 63$ cm

The balance condition is given by the relation,

$$\frac{E_1}{E_2} = \frac{l_1}{l_2}$$

$$E_2 = E_1 \times \frac{l_2}{l_1} = 1.25 \times \frac{63}{35} = 2.25 \text{ V}$$

Therefore, emf of the second cell is 2.25 V.

Q. 4. A silver wire has a resistance of 2.1 Ω at 27.5°C and a resistance of 2.7 Ω at 100°C. Determine the temperature coefficient of resistivity of silver. **[NCERT]**

Ans. Given,

Temperature, $T_1 = 27.5$ °C

Resistance of the silver wire at T_1, $R_1 = 2.1\ \Omega$

Temperature, $T_2 = 100$ ° C

Resistance of the silver wire at T_2, $R_2 = 2.7\ \Omega$

Temperature coefficient of silver = α

It is related with temperature and resistance as

$$\alpha = \frac{R_2 - R_1}{R_1 (T_2 - T_1)}$$

$$= \frac{2.7 - 2.1}{2.1\ (100 - 27.5)} = 0.0039°\text{C}^{-1}$$

Therefore, the temperature coefficient of silver is $0.0039°\text{C}^{-1}$.

Q. 5. A negligibly small current is passed through a wire of length 15 m and uniform cross-section 6.0×10^{-7} m^2, and its resistance is measured to be 5.0 Ω. What is the resistivity of the material at the temperature of the experiment ? **[NCERT]**

Ans. Given,

Length of the wire, $l = 15$ m

Area of cross-section of the wire, $a = 6.0 \times 10^{-7}$ m^2

Resistance of the material of the wire, R = 5.0 Ω

Resistivity of the material of the wire = ρ

Resistance is related with the resistivity as

$$R = \rho \frac{l}{A}$$

$$\rho = \frac{RA}{l}$$

$$= \frac{5 \times 6 \times 10^{-7}}{15} = 2 \times 10^{-7}\ \Omega \text{ m}$$

Therefore, the resistivity of the material is 2×10^{-7} Ω m.

Q. 6. A battery of emf 10 V and internal resistance 3 Ω is connected to a resistor. If the current in the circuit is 0.5 A, what is the resistance of the resistor ? What is the terminal voltage of the battery when the circuit is closed ?

Ans. Given,

Emf of the battery, E = 10 V

Internal resistance of the battery $(r) = 3\ \Omega$

Current in the circuit, I = 0.5 A

Let resistance of the resistor = R

From ohm's law

$$I = \frac{E}{R+r}$$

$$R + r = \frac{E}{I}$$

$$= \frac{10}{0.5} = 20\ \Omega$$

$$\therefore \quad R = 20 - 3 = 17\ \Omega$$

Consider terminal voltage of the resistor = V

According to Ohm's law,

$$V = IR$$

$$V = 0.5 \times 17$$

$$V = 8.5\ V$$

Therefore, the resistance of the resistor is 17 Ω and the terminal voltage is 8.5 V.

Q. 7. At room temperature (27.0 °C) the resistance of a heating element is 100 Ω. What is the temperature of the element if the resistance is found to be 117 Ω, given that the temperature coefficient of the material of the resistor is $1.70 \times 10^{-4}\ °C^{-1}$. **[NCERT]**

Ans. Given,

Room temperature, T = 27° C

Resistance of the heating element at T, R = 100 Ω

Assume T_1 is the increased temperature of the filament.

Resistance of the heating element at T_1, $R_1 = 117\ \Omega$

Temperature co-efficient of the material of the filament,

$$\alpha = 1.70 \times 10^{-4}\ °C^{-1}$$

α is given by the relation,

$$\alpha = \frac{R_1 - R}{R(T_1 - T)}$$

$$T_1 - T = \frac{R_1 - R}{R\alpha}$$

$$T_1 - 27 = \frac{117 - 100}{100\ (1.7 \times 10^{-4})}$$

$$T - 27 = 1000$$

$$T_1 = 1027\ °C$$

Therefore, at 1027°C, the resistance of the element is 117 Ω.

Q. 8. Two cells of emfs 1.5 V and 2.0 V having internal resistances 0.2 Ω and 0.3 Ω respectively are connected in parallel. Calculate the emf and internal resistance of the equivalent cell.★

Ans. Given,

$$E_1 = 1.5\ V,\ r_1 = 0.2\ \Omega$$

$$E_2 = 2\ V,\ r_2 = 0.3\ \Omega$$

Equivalent emf,
$$E = \frac{\frac{E_1}{r_1} + \frac{E_2}{r_2}}{\frac{1}{r_1} + \frac{1}{r_2}}$$

$$= \frac{(1.5 \times 0.3) + (2 \times 0.2)}{0.2 + 0.3} = \frac{0.45 + 0.4}{0.5}$$

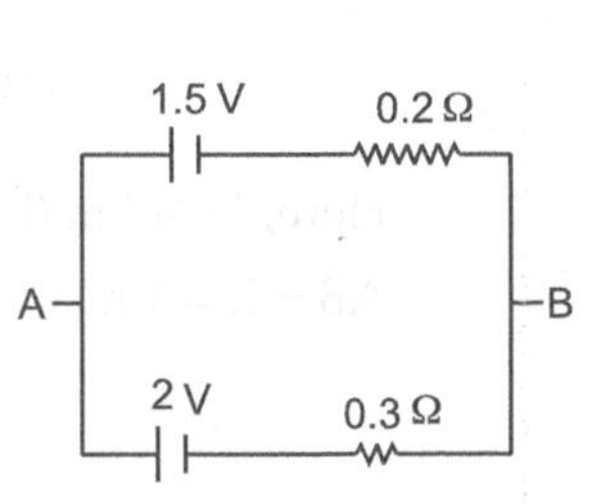

★ are board based questions from previous year

$$= \frac{0.85}{0.5} = 1.7 \text{ volt}$$

$$\text{Equivalent internal resistance} = \frac{r_1 r_2}{r_1 + r_2}$$

$$= \frac{0.2 \times 0.3}{0.2 + 0.3} = \frac{0.06}{0.5}$$

$$r_{eq} = 0.12\ \Omega$$

Q. 9. The number density of free electrons in a copper conductor estimated is $8.5 \times 10^{28}\ m^{-3}$. How long does an electron take to drift from one end of a wire 3.0 m long to its other end ? The area of cross-section of the wire is $2.0 \times 10^{-6}\ m^2$ and it is carrying a current of 3.0 A. **[NCERT]**

Ans. **Given :** Density of free electrons in a copper conductor, $n = 8.5 \times 10^{28}\ m^{-3}$

Length of copper wire, $l = 3.0$ m

Area of cross-section of the wire, $A = 2.0 \times 10^{-6}\ m^2$

Current carried by the wire, I = 3.0 A,

$$I = nAev_d$$

Where, e = Electric charge = 1.6×10^{-19} C

$$v_d = \text{Drift velocity} = \frac{\text{Length of the wire } (l)}{\text{Time taken to cover } (r)}$$

$$I = \frac{nAel}{t}$$

$$t = \frac{nAel}{I}$$

$$= \frac{3 \times 8.5 \times 10^{28} \times 2 \times 10^{-6} \times 1.6 \times 10^{-19}}{3.0}$$

$$= 2.7 \times 10^4 \text{ s}$$

Therefore, the time taken by an electron to drift from one end of the wire to the other is 2.7×10^4 s.

Q. 10. A potentiometer wire of length 1 m has a resistance of 10 Ω. It is connected to a 6 V battery in series with a resistance of 5 Ω. Determine the emf of the primary cell which gives a balance point at 40 cm.★

Ans. From the figure below :

Total resistance of the circuit,

$$R = (R_{AB} + 5) = 15\ \Omega$$

Current in the circuit,

$$I = \frac{V}{R} = \frac{6}{15} \text{ A}$$

$$V_{AB} = 6 - \frac{6}{15} \times 5$$

$$= 6 - 2 = 4 \text{ V}$$

∵ Voltage across AB,

$$\text{Emf of the cell, } e = \left(\frac{l}{L}\right) V_{AB}$$

Here, l = 40 m (balance point)

AB = L = 1 m = 100 cm (total length of the wire)

∴ $$e = \left(\frac{40}{100}\right) 4 = 1.6 \text{ V}$$

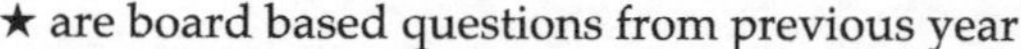

★ are board based questions from previous year

Q. 11. A storage battery of emf 8.0 V and internal resistance 0.5 Ω is being charged by a 120 V dc supply using a series resistor of 15.5 Ω. What is the terminal voltage of the battery during charging ? What is the purpose of having a series resistor in the charging circuit ? **[NCERT]**

Ans. Given,

Emf of the storage battery, E = 8.0 V

Internal resistance of the battery, $r = 0.5\ \Omega$

DC supply voltage, V = 120 V

Resistance of the resistor, R = 15.5 Ω

Effective voltage in the circuit = V′

R is connected to the storage battery in series. Hence, it can be written as

$$V' = V - E$$

$$V' = 120 - 8 = 112\text{ V}$$

Current flowing in the circuit = I, which is given by the relation,

$$I = \frac{V'}{R + r}$$

$$I = \frac{112}{15.5 + 0.5} = \frac{112}{16} = 7\text{ A}$$

Voltage across resistor R given by the product, IR = 7 × 15.5 = 108.5 V

DC supply voltage = Terminal voltage of battery + Voltage drop across R

Terminal voltage of battery = 120 – 108.5 = 11.5 V

A series resistor in a charging circuit limits the current drawn from the external source.

The current will be extremely high in its absence. This is very dangerous.

Q. 12. (i) The potential difference applied across a given resistor is altered so that the heat produced per second increases by a factor of 9. By what factor does the applied potential difference change ?★

(ii) In the figure shown, an ammeter A and a resistor of 4 Ω are connected to the terminals of the source. The emf of the source is 12 V having the internal resistance of 2 Ω.★

Calculate the voltmeter and ammeter readings.

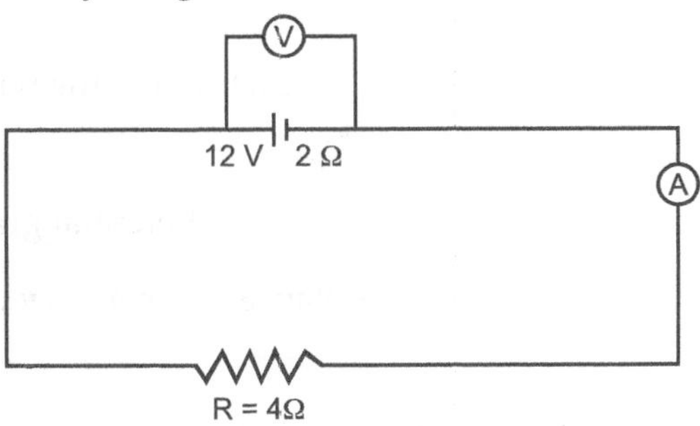

Ans. (i) Heat produced,

$$H = \frac{V^2}{R}t$$

$$\Rightarrow \quad H \propto V^2$$

$$\frac{H'}{H} = \frac{V'^2}{V^2}$$

$$\frac{9H}{H} = \frac{V'^2}{V^2} \qquad [\because H' = 9H]$$

$$V'^2 = 9V^2$$

$$V' = 3V$$

So, potential difference is increased by a factor of 3.

(ii) Current,
$$I = \frac{E}{R + r}$$

★ are board based questions from previous year

$$I = \frac{12}{4+2} = \frac{12}{6} = 2A$$

$$\text{Ammeter reading} = 2A$$

$$V = E - Ir$$

$$V = 12 - 2 \times 2$$

$$= 12 - 4$$

$$V = 8V$$

$$\text{Voltmeter reading} = 8V$$

Q. 13. In the following potentiometer circuit AB is a uniform wire of length 1 m and resistance 10 Ω. Calculate the potential gradient along the wire and balance length AO (= l).

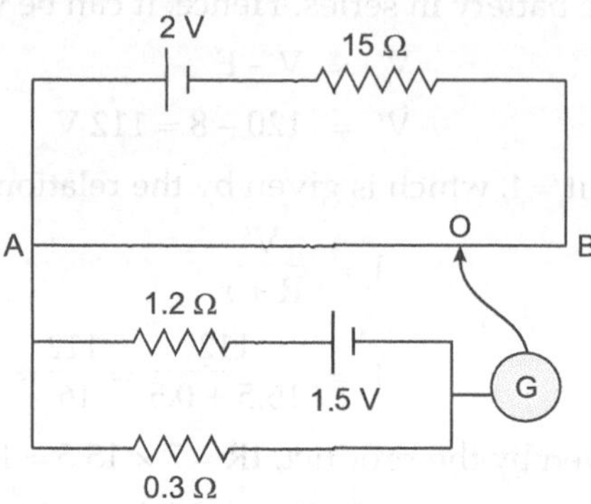

Ans. Total resistance of the primary circuit = 15 + 10 = 25 Ω.

$$\text{emf} = 2\text{ V}$$

∴ Current in the wire AB

$$I = \frac{2}{25} = 0.08\text{ A}$$

$$\text{P.D. across the wire AB} = \text{Current} \times \text{Resistance of wire AB}$$

$$= 0.08 \times 10 = 0.8\text{ V}$$

$$\text{Potential gradient} = \frac{\text{P.D.}}{\text{Length}} = \frac{0.8}{100} = 0.008\text{ V cm}^{-1}$$

$$\text{Resistance of secondary circuit} = 1.2 + 0.3 = 1.5\ \Omega$$

$$\text{emf} = 1.5\text{ V}$$

$$\text{Current in the secondary circuit} = \frac{1.5}{1.5} = 1.0\text{ A}$$

The same is the current in 0.3 Ω resistor.

P.D. between points A and O,

P.D. across 0.3 Ω resistor in the zero-deflection condition

$$= \text{Current} \times \text{Resistance}$$

$$= 1.0 \times 0.3 = 0.3\text{ V}$$

$$\text{Length AO} = \frac{\text{Potential difference}}{\text{Potential gradient}}$$

$$= \frac{0.3\text{ V}}{0.008\text{ V cm}^{-1}} = 37.5\text{ cm}$$

Q. 14. A heating element using nichrome connected to a 230 V supply draws an initial current of 3.2 A which settles after a few seconds to a steady value of 2.8 A. What is the steady temperature of the heating element if the room temperature is 27.0°C ? Temperature coefficient of resistance of nichrome averaged over the temperature range involved is $1.70 \times 10^{-4}\ °C^{-1}$. **[NCERT]**

Ans. Given,

Supply voltage, V = 230 V

Initial current drawn, $I_1 = 3.2$ A

Initial resistance = R_1, which is given by the relation,

$$R_1 = \frac{V}{I} = \frac{230}{3.2} = 71.87\ \Omega$$

Steady state value of the current, $I_2 = 2.8$ A

Resistance at the steady state = R_2, which is given as

$$R_2 = \frac{230}{2.8} = 82.14\ \Omega$$

Temperature co-efficient of nichrome, $\alpha = 1.70 \times 10^{-4}\ °C^{-1}$

Initial temperature of nichrome, $T_1 = 27.0°C$

Steady state temperature reached by nichrome = T_2

T_2 can be obtained by the relation for α,

$$\alpha = \frac{R_2 - R_1}{R_1 (T_2 - T_1)}$$

$$T_2 - 27°C = \frac{82.14 - 71.87}{71.87 \times 1.7 \times 10^{-4}} = 840.5$$

$$T_2 = 840.5 + 27 = 867.5\ °C$$

Therefore, the steady temperature of the heating element is 867.5°C.

Q. 15. In the given figure a long uniform potentiometer wire AB is having a constant potential gradient along its length. The null points for the two primary cells of emfs ε_1 and ε_2 connected in the manner shown are obtained at a distance of 120 cm and 300 cm from the end A. Find (i) $\frac{\varepsilon_1}{\varepsilon_2}$ and (ii) position of null point for the cell ε_1. How is the sensitivity of a potentiometer increased ?

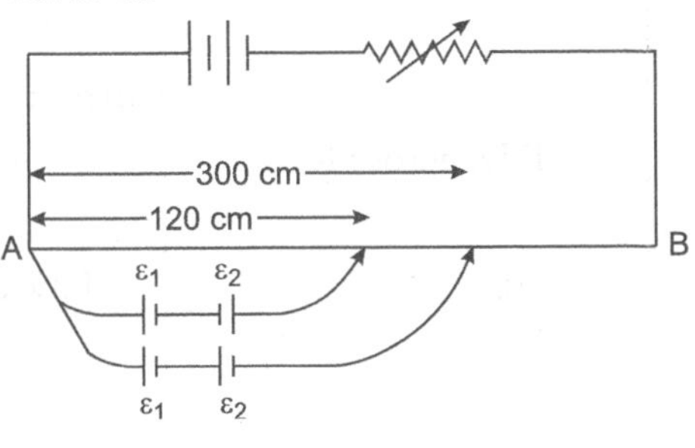

Ans. (i) Apply Kirchhoff's law in loop ACFGA :

$$k\,(120) = \varepsilon_1 - \varepsilon_2$$

k = potential drop per unit length

or $$\varepsilon_1 = \varepsilon_2 + k\,(120) \quad \ldots\text{(i)}$$

For loop AEHIA :

$$k\,(300) = \varepsilon_2 + \varepsilon_1$$

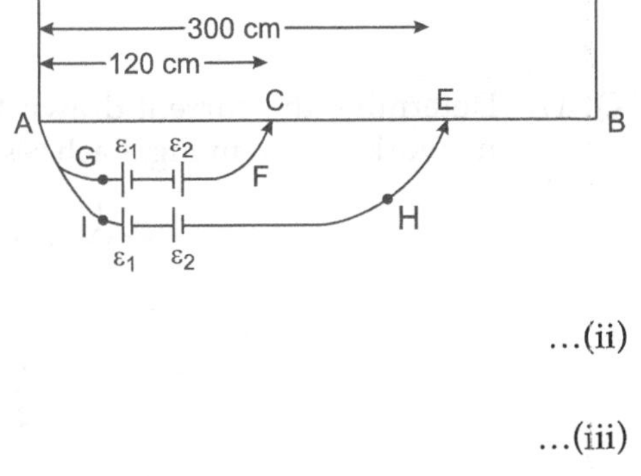

By substituting value of ε_1 from equation (i),

$$\varepsilon_2 + [\varepsilon_2 + k\,(120)] = k\,(300)$$

$$2\varepsilon_2 = k\,(300 - 120)$$

or $$\varepsilon_2 = 90\,k \quad \ldots\text{(ii)}$$

Thus, $$\varepsilon_1 = 90\,k + 120\,k$$

$$\varepsilon_1 = 210\,k \quad \ldots\text{(iii)}$$

Hence, $$\frac{\varepsilon_1}{\varepsilon_2} = \frac{210}{90} = \frac{7}{3}$$

(ii) As we know, $\varepsilon = kl$

Thus, from equations (ii) and (iii),

Null point for cell ε_2 is 90 cm

And for cell ε_1, it is 210 cm.

Sensitivity of the potentiometer can be increased by :

(a) Increasing the length of the potentiometer wire.

(b) Decreasing the resistance in the primary circuit.

Q. 16. In the circuit shown, $R_1 = 4\ \Omega$, $R_2 = R_3 = 15\ \Omega$, $R_4 = 30\ \Omega$ and $E = 10$ V. Calculate the equivalent resistance of the circuit and the current in each resistor.★

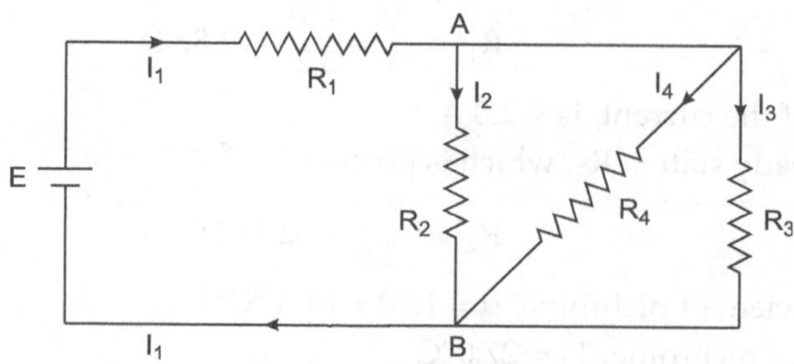

Ans. R_2, R_3, R_4 are in parallel combination.

$$\therefore \quad \frac{1}{R_{234}} = \frac{1}{R_2} + \frac{1}{R_3} + \frac{1}{R_4}$$

$$= \frac{1}{15} + \frac{1}{15} + \frac{1}{30} = \frac{2+2+1}{30} = \frac{5}{30}$$

$$\Rightarrow \quad R_{234} = 6\ \Omega$$

Now R_{234} is in series with R_1, so,

$$R_{eq} = 4\ \Omega + 6\ \Omega = 10\ \Omega$$

$$\therefore \quad I = \frac{E}{R_{eq}} = \frac{10}{10}\text{ A} = 1\text{ A}$$

$$\therefore \quad \text{Current through } R_1 = 1\text{ A}$$

P.D. across R_1,

$$V = IR_1$$

$$= 1 \times 4 = 4\text{V}$$

So,

$$\text{P.D. across } R_{234} = 6\text{ V}$$

$$\because \quad I_2R_2 = I_3R_3 = I_4R_4 = 6\text{ V}$$

$$I_2 = \frac{6}{15}\text{ A} = 0.4\text{ A}$$

$$I_3 = \frac{6}{15}\text{ A} = 0.4\text{ A}$$

$$I_4 = \frac{6}{30}\text{ A} = 0.2\text{ A}$$

Q. 17. Determine the current drawn from a 12 V supply with internal resistance 0.5 Ω by the infinite network shown in Fig. Each resistor has 1 Ω resistance. **[NCERT]**

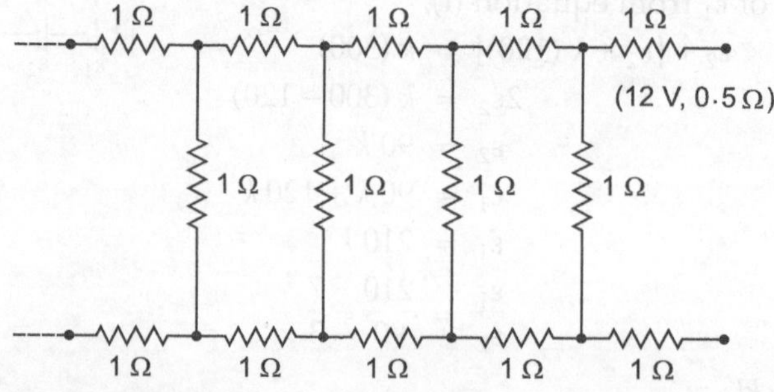

Ans. The resistance of each resistor connected in the given circuit, $R = 1\ \Omega$

Equivalent resistance of the given circuit = R′

The network is infinite. Hence, equivalent resistance is given by the relation,

Hence equivalent resistance can get from redrawn circuit as below

★ are board based questions from previous year

$$R' = 2 + R''$$

$$\therefore \quad R' = 2 + \frac{R'}{(R'+1)} \qquad \left[\because R'' = \frac{R'}{R'+1}\right]$$

$$(R')^2 - 2R' - 2 = 0$$

$$R' = \frac{2 \pm \sqrt{4+8}}{2}$$

$$= \frac{2 \pm \sqrt{12}}{2} = 1 \pm \sqrt{3}$$

$$R'' = \frac{R'}{R'+1}$$

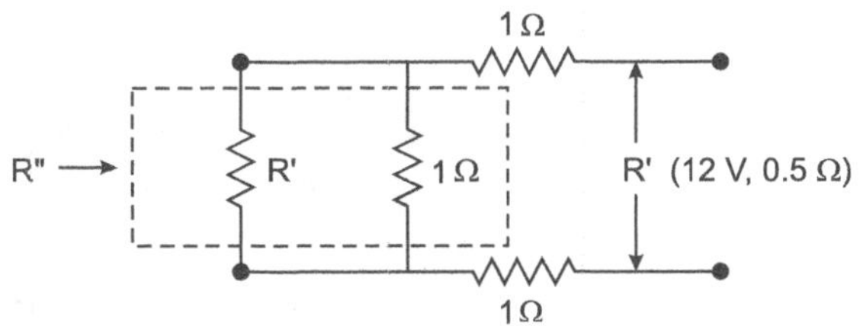

Negative value of R′ cannot be accepted. Hence, equivalent resistance,

$$R' = (1+\sqrt{3}) = 1 + 1.73 = 2.73\ \Omega$$

Internal resistance of the circuit, $r = 0.5\ \Omega$

Hence, total resistance of the given circuit = 2.73 + 0.5 = 3.23 Ω

Supply voltage, V = 12 V

According to Ohm's Law, current drawn from the source is given by the ratio,

$$I = \frac{V}{R} = \frac{12}{3.23} = 3.72\ A$$

Q. 18. An ammeter of resistance 0.6 Ω can measure current up to 1.0 A. Calculate :

(i) The shunt resistance required to enable the ammeter to measure current up to 5.0 A.

(ii) The combined resistance of the ammeter and the shunt.

Ans. We have, resistance of ammeter, $R_A = 0.60$ ohm.

Maximum current across ammeter, $I_A = 1.0$ A

So, voltage across ammeter,

$$V = IR = 1.0 \times 0.60$$

$$V = 0.6\ V$$

Let the value of shunt be x.

(i) Resistance of ammeter with shunt

$$R = \frac{R_A x}{R_A + x} = \frac{0.6x}{0.6+x}$$

$$\text{Current through ammeter (I)} = 5\ A$$

Now,

$$\left(\frac{0.6x}{0.6+x}\right) \times 5 = 0.6$$

$$0.6x \times 5 = 0.6\,(0.6+x)$$

$$3x = 0.36 + 0.6x$$

$$3x - 0.6x = 0.36$$

$$x = \frac{0.36}{2.4}$$

$$x = 0.15\ \Omega$$

Therefore, the shunt resistance is 0.15 Ω.

(ii) Combined resistance of the ammeter and the shunt is given by

$$R = \frac{0.6x}{0.6+x}$$

$$R = \frac{0.6 \times 0.15}{0.6 + 0.15}$$

$$R = 0.12\ \Omega$$

Q. 19. Determine the current in each branch of the network shown in figure. **[NCERT]**

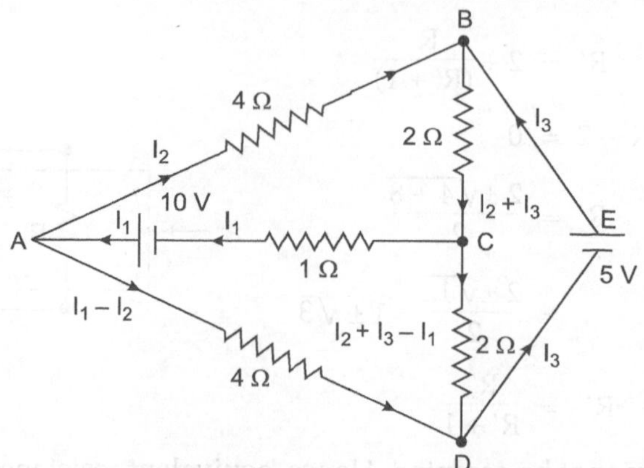

Ans. Consider current from 10 V battery in AC path as I_1 and current from 5 V battery in BD path as I_3 and In AB path I_2 Amp by use of Kirchhoff's current law, rest of the branch current can be assigned in terms of I_1, I_2, I_3.

Applying second rule of Kirchhoff to three different closed loops. Kirchhoff's second rule for the closed loop ADCA gives.

$$10 - 4(I_1 - I_2) + 2(I_2 + I_3 - I_1) - I_1 = 0$$

that is, $$7I_1 - 6I_2 - 2I_3 = 10 \quad \text{...(i)}$$

For the closed loop ABCA, we get

$$10 - 4I_2 - 2(I_2 + I_3) - I_1 = 0$$

that is, $$I_1 + 6I_2 + 2I_3 = 10 \quad \text{...(ii)}$$

For the closed loop BCDEB, we get

$$5 - 2(I_2 + I_3) - 2(I_2 + I_3 - I_1) = 0$$

that is, $$2I_1 - 4I_2 - 4I_3 = -5 \quad \text{...(iii)}$$

Equations (i), (ii), (iii) are three equations in three unknowns. After solving (i), (ii) and (iii)

$$I_1 = 2.5 \text{ A}, I_2 = \frac{5}{8} \text{ A}, I_3 = 1\frac{7}{8} \text{ A}$$

The currents in the various branches of the network are

$$AB = \frac{5}{8} \text{ A}, CA = 2\frac{1}{2} \text{ A}, DEB = 1\frac{7}{8} \text{ A}$$

$$AD = 1\frac{7}{8} \text{ A}, CD = 0\text{A}, BC = 2\frac{1}{2} \text{ A}$$

Q. 20. The four arms of a Wheatstone bridge (fig.) have the following resistances :

AB = 100 Ω, BC = 10 Ω, CD = 5 Ω, and DA = 60 Ω

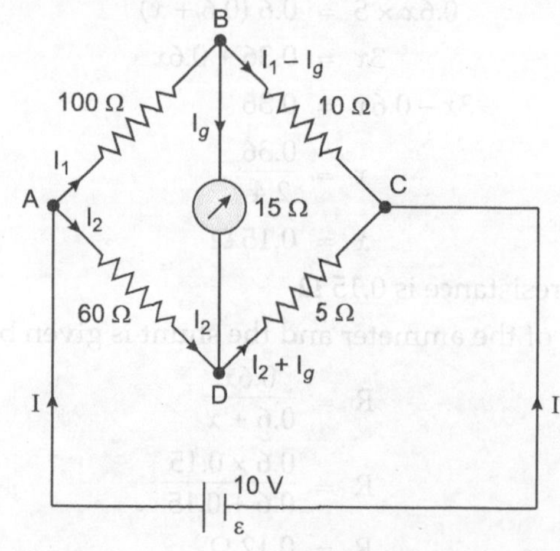

A galvanometer of 15 Ω resistance is connected across BD. Calculate the current through the galvanometer when a potential difference of 10 V is maintained across AC. **[NCERT]**

Ans. Applying Kirchhoff's voltage law in mesh BADB, we have

$$100\,I_1 + 15\,I_g - 60I_2 = 0$$

or
$$20I_1 + 3I_g - 12I_2 = 0 \quad \text{...(i)}$$

Considering KVL for mesh BCDB, we have

$$10\,(I_1 - I_g) - 15I_g - 5\,(I_2 + I_g) = 0$$
$$10I_1 - 30\,I_g - 5I_2 = 0$$
$$2I_1 - 6I_g - I_2 = 0 \quad \text{...(ii)}$$

Considering KVL for mesh ADCEA,

$$60I_2 + 5\,(I_2 + I_g) = 10$$
$$65I_2 + 5I_g = 10$$
$$13I_2 + I_g = 2 \quad \text{...(iii)}$$

Multiplying Equation (ii) by 10

$$20I_1 - 60I_g - 10I_2 = 0 \quad \text{...(iv)}$$

From equations (iv) and (i) we have

$$63I_g - 2I_2 = 0$$
$$I_2 = 31.5\,I_g$$

Substituting the value of I_2 into equation (iii), we get

$$13\,(31.5\,I_g) + I_g = 2$$
$$410.5\,I_g = 2$$
$$I_g = 4.87 \text{ mA.}$$

Q. 21. (i) Three resistors 2 Ω, 4 Ω and 5 Ω are combined in parallel. What is the total resistance of the combination ?

(ii) If the combination is connected to a battery of emf 20 V and negligible internal resistance, determine the current through each resistor and the total current drawn from the battery.

Ans. (i) There are three resistors of resistances,

$R_1 = 2\ \Omega$, $R_2 = 4\ \Omega$ and $R_3 = 5\ \Omega$

They are connected in parallel. Hence, total resistance (R) of the combination is given by :

$$\frac{1}{R} = \frac{1}{R_1} + \frac{1}{R_2} + \frac{1}{R_3}$$
$$\frac{1}{R} = \frac{1}{2} + \frac{1}{4} + \frac{1}{5}$$
$$\frac{1}{R} = \frac{10+5+4}{20} = \frac{19}{20}$$ 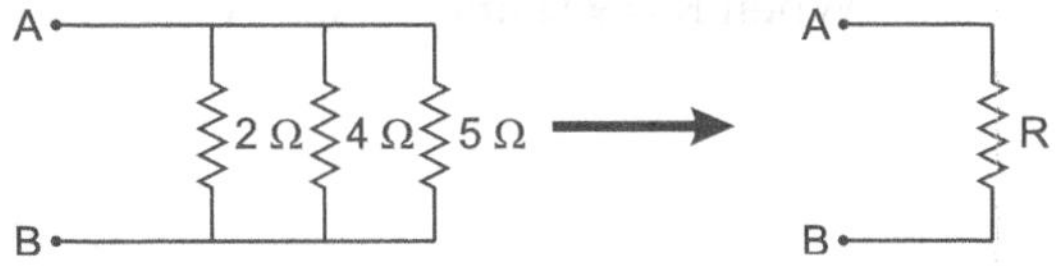
$$\therefore \quad R = \frac{20}{19}\ \Omega$$

A
2 Ω
4 Ω
5 Ω
B
A
R
B

Therefore, total resistance of the combination is $\frac{20}{19}$ Ω.

(ii) Emf of the battery, V = 20 V

Current (I_1) flowing through resistor R_1 is given by

$$I_1 = \frac{V}{R_1} = \frac{20}{2} = 10 \text{ A}$$

Current (I_2) flowing through resistor R_2 is given by,

$$I_2 = \frac{V}{R_2} = \frac{20}{4} = 5 \text{ A}$$

Current (I_3) flowing through resistor R_3 is given by,

$$I_3 = \frac{V}{R_3} = \frac{20}{5} = 4 \text{ A}$$

Total current , $I = I_1 + I_2 + I_3 = 10 + 5 + 4 = 19$ A

Therefore, the current through each resistor is 10 A, 5 A, and 4 A respectively and the total current is 19 A.

Q. 22. A cell of emf 'E' and internal resistance '*r*' is connected across a variable load resistor R. Draw the plots of the terminal voltage V versus (i) R and (ii) the current *i*. It is found that when R = 4 Ω, the current is 1 A when R is increased to 9 Ω, the current reduces to 0.5 A. Find the values of the emf E and internal resistance *r*.★

Ans. (i) Graph between terminal voltage (V) and resistance (R) :

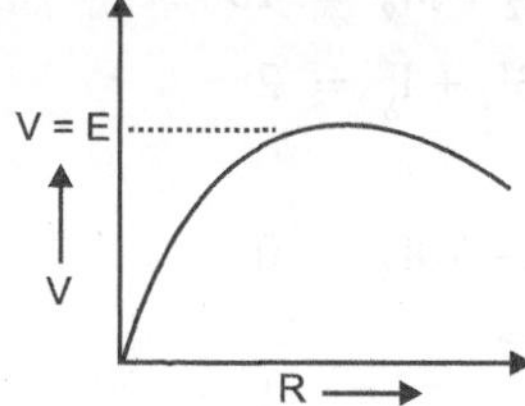

(ii) Graph between terminal voltage (V) and current (I) :

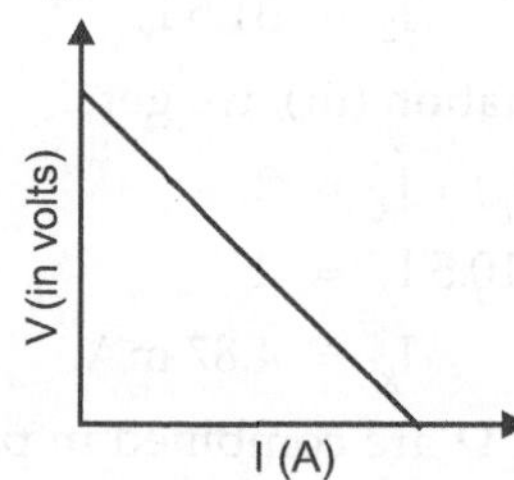

(iii) When R = 4 Ω and I = 1 A

We know

Terminal voltage, $V = E - Ir$

So, we have

$$V = IR = 4 = E - Ir$$

$$E - r = 4 \quad \text{...(i)}$$

When R = 9 Ω and I = 0.5 A

$$V = IR$$

$$= 0.5 \times 9 = E - 0.5r$$

$$E - 0.5r = 4.5 \quad \text{...(ii)}$$

Subtracting (i) from (ii), we get

$$E - 0.5r - E + r = 4.5 - 4$$

$$0.5r = 0.5$$

$$r = 1 \, \Omega$$

Substituting value of r in (i)

$$E - 1 = 4$$

$$E = 5 \text{ V}$$

Thus, $r = 1$ Ω and E = 5 V.

Q. 23. Use Kirchhoff's rules to determine the potential difference between the points A and D when no current flows in the arm BE of the electric network shown in the figure.★

★ are board based questions from previous year

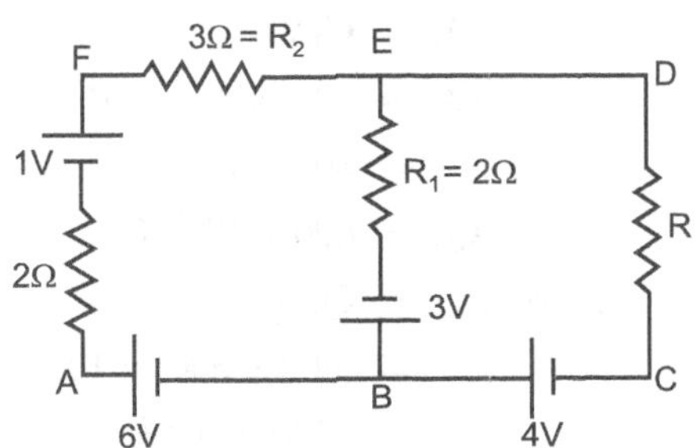

Ans.

According to Kirchhoff's junction law at B,

$$i_3 = i_1 + i_2$$

As $i_2 = 0$ (given)

$\therefore$ $i_3 = i_1$

Applying second law to loop AFEB,

$$\therefore \quad i_3 \times 2 + i_3 \times 3 + i_2 R_1 = 1 + 3 + 6$$

$$i_3 = i_1 = 2\text{ A}$$

The potential difference between A to D along the branch AFD,

$$\therefore \quad V_{AD} = 2i_3 - 1 + 3 \times i_3$$

$$= (4 - 1 + 6)\text{V} = 9\text{ V}$$

Q. 24. Two identical capacitors of 12 pF each are connected in series across a battery of 50 V. How much electrostatic energy is stored in the combination ? If these were connected in parallel across the same battery, how much energy will be stored in the combination now ? Also find the charge drawn from the battery in each case.★

Ans. **Case I :** When the capacitors are connected in series.

The equivalent capacitance is given by

$$C_{eq} = \frac{12 \times 12}{12 + 12} = \frac{12 \times 12}{24}$$

$$C_{eq} = 6\ pF$$

Electrostatic energy, $U = \frac{1}{2}CV^2$

$$= \frac{1}{2} \times 6 \times 10^{-12} \times 50 \times 50$$

$$U = 75 \times 10^{-10}\text{ J.}$$

As the capacitors are connected in series so, the charge remains the same.

$$\therefore \quad Q = C_{eq}V$$

$$Q = 6 \times 10^{-12} \times 50$$

$$Q = 300\ pC$$

Case II : When the capacitors are connected is parallel.

The equivalent capacitance is given by,

$$C_{eq} = 12 + 12$$

$$= 24\ pF$$

Electrostatic energy, $U = \frac{1}{2} \times 24 \times 10^{-12} \times (50)^2$

$$U = 3 \times 10^{-8}\text{ J}$$

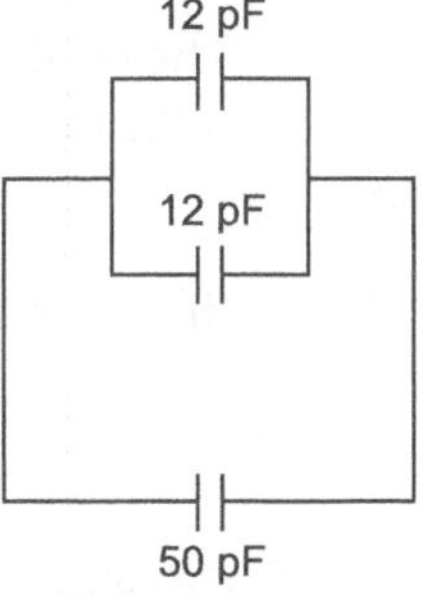

As the capacitors are connected in parallel so, the charge on each capacitor is different but potential difference remains the same.

Charge on C_1,

★ are board based questions from previous year

$$Q_1 = C_1V$$
$$Q_1 = 12 \times 10^{-12} \times 50$$
$$= 600 \times 10^{-12}\ C$$

Similarly,
$$Q_2 = 600 \times 10^{-12}\ C$$

Total charge,
$$Q = Q_1 + Q_2$$
$$= (600 + 600) \times 10^{-12}$$
$$= 1200 \times 10^{-12}$$
$$= 1200\ pC$$

Q. 25. A network of resistors is connected to a 16 V battery with internal resistance of 1 Ω, as shown in Fig. 1 :

(i) Compute the equivalent resistance of the network.

(ii) Obtain the current in each resistor.

(iii) Obtain the voltage drops V_{AB}, V_{BC} and V_{CD}. **[NCERT]**

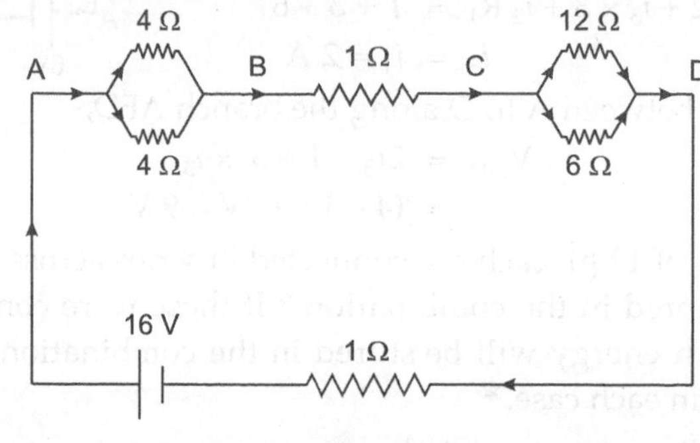

Fig. 1

Ans. (i) In network the two 4 Ω resistors in parallel are equivalent to a resistor

$$= [(4 \times 4)/(4 + 4)]\ \Omega = 2\ \Omega = R'.$$

In the same way, the 12 Ω and 6 Ω resistors in parallel are equivalent to a resistor of

$$\frac{[(12 \times 6)}{(12 + 6)]}\Omega = 4\ \Omega = R_1'$$

Fig. 2 **Fig. 2 and Fig. 3 shows equivalent circuit**

The equivalent resistance R of the network is obtained by combining these resistors (2 Ω and 4 Ω) with 1 Ω in series, that is

$$R = 2\ \Omega + 4\ \Omega + 1\ \Omega = 7\ \Omega = R_2'$$

(ii) The total current I in the circuit is

$$I = \frac{\varepsilon}{R + r} = \frac{16\ V}{(7 + 1)\ \Omega}$$
$$= 2\ A$$

Consider the resistors between A and B. If I_1 is the current in one of the 4 Ω resistors and I_2 the current in the other.

$$I_1 \times 4 = I_2 \times 4$$

(As both are in parallel so potential drop across them will be same)

$$I_1 = I_2 = 1\text{ A}$$

that is, current in each 4 Ω resistor is 1 A. Current in 1 Ω resistor between B and C would be 2 A. If I_3 is the current in the 12 Ω resistor, and I_4 in the 6 Ω resistor.

$$I_3 \times 12 = I_4 \times 6, \textit{ i.e., } I_4 = 2I_3$$

(Potential drop across both resistors must be same)

But, $$I_3 + I_4 = I = 2\text{ A}$$

Thus, $$I_3 = \left(\frac{2}{3}\right)\text{A}, I_4 = \left(\frac{4}{3}\right)\text{A}$$

that is, the current in the 12 Ω resistor is (2/3) A, while the current in the 6 Ω resistor is (4/3) A.

(iii) The voltage drop across AB is

$$V_{AB} = I_1 \times 4 = 1\text{ A} \times 4\ \Omega = 4\text{ V}.$$

The voltage drop across BC is

$$V_{BC} = 2\text{ A} \times 1\ \Omega$$
$$= 2\text{ V}$$

The voltage drop across CD is

$$V_{CD} = 12\ \Omega \times I_3$$
$$= 12\ \Omega \times \left(\frac{2}{3}\right)\text{A} = 8\text{ V}.$$

Q. 26. Determine the current in each branch of the network shown in figure : **[NCERT]**

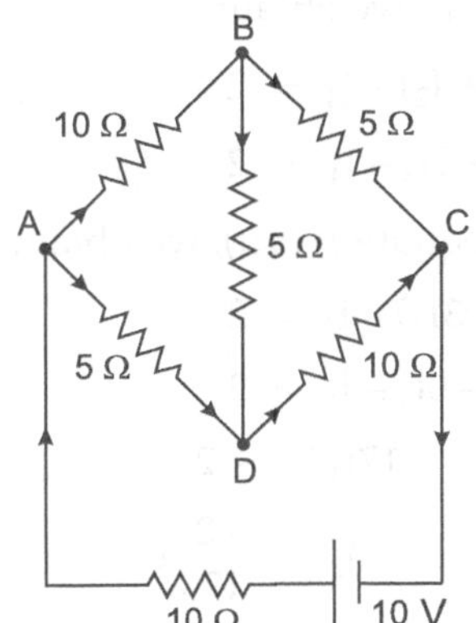

Ans. Current flowing through various branches of the circuit is represented in the given figure.

I_1 = Current flowing through the 10 V battery

I_2 = Current flowing through branch AB

I_3 = Current flowing through branch AD

$I_2 - I_4$ = Current flowing through branch BC

$I_3 + I_4$ = Current flowing through branch CD

I_4 = Current flowing through branch BD

For the closed circuit ABDA, potential is zero *i.e.,*

$$10I_2 + 5I_4 - 5I_3 = 0$$
$$2I_2 + I_4 - I_3 = 0$$
$$I_3 = 2I_2 + I_4 \quad \text{...(i)}$$

For the closed circuit BCDB, potential is zero *i.e.,*

$$5(I_2 - I_4) - 10(I_3 + I_4) - 5I_4 = 0$$
$$5I_2 + 5I_4 - 10I_3 - 10I_4 - 5I_4 = 0$$
$$5I_2 - 10I_3 - 20I_4 = 0$$

$$I_2 = 2I_3 + 4I_4 \quad \text{...(ii)}$$

For the closed circuit ABCFEA, potential is zero *i.e.*,

$$-10 + 10(I_1) + 10(I_2) + 5(I_2 - I_4) = 0$$

$$10 = 15I_2 + 10I_1 - 5I_4$$

$$3I_2 + 2I_1 - I_4 = 2 \quad \text{... (iii)}$$

From equations (i) and (ii), we obtain

$$I_3 = 2(2I_3 + 4I_4) + I_4$$

$$I_3 = 4I_3 + 8I_4 + I_4$$

$$-3I_3 = 9I_4$$

$$-3I_4 = +I_3 \quad \text{...(iv)}$$

Putting equation (iv) in equation (i), we obtain

$$I_3 = 2I_2 + I_4$$

$$-4I_4 = 2I_2$$

$$I_2 = -2I_4 \quad \text{...(v)}$$

It is evident from the given figure that,

$$I_1 = I_3 + I_2 \quad \text{...(vi)}$$

Putting equation (vi) in equation (i), we obtain

$$3I_2 + 2(I_3 + I_2) - I_4 = 2$$

$$5I_2 + 2I_3 - I_4 = 2 \quad \text{...(vii)}$$

Putting equations (iv) and (v) in equation (vii), we obtain

$$5(-2I_4) + 2(-3I_4) - I_4 = 2$$

$$-10I_4 - 6I_4 - I_4 = 2$$

$$17I_4 = -2$$

$$I_4 = \frac{-2}{17}\text{ A}$$

Equation (iv) reduces to

$$I_3 = -3(I_4)$$

$$I_3 = -3\left(\frac{-2}{17}\right) = \frac{6}{17}\text{ A}$$

$$I_2 = -2(I_4)$$

$$I_2 = -2\left(\frac{-2}{17}\right) = \frac{4}{17}\text{ A}$$

$$I_2 - I_1 = \frac{4}{17} - \left(\frac{-2}{17}\right) = \frac{6}{17}\text{ A}$$

$$I_3 + I_4 = \frac{6}{17} + \left(\frac{-2}{17}\right) = \frac{4}{17}\text{ A}$$

$$I_1 = I_3 + I_2$$

$$= \frac{6}{17} + \frac{4}{17} = \frac{10}{17}\text{ A}$$

Therefore, current in branch AB $= \frac{4}{17}$ A

$$\text{In branch BC} = \frac{6}{17}\text{ A}$$

$$\text{In branch CD} = \frac{-4}{17}\text{ A}$$

$$\text{In branch AD} = \frac{6}{17}\text{ A}$$

$$\text{In branch BD} = \left(\frac{-2}{17}\right)\text{ A}$$

$$\text{Total current} = \frac{4}{17}+\frac{6}{17}+\frac{-4}{17}+\frac{6}{17}+\frac{-2}{17}$$

$$= \frac{10}{17}\text{ A}$$

Q. 27. In a potentiometer arrangement for determining the emf of a cell, the balance point of the cell in open circuit is 350 cm. When a resistance of 9 Ω is used in the external circuit of the cell, the balance point shifts to 300 cm. Determine the internal resistance of the cell.★

Ans. Balance point in open circuit, $l_1 = 350$ cm, R = 9 Ω,

New balance point $l_2 = 300$ cm.

Internal resistance of a cell is given by,

$$r = R\left(\frac{l_1 - l_2}{l_2}\right)$$

$$= 9\left(\frac{350-300}{300}\right) = \frac{9 \times 50}{300} = \frac{3}{2} = 1.5\ \Omega$$

Q. 28. The plot of the variation of potential difference across a combination of three identical cells in series, versus current is shown below. What is the emf and internal resistance of each cell ?★

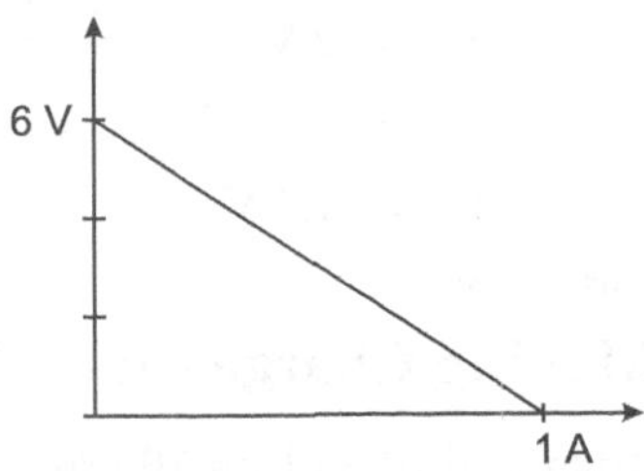

Ans. We know that,

$$V = E - Ir$$

Where E is the e.m.f. and r is the total internal resistance.

When $I = 0$,

$$\text{Total emf} = \text{Terminal voltage}$$

$$3E = 6\text{ V}$$

or $E = 2$ V

When $V = 0$

$$E = Ir$$

$$r = \frac{\text{Total e.m.f.}}{I}$$

$$= \frac{6}{1} = 6\ \Omega$$

As the cells are connected in series. So, the internal resistance of each cell = 2 Ω.

★ are board based questions from previous year

Q. 29. In a potentiometer, a standard source of emf 5V and negligible internal resistance maintains a steady current through the potentiometer wire of length 10 m. Two primary cells of emf E_1 and E_2 are joined together in a series with (i) same polarity and (ii) opposite polarity. The combination is connected to the potentiometer circuit in each case. The balancing length of the wire in the two cases are found to be 700 cm and 100 cm, respectively. Find the values of emf of the two cells.★

Ans.
$$\text{Emf, } V = 5\text{ V}$$

Length of potentiometer wire = 10 m = 1000 cm.
Let the emf of two cells be E_1 and E_2.
Balancing length in same polarity = 700 cm.
Balancing length in different polarity = 100 cm.

$$\text{Potential gradient} = \frac{V}{l}$$

$$= \frac{5\text{ V}}{1000\text{ cm}} = \frac{1}{200}\text{ C/cm}$$

$$E_1 + E_2 = Kl_1 \quad \text{(For same polarity)}$$

$$\Rightarrow \quad E_1 + E_2 = \frac{1}{200} \times 700$$

$$\Rightarrow \quad E_1 + E_2 = 3.5\text{ cm} \quad \text{...(i)}$$

and
$$E_1 - E_2 = Kl_2 \quad \text{(For different polarity)}$$

$$\Rightarrow \quad E_1 - E_2 = \frac{1}{200} \times 100$$

$$\Rightarrow \quad E_1 - E_2 = 0.5\text{ cm} \quad \text{...(ii)}$$

On adding (i) and (ii), we get,

$$2E_1 = 4$$

$$E_1 = 2\text{ V}$$

and
$$E_2 = 3.5 - 2$$

$$E_2 = 1.5\text{ V}$$

Hence, the emf of cells are 2 V and 1.5 V.

Chapter 4. Moving Charges and Magnetism

Q. 1. Consider a tightly wound 100 turn coil of radius 10 cm, carrying a current of 1 A. What is the magnitude of the magnetic field at the centre of the coil ? **[NCERT]**

Ans. Since the coil is tightly wound, we may take each circular element to have the same radius R = 10 cm = 0.1 m. The number of turns N = 100. The magnitude of the magnetic field is

$$B = \frac{\mu_0 NI}{2R}$$

$$= \frac{4\pi \times 10^{-7}\text{m} \times 10^2 \times 1}{2 \times 10^{-1}}$$

$$= 2\pi \times 10^{-4}$$

$$= 6.28 \times 10^{-4}\text{ T}$$

Q. 2. A solenoid of length 0.5 m has a radius of 1 cm and is made up of 500 turns. It carries a current of 5 A. What is the magnitude of the magnetic field inside the solenoid ? **[NCERT]**

Ans. The number of turns per unit length is,

$$n = \frac{500}{0.5} = 1000\text{ turns/m}$$

★ are board based questions from previous year

The length $l = 0.5$ m and radius $r = 0.01$ m. Thus $l/a = 50$ *i.e.*, $l >> a$.

Magnetic field inside the solenoid is given by

$$B = \mu_0 nI$$
$$= 4\pi \times 10^{-7} \times 10^3 \times 5$$
$$= 6.28 \times 10^{-3} \text{ T}$$

Q. 3. The wires which connect the battery of an automobile to its starting motor carry a current of 300 A (for a short time). What is the force per unit length between the wires if they are 70 cm long and 1.5 cm apart ? Is the force attractive or repulsive ? **[NCERT]**

Ans. Current in both wires (I) = 300 A

Distance between the wires (r) = 1.5 cm = 0.015 m

Length of the two wires (l) = 70 cm = 0.7 m

Force between the two wires is given by

$$F = \frac{\mu_0 I^2}{2\pi r}$$
$$= \frac{4\pi \times 10^{-7} \times (300)^2}{2\pi \times 0.015} = 1.2 \text{ N/m}$$

As the direction of the current in the wires is opposite, a repulsive force will exists between them.

Q. 4. A magnetic field of 100 G (1 G = 10^{-4} T) is required which is uniform in a region of linear dimension about 10 cm and area of cross-section about 10^{-3} m^2. The maximum current carrying capacity of a given coil of wire is 15 A and the number of turns per unit length that can be wound round a core is at most 1000 turns m^{-1}. Suggest some appropriate design particulars of a solenoid for the required purpose. Assume the core is not ferromagnetic. **[NCERT]**

Ans. Given,

Magnetic field strength, B = 100 G = 100×10^{-4} T

Number of turns per unit length, n = 1000 turns m^{-1}

Current through the coil (I) = 15 A

Permeability of free space (μ_0) = $4\pi \times 10^{-7}$ T m A^{-1}

Magnetic field due to current through the coil

$$B = \mu_0 nI$$
$$nI = \frac{B}{\mu_0} = \frac{100 \times 10^{-4}}{4\pi \times 10^{-7}} = 7957.74 \approx 8000 \text{ A m}^{-1}$$

If the length of the coil is 60 cm, radius 5 cm, number of turns 300 and current 20 A, then these values are not unique for this case. There is always a possibility of some adjustments with limits.

Q. 5. What is the magnitude of magnetic force per unit length on a wire carrying a current of 8 A and making an angle of 30° with the direction of a uniform magnetic field of 0.15 T ? **[NCERT]**

Ans. Given,

Current in the wire, I = 8 A

Magnitude of the uniform magnetic field, B = 0.15 T

Angle between the wire and magnetic field, $\theta = 30°$.

Magnetic force per unit length on the wire is given by

$$F = BIl \sin\theta$$
$$= 0.15 \times 8 \times 1 \times \sin 30°$$
$$= 0.6 \text{ N m}^{-1}$$

Q. 6. A 3.0 cm wire carrying a current of 10 A is placed inside a solenoid perpendicular to its axis. The magnetic field inside the solenoid is given to be 0.27 T. What is the magnetic force on the wire ? **[NCERT]**

Ans. Given,

Length of the wire, l = 3 cm = 0.03 m

Current flowing in the wire, I = 10 A

Magnetic field, B = 0.27 T

Angle between the current and magnetic field, $\theta = 90°$

Magnetic force exerted on the wire is given as

$$F = BIl \sin\theta$$
$$= 0.27 \times 10 \times 0.03 \sin 90°$$
$$= 8.1 \times 10^{-2} \text{ N}$$

Hence, the magnetic force on the wire is 8.1×10^{-2} N. The direction of the force can be obtained from Fleming's left hand rule.

Q. 7. A horizontal overhead power line carries a current of 90 A in east to west direction. What is the magnitude and direction of the magnetic field due to the current 1.5 m below the line ? **[NCERT]**

Ans. Given,

Current in the power line, I = 90 A

Point is located below the power line at distance, $r = 1.5$ m

Magnetic field at any point at r distance is given by the relation,

$$|\vec{B}| = \frac{\mu_0}{4\pi}\frac{2I}{r}$$

Where, μ_0 = Permeability of free space = $4\pi \times 10^{-7}$ T m A^{-1}

$$|\vec{B}| = \frac{4\pi \times 10^{-7}}{4\pi} \times \frac{2 \times 90}{1.5}$$
$$= 1.2 \times 10^{-5} \text{ T}$$

The current is flowing from east to west. The point is below the power line from Maxwell's right hand thumb rule, the direction of the magnetic field is towards the south.

Q. 8. A square coil of side 10 cm consists of 20 turns and carries a current of 12 A. The coil is suspended vertically and the normal to the plane of the coil makes an angle of 30° with the direction of a uniform horizontal magnetic field of magnitude 0.80 T. What is the magnitude of torque experienced by the coil ? **[NCERT]**

Ans. Given,

Length of a side of the square coil, I = 10 cm = 0.1 m

Current flowing in the coil, I = 12 A

Number of turns on the coil, $n = 20$

Magnitude of the magnetic torque experienced by the coil in the magnetic field

$$\tau = nBIA \sin\theta$$

Here, τ = torque, I = current

A = Area of the square coil

$= l \times l = 0.1 \times 0.1 = 0.01$ m^2

So,

$$\tau = 20 \times 0.8 \times 12 \times 0.01 \times \sin 30°$$
$$= 0.96 \text{ N m}$$

Q. 9. A long straight wire in the horizontal plane carries a current of 50 A in north to south direction. Give the magnitude and direction of B at a point 2.5 m east of the wire. **[NCERT]**

Ans. Given,

Current in the wire, I = 50 A

A point is 2.5 m away from the East of the wire.

∴ Magnitude of the distance of the point from the wire, $r = 2.5$ m

Magnitude of the magnetic field at any point at r distance is given by

$$|\vec{B}| = \frac{\mu_0}{4\pi}\frac{2I}{r}$$

Where, μ_0 = Permeability of free space = $4\pi \times 10^{-7}$ T m A^{-1}

$$|\vec{B}| = \frac{4\pi \times 10^{-7}}{4\pi} \times \frac{2 \times 50}{2.5}$$

$$= 4 \times 10^{-6} \text{ T}$$

The point is located normal to the wire length at a distance of 2.5 m. The direction of the current in the wire is vertically downward, in horizontal plane according to the Maxwell's right hand thumb rule, the direction of the magnetic field at the given point is vertically upward perpendicular to horizontal plane.

Q. 10. A galvanometer coil has a resistance of 12 Ω and the metre shows full scale deflection for a current of 3 mA. How will you convert the metre into a voltmeter of range 0 to 18 V ? **[NCERT]**

Ans. Given,

Resistance of galvanometer coil, G = 12 Ω

Current for full scale deflection, I_g = 3 mA = 3×10^{-3} A

Range of the voltmeter needs to be converted from 0 to 18 V.

$\therefore$ $V = 18$ V

Consider a resistor of resistance R is connected in series with the galvanometer to convert it into a voltmeter. This resistance is given by

$$R = \frac{V}{I_g} - G$$

$$= \frac{18}{3 \times 10^{-3}} - 12$$

$$= 5988 \ \Omega$$

Therefore, a resistor of resistance 5988 Ω is required to be connected in series with the galvanometer.

Q. 11. A magnetic needle free to rotate in a vertical plane parallel to the magnetic meridian has its north tip down in 60° with the horizontal. The horizontal component of the earth's magnetic field at the place is known to be 0.4 G. Determine the magnitude of the earth's magnetic field at the place.

Ans. Given,

Horizontal component of earth's magnetic field (B_H) = 0.4 G

Angle between needle and horizontal plane = Angle of dip = δ = 60°.

Earth's magnetic field strength = B

We can relate B and B_H as :

$$B_H = B \cos \theta$$

$\therefore$
$$B = \frac{B_H}{\cos \delta}$$

$$= \frac{0.4}{0.5} = 0.8 \text{ G}$$

The strength of earth's magnetic field at the given location is 0.8 G.

Q. 12. A galvanometer coil has a resistance of 15 Ω and the metre shows full scale deflection for a current of 4 mA. How will you convert the metre into an ammeter of range 0 to 6 A ? **[NCERT]**

Ans. Given,

Resistance of galvanometer coil, (G) = 15 Ω

Current for full scale deflection of galvanometer I_g = 4 mA = 4×10^{-3} A

Range of the ammeter is needs to be converted to 6 A from 0 A.

$\therefore$ Current (I) = 6 A

A shunt resistor of resistance R is to be connected in parallel with the galvanometer to convert it into an ammeter. The value of R is given as :

$$R_0 = \frac{I_g G}{I - I_g} = \frac{4 \times 10^{-3} \times 15}{6 - 4 \times 10^{-3}}$$

$$R_0 = \frac{6 \times 10^{-2}}{6 - 0.004}$$

$$= \frac{0.06}{5.996}$$

$$= 0.01\ \Omega = 10\ \text{m}\ \Omega$$

$$= 10\ \text{m}\ \Omega$$

Therefore a shunt resistor of 10 m Ω is to be connected in parallel with the galvanometer.

Q. 13. A toroid has a core (non-ferromagnetic) of inner radius 25 cm and outer radius 26 cm, around which 3500 turns of a wire are wound. If the current in the wire is 11 A, what is the magnetic field (i) outside the toroid, (ii) inside the core of the toroid, and (iii) in the empty space surrounded by the toroid ? **[NCERT]**

Ans. Given,

Inner radius of the toroid (r_1) = 25 cm = 0.25 m

Outer radius of the toroid (r_2) = 26 cm = 0.26 m

Number of turns of coil (N) = 3500

Current through coil (I) = 11 A

(i) Magnetic field outside of a toroid is zero. It is non-zero only inside the core of a toroid.

(ii) Magnetic field inside the core of a toroid will be

$$B = \frac{\mu_0 NI}{l}$$

Here, μ_0 = Permeability of free space = $4\pi \times 10^{-7}$ T m A^{-1}

l = length of toroid

$$l = 2\pi\left[\frac{r_1 + r_2}{2}\right]$$

$$l = \pi(0.25 + 0.26) = 0.51\pi$$

$$\therefore \quad B = \frac{4\pi \times 10^{-7} \times 3500 \times 11}{0.51\ \pi}$$

$$\approx 3.0 \times 10^{-2}\ \text{T}$$

(iii) Magnetic field in the empty space surrounded by the toroid will be zero.

Q. 14. Two long and parallel straight wires A and B carrying currents of 8.0 A and 5.0 A in the same direction are separated by a distance of 4.0 cm. Estimate the force on a 10 cm section of wire A. **[NCERT]**

Ans. Given,

Current flowing in wire A, $I_A = 8$.

Current flowing in wire B, $I_B = 5$.

As per question two wires can be represented as

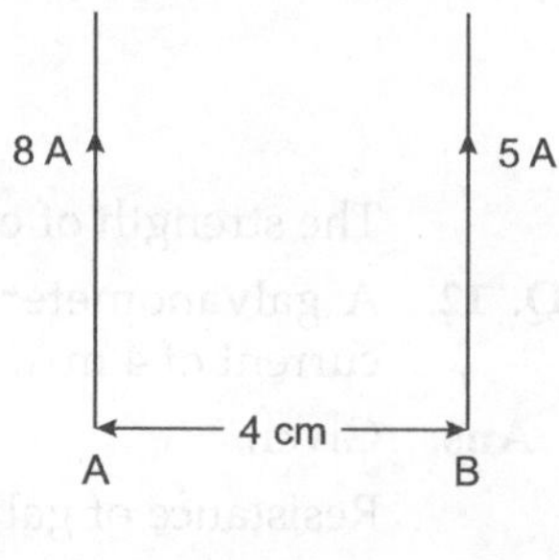

Distance between the two wires, r = 4.0 cm = 0.04 m

Length of a section of wire A, L = 10 cm = 0.1 m

Force exerted on length L due to the magnetic field is given as :

$$F = \frac{\mu_0 I_A I_B L}{2\pi r}$$

Where, μ_0 = Permeability of free space = $4\pi \times 10^{-7}$ T m A^{-1}

$$F = \frac{4\pi \times 10^{-7} \times 8 \times 5 \times 0.1}{2\pi \times 0.04}$$

$$= 2 \times 10^{-5}\ \text{N}$$

This is an attractive force normal to A towards B because the direction of the currents in the wires is the same.

Q. 15. A cyclotron's oscillator frequency is 10 MHz. What should be the operating magnetic field for accelerating protons ? If the radius of its 'dees' is 60 cm, what is the kinetic energy (in MeV) of the proton beam produced by the accelerator. **[NCERT]**

($e = 1.60 \times 10^{-19}$ C, $m_p = 1.67 \times 10^{-27}$ kg, 1 MeV = 1.6×10^{-13} J).

Ans. The oscillator frequency should be same as proton's cyclotron frequency.

As we know that,

$$B = \frac{2\pi m \nu}{q}$$

$$= \frac{2 \times 3.14 \times 1.67 \times 10^{-27} \times 10^{7}}{(1.6 \times 10^{-19})}$$

$$= 0.66 \text{ T}$$

Final velocity of protons are

$$v = r \times 2\pi\nu$$

$$= 0.6 \times 2 \times 3.14 \times 10^{7} = 3.78 \times 10^{7} \text{ m/s.}$$

$$E = \frac{1}{2} mv^2$$

$$= \frac{1}{2} \times 1.67 \times 10^{-27} \times (3.78 \times 10^{7})^2$$

$$= 7.4 \text{ MeV}$$

Q. 16. An element $\Delta l = \Delta x\hat{i}$ is placed at the origin and carries a large current I = 10 A (fig.). What is the magnetic field on the y-axis at a distance of 0.5 m. Δx = 1 cm. **[NCERT]**

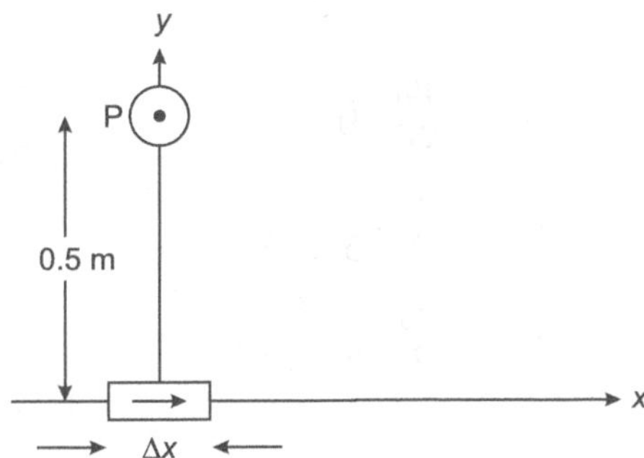

Ans. Magnetic field at a distance 'r' from a current carrying conductor is given by

$$|dB| = \frac{\mu_0}{4\pi} \frac{I dl \sin\theta}{r^2}$$

$$dl = \Delta x = 10^{-2} \text{ m}, I = 10 \text{ A}, r = 0.5 \text{ m} = y,$$

$$\frac{\mu_0}{4\pi} = 10^{-7} \text{ Tm/A},$$

$$\theta = 90°, \sin\theta = 1$$

$$|dB| = \frac{10^{-7} \times 10 \times 10^{-2}}{25 \times 10^{-2}}$$

$$= 4 \times 10^{-8} \text{ T}$$

The direction of the field is in the + z-direction. This is because

$$dl \times r = \Delta x\hat{i} \times y\hat{j} = y\Delta x\,(\hat{i} \times \hat{j}) = y\,\Delta x\,\hat{k}$$

Q. 17. What is the radius of the path of an electron (mass 9×10^{-31} kg and charge 1.6×10^{-19} C) moving at a speed of 3×10^{7} m/s in a magnetic field of 6×10^{-4} T perpendicular to it ? What is its frequency ? Calculate its energy in keV. (1 eV = 1.6×10^{-19} J). **[NCERT]**

Ans. The radius of the path of an electron is given by

$$r = \frac{mv}{(qB)}$$

$$= \frac{9 \times 10^{-31} \times 3 \times 10^{7}}{(1.6 \times 10^{-19} \times 6 \times 10^{-4})}$$

$$= 28 \times 10^{-2}\,\text{m} = 28\,\text{cm}$$

$$\text{Frequency, } \nu = \frac{v}{(2\pi r)}$$

$$= \frac{3 \times 10^{7}}{2 \times \pi \times 28 \times 10^{-2}}$$

$$= 1.7 \times 10^{7}\,\text{s}^{-1}$$

$$= 1.7 \times 10^{7}\,\text{Hz}$$

$$E = \frac{1}{2}mv^2 = \frac{1}{2} \times 9 \times 10^{-31} \times 9 \times 10^{14}$$

$$= 40.5 \times 10^{-17}\,\text{J}$$

$$= 4 \times 10^{-16}\,\text{J}$$

$$= 2.5\,\text{keV}$$

Q. 18. Two identical loops P and Q each of radius 5 cm are lying in perpendicular planes such that they have a common centre as shown in the figure. Find the magnitude and direction of the net magnetic field at the common centre of the two coils, if they carry currents equal to 3 A and 4 A respectively.★

Ans.

$$B_1 = \frac{\mu_0}{2R} \cdot I_1$$

$$B_1 = \frac{\mu_0}{2} \times \frac{3}{5 \times 10^{-2}}$$

Similarly,

$$B_2 = \frac{\mu_0}{2} \times \frac{4}{5 \times 10^{-2}}$$

Resultant magnetic field is,

$$B_2 = \sqrt{B_1^2 + B_2^2}$$

$$B = \sqrt{\left(\frac{\mu_0}{2} \times \frac{3}{5 \times 10^{-2}}\right)^2 + \left(\frac{\mu_0}{2} \times \frac{4}{5 \times 10^{-2}}\right)^2}$$

$$B = \frac{\mu_0}{2 \times 5 \times 10^{-2}} \times 5$$

$$B = \frac{\mu_0}{2} \times 100 = 50\,\mu_0$$

$$= 50 \times 4\pi \times 10^{-7}$$

$$= 62.83 \times 10^{-6}\,\text{T}$$

$$\tan\theta = \frac{B_1}{B_2} = \frac{3}{4}$$

$$\theta = \tan^{-1}\left(\frac{3}{4}\right)$$

Thus, direction of magnetic field makes an angle θ with the vertical.

★ are board based questions from previous year

Q. 19. A rectangular loop of wire of size 4 cm × 10 cm carries a steady current of 2 A. A straight long wire carrying 5 A current is kept near the loop as shown. If the loop and the wire are coplanar, find :★

(i) The magnitude and direction of the force on the loop due to the current carrying wire.

(ii) The torque acting on the loop.

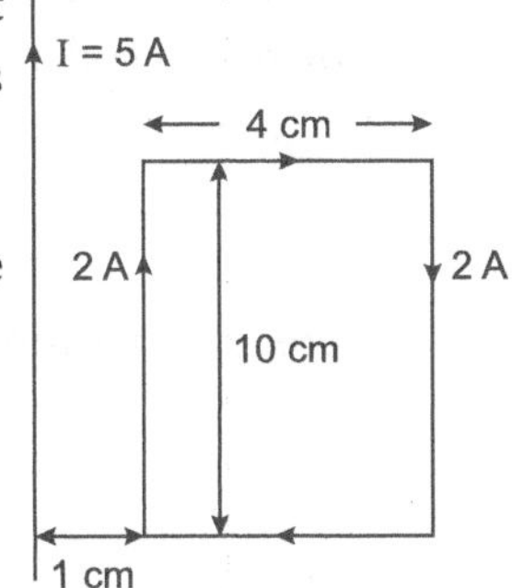

Ans.

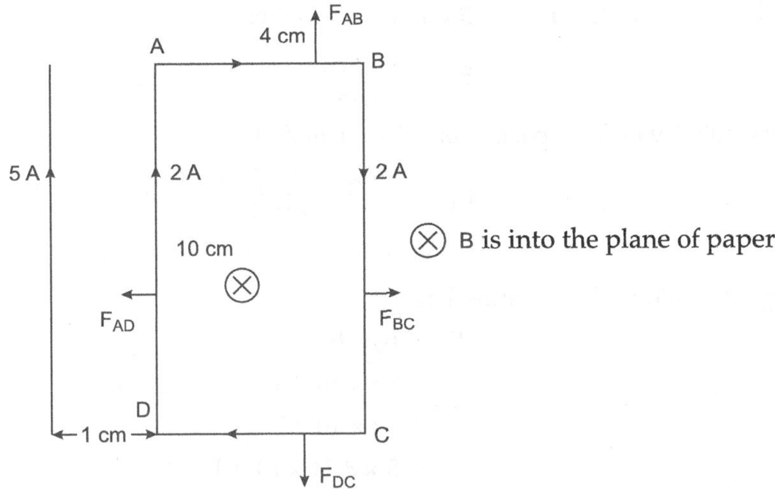

From the direction of current flow the force on each side of loop are as per fig. F_{AB} and F_{DC} are equal and opposite so they will cancel each other.

$$\Rightarrow \quad F_{AD} = \frac{\mu_0 I_a I_b}{2\pi \times d} \times L$$

$$= \frac{2 \times 10^{-7} \times 5 \times 2 \times (10 \times 10^{-2})}{(1 \times 10^{-2})} \text{(attractive) towards left}$$

$$= 2 \times 10^{-5} \text{ N (towards left)}$$

$$\Rightarrow \quad F_{BC} = \frac{(2 \times 10^{-7}) \times 5 \times 2 \times (10 \times 10^{-2})}{(1+4) \times 10^{-2}} \text{(repulsive) towards right}$$

$$= 4 \times 10^{-6} \text{ (towards right)}$$

$$= F_{net} = (F_{AD} - F_{DC}) = 2 \times 10^{-5} - 0.4 \times 10^{-5}$$

$$= 16 \times 10^{-6} \text{ N (towards right)}$$

The direction of net force is towards the straight wire *i.e.*, attractive.

(ii) Torque will be zero as all the forces are in the plane of loop.

Q. 20. Two concentric circular coils X and Y of radii 16 cm and 10 cm, respectively, lie in the same vertical plane containing the north to south direction. Coil X has 20 turns and carries a current of 16 A; coil Y has 25 turns and carries a current of 18 A. The sense of the current in X is anticlockwise, and clockwise in Y, for an observer looking at the coils facing west. Give the magnitude and direction of the net magnetic field due to the coils at their centre. **[NCERT]**

Ans. Given,

Radius of coil X, $(r_1) = 16 \text{ cm} = 0.16 \text{ m}$

Radius of coil Y $(r_2) = 10 \text{ cm} = 0.1 \text{ m}$

Number of turns of on coil X, $(n_1) = 20$

Number of turns of on coil Y, $(n_2) = 25$

★ are board based questions from previous year

Current in coil X (I_1) = 16 A
Current in coil Y (I_2) = 18 A
Magnetic field due to current in coil X at their centre is (B_1)

$$B_1 = \frac{\mu_0 N_1 I_1}{2\, r_1}$$

Here, μ_0 = Permeability of free space = $4\pi \times 10^{-7}$ T m A^{-1}

$$B_1 = \frac{4\pi \times 10^{-7} \times 20 \times 16}{2 \times 0.16}$$

$$= 4\pi \times 10^{-4} \text{ T in East direction.}$$

Magnetic field due to current coil Y at their centre is given by

$$B_2 = \frac{\mu_0 N_2 I_2}{2\, r_2}$$

Here, μ_0 = Permeability of free space = $4\pi \times 10^{-7}$ T m A^{-1}

$$B_1 = \frac{4\pi \times 10^{-7} \times 25 \times 18}{2 \times 0.10}$$

$$= 9\pi \times 10^{-4} \text{ T m A}^{-1} \text{ in West direction.}$$

Thus, net magnetic field will be obtained as :

$$B = B_2 - B_1$$
$$= 9\pi \times 10^{-4}\text{ T} - 4\pi \times 10^{-4}\text{ T}$$
$$= 5\pi \times 10^{-4}\text{ T}$$
$$= 5 \times 3.14 \times 10^{-4}\text{ T}$$
$$= 1.57 \times 10^{-3}\text{ T in West direction}$$

Q. 21. A uniform magnetic field of 1.5 T exists in a cylindrical region of radius 10.0 cm. Its direction parallel to the axis along east to west. A wire carrying current of 7.0 A in the north to south direction passes through this region. What is the magnitude and direction of the force on the wire if,

(i) the wire intersects the axis.

(ii) the wire is turned from N-S to northeast - northwest direction.

(iii) the wire in the N-S direction is lowered from the axis by a distance of 6.0 cm ? **[NCERT]**

Ans. Given,

Magnetic field strength (B) = 1.5 T
Radius of the cylindrical region (r) = 10 cm = 0.1 m
Current in the wire (I) = 7 A

(i) When wire intersects the axis, then the length of the wire is the diameter of the cylindrical region.

Thus, $l = 2r = 0.2$ m
Angle between magnetic field and current, $\theta = 90°$
Magnetic force acting on the wire is given by the relation,

$$F = BIl \sin\theta$$
$$= 1.5 \times 7 \times 0.2 \times \sin 90°$$
$$= 2.1 \text{ N}$$

Therefore, a force of 2.1N acts on the wire in a vertically downward direction.

(ii) New length of the wire after turning it to the northeast-northwest direction will be given as :

$$l_1 = \frac{l}{\sin\theta}$$

Angle between magnetic field and current, $\theta = 45°$

Force on the wire,

$$F = BIl_1 \sin\theta$$
$$= BI\frac{l}{\sin\theta}\sin\theta = BIl$$
$$= 1.5 \times 7 \times 0.2$$
$$= 2.1 \text{ N}$$

Hence, a force of 2.1 N acts vertically downward on the wire. This is independent of angle θ because $l \sin\theta$ is fixed.

(iii) The wire is lowered from the axis by distance, $d = 6.0$ cm

Let l_2 be the new length of the wire.

$$\therefore \quad \left(\frac{l_2}{2}\right)^2 = 4(d + r)$$
$$= 4(10 + 6) = 4(16)$$
$$\therefore \quad l_2 = 4 \times 4 = 16 \text{ cm} = 0.16 \text{ m}$$

Magnetic force exerted on the wire $(F_2) = BIl_2$

$$F_2 = BIl_2$$
$$= 1.5 \times 7 \times 0.16$$
$$= 1.68 \text{ N}$$

Hence, a force of 1.68 N acts in a vertically downward direction on the wire.

Q. 22. A uniform magnetic field of 3000 G is established along the positive *z*-direction. A rectangular loop of sides 10 cm and 5 cm carries a current of 12 A. What is the torque on the loop in the different cases shown in given figures ?

(i) What is the force on each case ?

(ii) Which case corresponds to stable equilibrium ? **[NCERT]**

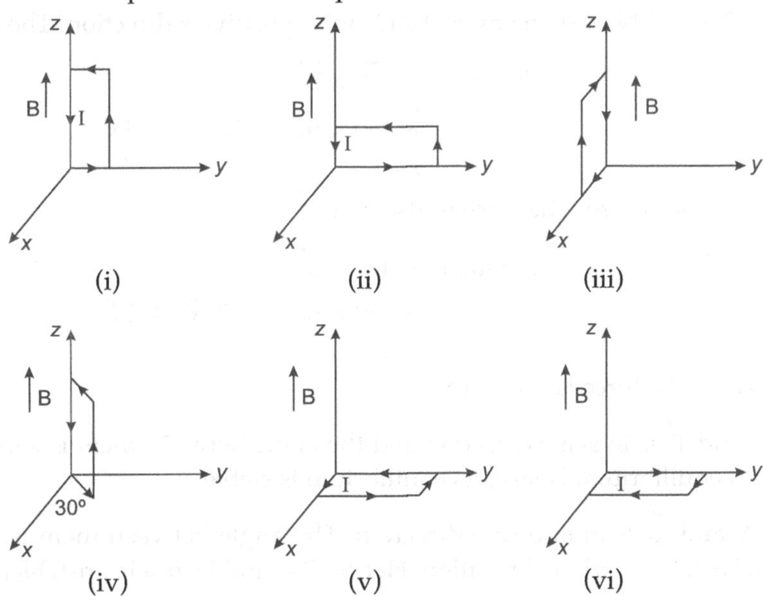

Ans. Given,

Magnetic field (B) = 3000 G = 3000×10^{-4} T = 0.3 T

Length of the rectangular loop (l) = 10 cm

Width of the rectangular loop (b) = 5 cm

Area of the rectangular loop

$$A = l \times b$$
$$= 10 \times 5 = 50 \text{ cm}^2$$
$$= 50 \times 10^{-4} \text{ m}^2$$

Current in the loop (I) = 12 A

Consider anti-clockwise direction of the current as positive and vice-versa :

(i) $$\text{Torque}\ (\vec{\tau}) = I(\vec{A}) \times (\vec{B})$$

From figure. Area vector $(\vec{A})$ is normal to the *y-z* plane and B is directed along the *z*-axis.

$$\therefore \quad \tau = 12 \times (50 \times 10^{-4})\ \hat{i} \times 0.3\ \hat{k}$$

$$= -1.8 \times 10^{-2}\ \hat{j}\ \text{N m}$$

Net torque is 1.8×10^{-2} N m along the negative *y*-direction.

The force on the loop is zero because the angle between A and B is zero.

(ii) Same as (i)

(iii) $$\text{Torque}\ (\vec{\tau}) = I(\vec{A}) \times (\vec{B})$$

From the given figure, it can be observed that A is normal to the *x-z* plane and B is directed along the z-axis.

$$\therefore \quad \tau = -12 \times (50 \times 10^{-4})\ \hat{j} \times 0.3\ \hat{k} \qquad [\because \hat{j} \times \hat{k} = \hat{i}\,]$$

$$= -1.8 \times 10^{-2}\ \hat{i}\ \text{N m}$$

Torque is 1.8×10^{-2} N m along the negative *x* direction and the force is zero.

(iv) Magnitude of torque is given as :

$$|\tau| = IAB$$

$$= 12 \times 50 \times 10^{-4} \times 0.3$$

$$= 1.8 \times 10^{-2}\ \text{N m}$$

The torque is 1.8×10^{-2} N m at an angle of 240° with positive *x* direction. The force is zero.

(v) $$\text{Torque}\ \tau = I\vec{A} \times \vec{B}$$

$$= (50 \times 10^{-4} \times 12)\ \hat{k} \times 0.3\ \hat{k} \qquad [\because \hat{k} \times \hat{k} = 0]$$

$$= 0$$

Therefore, the torque is zero. The force is also zero.

(vi) $$\text{Torque}\ \tau = I\vec{A} \times \vec{B}$$

$$= (50 \times 10^{-4} \times 12)\ \hat{k} \times 0.3\ \hat{k} \qquad [\because \hat{k} \times \hat{k} = 0]$$

$$= 0$$

The torque is zero. The force is also zero.

In case (v), $I\vec{A}$ and $\vec{B}$ is in same direction and the angle between them is zero. If displaced, they come back to an equilibrium. Hence, its equilibrium is stable.

In case (vi), $I\vec{A}$ and $\vec{B}$ is in opposite direction. The angle between them is 180°. If displaced it does not come back to its original position. Hence, its equilibrium is unstable.

Chapter 5. Magnetism and Matter

Q. 1. The earth's magnetic field at the equator is approximately 0.4 G. Estimate the earth's dipole moment. **[NCERT]**

Ans. The equatorial magnetic field is

$$B_E = \frac{\mu_0 m}{4\pi r^3}$$

We are given that $B_E \approx 0.4\ \text{G} = 4 \times 10^{-5}$ T. For *r*, we take the radius of the earth 6.4×10^6 m. Hence,

$$m = \frac{4 \times 10^{-5} \times (6.4 \times 10^6)^3}{\mu_0 / 4\pi}$$

$$= 4 \times 10^2 \times (6.4 \times 10^6)^3 \qquad \left(\frac{\mu_0}{4\pi} = 10^{-7}\right)$$

$$= 1.05 \times 10^{23} \text{ A m}^2$$

This is close to the value 8×10^{22} A m^2 quoted in geomagnetic texts.

Q. 2. An iron rod is subjected to a magnetising field of 1200 A/m. The susceptibility of iron is 599. Find the permeability and the magnetic field produced.

Ans. Given,

$$H = 1200 \text{ A/m}, \chi_m = 599$$

Since, we know that $\mu = \mu_0 (1 + \chi_m)$

$\Rightarrow$ $\mu = 4\pi \times 10^{-7} (1 + 599)$

$$\mu = 7.536 \times 10^{-4} \text{ Tm/A}$$

and $B = \mu H$

$$= 7.536 \times 10^{-4} \times 1200$$

$$B = 0.904 \text{ T}$$

Q. 3. In Figure, the magnetic needle has magnetic moment 6.7×10^{-2} A m^2 and moment of inertia $A = 7.5 \times 10^{-6}$ kg m^2. It performs 10 complete oscillations in 6.70 s. What is the magnitude of the magnetic field ? **[NCERT]**

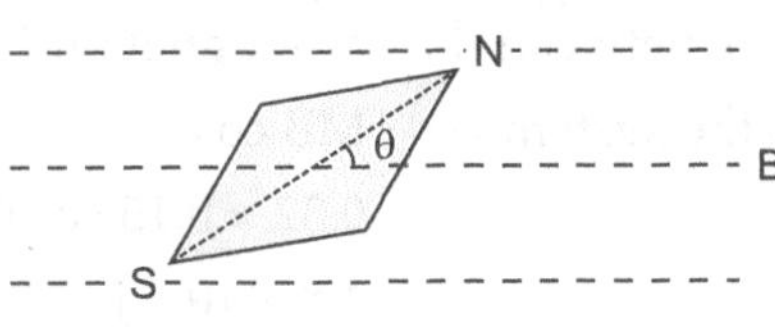

Ans. The time period of oscillation is

$$T = \frac{6.70}{10} = 0.67 \text{ s}$$

Relation between moment of inertia and magnetic field

$$B = \frac{4\pi^2 I}{mT^2}$$

$$= \frac{4 \times (3.14)^2 \times 7.5 \times 10^{-6}}{6.7 \times 10^{-2} \times (0.67)^2}$$

$$= 0.01 \text{ T}$$

Q. 4. What is the magnitude of the equatorial and axial fields due to a bar magnet of length 5.0 cm at a distance of 50 cm from its mid-point ? The magnetic moment of the bar magnet is 0.40 A m^2. **[NCERT]**

Ans. As we know that equatorial magnet field of bar magnet

$$(B_E) = \frac{\mu_0 m}{4\pi r^3}$$

$$= \frac{10^{-7} \times 0.4}{(0.5)^3} = \frac{10^{-7} \times 0.4}{0.125}$$

$$= 3.2 \times 10^{-7} \text{ T}$$

and Axial magnetic field $(B_A) = \dfrac{\mu_0 2m}{4\pi r^3}$

$$= 6.4 \times 10^{-7} \text{ T}$$

Q. 5. A magnetic needle free to rotate in a vertical plane parallel to the magnetic meridian has its north tip pointing down at 22° with the horizontal. The horizontal component of the earth's magnetic field at the place is known to be 0.35 G. Determine the magnitude of the earth's magnetic field at the place. **[NCERT]**

Ans. Given,

Horizontal component of earth's magnetic field, $B_H = 0.35$ G

Angle made by the needle with the horizontal plane = Angle of dip = $\delta = 22°$

Earth's magnetic field strength = B

B_H = Horizontal component of earth's magnetic field.

We can relate B and B_H as :

$$B_H = B\cos\theta$$

$$\therefore \quad B = \frac{B_H}{\cos\delta} = \frac{0.35}{\cos 22°} = 0.377 \text{ G}$$

Hence, the strength of earth's magnetic field at the given location is 0.377 G.

Q. 6. A short bar magnet of magnetic moment $m = 0.32$ J T^{-1} is placed in a uniform magnetic field of 0.15 T. If the bar is free to rotate in the plane of the field, which orientation would corresponds to its (i) stable, and (ii) unstable equilibrium ? What is the potential energy of the magnet in each case ? **[NCERT]**

Ans. Given,

Magnetic moment of the bar magnet, $M = 0.32$ J T^{-1}

Uniform external magnetic field, B = 0.15 T

(i) The bar magnet is aligned along the magnetic field. This system is considered in stable equilibrium. Hence, the angle θ, between the bar magnet and the magnetic field is 0°.

$$\begin{aligned}\text{Potential energy of the system} &= -MB\cos\theta \\ &= -0.32 \times 0.15 \cos 0° \\ &= -4.8 \times 10^{-2} \text{ J}\end{aligned}$$

(ii) The bar magnet is oriented 180° to the magnetic field. Hence, it is in unstable equilibrium,

$$\begin{aligned}\theta &= 180° \\ \text{Potential energy} &= -MB\cos\theta \\ &= -0.32 \times 0.15 \cos 180° \\ &= 4.8 \times 10^{-2} \text{ J}\end{aligned}$$

Q. 7. At a place, the horizontal component of earth's magnetic field is B and angle of dip is 60°. What is the value of horizontal component of the earth's magnetic field at equator ?★

Ans.

$$\begin{aligned}B_H &= B'\cos 60° \\ B &= B'\cos 60° \qquad [\text{given } B_H = B] \\ B &= B'\left(\frac{1}{2}\right) \\ B' &= 2B\end{aligned}$$

At equator,

$$\begin{aligned}\theta &= 0° \\ B_H &= B'\cos\theta \\ &= B'\cos 0' \\ &= 2B\end{aligned}$$

Q. 8. The magnetic field B and the magnetic intensity of a material are found to be 1.6 T and 1000 A/m respectively. Calculate the relative permeability (μ_r) and the susceptibility (χ) of the material.

Ans. Given,

B = 1.6 T, H = 1000 A/m

Since

$$\mu = \frac{B}{H}$$

★ are board based questions from previous year

$$\mu = \frac{1.6}{1000} = 1.6 \times 10^{-3} \text{ Tm/A}$$

As we know that $\mu_r = \dfrac{\mu}{\mu_0}$

$$\mu_r = \frac{1.6 \times 10^{-3}}{4\pi \times 10^{-7}}$$

$$\mu_r = 1273.23$$

Since, $\mu_r = (1 + \chi)$

$\Rightarrow$ $\chi = \mu_r - 1$

$$= 1273.23 - 1$$

$$= 1.272 \times 10^3$$

Q. 9. A closely wound solenoid of 2000 turns and area of cross-section 1.6×10^{-4} m^2, carrying a current of 4.0 A, is suspended through its centre allowing it to turn in a horizontal plane.

(i) What is the magnetic moment associated with the solenoid ?

(ii) What is the force and torque on the solenoid if a uniform horizontal magnetic field of 7.5×10^{-2} T is set-up at an angle of 30° with the axis of the solenoid ? **[NCERT]**

Ans. Given,

Number of turns on the solenoid (n) = 2000

Area of cross-section of the solenoid (A) = 1.6×10^{-4} m^2

Current through solenoid (I) = 4 A

(i) The magnetic moment along the axis of the solenoid is given by

$$M = nAI$$

$$= 2000 \times 1.6 \times 10^{-4} \times 4$$

$$= 1.28 \text{ Am}^2$$

(ii) Given,

Magnetic field (B) = 7.5×10^{-2} T

Angle between the magnetic field and the axis of the solenoid, $\theta = 30°$

$$\text{Torque } (\tau) = MB \sin\theta = 1.28 \times 7.5 \times 10^{-2} \sin 30°$$

$$= 4.8 \times 10^{-2} \text{ Nm}$$

As the magnetic field is uniform, the force on the solenoid is zero. The torque on the solenoid is 4.8×10^{-2} Nm.

Q. 10. A short bar magnet has a magnetic moment of 0.48 J T^{-1}. Give the direction and magnitude of the magnetic field produced by the magnet at a distance of 10 cm from the centre of the magnet on (i) the axis, (ii) the equatorial lines (normal bisector) of the magnet. **[NCERT]**

Ans. Given,

Magnetic moment (M) = 0.48 J T^{-1}

(i) Distance (d) = 10 cm = 0.1 m

The magnetic field at distance (d), from the centre of the magnet along the axis is given by

$$B = \frac{\mu_0}{4\pi} \frac{2M}{d^3}$$

Where, μ_0 = Permeability of free space = $4\pi \times 10^{-7}$ T m A^{-1}

$$\therefore \quad B = \frac{4\pi \times 10^{-7} \times 2 \times 0.48}{4\pi \times (0.1)^3}$$

$$= 0.96 \times 10^{-4} \text{ T} = 0.96 \text{ G}$$

The magnetic field is along the S–N direction.

(ii) The magnetic field at a distance of 10 cm along the equatorial line of the magnet is given as :

$$B = \frac{\mu_0 \times M}{4\pi \times d^3} = \frac{4\pi \times 10^{-7} \times 0.48}{4\pi \times (0.1)^3}$$

$$= 0.48 \text{ G}$$

The magnetic field is along the N–S direction.

Q. 11. A circular coil of 16 turns and radius 10 cm carrying a current of 0.75 A rests with its plane normal to an external field of magnitude 5.0×10^{-2} T. The coil is free to turn about an axis in its plane perpendicular to the field direction. When the coil is turned slightly and released, it oscillates about its stable equilibrium with a frequency of $2.0\ s^{-1}$.

What is the moment of inertia of the coil about its axis of rotation ? **[NCERT]**

Ans. Given,

Number of turns in the circular coil (N) = 16

Radius of the coil (r) = 10 cm = 0.1 m

Cross-section of the coil (A) = $\pi r^2 = \pi \times (0.1)^2\ m^2$

Current in the coil, I = 0.75 A

Magnetic field strength, $B = 5.0 \times 10^{-2}$ T

Frequency of oscillations of the coil (ν) = $2.0\ s^{-1}$

As Magnetic moment, $M = NIA = NI\pi r^2$

$$= 16 \times 0.75 \times \pi \times (0.1)^2$$

$$= 0.377\ J\ T^{-1}$$

Frequency is related with magnetic moment and field as

$$\nu = \frac{1}{2\pi}\sqrt{\frac{MB}{I}}$$

Where, I = Moment of inertia of the coil

$$I = \frac{MB}{4\pi^2\nu^2}$$

$$= \frac{0.377 \times 5 \times 10^{-2}}{4\pi^2 \times (2)^2}$$

$$= 1.19 \times 10^{-4}\ kg\ m^2$$

Q. 12. A domain in ferromagnetic iron is in the form of a cube of side length 1 μm. Estimate the number of iron atoms in the domain and the maximum possible dipole moment and magnetisation of the domain. The molecular mass of iron is 55 g/mole and its density is 7.9 g/cm³. Assume that each iron atom has a dipole moment of $9.27 \times 10^{-24}\ A\ m^2$. **[NCERT]**

Ans. The volume of the cubic domain is

$$V = (10^{-6}\ m)^3 = 10^{-18}\ m^3 = 10^{-12}\ cm^3$$

Its mass is volume × density = $7.9\ g\ cm^{-3} \times 10^{-12}\ cm^3 = 7.9 \times 10^{-12}$ g

It is given that Avogadro number (6.023×10^{23}) of iron atoms have a mass of 55 g. Hence, the number of atoms in the domain is

$$N = \frac{7.9 \times 10^{-12} \times 6.023 \times 10^{23}}{55}$$

$$= 8.65 \times 10^{10}\ \text{atoms}$$

The maximum possible dipole moment m_{max} is achieved for the (unrealistic) case when all the atomic moments are perfectly aligned. Thus,

$$m_{max} = (8.65 \times 10^{10}) \times (9.27 \times 10^{-24})$$

$$= 8.0 \times 10^{-13}\ A\ m^2$$

The consequent magnetisation is

$$M_{max} = \frac{m_{max}}{\text{Domain volume}}$$

$$= 8.0 \times 10^{-13}\ Am^2/10^{-18}\ m^3$$

$$= 8.0 \times 10^5\ Am^{-1}$$

Q. 13. A solenoid has a core of a material with relative permeability 400. The windings of the solenoid are insulated from the core and carry a current of 2A. If the number of turns is 1000 per metre, calculate (i) H, (ii) M, (iii) B and (iv) the magnetising current I_m. **[NCERT]**

Ans. (i) The field H is dependent of the material of the core, and is

$$H = nI$$

$$= 1000 \times 2.0 = 2 \times 10^3 \text{ A/m.}$$

(ii) The magnetic field B is given by

$$B = \mu_r \mu_0 H$$
$$= 400 \times 4\pi \times 10^{-7} \text{ (N/A}^2) \times 2 \times 10^3 \text{ (A/m)}$$
$$= 1.0 \text{ T}$$

(iii) Magnetisation is given by

$$M = \frac{(B - \mu_0 H)}{\mu_0} = \frac{(\mu_r \mu_0 H - \mu_0 H)}{\mu_0}$$
$$= (\mu_r - 1) H$$
$$= 399 \times H$$
$$\cong 8 \times 10^5 \text{ A/m}$$

(iv) The magnetising current I_m is the additional current that needs to be passed through the windings of the solenoid in the absence of the core which would give a B value as in the presence of the core. Thus,

$$B = \mu_r \mu_0 (I + I_m).$$

Using I = 2A, B = 1 T,

$$B = \mu_r \mu_0 \, nI$$
$$= \mu_0 \, n \, (I + I_m)$$
$$\mu_r I = I + I_m$$
$$\Rightarrow \quad I_m = (\mu_r - 1) I$$
$$= 798 \text{ A}$$

Q. 14. A compass needle free to turn in a horizontal plane is placed at the centre of circular coil of 30 turns and radius 12 cm. The coil is in a vertical plane making an angle of 45° with the magnetic meridian. When the current in the coil is 0.35 A, the needle points west to east.

(i) Determine the horizontal component of the earth's magnetic field at the location.

(ii) The current in the coil is reversed, and the coil is rotated about its vertical axis by an angle of 90° in the anticlockwise sense looking from above. Predict the direction of the needle. Take the magnetic declination at the places to be zero. **[NCERT]**

Ans. Given,

Number of turns in coil (N) = 30

Radius of the circular coil (r) = 12 cm = 0.12 m

Current through coil (I) = 0.35 A

Angle of dip (δ) = 45°

(i) The magnetic field due to current I, at a distance r, is given by

$$B = \frac{\mu_0 \, 2\pi NI}{4\pi r}$$

Where,

μ_0 = Permeability of free space = $4\pi \times 10^{-7}$

$$B = \frac{4\pi \times 10^{-7} \times 2\pi \times 30 \times 0.35}{4\pi \times 0.12}$$

$$B = 5.49 \times 10^{-5} \text{ T}$$

The compass needle points from west to East. Thus, the horizontal component of earth's magnetic field is given by

$$B_H = B \sin \delta$$
$$= 5.49 \times 10^{-5} \sin 45°$$
$$= 3.88 \times 10^{-5} \text{ T} = 0.388 \text{ G}$$

(ii) When the current through the coil is reversed and the coil gets rotated about its vertical axis by an angle of 90°, the needle will reverse its original direction. In this situation the needle will point from East to West.

Q. 15. A long straight horizontal cable carries a current of 2.5 A in the direction 10° south of west to 10° north of east. The magnetic meridian of the place happens to be 10° west of the geographic meridian. The earth's magnetic field at the location is 0.33 G and the angle of dip is zero. Locate the line of neutral points (ignore the thickness of the cable). (At neutral points, magnetic field due to a current-carrying cable is equal and opposite to the horizontal component of earth's magnetic field.) **[NCERT]**

Ans. Given,

Current in the wire, I = 2.5 A

Angle of dip at the given location on earth, $\delta = 10°$

Earth's magnetic field, $B_E = 0.33\ G = 0.33 \times 10^{-4}\ T$

The horizontal component of earth's magnetic field is given as :

$$B_H = B_E \cos\delta$$
$$= 0.33 \times 10^{-4} \times \cos 10°$$
$$\cong 0.33 \times 10^{-4}\ T$$

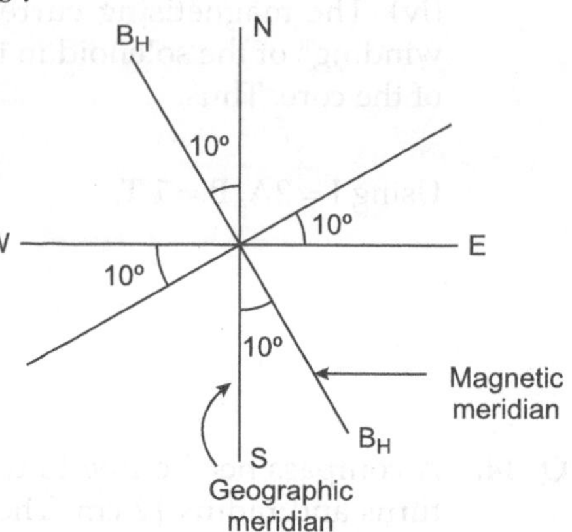

The magnetic field at the neutral point at a distance R from the cable is given by the relation :

$$\text{Neutral points} = B_{net} = 0°$$

$$B_H = \frac{\mu_0 I}{2\pi R}$$

Where,

μ_0 = Permeability of free space = $4\pi \times 10^{-7}$ T mA^{-1}

$$\therefore \quad R = \frac{\mu_0 I}{2\pi B_H}$$
$$= \frac{4\pi \times 10^{-7} \times 2.5}{2\pi \times 0.33 \times 10^{-4}}$$
$$= 15.15 \times 10^{-3}\ m$$
$$= 1.51\ cm$$

Hence, a set of neutral points parallel to and above the cable are located at a normal distance of 1.51 cm.

Q. 16. A short bar magnet of magnetic moment $5.25 \times 10^{-2}\ JT^{-1}$ is placed with its axis pependicular to the earth's field direction. At what distance from the centre of the magnet, the resultant field is inclined at 45° with earth's field on :

(i) its normal bisector, and (ii) its axis. Magnitude of the earth's field at the place is given to be 0.42 G. Ignore the length of the magnet in comparison to the distances involved.

Ans. Given,

Magnetic moment (M) = $5.25 \times 10^{-2}\ JT^{-1}$

Earth's magnetic field at a place (H) = $0.42\ G = 0.42 \times 10^{-4}\ T$

(i) The magnetic field at a distance R from the centre of the magnet on the normal bisector is given by the relation :

$$B = \frac{\mu_0 M}{4\pi R^3}$$

Where μ_0 = Permeability of free space = $4\pi \times 10^{-7}\ Tm\ A^{-1}$

If resultant field is inclined at 45° with earth's field, B = H

$$\frac{\mu_0 M}{4\pi R^3} = H = 0.42 \times 10^{-4}$$

$$R^3 = \frac{\mu_0 M}{0.42 \times 10^{-4} \times 4\pi}$$

$$= \frac{4\pi \times 10^{-7} \times 5.25 \times 10^{-2}}{4\pi \times 0.42 \times 10^{-4}}$$

$$= 12.5 \times 10^{-5}$$

$$\therefore \quad R = 0.05 \text{ m}$$

$$= 5 \text{ cm}$$

(ii) The magnetic field at a distance R′

from the centre of the magnet along its axis is given as :

$$B' = \frac{\mu_0 \, 2M}{4\pi R^3}$$

If resultant field is inclined at 45° with earth's field,

$$\therefore \quad B' = H$$

$$\frac{\mu_0 \, 2M}{4\pi (R')^3} = H$$

$$(R')^3 = \frac{\mu_0 \, 2M}{4\pi \times H}$$

$$= \frac{4\pi \times 10^{-7} \times 2 \times 5.25 \times 10^{-2}}{4\pi \times 0.42 \times 10^{-4}}$$

$$= 25 \times 10^{-3}$$

$$\therefore \quad R' = 0.063 \text{ m}$$

$$= 6.3 \text{ cm}$$

Q. 17. A sample of paramagnetic salt contains 2.0×10^{24} atomic dipoles each of dipole moment 1.5×10^{-23} J T^{-1}. The sample is placed under a homogeneous magnetic field of 0.64 T, and cooled to a temperature of 4.2 K. The degree of magnetic saturation achieved is equal to 15%. What is the total dipole moment of the sample for a magnetic field of 0.98 T and a temperature of 2.8 K ? (Assume Curie's law). **[NCERT]**

Ans. Given,

Number of atomic dipoles (n) = 2.0×10^{24}

Dipole moment of each atomic dipole (M) = 1.5×10^{-23} J T^{-1}

When the magnetic field, (B_1) = 0.64 T

The sample is cooled to a temperature (T_1) = 4.2 K

Total dipole moment of the atomic dipole

$$(M_{tot}) = n \times M$$

$$= 2 \times 10^{24} \times 1.5 \times 10^{-23}$$

$$= 30 \text{ J T}^{-1}$$

Magnetic saturation is achieved at 15%.

Hence, effective dipole moment,

$$M_1 = \frac{15}{100} \times 30$$

$$= 4.5 \text{ J T}^{-1}$$

When the magnetic field, $B_2 = 0.98$ T

Temperature, $T_2 = 2.8$ K

Its total dipole moment = M_2

According to Curie's law, we have the ratio of two magnetic dipoles as :

$$\frac{M_2}{M_1} = \frac{B_2}{B_1} \times \frac{T_1}{T_2}$$

$$\therefore \quad M_2 = \frac{B_2 T_1 M_1}{B_1 T_2}$$

$$= \frac{0.98 \times 4.2 \times 4.5}{0.64 \times 2.8}$$

$$= 10.336 \text{ J T}^{-1}$$

Therefore, 10.336 J T^{-1} is the total dipole moment of the sample for a magnetic field of 0.98 T and a temperature of 2.8 K.

Q. 18. A monoenergetic (18 keV) electron beam initially in the horizontal direction is subjected to a horizontal magnetic field of 0.04 G normal to the initial direction. Estimate the up or down deflection of the beam over a distance of 30 cm ($m_e = 9.11 \times 10^{-19}$ kg). [Note : Data in this exercise are so chosen that the answer will give you an idea of the effect of earth's magnetic field on the motion of the electron beam from the electron gun to the screen in a TV set.] **[NCERT]**

Ans. Given,

Energy of an electron beam, E = 18 keV = 18×10^3 eV

Charge on an electron, $e = 1.6 \times 10^{-19}$ C

F = $18 \times 10^3 \times 1.6 \times 10^{-19}$ T

Magnetic field (B) = 0.04 G

Mass of an electron (m_e) = 9.1×10^{-19} kg

Distance up to which the electron beam travels (d) = 30 cm = 0.3 m

The kinetic energy of the electron beam

$$E = \frac{1}{2} mv^2$$

$$v = \sqrt{\frac{2E}{m}}$$

$$= \sqrt{\frac{2 \times 18 \times 10^3 \times 1.6 \times 10^{-19} \times 10^{-15}}{9.11 \times 10^{-31}}}$$

$$= 0.795 \times 10^8 \text{ m/s}$$

The electron beam deflects along a circular path of radius, r

The force due to the magnetic field balances the centripetal force of the path. Thus,

$$Bev = \frac{mv^2}{r}$$

$$\therefore \quad r = \frac{mv}{Be}$$

$$= \frac{9.11 \times 10^{-31} \times 0.795 \times 10^8}{0.4 \times 10^{-4} \times 1.6 \times 10^{-19}}$$

$$= 11.3 \text{ m}$$

Let the up and down deflection of the electron beam be $x = r(1 - \cos\theta)$

Where, θ = Angle of declination

$$\sin\theta = \frac{d}{r} = \frac{0.3}{11.3}$$

$$\theta = \sin^{-1}\frac{0.3}{11.3} = 1.521°$$

and

$$x = 11.3\,(1 - \cos 1.521°)$$

$$= 0.0039 \text{ m}$$

$$= 3.9 \text{ mm}$$

Therefore, the up and down deflection of the beam is 3.9 mm.

Q. 19. A telephone cable at a place has four long straight horizontal wires carrying a current of 1.0 A in the same direction east to west. The earth's magnetic field at the place is 0.39 G, and the angle of dip is 35°. The magnetic declination is nearly zero. What are the resultant magnetic fields at points 4.0 cm below the cable ? **[NCERT]**

Ans. Given,

Number of horizontal wires in the telephone cable, $n = 4$

Current in each wire, I = 1.0 A

Earth's magnetic field at a location, $H = 0.39\ G = 0.39 \times 10^{-4}\ T$

Angle of dip at the location, $\delta = 35°$

Angle of declination, $\theta \sim 0°$

For a point 4 cm below the cable :

Distance, $r = 4\ cm = 0.04\ m$

The horizontal component of earth's magnetic field can be given by

$$H_h = H \cos \delta - B$$

Where,

B = Magnetic field at 4 cm due to current I through four wires

$$= 4 \times \frac{\mu_0 I}{2\pi r}$$

μ_0 = Permeability of free space = $4\pi \times 10^{-7}\ T\ mA^{-1}$

$$B = 4 \times \frac{4\pi \times 10^{-7} \times 1}{2\pi \times 0.04}$$

$$= 0.2 \times 10^{-4}\ T = 0.2\ G$$

$$\therefore \quad H_h = 0.39 \cos 35° - 0.2$$

$$= 0.39 \times 0.819 - 0.2 \approx 0.12\ G$$

The vertical component of earth's magnetic field is given by

$$H_V = H \sin \delta$$

$$= 0.39 \sin 35° = 0.22\ G$$

The angle between the field and its horizontal component is given by

$$\theta = \tan^{-1} \frac{Hv}{H_h}$$

$$= \tan^{-1} \frac{0.22}{0.12} = 61.39°$$

The resultant field at the point is given by

$$H = \sqrt{(Hv)^2 + (H_h)^2}$$

$$= \sqrt{(0.22)^2 + (0.12)^2}$$

$$= 0.25\ G$$

For a point 4 cm above the cable :

Horizontal component of earth's magnetic field :

$$H_h = H \cos \delta + B$$

$$= 0.39 \cos 35° + 0.2 = 0.52\ G$$

Vertical component of earth's magnetic field :

$$H_V = H \sin \delta$$

$$= 0.39 \sin 35° = 0.22\ G$$

$$\text{Angle, } \theta = \tan^{-1} \frac{H_v}{H_h} = \tan^{-1} \frac{0.22}{0.52} = 22.9°$$

And resultant field :

$$H = \sqrt{(H_v)^2 + (H_h)^2}$$

$$= \sqrt{(0.22)^2 + (0.52)^2} = 0.56\ T$$

Chapter 6. Electromagnetic Induction

Q. 1. Kamla peddles a stationary bicycle the pedals of the bicycle are attached to a 100 turn coil of area 0.10 m^2. The coil rotates at half a revolution per second and it is placed in a uniform magnetic field of 0.01 T perpendicular to the axis of rotation of the coil. What is the maximum voltage generated in the coil ? **[NCERT]**

Ans. Given,

$f = 0.5$ Hz, N = 100, A = 0.1 m^2 and B = 0.01 T.

As we know that

$$\varepsilon_0 = NBA\,(2\pi v)$$
$$= 100 \times 0.01 \times 0.01 \times 2 \times 3.14 \times 0.5$$
$$= 0.314 \text{ V}$$

The maximum voltage is 0.314 V.

Q. 2. Current in a circuit falls from 5.0 A to 0.0 A in 0.1 s. If an average emf of 200 V induced, give an estimate of the self-inductance of the circuit. **[NCERT]**

Ans. Given,

Initial current, $(I_1) = 5.0$ A

Final current, $(I_2) = 0.0$ A

Change in current, $(dI) = I_1 - I_2 = 5$ A

Time taken for the change, $(t) = 0.1$ s

Average emf, $(e) = 200$ V

$$e = L\frac{di}{dt}$$
$$L = \frac{e}{\left(\frac{di}{dt}\right)} = \frac{200}{\left(\frac{5}{0.1}\right)} = 4 \text{ H}$$

Hence, the self induction of the coil is 4 H.

Q. 3. A wheel with 10 metallic spokes each 0.5 m long is rotated with a speed of 120 rev/min in a plane normal to the horizontal component of earth's magnetic field H_E at a place. If $H_E = 0.4$ G at the place, what is the induced emf between the axle and the rim of the wheel ? Note that 1 G = 10^{-4} T. **[NCERT]**

Ans. Given,

R = 0.5 m, $H_E = 0.4$ G, B = $0{\cdot}4 \times 10^{-4}$ T

$$\text{Induced emf} = \left(\frac{1}{2}\right)\omega B R^2$$
$$= \left(\frac{1}{2}\right) \times 4\pi \times 0.4 \times 10^{-4} \times (0.5)^2$$
$$= 6.28 \times 10^{-5} \text{ V}$$

The number of spokes is immaterial because the emf's across the spokes are in parallel.

Q. 4. A jet plane is travelling towards west at a speed of 1800 km/h. What is the voltage difference developed between the ends of the wing having a span of 25 m, if the Earth's magnetic field at the location has a magnitude of 5 x 10^{-4} T and the dip angle is 30°. **[NCERT]**

Ans. Given,

Speed of the jet plane, (v) = 1800 km/h = 50 m/s

Wing span of jet plane, (l) = 25 m

Earth's magnetic field strength, (B) = 5.0 x 10^{-4} T

Angle of dip, $(\delta) = 30°$

Vertical component of Earth's magnetic field is given by

$$B_v = B \sin\delta$$
$$= 5 \times 10^{-4} \sin 30°$$
$$= 2.5 \times 10^{-4} \text{ T}$$

Voltage difference between the ends of the wing is given by

$$e = (B_v) \times l \times v$$
$$= 2.5 \times 10^{-4} \times 25 \times 500$$
$$= 3.125 \text{ V}$$

Therefore, the voltage difference arises between the ends of the wings is 3.125 V.

Q. 5. A pair of adjacent coils has a mutual indutance of 1.5 H. If the current in one coil changes from 0 to 20 A in 0.5 s, what is the change of flux linkage with the other coil ?★

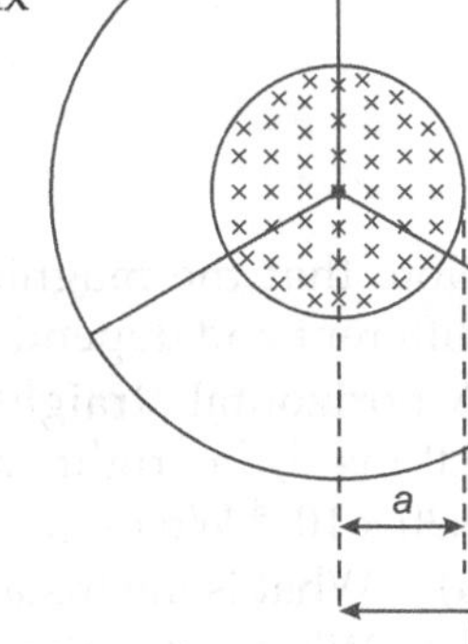

Ans.

$$M = 1.5 \text{ H}$$
$$I_i = 0\text{A}$$
$$I_b = 20 \text{ A}$$
$$dI = 20 \text{ A}, \Delta t = 0{\cdot}5 \text{ s}$$
$$e = \frac{-MdI}{dt} = -1.5 \times \frac{20}{0.5}$$
$$= -60 \text{ V}$$

So the flux linked with the other coil is given by

$$\Delta\phi = e\Delta t$$
$$= -60 \times 0.5$$
$$= -30 \text{ Wb}$$

Q. 6. A square loop of side 10 cm and resistance 0.5 Ω is placed vertically in the east-west plane. A uniform magnetic field of 0.10 T is set-up across the plane in the north-east direction. The magnetic field is decreased to zero in 0.70 s at a steady rate. Determine the magnitudes of induced emf and current during this time-interval. **[NCERT]**

Ans. The angle θ made by the area vector of the coil with the magnetic field is 45°. The initital magnetic flux is given by

$$\phi = BA \cos \theta$$
$$= \frac{0.1 \times 10^{-2}}{\sqrt{2}} \text{ Wb}$$

$$\text{Final flux, } (\phi_{min}) = 0$$

The change in flux is brought about in 0.70 s. The magnitude of the induced emf is given by

$$\varepsilon = \frac{|\Delta \phi_B|}{\Delta t} = \frac{|(\phi - 0)|}{\Delta t} = \frac{10^{-3}}{\sqrt{2} \times 0.7} = 1.0 \text{ mV}$$

And the magnitude of the current is

$$I = \frac{\varepsilon}{R} = \frac{10^{-3} \text{ V}}{0.5 \text{ Ω}} = 2 \text{ mA}$$

Note that the earth's magnetic field also produces a flux through the loop. But it is a steady field (which does not change within the time span of the experiment) and hence does not induce any emf.

Q. 7. A circular coil of radius 10 cm, 500 turns and resistance 2 Ω is placed with its plane perpendicular to the horizontal component of the earth's magnetic field. It is rotated about its vertical diameter through 180° in 0.25 s. Estimate the magnitudes of the emf and current induced in the coil. Horizontal component of the earth's magnetic field at the place is 3.0×10^{-5} T. **[NCERT]**

Ans. Initial flux through the coil.

$$\phi_{B(initial)} = BA \cos \theta$$
$$= 3.0 \times 10^{-5} \times (\pi \times 10^{-2}) \times \cos 0°$$
$$= 3\pi \times 10^{-7} \text{ Wb}$$

★ are board based questions from previous year

Final flux after the rotation,

$$\phi_{B\,(final)} = 3.0 \times 10^{-5} \times (\pi \times 10^{-2}) \times \cos 180°$$
$$= -3\,\pi \times 10^{-7}\text{ Wb}$$

Therefore, estimated value of the induced emf is,

$$\varepsilon = N\frac{\Delta\phi}{\Delta t} = \frac{500 \times (6\pi \times 10^{-7})}{0.25}$$
$$= 3.8 \times 10^{-3}\text{ V}$$
$$I = \frac{\varepsilon}{R} = \frac{3{\cdot}8 \times 10^{-3}}{2} = 1.9 \times 10^{-3}\text{ A}$$

Note that the magnitude of ε and I are the estimated values. Their instantaneous values are different and depend upon the speed of rotation at the particular instant.

Q. 8. A horizontal straight wire 10 m long extending from east to west is falling with a speed of 5.0 ms^{-1}, a right angles to the horizontal component of the earth's magnetic field, 0.30×10^{-4} Wb m^{-2}. **[NCERT]**

(i) What is the instantaneous value of the emf induced in the wire ?

(ii) What is the direction of the emf ?

(iii) Which end of the wire is at the higher electrical potential ?

Ans. Given,

Length of the wire (l) = 10 m

Falling speed of the wire (v) = 5.0 m/s

Magnetic field strength (B) = 0.3×10^{-4} Wb m^{-2}

(i) Emf induced in the wire is given by

$$e = Blv = 0.3 \times 10^{-4} \times 10 \times 5$$
$$= 1.5 \times 10^{-3}\text{ V}$$

(ii) From Fleming's right hand rule, it can be inferred that the induced emf is West to East.

(iii) The eastern end of the wire is at a higher potential.

Q. 9. A pair of adjacent coils has a mutual inductance of 1.5 H. If the current in one coil changes from 0 to 20 A in 0.5 s, what is the change of flux linkage with the other coil ? **[NCERT]**

Ans. Given,

Mutual inductance of a pair of coils (M) = 1.5 H

Initial current (I_1) = 0 A

Final current (I_2) = 20 A

Change in current, (dI) = $I_2 - I_1$ = 20 – 0 = 20 A

Time taken for the change (t) = 0.5 s

Induced emf is given by

$$e = \frac{d\phi}{dt}$$

Where $d\phi$ is the change in the flux linkage with the coil.

Emf is related with mutual inductance by

$$e = M\frac{dI}{dt}$$

Equating emf will give

$$\frac{d\phi}{dt} = M\frac{dI}{dt}$$
$$d\phi = 1.5 \times (20)$$
$$= 30\text{ Wb}$$

Q. 10. An aeroplane is flying horizontally from west to east with a velocity of 900 km/hour. Calculate the potential difference developed between the ends of its wings having a span of 20 m. The horizontal component of the Earth's magnetic field is 5×10^{-4} T and the angle of dip is 30°.

Ans. Given,

$$\text{Velocity } (v) = 900\text{ km/h}$$

$$= 900 \times \frac{5}{18} \text{ m/s}$$

$$= 250 \text{ ms}^{-1}$$

Wing span (l) = 20 m

Horiontal component of earth's field (B_H) = 5×10^{-4} T

Angle of dip (δ) = 30°

p.d. developed = ?

$$B_V = B_H \tan \delta$$

$$= 5 \times 10^{-4} \times \tan 30°$$

$$= 5 \times 10^{-4} \times \frac{1}{1.732}$$

$$\text{emf induced} = B_V l v$$

$$= \frac{5 \times 10^{-4}}{1.732} \times 20 \times 250 = 1.44 \text{ V}$$

Q. 11. An air-cored solenoid with length 30 cm, area of cross-section 25 cm^2 and number of turns 500, carries a current of 2.5 A. The current is suddenly switched off in a brief time of 10^{-3} s. How much is the average back emf induced across the ends of the open switch in the circuit ? Ignore the variation in magnetic field near the ends of the solenoid. **[NCERT]**

Ans. Given,

Length of the solenoid (l) = 30 cm = 0.3 m

Area of cross-section (A) = 25 cm^2 = 25×10^{-4} m^2

Number of turns on the solenoid (N) = 500

Current in the solenoid (I) = 2.5 A

Current flows for time (t) = 10^{-3} s

Average back emf, $$e = \frac{d\phi}{dt} \quad \text{...(i)}$$

Where, $d\phi$ = Change in flux

$$d\phi = NAB \quad \text{...(ii)}$$

Where,

B = Magnetic field strength

$$B = \mu_0 \frac{NI}{l} \quad \text{...(iii)}$$

Where, μ_0 = Permeability of free space = $4\pi \times 10^{-7}$ T mA^{-1}

From equation(i), (ii) and (iii)

$$e = \frac{\mu_0 N^2 IA}{l \times t}$$

$$= \frac{4\pi \times 10^{-7} \times (500)^2 \times 2.5 \times 25 \times 10^{-4}}{0.3 \times 10^{-3}}$$

$$= 6.5 \text{ V}$$

Hence, the average back emf induced in the solenoid is 6.5 V.

Q. 12. Suppose the rectangular loop of sides 8 cm and 2 cm is stationary but the current feeding the electromagnet that produces the magnetic field is gradually reduced so that the field decreases from its initial value of 0.3 T at the rate of 0.02 T s^{-1}. If the cut is joined and the loop has a resistance of 1.6 Ω how much power is dissipated by the loop as heat ? What is the source of this power ? **[NCERT]**

Ans. Given,

Sides of the rectangular loop are 8 cm and 2 cm.

Thus, area of the rectangular wire loop,

$$(A) = \text{length} \times \text{width}$$

$$= 8 \times 2 = 16 \text{ cm}^2 = 16 \times 10^{-4} \text{ m}^2$$

Initial value of the magnetic field, (B) = 0.3 T

Rate of decreases of the magnetic field, $\left(\frac{dB}{dt}\right) = 0.02$ T/s

Emf developed in the loop is given by

$$e = \frac{d\phi}{dt}$$

Where, $d\phi$ = Change in flux through the loop area

$$= AB$$

$$\therefore \quad e = \frac{d(AB)}{dt} = \frac{AdB}{dt}$$

$$= 16 \times 10^{-4} \times 0.02$$

$$= 0.32 \times 10^{-4} \text{ V}$$

Resistance of the loop, R = 1.6 Ω

The current induced in the loop is given by

$$i = \frac{e}{R} = \frac{0.32 \times 10^{-4}}{1.6} = 2 \times 10^{-5} \text{ A}$$

Power dissipated through loop in the form of heat is given by

$$P = i^2R$$

$$= (2 \times 10^{-5})^2 \times 1.6$$

$$= 6.4 \times 10^{-10} \text{ W}$$

The source of this heat loss is an external agent, which is responsible for changing the magnetic field with time.

Q. 13. A circular coil of radius 8.0 cm and 20 turns is rotated about its vertical diameter with an angular speed of 50 rad s^{-1} in a uniform horizontal magnetic field of magnitude 3.0×10^{-2} T. Obtain the maximum and average emf induced in the coil. If the coil forms a closed loop of resistance 10 Ω, calculate the maximum value of current in the coil. Calculate the average power loss due to Joule heating. Where does this power come from ? **[NCERT]**

Ans. Given,

Radius of the circular coil (r) = 8 cm = 0.08 m

Area of the coil (A) = $\pi r^2 = \pi \times (0.08)^2$ m²

Number of turns on the coil, (N) = 20

Angular speed (ω) = 50 rad/s

Magnetic field strength, (B) = 3×10^{-2} T

Resistance of the loop, (R) = 10 Ω

Maximum induced emf is given by

$$e = N\omega AB$$

$$= 20 \times 50 \times \pi \times (0.08)^2 \times 3 \times 10^{-2}$$

$$= 0.603 \text{ V}$$

The maximum emf induced in the coil is 0.603 V.

Over a full cycle, the average emf induced in the coil is zero.

Maximum current is given by

$$I = \frac{e}{R} = \frac{0.603}{10} = 0.0603 \text{ A}$$

Average power loss due to joule heating is given by

$$P = \frac{eI}{2} = \frac{0.603 \times 0.0603}{2} = 0.018 \text{ W}$$

The current induced in the coil produces a torque opposing the rotation of the coil. The rotor is an external agent. It must supply a torque to counter this torque in order to keep the coil rotating uniformly. Hence, dissipated power comes from the external rotor. (Power comes from external rotor)

Q. 14. A long solenoid with 15 turns per cm has a small loop of area 2.0 cm^2 placed inside the solenoid normal to its axis. If the current carried by the solenoid changes steadily from 2.0 A to 4.0 A in 0.1 s, what is the induced emf in the loop while the current is changing ? **[NCERT]**

Ans. Given,

Number of turns on the solenoid = 15 turns/cm = 1500 turns/m

Number of turns per unit length (n) = 1500 turns

Small loop of area (A) = 2.0 cm^2 = 2×10^{-4} m^2

Current through the solenoid changes from 2 A to 4 A.

∴ Change in current through solenoid (dI) = 4 – 2 = 2 A

Change in time (dt) = 0.1 s

Induced emf in the solenoid is given by Faraday's law as :

$$e = \frac{d\phi}{dt} \qquad \ldots\text{(i)}$$

Where, ϕ = Induced flux through the small loop

$$\phi = BA$$

$$B = \text{Magnetic field}$$

$$= \mu_0 ni \qquad \ldots\text{(ii)}$$

$$\mu_0 = \text{Permeability of free space}$$

$$= 4\pi \times 10^{-7} \text{ H/m}$$

Hence, equation (i) reduces to :

$$e = \frac{d}{dt}(BA)$$

$$= A\,\mu_0 n \times \left(\frac{di}{dt}\right)$$

$$= 2 \times 10^{-4} \times 4\pi \times 10^{-7} \times 1500 \times \frac{2}{0.1}$$

$$= 7.54 \times 10^{-6} \text{ V}$$

Hence, the induced voltage in the loop is 7.54×10^{-6} V

Q. 15. A rectangular wire loop of sides 8 cm and 2 cm with a small cut is moving out of a region of uniform magnetic field of magnitude 0.3 T directed normal to the loop. What is the emf developed across the cut if the velocity of the loop is 1 cm s^{-1} in a direction normal to the (i) longer side, (ii) shorter side of the loop ? For how long does the induced voltage last in each case ? **[NCERT]**

Ans. Given,

Length of the rectangular wire, l = 8 cm = 0.08 m

Width of the rectangular wire, b = 2 cm = 0.02 m

Hence, area of the rectangular loop,

$$A = lb = 0.08 \times 0.02$$

$$= 16 \times 10^{-4} \text{ m}^2$$

Magnetic field strength (B) = 0.3 T

Velocity of the loop (v) = 1 cm/s = 0.01 m/s

(i) Emf developed in the loop is given by

$$e = Blv$$

$$= 0.3 \times 0.08 \times 0.01 = 2.4 \times 10^{-4} \text{ V}$$

Time taken to travel along the width (t)

$$= \frac{\text{Distance travelled}}{\text{Velocity}} = \frac{b}{v}$$

$$= \frac{0.02}{0.01} = 2 \text{ s}$$

Hence, the induced voltage is 2.4×10^{-4} V which lasts for 2 s.

(ii)

$$\text{Emf developed, } (e) = Bbv$$
$$= 0.3 \times 0.02 \times 0.01$$
$$= 0.6 \times 10^{-4} \text{ V}$$

Time taken to travel along the length

$$t = \frac{\text{Distance travelled}}{\text{Velocity}} = \frac{l}{v}$$
$$= \frac{0.08}{0.01} = 8 \text{ s}$$

Hence, the induced voltage is 0.6×10^{-4} V which lasts for 8 s.

Q. 16. A square loop of side 12 cm with its sides parallel to X and Y axes is moved with a velocity of 8 cm s^{-1}m in the positive x-direction in an environment containing a magnetic field in the positive z-direction. The field is neither uniform in space nor constant in time. It has a gradient of 10^{-3} T cm^{-1} along the negative x-direction (that is it increases by 10^{-3} T cm^{-1} as one moves in the negative x-direction), and it is decreasing in time at the rate of 10^{-3} T s^{-1}. Determine the direction and magnitude of the induced current in the loop if its resistance is 4.50 mΩ. **[NCERT]**

Ans. Given, a square loop.

Side of the loop = 12 cm

Velocity with which the loop is moving = 8 cm s^{-1} = 8×10^{-2} m/s

Area of the loop, A = $(12 \times 10^{-2})^2 = 144 \times 10^{-4}$ m^2

Gradient in magnetic field,

$$\frac{dB}{dt} = 10^{-3} \text{ T/sec}$$

$$\frac{dB}{dx} = 10^{-3} \text{ T/cm} = 10^{-1} \text{ T/m}$$

Induced e.m.f. due to change of magnetic field B with time t is,

$$\varepsilon_1 = \frac{d\phi}{dt} = \frac{d(B \cdot A)}{dt} = \frac{A \cdot dB}{dt}$$
$$= 144 \times 10^{-4} \times 10^{-3}$$
$$\varepsilon_1 = 144 \times 10^{-7} \text{ V} \quad \text{...(i)}$$

Induced e.m.f. due to change of magnetic field B with distance (x) is

$$\varepsilon_2 = \frac{d\phi}{dt} = \frac{d}{dt} BA = A \cdot \frac{dB}{dt}$$
$$= A \cdot \frac{dB}{dx} \cdot \frac{dx}{dt}$$
$$= 144 \times 10^{-4} \times 10^{-1} \times 8 \times 10^{-2}$$
$$\varepsilon_2 = 1152 \times 10^{-7}$$

Total e.m.f.

$$\varepsilon_1 + \varepsilon_2 = 144 \times 10^{-7} + 1152 \times 10^{-7} = 1296 \times 10^{-7} \text{ V}$$
$$\varepsilon = 129.6 \times 10^{-6} \text{ V}$$

Resistance of the loop, R = 4.5 mΩ

Therefore,

$$\text{Induced current (I)} = \frac{\varepsilon}{R} = \frac{129.6 \times 10^{-6}}{4.5 \times 10^{-3}}$$
$$\cong 2.9 \times 10^{-2} \text{ A}$$

Thus, the direction of the induced current increase, the flux through the loop along positive z-direction.

Q. 17. Given figure shows a metal rod PQ resting on the smooth rails AB and positioned between the poles of a permanent magnet. The rails, the rod and the magnetic field are in three mutual pependicular directions. A galvanometer G connects the rails through a switch K. Length of the rod = 15 cm, B = 0.50 T, resistance of the closed loop containing the rod = 9.0 mΩ. Assume the field to be uniform.

(i) Suppose K is open and the rod is moved with a speed of 12 cm s^{-1} in the direction shown. Give the polarity and magnitude of the induced emf.

(ii) Is there an excess charge built up at the ends of the rods when K is open ? What if K is closed ?

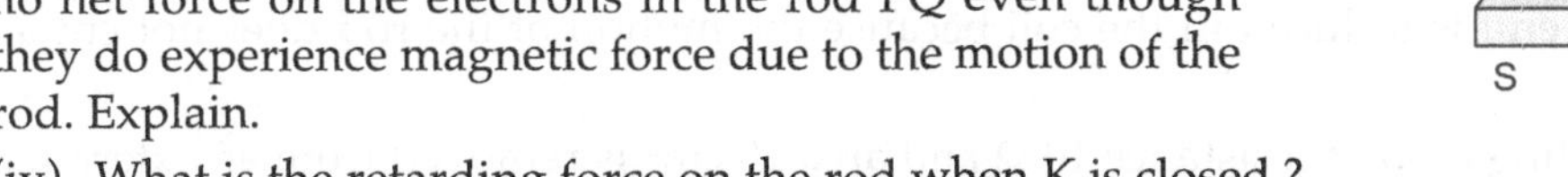

(iii) With K open and the rod moving uniformly, there is no net force on the electrons in the rod PQ even though they do experience magnetic force due to the motion of the rod. Explain.

(iv) What is the retarding force on the rod when K is closed ?

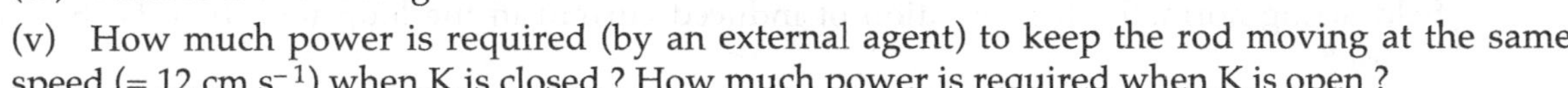

(v) How much power is required (by an external agent) to keep the rod moving at the same speed (= 12 cm s^{-1}) when K is closed ? How much power is required when K is open ?

(vi) How much power is dissipated as heat in the closed circuit ?

What is the source of this power ?

(vii) What is the induced emf in the moving rod if the magnetic field is parallel to the rails instead of being perpendicular ? **[NCERT]**

Ans. Given,

Length of the rod, (l) = 15 cm = 0.15 m

Magnetic field strength, (B) = 0.50 T

Resistance of the closed loop, (R) = 9 mΩ = 9×10^{-3} Ω

(i) Induced emf = 9 mV; polarity of the induced emf is makes end P shows positive while end Q negative ends. Speed of the rod (v) = 12 cm/s = 0.12 m/s

Induced emf is given by

$$\begin{aligned} e &= Bvl \\ &= 0.5 \times 0.12 \times 0.15 \\ &= 9 \times 10^{-3} V \\ &= 9 \text{ mV} \end{aligned}$$

The polarity of the induced emf is such that end P gives positive while end Q gives negative.

(ii) Yes; when key K is closed, excess charge is maintained by the continuous flow of current.

When key K is open, there is excess charge gets developed at both ends of the rods.

When key K is closed, excess charge is maintained by the continuous flow of current.

(iii) Magnetic force is cancelled by the electric force set-up due to the excess charge of opposite nature at both ends of the rod.

There is no net force on the electrons in rod PQ when key K is open and the rod is moving uniformly. This is because magnetic force is cancelled by the electric force set-up due to the excess charge of opposite nature at both ends of the rods.

(iv) Retarding force exerted on the rod, F = IBl

Where,

I = Current flowing through the rod,

$$\begin{aligned} F &= \frac{e}{R} \\ &= \frac{9 \times 10^{-3}}{9 \times 10^{-3}} = 1 \text{ A} \end{aligned}$$

$$\therefore \quad F = 1 \times 0.5 \times 0.15$$
$$= 75 \times 10^{-3} \text{ N}$$

(v) No power is expended when key K is open.
Speed of the rod, $v = 12$ cm/s = 0.12 m/s
Thus, power is given by

$$P = Fv = 75 \times 10^{-3} \times 0.12$$
$$= 9 \times 10^{-3} \text{ W} = 9 \text{ mW}$$

When key K is open, no power is expended.
(vi) Power is provided by an external agent.

$$\text{Power dissipated as heat} = I^2R$$
$$= (1)^2 \times 9 \times 10^{-3} = 9 \text{ mW}$$

The source of this power is an external agent.
(vii) Zero
In this case, no emf is induced in the coil because the motion of the rod does not cut across the field lines.

Q. 18. When a conducting loop of resistance 10 Ω and area 10 cm² is removed from an external magnetic field acting normally, the variation of induced current in the loop with time is shown in the figure.★

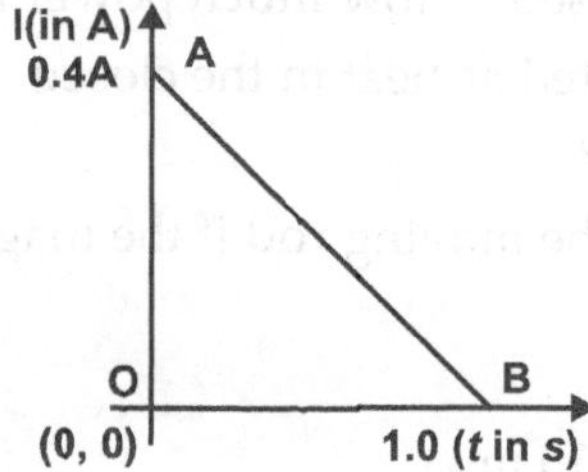

Find the :
(i) total charge passed through the loop.
(ii) change in magnetic flux through the loop.
(iii) Magnitude of the magnetic field applied.

Ans.

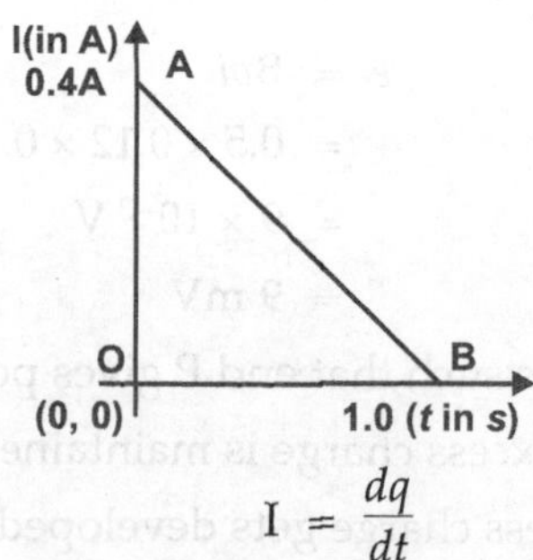

(i)
$$I = \frac{dq}{dt}$$
$$dq = Idt$$

Hence, the area under the I-t curve gives charge flow.

$$\text{Area of I-}t\text{ curve} = \frac{1}{2} \times OA \times OB$$
$$= \frac{1}{2} \times 0.4 \times 1 = 0.2 \text{ C}$$

$$\text{Total charge passed through loop } (\Delta q) = 0.2 \text{ C}$$

(ii) Now, we know

$$\Delta q = \frac{\Delta \phi}{R}$$

★ are board based questions from previous year

$$\therefore \quad \Delta\phi = \Delta q \times R$$
$$= 0.2 \times 10 = 2\ Wb$$

(iii) Now, change in flux, $\Delta\phi = B(\Delta A)$

$$2 = B\left(\frac{10}{100 \times 100}m\right)$$

$$B = \frac{2 \times 100 \times 100}{10}$$

$$B = 2000\ T$$

Hence, the magnitude of field applied = 2000 T

Chapter 7. Alternating Current

Q. 1. How much current is drawn by the primary of a transformer connected to 220 V supply when it delivers power to a 110 V– 550 W refrigerator ?

Ans. Given, Power (P) = 550 W

Supply voltage (V_S) = 220 V

$$\text{Power} = V_P I_P$$
$$550 = 220 \times I_P$$
$$\Rightarrow \quad I_P = \frac{5}{2} = 2.5\ A$$

Q. 2. A power transmission line feeds input power at 2200 V to a step-down transformer with its primary windings having 3000 turns. Find the number of turns in the secondary to get the power output at 220 V. ★

Ans. Given,

$$e_p = 2200\ V$$
$$N_p = 3000$$
$$N_s = ?$$
$$e_s = 220\ V$$
$$\therefore \quad \frac{e_s}{e_p} = \frac{N_s}{N_p}$$
$$\frac{220}{2200} = \frac{N_s}{3000}$$
$$\Rightarrow \quad N_s = 300$$

Q. 3. A light bulb is rated at 100 W for a 220 V supply. Find :

(i) the resistance of the bulb;

(ii) the peak voltage of the source;

(iii) the r.m.s. current through the bulb. **[NCERT]**

Ans. (i) We are given P = 100 W and V = 220 V. The resistance of the bulb is

$$R = \frac{V^2}{P}$$
$$= \frac{(220\ V)^2}{100\ W} = 484\ \Omega$$

(ii) The peak voltage of the source is

$$V_m = \sqrt{2}\,V = 311\ V$$

(iii) Since, $P = I\,V$

$$I = \frac{P}{V} = \frac{100\text{ W}}{220\text{ V}} = 0.450\text{ A}$$

Q. 4. A power transmission line feeds input power at 2300 V to a step-down transformer with its primary windings having 4000 turns. What should be the number of turns in the secondary in order to get output power at 230 V ?★ **[NCERT]**

Ans. Given,

Input voltage (V_1) = 2300V

Number of turns in primary coil (n_1) = 4000

Output voltage (V_2) = 230 V

Number of turns in secondary coil = n_2

Voltage is related to the number of turns as :

$$\frac{V_1}{V_2} = \frac{n_1}{n_2}$$

$$\frac{2300}{230} = \frac{4000}{n_2}$$

$$n_2 = \frac{4000 \times 230}{2300} = 400$$

Hence, there are 400 turns in the second winding.

Q. 5. At a hydroelectric power plant, the water pressure head is at a height of 300 m and the water flow available is 100 $m^3\ s^{-1}$. If the turbine generator efficiency is 60%, estimate the electric power available from the plant ($g = 9.8\ m/s^{-2}$).★

Ans. Given,

Height of water pressure head (h) = 300 m

Volume of water flow per second (V) = 100 m^3/s

Efficiency of turbine generator (η) = 60% = 0.6

Acceleration due to gravity (g) = 9.8 m/s^2

Density of water (ρ) = 10^3 kg/m^3

Electric power available from the plant

$$= \eta \times h\rho g V$$

$$= 0.6 \times 300 \times 10^3 \times 9.8 \times 100$$

$$= 176.4 \times 10^6\text{ W}$$

$$= 176.4\text{ MW}$$

Q. 6. Assume the initial charge on the capacitor in previous question is 6 mC. What is the total energy stored in the circuit initially ? What is the total energy at later time ? **[NCERT]**

Ans. Given,

Capacitance of capacitor (C) = 30 μF = 30×10^{-6} F, inductance of the inductor, L = 27 mH = 27×10^{-3} H, charge on the capacitor, Q = 6 mC = 6×10^{-3} C

Total energy stored in the capacitor

$$E = \frac{1}{2}\frac{Q^2}{C}$$

$$= \frac{1}{2}\frac{(6 \times 10^{-3})^2}{30 \times 10^{-6}} = \frac{6}{10} = 0.6\text{ J}$$

Total energy at a latter time will remain the same because energy is shared between the capacitor and the inductor.

Q. 7. Obtain the resonant frequency ω_r of a series LCR circuit with L = 2.0 H, C = 32 μF and R = 10 Ω. What is the Q-value of this circuit ? **[NCERT]**

Ans. Given,

★ are board based questions from previous year

Inductance (L) = 2.0 H, capacitance (C) = 32 μF = 32×10^{-6} F, resistance (R) = 10 Ω

Resonant frequency is given by the relation,

$$(\omega_r) = \frac{1}{\sqrt{LC}} = \frac{1}{(2 \times 32 \times 10^{-6})^{\frac{1}{2}}}$$

$$= \frac{1}{8 \times 10^{-3}} = 125 \text{ rad/s.}$$

Now Q-value of the circuit is given by

$$Q = \frac{1}{R}\sqrt{\frac{L}{C}} = \frac{1}{10}\sqrt{\frac{2}{32 \times 10^{-6}}}$$

$$= \frac{1}{10 \times 4 \times 10^{-3}} = 25$$

Hence, the Q-value of this circuit is 25.

Q. 8. A 44 mH inductor is connected to 220 V, 50 Hz a.c. supply. Determine the r.m.s. value of the current in the circuit. **[NCERT]**

Ans. Given,

Inductance of inductor (L) = 44 mH = 44×10^{-3} H, supply voltage (V) = 220 V, frequency (ν) = 50 Hz, angular frequency (ω) = 2πν

$$\text{Inductive reactance } (X_L) = \omega L = 2\pi\nu L$$

$$= 2\pi \times 50 \times 44 \times 10^{-3}\ \Omega$$

r.m.s. value of current is given by

$$(I) = \frac{V}{X_L} = \frac{220}{2\pi \times 50 \times 44 \times 10^{-3}}$$

$$= 15.92 \text{ A}$$

Hence, the r.m.s. value of current in the circuit is 15.92 A.

Q. 9. A 15.0 μF capacitor is connected to a 220 V, 50 Hz source. Find the capacitive reactance and the current (r.m.s. and peak) in the circuit. If the frequency is doubled, what happens to the capacitive reactance and the current ? **[NCERT]**

Ans. The capacitive reactance is

$$X_C = \frac{1}{2\pi\nu C}$$

$$= \frac{1}{2\pi\,(50 \text{ Hz})\,(15.0 \times 10^{-6} \text{ F})}$$

$$= 212\ \Omega$$

The r.m.s. current is

$$I = \frac{V}{X_C} = \frac{220 \text{ V}}{212\ \Omega} = 1.04 \text{ A}$$

The peak current is

$$i_m = \sqrt{2}\,I = (1.41)\,(1.04 \text{ A}) = 1.47 \text{ A}$$

This current oscillates between + 1.47 A and – 1.47 A, and is ahead of the voltage by $\frac{\pi}{2}$.

If the frequency is doubled, the capacitive reactance is halved and consequently, the current is doubled.

Q. 10. A sinusoidal voltage of peak value 283 V and frequency 50 Hz is applied to a series LCR circuit in which R = 3 Ω, L = 25.48 mH, and C = 796 μF. Find : (i) the impedance of the circuit. (ii) the phase difference between the voltage across the source and the current. (iii) the power dissipated in the circuit, and (iv) the power factor.

Ans. (i) To find the impedance of the circuit, we first calculate X_L and X_C.

$$X_L = 2\pi\nu L$$

$$= 2 \times 3.14 \times 50 \times 25.48 \times 10^{-3}\ \Omega = 8\ \Omega$$

$$X_C = \frac{1}{2\pi\nu C}$$

$$= \frac{1}{2 \times 3.14 \times 50 \times 796 \times 10^{-6}} = 4\ \Omega$$

Therefore,

$$Z = \sqrt{R^2 + (X_L - X_C)^2} = \sqrt{3^2 + (8-4)^2}$$

$$= 5\ \Omega$$

(ii) Phase difference, $\phi = \tan^{-1}\dfrac{X_C - X_L}{R}$

$$= \tan^{-1}\left(\frac{4-8}{3}\right) = -53.1°$$

Since ϕ is negative, the current in the circuit lags the voltage across the source.

(iii) The power dissipated in the circuit is

$$P = I^2R$$

Now,

$$I = \frac{i_m}{\sqrt{2}}$$

$$= \frac{1}{\sqrt{2}}\left(\frac{283}{5}\right) = 40\text{ A}$$

Therefore,

$$P = (40\text{ A})^2 \times 3\ \Omega$$

$$= 4800\text{ W}$$

(iv) Power factor $= \cos\phi = \cos 53.1° = 0.6$

Q. 11. A series LCR circuit with R = 20 Ω, L = 1.5 H and C = 35 μF is connected to a variable frequency 200 V a.c. supply. When the frequency of the supply equals the natural frequency of the circuit, what is the average power transferred to the circuit in one complete cycle. **[NCERT]**

Ans. At resonance, the frequency of the supply power equals the natural frequency of the given LCR circuit.

Given,

Resistance, R = 20 Ω, inductance, L = 1.5 H, capacitance, C = 35 μF = 35×10^{-6} F

AC supply voltage to the LCR circuit, V = 200 V

Impedance of the circuit is given by the relation,

$$Z = \sqrt{R^2 + (X_L - X_C)^2}$$

At resonance, $X_L = X_C$

$$\therefore \quad Z = R = 20\ \Omega$$

Current in the circuit can be calculated as :

$$I = \frac{V}{Z} = \frac{200}{20} = 10\text{ A}$$

Thus, the average power transferred to the circuit is one complete cycle :

$$P = VI = 200 \times 10 = 2000\text{ W}$$

Q. 12. Obtain the resonant frequency and Q-factor of a series LCR circuit with L = 3.0 H, C = 27 μF, and R = 7.4 Ω. It is desired to improve the sharpness of the resonance of the circuit by reducing its full width at half maximum by a factor of 2. Suggest a suitable way.★

★ are board based questions from previous year

Ans. Given,

Inductance (L) = 3.0 H

Capacitance (C) = 27 μF = 27×10^{-6} F

Resistance (R) = 7.4 Ω

At resonance, angular frequency of the source for the given LCR series circuit is given by

$$\omega_r = \frac{1}{\sqrt{LC}}$$

$$= \frac{1}{\sqrt{3 \times 27 \times 10^{-6}}} = \frac{10^3}{9} = 111.11 \text{ rad s}^{-1}$$

Q-factor of the series :

$$Q = \frac{\omega_r L}{R} = \frac{111.11 \times 3}{7.4} = 45.0446$$

To improve the sharpness of the resonance by reducing its 'full width at half maximum' by a factor of 2 without changing (ω_r) we need to reduce R to half *i.e.*, resistance $= \frac{R}{2} = \frac{7.4}{2} = 3.7$ Ω.

Q. 13. (i) Find the value of the phase difference between the current and the voltage in the series LCR circuit shown below. Which one leads in phase : current or voltage ?★

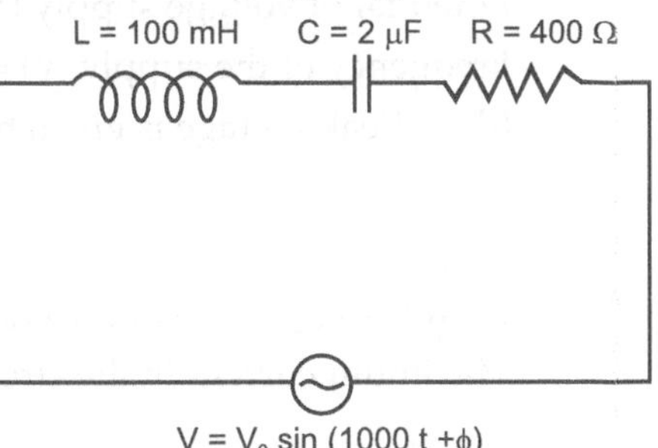

(ii) Without making any other change, find the value of the additional capacitor C_1, to be connected in parallel with capacitor C, in order to make the power factor of the circuit unity.★

Ans. (i) Given,

$$L = 100 \times 10^{-3} \text{ H}$$
$$C = 2 \times 10^{-6} \text{ F}$$
$$R = 400 \, \Omega$$
$$\omega = 1000$$

Now,

$$X_L = \omega L$$
$$X_L = 1000 \times 100 \times 10^{-3}$$
$$X_L = 100 \, \Omega$$

And

$$X_C = \frac{1}{\omega C} = \frac{1}{1000 \times 2 \times 10^{-6}} = \frac{10^6}{2 \times 10^{-3}}$$

$$X_C = \frac{1000}{2} = 500 \, \Omega$$

Here $X_C > X_L$ so, the current will lead.

$$\tan \phi = \frac{X_L - X_C}{R} = \frac{100 - 500}{400}$$

$$\tan \phi = -\frac{400}{400}$$

$$\tan \phi = -1$$

$$\Rightarrow \quad \phi = -45°$$

(ii) For unity power factor,

$$X_L = X_C$$

$$\omega L = \frac{1}{LC_{eq}}$$

$$\omega^2 = \frac{1}{LC_{eq}}$$

★ are board based questions from previous year

$$C_{eq} = \frac{1}{\omega^2 L}$$

$$C_{eq} = \frac{1}{(1000)^2 \times 100 \times 10^{-3}} = 10\ \mu F$$

Now,

$$C_{eq} = C + C_1$$
$$10 = 2 + C_1$$
$$10 - 2 = C_1$$
$$C_1 = 8\mu F$$

Q. 14. A coil of inductance 0.50 H and resistance 100 Ω is connected to a 240 V, 50 Hz a.c. supply.

(i) What is the maximum current in the coil ?

(ii) What is the time lag between the voltage maximum and the current maximum ? **[NCERT]**

Ans. Given,

Inductance of inductor (L) = 0.50 H

Resistance of resistor (R) = 100 Ω

Potential of voltage supply (V) = 240 V

Frequency of the supply (ν) = 50 Hz

(i) Peak voltage is given by

$$(V_0) = \sqrt{2}\,V$$
$$= \sqrt{2} \times 240 = 339.41\ V$$

Angular frequency (ω) = 2πν = 2π × 50 = 100 π rad/s

Maximum current in the circuit is given by

$$I_0 = \frac{V_0}{\sqrt{R^2 + \omega^2 L^2}}$$
$$= \frac{339.41}{\sqrt{(100)^2 + (100\pi)^2\,(0.50)^2}} = 1.82\ A$$

(ii) Equation for voltage is given by

$$V = V_0 \cos \omega t$$

Equation for current is given by

$$I = I_0 \cos (\omega t - \phi)$$

Here, φ = Phase difference between voltage and current. At time $t = 0$.

$V = V_0$ (voltage is maximum)

For $\omega t - \phi = 0$ *i.e*, at time $(t) = \frac{\phi}{\omega}$.

$I - I_0 =$ (current is maximum)

Thus, the time lag between maximum voltage and maximum current is $\frac{\phi}{\omega}$.

Now, phase angle φ is given by the relation,

$$\tan \phi = \frac{\omega L}{R} = \frac{2\pi \times 50 \times 0.5}{100}$$
$$= 1.57$$
$$\phi = 57.5° = \frac{57.5\,\pi}{180}\ \text{rad}$$
$$\omega t = \frac{57.5\,\pi}{180}$$
$$t = \frac{57.5\,\pi}{180 \times 2\pi \times 50}$$
$$= 3.19 \times 10^{-3}\ s$$
$$= 3.2\ ms.$$

Therefore, the time lag between maximum voltage and maximum current is 3.2.

Q. 15. Obtain the answers (i) and (ii) in previous exercise if the circuit is connected to a high frequency supply (240 V, 10 kHz). Hence, explain the statement that at very high frequency, an inductor in a circuit nearly amounts to an open circuit. How does an inductor behave in a d.c. circuit after the steady state ? **[NCERT]**

Ans. Given,

Inductance of the inductor (L) = 0.5 H

Resistance of the resistor (R) = 100 Ω

Potential of the supply voltages (V) = 240 V

Frequency of the supply (υ) = 10 kHz = 10^4 Hz,

Angular frequency (ω) = $2\pi\upsilon = 2\pi \times 10^4$ rad/s

(i)

$$\text{Peak voltage } (V_0) = \sqrt{2} \times V = 240\sqrt{2} \text{ V}$$

$$\text{Maximum current } (I_0) = \frac{V_0}{\sqrt{R^2 + \omega^2 L^2}}$$

$$= \frac{240\sqrt{2}}{\sqrt{(100)^2 + (2\pi \times 10^4)^2 \times (0.50)^2}}$$

$$= 1.1 \times 10^{-2} \text{ A}$$

(ii) For phase difference ϕ, we have the relation :

$$\tan\phi = \frac{\omega L}{R} = \frac{2\pi \times 10^4 \times 0.5}{100} = 100\pi$$

$$\phi = 89.82° = \frac{89.82\pi}{180} \text{ rad}$$

$$\omega t = \frac{89.82\pi}{180}$$

$$t = \frac{89.82\pi}{180 \times 2\pi \times 10^4} = 25\ \mu\text{s}$$

It can be observed that I_0 is very small in this case. Thus, at high frequencies, the inductor amounts to an open circuit.

In a d.c. circuit, after a steady state is achieved, (ω = 0). Thus, inductor L behaves like a pure conducting object.

Q. 16. The primary coil of an ideal step up transformer has 100 turns and transformation ratio is also 100. The input voltage and power are respectively 220 V and 1100 W. Calculate :

(i) number of turns in secondary (ii) current in primary

(iii) voltage across secondary (iv) current in secondary

(v) power in secondary

Ans. Given,

$$N_1 = 100$$

$$K = 100$$

$$V_1 = 220 \text{ V}$$

$$P_1 = 1100 \text{ W}$$

(i) As,

$$K = \frac{N_2}{N_1}$$

$$N_2 = KN_1 = 100 \times 100$$

$$N_2 = 10000$$

(ii)

$$P_1 = V_1 I_1$$

$$I_1 = \frac{P_1}{V_1} = \frac{1100}{220} = 5\text{A}$$

(iii)

$$\frac{V_2}{V_1} = K$$

$$V_2 = KV_1;\ V_2 = 100 \times 220$$

$$V_2 = 22000 \text{ V}$$

(iv)
$$\frac{I_1}{I_2} = K$$
$$I_2 = \frac{I_1}{K} = \frac{5}{100}$$
$$I_2 = 0.05 \text{ A}$$

(v)
$$P_2 = V_2 I_2$$
$$P_2 = 22000 \times \frac{5}{100}; P_2 = 1100 \text{ W}$$

Q. 17. (i) Determine the value of phase difference between the current and the voltage in the given series LCR circuit.

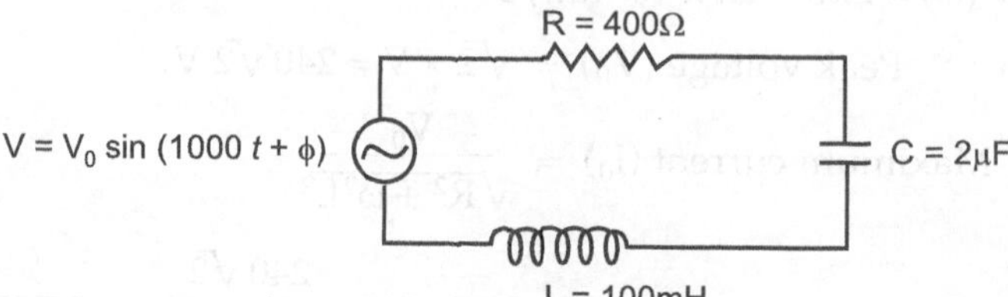

(ii) Calculate the value of the additional capacitor which may be joined suitably to the capacitor C that would make the power factor of the circuit unity.

Ans. (i)
$$V = V_0 \sin(1000\,t + \phi)$$

Comparing this with
$$V = V_0 \sin(\omega t + \phi)$$

we have,
$$\omega = 1000$$

Given,

$L = 100 \times 10^{-3}$ H, R = 400 Ω, C = 2 μF

$$X_L = \omega L$$
$$X_L = 1000 \times 100 \times 10^{-3} = 10^2 = 100\ \Omega$$
$$X_C = \frac{1}{\omega C} = \frac{1}{1000 \times 2 \times 10^{-6}} = \frac{1}{2} \times 10^3$$
$$= 500\ \Omega$$

Phase difference,
$$\tan\phi = \frac{X_L - X_C}{R}$$
$$\tan\phi = \frac{X_C - X_L}{R}, X_C > X_L$$

we have,
$$R = 400\ \Omega$$
$$\tan\phi = \frac{500 - 100}{400} = \frac{400}{400} = 1$$
$$\tan\phi = 1$$

∴
$$\tan\phi = \tan 45°$$
$$\phi = 45°$$

(ii) The power factor of the circuit is unity. It means that the given circuit is in resonance. It is possible, if another capacitor C′ is used in the circuit.

$$X_{C'} = X_L$$
$$\frac{1}{\omega C'} = \omega L$$
$$C' = \frac{1}{\omega^2 L} = \frac{1}{(1000)^2 \times 100 \times 10^{-3}}$$
$$C' = 10\ \mu F$$

Since C′ > C, so an additional capacitor of 8 μF would be connected in parallel to the capacitor of C = 2 μF.

Q. 18. A circuit containing a 80 mH inductor and a 60 μF capacitor in series is connected to a 230 V, 50 Hz supply. The resistance of the circuit is negligible. **[NCERT]**

(i) Obtain the current amplitude and r.m.s. values.

(ii) Obtain the r.m.s. values of potential drops across each element.

(iii) What is the average power transferred to the inductor ?

(iv) What is the average power transferred to the capacitor ?

(v) What is the total average power absorbed by the circuit ?

['Average' impies 'averaged over one cycle'.]

Ans. Given,

Inductance (L) = 80 mH = 80×10^{-3} H

Capacitance (C) = 60 μF = 60×10^{-6} F

Supply voltage (V) = 230 V

Frequency (v) = 50 Hz

Angular frequency (ω) = 2πv = 100 π rad/s

Peak voltage $(V_0) = V\sqrt{2} = 230\sqrt{2}$ V

(i) Maximum current is given by

Hence, r.m.s. values of current,

$$I_0 = \frac{V_0}{\left(\omega L - \frac{1}{\omega C}\right)}$$

$$= \frac{230\sqrt{3}}{\left(100\,\pi \times 80 \times 10^{-3} - \frac{1}{100\,\pi \times 60 \times 10^{-6}}\right)}$$

$$= \frac{230\sqrt{2}}{\left(8\,\pi - \frac{1000}{6\pi}\right)} = -\,11.63\text{ A}$$

The negative sign appears because $\omega L < \frac{1}{\omega C}$

Amplitude of maximum current, $|I_0| = 11.63$ A

$$I = \frac{I_0}{\sqrt{2}} = \frac{-\,11.63}{\sqrt{2}} = -\,8.22\text{ A}$$

(ii) Potential difference across the inductor,

$$(V_L) = I \times \omega L$$

$$= 8.22 \times 100\,\pi \times 80 \times 10^{-3}$$

$$= 206.61\text{ V}$$

Potential difference across the capacitor,

$$(V_C) = I \times \frac{1}{\omega C}$$

$$= 8.22 \times \frac{1}{100\pi \times 60 \times 10^{-6}}$$

$$= 436.3\text{ V}$$

(iii) Average power consumed by the inductor is zero as actual voltage leads the current by $\frac{\pi}{2}$.

(iv) Average power consumed by the capacitor is zero as voltage lags current by $\frac{\pi}{2}$.

(v) The total power absorbed (averaged over one cycle) is zero.

Q. 19. Suppose the circuit in previous question has a resistance of 15 Ω. Obtain the average power transferred to each element of the circuit, and the total power absorbed. **[NCERT]**

Ans. Given,

Inductance of inductor (L) = 80 mH = 80×10^{-3} H

Capacitance of capacitor (C) = 60 μF = 60×10^{-6} F

Resistance of resistor (R) = 15 Ω

Potential of voltage supply (V) = 230 V

Frequency of signal (ν) = 50 Hz

Angular frequency of signal (ω) = 2πν = 2π × (50) = 100 π rad/s

The elements are connected in series. Thus, impedance of the circuit is given by

$$\text{Current flowing in the circuit, } Z = \sqrt{R^2 + \left(\omega L - \frac{1}{\omega C}\right)^2}$$

$$= \sqrt{(15)^2 + \left(100\pi\,(80 \times 10^{-3}) - \frac{1}{(100\,\pi \times 60 \times 10^{-6})}\right)^2}$$

$$= \sqrt{(15)^2 + (25.12 - 53.08)^2}$$

$$= 31.728\ \Omega$$

$$I = \frac{V}{Z} = \frac{230}{31.728} = 7.25\text{ A}$$

Average power transferred to resistance is given by

$$P_R = I^2R$$

$$= (7.25)^2 \times 15 = 788.44\text{ W}$$

Average power transferred to capacitor, (P_C)

$$= \text{Average power transferred to inductor, } (P_L) = 0$$

$$\text{Total power absorbed by the circuit} = P_R + P_C + P_L$$

$$= 788.44 + 0 + 0 = 788.44\text{ W}$$

Thus, the total power absorbed by the circuit is 788.44 W.

Q. 20. Figure shows a series LCR circuit connected to a variable frequency 230 V source, L = 5.0 H, C = 80 μF, R = 40 Ω

(i) Determine the source frequency which drives the circuit in resonance.

(ii) Obtain the impedance of the circuit and the amplitude of current at the resonating frequency.

(iii) Determine the r.m.s. potential drops across the three elements of the circuit. Show that the potential drop across the LC combination is zero at the resonating frequency.

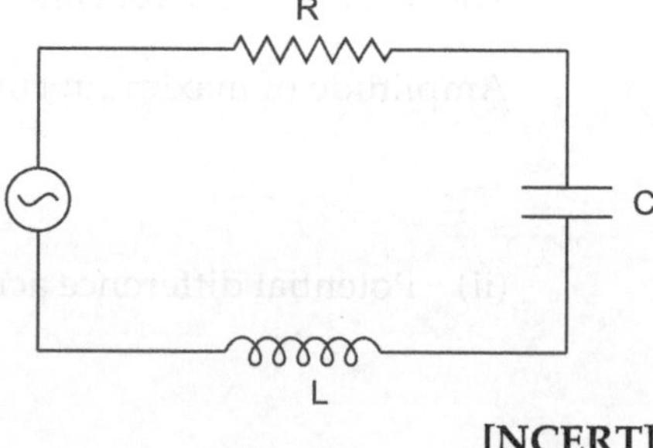

[NCERT]

Ans. Given,

Inductance of the inductor (L) = 5.0 H

Capacitance of the capacitor (C) = 80 μH = 80×10^{-6} F

Resistance of the resistor (R) = 40 Ω

Potential of the variable voltage source (V) = 230 V

(i) Resonance angular frequency is given as :

$$(\omega_r) = \frac{1}{\sqrt{LC}} = \frac{1}{\sqrt{5 \times 80 \times 10^{-6}}}$$

$$= \frac{10^3}{20} = 50\text{ rad/s}$$

Hence, the circuit will come in resonance for a source frequency of 50 rad/s.

(ii) Impedance of the circuit is given by the relation :

$$Z = \sqrt{R^2 + (X_L - X_C)^2}$$

At resonance, $X_L = X_C \Rightarrow Z = R = 40\ \Omega$

Amplitude of the current at the resonating frequency is given as :

$$I_0 = \frac{V_0}{Z}$$

Where, V_0 = Peak voltage = $\sqrt{2}$ V

$$\therefore \quad I_0 = \frac{\sqrt{2}\,V}{Z}$$

$$= \frac{\sqrt{2} \times 230}{40} = 8.13 \text{ A}$$

Hence, at resonance, the impedance of the circuit is 40 Ω and the amplitude of the current is 8.13 A.

(iii) r.m.s. potential drop across the inductor,

$$(V_L)_{rms} = I \times \omega_r L$$

Where, $$I_{rms} = \frac{I_o}{\sqrt{2}} = \frac{\sqrt{2}\,V}{\sqrt{2}\,Z} = \frac{230}{40} = \frac{23}{4} \text{ A}$$

$$\therefore \quad (V_L)_{rms} = \frac{23}{4} \times 50 \times 5 = 1437.5 \text{ V}$$

Potential drop across the capacitor :

$$\therefore \quad (V_C)_{rms} = I \times \frac{1}{\omega_r C}$$

$$= \frac{23}{4} \times \frac{1}{50 \times 80 \times 10^{-6}} = 1437.5 \text{ V}$$

Potential drop across the resistor :

$$(V_R)_{rms} = IR$$

$$= \frac{23}{4} \times 40 = 230 \text{ V}$$

Potential drop across the LC combination :

$$V_{LC} = I\,(X_L - X_C)$$

At resonance, $X_L = X_C \Rightarrow X_{LC} = 0$

Hence, it is proved that the potential drop across the LC combination is zero at resonating frequency.

Q. 21. A series LCR circuit with L = 0.12 H, C = 480 nF, R = 23 Ω is connected to a 230 variable frequency supply.

(i) What is the source frequency for which current amplitude is maximum ? Obtain this maximum value.

(ii) What is the source frequency for which average power absorbed by the circuit is maximum ? Obtain the value of this maximum power.

(iii) For which frequencies of the source is the power transferred to the circuit half the power at resonant frequency ? What is the current amplitude at these frequencies ?

(iv) What is the Q-factor of the given circuit ? **[NCERT]**

Ans. Given,

Inductance (L) = 0.12 H

Capacitance (C) = 480 nF = 480×10^{-9} F

Resistance (R) = 23 Ω

Supply voltage (V) = 230 V

Peak voltages is given by

$$(V_0) = \sqrt{2} \times 230 = 325.22 \text{ V}$$

(i) The current flowing through the circuit is given by

$$I_0 = \frac{V_0}{\sqrt{R^2 + \left(\omega L - \frac{1}{\omega C}\right)^2}}$$

Here, I_0 = maximum at resonance.

At resonance, we have

$$\omega_R L - \frac{1}{\omega_R C} = 0$$

Here, ω_R = Resonance angular frequency

$$\therefore \quad \omega_R = \frac{1}{\sqrt{LC}}$$

$$= \frac{1}{\sqrt{0.12 \times 480 \times 10^{-9}}}$$

$$= 4166.67 \text{ rad/s}$$

$$\therefore \quad \text{Resonant frequency } (\nu_R) = \frac{\omega_R}{2\pi}$$

$$= \frac{4166.67}{2 \times 3.14} = 663.48 \text{ Hz}$$

and Maximum current $(I_0)_{max} = \frac{V_0}{R} = \frac{325.22}{23} = 14.14 \text{ A}$

(ii) Maximum average power absorbed by the circuit is given as :

$$(P_{av})_{max} = \frac{1}{2}(I_0)^2_{max} R$$

$$= \frac{1}{2} \times (14.14)^2 \times 23 = 2299.3 \text{ W}$$

(iii) The power transferred to the circuit is half the power at resonant frequency.

Frequencies for half transferred power $= \omega_R \pm \Delta\omega$

$$= 2\pi(\nu_R \pm \Delta\nu)$$

Here, $\Delta\omega = \frac{R}{2L}$

$$= \frac{23}{2 \times 0.12} = 95.83 \text{ rad/s}$$

Thus change in frequency, $\Delta\nu = \frac{1}{2\pi}\Delta\omega$

$$= \frac{95.83}{2\pi} = 15.26 \text{ Hz}$$

$$\therefore \quad \nu_R + \Delta\nu = 663.48 + 15.26$$

$$= 678.74 \text{ Hz}$$

and $\nu_R - \Delta\nu = 663.48 - 15.26 = 648.22 \text{ Hz}$

Hence, at 648.22 Hz and 678.74 Hz frequencies, the power transferred is half.

At these frequencies, current amplitude can be given as :

$$I' = \frac{1}{\sqrt{2}} \times (I_0)_{max}$$

$$= \frac{14.14}{\sqrt{2}} = 10 \text{ A}$$

(iv) Q-factor of the given circuit can be obtained using the relation,

$$Q = \frac{\omega_R L}{R} = \frac{4166.67 \times 0.12}{23}$$

$$= 21.74$$

Hence, the Q-factor of the given circuit is 21.74.

Chapter 8. Electromagnetic Waves

Q. 1. The magnetic field in a travelling electromagnetic wave has a peak value of 20nT. The peak value of electric field strength is :

Ans. Since the medium is not mentioned, we consider it to be vacuum.

The peak value of electric field vector is

$$E_0 = cB_0$$

$$c = 3 \times 10^8 \text{ m/s}$$

$$B_0 = 20 \text{ nT} = 20 \times 10^{-9} \text{ T}$$

$$E_0 = 3 \times 10^8 \times 20 \times 10^{-9} \text{ V/m}$$

$$= 6 \text{ V/m}$$

Q. 2. A plane electromagnetic wave of frequency 25 MHz travels in free space along the x-direction. At a particular point in space and time, $E = 6.3\ \hat{j}$ V/m. What is B at this point ?★

Ans. The magnitude of B is given by

$$B = \frac{E}{c} = \frac{6.3 \text{ V/m}}{3 \times 10^8 \text{ m/s}} = 2.1 \times 10^{-8} \text{ T}$$

E is along y-direction and the wave propagates along X-axis. Therefore, B should be in a direction perpendicular to both X and Y-axis. Using vector algebra. E × B should be along X-direction. Since, $(+\hat{j}) \times (+\hat{k}) = \hat{i}$, B is along the z-direction.

Thus, $$B = 2.1 \times 10^{-8}\ \hat{k} \text{ T}$$

Q. 3. The amplitude of the magnetic field part of a harmonic electromagnetic wave in vacuum is B_0 = 510 nT. What is the amplitude of the electric field part of the wave ?

Ans. Amplitude of magnetic field of an electromagnetic wave in a vacuum,

$$B_0 = 510 \text{ nT} = 510 \times 10^{-9} \text{ T}$$

Speed of light in vacuum, $$c = 3 \times 10^8 \text{ m/s}$$

Amplitude of electric field of the electromagnetic wave is given by the relation,

$$E = cB_0$$

$$= 3 \times 10^8 \times 510 \times 10^{-9} = 153 \text{ N/C}$$

Therefore, the electric field part of the wave is 153 N/C.

Q. 4. Calculate the velocity of light in free space if the permittivity of free space is given an 8.85×10^{-12} C^2/Nm^2 and permittivity is $4\pi \times 10^{-7}$ N/Cs^2.

Ans.

$$\varepsilon_0 = 8.85 \times 10^{-12} \text{ C}^2/\text{Nm}^2$$

$$\mu_0 = 4\pi \times 10^{-7} \text{ C}^2 \text{ NS/C}^2$$

$$c = \frac{1}{\sqrt{\mu_0 \varepsilon_0}}$$

$$= \frac{1}{\sqrt{4\pi \times 10^{-7} \times 8.85 \times 10^{-12}}}$$

$$= 3 \times 10^8 \text{ m/s}$$

★ are board based questions from previous year

Q. 5. A plane electromagnetic wave travels in vacuum along z-direction. What can you say about the directions of its electric and magnetic field vectors ? If the frequency of the wave is 30 MHz, what is its wavelength ?

Ans. The EM wave travels in a vacuum along Z-direction. The electric field (E) and the magnetic field (H) are in X-Y plane. They are mutually perpendicular. Frequency of the wave, $\nu = 30\text{ MHz} = 30 \times 10^6\text{ s}^{-1}$

Speed of light in a vacuum, $c = 3 \times 10^8\text{ m/s}$

Wavelength is given by

$$\lambda = \frac{c}{\nu}$$

$$= \frac{3 \times 10^8}{30 \times 10^6} = 10\text{ m.}$$

Q. 6. Light with an energy flux of 18 W/cm² falls on a non-reflecting surface at normal incidence. If the surface has an area of 20 cm², find the average force exerted on the surface during a 30 minute time span. **[NCERT]**

Ans. The total energy falling at the surface

$$U = (18\text{ W/cm}^2) \times (20\text{ cm}^2) \times (30 \times 60)$$

$$= 6.48 \times 10^5\text{ J}$$

Thus, the total momentum delivered (for complete absorption) will be

$$p = \frac{U}{c} = \frac{6.48 \times 10^5\text{ J}}{3 \times 10^8\text{ m/s}} = 2.16 \times 10^{-3}\text{ kg m/s}$$

The average force exerted on the surface will be

$$F = \frac{p}{t} = \frac{2.16 \times 10^{-3}}{0.18 \times 10^4} = 1.2 \times 10^{-6}\text{ N}$$

Q. 7. An electromagnetic wave has a wavelength of 3×10^{-3} m and electric field associated with it has an amplitude of 30 V/m. Calculate the amplitude and frequency of oscillation of the magnetic field in vacuum.

Ans. Velocity of propagation (c) = 3×10^8 m/s

The amplitudes of electric and magnetic fields are related as

$$\frac{E_0}{B_0} = c$$

$$B_0 = E_0/c$$

$$= \frac{30}{3 \times 10^8\text{ T}} = 1 \times 10^{-9}\text{ T}$$

Frequency of oscillations of magnetic field and frequency of the wave are equal.

Therefore $\lambda = 3 \times 10^{-3}$

$$f = \frac{c}{\lambda} = \frac{3 \times 10^8}{3 \times 10^{-3}}$$

$$= 1 \times 10^{11}\text{ Hz}$$

Q. 8. A parallel plate capacitor with circular plates of radius 1 m has a capacitance of 1 nF. At $t = 0$. It is connected for charging in series with a resistor R = 1 M Ω across a 2V battery (Fig.). Calculate the magnetic field at a point P, half-way between the centre and the periphery of the plates, after $t = 10^{-3}$ s. (The charge on the capacitor at time t is $q(t) = CV[1 - \exp(-t/\tau)]$. Where the time constant τ is equal to CR.) **[NCERT]**

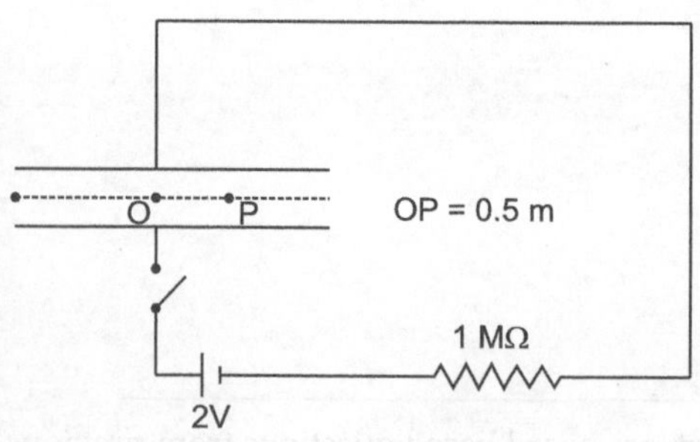

Ans. The time constant of the CR circuit is $\tau = CR = 10^{-3}$ s.

$$\therefore \quad q(t) = CV\,[1 - \exp.(-t/\tau)]$$
$$= 2 \times 10^{-9}\,[1 - \exp.(-t/10^{-3})]$$

The electric field in between the plates at any time (t) is

$$E = \frac{q(t)}{\varepsilon_0 A} = \frac{q}{\pi\varepsilon_0};\ A = \pi\,(1)^2\ m^2 = \text{area of the plates.}$$

Consider now a circular loop of radius (1/2) m parallel to the plates passing through P. The magnetic field B at all points on the loop is along the loop and of the same value.

The flux ϕ_E through this loop is

$$\phi_E = E \times \text{Area of the loop}$$
$$= E \times \pi \times \left(\frac{1}{2}\right)^2 = \frac{\pi E}{4} = \frac{q}{4\varepsilon_0}$$

The displacement current

$$i_d = \varepsilon_0 \frac{d\phi_E}{dt} = \frac{1}{4}\frac{dq}{dt} = 0.5 \times 10^{-6} \exp.(-1)$$

at $t = 10^{-3}$ s. Ampere-Maxwell law to the loop. We will get

$$B \times 2\pi \times \left(\frac{1}{2}\right) = \mu_0\,(i_c + i_d)$$
$$= \mu_0\,(0 + i_d)$$
$$= 0.5 \times 10^{-6}\,\mu_0 \exp.(-1)$$

or $$B = 0.74 \times 10^{-13}\ T$$

Q. 9. The magnetic field in a plane electromagnetic wave is given by

$$B_y = 2 \times 10^{-7} \sin (0.5 \times 10^3 x + 1.5 \times 10^{11} t)\ T$$

(i) What is the wavelength and frequency of the wave ?

(ii) Write an expression for the electric field. **[NCERT]**

Ans. (i) Comparing the given equation with

$$B_y = B_0 \sin \left[2\pi\left(\frac{x}{\lambda} + \frac{t}{T}\right)\right]$$

It will give, $$\lambda = \frac{2\pi}{0.5 \times 10^3}\ m = 1.26\ cm$$

and $$\frac{1}{T} = \nu = \frac{(1.5 \times 10^{11})}{2\pi} = 23.9\ GHz$$

(ii) $$E_0 = B_0 c = 2 \times 10^{-7}\ T \times 3 \times 10^8\ m/s$$
$$= 60\ V/m$$

The electric field component is perpendicular to the direction of propagation and the direction of magnetic field. Therefore, the electric field component along the Z-axis is as follows

$$E_z = 60 \sin (0.5 \times 10^3 x + 1.5 \times 10^{11} t) V/m$$

Q. 10. About 5% of the power of a 100 W light bulb is converted to visible radiation. What is the average intensity of visible radiation :

(i) At a distance of 1 m from the bulb ?

(ii) At a distance of 10 m ?

Assume that the radiation is emitted isotropically and neglect reflection. **[NCERT]**

Ans. Power rating of bulb, P = 100 W

It is given that about 5% of its power is converted into visible radiation.

∴ Power of visible radiation,

$$P' = \frac{5}{100} \times 100 = 5\ W$$

Thus, the power of visible radiation is 5 W.

(i) Distance of a point from the bulb (d) = 1 m.

Thus, intensity of radiation at the point is given by

$$I = \frac{P'}{4\pi d^2} = \frac{5}{4\pi(1)^2} = 0.398\ W/m^2$$

(ii) Distance of a point from the bulb $(d_1) = 10$ m

Thus, intensity of radiation at that point is given by

$$I = \frac{P'}{4\pi(d_1)^2} = \frac{5}{4\pi(10)^2} = 0.00398\ W/m^2$$

Q. 11. The terminology of different parts of the electromagnetic spectrum is given in the text. Use the formula $E = h\nu$ (for energy of a quantum of radiation : photon) and obtain the photon energy in units of eV for different parts of the electromagnetic spectrum. In what way are the different scales of photon energies that you obtain related to the sources of electromagnetic radiation ?

[NCERT]

Ans. Energy of photon is given as :

$$E = h\nu = \frac{hc}{\lambda}$$

where, h = Planck's constant $= 6.6 \times 10^{-34}$ Js

c = Speed of light $= 3 \times 10^8$ m/s

λ = Wavelength of radiation

$$\therefore \quad E = \frac{6.6 \times 10^{-34} \times 3 \times 10^8}{\lambda}$$

$$= \frac{19.8 \times 10^{-26}}{\lambda} J$$

$$= \frac{19.8 \times 10^{-26}}{\lambda \times 1.6 \times 10^{-19}} eV = \frac{12.375 \times 10^{-7}}{\lambda} eV$$

The given table lists the photon energies for different parts of an electromagnetic spectrum for different λ.

λ (m)	10^3	1	10^{-3}	10^{-6}	10^{-8}	10^{-10}	10^{-12}
E (eV)	12.375×10^{-10}	12.375×10^{-7}	12.375×10^{-4}	12.375×10^{-1}	12.375×10^{1}	12.375×10^{3}	12.375×10^{5}

The photon energies for the different parts of the spectrum of a source indicate the spacing of the relevant energy levels of the source.

Q. 12. Calculate the electric and magnetic fields produced by the radiation coming from a 100 W bulb at a distance of 3 m. Assume that the efficiency of the bulb is 2.5% and it is a point source. **[NCERT]**

Ans. The bulb (point source) radiates light in all directions uniformly. At distance of 3 m, the surface area of the surrounding sphere will be

$$A = 4\pi r^2 = 4\pi\,(3)^2 = 113\ m^2$$

The intensity at this distance will be

$$I = \frac{\text{Power}}{\text{Area}} = \frac{100\ W \times 2.5\%}{113\ m^2}$$

$$= 0.022\ W/m^2$$

Half of this intensity is provided by the electric field and half by the magnetic field.

$$\frac{1}{2} I = \frac{1}{2}(\varepsilon_0 E_{rms}^2\, c)$$

$$= \frac{1}{2}(0.022\ W/m^2)$$

$$E_{rms} = \sqrt{\frac{0.022}{(8.85 \times 10^{-12})\,(3 \times 10^8)}}\ V/m$$

$$= 2.9\ V/m$$

The peak electric field, E_0 is,

$$E_0 = \sqrt{2}E_{rms}$$
$$= \sqrt{2} \times 2.9 \text{ V/m} = 4.07 \text{ V/m}$$

Compare it with electric field strength of TV or FM waves, which is of the order of a few microvolts per metre.

Let us calculate the strength of the magnetic field. It is

$$B_{rms} = \frac{E_{rms}}{c} = \frac{2.9 \times \text{Vm}^{-1}}{3 \times 10^8 \text{ ms}^{-1}}$$
$$= 9.6 \times 10^{-9} \text{ T}$$

The peak magnetic field is $B_0 = \sqrt{2}\, B_{rms} = 1.4 \times 10^{-8}$ T. Note that although the energy in the magnetic field is equal to the energy in the electric field, the magnetic field strength is evidently very weak.

Q. 13. Given figures shows a capacitor made of two circular plates each of radius 12 cm, and separated by 5.0 cm. The capacitor is being charged by an external source (not shown in the figure). The charging current is constant and equal to 0.15 A.

(i) Calculate the capacitance and the rate of charge of poten-tial difference between the plates.

(ii) Obtain the displacement current across the plates.

(iii) Is Kirchhoff's first rule (junction rule) valid at each plate of the capacitor ? Explain.

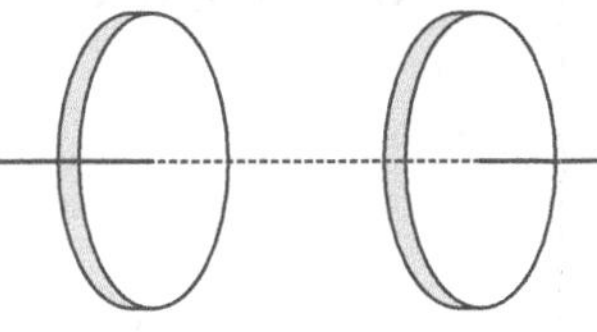

[NCERT]

Ans. Given,

Radius of each circular plate (r) = 12 cm = 0.12 m

Distance between the plates (d) = 5 cm = 0.05 m

Charging current (I) = 0.15 A

Permittivity of free space (ε_0) = 8.85×10^{-12} C^{2}N^{-1} m^{-2}

(i) Capacitance between the two plates is given by

$$C = \frac{\varepsilon_0 A}{d}$$

where, A = Area of each plate = πr^2

$$C = \frac{\varepsilon_0 \pi r^2}{d}$$
$$= \frac{8.85 \times 10^{-12} \times \pi \times (0.12)^2}{0.05}$$
$$= 8.0032 \times 10^{-12} \text{ F} = 80.032 \text{ pF.}$$

Charge on each plate (q) = CV

where, V = Potential difference across the plates

Taking differentiation on both sides with respect to time (t)

$$\frac{dq}{dt} = C\frac{dV}{dt}$$

But, $\frac{dq}{dt}$ = current (I)

$$\therefore \quad \frac{dV}{dt} = \frac{I}{C}$$

$$\Rightarrow \quad \frac{0.15}{80.032 \times 10^{-12}} = 1.87 \times 10^9 \text{ V/s}$$

Therefore, the change in potential difference between the plates is 1.87×10^9 V/s.

(ii) The displacement current through the plates is the same as the conduction current. Hence, the displacement current, i_d is 0.15 A.

(iii) Yes.

Kirchhoff's first rule is valid at each plate of the capacitor if we consider the sum of conduction and displacement for current.

Q. 14. Suppose that the electric field amplitude of an electromagnetic wave is $E_0 = 120$ N/C and that its frequency is $\upsilon = 50.0$ MHz.

(i) Determine, B_0, ω, k and λ.

(ii) Find expressions for E and B. **[NCERT]**

Ans. Given,

Electric field amplitude, $E_0 = 120$ N/C

Frequency of source, $\upsilon = 50.0$ MHz $= 50 \times 10^6$ Hz

Speed of light, $c = 3 \times 10^8$ m/s

(i) Magnitude of magnetic field strength is given as :

$$B_0 = \frac{E_0}{c} = \frac{120}{3 \times 10^8}$$

$$= 4 \times 10^{-7} \text{ T} = 400 \text{ nT}$$

Angular frequency of source is given as :

$$\omega = 2\pi\nu = 2\pi \times 50 \times 10^6$$

$$= 3.14 \times 10^8 \text{ rad/s}$$

Propagation constant is given as :

$$k = \frac{\omega}{c} = \frac{3.14 \times 10^8}{3 \times 10^8}$$

$$= 1.05 \text{ rad/m}$$

Wavelength of wave is given as :

$$\lambda = \frac{c}{\nu}$$

$$= \frac{3 \times 10^8}{50 \times 10^6} = 6.0 \text{ m}$$

(ii) Suppose the wave is propagating in the positive X-direction. Then, the electric field vector will be in the positive y-direction and the magnetic field vector will be in the positive z-direction. This is because all three vectors are mutually perpendicular.

Equation of electric field vector is given as :

$$\vec{E} = E_0 \sin (kx - \omega t) \hat{j}$$

$$= 120 \sin [1.05\, x - 3.14 \times 10^8\, t] \hat{j}$$

and, magnetic field vector is given as :

$$\vec{B} = B_0 \sin (kx - \omega t) \hat{k}$$

$$\vec{B} = (4 \times 10^{-7}) \sin [1.05x - 3.14 \times 10^8 t] \hat{k}.$$

Q. 15. A parallel plate capacitor (Fig.) made of circular plates each of radius R = 6.0 cm has a capacitance C = 100 pF. The capacitor is connected to a 230 V a.c. supply with a (angular) frequency of 300 rad s^{-1}.

(i) What is the r.m.s. value of the conduction current ?

(ii) Is the conduction current equal to the displacement current ?

(iii) Determine the amplitude of B at a point 3.0 cm from the axis between the plates. **[NCERT]**

Ans. Given,

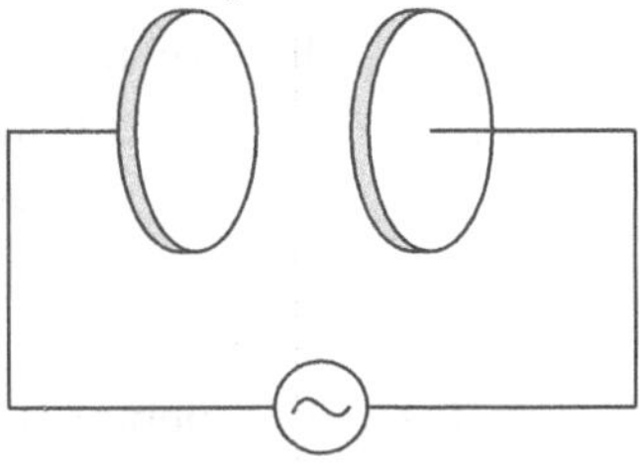

Radius of each circular plate (R) = 6.0 cm = 0.06 cm

Capacitance of a parallel plate capacitor (C)

$= 100 \text{ pF} = 100 \times 10^{-12}\text{F}$

Supply voltage (V) = 230 V

Angular frequency (ω) = 300 rad s^{-1}

(i) R.m.s. value of conduction current,

$$I = V/X_C$$

where, X_C = Capacitive reactance

$$= \frac{1}{\omega C}$$

$$\therefore \quad I = V \times \omega C$$

$$= 230 \times 300 \times 100 \times 10^{-12}$$

$$= 6.9 \times 10^{-6} \text{ A}$$

$$= 6.9 \text{ μA}$$

Hence, the r.m.s. value of conduction current is 6.9 μA.

(ii) Yes, conduction current is equal to displacement current.

(iii) Magnetic field is given by

$$B = \frac{\mu_0 r}{2\pi R^2} I_0$$

Where, μ_0 = Free space permeability

$= 4\pi \times 10^{-7}\text{ NA}^{-2}$.

I_0 = Maximum value of current = $\sqrt{2}I$

r = Distance between the plates from the axis

= 3.0 cm = 0.03 m

$$\therefore \quad B = \frac{4\pi \times 10^{-7} \times 0.03 \times \sqrt{2} \times 6.9 \times 10^{-6}}{2\pi \times (0.06)^2}$$

$$= 1.63 \times 10^{-11} \text{ T}$$

Thus, the magnetic field at that point is 1.63×10^{-11} T.

Q. 16. Suppose that the electric field part of an electromagnetic wave in vacuum is E = {(3.1 N/C) cos [(1.8 rad/m) y + (5.4×10^6 rad/s)t]}$\hat{i}$.

(i) What is the direction of propagation ?

(ii) What is the wavelength λ ?

(iii) What is the frequency υ ?

(iv) What is the amplitude of the magnetic field part of the wave ?

(v) Write an expression for the magnetic field part of the wave. **[NCERT]**

Ans. (i) From the given electric field vector, it can be understood that the electric field is directed along the negative x-direction. Hence, the direction of motion is along the negative y-direction *i.e.*, $(-\hat{j})$.

(ii) From question

$$\vec{E} = 3.1 \cos [(1.8 \text{ rad/m})y + (5.4 \times 10^8 \text{ rad/s})t]\hat{i} \quad ...(i)$$

The general equation for the electric field vector in the positive x-direction is given by

$$\vec{E} = E_0 \sin(kx - \omega t)\hat{i}. \quad ...(ii)$$

From equations (i) and (ii)

Electric field amplitude, $E_0 = 3.1$ N/C

Angular frequency, $\omega = 5.4 \times 10^8$ rad/s

Wave number, $k = 1.8$ rad/m

$$\text{Wavelength, } \lambda = \frac{2\pi}{1.8} = 3.490 \text{ m}$$

(iii) Frequency of wave is given as :

$$\nu = \frac{\omega}{2\pi} = \frac{5.4 \times 10^8}{2\pi}$$

$$= 8.6 \times 10^7 \text{ Hz}$$

(iv) Magnetic field strength is given by

$$B_0 = \frac{E_0}{c}$$

where, c = Speed of light = 3×10^8 m/s

$$\therefore \quad B_0 = \frac{3.1}{3 \times 10^8} = 1.03 \times 10^{-7} \text{ T}$$

(v) From given vector field, it can be inferred that the magnetic field vector is directed along the negative z-direction. Hence, the general equation for the magnetic field vector is written as :

$$\vec{B} = B_0 \cos(ky + \omega t)\hat{k}$$

$$= \{(1.03 \times 10^{-7} \text{ T}) \cos[(1.8 \text{ rad/m})\, y + (5.4 \times 10^6 \text{ rad/s}]\}\hat{k}$$

Q. 17. In a plane electromagnetic wave, the electric field oscillates sinusoidally at a frequency of 2.0×10^{10} Hz and amplitude 48 Vm^{-1}.

(i) What is the wavelength of the wave ?

(ii) What is the amplitude of the oscillating magnetic field ?

(iii) Show that the average energy density of the E field equals the average energy density of the B field. [$c = 3 \times 10^9$ ms^{-1}]. **[NCERT]**

Ans. Given,

Frequency of the electromagnetic wave (ν)

$$= 2.0 \times 10^{10} \text{ Hz}$$

Electric field amplitude (E_0) = 48 Vm^{-1}

Speed of light, (c) = 3×10^8 m/s

(i) Wavelength of a wave is given by

$$\lambda = \frac{c}{\nu} = \frac{3 \times 10^8}{2 \times 10^{10}}$$

$$= 0.015 \text{ m}$$

(ii) Magnetic field strength is given by

$$B_0 = \frac{E_0}{c} = \frac{48}{3 \times 10^8}$$

$$= 1.6 \times 10^{-7} \text{ T}$$

(iii) Energy density of the electric field is given by

$$U_E = \frac{1}{2}\varepsilon_0 E^2$$

and, energy density of the magnetic field is given by

$$U_B = \frac{1}{2\mu_0} B^2$$

where, ε_0 = Permittivity of free space

μ_0 = Permeability of free space

The relation between E and B is

$$E = cB \quad \ldots(i)$$

where,

$$c = \frac{1}{\sqrt{\varepsilon_0\mu_0}} \quad \ldots(ii)$$

From equations (ii) and (i), we get

$$E = \frac{1}{\sqrt{\varepsilon_0\mu_0}} B$$

Squaring both sides will give

$$E^2 = \frac{1}{\varepsilon_0\mu_0} B^2$$

$$\varepsilon_0 E^2 = \frac{B^2}{\mu_0}$$

$$\frac{1}{2}\varepsilon_0 E^2 = \frac{1}{2}\frac{B^2}{\mu_0}$$

$$U_E = U_B$$

Q. 18. A 1000 W bulb is kept at the centre of a spherical surface and is at a distance of 10 m from the surface. Calculate the force acting on the surface of the sphere by the electromagnetic waves, along with E_0, B_0 and intensity I. Take the working efficiency of the bulb to be 2.5% and consider it as a point source, $\varepsilon_0 = 8.85 \times 10^{-12}$ SI and $c = 3.0 \times 10^8$ ms^{-1}. Also calculate the energy density on the surface.

Ans. The energy consumed every second by a 1000 W bulb = 1000 J.

The working efficiency of the bulb is equal to 2.5%, the energy radiated by the bulb per second

$$\Delta U = 1000 \times \frac{2.5}{100}$$

$$\therefore \quad \Delta U = 25 \text{ Js}^{-1}$$

Considering, the bulb at the centre of the sphere, surface Area of the sphere.

$$A = 4\pi R^2$$

$$= (4)(3.14)(10^2) = 1256 \text{ m}^2$$

$$\text{Intensity, } I = \frac{\text{The energy of the incident radiation per second}}{\text{Area}}$$

$$= \frac{25}{1256} = 0.02 \text{ Wm}^{-2}$$

Now,

$$I = \varepsilon_0 c E_{rms}^2 = 0.02$$

$$\therefore \quad E_{rms} = \left[\frac{0.02}{8.85 \times 10^{-12} \times 3.0 \times 10^8}\right]^{1/2}$$

$$= 2.74 \text{ Vm}^{-1}$$

Now,

$$B_{rms} = \frac{E_{rms}}{c} = \frac{2.74}{3.0 \times 10^8}$$

$$= 9.13 \times 10^{-9} \text{ T}$$

$$E_0 = \sqrt{2} E_{rms}$$

$$= 3.87 \text{ Vm}^{-1}$$

and

$$B_0 = \sqrt{2} B_{rms} = 1.29 \times 10^{-8} \text{ T}$$

The total energy incident on the surface = 25 J

$\therefore$ The momentum (Δp) imparted to the surface in one second (= force).

$$\Delta p = \frac{\Delta U}{c} = F = \frac{25}{3 \times 10^8} = 8.33 \times 10^{-8} \text{ N}$$

From
$$I = \rho c, \text{ (energy density} = \rho)$$

$$\rho = \frac{I}{c} = \frac{0.02}{3 \times 10^8} = 6.67 \times 10^{-11} \text{ Jm}^{-3}.$$

Q. 19. A radio can tune into any station in the 7.5 MHz to 12 MHz band. What is the corresponding wavelength band ?

Ans. Given, A radio can tune to minimum frequency (ν_1)

$$= 7.5 \text{ MHz} = 7.5 \times 10^6 \text{ Hz}$$

$$\text{Maximum frequency } (\nu_2) = 12 \text{ MHz} = 12 \times 10^6 \text{ Hz}$$

$$\text{Speed of light, } (c) = 3 \times 10^8 \text{ m/s}$$

Corresponding wavelength for (ν_1) can be calculated as :

$$\lambda_1 = \frac{c}{\nu_1}$$

$$= \frac{3 \times 10^8}{7.5 \times 10^6} = 40 \text{ m}$$

Corresponding wavelength for ν_2 can be calculated as :

$$\lambda_2 = \frac{c}{\nu_2}$$

$$= \frac{3 \times 10^8}{12 \times 10^6} = 25 \text{ m}$$

Hence, the wavelength band of the radio is 40 m to 25 m.

Chapter 9. Ray Optics and Optical Instruments

Q. 1. (i) Monochromatic light of wavelength 589 nm is incident from air on a water surface. If μ for water is 1.33, find the wavelength, frequency and speed of the refracted light.

(ii) A double convex lens is made of a glass of refractive index 1.55, with both faces of the same radius of curvature.

Find the radius of curvature required, if the focal length is 20 cm.★

Ans. (i) Given,

$$\mu = 1.33$$

$$\lambda_a = 589 \text{ nm}$$

We know that,
$$\mu = \frac{c}{v}$$

$$v = \frac{c}{\mu} = \frac{3 \times 10^8}{1.33}$$

$$\text{Speed, } v = 2.26 \times 10^8 \text{ m/s}$$

Frequency remains same.

$$\therefore \quad \nu = \frac{c}{\lambda_a} = \frac{3 \times 10^8}{589 \times 10^{-9}}$$

$$\text{Frequency, } \nu = 5.09 \times 10^{14} \text{ Hz}$$

$$\text{Wavelength, } \lambda_w = \frac{v}{\nu} = \frac{2.26 \times 10^8}{5.09 \times 10^{14}}$$

$$= 444 \text{ nm}$$

(ii) Given,

$\mu = 1.55, f = 20$ cm

★ are board based questions from previous year

We know that,
$$\frac{1}{f} = (\mu - 1)\left(\frac{1}{R_1} - \frac{1}{R_2}\right)$$
$$\frac{1}{20} = (1.55 - 1)\left(\frac{1}{R} + \frac{1}{R}\right)$$
$$\frac{1}{20} = 0.55 \times \frac{2}{R}$$
$$R = 20 \times 0.55 \times 2$$
$$R = 22 \text{ cm}$$

Q. 2. An equiconvex lens of focal length 15 cm is cut into two equal halves in thickness. What is the focal length of each half ?

Ans. From
$$\frac{1}{f} = (r - 1)\left(\frac{1}{R_1} - \frac{1}{R_2}\right)$$
Taking
$$R_1 = R \text{ and } R_2 = -R$$
$$\frac{1}{15} = (r - 1)\left(\frac{1}{R} + \frac{1}{R}\right) = \frac{2}{R}(r - 1)$$
For each half (plano-convex) lens $R_1 = R$, $R_2 = \infty$
$$\therefore \quad \frac{1}{f'} = (r - 1)\left(\frac{1}{R} - \frac{1}{\infty}\right) = \frac{r - 1}{R} = \frac{1}{30}$$
$$\therefore \quad f' = 30 \text{ cm}$$

Q. 3. You are given two converging lenses of focal lengths 1.25 cm and 5 cm to design a compound microscope. If it is desired to have a magnification of 30, find out the separation between the objective and the eyepiece. ★

Ans. Given,

$f_e = 5$ cm; $f_o = 1.25$ cm and $M = -30$.

Let L be the tube length (distance between the objective and the eyepiece)
$$\therefore \quad M = \frac{-L}{f_o}\left(1 + \frac{d}{f_e}\right)$$
[d is a constant and equals to the normal distance of clear vision of the human eye]

Hence,
$$-30 = -\frac{L}{1.25}\left(1 + \frac{25}{5}\right)$$
$$\therefore \quad L = \frac{30 \times 1.25}{6} = 6.25 \text{ cm}$$
Hence, the tube length = 6.25 cm

Q. 4. A ray PQ incident normally on the refracting face BA is refracted in the prism BAC made of material of refractive index 1.5. Complete the path of ray through the prism. From which face will the ray emerge ? Justify your answer.★

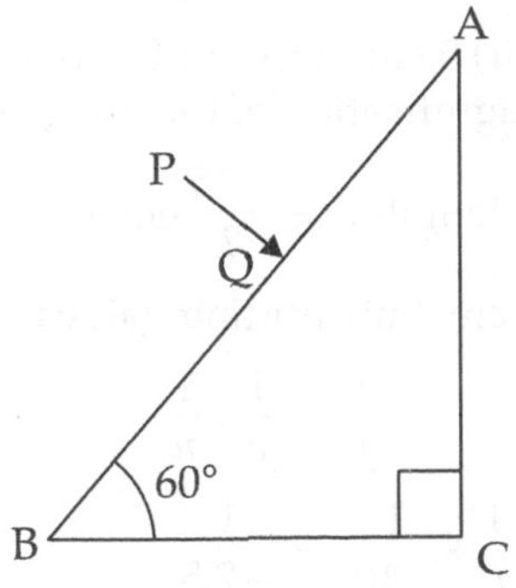

★ are board based questions from previous year

Ans. We know that,

$$n = \frac{1}{\sin C}$$

Where,
$$n = \text{refractive index}$$
$$C = \text{Critical angle}$$

So,
$$\sin C = 1/n = 1/1.5 = 0.667$$
$$C = 41° \, 49'$$
$$C > 30°$$

Thus, the light ray PQ will emerge out from face AC.

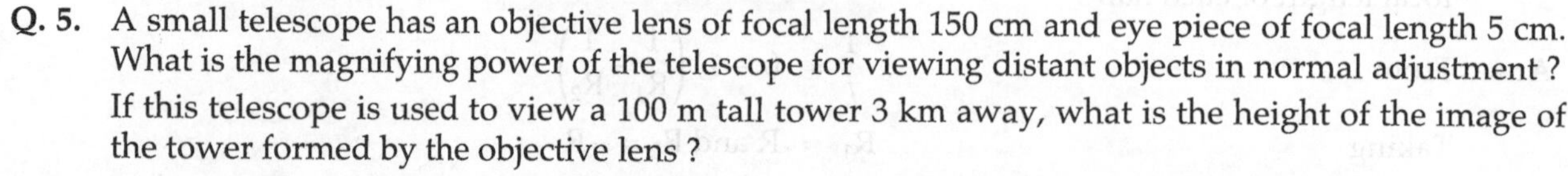

Q. 5. A small telescope has an objective lens of focal length 150 cm and eye piece of focal length 5 cm. What is the magnifying power of the telescope for viewing distant objects in normal adjustment ? If this telescope is used to view a 100 m tall tower 3 km away, what is the height of the image of the tower formed by the objective lens ?

Ans. If the telescope is in normal adjustment, *i.e.*, the final image is at infinity.

$$M = \frac{f_o}{f_e}$$

Since $f_o = 150$ cm, $f_e = 5$ cm

$$M = \frac{150}{5} = 30$$

If tall tower is at distance 3 km from the objective lens of focal length 150 cm. It will form its image at distance v_0. So,

$$\frac{1}{f_o} = \frac{1}{v_o} + \frac{1}{u_o}$$

$$\frac{1}{150 \text{ cm}} = \frac{1}{v_o} + \frac{1}{(-3 \text{ km})}$$

$$\frac{1}{v_o} = \frac{1}{1.5 \text{ m}} - \frac{1}{3000 \text{ m}}$$

$$v_o = \frac{3000 \times 1.5}{3000 - 1.5} = \frac{4500}{2998.5} = 1.5 \text{ m}$$

Magnification,
$$m_o = \frac{I}{O} = \frac{h_i}{h_o} = \frac{v_o}{u_o}$$

$$\frac{h_i}{100 \text{ m}} = \frac{1.5 \text{ m}}{3 \text{ km}} = \frac{1.5}{3000}$$

$$h_i = \frac{1.5 \times 100}{3000} = \frac{1}{20} \text{ m}$$

$$h_i = 0.05 \text{ m}$$

Q. 6. An object is placed at (i) 10 cm, (ii) 5 cm in front of a concave mirror of radius of curvature 15 cm. Find the position, nature, and magnification of the image in each case. **[NCERT]**

Ans. The focal length $f = \frac{-15}{2}$ cm $= -7.5$ m

(i) The object distance $u = -10$ cm. Substituting values in mirror equation

$$\frac{1}{f} = \frac{1}{v} + \frac{1}{u}$$

$$\frac{1}{v} + \frac{1}{-10} = \frac{1}{-7.5}$$

or
$$v = \frac{10 \times 7.5}{-2.5} = -30 \text{ cm}$$

The image is 30 cm from the mirror on the same side as the object.

Also, magnification $m = -\frac{v}{u} = -\frac{(-30)}{(-10)} = +3$

The image is magnified, real and inverted.

(ii) The object distance $u = -5$ cm. Substituting values in mirror equation

$$\frac{1}{f} = \frac{1}{v} + \frac{1}{u}$$

$$\frac{1}{v} + \frac{1}{-5} = \frac{1}{-7.5}$$

or

$$v = \frac{5 \times 7.5}{(7.5 - 5)}$$

$$= 15 \text{ cm}$$

This image is formed at 15 cm behind the mirror. It is a virtual image.

$$\text{Magnification, } m = -\frac{v}{u} = -\frac{15}{(-5)} = 3$$

The image is magnified, virtual and erect.

Q. 7. (i) If $f = 0.5$ m for a glass lens, what is the power of the lens ?

(ii) The radii of curvature of the faces of a double convex lens are 10 cm and 15 cm. Its focal length is 12 cm. What is the refractive index of glass ?

(iii) A convex lens has 20 cm focal length in air. What is focal length in water ? (Refractive index of air-water = 1.33, refractive index for air-glass = 1.5) **[NCERT]**

Ans. (i) Power = + 2 dioptre.

(ii) Given,

$f = +12$ cm, $R_1 = +10$ cm, $R_2 = -15$ cm.

Refractive index of air is taken as unity.

From lens formula as per sign convention has to be applied for f, R_1 and R_2.

Substituting the values will give

$$\frac{1}{12} = (n-1)\left(\frac{1}{10} - \frac{1}{-15}\right)$$

This gives $n = 1.5$.

(iii) For a glass lens in air, $n_2 = 1.5$, $n_1 = 1$, $f = +20$ cm. Thus, the lens formula gives

$$\frac{1}{20} = (1{\cdot}5 - 1)\left(\frac{1}{R_1} - \frac{1}{R_2}\right)$$

For the same glass lens in water, $n_2 = 1.5$, $n_1 = 1.33$. Therefore,

$$\frac{1.33}{f} = (1.5 - 1.33)\left(\frac{1}{R_1} - \frac{1}{R_2}\right)$$

Combining these two equations, we find $f = +78.2$ cm.

Q. 8. A small bulb is placed at the bottom of a tank containing water to a depth of 80 cm. What is the area of the surface of water through which light from the bulb can emerge out ? Refractive index of water is 1.33. (Consider the bulb to be a point source). **[NCERT]**

Ans. Given,

Actual depth of the bulb in water, $d_1 = 80$ cm $= 0.8$ m

Refractive index of water, $\mu = 1.33$

The given situation is shown in the following figure :

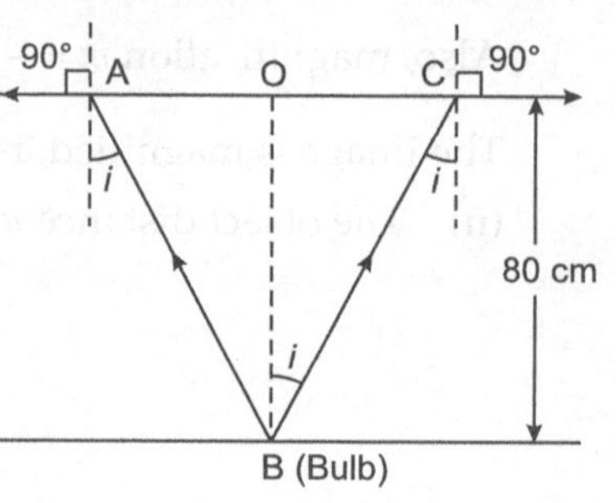

where, i = angle of incidence,

$$r = \text{Angle of refraction} = 90°.$$

As the bulb is a point source, the emergent light can be considered as a circle of radius,

$$R = \frac{AC}{2} = AO = OB$$

From Snell's law, we can write the relation for the refractive index of water by

$$\mu = \frac{\sin r}{\sin i}$$

$$1.33 = \frac{\sin 90°}{\sin i}$$

$$\therefore \quad i = \sin^{-1}\left(\frac{1}{1.33}\right) = 48.75°$$

From given figure,

$$\tan i = \frac{OC}{OB} = \frac{R}{d_1}$$

$$\therefore \quad R = \tan 48{\cdot}75° \times 0.8 = 0.91 \text{ m}$$

Area of the surface of water $= \pi R^2 = \pi\,(0.91)^2 = 2.61 \text{ m}^2$.

Hence, the area of the surface of water through which the light from the bulb can emerge is approximately 2.61 m^2.

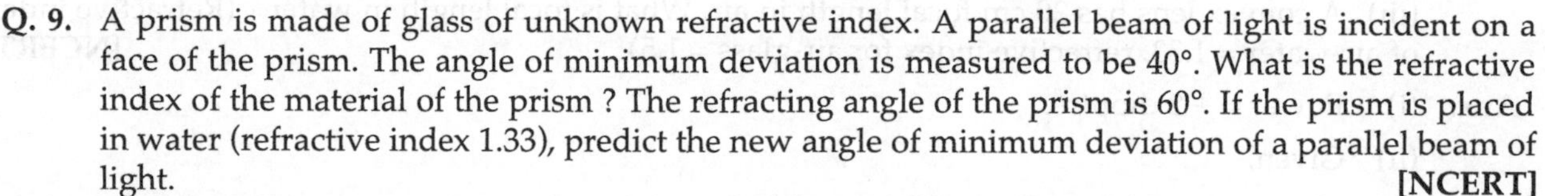

Q. 9. A prism is made of glass of unknown refractive index. A parallel beam of light is incident on a face of the prism. The angle of minimum deviation is measured to be 40°. What is the refractive index of the material of the prism ? The refracting angle of the prism is 60°. If the prism is placed in water (refractive index 1.33), predict the new angle of minimum deviation of a parallel beam of light. **[NCERT]**

Ans. Given,

Angle of minimum deviation, $\delta_m = 40°$

Angle of the prism, A = 60°

Refractive index of water, $\mu = 1.33$

Refractive index of the material of the prism = μ'

The angle of deviation is related to refractive index (μ') as :

$$\mu' = \frac{\sin\frac{(A+\delta_m)}{2}}{\sin\frac{A}{2}} = \frac{\sin\frac{(60°+40°)}{2}}{\sin\frac{60°}{2}} = \frac{\sin 50°}{\sin 30°} = 1.532$$

Hence, the refractive index of the material of the prism is 1.532.

Since the prism is placed in water, let δ'_m be the new angle of minimum deviation for the same prism.

The refractive index of glass with respect to water is given by the relation :

$${}^{w}g_{g} = \frac{\mu'}{\mu} = \frac{\sin\frac{(A+\delta'_m)}{2}}{\sin\frac{A}{2}}$$

$$\sin\frac{(A+\delta'_m)}{2} = \frac{\mu'}{\mu} = \sin\frac{A}{2}$$

$$\sin\frac{(A+\delta'_m)}{2} = \frac{1.532}{1.33} \times \sin\frac{60°}{2} = 0.5759$$

$$\frac{(A+\delta'_m)}{2} = \sin^{-1} 0.5759 = 35.16°$$

$$60° + \delta'_m = 70.32°$$

$$\therefore \quad \delta'_m = 70.32° - 60° = 10.32°$$

Hence, the new minimum angle of deviation is 10.32°.

Q. 10. A card sheet divided into square each of size 1 mm^2 is being viewed at a distance of 9 cm through a magnifying glass (a converging lens of focal length 9 cm) held close to the eye.

(i) What is the magnification produced by the lens ? How much is the area of each square in the virtual image ?

(ii) What is the angular magnification (magnifying power) of the lens ?

(iii) Is the magnification in (i) equal to the magnifying power in (ii) ? **[NCERT]**

Ans. (i) Given,

Area of each square (A) = 1 mm^2

Object distance (u) = – 9 cm

Focal length of a converging lens (f) = 10 cm

For image distance, v from lens formula

$$\frac{1}{f} = \frac{1}{v} - \frac{1}{u}$$

$$\frac{1}{10} = \frac{1}{v} + \frac{1}{9}$$

$$\frac{1}{v} = -\frac{1}{90}$$

$$\therefore \quad v = -90 \text{ cm}$$

$$\text{Magnification, } m = \frac{v}{u} = \frac{-90}{-9} = 10$$

$$\text{Area of each square in the virtual image} = (10)^2 \text{ A}$$

$$= 10^2 \times 1 = 100 \text{ mm}^2$$

$$= 1 \text{ cm}^2$$

(ii) $$\text{Magnifying power of the lens} = \frac{d}{|u|} = \frac{25}{9} = 2.8$$

(iii) Magnification of part (i) is not the same as the magnifying power in (ii).

The magnification magnitude is $\left(\frac{|v|}{|u|}\right)$ and the magnifying power is $\left(\frac{d}{|u|}\right)$

The two quantities will be equal when the image is formed at the near point (25 cm).

Q. 11. A convex lens of focal length 20 cm is placed coaxially with a concave mirror of focal length 10 cm at a distance of 50 cm apart from each other. A beam of light coming parallel to the principal axis is incident on the convex lens. Find the position of the final image formed by this combination. Draw the ray diagram showing the formation of the image.

Ans. The beam incident on lens L is parallel to principal axis. Hence the lens forms an image I_1 at its focus. *i.e.*, at a distance OI_1 (= 20 cm) from the lens.

The image I_1 is formed in front of the mirror and hence, acts as a real source for the mirror. The concave mirror forms the image I_2, whose distance from the mirror can be calculated as;

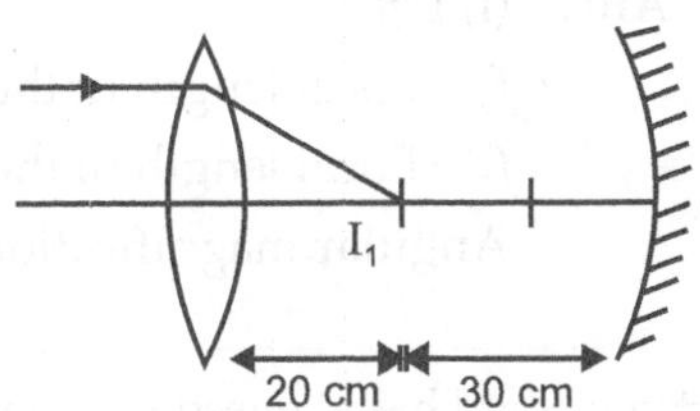

$$\frac{1}{v} + \frac{1}{u} = \frac{1}{f}$$

Here : u = – 30 cm, and f = – 10 cm

$$\Rightarrow \quad \frac{1}{v} = \frac{1}{f} - \frac{1}{u}$$

$$\Rightarrow \quad \frac{1}{v} = -\frac{1}{10} + \frac{1}{30}$$

$$\Rightarrow \quad \frac{1}{v} = \frac{1-3}{30} = -\frac{2}{30}$$

$$\Rightarrow \quad v = -15 \text{ cm}$$

Hence, the final image is formed at a distance of 15 cm from the concave mirror, as shown in the following figure.

Q. 12. A convex lens of focal length 20 cm is placed coaxially with a convex mirror of radius of curvature 20 cm. The two are kept at 15 cm apart from each other. A point object lies 60 cm in front of the convex lens. Draw a ray diagram to show the formation of the image by the combination. Determine the nature and position of the image formed.★

Ans. Let us first locate the image of the point object S formed by the convex lens.

Here, $u = -60$ cm and $f = 20$ cm

From the lens formula, we have :

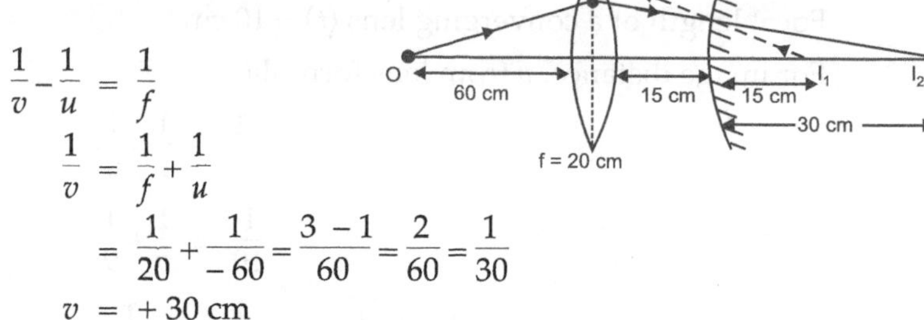

$$\frac{1}{v} - \frac{1}{u} = \frac{1}{f}$$

$$\frac{1}{v} = \frac{1}{f} + \frac{1}{u}$$

$$= \frac{1}{20} + \frac{1}{-60} = \frac{3-1}{60} = \frac{2}{60} = \frac{1}{30}$$

$$v = +30 \text{ cm}$$

The positive sign indicates that the image is formed at the right of the lens.

The image I_1 is formed behind the mirror and acts as a virtual source for the mirror. The convex mirror forms the image I_2, whose distance from the mirror can be determined as :

$$\frac{1}{v} + \frac{1}{u} = \frac{1}{f}$$

Here, $u = 15$ cm

and, $\frac{R}{2} = 10$ cm

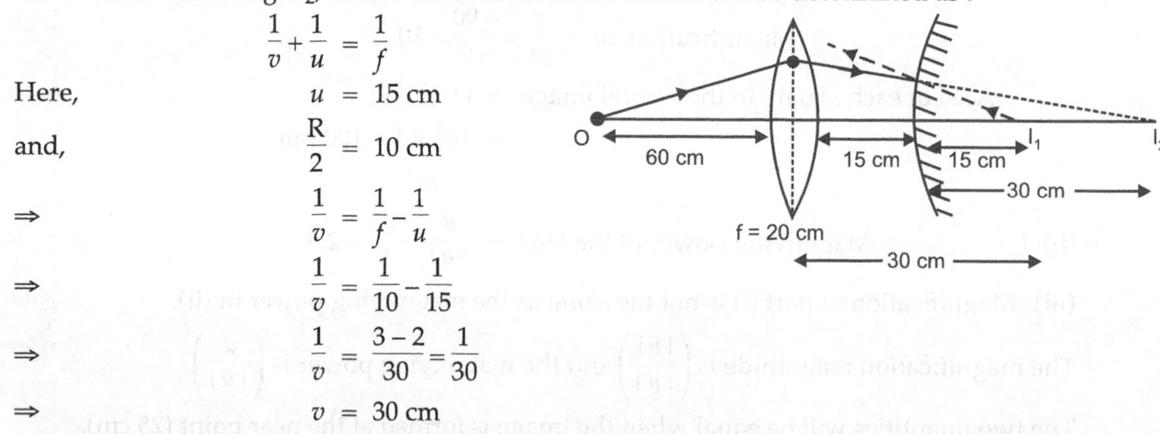

$$\Rightarrow \quad \frac{1}{v} = \frac{1}{f} - \frac{1}{u}$$

$$\Rightarrow \quad \frac{1}{v} = \frac{1}{10} - \frac{1}{15}$$

$$\Rightarrow \quad \frac{1}{v} = \frac{3-2}{30} = \frac{1}{30}$$

$$\Rightarrow \quad v = 30 \text{ cm}$$

Hence, the final virtual image is formed at a distance of 30 cm from the convex mirror.

Q. 13. (i) A giant refracting telescope has an objective lens of focal length 15 m. If an eye-piece of focal length 1.0 cm is used. What is the angular magnification of the telescope ?★

(ii) If this telescope is used to view the moon. What is the diameter of the image of the moon formed by the objective lens ? The diameter of the moon is 3.48×10^6 m and the radius of lunar orbit is 3.8×10^8 cm.★

Ans. (i) Let

f_o = Focal length of the objective lens = 15 m = 1500 cm

f_e = Focal length of the eye lens = 1.0 cm

Angular magnification of the giant refracting telescope is given by

★ are board based questions from previous year

$$m_0 = \left|\frac{f_o}{f_e}\right|$$

$$\therefore \quad m_0 = \left|\frac{1500}{1}\right| = 1500$$

(ii) Diameter of the image of the moon formed by the objective lens, $d = \alpha f_o$

$$\Rightarrow \quad d = \frac{\text{Diameter of the moon}}{\text{Radius of the lunar orbit}} \times f_o$$

$$\Rightarrow \quad d = \frac{3.48 \times 10^6}{3.8 \times 10^8} \times 1500 \times 10^{-2}$$

$$\Rightarrow \quad d = 0.1373 \text{ m}$$

$$\Rightarrow \quad d = 13.73 \text{ cm}$$

Q. 14. Find out the relation between the refractive index (μ) of the glass prism and $\angle$A for the case when the angle of prism (A) is equal to the angle of minimum deviation (δ_m). Hence obtain the value of the refractive index for angle of prism A = 60°.★

Ans.

$$\mu = \frac{\sin\left(\frac{A + \delta_{min}}{2}\right)}{\sin\frac{A}{2}}$$

But

$$A = \delta_{min}$$

$$\therefore \quad \mu = \frac{\sin\left(\frac{A + A}{2}\right)}{\sin\frac{A}{2}}$$

$$= \frac{\sin A}{\sin\frac{A}{2}} = \frac{2\sin\frac{A}{2}\cdot\cos\frac{A}{2}}{\sin\frac{A}{2}} = 2\cos\frac{A}{2}$$

If

$$A = 60°,$$

$$\mu = 2\cos\frac{60°}{2} = 2\cos 30°$$

$$\mu = 2 \times \frac{\sqrt{3}}{2} = \sqrt{3}$$

$$\therefore \quad \mu = \sqrt{3}$$

Q. 15. (i) Calculate the distance of an object of height h from a concave mirror of radius of curvature 20 cm, so as to obtain a real image of magnification 2. Find the location of image also.★

(ii) Using mirror formula, explain why does a convex mirror always produce a virtual image.★

Ans. (i) Given,

Height of object = h

Radius of curvature = – 20 cm

Magnification, m = – 2

Object distance, u = ?

Image distance, v = ?

$$\text{Magnification, M} = \frac{-v}{u} = \frac{h_i}{h_o}$$

★ are board based questions from previous year

$$-2 = -\frac{v}{u}$$

$$v = 2u \qquad \text{...(i)}$$

Using mirror formula,

$$\frac{1}{v}+\frac{1}{u} = \frac{1}{f}$$

$$\frac{1}{2u}+\frac{1}{u} = \frac{-2}{20}$$

$$\frac{+3}{2u} = \frac{-2}{20}$$

$$u = -\frac{60}{4} = -15 \text{ cm}$$

Putting in (i), we get

$$v = 2 \times -15 \text{ cm} = -30 \text{ cm}$$

$$\frac{h_i}{h_o} = \left|\frac{v}{u}\right|$$

$$\frac{h_i}{h_o} = \left|\frac{2u}{u}\right|$$

Height of image, $h_i = 2h$

(ii) For convex mirror, $f = +\text{ve (always)}$

Mirror formula,

$$\frac{1}{v}+\frac{1}{u} = \frac{1}{f}$$

$$\frac{1}{v} = \frac{1}{f}-\frac{1}{u}$$

As, $u = -ve$ (for real object)

$$\frac{1}{v} = \frac{1}{f}-\left(\frac{1}{-u}\right)$$

$$\frac{1}{v} = \frac{1}{f}+\frac{1}{u}$$

$$v = +ve$$

Hence, it will form virtual image.

Q. 16. (i) A ray of light incident on face AB of an equilateral glass prism, shows minimum deviation of 30°. Calculate the speed of light through the prism.★

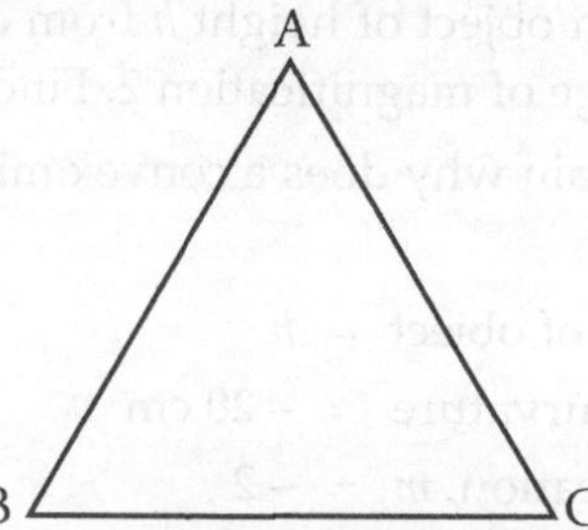

(ii) Find the angle of incidence at face AB so that the emergent ray grazes along the face AC.

Ans. (i) Given,

$$\delta_m = 30°$$

$$A = 60°$$

★ are board based questions from previous year

Refractive index of the prism,

$$\mu = \frac{\sin\left(\frac{\delta_m + A}{2}\right)}{\sin (A/2)}$$

$$\mu = \frac{\sin\left(\frac{30° + 60°}{2}\right)}{\sin\left(\frac{60°}{2}\right)}$$

$$\mu = \frac{\sin (45°)}{\sin (30°)}$$

$$\mu = \frac{1\sqrt{2}}{1/2} = \frac{2}{\sqrt{2}} = \sqrt{2}$$

Now,
$$\mu = \frac{c}{v}$$

$$v = \frac{c}{\mu}$$

$$v = \frac{3 \times 10^8}{\sqrt{2}} = \frac{3 \times 10^8}{1.414}$$

$$v = 2.12 \times 10^8 \text{ m/s}$$

(ii) Let c be the critical angle, then

$$\sin c = \frac{1}{\mu}$$

$$\sin c = \frac{1}{\sqrt{2}}$$

$$\sin c = \sin 45°$$

$$c = 45°$$

∴
$$r_2 = c = 45°$$

$$r_1 + r_2 = A$$

$$r_1 = A - r_2 = 60° - 45°$$

$$r_1 = 15°$$

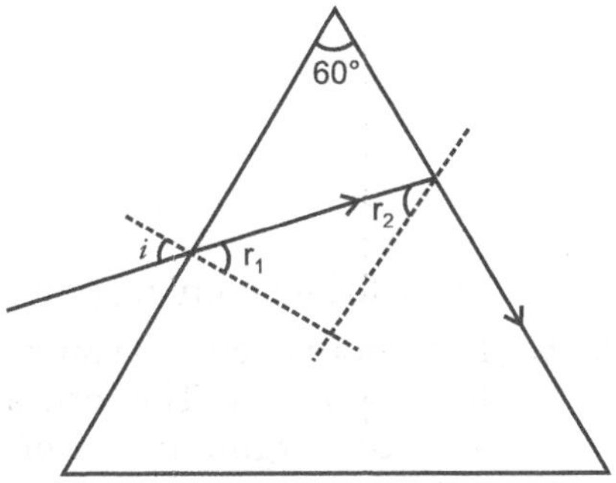

Now,
$$\mu = \frac{\sin i}{\sin r_1}$$

$$\sin i = \mu \sin r_1 = \sqrt{2} \sin (15°) = 1.414 \times 0.2588$$

$$\sin i = 0.366$$

$$i = 21° \text{ (approx).}$$

Q. 17. A person with a normal near point (25 cm) using a compound microscope with objective of focal length 8.0 mm and an eyepiece of focal length 2.5 cm can bring an object placed at 9.0 mm from the objective in sharp focus. What is the separation between the two lenses ? Calculate the magnifying power of the microscope. **[NCERT]**

Ans. Given,

Focal length of the objective lens (f_o) = 8 mm = 0.8 cm

Focal length of the eyepiece (f_e) = 2.5 cm

Object distance for the objective lens (u_o) = – 9.0 mm = – 0.9 cm

Least distance of distant vision (d) = 25 cm

Image distance for the eyepiece (v_e) = – d = – 25 cm

Let object distance for the eyepiece = u_e

From the lens formula, the value of u_e as :

$$\frac{1}{v_e} - \frac{1}{u_e} = \frac{1}{f_e}$$

$$\frac{1}{u_e} = \frac{1}{v_e} - \frac{1}{f_e}$$

$$= \frac{1}{-25} - \frac{1}{2.5} = \frac{-1-10}{25} = \frac{-11}{25}$$

$$\therefore \quad u_e = -\frac{25}{11} = -2.27 \text{ cm}$$

The value of the image distance for the objective lens (v_o) using the lens formula,

$$\frac{1}{v_o} - \frac{1}{u_o} = \frac{1}{f_o}$$

$$\frac{1}{v_o} = \frac{1}{f_o} + \frac{1}{u_o}$$

$$= \frac{1}{0.8} - \frac{1}{0.9} = \frac{0.9 - 0.8}{0.72} = \frac{0.1}{0.72}$$

$$\therefore \quad v_o = 7.2 \text{ cm}$$

The distance between the objective lens and the eyepiece = $|u_e| + v_o$

$$= 2.27 + 7.2 = 9.47 \text{ cm}$$

The magnifying power of the microscope is calculate as :

$$\frac{v_o}{|u_o|}\left(1 + \frac{d}{f_e}\right)$$

$$= \frac{7.2}{0.9}\left(1 + \frac{25}{2.5}\right) = 8\,(1 + 10) = 88$$

Thus, the magnifying power of the microscope is 88.

Q. 18. For a normal eye, the far point is at infinity and the near point of distinct vision is about 25 cm in front of the eye. The cornea of the eye provides a converging power of about 40 dioptres, and the least converging power of the eye-lens behind the cornea is about 20 dioptres. From this rough data estimate the range of accommodation (*i.e.*, the range of converging power of the eye-lens) of a normal eye. **[NCERT]**

Ans. As we know,

Least distance of distinct vision (d) = 25 cm

Far point of a normal eye (d) = ∞

Converging power of the cornea, (P_c) = 40 D

Least converging power of the eye-lens, P_e = 20D

To see the objects at infinity, the eye uses its least converging power,

Power of the eye-lens, $P = P_c + P_e = 40 + 20 = 60$ D

Power of the eye-lens is given by

$$P = \frac{1}{\text{Focal length of the eye lens } (f)}$$

$$f = \frac{1}{P} = \frac{1}{60\text{ D}} = \frac{100}{60} = \frac{5}{3} \text{ cm}$$

The focus an object at the near point, object distance (u) = – d = – 25 cm.

Focal length of the eye-lens = distance between the cornea and retina = image distance

Thus, image distance, $v = \frac{5}{3}$ cm

From lens formula, $\frac{1}{f'} = \frac{1}{v} - \frac{1}{u}$

Where, f' = focal length

$$\frac{1}{f'} = \frac{3}{5} + \frac{1}{25} = \frac{15+1}{25} = \frac{16}{25} \text{ cm}$$

Power, $$P' = \frac{1}{f'} \times 100 = \frac{16}{25} \times 100 = 64 \text{ D}$$

$\therefore$ Power of the eye-lens = 64 – 40 = 24D

Thus, the range of accommodation of the eye-lens is from 20 D to 24 D.

Q. 19. A convex lens of focal length 20 cm is placed coaxially with a convex mirror of radius of curvature 20 cm. The two are kept at 15 cm from each other. A point object placed 40 cm in front of the convex lens. Find the position of the image formed by this combination. Draw a ray diagram to show the formation.★

Ans. Given,

$n = -40$ cm and, $f = 20$ cm

From the lens formula, we have :

$$\frac{1}{v} - \frac{1}{u} = \frac{1}{f}$$

$$\Rightarrow \quad \frac{1}{v} = \frac{1}{f} + \frac{1}{u}$$

$$\Rightarrow \quad \frac{1}{v} = \frac{1}{20} + \frac{1}{(-40)}$$

$$\Rightarrow \quad \frac{1}{v} = \frac{2-1}{40} = \frac{1}{40}$$

$$\Rightarrow \quad v = 40 \text{ cm}$$

The positive sign describes that the image is formed to the right of the lens.

The image I_1 is formed behind the mirror and thus acts as a virtual source for the mirror. The convex mirror forms the image I_2, whose distance from the mirror is given by :

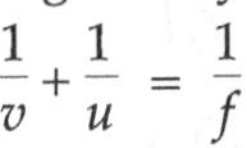

$$\frac{1}{v} + \frac{1}{u} = \frac{1}{f}$$

Here : $u = 25$ cm

$$f = \frac{R}{2} = 10 \text{ cm}$$

$$\frac{1}{v} + \frac{1}{u} = \frac{1}{f}$$

$$\Rightarrow \quad \frac{1}{v} = \frac{1}{f} - \frac{1}{u}$$

$$\Rightarrow \quad \frac{1}{v} = \frac{1}{10} - \frac{1}{25}$$

$$\Rightarrow \quad \frac{1}{v} = \frac{5-2}{50} = \frac{3}{50}$$

$$\Rightarrow \quad v = +16.67 \text{ cm}$$

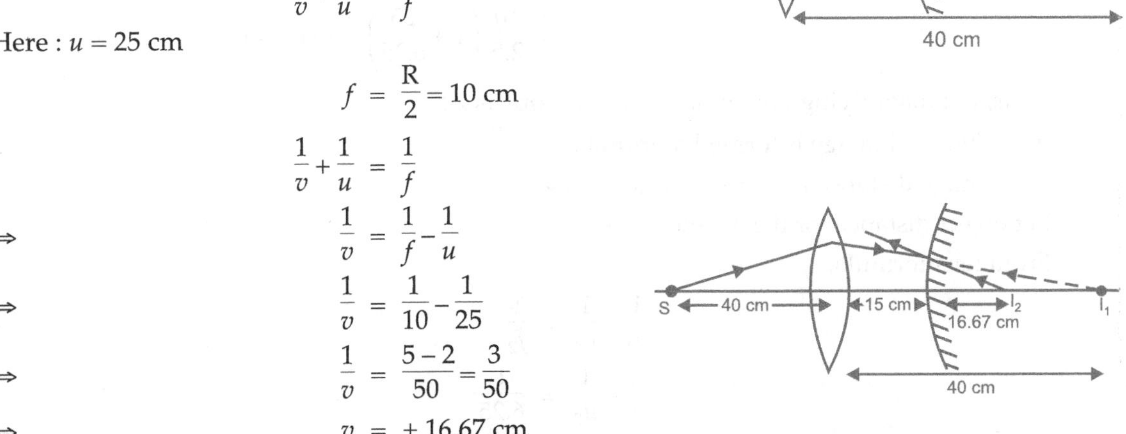

Hence, the final image is formed at a distance of 16.67 cm behind the convex mirror.

Q. 20. A compound microscope consists of an objective lens of focal length 2.0 cm and an eyepiece of focal length 6.25 cm separated by a distance of 15 cm. How far from the objective should an object

---

★ are board based questions from previous year

be placed in order to obtain the final image at (i) the least distance of distinct vision (25 cm), and (ii) at infinity ? What is the magnifying power of the microscope in each case ? **[NCERT]**

Ans. Given,

Focal length of the objective lens (f_1) = 2.0 cm

Focal length of the eyepiece (f_2) = 6.25 cm

Distance between eyepiece and the objective lens (d) = 15 cm

(i) Least distance of distinct vision (d') = 25 cm

∴ Image distance for the eyepiece, (v_2) = – 25 cm

Let object distance for the eyepiece = u_2

From lens formula,
$$\frac{1}{v_2} - \frac{1}{u_2} = \frac{1}{f_2}$$
$$\frac{1}{u_2} = \frac{1}{v_2} - \frac{1}{f_2}$$
$$= \frac{1}{-25} - \frac{1}{6.25} = \frac{-1-4}{25} = \frac{-5}{25}$$
$$\therefore \quad u_2 = -5 \text{ cm}$$

Image distance for the objective lens, (v_1) = $d + u_2$ = 15 – 5 = 10 cm

Let object distance for the objective lens = u_1

From lens formula,
$$\frac{1}{v_1} - \frac{1}{u_1} = \frac{1}{f_1}$$
$$\frac{1}{u_1} = \frac{1}{v_1} - \frac{1}{f_1}$$
$$= \frac{1}{10} - \frac{1}{2} = \frac{1-5}{10} = \frac{-4}{10}$$
$$\therefore \quad u_1 = -2.5 \text{ cm}$$

Magnitude of the object distance $|u_1|$ = 2.5 m.

The magnifying power of compound microscope is given by the relation :
$$m = \frac{v_1}{|u_1|}\left(1 + \frac{d'}{f_2}\right)$$
$$= \frac{10}{2.5}\left(1 + \frac{25}{6.25}\right) = 4\,(1 + 4) = 20$$

Thus, the magnifying power of the microscope is 20.

(ii) The final image is formed at infinity.

∴ Image distance for the eyepiece, $v_2 = \infty$

Let object distance for the eyepiece = u_2

From lens formula,
$$\frac{1}{v_2} - \frac{1}{u_2} = \frac{1}{f_2}$$
$$\frac{1}{\infty} - \frac{1}{u_2} = \frac{1}{6.25}$$
$$\therefore \quad u_2 = -6.25 \text{ cm}$$

Image distance for the objective lens, (v_1) = $d + u_2$ = 15 – 6.25 = 8.75 cm

Object distance for the objective lens = u_1

From lens formula,
$$\frac{1}{v_1} - \frac{1}{u_1} = \frac{1}{f_1}$$
$$\frac{1}{u_1} = \frac{1}{v_1} - \frac{1}{f_1}$$

$$= \frac{1}{8.75} - \frac{1}{2.0} = \frac{2 - 8.75}{17.5}$$

$$\therefore \qquad u_1 = -\frac{17.5}{6.75} = -2.59$$

Magnitude of the object distance $|u_1| = 2.59$ cm

The magnifying power of compound microscope is given by :

$$m = \frac{v_1}{|u_1|}\left(\frac{d'}{|u_2|}\right)$$

$$= \frac{8.75}{2.59} \times \frac{25}{6.25} = 13.51$$

Hence, the magnifying power of the microscope is 13.51.

Q. 21. (i) Draw a labelled ray diagram showing the formation of a final image by a compound microscope at least distance of distinct vision.★

(ii) The total magnification produced by a compound microscope is 20. The magnification produced by the eye piece is 5. The microscope is focussed on a certain object. The distance between the objective and eye-piece is observed to be 14 cm. If least distance of distinct vision is 20 cm, calculate the focal length of the objective and the eye-piece.★

Ans. (i)

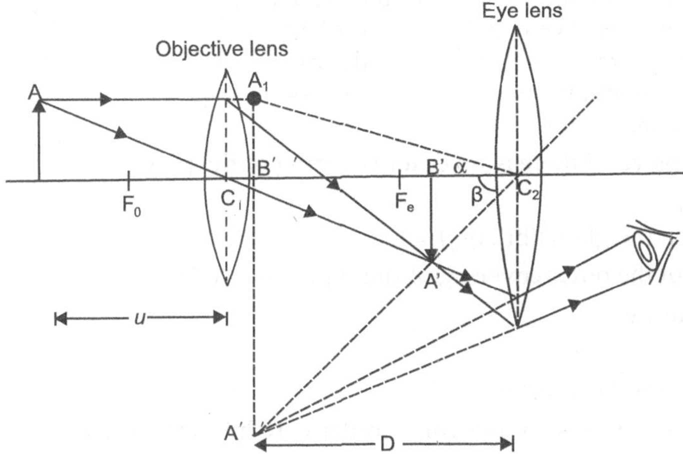

(ii) For the least distance of clear vision, the total magnification is given by :

$$m = \frac{L}{f_o}\left(1 + \frac{D}{f_e}\right) = m_o m_e \qquad \text{...(i)}$$

where,

L is the separation between the eyepiece and the objective

f_o is the focal length of the objective

f_e is the focal length of the eyepiece.

D is the least distance for clear vision.

Also, the given magnification for the eyepiece :

$$m_e = 5 = 1 + \frac{D}{f_e}$$

$$\Rightarrow \qquad 5 = 1 + \frac{20}{f_e}$$

$$\Rightarrow \qquad f_e = 5 \text{ cm}$$

★ are board based questions from previous year

Substituting the value of m and m_e in equation (i), we get :

$$m = m_o m_e$$

$$20 = m_o \times 5$$

$$m_o = 4$$

Now, we have

$$|u_e| + |v_o| = 14$$

(*i. e.*, distance between objective and eye piece)

$$u + v_o = 14$$

$$\Rightarrow \quad v_o = 10 \text{ cm}$$

Now, $$m_o = 1 - \frac{v_o}{f_o}$$

$$\Rightarrow \quad -4 = 1 - \frac{10}{f_o}$$

$$f_o = 2 \text{ cm}$$

Q. 22. (i) Given Figure shows a cross-section of a 'light pipe' made of a glass fibre of refractive index 1.68. The outer covering of the pipe is made of a material of refractive index 1.44. What is the range of the angles of the incident rays with the axis of the pipe for which total reflections inside the pipe take place, as shown in the figure.

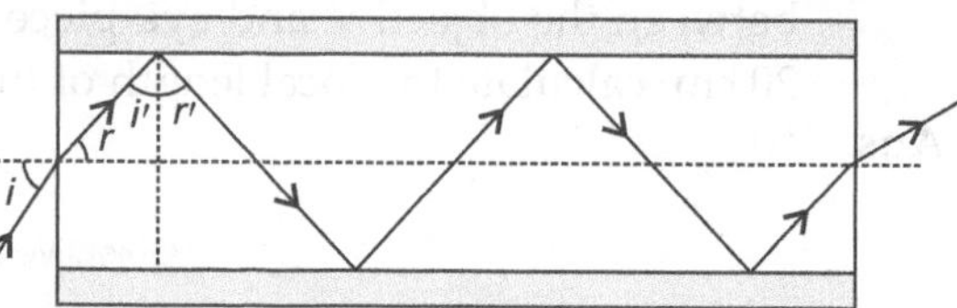

(ii) What is the answer if there is no outer covering of the pipe ? **[NCERT]**

Ans. (i) Given,

Refractive index of the glass fibre (μ_2) = 1.68

Refractive index of the outer covering of the pipe (μ_1) = 1.44

Let angle of incidence = i

and angle of refraction = r

Angle of incidence at the interface = i'

The refractive index (μ) of the inner core – outer core interface is given by

$$\mu = \frac{\mu_2}{\mu_1} = \frac{1}{\sin i'}$$

$$\sin i' = \frac{\mu_1}{\mu_2}$$

$$= \frac{1.44}{1.68} = 0.8571$$

$$\therefore \quad i' = 59°$$

For the critical angle, total internal reflection (TIR) takes place only when $i > i'$ *i.e.,* $i > 59°$.

Maximum angle of reflection, $r_{max} = 90° - i' = 90° - 59° = 31°$

Let $i_{max.}$ be the maximum angle of incidence.

The refractive index at the air – glass interface,

$$\mu_1 = 1.68$$

$$\mu_1 = \frac{\sin i_{max}}{\sin r_{max}}$$

$$\sin i_{max} = \mu_1 \sin r_{max}$$

$$= 1.68 \sin 31°$$

$$= 1.68 \times 0.5150$$

$$= 0.8652$$

$$\therefore \quad i_{max} = \sin^{-1} 0.8652 = 60°$$

Thus, all the rays incident at angles lying in the range $0 < i < 60°$ will suffer total internal reflection.

(ii) If the outer covering of the pipe is not present, then :

Refractive index of the outer pipe (μ_1) = Refractive index of air = 1

For the angle of incidence $i = 90°$, we can write Snell's law at the air – Pipe interface as :

$$\frac{\sin i}{\sin r} = (\mu_2) = 1.68$$

$$\sin r = \frac{\sin 90°}{1.68} = \frac{1}{1.68}$$

$$r = \sin^{-1}(0.5952)$$

$$= 36.5°$$

$$\therefore \quad i' = 90° - 36.5° = 53{\cdot}5°$$

Since $i' > r$, all incident rays will suffer total internal reflection.

Q. 23. An angular magnification (magnifying power) of 30X is desired using an objective of focal length 1.25 cm and an eyepiece of focal length 5 cm. How will you set up the compound microscope ? **[NCERT]**

Ans. Given,

Focal length of the objective lens (f_o) = 1.25 cm

Focal length of the eyepiece (f_e) = 5 cm

Least distance of distinct vision (d) = 25 cm

Angular magnification of the compound microscope = 3X

Total magnifying power of the compound microscope (m) = 30

The angular magnification of the eyepiece is given by

$$m_e = \left(1 + \frac{d}{f_e}\right)$$

$$= \left(1 + \frac{25}{5}\right) = 6$$

The angular magnification of the objective lens (m_o) is related to m_e by

$$m_o m_e = m$$

$$m_o = \frac{m}{m_e} = \frac{30}{6} = 5$$

We also have the relation

$$m_o = \frac{\text{Image distance for the objective lens } (v_0)}{\text{Object distance for the objective lens } (-u_e)}$$

$$5 = \frac{v_o}{-u_o}$$

$$\therefore \quad v_o = -5u_o \quad \text{...(i)}$$

From lens formula for the objective lens :

$$\frac{1}{f_o} = \frac{1}{v_o} - \frac{1}{u_o}$$

$$\frac{1}{1.25} = \frac{1}{-5u_o} - \frac{1}{u_o} = \frac{-6}{5u_o}$$

$$\therefore \quad u_o = \frac{-6}{5} \times 1.25 = -1.5 \text{ cm}$$

and

$$v_o = -5u_o$$

$$= -5 \times (-1.5) = 7.5 \text{ m}$$

The object should be placed 1.5cm away from the objective lens to obtain the desired magnification.

Applying the lens formula for the eyepiece :

$$\frac{1}{v_e}-\frac{1}{u_e}=\frac{1}{f_e}$$

Where, v_e = Image distance for the eyepiece = – d = – 25 cm

u_e = Object distance for the eyepiece

$$\frac{1}{u_e}=\frac{1}{v_e}-\frac{1}{f_e}=\frac{-1}{25}-\frac{1}{5}=-\frac{6}{25}$$

$\therefore \quad u_e = -4.17$ cm

Separation between the objective lens and the eyepiece = $|u_e| + |v_o|$

$= 4.17 + 7.5$

$= 11.67$ cm

Therefore, the separation between the objective lens and the eyepiece given by 11.67 cm.

Q. 24. (i) Determine the 'effective focal length' of the combination of the two lenses in which have a convex lens of focal length 30 cm in contact with a concave lens of focal length 20 cm. If they are placed 8.0 cm apart with their principal axes coincident. Does the answer depend on which side of combination a beam of Parallel light is incident ? Is the notion of effective focal length of this system useful at all ?

(ii) An object 1.5 cm in size is placed on the side of the convex lens in the arrangement (i) above. The distane between the object and the convex lens is 40 cm. Determine the magnification produced by the two-lens system, and the size of the image. **[NCERT]**

Ans. (i) Given,

Focal length of the convex lens (f_1) = 30 cm

Focal length of the concave lens (f_2) = – 20 cm

Distance between the two lenses (d) = 8.0 cm

(a) When the parallel beam of light is incident on the convex lens first :

From lens formula,

$$\frac{1}{v_1}-\frac{1}{u_1}=\frac{1}{f_1}$$

Here, u_1 = Object distance = ∞

v_1 = Image distance

$$\frac{1}{v_1}=\frac{1}{30}-\frac{1}{\infty}=\frac{1}{30}$$

$\therefore \quad v_1 = 30$ cm

The image will act as a virtual object for the concave lens.

From lens formula for concave lens,

$$\frac{1}{v_2}-\frac{1}{u_2}=\frac{1}{f_2}$$

Here, u_2 = Object distance

= $(30 - d) = 30 - 8 = 22$ cm

v_2 = Image distance

$$\frac{1}{v_2}=\frac{1}{22}-\frac{1}{20}=\frac{10-11}{220}=\frac{-1}{220}$$

$\therefore \quad v_2 = -220$ cm

The parallel incident beam appears to diverge from a point that is $\left(220-\frac{d}{2}=220-4\right)$ 216 cm from the centre of the combination of the two lenses.

(b) When the parallel beam of light is incident, from the left, on the concave lens first :

From lens formula, we have :

$$\frac{1}{v_2} - \frac{1}{u_2} = \frac{1}{f_2}$$

$$\frac{1}{v_2} = \frac{1}{f_2} + \frac{1}{u_2}$$

Where, u_2 = Object distance = $-\infty$

v_2 = Image distance

$$\frac{1}{v_2} = \frac{1}{-20} + \frac{1}{-\infty} = \frac{-1}{20}$$

$$\therefore \quad v_2 = -20 \text{ cm}$$

The image will act as a real object for the convex lens.

Applying lens formula to the convex lens, we have :

$$\frac{1}{v_1} - \frac{1}{u_1} = \frac{1}{f_1}$$

Here, u_1 = Object distance

$= -(20 + d) = -(20 + 8) = -28$ cm

v_1 = Image distance

$$\frac{1}{v_1} = \frac{1}{30} + \frac{1}{-28} = \frac{14-15}{420} = \frac{-1}{420}$$

$$\therefore \quad v_2 = -420 \text{ cm}$$

Thus parallel incident beam seems to diverge from a point that is (420 – 4) 416 cm from the left of the centre of the combination of the two lenses.

The answer does depend on the side of the combination at which the parallel beam of light is incident. The notion of effective focal length does not seem to be useful for this combination.

(ii) Height of the image (h_1) = 1.5 cm

Object distance from the side of the convex lens (u_1) = – 40 cm

$$|u_1| = 40 \text{ cm}$$

Lens formula—

$$\frac{1}{v_1} - \frac{1}{u_1} = \frac{1}{f_1}$$

Here, v_1 = Image distance

$$\frac{1}{v_1} = \frac{1}{30} + \frac{1}{-40}$$

$$= \frac{4-3}{120} = \frac{1}{120}$$

$$\therefore \quad v_1 = 120 \text{ cm}$$

Magnification, $$m = \frac{v_1}{|u_1|} = \frac{120}{40} = 3$$

Therefore, the magnification due to the convex lens is 3.

The image formed by the convex lens behaves as an object for the concave lens.

From lens formula :

$$\frac{1}{v_2} - \frac{1}{u_2} = \frac{1}{f_2}$$

where, u_2 = Object distance

$= +(120 - 8) = 112$ cm.

v_2 = Image distance

$$\frac{1}{v_2} = \frac{1}{-20} - \frac{1}{112}$$

$$= \frac{-112+20}{2240} = \frac{-92}{2240}$$

$$\therefore \quad v_2 = -\frac{2240}{92} \text{ cm}$$

$$\text{Magnification } (m') = \left|\frac{v_2}{u_2}\right| = \frac{2240}{92} \times \frac{1}{112} = \frac{20}{92}$$

Hence, the magnification due to the concave lens is $\frac{20}{92}$.

The magnification by the combination of the two lenses will be

$$= m \times m'$$

$$= 3 \times \frac{20}{92} = \frac{60}{92} = 0.652$$

The magnification of the combination is given by

$$\frac{h_2}{h_1} = 0.652$$

$$h_2 = 0.652 \times h_1$$

Here, h_1 = Object size = 1.5 cm

h_2 = Size of the image

$\therefore \quad h_2 = 0.652 \times 1.5 = 0.98$ cm

Therefore, the height of the image is 0.98 cm.

Q. 25. Find the position of the image formed by the lens combination given in the fig. **[NCERT]**

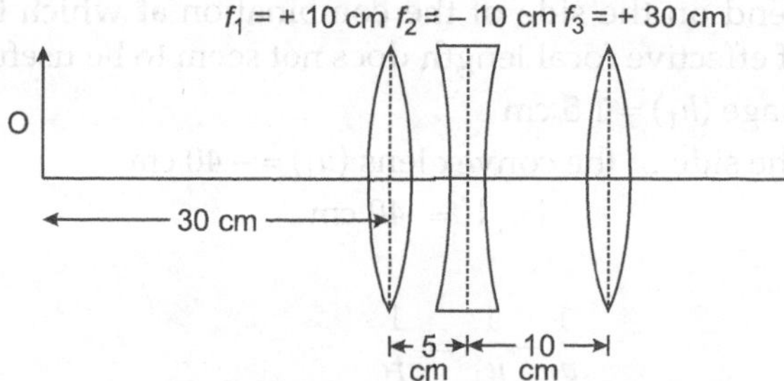

Ans. Image formed by the first lens (given by lens formula)

$$\frac{1}{v_1} - \frac{1}{u_1} = \frac{1}{f_1}$$

$$\frac{1}{v_1} - \frac{1}{-30} = \frac{1}{10}$$

or $\quad v_1 = 15$ cm

The image formed by the first lens serves as the object for the second. This is at a distance of (15 – 5) cm = 10 cm to the right of the second lens. Though the image of real, it serves as a virtual object for the second lens, which means that the rays appear to come from it for the second lens.

$$\frac{1}{v_2} - \frac{1}{10} = \frac{1}{-10}$$

or $\quad v_2 = \infty$

The virtual image is formed at an infinite distance to the left of the second lens. This acts as an object for the third lens.

$$\frac{1}{v_3} - \frac{1}{u_3} = \frac{1}{f_3}$$

$$\text{or} \qquad \frac{1}{v_3} = \frac{1}{\infty} + \frac{1}{30}$$

$$\text{or} \qquad v_3 = 30 \text{ cm}$$

The final image is formed 30 cm to the right of the third lens.

Q. 26. Figures (a) and (b) show refraction of a ray in air incident at 60° with the normal to a glass-air and water-air interface, respectively. Predict the angle of refraction in glass when the angle of incidence in water is 45° with the normal to a water-glass interface [fig. (c)] **[NCERT]**

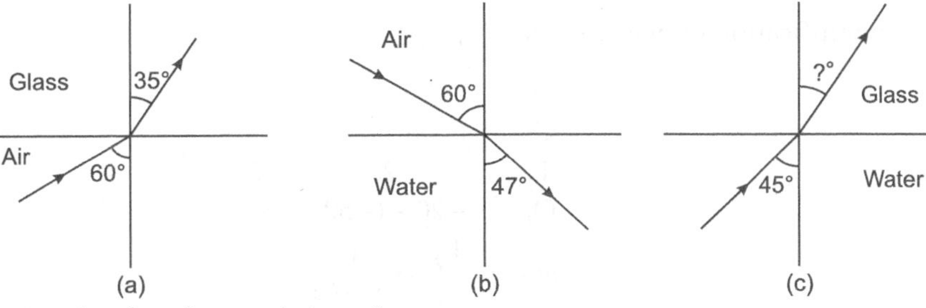

(a) (b) (c)

Ans. From question for the glass—air interface :

Angle of incidence, $i = 60°$

Angle of refraction, $r = 35°$

The relative refractive index of glass with respect to air is given by

$$^{a}\mu_{g} = \frac{\sin i}{\sin r} = \frac{\sin 60°}{\sin 35°} = \frac{0.8660}{0.5736} = 1.51 \qquad \text{...(i)}$$

For the air—water interface :

Angle of incidence, $i = 60°$

Angle of refraction, $r = 47°$

The relative refractive index of water with respect to air is given by Snell's law as :

$$^{a}\mu_{w} = \frac{\sin i}{\sin r} = \frac{\sin 60°}{\sin 47°} = \frac{0.8660}{0.7314} = 1.184 \qquad \text{...(ii)}$$

From equation (i) and (ii), the relative refractive index of glass with respect to water can be obtained as :

$$^{w}\mu_{g} = \frac{^{a}\mu_{g}}{^{a}\mu_{w}} = \frac{1.51}{1.184} = 1.275$$

The following figure shows the situation involving the glass—water interface.

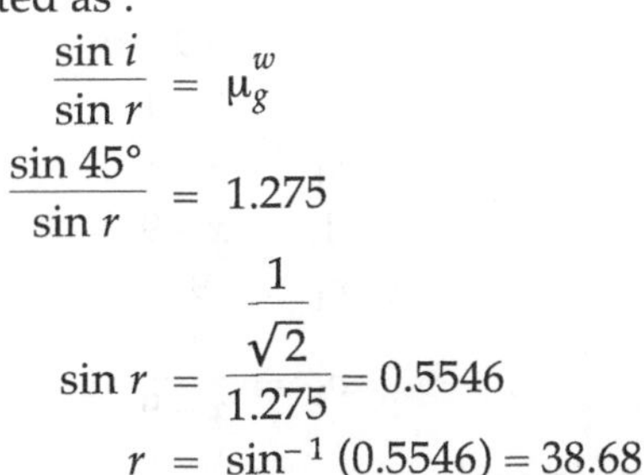

Angle of incidence, $i = 45°$

Angle of refraction $= r$

From Snell's law, r can be calculated as :

$$\frac{\sin i}{\sin r} = {^{w}\mu_{g}}$$

$$\frac{\sin 45°}{\sin r} = 1.275$$

$$\sin r = \frac{\frac{1}{\sqrt{2}}}{1.275} = 0.5546$$

$$\therefore \qquad r = \sin^{-1}(0.5546) = 38.68°$$

Thus, the angle of refraction at the water—glass interface is 38.68°.

Q. 27. Two objects P and Q when placed at different positions in front of a concave mirror of focal length 20 cm, form real images of equal size. Size of object P is three times size of object Q. If the distance of P is 50 cm from the mirro, find the distance of Q from the mirror.★

★ are board based questions from previous year

Ans. Here, focal length of concave mirror, $f = -20$ cm.

Let, I_P and I_Q be the image of P and Q and O_P and O_Q be the object size of P and Q.

Given, $I_P = I_Q$

and $O_P = 3O_Q$

$u_P = -50$ cm

$u_Q = ?$

Now, magnification of object P, $m_P = \dfrac{f}{f-u_P}$

$$\Rightarrow \frac{I_P}{O_P} = \frac{f}{f-u_P}$$

$$\Rightarrow \frac{I_P}{O_P} = \frac{-20}{-20-(-50)} = \frac{-2}{3}$$

Now, $m_Q = \dfrac{I_Q}{O_Q} = \dfrac{I_P}{O_P/3}$

$$= \frac{3\,I_P}{O_P} = 3 \times \left(\frac{-2}{3}\right) = -2$$

But, $m_Q = \dfrac{f}{f-u_Q}$

$$-2 = \frac{-20}{-20-u_Q}$$

$$\Rightarrow 40 + 2u_Q = -20$$

$$\Rightarrow u_Q = -30 \text{ cm.}$$

Hence, the distance of object Q from the mirror is 30 cm.

Q. 28. A concave mirror forms a real image of an object kept at a distance 9 cm from it. If the object is taken away from the mirror by 6 cm, the image size reduces to $\frac{1}{4}$th of its previous size. Find the focal length of the mirror.★

Ans. Let, the focal length of concave mirror be $-x$.

First object distance, $u_1 = -9$ cm

Second object distance, $u_2 = -15$ cm

I case, using mirror formula,

$$\frac{1}{f} = \frac{1}{u_1} + \frac{1}{v_1}$$

$$\Rightarrow \frac{1}{-x} = \frac{1}{-9} + \frac{1}{v_1}$$

$$\Rightarrow \frac{1}{v_1} = \frac{1}{9} - \frac{1}{x}$$

$$\Rightarrow \frac{1}{v_1} = \frac{x-9}{9x}$$

$$\Rightarrow v_1 = \frac{9x}{x-9} \quad \text{...(i)}$$

II case, using mirror formula again

$$\Rightarrow \frac{1}{f} = \frac{1}{u_2} + \frac{1}{v_2}$$

$$\Rightarrow \frac{1}{-x} = \frac{1}{-15} + \frac{1}{v_2}$$

★ are board based questions from previous year

$$\Rightarrow \quad \frac{1}{v_2} = \frac{1}{15} - \frac{1}{x} = \frac{x-15}{15x}$$

$$\Rightarrow \quad v_2 = \frac{15x}{x-15} \qquad \text{...(ii)}$$

Now, magnification for I cast,

$$m_1 = \frac{v_1}{u_1} = \frac{9x}{x-9} \times \frac{1}{(-9)} = \frac{I_1}{O}$$

Now, magnification for II case,

$$m_2 = \frac{v_2}{u_2} = \frac{15x}{x-15} \times \frac{1}{(-15)} = \frac{I_2}{O}$$

According to question,

$$I_2 = \frac{1}{4} I_1 \qquad \text{[Object remains of the same size]}$$

$$\frac{x}{-(x-15)} = \frac{1}{4} \times \frac{x}{-(x-9)}$$

$$\Rightarrow \quad \frac{1}{x-15} = \frac{1}{4(x-9)}$$

$$\Rightarrow \quad 4x - 36 = x - 15$$

$$\Rightarrow \quad 3x = 21$$

$$\Rightarrow \quad x = 7 \text{ cm}$$

Hence, the focal length of mirror is 7 cm.

Q. 29. A screen is placed 80 cm from an object. The image of the object on the screen is formed by a convex lens placed between them at two different locations separated by a distance 20 cm. Determine the focal length of the lens.★

Ans. Case 1

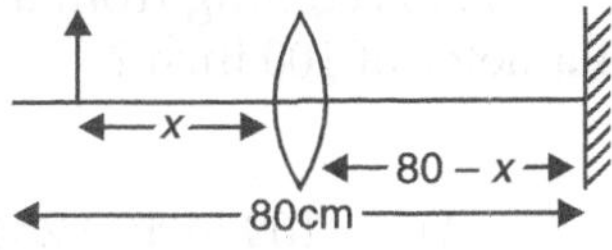

Let

Object distance, $u = x$

Image distance, $v = 80 - x$

Focal length $= f$

According to lens formuals,

$$\frac{1}{f} = \frac{1}{v} + \frac{1}{u}$$

$$\frac{1}{f} = \frac{1}{80-x} + \frac{1}{x} \qquad \text{...(i)}$$

Similarly,

Case 2

Now,

$$u = x + 20$$

$$v = 60 - x$$

★ are board based questions from previous year

$$\therefore \qquad \frac{1}{f} = \frac{1}{60-x} + \frac{1}{20+x} \qquad \ldots(ii)$$

On comparing (i) and (ii) we get

$$\frac{1}{80-x} + \frac{1}{x} = \frac{1}{60-x} + \frac{1}{20+x}$$

$$\Rightarrow \qquad \frac{80}{(80-x)x} = \frac{80}{(60-x)(20+x)}$$

$$80x - x^2 = 1200 + 40x - x^2$$

$$\Rightarrow \qquad 40x = 1200$$

$$\Rightarrow \qquad x = 30 \text{ cm}$$

Put the value of x in equation (i)

$$\frac{1}{f} = \frac{1}{80-30} + \frac{1}{30}$$

$$= \frac{1}{50} + \frac{1}{30}$$

$$\Rightarrow \qquad f = 18.75 \text{ cm}$$

Hence, the focal length of lens is 18.75 cm.

Chapter 10. Wave Optics

Q. 1. For what distance is ray optics a good approximation when the aperture is 3 mm wide and the wavelength is 500 nm ? **[NCERT]**

Ans.

$$Z_F = \frac{a^2}{\lambda} = \frac{(3 \times 10^{-3})^2}{5 \times 10^{-7}} = 18 \text{ m}$$

This example shows that even with a small aperture, diffraction spreading can be neglected for rays many metres in lengths. Thus, ray optics is valid in many common situations.

Q. 2. Assume that light of wavelength 6000 Å is coming from a star. What is the limit of resolution of a telescope whose objective has a diameter of 100 inch ? **[NCERT]**

Ans. A 100 inch telescope implies that

$$2a = 100 \text{ inch} = 254 \text{ cm.}$$

Thus if,

$$\lambda \approx 6000 \text{ Å} = 6 \times 10^{-5} \text{ cm}$$

Then

$$\Delta\theta \approx \frac{0.61 \times 6 \times 10^{-5}}{127}$$

$$\approx 2.9 \times 10^{-7} \text{ radians}$$

Q. 3. (i) Two slits are made one milimetre apart and the screen is placed one metre away. What is the fringe separation when blue-green light of wavelength 500 nm is used ?

(ii) In previous part, what should the width of each slit be to obtain 10 maxima of the double slit pattern within the central maximum of the single slit pattern ?

Ans. (i)

$$\text{Fringe spacing} = \frac{D\lambda}{d} = \frac{1 \times 5 \times 10^{-7}}{1 \times 10^{-3}} \text{ m}$$

$$= 5 \times 10^{-4} \text{ m} = 0.5 \text{ mm}$$

(ii) We want

$$2\frac{\lambda D}{a} = 10\left(\frac{\lambda D}{d}\right)$$

$$a = \frac{2d}{10} = \frac{2 \times 10^{-3}}{10} = 0.2 \text{ mm}$$

Notice that the wavelength of light and distance of the screen do not enter in the calculation of a.

Q. 4. Estimate the distance for which ray optics is good approximation for an aperture of 4 mm and wavelength 400 nm. **[NCERT]**

Ans. Fresnel's distance (Z_F) is the distance for which the ray optics is a good approximation. It is given by the relation.

$$Z_F = \frac{a^2}{\lambda}$$

where,

$$\text{Aperture width, } a = 4 \text{ mm} = 4 \times 10^{-3} \text{ m}$$

$$\text{Wavelength of light, } \lambda = 400 \text{ nm} = 400 \times 10^{-9} \text{ m}$$

$$Z_F = \frac{(4 \times 10^{-3})^2}{400 \times 10^{-9}} = 40 \text{ m}$$

Therefore, the distance for which the ray optics is a good approximation is 40 m.

Q. 5. In double slit experiment using light of wavelength 600 nm, the angular width of a fringe formed on a distance screen is 0.1°. What is the spacing between the two slits ? **[NCERT]**

Ans. Wavelength of light used, λ = 6000 nm = 600×10^{-9} m

Angular width of fringe,

$$\theta = 0.1° = 0.1 \times \frac{\pi}{180} = \frac{3.14}{1800} \text{ rad}$$

Angular width of a fringe is related to slit spacing (d) as :

$$\theta = \frac{\lambda}{d}$$

$$d = \frac{\lambda}{\theta} = \frac{600 \times 10^{-9}}{\frac{3.14}{1800}} = 3.44 \times 10^{-4} \text{ m}$$

Therefore, the spacing between the slits is 3.44×10^{-4} m.

Q. 6. What is the Brewster angle for air to glass transition ? (Refractive index of glass = 1.5). **[NCERT]**

Ans. Given,

Refractive index of glass, (μ) = 1.5
Let Brewster angle = θ
Brewster angle is related to refractive index by

$$\tan \theta = \mu$$

$$\theta = \tan^{-1}(1.5) = 56.31°$$

Hence, the Brewster angle for air to glass transition is 56.31°.

Q. 7. What speed should a galaxy move with respect to us so that the sodium line at 589.0 nm is observed at 589.6 nm ? **[NCERT]**

Ans. As

$$c = v\lambda,$$

$$\frac{\Delta v}{v} = -\frac{\Delta \lambda}{\lambda} \text{ (for small changes in } v \text{ and } \lambda\text{). For}$$

$$\Delta\lambda = 589.6 - 589.0 = +0.6 \text{ nm}$$

$$\frac{\Delta v}{v} = -\frac{\Delta \lambda}{\lambda} = -\frac{v_{radial}}{c}$$

or

$$v_{radial} \cong +c\left(\frac{0.6}{589.0}\right)$$

$$= +3.06 \times 10^5 \text{ ms}^{-1}$$

$$= 306 \text{ km/s}$$

Therefore, the galaxy is moving away from us.

Q. 8. The 6563 Å H_α line emitted by hydrogen in a star is found to be red shifted by 15 Å. Estimate the speed with which the star is receding from the earth. **[NCERT]**

Ans. Given : Wavelength or H_α line emitted by

$$\text{Hydrogen, } (\lambda) = 6563 \text{ Å}$$

$$= 6563 \times 10^{-10} \text{ m.}$$

$$\text{Star's red-shift } (\lambda' - \lambda) = 15 \text{ Å} = 15 \times 10^{-10} \text{ m}$$

$$\text{Speed of light } (c) = 3 \times 10^{8} \text{ m/s}$$

Let the velocity of the star receding away from the Earth be v.

The red shift is related with velocity by

$$\lambda' - \lambda = \frac{v}{c}\lambda$$

$$v = \frac{c}{\lambda} \times (\lambda' - \lambda) = \frac{3 \times 10^{8} \times 15 \times 10^{-10}}{6563 \times 10^{-10}}$$

$$= 6.87 \times 10^{5} \text{ m/s}$$

Hence, the speed with which the star is receding away from the earth is 6.87×10^{5} m/s.

Q. 9. In a double slit experiment the angular width of a fringe is found to be 0.2° on a screen placed 1 m away. The wavelength of light used is 600 nm. What will be the angular width of the fringe. If the entire experimental apparatus is immersed in water ? Take refractive index of water to be 4/3. **[NCERT]**

Ans. Given,

Distance of the screen from the slits (D) = 1 m

Wavelength of light used (λ_1) = 600 nm

Angular width of the fringe in air (θ_1) = 0.2°

Angular width of the fringe in water (θ_2) = ?

Refractive index of water (μ) $= \frac{4}{3}$

Refractive index is related to angular width by

$$\mu = \frac{\theta_1}{\theta_2}$$

$$\theta_2 = \frac{3}{4}\theta_1 = \frac{3}{4} \times 0.2 = 0.15$$

Hence, the angular width of the fringe in water will reduce to 0.15°.

Q. 10. Two towers on top of two hills are 40 km apart. The line joining them passes 50 m above a hill halfway between the towers. What is the longest wavelength of radio waves, which can be sent between the towers without appreciable diffraction effects ? **[NCERT]**

Ans. Given,

Distance between the towers (d) = 40 km

Height of the line joining the hills, (d') = 50 m.

Thus, the radial spread of the radio waves should not exceed 50 km.

Since the hill is located halfway between the towers. Fresnel's distance can be obtained as :

$$(Z_P) = 20 \text{ km} = 2 \times 10^{4} \text{ m}$$

Aperture can be taken by

$$a = d = 50 \text{ m}$$

Fresnel's distance is given by $Z_F = \frac{a^2}{\lambda}$

where, (λ) = Wavelength of radio waves

$$\therefore \quad \lambda = \frac{a^2}{Z_F}$$

$$= \frac{(50)^2}{2 \times 10^{4}}$$

$$= 1250 \times 10^{-4} = 0.1250 \text{ m} = 12.5 \text{ cm}$$

Therefore, the wavelength of the radiowaves is 12.5 cm.

Q. 11. In a Young's double-slit experiment, the slits are separated by 0.28 mm and the screen is placed 1.4 m away. The distance between the central bright fringe and the fourth bright fringe is measured to be 1.2 cm. Determine the wavelength of light used in the experiment. **[NCERT]**

Ans. Given,

Distance between the slits, $(d) = 0.28 \text{ mm} = 0.28 \times 10^{-3}$ m

Distance between the slit and the screen (D) = 1.4 m

Distance between the central fringe and the fourth ($n = 4$) fringe,

$$(u) = 1.2 \text{ cm} = 1.2 \times 10^{-2} \text{ m}$$

In case of a constructive interference, the distance between the two fringes is

$$u = n\lambda \frac{D}{d}$$

where, n = Order of fringes = 4

λ = Wavelength of light

$$\therefore \quad \lambda = \frac{ud}{nD} = \frac{1.2 \times 10^{-2} \times 0.28 \times 10^{-3}}{4 \times 1.4}$$

$$= 6 \times 10^{-7}$$

$$= 600 \text{ nm}$$

Thus, the wavelength of the light is 600 nm.

Q. 12. A parallel beam of light of wavelength 500 nm falls on a narrow slit and the resulting diffraction pattern is observed on a screen 1 m away. It is observed that the first minimum is at a distance of 2.5 mm from the centre of the screen. Find the width of the slit. **[NCERT]**

Ans. Wavelength of light beam, $\lambda = 500 \text{ nm} = 500 \times 10^{-9}$ m

Distance of screen from the slit, D = 1 m

For first minima, $n = 1$

Distance between the slits = d

Distance of the first minimum from the centre of the screen can be obtained as :

$$x = 2.5 \text{ mm} = 2.5 \times 10^{-3} \text{ m}$$

It is related to the order of minima as :

$$n\lambda = x\frac{d}{D}$$

$$d = \frac{n\lambda D}{x}$$

$$= \frac{1 \times 500 \times 10^{-9} \times 1}{2.5 \times 10^{-3}}$$

$$= 2 \times 10^{-4} \text{ m} = 0.2 \text{ mm}$$

Therefore, the width of the slits is 0.2 mm.

Q. 13. Two wavelength of sodium light 590 nm and 596 nm are used, in turn, to study the diffraction taking place due to a single slit of aperture 1×10^{-4} m. The distance between the slit and the screen is 1.8 m. Calculate the separation between the positions of the first maxima of the diffraction pattern obtained in the two cases.

Ans. Wavelength of the light beam, $\lambda_1 = 590 \text{ nm} = 5.9 \times 10^{-7}$ m

Wavelength of another light beam, $\lambda_2 = 596 \text{ nm} = 5.96 \times 10^{-7}$ m

Distance of the slits from the screen = D = 1.8 m

Distance between the two slits = 1×10^{-4} m

For the first secondary maxima,

$$\sin\theta = \frac{3\lambda_1}{2a} = \frac{x_1}{D}$$

$$x_1 = \frac{3\lambda_1 D}{2a}$$

and
$$x_2 = \frac{3\lambda_2 D}{2a}$$

∴ Spacing between the position of first secondary maxima of two sodium ines

$$= x_2 - x_1$$

$$= \frac{3D}{2a}(\lambda_2 - \lambda_1) = 1.62 \times 10^{-4} \text{ m.}$$

Q. 14. If one of two identical slits producing interference in Young's experiment is covered with glass, so that the light intensity passing through it is reduced to 50%, find the ratio of the maximum and minimum intensity of the fringe in the interference pattern.★

Ans. The resultant intensity in Young's experiment is given by

$$I_R = I_1 + I_2 + 2\sqrt{I_1 I_2} \cos\phi$$

When slit is not covered, then I_0 is the intensity from each slit.

Maximum intensity (I_{max}) occurs when $\phi = 0°$.

Minimum intensity (I_{min}) occurs when $\phi = 180°$.

If one slit is covered with glass to reduce its intensity by 50%, then

$$I_{max} = I_0 + \frac{I_0}{2} + 2\sqrt{I_0 \times \frac{I_0}{2}} \cos 0°$$

$$= 1.5\, I_0 + 2 \times 0.707\, I_0 = 2.914\, I_0$$

$$I_{min} = I_0 + \frac{I_0}{2} + 2\sqrt{I_0 \times \frac{I_0}{2}} \cos 180°$$

$$= 1.5\, I_0 - 2 \times 0.707\, I_0 = 0.086\, I_0$$

$$\text{Ratio} = \frac{I_{max}}{I_{min}} = \frac{2.914\, I_0}{0.086\, I_0} = 33.884$$

Q. 15. Light of wavelength 5000 Å falls on a plane reflecting surface. What are the wavelength and frequency of the reflected light ? For what angle of incidence is the reflected ray normal to the incident ray ? **[NCERT]**

Ans. Given,

Wavelength of incident light (λ) = 5000 Å

$= 5000 \times 10^{-10}$ m

Speed of light (c) = 3×10^8 m/s

Frequency of incident light is given by

$$\nu = \frac{c}{\lambda}$$

$$= \frac{3 \times 10^8}{5000 \times 10^{-10}}$$

$$= 6 \times 10^{14} \text{ Hz}$$

The wavelength and frequency of incident light is the same as that of reflected ray. Thus, the wavelength of reflected light is 5000 Å and its frequency is 6×10^{14} Hz. When reflected ray is normal to incident ray, the sum of the angle of incidence, $\angle i$ and angle of reflection $\angle r$ is 90°.

From law of reflection, the angle of incidence is always equal to the angle of reflection.

$$\angle i + \angle r = 90°$$

★ are board based questions from previous year

$$\angle i + \angle i = 90°$$

$$\angle i = \frac{90°}{2} = 45°$$

Therefore, the angle of incidence for the given condition is 45°.

Q. 16. A beam of light consisting of two wavelengths, 650 nm and 520 nm, is used to obtain interference fringes in a Young's double slit experiment.

(i) Find the distance of the third bright fringe on the screen from the central maximum for wavelength 650 nm.

(ii) What is the least distance from the central maximum where the bright fringes due to both the wavelength coincide ? **[NCERT]**

Ans. Given,

Wavelength of the light beam (λ_1) = 650 nm

Wavelength of another light beam (λ_2) = 520 nm

Let distance of the slits from the screen = D

and distance between the two slits = d

(i) Distance of the nth bright fringe on the screen from the central maximum is given by

$$x = n\lambda\left(\frac{D}{d}\right)$$

For third bright fringe, $n = 3$

$$\therefore \quad x = 3 \times 650\frac{D}{d} = 1950\left(\frac{D}{d}\right) \text{nm}$$

(ii) Let the nth bright fringe due to wavelength λ_2 and $(n-1)^{th}$ bright fringe due to wavelength λ_2 coincide on the screen. On equating the conditions for bright fringes as

$$n\lambda_2 = (n-1)\lambda$$

$$520n = 650n - 650$$

$$650 = 130n$$

$$\therefore \quad n = 5$$

Thus, the least distance from the central maximum can be obtained by the relation :

$$x = n\lambda_2\frac{D}{d}$$

$$= 5 \times 520\frac{D}{d} = 2600\frac{D}{d} \text{ nm}$$

Q. 17. The ratio of the intensities at minima to the maxima in the Young's double slit experiment is 9 : 25. Find the ratio of the widths of the two slits.★

Ans. Let w, a and I represent the slit width, amplitude and intensity respectively.

$$\frac{I_{min}}{I_{max}} = \frac{(a_1 - a_2)^2}{(a_1 + a_2)^2} = \frac{9}{25}$$

$$\frac{(a_1 - a_2)}{(a_1 + a_2)} = \frac{3}{5}$$

$$\frac{(a_1 - a_2) + (a_1 + a_2)}{(a_1 + a_2) - (a_1 - a_2)} = \frac{3+5}{5-3}$$

$$\frac{2a_1}{2a_2} = \frac{8}{2}$$

or $$\frac{a_1}{a_2} = \frac{4}{1}$$

and $$\frac{w_1}{w_2} = \frac{(a_1)^2}{(a_2)^2} = \frac{16}{1}$$

★ are board based questions from previous year

Q. 18. (i) In what way is diffraction from each slit related to the interference pattern in a double slit experiment.★

(ii) Two wavelength of sodium light 590 nm and 596 nm are used, in turn to study the diffraction taking place at a single slit of aperture 2×10^{-4} m. The distance between the slit and the screen is 1.5 m. Calculate the separation between the positions of the first maxima of the diffraction pattern obtained in the two cases.★

Ans. (i) If the width of each slit is comparable to the wavelength of light used the interference pattern thus obtained in the double-slit experiment is modified by diffraction from each of the two slits.

(ii) Given,

Wavelength of the light beam,

$$\lambda_1 = 590 \text{ nm} = 5.9 \times 10^{-7} \text{ m}$$

Wavelength of another light beam,

$$\lambda_2 = 596 \text{ nm} = 5.96 \times 10^{-7} \text{ m}$$

Distance of the slits from the screen = D = 1.5 m

Distance between the two slits, $a = 2 \times 10^{-4}$ m

For the first secondary maxima, $\sin\theta = \dfrac{3\lambda_1}{2a} = \dfrac{x_1}{D}$

$$x_1 = \frac{3\lambda_1 D}{2a} \text{ and } x_2 = \frac{3\lambda_2 D}{2a}$$

$$x_1 = \frac{3 \times 590 \times 10^{-9} \times 15}{2 \times 2 \times 10^{-4}},\ x_2 = \frac{3 \times 596 \times 10^{-9} \times 1.5}{2 \times 2 \times 10^{-4}}$$

$$x_1 = 663.75 \times 10^{-5},\ x_2 = 670.50 \times 10^{-5}$$

∴ Spacing between the positions of first secondary maxima of two sodium lines

$$x_2 - x_1 = 6.75 \times 10^{-5} \text{ m} = 0.0675 \text{ mm}$$

Q. 19. A parallel beam of light of 600 nm falls on a narrow slit and the resulting diffraction pattern is observed on a screen 1.2 m away. It is observed that the first minimum is at a distance of 3 mm from the centre of the screen. Calculate the width of the slit.

Ans. The distance of the nth minimum from the centre of the screen is,

$$x_n = \frac{nD\lambda}{a}$$

where, D = distance of slit from screen,

λ = wavelength of the light,

a = width of the slit.

For first minimum, $n = 1$

$x_n = 3 \times 10^{-3}$ m,

$\lambda = 600 \times 10^{-9}$ m,

D = 1.2 m

$$3 \times 10^{-3} = \frac{1 \times 1.2 \times 600 \times 10^{-9}}{a}$$

$$a = \frac{1 \times 1.2 \times 600 \times 10^{-9}}{3 \times 10^{-3}}$$

$$= \frac{1.2 \times 6 \times 10^{-4}}{3}$$

$$= 2.4 \times 10^{-4} \text{ m}$$

$$= 0.24 \times 10^{-3} \text{ m}$$

$$a = 0.24 \text{ mm}$$

★ are board based questions from previous year

Q. 20. Two wavelengths of sodium light of 590 nm and 596 nm are used in turn to study the diffraction taking place at a single slit of aperture 2×10^{-6} m. The distance between the slit and the screen is 1.5 m. Calculate the separation between the positions of first maxima of the diffraction pattern obtained in the two cases.

Ans. For first maxima of the diffraction pattern we know

$$\sin\theta = \frac{3\lambda}{2a}$$

where a is aperture of slit.

From small value of θ, $\sin\theta \approx \tan\theta = \frac{y}{D}$

Where y is the distance of first minima from central line and D is the distance between the slit and the screen.

So, $$y = \frac{3\lambda}{2a}D$$

When $\lambda = 590$ nm,

$$y_1 = \frac{3 \times 590 \times 10^{-9}}{2 \times 2 \times 10^{-6}} \times 1.5$$

$$y_1 = 0.66375 \text{ m}$$

When $\lambda = 596$ nm,

$$y_2 = \frac{3 \times 596 \times 10^{-9}}{2 \times 2 \times 10^{-6}} \times 1.5$$

$$y_2 = 0.6705 \text{ m}$$

Separation between the positions of first maxima $= y_2 - y_1 = 0.00675$ m or 6.75 mm

Q. 21. Answer the following questions :★

(i) In a double slit experiment using light of wavelength 600 nm, the angular width of the fringe formed on a distant screen is 0.1°. Find the spacing between the two slits.

(ii) Light of wavelength 5000 Å propagating in air gets partly reflected from the surface of water. How will the wavelengths and frequencies of the reflected and refracted light be affected ?

Ans. (i) Angular width (θ) of fringe in double-slit experiment is given by $\theta = \frac{\lambda}{d}$

Where, d = Spacing between the slits.
Given, Wavelength of light, $\lambda = 600$ nm
Angular width of fringe,

$$\theta = 0.1° = 0.1 \times \frac{\pi}{180} \text{ rad} = 0.0018 \text{ rad}$$

$$\therefore \quad d = \frac{\lambda}{\theta}$$

$$d = \frac{600 \times 10^{-9}}{18 \times 10^{-1}}$$

$$\Rightarrow \quad d = 0.33 \times 10^{-3} \text{ m}$$

(ii) The frequency and wavelength of reflected wave will not change.
The refracted wave will have same frequency, only wavelength will change.
The velocity of light in water is given by

$$v = \lambda f$$

where, v = Velocity of light
f = Frequency of light
λ = Wavelength of light

As light ray in travelling from rarer (air) medium to denser medium, its speed will decrease. Hence wavelength (λ) will also decrease.

★ are board based questions from previous year

Chapter 11. Dual Nature of Radiation and Matter

Q. 1. The photoelectric cut-off voltage in a certain experiment is 1.5 V. What is the maximum kinetic energy of photoelectrons emitted ? **[NCERT]**

Ans. Given,

Photoelectric cut-off voltage $(V_0) = 1.5$ V

The maximum kinetic energy of the emitted photoelectrons is given by

$$K_{max} = eV_0$$

Here, charge on an electron $(e) = 1.6 \times 10^{-19}$ C

$$\therefore \quad K_{max} = 1.6 \times 10^{-19} \times 1.5$$
$$= 2.4 \times 10^{-19} \text{ J}$$

Hence, the maximum kinetic energy of the photoelectrons emitted in the given experiment is 2.4×10^{-19} J.

Q. 2. Electrons initially at rest are passed through a potential difference of 2 kV. Calculate their :

(i) Final velocity, and (ii) de-Broglie wavelength.

Ans. (i) We know that,

$$\frac{1}{2} mv^2 = eV$$

$$\frac{1}{2} \times 9.1 \times 10^{-31} \times v^2 = 1.6 \times 10^{-19} \times 2 \times 10^3$$

$$v = 2.65 \times 10^7 \text{ m/sec.}$$

(ii)
$$\lambda = \frac{12.27}{\sqrt{V}} \text{ Å}$$

$$= \frac{12.27}{\sqrt{2 \times 10^3}} = 0.274 \text{ Å}$$

Q. 3. The electron is 2000 times lighter than the proton. An electron and a proton are moving so that their matter waves have a length 1 Å. Find the ratio of their kinetic energies.

Ans. We know,
$$\lambda = \frac{h}{mv}$$

Since, $\lambda_p = \lambda_e$ therefore
$$mv = \text{constant}$$
$$m_p v_p = m_e v_e$$
$$\frac{m_e}{m_p} = \frac{1}{2000} = \frac{v_p}{v_e}$$
$$\frac{E_{k_e}}{E_{k_p}} = \frac{1/2\, m_e \times v_e^2}{1/2\, m_p \times v_p^2}$$
$$= \frac{1 \times 2000 \times 2000}{2000 \times 1 \times 1} = \frac{2000}{1}$$

Q. 4. Monochromatic light of wavelength 198 nm is incident on the surface of a metal, whose work function is 2.5 eV. Calculate the stopping potential.

Ans. Given,

$\lambda = 198$ nm, $W_o = 2.5$ eV, $V_0 = ?$

We know,
$$\frac{hc}{\lambda} = W_o + E_k$$

$$\frac{6.6 \times 10^{-34} \times 3 \times 10^8}{198 \times 10^{-9}} = 2.5 \times 1.6 \times 10^{-19} + E_k$$
$$E_k = 6 \times 10^{-19}$$
$$E_k = eV_o$$
$$6 \times 10^{-19} = 1.6 \times 10^{-19} \times V_o$$
$$V_o = 3.75 \text{ volt.}$$

Q. 5. In an experiment on photoelectric effect, the slope of the cut off voltage versus frequency of incident light is found to be 4.12×10^{-15} Vs. Calculate the value of Planck's constant. **[NCERT]**

Ans. The slope of the cut-off voltage (V) versus frequency (ν) of an incident light is given by

$$\frac{V}{\nu} = 4.12 \times 10^{-15} \text{ Vs}$$

V is related to frequency by

$$h\nu = eV$$

Here, Charge on an electron $(e) = 1.6 \times 10^{-19}$ C

Planck's constant $(h) = ?$

$$\therefore \quad h = e \times \frac{V}{\nu}$$

$$= 1.6 \times 10^{-19} \times 4.12 \times 10^{-15}$$

$$= 6.592 \times 10^{-34} \text{ Js}$$

Hence, the value of Planck's constant is 6.592×10^{-34} Js.

Q. 6. What is the de-Broglie wavelength associated with (i) an electron moving with a speed of 5.4×10^6 m/s and (ii) a ball of mass 150 g travelling at 30.0 m/s ? **[NCERT]**

Ans. (i) **For the electron :**

Mass $(m) = 9.11 \times 10^{-31}$ kg,

speed $(v) = 5.4 \times 10^6$ m/s.

Then, momentum $p = mv = 9.11 \times 10^{-31}$ kg $\times 5.4 \times 10^6$ m/s,

$p = 4.92 \times 10^{-24}$ kg m/s

$$\text{de Broglie wavelength } (\lambda) = \frac{h}{p} = \frac{6.63 \times 10^{-34} \text{ Js}}{4.92 \times 10^{-24} \text{ kg m/s}}$$

$$\lambda = 0.135 \text{ nm}$$

(ii) **For the ball :**

Mass $(m') = 0.150$ kg, speed $(v') = 30.0$ m/s.

Then momentum $p' = m'v' = 0.150$ kg $\times$ 30.0 m/s, $p' = 4.50$ kg m/s

$$\text{de-Broglie wavelength } \lambda' = \frac{h}{p'} = \frac{6.63 \times 10^{-34} \text{ Js}}{4.50 \text{ kg m/s}}$$

$$\lambda' = 1.47 \times 10^{-34} \text{ m}$$

The de-Broglie wavelength of electron is comparable with X-ray wavelength. However, for the ball it is about 10^{-19} times the size of the proton, quite beyond experimental measurement.

Q. 7. Find the de-Broglie wavelength associated with an electron moving with a velocity of $0.5\,c$ and rest mass 9.1×10^{-31} kg. [Given, $h = 6.6 \times 10^{-34}$ Js and $c = 3 \times 10^8$ ms^{-1}.]

Ans. Given,

Velocity of electron, $v = 0.5\,c = 0.5 \times 3 \times 10^8 = 1.5 \times 10^8$ ms^{-1}.

$$\text{Mass of electron} = \frac{m_0}{\sqrt{1 - \frac{v^2}{c^2}}}$$

$$= \frac{9.1 \times 10^{-31}}{\sqrt{1 - (0.5)^2}}$$

$$= \frac{9.1 \times 10^{-31}}{\sqrt{0.75}} = 10.51 \times 10^{-31} \text{ kg.}$$

$$\text{de-Broglie wavelength, } \lambda = \frac{h}{mv}$$

$$= \frac{6.6 \times 10^{-34}}{10.51 \times 10^{-31} \times 1.5 \times 10^8}$$

$$= 0.4187 \times 10^{-11} \text{ m} = 0.04187 \text{ Å}.$$

Q. 8. Light of frequency 7.21×10^{14} Hz is incident on a metal surface. Electrons with a maximum speed of 6.0×10^5 m/s are ejected from the surface. What is the threshold frequency for photoemission of electrons ? **[NCERT]**

Ans. Given,

Frequency of the incident photon (ν) = $7{\cdot}21 \times 10^{14}$ Hz

Maximum speed of the electrons (v) = 6.0×10^5 m/s

Planck's constant (h) = 6.626×10^{-34} Js

Mass of an electron (m) = 9.1×10^{-31} kg

For threshold frequency (ν_0) kinetic energy is given by

$$\frac{1}{2} mv^2 = h(\nu - \nu_0)$$

$$\nu_0 = \nu - \frac{mv^2}{2h}$$

$$= 7.21 \times 10^{14} - \frac{(9.1 \times 10^{-31}) \times (6 \times 10^5)^2}{2 \times (6.626 \times 10^{-34})}$$

$$= 7.21 \times 10^{14} - 2.472 \times 10^{14}$$

$$= 4.738 \times 10^{14} \text{ Hz}$$

Hence, the threshold frequency for the photo-emission of electrons is 4.738×10^{14} Hz.

Q. 9. The threshold frequency for a certain metal is 3.3×10^{14} Hz. If light of frequency 8.2×10^{14} Hz is incident on the metal, predict the cut-off voltage for the photoelectric emission. **[NCERT]**

Ans. Given,

Threshold frequency of the metal (ν_0) = 3.3×10^{14} Hz

Frequency of light incident on the metal (ν) = 8.2×10^{14} Hz

Charge on an electron (e) = 1.6×10^{-19} C

Planck's constant, $h = 6.626 \times 10^{-34}$ Js

Cut-off voltage for the photoelectric emission from the metal = V_0

The equation for the cut-off energy is given by

$$eV_0 = h(\nu - \nu_0)$$

$$V_0 = \frac{h(\nu - \nu_0)}{e}$$

$$= \frac{6.626 \times 10^{-34} \times (8.2 \times 10^{14} - 3.3 \times 10^{14})}{1.6 \times 10^{-19}}$$

$$= 2.0292 \text{ V}$$

Therefore, the cut-off voltage for the photoelectric emission is 2.0292 V.

Q. 10. In the study of a photoelectric effect the graph between the stopping potential V and frequency ν of the incident radiation on two different metals P and Q is shown below :★

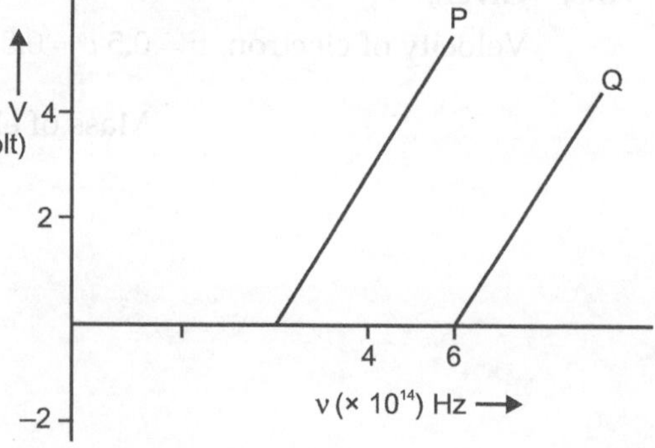

(i) Which one of the two metals has higher threshold frequency ?

(ii) Determine the work function of the metal which has greater value.

(iii) Find the maximum kinetic energy of electron emitted by light of frequency 8×10^{14} Hz for this metal.

Ans. (i) Q has higher threshold frequency.

(ii) Work function, $\phi_0 = h\nu_0$

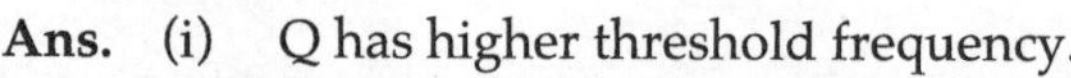

★ are board based questions from previous year

$$h\nu_0 = (6.6 \times 10^{-34}) \times \frac{6 \times 10^{14}}{1.6 \times 10^{-19}} \text{ eV} = 2.5 \text{ eV}$$

(iii)
$$K_{max} = h(\nu - \nu_0)$$
$$= \frac{6.6 \times 10^{-34} (8 \times 10^{14} - 6 \times 10^{14})}{1.6 \times 10^{-19}} \text{eV}$$
$$= \frac{6.6 \times 10^{-34} \times 2 \times 10^{14}}{1.6 \times 10^{-19}} \text{eV}$$
$$K_{max} = 0.83 \text{ eV}$$

Q. 11. Light of wavelength 488 nm is produced by an argon laser which is used in the photoelectric effect. When light from this spectral line is incident on the emitter, the stopping (cut-off) potential of photo-electrons is 0.38 V. Find the work function of the material from which the emitter is made. **[NCERT]**

Ans. Given,

Wavelength of light produced by the argon laser (λ) = 488 nm = 488×10^{-9} m

Stopping potential of the photoelectrons (V_0) = 0.38 V

$$1 \text{ eV} = 1.6 \times 10^{-19} \text{ J}$$

$$\therefore \quad (V_0) = \frac{0.38}{1.6 \times 10^{-19}} \text{ eV}$$

Planck's constant (h) = 6.6×10^{-34} Js

Charge on an electron (e) = 1.6×10^{-19} C

Speed of light (c) = 3×10^8 m/s

From Einstein's photoelectric effect, we have the relation involving the work function ϕ_0 of the material of the emitter

$$eV_0 = \frac{hc}{\lambda} - \phi_0$$

$$\phi_0 = \frac{hc}{\lambda} - eV_0$$

$$= \frac{6.6 \times 10^{-34} \times 3 \times 10^8}{1.6 \times 10^{-19} \times 488 \times 10^{-9}} - \frac{1.6 \times 10^{-19} \times 0.38}{1.6 \times 10^{-19}}$$

$$= 2.54 - 0.38 = 2.16 \text{ eV}$$

Therefore, the material with which the emitter is made has the work function of 2.16 eV.

Q. 12. A 100 W sodium lamp radiates energy uniformly in all directions. The lamp is located at the centre of a large sphere that absorbs all the sodium light which is incident on it. The wavelength of the sodium light is 589 nm.

(i) What is the energy per photon associated with the sodium light ?

(ii) At what rate are the photons delivered to the sphere ? **[NCERT]**

Ans. Power of the sodium lamp (P) = 100 W

Wavelength of the emitted sodium light (λ) = 589 nm = 589×10^{-9} m

Planck's constant (h) = 6.26×10^{-34} Js

Speed of light (c) = 3×10^8 m/s

(i) The energy per photon associated with the sodium light is given by

$$E = \frac{hc}{\lambda}$$

$$= \frac{6.626 \times 10^{-34} \times 3 \times 10^8}{589 \times 10^{-9}}$$

$$= 3.37 \times 10^{-19} \text{ J}$$

$$= \frac{3.37 \times 10^{-19}}{1.6 \times 10^{-19}} = 2.11 \text{ eV}$$

(ii) Number of photons delivered to the sphere = n

The equation for power can be written by

$$P = nE$$

$$\therefore \quad n = \frac{P}{E}$$

$$= \frac{100}{3.37 \times 10^{-19}} = 2.96 \times 10^{20} \text{ photons/s}$$

Therefore, every second, 2.96×10^{20} photons are delivered to the sphere.

Q. 13. Find the :

(i) Maximum frequency, and

(ii) Minimum wavelength of X-rays produced by 30 kV electrons. **[NCERT]**

Ans. Given,

Potential of the electrons (V) = 30 kV = 3×10^4 V

Hence, energy of the electrons (E) = 3×10^4 eV

Here, Charge on an electron (e) = 1.6×10^{-19} C

(i) Maximum frequency produced by the X-rays = ν

The energy of the electrons is given by

$$(E) = h\nu$$

Here, h = Planck's constant = 6.626×10^{-34} Js

$$\therefore \quad \nu = \frac{E}{h}$$

$$= \frac{1.6 \times 10^{-19} \times 3 \times 10^4}{6.626 \times 10^{-34}}$$

$$\approx 7.24 \times 10^{18} \text{ Hz.}$$

The maximum frequency of X-rays produced is 7.24×10^{18} Hz.

(ii) The minimum wavelength produced by the X-rays is given by

$$\lambda = \frac{c}{v} = \frac{3 \times 10^8}{7.24 \times 10^{18}}$$

$$= 4.14 \times 10^{-11} \text{ m}$$

$$= 0.0414 \text{ nm}$$

Hence, the minimum wavelength of X-rays produced is 0.0414 nm.

Q. 14. The wavelength of a probe is roughly a measure of the size of a structure that it can probe in some detail. The quark structure of protons and neutrons appears at the minute length-scale of 10^{-15} m or less. This structure was first probed in early 1970's using high energy electron beams produced by a linear accelerator at Stanford, USA. Guess what might have been the order of energy of these electron beams. (Rest mass energy of electron = 0.511 MeV.)

Ans. Given,

Wavelength of a proton or a neutron (λ) $\approx 10^{-15}$ m

Rest mass energy of an electron (m_0) = ?

$$m_0c^2 = 0.511 \text{ MeV}$$

$$= 0.511 \times 10^6 \times 1.6 \times 10^{-19}$$

$$= 0.8176 \times 10^{-13} \text{ J}$$

$$\text{Planck's constant } (h) = 6.6 \times 10^{-34} \text{ Js}$$

$$\text{Speed of light } (c) = 3 \times 10^8 \text{ m/s}$$

The momentum of a proton or a neutron is given by

$$(p) = \frac{h}{\lambda} = \frac{6.6 \times 10^{-34}}{10^{-15}}$$

$$= 6.6 \times 10^{-19} \text{ kg m/s}$$

The relativistic relation for energy (E) is given as :

$$E^2 = p^2c^2 + m^2c^4$$

$$= (6.6 \times 10^{-19} \times 3 \times 10^8)^2 + (0.8176 \times 10^{-13})^2$$
$$= 392.04 \times 10^{-22} + 0.6685 \times 10^{-26}$$
$$= 392.04 \times 10^{-22}$$
$$\therefore \quad E = 1.98 \times 10^{-10} \text{ J}$$
$$= \frac{1.98 \times 10^{-10}}{1.6 \times 10^{-19}}$$
$$= 1.24 \times 10^9 \text{ eV} = 1.24 \text{ GeV}$$

Thus, the electron energy emitted from the accelerator at Stanford, USA might be of the order of 1.24 BeV.

Q. 15. Ultraviolet light of wavelength 2271 Å from a 100 W mercury source irradiates a photocell made of molybdenum metal. If the stopping potential is – 1.3 V, estimate the work function of the metal. How would the photo-cell respond to a high intensity (~ 10^5 W m^{-2}) red light of wavelength 6328 Å produced by a He-Ne laser ? **[NCERT]**

Ans. Given,

Wavelength of ultraviolet light (λ) = 2271 Å = 2271×10^{-10} m

Stopping potential of the metal (V_0) = 1.3 V

Planck's constant (h) = 6.6×10^{-34} J

Charge on an element (e) = 1.6×10^{-19} C

Let work function of the metal = ϕ_0

Frequency of light = ν

The photo-energy relation from the photoelectric effect is

$$\phi_0 = h\nu - eV_0$$
$$= \frac{hc}{\lambda} - eV_0 = \frac{6.6 \times 10^{-34} \times 3 \times 10^8}{2271 \times 10^{-10}} - 1.6 \times 10^{-19} \times 1.3$$
$$= 8.72 \times 10^{-19} - 2.08 \times 10^{-19}$$
$$= 6.64 \times 10^{-19} \text{ J}$$
$$= \frac{6.64 \times 10^{-19}}{1.6 \times 10^{-19}} = 4.15 \text{ eV}$$

Let ν_0 be the threshold frequency of the metal.

$$\therefore \quad \phi_0 = h\nu_0$$
$$\nu_0 = \frac{\phi_0}{h}$$
$$= \frac{6.64 \times 10^{-19}}{6.6 \times 10^{-34}} = 1.006 \times 10^{15} \text{ Hz}$$

$$\text{Wavelength of red light } (\lambda_r) = 6328 \text{ Å}$$
$$\therefore \quad \text{Frequency of red light, } \nu_r = \frac{c}{\lambda_r}$$
$$= \frac{3 \times 10^8}{6328 \times 10^{-10}} = 4.74 \times 10^{14} \text{ Hz}$$

Since $\nu_0 > \nu_r$, the photocell will not respond to the red light produced by the laser.

Q. 16. An electron gun with its collector at a potential of 100 V fires out electrons in a spherical bulb containing hydrogen gas at low pressure (~ 10^2 mm of Hg). A magnetic field of 2.83×10^{-4} T curves the path of the electrons in a circular orbit of radius 12.0 cm. (The path can be viewed because the gas ions in the path focus the beam by attracting electrons, and emitting light by electron capture. This method is known as the 'fine beam tube' method. Determine e/m from the data. **[NCERT]**

Ans. Potential of an anode (V) = 100 V

Magnetic field experienced by the electrons (B) = 2.83×10^{-4} T

Radius of the circular orbit (r) = 12.0 cm = 12.0×10^{-2} m

Let mass of each electron = m

Charge on each electron = e
Velocity of each electron= v
The energy of each electron is equal to its kinetic energy, *i.e.*,

$$\frac{1}{2}mv^2 = eV$$

$$v^2 = \frac{2\,eV}{m} \quad \text{...(i)}$$

It is the magnetic field, due to its bending nature, that provides the centripetal force

$$F = \frac{mv^2}{r} \text{ for the beam.}$$

Hence, Centripetal force = Magnetic force

$$\frac{mv^2}{r} = evB$$

$$eB = \frac{mv}{r}$$

$$v = \frac{eBr}{m} \quad \text{...(ii)}$$

Substituting the value of v in equation (i), will give

$$\frac{2eV}{m} = \frac{e^2B^2r^2}{m^2}$$

$$\frac{e}{m} = \frac{2V}{B^2r^2}$$

$$= \frac{2 \times 100}{(2.83 \times 10^{-4})^2 \times (12 \times 10^{-2})^2}$$

$$= 1.73 \times 10^{11}\ \text{C kg}^{-1}$$

Therefore, the specific charge ratio (e/m) is 1.73×10^{11} C kg^{-1}.

Q. 17. In an accelerator experiment on high-energy collisions of electrons with positrons, a certain event is interpreted as annihilation of an electron-positron pair of total energy 10.2 BeV into two γ-rays of equal energy. What is the wavelength associated with each γ-ray ? (1 BeV = 10^9 eV).

[NCERT]

Ans. From question, total energy of two γ-rays :

$$(E) = 10.2\ \text{BeV}$$

$$= 10.2 \times 10^9\ \text{eV}$$

$$= 10.2 \times 10^9 \times 1.6 \times 10^{-19}\ \text{J}$$

Hence, the energy of each γ-ray :

$$E' = \frac{E}{2}$$

$$= \frac{10.2 \times 10^9 \times 1.6 \times 10^{-19}}{2} = 8.16 \times 10^{-10}\ \text{J}$$

Planck's constant (h) = 6.626×10^{-34} Js
Speed of light (c) = 3×10^8 m/s
Energy is related to wavelength by

$$E' = \frac{hc}{\lambda}$$

$$\therefore \quad \lambda = \frac{hc}{E'}$$

$$= \frac{6.626 \times 10^{-34} \times 3 \times 10^{8}}{8.16 \times 10^{-10}}$$

$$= 2.436 \times 10^{-16} \text{ m}$$

Therefore, the wavelength associated with each γ-ray is 2.436×10^{-16} m.

Q. 18. The work function of caesium is 2.14 eV. Find :

(i) the threshold frequency for caesium, and

(ii) the wavelength of the incident light if the photocurrent is brought to zero by a stopping potential of 0.60 V. **[NCERT]**

Ans. (i) For the cut-off or threshold frequency, the energy ($h\nu_0$) of the incident radiation must be equal to work function (ϕ_0), so that

$$\nu_0 = \frac{\phi_0}{h}$$

$$= \frac{2.14 \text{ eV}}{6.63 \times 10^{-34} \text{ Js}}$$

$$= \frac{2.14 \times 1.6 \times 10^{-19} \text{ J}}{6.63 \times 10^{-34} \text{ Js}}$$

$$= 5.16 \times 10^{14} \text{ Hz}$$

Thus, for frequencies less than this threshold frequency, no photoelectrons are ejected.

(ii) Photocurrent reduces to zero, when maximum kinetic energy of the emitted photoelectrons equals the potential energy (eV_0) by the retarding potential (V_0). Einstein's photoelectric equation is

$$eV_0 = h\nu - \phi_0$$

$$= \frac{hc}{\lambda} - \phi_0$$

or

$$\lambda = \frac{hc}{(eV_0 + \phi_0)}$$

$$= \frac{(6.63 \times 10^{-34} \text{ Js}) \times (3 \times 10^{8} \text{ m/s})}{(0.60 \text{ eV} + 2.14 \text{ eV})}$$

$$= \frac{19.89 \times 10^{-26} \text{ Jm}}{(2.74 \text{ eV})}$$

$$\lambda = \frac{19.89 \times 10^{-26} \text{ Jm}}{2.74 \times 1.6 \times 10^{-19} \text{ J}}$$

$$\lambda = 454 \text{ nm}$$

Q. 19. A proton and an α-particle have the same de-Broglie wavelength. Determine the ratio of (i) their accelerating potentials (ii) their speeds.★

Ans. (i) The de-Broglie wavelength of a particle is given by

$$\lambda = \frac{h}{\sqrt{2mqV}}$$

where, V is the accelerating potential of the particle.

It is given that

$$\lambda_{proton} = \lambda_{alpha}$$

$$\Rightarrow \quad \frac{h}{\sqrt{2m_p q_p V_p}} = \frac{h}{\sqrt{2m_\alpha q_\alpha V_\alpha}}$$

$$\sqrt{2m_\alpha q_\alpha V_\alpha} = \sqrt{2m_p q_p V_p}$$

On squaring both sides, we get

$$2m_\alpha q_\alpha V_\alpha = 2m_p q_p V_p$$

★ are board based questions from previous year

$$\Rightarrow \quad \frac{V_p}{V_\alpha} = \frac{m_\alpha q_\alpha}{m_p q_p} = \frac{4m_p \times 2q_p}{m_p q_p} \qquad \left[\because m_\alpha = 4m_p \text{ and } q_\alpha = 1q_p\right]$$

$$= \frac{8}{1}$$

(ii) We can also write de-Broglie wavelength as $\lambda = \frac{h}{mv}$

where h is Planck's constant, m is mass of the particle and v is speed of the particle.

It is given that

$$\lambda_{proton} = \lambda_{alpha}$$

We know

$$m_{alpha} = 4m_{proton}$$

$$\therefore \quad \lambda_{alpha} = \frac{h}{4m_{proton}v_{alpha}}$$

$$\Rightarrow \quad \frac{h}{m_{proton}v_{proton}} = \frac{h}{4m_{proton}v_{alpha}}$$

$$\frac{v_{proton}}{v_{alpha}} = 4$$

Q. 20. What is the de-Broglie wavelength of a nitrogen molecule in air at 300 K ? Assume that the molecule is moving with the root-mean square speed of molecule at this temperature. (Atomic mass of nitrogen = 14.0076 u). **[NCERT]**

Ans. Given,

Temperature of the nitrogen molecule (T) = 300 K

Atomic mass of nitrogen = 14.0076 u

Hence, mass of the nitrogen molecule (m) = 2 × 14.0076 = 28.0152 u

But 1 u = 1.66×10^{-27} kg

$$\therefore \quad m = 28.0152 \times 1.66 \times 10^{-27} \text{ kg}$$

$$\text{Planck's constant } (h) = 6.63 \times 10^{-34} \text{ Js}$$

$$\text{Boltzmann constant } (k) = 1.38 \times 10^{-23} \text{ JK}^{-1}$$

The mean kinetic energy of the neutron molecule $\left(\frac{3}{2}kT\right)$ is related with root mean square speed (v_{rms}) by

$$\frac{1}{2}mv_{rms}^2 = \frac{3}{2}kT$$

$$v_{rms} = \sqrt{\frac{3kT}{m}}$$

Therefore, the de-Broglie wavelength of the nitrogen molecule is given by

$$\lambda = \frac{h}{mv_{rms}} = \frac{h}{\sqrt{3mkT}}$$

$$= \frac{6.63 \times 10^{-34}}{\sqrt{3 \times 28.0152 \times 1.66 \times 10^{-27} \times 1.38 \times 10^{-23} \times 300}}$$

$$= 0.028 \times 10^{-9} \text{ m}$$

$$= 0.028 \text{ nm}$$

Hence, the de-Broglie wavelength of the nitrogen molecule is 0.028 nm.

Q. 21. An electron microscope uses electrons accelerated by a voltage of 50 kV. Determine the de-Broglie wavelength associated with the electrons. Taking other factors, such as numerical aperture etc., to be same, how does the resolving power of an electron microscope compare with that of an optical microscope which used yellow light ?★

★ are board based questions from previous year

Ans. The de-Broglie wavelength of the electrons is given by :

$$\lambda = \frac{h}{\sqrt{2meV}}$$

Here,

m = mass of the electron = 9.1×10^{-31} kg

e = charge on the electron = 1.6×10^{-19} C

V = accelerating potential = 50 kV

h = Planck's constant = 6.626×10^{-34} Js

$$\Rightarrow \qquad \lambda = \frac{6.626 \times 10^{-34}}{\sqrt{2\,(9.1 \times 10^{-31})\,(1.6 \times 10^{-19})\,(50 \times 10^{3})}}$$

$$\Rightarrow \qquad \lambda = 0.0549 \text{ Å}$$

Resolving power of a microscope, $R = \dfrac{2\mu \sin\theta}{\lambda}$

This formula shows that to enhance resolution, we have to use shorter wavelength and media with large indices of refraction.

For an electron microscope, μ is equal to 1 (vacuum).

For an electron microscope, the electrons are accelerated through a 60,000 V potential difference. Thus, the wavelength of electrons is given by,

$$\lambda = \frac{12.27}{\sqrt{V}} = \frac{12.27}{\sqrt{60000}} = 0.05 \text{ Å}$$

As, λ is very little (roughly 10^{-5} times smaller) for electron microscope than an optical microscope which uses yellow light of wavelength (5700 Å to 5900 Å). So, the resolving power of electron microscope is about 10^5 greater than that of optical microscope.

Q. 22. Crystal diffraction experiments can be performed using X-rays, or electrons accelerated through appropriate voltage. Which probe has greater energy ? (For quantitative comparison, take the wavelength of the probe equal to 1 Å, which is of the order of inter-atomic spacing in the lattice) ($m_e = 9.11 \times 10^{-31}$ kg). **[NCERT]**

Ans. An X-ray probe has a greater energy than an electron probe for the same wavelength.

Wavelength of light emitted from the probe (λ) = 1 Å = 10^{-10} m

Mass of an electron (m_e) = 9.11×10^{-31} kg

Planck's constant (h) = 6.6×10^{-34} Js

Charge on an electron (e) = 1.6×10^{-19} C

The kinetic energy of the electron is given by

$$E = \frac{1}{2} m_e v^2$$

$$m_e v = \sqrt{2\,E m_e}$$

Here, v = Velocity of the electron

$m_e v$ = Momentum (p) of the electron

According to the de-Broglie principle, the de-Broglie wavelength is given by

$$\lambda = \frac{h}{p} = \frac{h}{m_e v} = \frac{h}{\sqrt{2\,E m_e}}$$

$$\therefore \qquad E = \frac{h^2}{2\lambda^2 m_e}$$

$$= \frac{(6.6 \times 10^{-34})^2}{2 \times (10^{-10})^2 \times 9.11 \times 10^{-31}}$$

$$= 2.39 \times 10^{-17} \text{ J}$$

$$= \frac{2.39 \times 10^{-17}}{1.6 \times 10^{-19}} = 149.375 \text{ eV}$$

$$\text{Energy of a photon (E)} = \frac{hc}{\lambda_0} \text{ eV}$$

$$= \frac{6.6 \times 10^{-34} \times 3 \times 10^8}{10^{-10} \times 1.6 \times 10^{-19}}$$

$$= 12.375 \times 10^3 \text{ eV}$$

$$= 12.375 \text{ keV}$$

Hence, a photon has a greater energy than an electron for the same wavelength.

Q. 23. (i) Obtain the de-Broglie wavelength of a neutron of kinetic energy 150 eV. As you have seen in previous question an electron beam of this energy is suitable for crystal diffraction experiments. Would a neutron beam of the same energy be equally suitable ? Explain. ($m_n = 1.675 \times 10^{-27}$ kg)

(ii) Obtain the de-Broglie wavelength associated with thermal neutrons at room temperature (27°C). Hence explain why a fast neutron beam needs to be thermalised with the environment before it can be used for neutron diffraction experiments. **[NCERT]**

Ans. (i) **Given,**

de-Broglie wavelength = 2.327×10^{-12} m; neutron is not suitable for the diffraction experiment.

$$\text{Kinetic energy of the neutron (K)} = 150 \text{ eV}$$

$$= 150 \times 1.6 \times 10^{-19}$$

$$= 2.4 \times 10^{-17} \text{ J}$$

$$\text{Mass of a neutron } (m_n) = 1.675 \times 10^{-27} \text{ kg}$$

The kinetic energy of the neutron is given by

$$K = \frac{1}{2} m_n v^2$$

$$m_n v = \sqrt{2 K m_n}$$

Here, v = Velocity of the neutron

$m_n v$ = Momentum of the neutron

de-Broglie wavelength of the neutron is given as :

$$\lambda = \frac{h}{m_n v} = \frac{h}{\sqrt{2 K m_n}}$$

As wavelength is inversely proportional to the square root of mass.

Thus, wavelength decreases with increase in mass and vice-versa.

$$\lambda = \frac{6.6 \times 10^{-34}}{\sqrt{2 \times 2.4 \times 10^{-17} \times 1.675 \times 10^{-27}}}$$

$$= 2.327 \times 10^{-12} \text{ m}$$

As inter-atomic spacing of a crystal is about 1 Å, *i.e.*, 10^{-10} m. Thus, the inter-atomic spacing is about a hundred times greater. Hence, a neutron beam of energy 150 eV is not suitable for diffraction experiments.

(ii)
$$\text{de-Broglie wavelength} = 1.447 \times 10^{-10} \text{ m}$$

$$\text{Room temperature (T)} = 27° \text{ C} = 27 + 273 = 300 \text{ K}$$

The average kinetic energy of the neutron is given by

$$E = \frac{3}{2} kT$$

Here, (k) = Boltzmann constant = 1.38×10^{-23} J mol^{-1} K^{-1}

The wavelength of the neutron is given by

$$\lambda = \frac{h}{\sqrt{2 m_n E}} = \frac{h}{\sqrt{3 m_n kT}}$$

$$= \frac{6.6 \times 10^{-34}}{\sqrt{3 \times 1.675 \times 10^{-27} \times 1.38 \times 10^{-23} \times 300}}$$
$$= 1.447 \times 10^{-10} \text{ m}$$

This wavelength is comparable to the inter-atomic spacing of a crystal. Hence, the high energy neutron beam should first be thermalised, before using it for diffraction.

Q. 24. (i) For what kinetic energy of a neutron will the associated de-Broglie wavelength be 1.40×10^{-10} m ?

(ii) Also find the de-Broglie wavelength of a neutron, in thermal equilibrium with matter, having an average kinetic energy of (3/2) kT at 300 K. **[NCERT]**

Ans. (i) Given,

de-Broglie wavelength of the neutron (λ) = 1.40×10^{-10} m

Mass of a neutron (m_n) = 1.66×10^{-27} kg

Planck's constant (h) = 6.6×10^{-34} Js

Kinetic energy (K) is given by

$$K = \frac{1}{2} m_n v^2 \quad \text{...(i)}$$

de-Broglie wavelength (λ) and velocity (v) are related as :

$$\lambda = \frac{n}{m_n v} \quad \text{...(ii)}$$

Using equation (ii) in equation (i), we get

$$K = \frac{1}{2} \frac{m_n h^2}{\lambda^2 m_n^2} = \frac{h^2}{2\lambda^2 m_n}$$
$$= \frac{(6.63 \times 10^{-34})^2}{2 \times (1.40 \times 10^{-10})^2 \times 1.66 \times 10^{-27}}$$
$$= 6.75 \times 10^{-21} \text{ J}$$
$$= \frac{6.75 \times 10^{-21}}{1.6 \times 10^{-19}}$$
$$= 4.219 \times 10^{-2} \text{ eV}$$

Thus, the kinetic energy of the neutron is 6.75×10^{-21} J or 4.219×10^{-2} eV.

(ii) Temperature of the neutron (T) = 300 K

Boltzmann constant (k)= 1.38×10^{-23} kg m^2 s^{-2} K^{-1}

Average kinetic energy of the neutron :

$$K' = \frac{3}{2} kT = \frac{3}{2} \times 1.38 \times 10^{-23} \times 300$$
$$= 6.21 \times 10^{-21} \text{ J}$$

The de-Broglie wavelength is given by

$$\lambda' = \frac{h}{\sqrt{2k' m_n}}$$

Here, $m_n = 1.66 \times 10^{-27}$ kg

$h = 6.63 \times 10^{-34}$ Js

$K' = 6.21 \times 10^{-21}$ J

$$\therefore \quad \lambda' = \frac{6.63 \times 10^{-34}}{\sqrt{2 \times 6.21 \times 10^{-21} \times 1.66 \times 10^{-27}}}$$
$$= 1.46 \times 10^{-10} \text{ m} = 0.146 \text{ nm}$$

Hence, the de-Broglie wavelength of the neutron is 0.146 nm.

Q. 25. Light of intensity 10^{-5} Wm^{-2} falls on a sodium photo-cell of surface area 2 cm^2. Assuming that the top 5 layers of sodium absorb the incident energy, estimate time required for photoelectric emission in the wave-picture of radiation. The work function for the metal is given to be about 2 eV. What is the implication of your answer ? **[NCERT]**

Ans. Intensity of incident light (I) = 10^{-5} W m^{-2}

$$\text{Surface area of a sodium photocell (A)} = 2\text{ cm}^2 = 2 \times 10^{-4}\text{ m}^2$$

$$\text{Incident power of the light (P)} = \text{I} \times \text{A}$$
$$= 10^{-5} \times 2 \times 10^{-4}$$
$$= 2 \times 10^{-9}\text{ W}$$

$$\text{Work function of the metal } (\phi) = 2\text{ eV} \quad \text{[from question]}$$
$$= 2 \times 1.6 \times 10^{-19}$$
$$= 3.2 \times 10^{-19}\text{ J}$$

Number of layers of sodium that absorbs the incident energy, $n = 5$

We know that the effective atomic area of a sodium atom, A_0 is 10^{-20} m^2.

Hence, the number of conduction electrons in n layers is given by

$$n' = n \times \frac{A}{A_0}$$
$$= 5 \times \frac{2 \times 10^{-4}}{10^{-20}} = 10^{17}$$

The incident power is uniformly absorbed by all the electrons continuously. Hence, the amount of energy absorbed per second per electron is :

$$E = \frac{P}{n'}$$
$$= \frac{2 \times 10^{-9}}{10^{17}} = 2 \times 10^{-26}\text{ J/s}$$

Time required for photoelectric emission :

$$t = \frac{\phi_0}{E}$$
$$= \frac{3.2 \times 10^{-19}}{2 \times 10^{-26}}$$
$$= 1.6 \times 10^7\text{ s} \approx 0.507\text{ years}$$

The time required for the photoelectric emission is nearly half a year, which is not practical. Hence, the wave picture is in disagreement with the given experiment.

Q. 26. Find the typical de-Broglie wavelength associated with a He atom in helium gas at room temperature (27° C) and 1 atm pressure; and compare it with the mean separation between two atoms under these conditions. **[NCERT]**

Ans. Given,

de-Broglie wavelength associated with the He atom = 0.7268×10^{-10} m

Room temperature, (T) = 27°C = 27 + 273 = 300 K

Atmospheric pressure, (P) = 1 atm = 1.01×10^5 Pa

Atomic weight of a He atom = 4

Avogadro's number (N_A) = 6.023×10^{23}

Boltzmann constant (k) = 1.38×10^{-23} J mol^{-1} K^{-1}

Average energy of a gas at temperature T, is given by

$$E = \frac{3}{2}kT$$

de-Broglie wavelength is given by

$$\lambda = \frac{h}{\sqrt{2mE}}$$

Here, m = Mass of a He atom

$$= \frac{\text{Atomic weight}}{N_A}$$

$$= \frac{4}{6.023 \times 10^{23}}$$

$$= 6.64 \times 10^{-24}\ \text{g} = 6.64 \times 10^{-27}\ \text{kg}$$

$$\therefore \qquad \lambda = \frac{h}{\sqrt{3mkT}}$$

$$= \frac{6.6 \times 10^{-34}}{\sqrt{3 \times 6.64 \times 10^{-27} \times 1.38 \times 10^{-23} \times 300}}$$

$$= 0.7268 \times 10^{-10}\ \text{m}$$

From ideal gas formula :

$$PV = RT$$

$$PV = kNT$$

$$\frac{V}{N} = \frac{kT}{P}$$

Here, V = Volume of the gas

N = Number of moles of the gas

Mean separation between two atoms of the gas is given by

$$r = \left(\frac{V}{N}\right)^{\frac{1}{3}} = \left(\frac{kT}{P}\right)^{\frac{1}{3}}$$

$$= \left[\frac{1.38 \times 10^{-23} \times 300}{1.01 \times 10^{5}}\right]^{\frac{1}{3}}$$

$$= 3.35 \times 10^{-9}\ \text{m}$$

Thus, the mean separation between the atoms is much greater than the de-Broglie wavelength.

Q. 27. Monochromatic light of wavelength 632.8 nm is produced by a helium-neon laser. The power emitted is 9.42 mW.

(i) Find the energy and momentum of each photon in the light beam.

(ii) How many photons per second, on the average, arrive at a target irradiated by this beam ? (Assume the beam to have uniform cross-section which is less than the target area).

(iii) How fast does a hydrogen atom have to travel in order to have the same momentum as that of the photon ? **[NCERT]**

Ans. Given,

Wavelength of the monochromatic light (λ) = 632.8 nm = 632.8×10^{-9} m

Power emitted by the laser (P) = 9.42 mW = 9.42×10^{-3} W

Planck's constant (h) = 6.626×10^{-34} Js

Speed of light (c) = 3×10^{8} m/s

Mass of a hydrogen atom, $m = 1.66 \times 10^{-27}$ kg

(i) The energy of each photon is given by

$$E = \frac{hc}{\lambda} = \frac{6.626 \times 10^{-34} \times 3 \times 10^{8}}{632.8 \times 10^{-9}}$$

$$= 3.141 \times 10^{-19}\ \text{J}$$

The momentum of each photon is given by

$$p = \frac{h}{\lambda} = \frac{6.626 \times 10^{-34}}{632.8}$$

$$= 1.047 \times 10^{-27}\ \text{kg ms}^{-1}$$

(ii) Number of photons arriving per second, at a target irradiated by the beam = n

Assume that the beam has a uniform cross-section that is less than the target area.

Hence, the equation for power can be written by

$$P = nE$$

$$\therefore \quad n = \frac{P}{E}$$

$$= \frac{9.42 \times 10^{-3}}{3.141 \times 10^{-19}} \approx 3 \times 10^{16} \text{ photons}$$

(iii) Momentum of the hydrogen atom is the same as the momentum of the photon,

$$p = 1.047 \times 10^{-27} \text{ kg ms}^{-1}$$

Momentum is given by

$$p = mv$$

Here,

v = Speed of the hydrogen atom

$$\therefore \quad v = \frac{p}{m} = \frac{1.047 \times 10^{-27}}{1.66 \times 10^{-27}} = 0.621 \text{ m/s}$$

Q. 28. (i) Estimate the speed with which electrons emitted from a heated emitter of an evacuated tube impinge on the collector maintained at a potential difference of 500 V with respect to the emitter. Ignore the small initial speeds of the electrons. The specific charge of the electron, *i.e.*, its e/m is given to be 1.76×10^{11} C kg^{-1}.

(ii) Use the same formula you employ in (i) to obtain electron speed for an collector potential of 10 MV. Do you see what is wrong ? In what way is the formula be modified ?

Ans. (i) Given,

Potential difference of the evacuated tube (V) = 500 V

Specific charge of an electron (e/m) = 1.76×10^{11} C kg^{-1}.

The speed of each emitted electron is given in terms of kinetic energy as follows :

$$KE = \frac{1}{2}mv^2 = eV$$

$$\therefore \quad v = \left(\frac{2eV}{m}\right)^{\frac{1}{2}} = \left(2V \times \frac{e}{m}\right)^{\frac{1}{2}} \text{m}$$

$$= (2 \times 500 \times 1.76 \times 10^{11})^{\frac{1}{2}}$$

$$= 1.327 \times 10^{7} \text{ m/s}$$

(ii) Potential of the anode (V) = 10 MV = 10×10^{6} V

The speed of each electron is given by

$$v = \left(2V \times \frac{e}{m}\right)^{\frac{1}{2}}$$

$$= (2 \times 10^{7} \times 1.76 \times 10^{11})^{\frac{1}{2}}$$

$$= 1.88 \times 10^{9} \text{ m/s}$$

This result is wrong because nothing can move faster than light. In the above formula, the expression ($mv^2/2$) for energy can only be used in the non-relativistic limit, *i.e.*, for $v << c$.

For very high speed problems, relativistic equations must be considered for solving them.

In the relativistic limit, the total energy is given by

$$E = mc^2$$

Where, m = Relativistic mass

$$= m_0\left(1 - \frac{v^2}{c^2}\right)^{\frac{1}{2}}$$

m_0 = Mass of the particle at rest

Kinetic energy is given as :

$$K = mc^2 - m_0c^2$$

Q. 29. Estimating the following two numbers should be interesting. The first number will tell you why radio engineers do not need to worry much about photons! The second number tells you why our eye can never 'count photons', even in barely detectable light.

(i) The number of photons emitted per second by a Medium wave transmitter of 10 kW power, emitting radiowaves of wavelength 500 m.

(ii) The number of photons entering the pupil of our eye per second corresponding to the minimum intensity of white light that we humans can perceive ($\sim 10^{-10}$ W m^{-2}). Take the area of the pupil to be about 0.4 cm^2, and the average frequency of white light to be about 6×10^{14} Hz.

[NCERT]

Ans. (i) Given,

Power of the medium wave transmitter (P) = 10 kW = 10^4 W = 10^4 J/s

Hence, energy emitted by the transmitter per second (E) = 10^4

Wavelength of the radio wave (λ) = 500 m

The energy of the wave is given by

$$E_1 = \frac{hc}{\lambda}$$

Here, Planck's constant (h) = 6.6×10^{-34} Js

Speed of light (c) = 3×10^8 m/s

$$\therefore \quad E_1 = \frac{6.6 \times 10^{-34} \times 3 \times 10^8}{500}$$

$$= 3.96 \times 10^{-28} \text{ J}$$

Let n be the number of photons emitted by the transmitter.

$$\therefore \quad nE_1 = E$$

$$n = \frac{E}{E_1} = \frac{10^4}{3.96 \times 10^{-28}}$$

$$= 2.525 \times 10^{31} \approx 3 \times 10^{31}$$

The energy (E_1) of a radio photon is very less, but the number of photons (n) emitted per second in a radiowave is very large.

The existence of a minimum quantum of energy can be ignored and the total energy of a radiowave can be treated as being continuous.

(ii) Intensity of light perceived by the human eye (I) = 10^{-10} W m^{-2}

Area of a pupil (A) = 0.4 cm^2 = 0.4×10^{-4} m^2

Frequency of white light (ν) = 6×10^{14} Hz

The energy emitted by a photon is given by

$$E = h\nu$$

Here, Planck's constant = 6.6×10^{-34} Js

$$\therefore \quad E = 6.6 \times 10^{-34} \times 6 \times 10^{14}$$

$$= 3.96 \times 10^{-19} \text{ J}$$

Let n be the total number of photons falling per second, per unit area of the pupil.

The total energy per unit for n falling photons is given as :

$$E = n \times 3.96 \times 10^{-19} \text{ J s}^{-1} \text{ m}^{-2}$$

The energy per unit area per second is the intensity of light.

$$\therefore \quad E = I$$

$$n \times 3.96 \times 10^{-19} = 10^{-10}$$

$$n = \frac{10^{-10}}{3.96 \times 10^{-19}} = 2.52 \times 10^{8}\ m^2\ s^{-1}$$

The total number of photons entering the pupil per second is given as :

$$n_A = n \times A$$
$$= 2.52 \times 10^{8} \times 0.4 \times 10^{-4}$$
$$= 1.008 \times 10^{4}\ s^{-1}$$

This number is not as large as the one found in problem (i), but it is large enough for the human eye to never see the individual photons.

Chapter 12. Atoms

Q. 1. Find angular momentum of an electron when it is in the second Bohr orbit of hydrogen atom.

Ans. Angular momentum $= mvr$

$$= \frac{nh}{2\pi} = \frac{2h}{2\pi} = \frac{h}{\pi} = \frac{6.6 \times 10^{-34}}{3.14}$$

$$= 2.10 \times 10^{-34}\ kg\ m^2/s.$$

Q. 2. The radius of the first orbit of the electron of a hydrogen atom is 5.3×10^{-11} m. What is the radius of its second orbit ?

Ans. Given,

$r_1 = 5.3 \times 10^{-11}$ m

We know that, $r = \frac{n^2 h^2 \varepsilon_0}{\pi m Z e^2}$

For first orbit, $n = 1$,

$$r_1 = \frac{h^2 \varepsilon_0}{\pi m Z e^2} = 5.3 \times 10^{-11}\ m$$

For second orbit, $n = 2$,

$$\therefore \quad r_2 = (2)^2 \times \frac{h^2 \varepsilon_0}{\pi m Z e^2}$$
$$= 4 \times 5.3 \times 10^{-11}\ m$$
$$= 2.12 \times 10^{-10}\ m.$$

Q. 3. Find out the wavelength of the electron orbiting in the ground state of hydrogen atom.

Ans. We know that,

$$2\pi r = n\lambda$$

For first orbit, $n = 1$

$$2\pi r = n\lambda$$
$$\lambda = 2\pi \times 0.53\ Å \qquad [\because r = 0.53\ Å]$$
$$\lambda = 3.331\ Å$$

Q. 4. Calculate the shortest wavelength of the spectral lines emitted in Balmer series. [Given Rydberg constant, $R = 10^7\ m^{-1}$].★

Ans. For Balmer series, wavelength is given by :

$$\frac{1}{\lambda} = R\left[\frac{1}{2^2} - \frac{1}{n^2}\right]$$

Where, $n = 3, 4, 5$

For shortest wavelength, $n = \infty$

★ are board based questions from previous year

So, $$\frac{1}{\lambda} = R\left[\frac{1}{2^2} - \frac{1}{\infty^2}\right]$$

$$\frac{1}{\lambda} = 10^7\left[\frac{1}{4}\right]$$

$$\lambda = 4 \times 10^{-7}\,\text{m}$$

Q. 5. A difference of 2.3 eV separates two energy levels in an atom. What is the frequency of radiation emitted when the atom makes a transition from the upper level to the lower level ? **[NCERT]**

Ans. Separation of two energy levels in an atom,

$$E = 2.3\text{ eV}$$
$$= 2.3 \times 1.6 \times 10^{-19}$$
$$= 3.68 \times 10^{-19}\text{ J}$$

If ν be the frequency of radiation emitted when the atom transits from the upper level to the lower level.

Energy associated with it is

$$E = h\nu$$

Here, h = Planck's constant = 6.62×10^{-34} Js

∴ $$\nu = \frac{E}{h} = \frac{3.68 \times 10^{-19}}{6.62 \times 10^{-34}}$$

$$= 5.55 \times 10^{14}\text{ Hz}$$

Thus, the frequency of the radiation is 5.6×10^{14} Hz.

Q. 6. A 10 kg satellite circles earth once every 2 h in an orbit having a radius of 8000 km. Assuming that Bohr's angular momentum postulate applies to satellites just as it does to an electron in the hydrogen atom. Find the quantum number of the orbit of the satellite. **[NCERT]**

Ans. As we know

$$mv_nr_n = \frac{nh}{2\pi}$$

Here m = 10 kg

and $r_n = 8 \times 10^6$ m.

We have the time period T of the circling satellite as 2 h.

That is T = 7200 s.

Thus the velocity $$v_n = \frac{2\pi\, r_n}{T}$$

The quantum number of the orbit of satellite

$$n = \frac{(2\pi\, r_n)^2 \times m}{(T \times h)}$$

Substituting the values $$n = \frac{(2\pi \times 8 \times 10^6\text{ m})^2 \times 10}{(7200\text{ s} \times 6.64 \times 10^{-34})}\text{J s}$$

$$= 5.3 \times 10^{45}$$

Note that the quantum number for the satellite motion is extremely large. In fact for such large quantum numbers the results of quantisation conditions tend to those of classical physics.

Q. 7. In a hydrogen atom a transition of electrons takes place from 3rd orbit to 2nd orbit. Calculate the wavelength of the emitted photon. To which spectral series will this photon belong ?

Ans. The wavelength of the emitted photon is given by :

$$\frac{1}{\lambda} = R\left(\frac{1}{n_1^2} - \frac{1}{n_2^2}\right)$$

Here, R = 1.097×10^7, $n_1 = 1$ and $n_3 = 3$.

$$\frac{1}{\lambda} = 1.097 \times 10^7 \left(\frac{1}{2^2} - \frac{1}{3^2}\right)$$

$$= 1.097 \times 10^7 \left(\frac{5}{36}\right)$$

$$\lambda = 6.563 \times 10^{-7} \text{ m}$$

$$= 6563 \text{ Å}.$$

Q. 8. It is found experimentally that 13.6 eV energy is required to separate a hydrogen atom into a proton and an electron. Compute the orbital radius and the velocity of the electron in a hydrogen atom. **[NCERT]**

Ans. Total energy of the electron in hydrogen atom is

$$-13.6 \text{ eV} = -13.6 \times 1.6 \times 10^{-19} \text{ J}$$

$$= 2.2 \times 10^{-18} \text{ J}.$$

Thus from we have

$$-\frac{e^2}{8\pi\varepsilon_0 r} = -2.2 \times 10^{-18} \text{ J}.$$

This gives the orbital radius,

$$r = -\frac{e^2}{8\pi\varepsilon_0 E}$$

$$= \frac{-(9 \times 10^9 \text{ N m}^2/\text{C}^2)(1.6 \times 10^{-19} \text{ C})^2}{(2)(-2.2 \times 10^{-18} \text{ J})}$$

$$= 5.3 \times 10^{-11} \text{ m}.$$

The velocity of the revolving electron can be computed

$$v = \frac{e}{\sqrt{4\pi\varepsilon_0 mr}}$$

$$= 2.2 \times 10^6 \text{ m/s}$$

Q. 9. What is the shortest wavelength present in the Paschen series of spectral lines ? **[NCERT]**

Ans. For calculating wavelength, formula is,

$$\frac{hc}{\lambda} = 21.76 \times 10^{-19} \left[\frac{1}{n_1^2} - \frac{1}{n_2^2}\right]$$

Here, h = Planck's constant $= 6.6 \times 10^{-34}$ Js
c = Speed of light $= 3 \times 10^8$ m/s
(n_1 and n_2 are integers)

The shortest wavelength present in the Paschen series of the spectral lines is given by values $n_1 = 3$ and $n_2 = \infty$.

$$\therefore \frac{hc}{\lambda} = 21.76 \times 10^{-19} \left[\frac{1}{(3)^2} - \frac{1}{(\infty)^2}\right]$$

$$\lambda = \frac{6.6 \times 10^{-34} \times 3 \times 10^8 \times 9}{21.76 \times 10^{-19}}$$

$$= 8.189 \times 10^{-7} \text{ m}$$

$$= 818.9 \text{ nm}.$$

Q. 10. Using the Rydberg formula, calculate the wavelengths of the first four spectral lines in the Lyman series of the hydrogen spectrum. **[NCERT]**

Ans. The Rydberg formula is

$$\frac{hc}{\lambda_{if}} = \frac{me^4}{8\varepsilon_0^2 h^2}\left(\frac{1}{n_f^2} - \frac{1}{n_i^2}\right)$$

The wavelengths of the first four lines in the Lyman series correspond to transitions from $n_i = 2, 3, 4, 5$ to $n_f = 1$. We know that

$$\frac{me^4}{8\varepsilon_0^2 h^2} = 13.6 \text{ eV} = 21.76 \times 10^{-19} \text{ J}$$

Therefore,

$$\lambda_{i1} = \frac{hc}{21.76 \times 10^{-19}\left(\frac{1}{1} - \frac{1}{n_i^2}\right)} \text{ m}$$

$$= \frac{6.625 \times 10^{-34} \times 3 \times 10^8 \times n_i^2}{21.76 \times 10^{-19} \times (n_i^2 - 1)} \text{ m}$$

$$= \frac{0.9134\, n_i^2}{(n_i^2 - 1)} \times 10^{-7} \text{ m}$$

$$= \frac{913.4\, n_i^2}{(n_i^2 - 1)} \text{ Å}$$

Substituting $n_i = 2, 3, 4, 5$, we get $\lambda_{21} = 1218$ Å, $\lambda_{31} = 1028$ Å, $\lambda_{41} = 974.3$ Å, and $\lambda_{51} = 951.4$ Å.

Q. 11. The radius of the innermost electron orbit of a hydrogen atom is 5.3×10^{-11}m. What are the radii of the $n = 2$ and $n = 3$ orbits ? **[NCERT]**

Ans. Given,

The radius of the innermost orbit of a hydrogen atom,

$$r_1 = 5.3 \times 10^{-11} \text{ m.}$$

Consider r_2 be the radius of the orbit at $n = 2$. It is related to the radius of the innermost orbit.

$$r_2 = (n)^2 r_1$$

$$= 4 \times 5.3 \times 10^{-11}$$

$$= 2.12 \times 10^{-10} \text{ m}$$

For $n = 3$,

$$r_3 = (n)^2 r_1$$

$$= 9 \times 5.3 \times 10^{-11}$$

$$= 4.77 \times 10^{-10} \text{ m}$$

Thus, the radii of an electron for $n = 2$ and $n = 3$ orbits are 2.12×10^{-10} m and 4.77×10^{-10} m respectively.

Q. 12. In accordance with the Bohr's model, find the quantum number that characterises the earth's revolution around the sun in an orbit of radius 1.5×10^{11} m with orbital speed 3×10^4 m/s. (Mass of earth = 6.0×10^{24} kg.) **[NCERT]**

Ans. Given,

Radius of the orbit of the Earth around the Sun $(r) = 1.5 \times 10^{11}$ m

Orbital speed of the Earth $(v) = 3 \times 10^{24}$ m/s

Mass of the Earth $(m) = 6.0 \times 10^{24}$ kg

From Bohr's model, angular momentum is quantised and given by

$$mvr = \frac{nh}{2\pi}$$

Here, h = Planck's constant = 6.62×10^{-34} Js

n = Quantum number

$$\therefore \quad n = \frac{mvr \times 2\pi}{h} = \frac{2\pi \times 6 \times 10^{24} \times 3 \times 10^4 \times 1.5 \times 10^{11}}{6.62 \times 10^{-34}}$$

$$= 25.61 \times 10^{73}$$

$$= 2.6 \times 10^{74}$$

Therefore, the quanta number that characterises that Earth's revolution is 2.6×10^{74}.

Q. 13. A dust particle of mass 1.0 μg is confined to move in a line between two walls separated by 0.1 mm. The speed of the particle is 10^{-6} ms^{-1}. Does this system belong to the realm of quantum physics ? Give the reasons for your answer.

Ans.
$$m = 1.0\ \mu g$$
$$= 1.0 \times 10^{-9}\ kg$$
$$l = 1 \times 10^{-4}\ m$$
$$v = 10^{-6}\ ms^{-1}$$
$$E_n = \frac{1}{2} mv^2$$
$$= \frac{1}{2} \times 10^{-9} \times (10^{-6})^2 = 5 \times 10^{-22}\ J$$
$$E_n = \frac{\eta^2 h^2}{8ml^2}$$
$$\eta = \sqrt{8mE_n}\,(l/h)$$
$$= \sqrt{8 \times 10^{-9} \times 5 \times 10^{-22}}\left(\frac{10^{-4}}{6.63 \times 10^{-34}}\right)$$
$$\approx 3 \times 10^{14}$$

This is a very large value and no experiment will be able to distinguish between n and $(n + 1)$ energy states. This problem, in fact, does not belong to the realm of quantum physics but rather to that of classical physics. It is for this reason that the motion of planets round the sun in not quantised.

Q. 14. An electron of energy 20 eV collides with to hydrogen atom in the ground state. As a result of this collision, the atom is excited a higher energy state and the electron is scattered with reduced velocity. The atom subsequently returns to its ground state with the emission of radiation of wavelength 1.216×10^{-7} m. Find the velocity of the scattered electron.

Ans. Energy lost by the electron in exciting the hydrogen atom
$$= \frac{hc}{\lambda} = \frac{6.63 \times 10^{-34} \times 3 \times 10^8}{1.216 \times 10^{-7}}$$
$$= 16.36 \times 10^{-19}\ J$$

Initial energy of the electron $= 20\ eV = 20 \times 1.6 \times 10^{-19}\ J = 32 \times 10^{-19}\ J$

∴ Kinetic energy of the scattered electron,
$$E_k = 32 \times 10^{-19} - 16.36 \times 10^{-19}$$
$$= 15.64 \times 10^{-19}\ J$$

But
$$E_k = \frac{1}{2} mv^2$$

⇒
$$v = \sqrt{\frac{2E_k}{m}} = \sqrt{\frac{2 \times 15.64 \times 10^{-19}}{9.11 \times 10^{-31}}}$$
$$= 1.86 \times 10^6\ ms^{-1}$$

Q. 15. The energy of the lowest level of hydrogen atom is – 13.6 eV. Calculate the energy, frequency and wavelength of the photon emitted by transition from $n = 2$ to $n = 4$.

Ans. We have,
$$E = \frac{-13.6}{n^2}\ eV$$
$$E_4 = \frac{-13.6}{4^2}\ eV = \frac{-13.6}{16}\ eV$$

and
$$E_2 = \frac{-13.6}{2^2} = \frac{-13.6}{4}\ eV$$

∴
$$\Delta E = E_4 - E_2$$

$$= \frac{-13.6}{16} - \left(\frac{-13.6}{4}\right)$$

$$= \frac{-13.6}{16} + \frac{13.6}{4}$$

$$= \frac{-13.6 + 13.6 \times 4}{16}$$

or $$\Delta E = \frac{40.8}{16} = 2.55 \text{ eV}$$

We know, $$\Delta E = \frac{hc}{\lambda}$$

or $$\lambda = \frac{hc}{\Delta E} = \frac{6.6 \times 10^{-34} \times 3 \times 10^{8}}{2.55 \times 1.6 \times 10^{-19}}$$

$$\lambda = \frac{12375 \times 10^{-10}}{2.55} = 4852.9 \text{ Å}$$

$$\Delta E = h\nu$$

or $$\nu = \frac{\Delta E}{h} = \frac{2.55 \times 1.6 \times 10^{-19}}{6.6 \times 10^{-34}} \text{ Hz}$$

$$\nu = 6.18 \times 10^{15} \text{ Hz.}$$

Q. 16. Calculate the de-Broglie wavelength of the electron orbitting in the $n = 2$ stage of hydrogen atom.★

Ans. The velocity of an orbitting electron,

$$v = \frac{Zc}{137n}$$

Where, c = Speed of light in vacuum.

So, $$v = \frac{1 \times 3 \times 10^{8}}{137 \times 2} = 1.09 \times 10^{6} \text{ m/s}$$

The de-Broglie wavelength, $\lambda = h/mv$

Where h is Planck's constant = 6.6×10^{-34} Js

and mass of electron = 9.1×10^{-31} kg

So, $$\lambda = \frac{6.6 \times 10^{-34}}{(9.1 \times 10^{-31} \times 1.09 \times 10^{6})}$$

$$= 6.65 \times 10^{-10} \text{ m}$$

Q. 17. In a Geiger-Marsden experiment, what is the distance of closest approach to the nucleus of a 7.7 MeV α-particle before it comes momentarily to rest and reverses its direction ? **[NCERT]**

Ans. The key idea here is that throughout the scattering process, the total mechanical energy of the system consisting of an α-particle and a gold nucleus is conserved. Let system's initial mechanical energy is E_i, before the particle and nucleus interact, and it is equal to its mechanical energy E_f when the α-particle momentarily stops. The initial energy E_i is just the kinetic energy K of the incoming α-particle. The final energy E_f is just the electric potential energy U of the system.

Let d be the centre-to-centre distance between the α-particle and the gold nucleus when the α-particle is at its stopping point. Then we can write the conservation of energy $E_i = E_f$ as

$$K = \frac{1}{4\pi\varepsilon_0} \frac{(2e)(Ze)}{d} = \frac{2Ze^2}{4\pi\varepsilon_0 d}$$

Thus, the distance of closest approach d is given by

$$d = \frac{2Ze^2}{4\pi\varepsilon_0 K}$$

The maximum kinetic energy found in α-particles of natural origin is 7.7 MeV or 1.2×10^{-12} J. Since $1/4\pi\varepsilon_0 = 9.0 \times 10^{9}$ Nm^2/C^2. Therefore ,with $e = 1.6 \times 10^{-19}$ C, we have,

★ are board based questions from previous year

$$d = \frac{(2)(9.0 \times 10^9 \text{ Nm}^2/\text{C}^2)(1.6 \times 10^{-19}\text{C})^2 \, Z}{1.2 \times 10^{-12}\text{ J}}$$

$$= 3.84 \times 10^{-16} \text{ Z m}$$

The atomic number of foil material gold is Z = 79, so that

$$d \text{ (Au)} = 3.0 \times 10^{-14} \text{ m}$$

$$= 30 \text{ fm. (1 fm (\textit{i.e.}, fermi)} = 10^{-15} \text{ m)}$$

The radius of gold nucleus is, therefore, than 3.0×10^{-14} m. This is not in very good agreement with the observed result as the actual radius of gold nucleus is 6 fm. The cause of discrepancy is that the distance of closest approach is considerably larger than the sum of the radii of the gold nucleus and the α-particle. Thus, the α-particle reverses its motion without ever actually touching the gold nucleus.

Q. 18. A 12.9 eV beam of electronic is used to bombard gaseous hydrogen at room temperature. Upto which energy level the hydrogen atoms would be excited ? Calculate the wavelength of the first member of Paschen series and first member of Balmer series.★

Ans. Energy of the electron in the nth state of an atom

$$= \frac{-13.6 \, Z^2}{n^2} \text{ eV}$$

Here, Z is the atomic number of the atom.

For hydrogen atom, Z = 1

Energy required to excite an atom from initial state (n_i) to final state (n_f),

$$E = -13.6 \left(\frac{1}{n_f^2} - \frac{1}{n_i^2}\right) \text{eV}$$

$$\Rightarrow \quad \frac{-13.6}{n_f^2} + \frac{13.6}{n_i^2} = 12.9$$

This energy must be equal to or less than the energy of the incident electron beam.

$$\Rightarrow \quad 13.6 - 12.9 = \frac{13.6}{n_f^2}$$

$$n_f^2 = \frac{13.60}{0.7} = 19.43$$

$$\Rightarrow \quad n_f = 4.4$$

State cannot be a fraction number.

$$\Rightarrow \quad n_f = 4$$

Hence, the hydrogen atom would be excited up to 4th energy level.

Rydberg's formula for the spectrum of the hydrogen atom is given by :

$$\frac{1}{\lambda} = R\left(\frac{1}{n_1^2} - \frac{1}{n_2^2}\right)$$

Here, λ is the wavelength.

Rydberg's constant, $R = 1.097 \times 10^7 \text{ m}^{-1}$

For the first member of the Paschen series

$$n_1 = 3, n_2 = 4$$

$$\frac{1}{\lambda} = 1.097 \times 10^7 \left(\frac{1}{3^2} - \frac{1}{4^2}\right)$$

$$\lambda = 18752.4 \text{ Å}$$

For the first member of Balmer series

$$n_1 = 2, n_2 = 3$$

★ are board based questions from previous year

$$\frac{1}{\lambda} = 1.097 \times 10^7 \left(\frac{1}{2^2} - \frac{1}{3^2}\right)$$

$$\lambda = 6563.3 \text{ Å}$$

Q. 19. Classically, an electron can be in any orbit around the nucleus of an atom. Then what determines the typical atomic size ? Why is an atom not, say, thousand times bigger than its typical size ? The question had greatly puzzled Bohr before he arrived at his famous model of the atom that you have learnt in the text. To simulate what he might well have done before his discovery, let us play as follows with the basic constants of nature and see if we can get a quantity with the dimensions of length that is roughly equal to the known size of an atom (~ 10^{-10} m).

(i) Construct a quantity with the dimensions of length from the fundamental constants e, m_e and c. Determine its numerical value.

(ii) You will find that the length obtained in (i) is many orders of magnitude smaller than the atomic dimensions. Further, it involves c. But energies of atoms are mostly in non-relativistic domain where c is not expected to play any role. This is what may have suggested Bohr to discard c and look for 'something else' to get the right atomic size. Now, the Planck's constant h had already made its appearance elsewhere. Bohr's great insight lay in recognising that h, m_e and e will yield the right atomic size. Construct a quantity with the dimension of length from h, m_e, and e and confirm that its numerical value has indeed the correct order of magnitude. **[NCERT]**

Ans. (i) Given,

Charge on an electron (e) = 1.6×10^{-19} C

Mass of an electron (m_e) = 9.1×10^{-31} kg

Speed of light (c) = 3×10^8 m/s

Consider a quantity involving the given quantities as $\left(\frac{e^2}{4\pi \varepsilon_0 m_e c^2}\right)$

Here, ε_0 = Permittivity of free space

And, $\frac{1}{4\pi\varepsilon_0} = 9 \times 10^9 \text{N m}^2 \text{C}^{-2}$

The numerical value of the taken quantity will be :

$$\frac{1}{4\pi\varepsilon_0} \times \frac{e^2}{m_e c^2} = 9 \times 10^9 \times \frac{(1.6 \times 10^{-19})^2}{9.1 \times 10^{-31} \times (3 \times 10^8)^2}$$

$$= 2.81 \times 10^{-15} \text{ m}$$

Therefore, the numerical value of the taken quantity is much smaller than the typical size of an atom.

(ii) Charge on an electron (e) = 1.6×10^{-19} C

Mass of an electron (m_e) = 9.1×10^{-31} kg

Planck's constant (h) = 6.63×10^{-34} Js

Consider a quantity involving the given quantities as

$$= \frac{4\pi\varepsilon_0 \left(\frac{h}{2\pi}\right)^2}{m_e e^2}$$

Here, ε_0 = Permittivity of free space

and $\frac{1}{4\pi\varepsilon_0} = 9 \times 10^9 \text{ Nm}^2\text{C}^{-2}$

The numerical value of the taken quantity will be :

$$4\pi\varepsilon_0 \times \frac{\left(\frac{h}{2\pi}\right)^2}{m_e e^2} = \frac{1}{9 \times 10^9} \times \frac{\left(\frac{6.63 \times 10^{-34}}{2 \times 3.14}\right)^2}{9.1 \times 10^{-31} \times (1.6 \times 10^{-19})^2}$$

$$= 0.53 \times 10^{-10} \text{ m}$$

Therefore, the value of the quantity taken is of the order of the atomic size.

Q. 20. A 12.5 eV electron beam is used to excite a gaseous hydrogen atom at room temperature. Determine the wavelengths and the corresponding series of the lines emitted.

Ans. Given,

$$\Delta E = 12.5 \text{ eV}$$

Let the electron jump from $n = 1$ to $n = n$ level.

$$\Delta E = E_n - E_1$$

$$\therefore \quad 12.5 = -\frac{13.6}{n^2} - \left(\frac{-13.6}{1^2}\right)$$

$$12.5 = 13.6\left(1 - \frac{1}{n^2}\right)$$

$$1 - \frac{12.5}{13.6} = \frac{1}{n^2}$$

$$\frac{1.1}{13.6} = \frac{1}{n^2}$$

$$\frac{13.6}{1.1} = n^2$$

$$12.36 = n^2$$

$$n = 3.5$$

$$\therefore \quad n = 3^{rd}$$

For $n = 3$ to $n = 1$,

$$\frac{1}{\lambda} = R\left(\frac{1}{1^2} - \frac{1}{n^2}\right)$$

$$\frac{1}{\lambda} = R\left(\frac{1}{1^2} - \frac{1}{3^2}\right)$$

$$\frac{1}{\lambda} = 1.097 \times 10^7\left(1 - \frac{1}{9}\right)$$

$$\frac{1}{\lambda} = 1.097 \times 10^7 \times \frac{8}{9}$$

$$\lambda = 102.55 \text{ nm}$$

For $n = 2$ to $n = 1$,

$$\frac{1}{\lambda} = 1.097 \times 10^7\left(\frac{1}{1^2} - \frac{1}{2^2}\right)$$

$$\frac{1}{\lambda} = 1.097 \times 10^7\left(\frac{3}{4}\right)$$

$$\lambda = 121.54 \text{ nm}$$

For $n = 3$ to $n = 2$,

$$\frac{1}{\lambda} = 1.097 \times 10^7\left(\frac{1}{2^2} - \frac{1}{3^2}\right)$$

$$\frac{1}{\lambda} = 1.097 \times 10^7\left(\frac{1}{4} - \frac{1}{9}\right)$$

$$\frac{1}{\lambda} = 1.097 \times 10^7 \times \frac{5}{36}$$

$$\lambda = 656.33 \text{ nm}$$

Hence, $\lambda = 102.5$ nm and 121.5 nm → Lyman series and $\lambda = 656.33$ nm → Balmer series

Q. 21. A 12.5 eV electron beam is used to bombard gaseous hydrogen at room temperature. Upto which energy level the hydrogen atoms would be excited ? Calculate the wavelength of the first member of Lyman and first member of Balmer series.★

★ are board based questions from previous year

Ans. Energy of the electron in the nth state of an atom

$$= \frac{-13.6}{n^2} \times Z^2 \text{ eV}$$

Here, Z is the atomic number of the atom. For hydrogen atom, Z is equal to 1. Energy required to excite an atom from the initial state (n_i) to the final state (n_f).

$$E_f = \frac{-13.6}{n_f^2} + \frac{13.6}{n_i^2} \text{ eV}$$

This energy must be equal to or less than the energy of the incident electron beam.

$$\Rightarrow \quad \frac{-13.6}{n_f^2} + \frac{13.6}{n_i^2} = 12.5$$

Energy of the electron in the ground state

$$= 13.612$$

$$= -13.6 \text{ eV}$$

$$\Rightarrow \quad \frac{-13.6}{n_f^2} + 13.6 = 12.5$$

$$\Rightarrow \quad 13.6 - 12.5 = \frac{-13.6}{n_f^2}$$

$$\Rightarrow \quad n_f^2 = \frac{13.6}{1.1}$$

$$\Rightarrow \quad n_f^2 = 12.36$$

$$\Rightarrow \quad n_f = 3.5$$

State cannot be a fraction number $n_f = 3$

Hence, hydrogen atom would be excited up to 3rd energy level.

Rydberg formula for the spectrum of the hydrogen atom is given below.

$$\frac{1}{\lambda} = R\left(\frac{1}{n_f^2} - \frac{1}{n_i^2}\right)$$

Here, λ is the wavelength and R is the Rydberg constant.

$$R = 1.097 \times 10^7 \text{ m}^{-1}$$

For the first member of the Lyman series : $n_f = 1$ and $n_i = 2$

$$\frac{1}{\lambda} = 1.097 \times 10^7 \left(\frac{1}{1^2} - \frac{1}{2^2}\right)$$

$$\lambda = 1215 \text{ (Å)}$$

For the first member of Balmer series : $n_f = 2$ and $n_i = 3$

$$\frac{1}{\lambda} = 1.097 \times 10^7 \left[\frac{1}{2^2} - \frac{1}{3^2}\right]$$

$$\Rightarrow \quad \lambda = 6563 \text{ Å}$$

Q. 22. (i) Using the Bohr's model calculate the speed of the electron in a hydrogen atom in the $n = 1, 2$ and 3 levels.

(ii) Calculate the orbital period in each of these levels. **[NCERT]**

Ans. (i) Consider v_1 be the oribtal speed of the electron in a hydrogen atom in the ground state level, $n_1 = 1$. For charge (e) of an electron, v_1 is given by

$$v_1 = \frac{e^2}{n_1 4\pi\,\varepsilon_0 \left(\frac{h}{2\pi}\right)} = \frac{e^2}{2\varepsilon_0\, h}$$

where, $e = 1.6 \times 10^{-19}$ C

ε_0 = Permittivity of free space = 8.85×10^{-12} N^{-1} C^2 m^{-2}

h = Planck's constant = 6.62×10^{-34} Js

$$\therefore \quad v_1 = \frac{(1.6 \times 10^{-19})^2}{2 \times 8.85 \times 10^{-12} \times 6.62 \times 10^{-34}}$$

$$= 0.0218 \times 10^8$$

$$= 2.18 \times 10^6 \text{ m/s}$$

For level $n_2 = 2$, we can write the relation for the corresponding orbital speed as :

$$v_2 = \frac{e^2}{2n_2\,\varepsilon_0 h}$$

$$= \frac{(1.6 \times 10^{-19})^2}{2 \times 2 \times 8.85 \times 10^{-12} \times 6.62 \times 10^{-34}}$$

$$= 1.09 \times 10^6 \text{ m/s}$$

And, for $n_3 = 3$, we can write the relation for the corresponding orbital speed as :

$$v_3 = \frac{e^2}{n_3\, 2\varepsilon_0 h}$$

$$= \frac{(1.6 \times 10^{-19})^2}{3 \times 2 \times 8.85 \times 10^{-12} \times 6.62 \times 10^{-34}} = 7.27 \times 10^5 \text{ m/s}$$

Hence, the speed of the electron in a hydrogen atom in $n = 1$, $n = 2$, and $n = 3$ is 2.18×10^6 m/s, 1.09×10^6 m/s, 7.27×10^5 m/s respectively.

(ii) Let T_1 be the orbital period of the electron when it is in level $n_1 = 1$. orbital period is related to orbital speed as follows

$$T_1 = \frac{2\pi r_1}{v_1}$$

Here, r_1 = Radius of orbit = $\dfrac{n_1^2 h^2\,\varepsilon_0}{\pi m e^2}$

h = Planck's constant = 6.62×10^{-34} Js

e = Charge on an electron = 1.6×10^{-19} C

ε_0 = Permittivity of free space = 8.85×10^{-12} N^{-1} C^2 m^{-2}

m = Mass of an electron = 9.1×10^{-31} kg

$$\therefore \quad T_1 = \frac{2\pi r_1}{v_1}$$

$$= \frac{2\pi \times (1)^2 \times (6.62 \times 10^{-34})^2 \times 8.85 \times 10^{-12}}{2.18 \times 10^6 \times \pi \times 9.1 \times 10^{-31} \times (1.6 \times 10^{-19})^2}$$

$$= 15.27 \times 10^{-17} = 1.527 \times 10^{-16} \text{ s}$$

For level $n_2 = 2$, we can write the period as :

$$T_2 = \frac{2\pi r_2}{v_2}$$

Where, r_2 = Radius of the electron in $n_2 = 2$

$$= \frac{(n_2)^2 h^2 \varepsilon_0}{\pi m e^2}$$

$$\therefore \quad T_2 = \frac{2\pi r_2}{v_2}$$

$$= \frac{2\pi \times (2)^2 \times (6.62 \times 10^{-34})^2 \times 8.85 \times 10^{-12}}{1.09 \times 10^6 \times \pi \times 9.1 \times 10^{-31} \times (1.6 \times 10^{-19})^2}$$

$$= 1.22 \times 10^{-15} \text{ s}$$

And, for level $n_3 = 3$,

$$T_3 = \frac{2\pi r_3}{v_3}$$

Here, r_3 = Radius of the electron in $n_3 = 3$

$$= \frac{(n_3)^2 h^2 \varepsilon_0}{\pi m e^2}$$

$$\therefore \quad T_3 = \frac{2\pi r_3}{v_3}$$

$$= \frac{2\pi \times (3)^2 \times (6.62 \times 10^{-34})^2 \times 8.85 \times 10^{-12}}{7.27 \times 10^5 \times \pi \times 9.1 \times 10^{-31} \times (1.6 \times 10^{-19})^2}$$

$$= 4.12 \times 10^{-15} \text{ s}$$

Thus, the orbital period in each of these levels is 1.52×10^{-16} s, 1.22×10^{-15} s, and 4.12×10^{-15} s respectively.

Q. 23. A hydrogen atom initially in the ground level absorbs a photon, which excites it to the $n = 4$ level. Determine the wavelength and frequency of the photon. **[NCERT]**

Ans. For ground level, $n_1 = 1$

Consider E_1 be the energy of this level. It is known that E_1 is related with n_1 by

$$E_1 = \frac{-13.6}{n_1^2} \text{ eV} = \frac{-13.6}{1^2} = -13.6 \text{ eV}$$

The atom is excited to a higher level, $n_2 = 4$.

Consider E_2 be the energy of this level.

$$\therefore \quad E_2 = \frac{-13.6}{n_2^2} \text{ eV}$$

$$= \frac{-13.6}{4^2} = -\frac{13.6}{16} \text{ eV}$$

The amount of energy absorbed by the photon is given by

$$E = E_2 - E_1$$

$$= \frac{-13.6}{16} - \left(-\frac{13.6}{1}\right)$$

$$= \frac{13.6 \times 15}{16} \text{ eV}$$

$$= \frac{13.6 \times 15}{16} \times 1.6 \times 10^{-19}$$

$$= 2.04 \times 10^{-18} \text{ J}$$

For a photon of wavelength 'λ' is given by

$$E = \frac{hc}{\lambda}$$

Here, h = Planck's constant = 6.6×10^{-34} Js

c = Speed of light = 3×10^8 m/s

$$\therefore \quad \lambda = \frac{hc}{E} = \frac{6.6 \times 10^{-34} \times 3 \times 10^8}{2.04 \times 10^{-18}}$$

$$= 9.7 \times 10^{-8} \text{ m} = 97 \text{ nm}$$

and frequency of a photon is given by the relation,

$$\nu = \frac{c}{\lambda} = \frac{3 \times 10^8}{9.7 \times 10^{-8}} \approx 3.1 \times 10^{15} \text{ Hz}$$

Hence, the wavelength of the photon is 97 nm while the frequency is 3.1×10^{15} Hz.

Chapter 13. Nuclei

Q. 1. A 1 MeV positron collides head on with a 1 MeV electron and they annihilated, giving rise to gamma rays. What is the total energy of resulting gamma rays ?

Ans. $$\text{Total energy} = 2\,[m_0c^2 + E_k]$$

where m_0 is the mass of the electron, m_0c^2 denotes the rest mass energy and E_k is the kinetic energy.

or
$$E_\gamma = 2 \times 0.000549 \times 931 \text{ MeV} + 2 \text{ MeV}$$
$$= 1.022 \text{ MeV} + 2 \text{ MeV}$$
$$= 3.022 \text{ MeV}$$

Q. 2. Calculate the energy in fusion reaction :★

$^2_1H + ^2_1H \longrightarrow ^3_2H + n$, where BE of $^2_1H = 2.23$ MeV and of $^3_2He = 7.73$ MeV.

Ans. $^2_1H + ^2_1H \longrightarrow ^3_2H + n$, where BE of $^2_1H = 2.23$ MeV and of $^3_2He = 7.73$ MeV.

The energy released in the fusion reaction is
$$DE = (7.73) - 2\,(2.23)$$
$$= 7.73 - 4.46$$
$$= 3.27 \text{ MeV}$$

Q. 3. A radioactive sample has 6.0×10^{18} active nuclei at a certain instant. How many of these nuclei will still be in the same active state after two half-lives ?

Ans. In one half-life, the number of active nuclei reduces to half the original number. Thus, in two half-life the number is reduced to (1/2) of (1/2) of the original number.

$$\text{Number of remaining active nuclei} = 6.0 \times 10^{18} \times \left(\frac{1}{2}\right) \times \left(\frac{1}{2}\right) = 1.5 \times 10^{18}$$

Q. 4. Calculate the energy equivalent of 1 g of substance. **[NCERT]**

Ans. Energy,
$$E = 10^{-3} \times (3 \times 10^8)^2 \text{ J}$$
$$E = 10^{-3} \times 9 \times 10^{16}$$
$$= 9 \times 10^{13} \text{ J}$$

Thus, if one gram of matter is converted to energy, there is a release of enormous amount of energy.

Q. 5. Find out the binding energy per nucleon of an α-particle in MeV. It is given that the masses of α-particle, proton and neutron are respectively 4.00150 a.m.u., 1.00728 a.m.u. and 1.00867 a.m.u.

Ans. Given that,

$$\text{The mass of an } \alpha\text{-particle, } m_\alpha = 4.00150 \text{ a.m.u.}$$
$$\text{The mass of the proton, } m_p = 1.00728 \text{ a.m.u.}$$
$$\text{The mass of the neutron, } m_n = 1.00867 \text{ a.m.u.}$$
$$\text{Mass Defect, } \Delta m = 2m_p + 2\,m_n - m_\alpha$$
$$= 2 \times 1.00728 + 2 \times 1.00867 - 4.00150$$
$$= 0.0304 \text{ a.m.u.}$$

Since, $$1 \text{ a.m.u.} = 931.5 \text{ MeV}$$

Thus, total binding energy will be $$\Delta E = 0.0304 \times 931.5 \text{ MeV} = 28.31 \text{ MeV}$$

Q. 6. Given the mass of iron nucleus as 55.85u and A = 56. Find the nuclear density ? **[NCERT]**

Ans. $m_{Fe} = 55.85$

$u = 9.27 \times 10^{-26}$ kg

$$\text{Nuclear density} = \frac{\text{Mass}}{\text{Volume}}$$
$$= \frac{9.27 \times 10^{-26}}{(4\pi/3)(1.2 \times 10^{-15})^3} \times \frac{1}{56}$$
$$= 2.29 \times 10^{17} \text{ kg m}^{-3}$$

★ are board based questions from previous year

The density of matter in neutron stars (an astrophysical object) is comparable to this density. This shows that matter in these objects has been compressed to such an extent that they resemble a big nucleus.

Q. 7. Find the energy equivalent of one atomic mass unit, first in Joules and then in MeV. Using this, express the mass defect of ${}^{16}_{8}O$ in MeV/c². **[NCERT]**

Ans. Given,

$$1u = 1.6605 \times 10^{-27} \text{ kg}$$

To convert it into energy units, we multiple it by c^2 and find that energy equivalent

$$= 1.6605 \times 10^{-27} \times (2.9979 \times 10^8)^2 \text{ kg m}^2/\text{s}^2$$
$$= 1.4924 \times 10^{-10} \text{ J}$$
$$= \frac{1.4924 \times 10^{-10}}{1.602 \times 10^{-19}} \text{ eV}$$
$$= 0.9315 \times 10^9 \text{ eV}$$
$$= 931.5 \text{ MeV}$$

or, $$1u = 931.5 \text{ MeV}/c^2$$

For $${}^{16}_{8}O, \Delta M = 0.13691 \text{ u} = 0.13691 \times 931.5 \text{ MeV}/c^2$$
$$= 127.5 \text{ MeV}/c^2$$

The energy needed to separate ${}^{16}_{8}O$ into its constituents is thus 127.5 MeV/c².

Q. 8. The half-life of ${}^{90}_{38}Sn$ is 28 years. What is the disintegration rate of 15 mg of this isotope ? **[NCERT]**

Ans. We have $$\frac{dN}{dt} = \lambda N \quad \text{...(i)}$$

Disintegration constant $$\lambda = \frac{0.6931}{T} \quad \text{...(ii)}$$

Here T = 28 years = 28 × 3.154 × 10⁷ seconds

90 g of ^{90}Sr contain 6.023 × 10²³ atoms

∴ Number of Sr-90 atoms in 15 mg (= 15 × 10⁻³ g)

$$N = \frac{15 \times 10^{-3}}{90} \times 6.023 \times 10^{23} = 1.00 \times 10^{20}$$

∴ Disintegration rate, $$\frac{dN}{dt} = \left(\frac{0.6931}{28 \times 3.154 \times 10^7}\right) \times 1.00 \times 10^{20}$$
$$= 7.85 \times 10^{10} \text{ Bq}$$

Q. 9. A 1000 MW fission reactor consumes half of its fuel in 5 years. How much ${}^{235}_{92}U$ did it contain initially ? Assume that the reactor operates 80% of the time, that all the energy generated arises from the fission of ${}^{23}_{92}U$ and that this nuclide is consumed only by the fission process. **[NCERT]**

Ans. Half-life of the fuel of the fission reactor, $T_{\frac{1}{2}}$ = 5 years

$$= 5 \times 365 \times 24 \times 60 \times 60 \text{ s}$$

We know that in the fission of 1 g of ${}^{235}_{92}U$ nucleus, the energy released is equal to 200 MeV.

1 mole, *i.e.*, 235 g of ${}^{235}_{92}U$ contains 6.023 × 10²³ atoms.

∴ $$1\text{g } {}^{235}_{92}U \text{ contains} = \frac{6.023 \times 10^{23}}{235} \text{ atoms}$$

The total energy generated per gram of ${}^{235}_{92}U$ is calculated as :

$$E = \frac{6.023 \times 10^{23}}{235} \times 200 \text{ MeV/g}$$
$$= \frac{200 \times 6.023 \times 10^{23} \times 10^{-19} \times 10^6}{235}$$
$$= 820 \times 10^{19} \text{ J/g}$$

The reactor operates only 80% of the time.

Hence, the amount of ${}^{235}_{92}U$ consumed in 5 years by the 1000 MW fission reactor is calculated by

$$= \frac{5 \times 80 \times 60 \times 365 \times 24 \times 1000 \times 10^6}{100 \times 8.20 \times 10^{19}} \text{ g}$$

$$= 1538 \text{ kg}$$

$\therefore$ Initial amount of ${}^{235}_{92}U$ = 2 × 1538 = 3076 kg

Q. 10. Obtain approximately the ratio of the nuclear radii of the gold isotope ${}^{197}_{79}Au$ and the silver isotope ${}^{107}_{47}Ag$. **[NCERT]**

Ans. **Given,**

Nuclear radius of the gold isotope

$${}_{79}Au^{197} = R_{Au}$$

Nuclear radius of the silver isotope ${}_{47}Ag^{107} = R_{Ag}$

Mass number of gold (A_{Au}) = 197

Mass number of silver (A_{Ag}) = 107

The ratio of the radii of the two nuclei is related with their mass numbers by

$$\frac{R_{Au}}{R_{Ag}} = \left(\frac{A_{Au}}{A_{Ag}}\right)^{\frac{1}{2}}$$

$$= \left(\frac{197}{107}\right)^{\frac{1}{2}}$$

$$= 1.2256$$

Therefore, the ratio of the nuclear radii of the gold and silver isotopes is about 1.23.

Q. 11. A radioactive sample has a half-life of 1.42×10^{17} second and a mass number of 238. Calculate the number of disintegrations per second in 1g of this sample. Avogadro's number is 6.023×10^{23} mol^{-1}.

Ans. The number of atoms in 1 g of the sample is,

$$N = \frac{6.023 \times 10^{23}}{238} = 2.53 \times 10^{21}$$

The half-life T of the radioactive sample is 1.42×10^{17} s.

$$\therefore \quad \text{Decay constant, } \lambda = \frac{0.6931}{T} = \frac{0.6931}{1.42 \times 10^{17}}$$

$$= 0.488 \times 10^{-17} \text{ s}^{-1}$$

The number of disintegrations per second in 1 g of the sample is

$$\frac{-dN}{dt} = \lambda N$$

$$= 0.488 \times 10^{-17} \times 2.53 \times 10^{21}$$

$$= 1.23 \times 10^4 \text{ s}^{-1}$$

Q. 12. In a sphere of 10^2 m radius, radioactive material emits β^- particles at the rate of 5×10^7 s^{-1}. If 40% of these emitted β^--particles escape from the sphere, how long would it take to raise the potential of the sphere from 0 to 16 V ? (Take $k = 9 \times 10^9$ S.I. unit)

Ans. Out of emitted β^--particles the number of β^-particle leaving the sphere in one second

$$n = (0.4)(5 \times 10^7)$$

$$= (2 \times 10^7)\text{ S}^{-1}$$

$\therefore$ The number of β^--particles leaving the sphere in t seconds = $n \times t$

$\therefore$ In t seconds, the charge acquired by the sphere

$$Q = n \times t \times e \ldots \text{(positive)}$$

$$V = \frac{kQ}{R} = \frac{k(n \times t \times e)}{R}$$

$$\therefore \quad 16 = 9 \times 10^9 \frac{(2 \times 10^7)(t)(1.6 \times 10^{-19})}{10^2}$$

$$\therefore \quad t = \frac{16 \times 10^2}{9 \times 2 \times 1.6 \times 10^{-3}}$$

$$= 55578 \text{ s} = 15.438 \text{ hr.}$$

Q. 13. Suppose, we think of fission of a ${}^{56}_{26}Fe$ nucleus into two equal fragments, ${}^{28}_{13}Al$. Is the fission energetically possible ? Argue by working out Q of the process. Given :

$$m\left({}^{56}_{26}Fe\right) = 55.93494 \text{ u and } m\left({}^{28}_{13}Al\right) = 27.981914$$ **[NCERT]**

Ans. The fission of ${}^{56}_{26}Fe$ can be given by

$${}^{56}_{26}Fe \longrightarrow 2\,{}^{28}_{13}Al$$

From data given in question

$$\text{Atomic mass of } m\left({}^{56}_{26}Fe\right) = 55.93494 \text{ u}$$

$$\text{Atomic mass of } m\left({}^{28}_{13}Al\right) = 27.98191 \text{ u}$$

The Q-value of this nuclear reaction is given by

$$Q = \left[m\left({}^{56}_{26}Fe\right) - 2m\left({}^{28}_{13}Al\right)\right] c^2$$

$$= [55.93494 - 2 \times 27.98191]\, c^2 = (-0.02888\, c^2) \text{ u}$$

But $\quad 1\text{u} = 931.5 \text{ MeV}/c^2$

$$\therefore \quad Q = -0.02888 \times 931.5 = 26.902 \text{ MeV}$$

The Q-value of the fission is negative. Therefore, the fission is not possible energetically. For an energetically-possible fission reaction, the Q-value must be positive.

Q. 14. The three stable isotopes of neon : ${}^{20}_{10}Ne$, ${}^{21}_{10}Ne$ and ${}^{22}_{10}Ne$ have respective abundances of 90.51%, 0.27% and 9.22%. The atomic masses of the three isotopes are 19.99 u, 20.99 u and 21.99 u, respectively. Obtain the average atomic mass of neon. **[NCERT]**

Ans.

$$\text{Atomic mass of } {}^{20}_{10}Ne,\ m_1 = 19.99 \text{ u}$$

$$\text{Abundance of } {}^{20}_{10}Ne,\ \eta_1 = 90.51\%$$

$$\text{Atomic mass of } {}^{21}_{10}Ne,\ m_2 = 20.99 \text{ u}$$

$$\text{Abundance of } {}^{21}_{10}Ne,\ \eta_2 = 0.27\%$$

$$\text{Atomic mass of } {}^{22}_{10}Ne,\ m_3 = 21.99 \text{ u}$$

$$\text{Abundance of } {}^{22}_{10}Ne,\ \eta_3 = 9.22\%$$

The average atomic mass of neon is

$$m = \frac{m_1\eta_1 + m_2\eta_2 + m_3\eta_3}{\eta_1 + \eta_2 + \eta_3}$$

$$= \frac{19.99 \times 90.51 + 20.99 \times 0.27 + 21.99 \times 9.22}{90.51 + 0.27 + 9.22}$$

$$= 20.1771 \text{ u}$$

Q. 15. The half-life of ${}^{238}_{92}U$ undergoing α-decay is 4.5×10^9 years. What is the activity of 1 g sample of ${}^{238}_{92}U$? **[NCERT]**

Ans. $$T_{\frac{1}{2}} = 4.5 \times 10^9 \text{ y}$$
$$= 4.5 \times 10^9 \text{ y} \times 3.16 \times 10^7 \text{ s/y}$$
$$= 1.42 \times 10^{17} \text{ s}$$

One kmol of any isotope contains Avogadro's number of atoms and so 1g of ${}^{238}_{92}U$ contains

$$= \frac{1}{238 \times 10^{-3}} \text{kmol} \times 6.025 \times 10^{26} \text{ atoms/kmol}$$
$$= 25.3 \times 10^{20} \text{ atoms.}$$

The decay rate R is $$R = \lambda N$$
$$= \frac{0.693}{T_{1/2}} N$$
$$= \frac{0.693 \times 25.3 \times 10^{20}}{1.42 \times 10^{17}} \text{ s}^{-1}$$
$$= 1.23 \times 10^4 \text{ s}^{-1}$$
$$= 1.23 \times 10^4 \text{ Bq}$$

Q. 16. The half-life of a radioactive substance is 4 hours. In how much time will 7/8th of the material decay ?

Ans. Given,

T = 4 hours,

We known that, $$T = \frac{0.693}{\lambda}$$

$$\therefore \quad \lambda = \frac{0.693}{4}$$
$$= 0.17325 \text{ per hour}$$

$\left(\frac{7}{8}\right)^{th}$ of the material decays. So, the remaining material

$$N = N_0 - \left(\frac{7}{8}\right) N_0 = \frac{N_0}{8}$$

$$\therefore \quad \frac{N_0}{N} = 8$$

$$\text{Time taken, } t = \frac{1}{\lambda} \log_e \frac{N_0}{N}$$
$$= \frac{1}{0.17325} \times \log_e 8$$
$$= 12.002 \text{ hours.}$$

Q. 17. The atomic mass of ${}_8O^{16}$ is 16.000 amu. Find out its binding energy per nucleon. The mass of electron = 0.00055 amu, mass of proton = 1.007593 amu and mass of neutron = 1.008982 amu and 1 amu = 931.5 MeV.

Ans. ${}_8O^{16}$ has 8 protons, 8 neutrons and 8 electrons.

∴ Mass of 8 (electron + protons + neutrons)

$$= 8\,[0.00055 + 1.007593 + 1.008982] \text{ amu}$$
$$= 16.1370 \text{ amu}$$

$$\therefore \quad \text{Mass defect, } \Delta m = 16.1370 \text{ amu} - 16.0000 \text{ amu}$$
$$= 0.1370 \text{ amu}$$

$$\text{Binding energy} = 0.1370 \times 931.5 \text{ MeV} = 127.615 \text{ MeV}$$

$$\therefore \quad \text{Binding energy/nucleon} = \frac{127.615}{16} = 7.97 \text{ MeV}$$

Q. 18. In a sample of radioactive element, after $\frac{1}{\lambda}$ time (where λ = decay constant) :

(i) What percent of initial amount remains undisintegrated ?

(ii) What percent is disintegrated ?

Ans. (i) Let number of undisintegrated nuclei is N at time and it is given by

$$N = N_0 e^{-\lambda t} = N_0 e^{-\lambda\left(\frac{1}{\lambda}\right)} \qquad \left(\because t = \frac{1}{\lambda}\right)$$

$$= N_0 e^{-1} = \frac{N_0}{e}$$

At this time, the fraction of initial amount remaining undisintegrated will be

$$\frac{N}{N_0} = \frac{1}{e} = \frac{1}{2.718} = 0.368$$

$\therefore$ The percentage remaining undisintegrated is

$$= \frac{N}{N_0} \times 100 = 0.368 \times 100 = 36.8\%$$

(ii) At this time if number of nuclei disintegrated is N', Then

$$N' = N_0 - N$$

$$= N_0 - \frac{N_0}{e} = N_0\left(\frac{e-1}{e}\right)$$

$$= N_0\left(\frac{1718}{2718}\right) = N_0\,(0.632)$$

$\therefore$ At this time, fraction of initial amount disintegrated

$$= \frac{N'}{N_0} = 0.632$$

$\therefore$ At this time, percentage disintegrated

$$= \frac{N'}{N_0} \times 100 = 63.2\%$$

Q. 19. Suppose India had a target of producing by 2020 AD, 200,000 MW of electric power, ten percent of which was to be obtained from nuclear power plants. Suppose we are given that, on an average, the efficiency of utilisation (*i.e.*, conversion to electric energy) of thermal energy produced in a reactor was 25%. How much amount of fissionable uranium would our country need per year by 2020 ? Take the heat energy per fission of ^{235}U to be about 200 MeV. **[NCERT]**

Ans. Given,

Amount of electric power to be generated,

$$P = 2 \times 10^5 \text{ MW}$$

10% of this amount has to be obtained from nuclear power plants.

$$P = \frac{10}{100} \times 2 \times 10^5$$

$\therefore$ Amount of nuclear power,

$$= 2 \times 10^4 \text{ MW}$$

$$= 2 \times 10^4 \times 10^6 \text{ J/s}$$

$$= 2 \times 10^{10} \times 60 \times 60 \times 24 \times 365 \text{ J/y}$$

Heat energy released per fission of a ^{235}U nucleus, E = 200 MeV

Efficiency of a reactor = 25%

Hence, the amount of energy converted into the electrical energy per fission is calculated by

$$\frac{25}{100} \times 200 = 50 \text{ MeV}$$

$$= 50 \times 1.6 \times 10^{-19} \times 10^6 = 8 \times 10^{-12} \text{ J}$$

Number of atoms required for fission per year :

$$\frac{2 \times 10^{10} \times 60 \times 60 \times 24 \times 365}{8 \times 10^{-12}} = 78840 \times 10^{24} \text{ atoms}$$

1 mole, *i.e.*, 235 g of U^{235} contains 6.023×10^{23} atoms.

$\therefore$ Mass of 6.023×10^{23} atoms of U^{235} $= 235 \text{ g} = 235 \times 10^{-3}$ kg

$\therefore$ Mass of 78840×10^{24} atoms of U^{235} $= \dfrac{235 \times 10^{-3}}{6.023 \times 10^{23}} \times 78840 \times 10^{24}$

$= 3.076 \times 10^{4}$ kg

Thus, the mass of uranium needed per year is 3.076×10^4 kg.

Q. 20. An electron and its antiparticle are annihilated after collision. How much energy will be obtained ? What will be the wavelength of the emitted radiation (energy) ? (Mass of electron = 9.1×10^{-31} kg, speed of light = 3.0×10^8 ms^{-1} and Planck's constant = 6.6×10^{-34} Js)

Ans. The positron is antiparticle of electron. In electron-positron collision, both are annihilated and the total mass is converted into two γ-photons (energy).

$$e^- + e^+ \rightarrow 2\gamma$$

The mass-loss in this process is $\Delta m = 2 \times (9.1 \times 10^{-31}) = 18.2 \times 10^{-31}$ kg.

According to mass-energy relation, the energy released is,

$$\Delta E = \Delta mc^2$$

$$= 18.2 \times 10^{-31} \times (3 \times 10^8)^2$$

$$= 1.64 \times 10^{-13} \text{ joule}$$

$$= \frac{1.64 \times 10^{-13}}{1.6 \times 10^{-19}} \text{ eV} = 1.025 \times 10^6 \text{ eV}.$$

Two γ-photons are produced in this process, hence energy of each photon is,

$$E = \frac{\Delta E}{2} = \frac{1.64 \times 10^{-13}}{2}$$

$$= 0.82 \times 10^{-13} \text{ joule.}$$

The wavelength corresponding to it is,

$$\lambda = \frac{hc}{E} = \frac{6.6 \times 10^{-34} \times 3 \times 10^8}{0.82 \times 10^{-13}}$$

$$= 2.4 \times 10^{-12} \text{ m}$$

$$= 0.024 \text{ Å}$$

Q. 21. How long can an electric lamp of 100 W be kept glowing by fusion of 2.0 kg of deuterium ? Take the fusion reaction as

$${}_1^2H + {}_1^2H \longrightarrow {}_2^3He + n + 3.27 \text{ MeV}$$ **[NCERT]**

Ans. The given fusion reaction is :

$${}_1^2H + {}_1^2H \longrightarrow {}_2^3He + n + 3.27 \text{ MeV}$$

Amount of deuterium, $m = 2$ kg

1 mole, *i.e.*, 2 g of deuterium contains 6.023×10^{23} atoms.

$\therefore$ 2.0 kg of deuterium contains $= \dfrac{6.023 \times 10^{23}}{2} \times 2000$

$= 6.023 \times 10^{26}$ atoms

When two atoms of deuterium fuse, 3.27 MeV energy is released.

$\therefore$ Total energy per nucleus released in the fusion reaction :

$$E = \frac{3.27}{2} \times 6.023 \times 10^{26} \text{ MeV}$$

$$= \frac{3.27}{2} \times 6.023 \times 10^{26} \times 1.6 \times 10^{-19} \times 10^{6}$$

$$= 1.576 \times 10^{14} \text{ J}$$

Power of the electric lamp, P = 100 W = 100 J/s

Hence, the energy consumed by the lamp per second = 100 J

The total time for which the electric lamp will glow is calculated by

$$= \frac{1.576 \times 10^{14}}{100} \text{ s}$$

$$= \frac{1.576 \times 10^{14}}{100 \times 60 \times 60 \times 24 \times 365} \approx 4.9 \times 10^{4} \text{ years.}$$

Q. 22. The fission properties of $^{239}_{94}Pu$ are very similar to those of $^{235}_{92}U$.

The average energy released per fission is 180 MeV. How much energy, in MeV, is released. If all the atoms in 1 kg of pure $^{239}_{94}Pu$ undergo fission ? **[NCERT]**

Ans. Given,

Average energy released per fission of $^{239}_{94}Pu$, E_{av}= 180 MeV.

Amount of pure $_{94}Pu^{239}$ (m) = 1 kg = 1000 g

Avogadro number (N_A) = 6.023×10^{23}

Mass number of $^{239}_{94}Pu$ = 239 g

1 mole of $_{94}Pu^{239}$ contains N_A atoms.

$\therefore$ Mg of $_{94}Pu^{239}$ contains $\left(\frac{N_A}{\text{Mass number}} \times m\right)$ atom

$$= \frac{6.023 \times 10^{23}}{239} \times 1000$$

$$= 2.52 \times 10^{24} \text{ atoms}$$

$\therefore$ Total energy released during the fission of 1 kg of $^{239}_{94}Pu$ is given by

$$E = E_{av} \times 2.52 \times 10^{24}$$

$$= 180 \times 2.52 \times 10^{24}$$

$$= 4.536 \times 10^{26} \text{ MeV}$$

Therefore 4.536×10^{26} MeV is released if all the atoms in 1 kg of pure $_{94}Pu^{239}$ undergo fission.

Q. 23. (i) Find the binding energy per nucleon for $^{56}_{26}Fe$ nucleus from the data given below.

(ii) If the proton with least binding in this nucleus is emitted. $^{55}_{25}Mn$ nucleus is formed. Find the binding energy of this proton. [Mass of proton m_p = 1.007825 u, mass of neutron. m_n = 1.008665 u, mass of Fe nucleus M_{Fe} = 55.934939 u, mass of Mn nucleus m = 54.938046 u, 1u = 931.494 MeV.] **[NCERT]**

Ans. (i) $^{56}_{26}Fe$ nucleus has 26 protons and 30 neutrons. When the protons and neutrons are in free state their total mass

$$= Zm_p + Nm_n = 26\, m_p + 30 m_n$$

$\therefore$ Mass defect, $\Delta m = (Zm_p + Nm_n) - (M_{Fe})$

$$= (26 \times 1.007825 + 30 \times 1.008665) - (55.934939)$$

$$= 0.528461 \text{ u}$$

$\therefore$ Binding energy E_b = energy equivalent to Δm

$$= 0.528461 \times 931.494$$

$$= 492.258 \text{ MeV}$$

$\therefore$ Binding energy per nucleon $E_{bn} = \frac{E_p}{A} = \frac{492.258}{56}$

$\therefore$ $E_{bn} = 8.79$ MeV / Nucleon

(ii) If the proton is separated from $^{56}_{25}Fe$, the following reaction takes place

$$^{56}_{26}Fe \longrightarrow {}^{55}_{25}Mn + {}^{1}_{1}p$$

Total mass of M_n and p = 54.938046 + 1.007825
= 55.945871 u

and Mass of Fe is M_{Fe} = 55.934949 u

Thus in this reaction mass is increased. It shows that this reaction cannot occur spontaneously (naturally by itself) but if energy is suplied from outside, then only it can occur

$\therefore$ Proton's binding energy = Energy to be supplied
= (energy equivalent to increase in mass)
= (55.945871 – 55.934939) (931.494) MeV
= 0.010932 × 931.494
= 10.18 MeV

The energy to be supplied to separate a nucleon from a nucleus is called separation energy.

Q. 24. Under certain circumstances, a nucleus can decay be emitting a particle more massive than an α-particle. Consider the following decay processes :

$$^{223}_{88}Ra \longrightarrow {}^{209}_{82}Pb + {}^{14}_{6}C$$

$$^{223}_{88}Ra \longrightarrow {}^{219}_{86}Rn + {}^{4}_{2}He$$

Calculate the Q-values for these decays and determine that both are energetically allowed.

[NCERT]

Ans. Take a $^{14}_{6}C$ emission nuclear reaction :

$$^{223}_{88}Ra \longrightarrow {}^{209}_{82}Pb + {}^{14}_{6}C$$

Given,

Mass of $^{223}_{88}Ra$, m_1 = 223.01850 u

Mass of $^{209}_{82}Pb$, m_2 = 208.98107 u

Mass of $^{14}_{6}C$, m_3 = 14.00324 u

Hence, the Q-value of the reaction is given as :

$Q = (m_1 - m_2 - m_3)\, c^2$
= $(223.01850 - 208.98107 - 14.00324)\, c^2$
= $(0.03419\, c^2)$ u

But 1u = 931.5 MeV / c^2

$\therefore$ Q = 0.03419 × 931.5
= 31.848 MeV

Hence, the Q-value of the nuclear reaction is 31.848 MeV. Since the value is positive, the reaction is energetically allowed.

Now take a $^{4}_{2}He$ emission nuclear reaction :

$$^{223}_{88}Ra \longrightarrow {}^{219}_{86}Rn + {}^{4}_{2}He$$

We know that :

Mass of $^{223}_{88}Ra$, m_1 = 223.01850

Mass of $^{219}_{86}Rn$, m_2 = 219.00948

Mass of $^{4}_{2}He$, m_3 = 4.00260

Q-value of this nuclear reaction is given as :

$$\begin{aligned} Q &= (m_1 - m_2 - m_3)\,c^2 \\ &= (223.01850 - 219.00948 - 4.00260)\,c^2 \\ &= (0.00642\,c^2)\,u \\ &= 0.00642 \times 931.5 = 5.98 \text{ MeV} \end{aligned}$$

Thus, the Q value of the second nuclear reaction is 5.98 MeV. Since the value is positive, the reaction is energetically allowed.

Q. 25. A given coin has a mass of 3.0 g. Calculate the nuclear energy that would be required to separate all the neutrons and protons from each other. For simplicity assume that the coin is entirely made of ${}^{63}_{29}Cu$ atoms (of mass 62.92960 u). **[NCERT]**

Ans.

$$\begin{aligned} \text{Mass of a copper coin } (m') &= 3g \\ \text{Atomic mass of } {}_{29}Cu^{63} \text{ atom } (m) &= 62.92960 \text{ u} \end{aligned}$$

The total number of ${}_{29}Cu^{63}$ atoms in the coin (N)

$$= \frac{N_A \times m'}{\text{Mass number}}$$

$$\begin{aligned} N_A &= \text{Avogadro's number} \\ &= 6.023 \times 10^{23} \text{ atoms/g} \\ \text{Mass number} &= 63 \text{ g} \end{aligned}$$

$$\therefore \quad N = \frac{6.023 \times 10^{23} \times 3}{63}$$

$$= 2.868 \times 10^{22} \text{ atoms}$$

${}_{29}Cu^{63}$ nucleus has 29 protons and (63 – 29)

$$= 34 \text{ neutrons}$$

$$\therefore \quad \text{Mass defect of this nucleus, } \Delta m' = 29 \times m_p + 34 \times m_n - m$$

Where,

$$\begin{aligned} \text{Mass of a proton, } (m_p) &= 1.007825 \text{ u} \\ \text{Mass of neutron, } (m_n) &= 1.008665 \text{ u} \end{aligned}$$

$$\therefore \quad \Delta m' = 29 \times 1.007825 + 34 \times 1.008665 - 62.9296$$

$$= 0.591935 \text{ u}$$

Mass defect of all the atoms present in the coin (Δm)

$$\begin{aligned} &= 0.591935 \times 2.868 \times 10^{22} \\ &= 1.69766958 \times 10^{22} \text{ u} \end{aligned}$$

But

$$1u = 931.5 \text{ MeV}/c^2$$

$$\therefore \quad \Delta m = 1.69766958 \times 10^{22} \times 931.5 \text{ MeV}/c^2$$

Therefore, the binding energy of the nuclei of the coin is given by

$$\begin{aligned} E_b &= \Delta m c^2 \\ &= 1.69766958 \times 10^{22} \times 931.5 \left(\frac{\text{MeV}}{c^2}\right) \times c^2 \\ &= 1.581 \times 10^{25} \text{ MeV} \end{aligned}$$

But

$$\begin{aligned} 1 \text{ MeV} &= 1.6 \times 10^{-13} \text{ J} \\ E_b &= 1.581 \times 10^{25} \times 1.6 \times 10^{-13} \\ &= 2.5296 \times 10^{12} \text{ J} \end{aligned}$$

This much energy is required to separate all the neutrons and protons from the given coin.

Q. 26. Obtain the binding energy of the nuclei ${}^{56}_{26}Fe$ and ${}^{209}_{83}Bi$ in units of MeV from the following data :

$$m\left({}^{56}_{26}Fe\right) = 55.934939\,u;\ m\left({}^{209}_{83}Bi\right) = 208.980388\,u$$ **[NCERT]**

Ans. From question : Atomic mass of ${}^{56}_{26}Fe$, $m_1 = 55.934939$ u

${}^{56}_{26}Fe$ nucleus has 26 protons and (56 – 26) = 30 neutrons

Hence, the mass defect of the nucleus,

$$\Delta m = 26 \times m_H + 30 \times m_n - m_1$$

Where, Mass of proton, $m_H = 1.007825$ u

Mass of neutron, $m_n = 1.008665$ u

$$\therefore \quad \Delta m = 26 \times 1.007825 + 30 \times 1.008665 - 55.934939$$
$$= 26.20345 + 30.25995 - 55.934939$$
$$= 0.528461 \text{ u}$$

But $1\text{u} = 931.5 \text{ MeV}/c^2$

$$\therefore \quad \Delta m = 0.528461 \times 931.5 \text{ MeV}/c^2$$

The binding energy of this nucleus is

$$E_{b1} = \Delta m c^2$$

Where, c = Speed of light

$$\therefore \quad E_{b1} = 0.528461 \times 931.5 \left(\frac{\text{MeV}}{c^2}\right) \times c^2$$
$$= 492.26 \text{ MeV}$$

$$\text{Average binding energy per nucleon} = \frac{492.26}{56}$$
$$= 8.79 \text{ MeV}$$

Atomic mass of ${}^{209}_{83}\text{Bi}$, $m_2 = 208.980388$ u

${}^{209}_{83}\text{Bi}$ nucleus has 83 protons and (209 – 83) = 126 neutrons.

Hence, the mass defect of this nucleus is given as

$$\Delta m' = 83 \times m_H + 126 \times m_n - m_2$$

Here, Mass of proton, $m_H = 1.007825$ u

Mass of a neutron, $m_n = 1.008665$ u

$$\therefore \quad \Delta m' = 83 \times 1.007825 + 126 + 1.008665 - 208.980388$$
$$= 83.649475 + 127.091790 - 208.980388$$
$$= 1.760877 \text{ u}$$

But $1u = 931.5 \text{ MeV}/c^2$

$$\therefore \quad \Delta m' = 1.760877 \times 931.5 \text{ MeV}/c^2$$

Thus the binding energy of this nucleus is given by

$$E_{b_2} = \Delta m' c^2$$
$$= 1.760877 \times 931.5 \left(\frac{\text{MeV}}{c^2}\right) \times c^2$$
$$= 1640.26 \text{ MeV}$$

$$\text{Average binding energy per nucleon} = \frac{1640.26}{209} = 7.848 \text{ MeV}$$

Q. 27. Consider the D–T reaction (deuterium-tritium fusion)

$${}^{2}_{1}\text{H} + {}^{3}_{1}\text{H} \longrightarrow {}^{4}_{2}\text{He} + n$$

(i) Calculate the energy released in MeV in this reaction from the data :

$$m\left({}^{2}_{1}\text{H}\right) = 2.014102 \text{ u}$$
$$m\left({}^{3}_{1}\text{H}\right) = 3.016049 \text{ u}$$

(ii) Consider the radius of both deuterium and tritium to be approximately 2.0 fm. What is the kinetic energy needed to overcome the coulomb repulsion between the two nuclei ? To what temperature must the gas be heated to initiate the reaction ?

(**Hint** : Kinetic energy required for one fusion event = average thermal kinetic energy available with the interacting particles = $2(3kT/2)$; k = Boltzman's constant, T = absolute temperature.).

[NCERT]

Ans. (i) Take the D-T nuclear reaction :

$$^2_1H + ^3_1H \longrightarrow ^4_2He + n$$

Data given in question is as follows

$$\text{Mass of } ^2_1H,\ m_1 = 2.014102 \text{ u}$$
$$\text{Mass of } ^3_1H,\ m_2 = 3.016049 \text{ u}$$
$$\text{Mass of } ^4_2He,\ m_3 = 4.002603 \text{ u}$$
$$\text{Mass of } ^1_0n,\ m_4 = 1.008665 \text{ u}$$

Q-value of the given reaction is :

$$Q = [m_1 + m_2 - m_3 - m_4]\ c^2$$
$$= [2.014102 + 3.016049 - 4.002603 = 1.008665]\ c^2$$
$$= [0.01883\ c^2]\ u$$

But $1u = 931.5 \text{ MeV}/c^2$

$\therefore$ $Q = 0.018883 \times 931.5 = 17.59 \text{ MeV}$

(ii) Radius of deuterium and tritium,

$$r \approx 2.0\ fm = 2 \times 10^{-15} \text{ m}$$

Distance between the two nuclei at the moment when they touch each other,

$$d = A + A = 4 \times 10^{-15} \text{ m}$$

Charge on the deuterium nucleus = e

Charge on the tritium nucleus = e

Hence, the repulsive potential energy between the two nuclei is given by

$$V = \frac{e^2}{4\pi\varepsilon_0\ (d)}$$

Here, ε_0 = Permittivity of free space

$$\frac{1}{4\pi\varepsilon_0} = 9 \times 10^9 \text{ Nm}^2\text{ C}^{-2}$$

$\therefore$
$$V = \frac{9 \times 10^9 \times (1.6 \times 10^{-19})^2}{4 \times 10^{-15}} = 5.76 \times 10^{-14} \text{ J}$$
$$= \frac{5.16 \times 10^{-14}}{1.6 \times 10^{-19}} = 3.6 \times 10^5 \text{ eV} = 360 \text{ keV}$$

Therefore, 5.76×10^{-14} J or 360 keV of kinetic energy (KE) is needed to overcome the Coulomb repulsion between the two nuclei.

$$\text{K.E.} = 2 \times \frac{3}{2}\text{kT}$$

Here, k = Boltzmann constant

$= 1.38 \times 10^{-23} \text{ m}^2 \text{ kg s}^{-2} \text{ k}^{-1}$

T = Temperature required for triggering the reaction

$\therefore$
$$T = \frac{KE}{3K}$$
$$= \frac{5.76 \times 10^{-14}}{3 \times 1.34 \times 10^{-23}}$$
$$= 1.39 \times 10^9 \text{ K}$$

Thus, the gas must be heated to a temperature of 1.39×10^9 K to initiate the reaction.

Q. 28. A source contains two phosphorous radio nuclide $^{32}_{15}P$ ($T_{1/2} = 14.3\ d$) and $^{33}_{15}P$ ($T_{1/2} = 25.3\ d$). Initially, 10% of the decays come from $^{33}_{15}P$. How long one must wait until 90% do so ?

Ans. Given,

$$\text{Half-life of } ^{32}_{15}P,\ T_{1/2} = 14.3 \text{ days}$$
$$\text{Half-life of } ^{33}_{15}P,\ T_{1/2} = 25.3 \text{ days}$$

$^{33}_{15}P$ nucleus decay is 10% of the total amount of decay.

The source has initially 10% of $^{33}_{15}P$ nucleus and 90% of $^{32}_{15}P$ nucleus.

Consider after *t*-days, the source has 10% of $^{32}_{15}P$ nucleus and 90% of $^{33}_{15}P$ nucleus.

Initially :

$$\text{Number of } ^{33}_{15}P \text{ nucleus} = N$$
$$\text{Number of } ^{32}_{15}P \text{ nucleus} = 9\,N$$

Finally :

$$\text{Number of } ^{33}_{15}P \text{ nucleus} = 9N'$$
$$\text{Number of } ^{32}_{15}P \text{ nucleus} = N'$$

For $^{32}_{15}P$ nucleus, we can write the number ratio as :

$$\frac{N'}{9N} = \left(\frac{1}{2}\right)^{\frac{t}{T_{1/2}}}$$
$$N' = 9N\,(2)^{\frac{-t}{14.3}} \qquad \text{...(i)}$$

For $^{33}_{15}P$, we can write the number ratio as :

$$\frac{9N'}{N} = \left(\frac{1}{2}\right)^{\frac{t}{T_{1/2}}}$$
$$9N' = N\,(2)^{\frac{-t}{25.3}} \qquad \text{...(ii)}$$

On dividing equation (i) by equation (ii), we get :

$$\frac{1}{9} = 9 \times 2^{\left(\frac{t}{25.3} - \frac{t}{14.3}\right)}$$
$$\frac{1}{81} = 2^{\left(\frac{11t}{25.3 \times 14.3}\right)}$$
$$\log 1 - \log 81 = \frac{-11t}{25.3 \times 14.3} \log 2$$
$$\frac{-11t}{25.3 \times 14.3} = \frac{0 - 1.908}{0.301}$$
$$t = \frac{25.3 \times 14.3 \times 1.908}{11 \times 0.301} = 208.5 \text{ days.}$$

Thus, it will take about 208.5 days for 90% decay of $_{15}P^{33}$.

Q. 29. The neutron separation energy is defined as the energy required to remove a neutron from the nucleus. Obtain the neutron separation energies of the nuclei $^{41}_{20}Ca$ and $^{27}_{13}Al$ from the following data :

$$m\left(^{40}_{20}Ca\right) = 39.962591 \text{ u}$$
$$m\left(^{41}_{20}Ca\right) = 40.962278 \text{ u}$$
$$m\left(^{26}_{13}Al\right) = 25.986895 \text{ u}$$
$$m\left(^{27}_{13}Al\right) = 26.981541 \text{ u}$$

[NCERT]

Ans. For $^{41}_{20}\text{Ca}$: Separation energy $=$ 8.363007 MeV

$^{27}_{13}\text{Al}$: Separation energy $=$ 13.059 MeV

$({}_0n^1)$ is removed from a $^{41}_{20}\text{Ca}$

The required nuclear reaction can be written as :

$$^{41}_{20}\text{Ca} \longrightarrow {}^{40}_{20}\text{Ca} + {}^{1}_{0}n$$

It is given that :

$$m\left({}^{40}_{20}\text{Ca}\right) = 39.962591\ u$$

$$m\left({}^{41}_{20}\text{Ca}\right) = 40.962278\ u$$

$$m\,({}_0n^1) = 1.008665\ u$$

The mass defect of this reaction is given by

$$\Delta m = m\left({}^{40}_{20}\text{Ca}\right) + \left({}^{1}_{0}n\right) - m\left({}^{41}_{20}\text{Ca}\right)$$

$$= 39.962591 + 1.008665 - 40.962278 = 0.008978 \text{ u}$$

But $\quad 1\text{u} = 931.5 \text{ MeV}/c^2$

$\therefore \quad \Delta m = 0.008978 \times 931.5 \text{ MeV}/c^2$

Therefore, the energy required for neutron removal is

$$\text{E} = \Delta mc^2$$

$$= 0.008978 \times 931.5$$

$$= 8.363007 \text{ MeV}$$

For $^{27}_{13}\text{Al}$, the neutron removal reaction can be written as :

$$^{27}_{13}\text{Al} \longrightarrow {}^{26}_{13}\text{Al} + {}^{1}_{0}n$$

It is given that :

$$m\left({}^{27}_{13}\text{Al}\right) = 26.981541 \text{ u}$$

$$m\left({}^{26}_{13}\text{Al}\right) = 25.986895 \text{ u}$$

The mass defect of this reaction is given by

$$\Delta m = m\left({}^{27}_{13}\text{Al}\right) + m\left({}^{1}_{0}n\right) - m\left({}^{27}_{13}\text{Al}\right)$$

$$= 25.986895 + 1.008665 - 26.981541$$

$$= 0.014019 \text{ u} = 0.014019 \times 931.5 \text{ MeV}/c^2$$

Therefore, the energy required for neutron removal is calculated by

$$\text{E} = \Delta mc^2$$

$$= 0.014019 \times 931.5 = 13.059 \text{ MeV.}$$

Q. 30. Obtain the maximum kinetic energy of β-particles, and the radiation frequencies of γ decays in the decay scheme shown in given figure. You are given that $m\,(^{198}\text{Au}) = 197.968233\ u$ $m\,(^{198}\text{Hg}) = 197.966760\ u$. **[NCERT]**

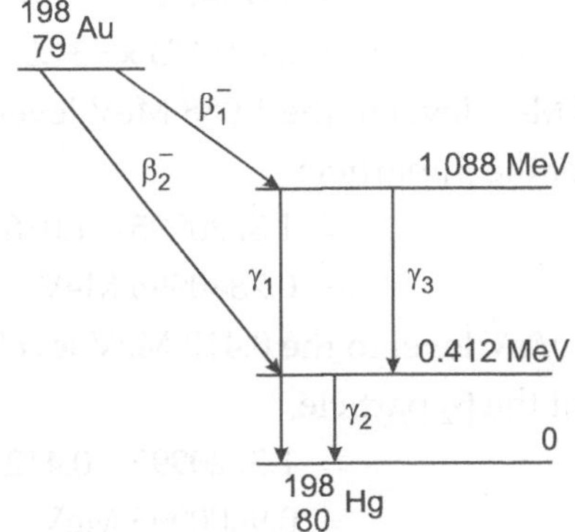

Ans. From the given γ-decay diagram it can be observed that γ_1 decays from the 1.088 MeV energy level to the 0 MeV energy level.

Therefore, the energy corresponding to γ_1-decay is given by

$$E_1 = 1.088 - 0 = 1.088 \text{ MeV}$$

$$h\nu_1 = 1.088 \times 1.6 \times 10^{-19} \times 10^6 \text{ J}$$

Here,

h = Planck's constant = 6.6×10^{-34} Js

ν_1 = Frequency of radiation radiated by γ_1-decay

$$\therefore \quad \nu_1 = \frac{E_1}{h} = \frac{1.088 \times 1.6 \times 10^{-19} \times 10^6}{6.6 \times 10^{-34}}$$

$$= 2.637 \times 10^{20} \text{ Hz}$$

It can be observed from the given γ-decay diagram that γ_2 decays from the 0.412 MeV energy level to the 0 MeV energy level.

Thus, the energy corresponding to γ_2-decay is given by

$$E_2 = 0.412 - 0 = 0.412 \text{ MeV}$$

$$h\nu_2 = 0.412 \times 1.6 \times 10^{-19} \times 10^6 \text{ J}$$

Here,

ν_2 = Frequency of radiation radiated by γ_2-decay

$$\therefore \quad \nu_2 = \frac{E_2}{h} = \frac{0.412 \times 1.6 \times 10^{-19} \times 10^6}{6.6 \times 10^{-34}} = 9.988 \times 10^{19} \text{ Hz}$$

From the given γ-decay diagram it is clear that γ_3 decays from the 1.088 MeV energy level to the 0.412 MeV energy level.

Therefore, the energy corresponding to γ_3-decay is given by

$$E_3 = 1.088 - 0.412 = 0.676 \text{ MeV}$$

$$h\nu_3 = 0.676 \times 10^{-19} \times 10^6 \text{ J}$$

Here,

ν_3 = Frequency of radiation radiated by γ_3-decay

$$\therefore \quad \nu_3 = \frac{E_3}{h} = \frac{0.676 \times 1.6 \times 10^{-19} \times 10^6}{6.6 \times 10^{-34}}$$

$$= 1.639 \times 10^{20} \text{ Hz}$$

$$\text{Mass of } m\left({}^{198}_{78}\text{Au}\right) = 197.968233 \text{ u}$$

$$\text{Mass of } m\left({}^{198}_{80}\text{Hg}\right) = 197.966760 \text{ u}$$

$$1\text{u} = 931.5 \text{ MeV}/c^2$$

Energy of the highest level is given by

$$E = \left[m\left({}^{198}_{78}\text{Au}\right) - m\left({}^{198}_{80}\text{Hg}\right)\right]$$

$$= 197.968233 - 197.966760 = 0.001473 \text{ u}$$

$$= 0.001473 \times 931.5 = 1.3720995 \text{ MeV}$$

β_1 decays from the 1.3720995 MeV level to the 1.088 MeV level

∴ Maximum kinetic energy of the β_1 particle

$$= 1.3720995 - 1.088$$

$$= 0.2840995 \text{ MeV}$$

β_2 decays from the 1.3720995 MeV level to the 0.412 MeV level

∴ Maximum kinetic energy of the β_2 particle

$$= 1.3720995 - 0.412$$

$$= 0.9600995 \text{ MeV}$$

Q. 31. Calculate for how many years will the fusion of 2.0 kg deuterium keep 800 W electric lamp glowing. Take the fusion reaction as : ★

$$^{2}_{1}H + ^{2}_{1}H \longrightarrow ^{3}_{1}H + ^{1}_{0}n + 3.27 \text{ MeV}$$

Ans. 2 deuterium nuclei fuse to produce 3.27 MeV

$$= 3.27 \times 1.6 \times 10^{-13} \text{ J} = 5.238 \times 10^{-13} \text{ J}$$

Energy produced per nuclei

$$= \frac{5.238 \times 10^{-13}}{2} = 2.61 \times 10^{-13} \text{ J}$$

2 kg deuterium atoms will have

$$= \frac{6.023 \times 10^{23}}{2} \times 2000 = 6.023 \times 10^{26} \text{ deuterium atom}$$

6.022×10^{26} deuterium atoms will produce

$$= 2.61 \times 10^{-13} \times 6.023 \times 10^{26} \text{ J}$$

$$= 15.77 \times 10^{13} \text{ J}$$

$$\because \quad \text{Power} = \frac{E}{t}$$

$$800 = \frac{15.77 \times 10^{13}}{t}$$

$$\Rightarrow \quad t = \frac{15.77 \times 10^{13}}{800} = 1.971 \times 10^{11} \text{ seconds}$$

$$t = \frac{1.971 \times 10^{11}}{365 \times 24 \times 60 \times 60} = 6.24 \times 10^{3} \text{ years}$$

Hence, electric lamp will glow for 6.24×10^3 years.

Chapter 14. Semiconductor Electronics : Materials, Devices and Simple Circuits

Q. 1. Suppose a pure Si crystal has 5×10^{28} atoms m^{-3}. It is doped by 1 ppm concentration of pentavalent As. Calculate the number of electrons and holes. [Given that $n_i = 1.5 \times 10^{16}$ m^{-3}.]

[NCERT]

Ans. Note that thermally generated electrons ($n_i \approx 10^{16}$ m^{-3}) are negligibly small as compared to those produced by doping.

Therefore, $n_e \approx N_D$.

Since $n_e n_h = n_i^2$

The number of holes $n_h = \dfrac{(2.25 \times 10^{32})}{(5 \times 10^{22})} \approx 4.5 \times 10^9$ m^{-3}

Q. 2. Two amplifier are connected one after the other in series (cascaded). The first amplifier has a voltage gain of 10 and the second has a voltage gain of 20. If the input signal is 0.01 volt, calculate the output ac signal. **[NCERT]**

Ans. Given,

Voltage gain of the first amplifier (V_1) = 10

Voltage gain of the second amplifier (V_2) = 20

Input signal voltage (V_i) =0.01 V

Let output a.c. signal voltage = V_o

The total voltage gain of a two stage cascaded amplifier is given by the product of voltage gains of both the stages, *i.e.*,

★ are board based questions from previous year

$$V = V_1 \times V_2 = 10 \times 20 = 200$$

We have the relation :

$$V = \frac{V_o}{V_i}$$

$$V_o = V \times V_i = 200 \times 0.01 = 2V$$

Therefore, the output a.c. signal of the given amplifier is 2V.

Q. 3. In a zener regulated power supply a Zener diode with $V_Z = 6.0$ V is used for regulation. The load current is to be 4.0 mA and the unregulated input is 10.0 V. What should be the value of series resistor R_S ? **[NCERT]**

Ans. The value of R_S should be such that the current through the zener diode is much larger than the load current. This is to have good load regulation. Choose zener current as five times the load current *i.e.,* $I_Z = 20$ mA. The total current through R_S is, therefore, 24 mA. The voltage drop across R_S is 10.0 – 6.0 = 4.0 V. This gives

$R_S = \frac{4.0 \text{ V}}{(24 \times 10^{-3})}$ A = 167 Ω. The nearest value of carbon resistor is 150 Ω. So, a series resistor of 150 Ω is appropriate.

Q. 4. The V-I characteristic of a silicon diode is shown in the Fig. Calculate the resistance of the diode at :

(i) $I_D = 15$ mA and

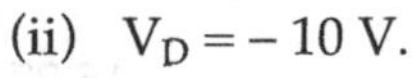

(ii) $V_D = -10$ V. **[NCERT]**

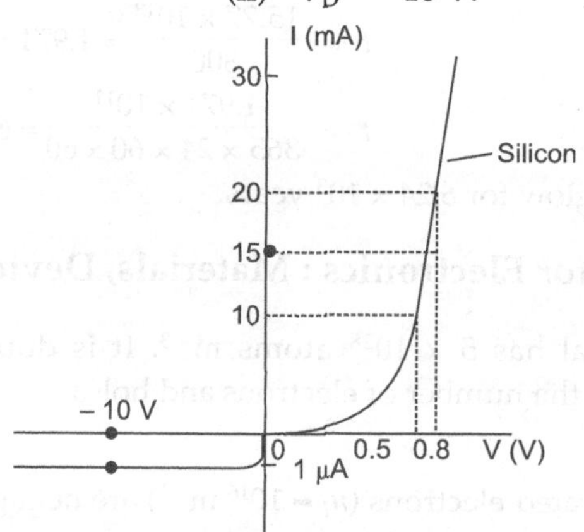

Ans. Considering the diode characteristics as a straight line between I = 10 mA to I = 20 mA passing through the origin, from Ohm's law, value of resistance can be calculated :

(i) From the curve, at I = 20 mA, V = 0.8 V and I = 10 mA, V = 0.7 V,

$$r_{fb} = \frac{\Delta V}{\Delta I} = \frac{0.1V}{10 \text{ mA}} = 10 \ \Omega$$

(ii) From the curve at V = – 10 V, I = – 1 μA,

$$\therefore \qquad r_{rb} = \frac{10 \text{ V}}{1 \mu\text{A}} = 1.0 \times 10^7 \ \Omega$$

Q. 5. A *p-n* photodiode is fabricated from a semiconductor with band gap of 2.8 eV. Can it detect a wavelength of 6000 nm ? **[NCERT]**

Ans. Given,

Energy band gap of the given photodiode (E_b) = 2.8 eV

Wavelength (λ) = 6000 nm = 6000×10^{-9} m

The energy of a signal is given by

$$E = \frac{hc}{\lambda}$$

Here, h = Planck's constant = 6.626×10^{-34} Js

c = Speed of light = 3×10^8 m/s

$$E = \frac{6.626 \times 10^{-34} \times 3 \times 10^8}{6000 \times 10^{-9}} = 3.313 \times 10^{-20} \text{ J}$$

But $\quad 1.6 \times 10^{-19} \text{ J} = 1 \text{ eV}$

$$\therefore \quad E = 3.313 \times 10^{-20} \text{ J}$$

$$= \frac{3.313 \times 10^{-20}}{1.6 \times 10^{-19}} = 0.207 \text{ eV}$$

Energy of a signal of wavelength 6000 nm is 0.207 eV, which is less than 2.8 eV the energy band gap of a photodiode. Therefore, the photodiode cannot detect the signal.

Q. 6. The number of silicon atoms per m^3 is 5×10^{23}. This is doped simultaneously with 5×10^{23} atoms per m^3 of Arsenic and 5×10^{20} per m^3 atoms of Indium. Calculate the number of electrons and holes. Given that $n_i = 1.5 \times 10^{16}$ m^{-3}. Is the material n-type or p-type. **[NCERT]**

Ans. Given,

Number of silicon atoms (N) = 5×10^{23} atoms/m^3

Number of arsenic atoms (n_{As}) = 5×10^{22} atoms/m^3

Number of indium atoms (n_{In}) = 5×10^{20} atoms/m^3

Number of thermally-generated electrons (n_i) = 1.5×10^{16} electrons/m^3

Number of electrons (n_e) = $5 \times 10^{22} - 1.5 \times 10^{16} \approx 4.99 \times 10^{22}$

Let number of holes = n_h

In thermal equilibrium, the concentrations of electrons and holes in a semiconductor are related as

$$n_e n_h = n_i^2$$

$$\therefore \quad n_h = \frac{n_i^2}{n_e} = \frac{(1.5 \times 10^{16})^2}{4.99 \times 10^{22}} \approx 4.51 \times 10^9$$

The number of electrons is a approximately 4.99×10^{22} and the number of holes are 4.51×10^9. Since the number of electrons is more than the number of holes, the material is an n-type semiconductor.

Q. 7. In half-wave rectification, what is the output frequency if the input frequency is 50 Hz. What is the output frequency of a full-wave rectifier for the same input frequency. **[NCERT]**

Ans. Given,

Input frequency = 50 Hz

For a half-wave recifier, the output frequency is equal to the input frequency.

$\therefore$ Output frequency = 50 Hz

For a full-wave recifier, the output frequency is twice the input frequency.

$\therefore$ Output frequency = $2 \times 50 = 100$ Hz

Q. 8. In a p-n junction diode, the current I can be expressed as

$$I = I_0 \exp.\left(\frac{eV}{2k_B T} - 1\right)$$

where T_0 is called the reverse saturation current, V is the voltage across the diode and is positive for forward bias and negative for reverse bias, and I is the current through the diode, k_B is the Boltzmann constant (8.6×10^{-5} eV/K) and T is the absolute temperature. If for a given diode $I_0 = 5 \times 10^{-12}$ and T = 300 K, then

(i)What will be the forward current at a forward voltage of 0.6 V ?

(ii)What will be the increase in the current it the voltage across the diode is increased to 0.7 V ?

(iii) What is the dynamic resistance ?

(iv)What will be the current if reverse bias voltage changes from 1 V to 2 V ? **[NCERT]**

Ans. In a *p-n* junction diode, the expression for current is given by

$$I = I_0 \exp.\left(\frac{eV}{2k_BT} - 1\right)$$

Here, I_0 = Reverse saturation current = 5×10^{-12} A

T = Absolute temperature = 300 K

k_B = Boltzmann constant = 8.6×10^{-5} eV/K = 1.376×10^{-23} J K^{-1}

V = Voltage across the diode

(i) Forward voltage (V) = 0.6 V

$$\therefore \quad \text{Current (I)} = 5 \times 10^{-12}\left[\exp\left(\frac{1.6 \times 10^{-19} \times 0.6}{1.376 \times 10^{-23} \times 300}\right) - 1\right]$$

$$= 5 \times 10^{-12} \times \exp[22.36] = 0.0256 \text{ A}$$

Thus, the forward current is about 0.0256 A.

(ii) For forward voltage (V) = 0.7 V, we can write :

$$I' = 5 \times 10^{-12}\left[\exp\left(\frac{1.6 \times 10^{-19} \times 0.7}{1.376 \times 10^{-23} \times 300} - 1\right)\right]$$

$$= 5 \times 10^{-12} \times \exp[26.25] = 1.257 \text{ A}$$

Therefore, the increases in current,

$$\Delta I = I' - I$$

$$= 1.257 - 0.0256 = 1.23 \text{ A}$$

(iii) $$\text{Dynamic resistance} = \frac{\text{Change in voltage}}{\text{Change in current}}$$

$$= \frac{0.7 - 0.6}{1.23} = \frac{0.1}{1.23} = 0.081\ \Omega$$

(iv) If the reverse bias voltage changes from 1 V to 2 V, then the current (I) will almost remain equal to I_0 in both cases. Therefore, the dynamic resistance in the reverse bias will be infinite.

□□

Topper's Answers - 2019

SECTION – A

1. Define the term "threshold frequency", in the context of photoelectric emission.

OR

Define the term "Intensity" in photon picture of electromagnetic radiation.

1. Threshold frequency is defined as that minimum frequency that an electromagnetic radiation should have, so that is able to eject electrons (with zero kinetic energy) from the surface of the metal on which it is incident.

Mathematically, threshold frequency, ν_0

$$\nu_0 = \frac{\phi_0}{h}$$

where ϕ_0 is the work function of the corresponding metal.

Or in other words, it is the minimum frequency of incident radiation for which photoelectric effect is observed.

2. When unpolarised light is incident on the interface separating the rarer medium and the denser medium, Brewster angle is found to be 60°. Determine the refractive index of the denser medium.

2. Brewster angle, $i_B = 60°$

Therefore, from Brewster's law, refractive index of denser medium w.r.t rarer medium, $n = \tan i_B = \tan 60° = 1.73$

Hence, refractive index = 1.73.

3. In sky wave mode of propagation, why is the frequency range of transmitting signals restricted to less than 30 MHz ?

OR

On what factors does the range of coverage in ground wave propagation depend ?

3. Signals having frequency greater than 30 MHz are sufficiently energetic that they are able to penetrate the ionosphere of the atmosphere. Hence, these are no longer reflected by the ionosphere. Thus, these are not used for sky wave propagation (which makes use of ionospheric reflection). Hence, frequency is kept less than 30 MHz

5

4. When a potential difference is applied across the ends of a conductor, how is the drift velocity of the electrons related to the relaxation time ?

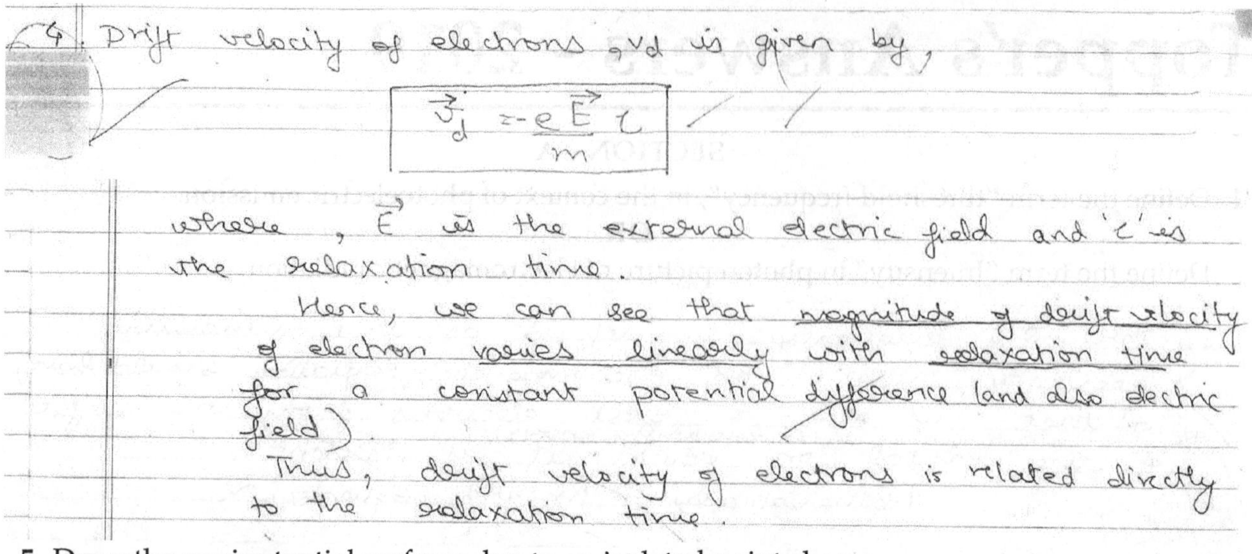

4. Drift velocity of electrons, v_d is given by,

$$\vec{v}_d = \frac{-e\vec{E}\tau}{m}$$

where, $\vec{E}$ is the external electric field and 'τ' is the relaxation time.

Hence, we can see that magnitude of drift velocity of electron varies linearly with relaxation time for a constant potential difference (and also electric field).

Thus, drift velocity of electrons is related directly to the relaxation time.

5. Draw the equipotential surfaces due to an isolated point charge.

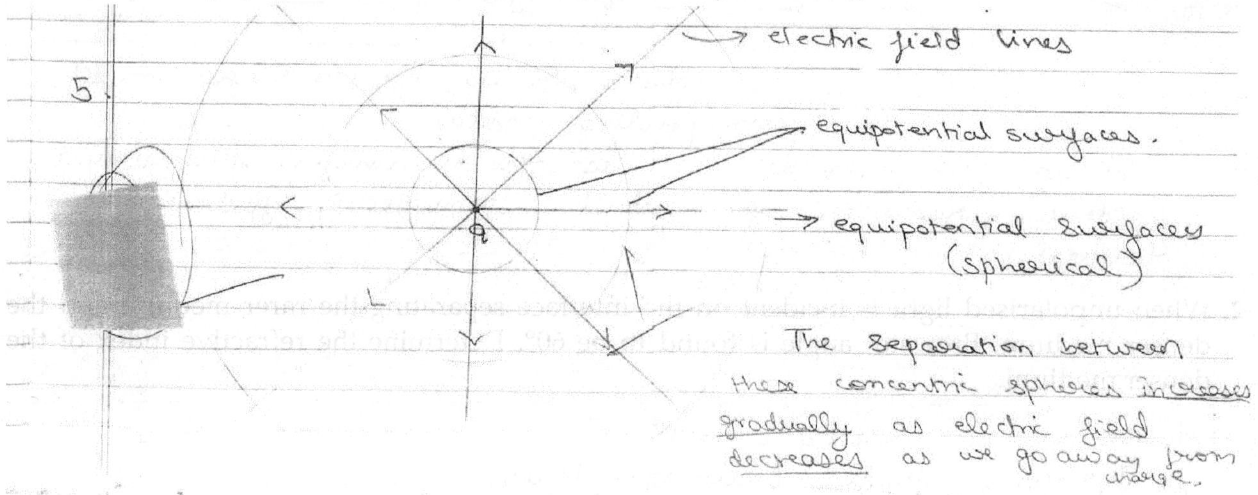

SECTION – B

6. Explain with the help of Einstein's photo electric equation any two observed features in photoelectric effect which cannot be explained by wave theory.

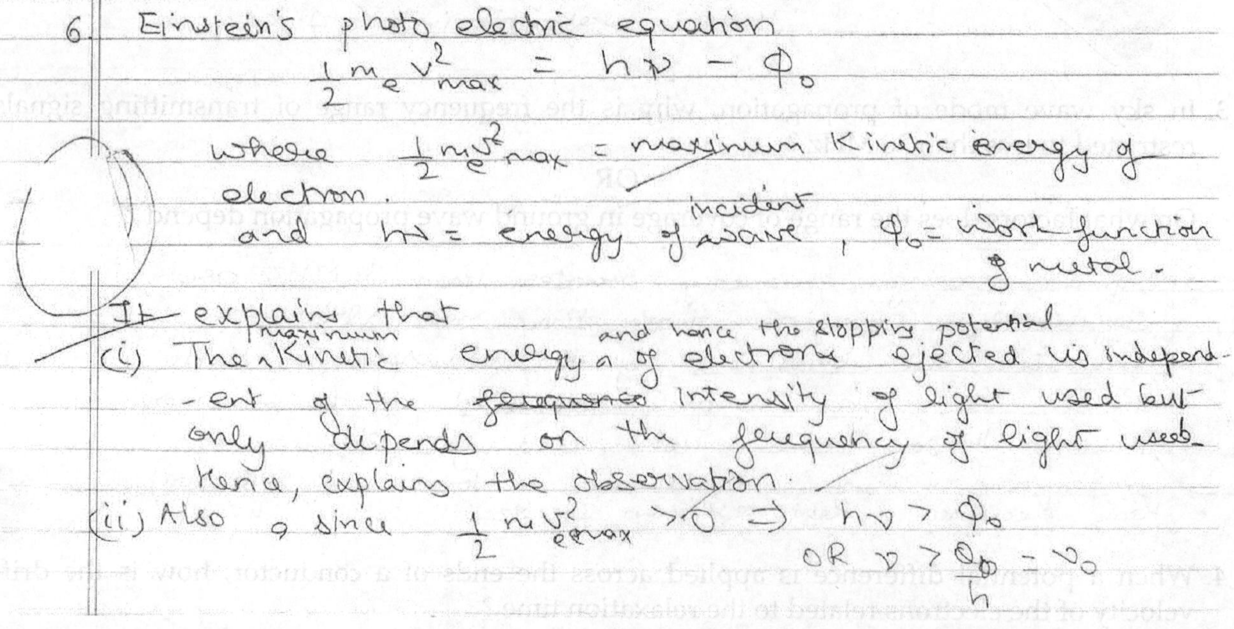

6. Einstein's photo electric equation

$$\frac{1}{2} m_e v^2_{max} = h\nu - \phi_0$$

where $\frac{1}{2} m v^2_{e\,max}$ = maximum Kinetic energy of electron.

and $h\nu$ = energy of incident wave, ϕ_0 = work function of metal.

It explains that

(i) The maximum Kinetic energy and hence the stopping potential of electrons ejected is independent of the intensity of light used but only depends on the frequency of light used. Hence, explains the observation.

(ii) Also, since $\frac{1}{2} m v^2_{e\,max} > 0 \Rightarrow h\nu > \phi_0$

OR $\nu > \frac{\phi_0}{h} = \nu_0$

where $\nu_0 = \frac{\phi_0}{h}$ = threshold frequency

Hence, a minimum frequency of light is required to observe photoelectric effect, characteristic to the metal. Thus, explains the observations and rules out the fact that intensity has got to do anything with ejection of electron.

7. A deuteron and an alpha particle having same momentum are in turn allowed to pass through a magnetic field , acting normal to the direction of motion of the particles. Calculate the ratio of the radii of the circular paths described by them.

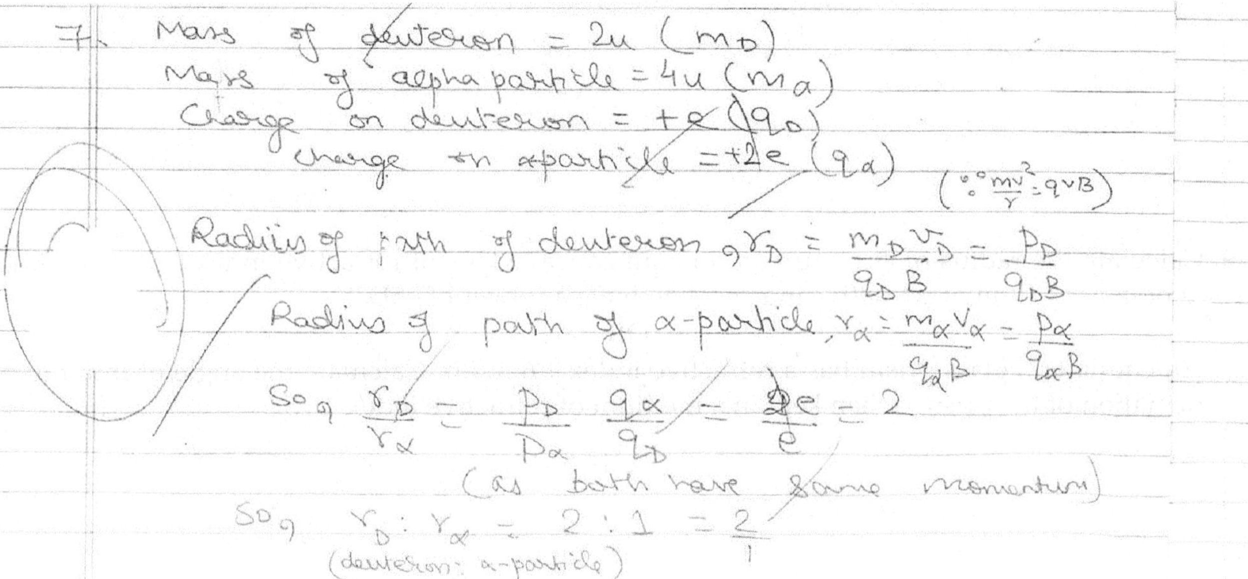

7. Mass of deuteron = 2u (m_D)
Mass of alpha particle = 4u (m_α)
Charge on deuteron = +e (q_D)
Charge on α-particle = +2e (q_α)

$\left(\because \frac{mv^2}{r} = qvB\right)$

Radius of path of deuteron, $r_D = \frac{m_D v_D}{q_D B} = \frac{p_D}{q_D B}$

Radius of path of α-particle, $r_\alpha = \frac{m_\alpha v_\alpha}{q_\alpha B} = \frac{p_\alpha}{q_\alpha B}$

So, $\frac{r_D}{r_\alpha} = \frac{p_D}{p_\alpha} \frac{q_\alpha}{q_D} = \frac{2e}{e} = 2$

(as both have same momentum)

So, $r_D : r_\alpha = 2 : 1 = \frac{2}{1}$
(deuteron : α-particle)

8. Two bulbs are rated (P1, V) and (P2, V). If they are connected (i) in series and (ii) in parallel across a supply V, find the power dissipated in the two combinations in terms of P1 and P2.

8. Power of bulb 1 = P_1
Voltage rating = V
Resistance = $\frac{V^2}{P_1} = R_1$ $\left(\because P_1 = \frac{V^2}{R}\right)$

Power of bulb 2 = P_2
Voltage rating = V
Resistance = $\frac{V^2}{P_2} = R_2$

(i) When connected in series,
total Resistance, $R_S = R_1 + R_2 = V^2\left[\frac{P_1 + P_2}{P_1 P_2}\right]$

Total current flowing through the circuit = $\frac{V}{R_S}$ (∵ they are connected to a supply of voltage V)

$I_T = \frac{V}{V^2}\left[\frac{P_1 P_2}{P_1 + P_2}\right] = \frac{1}{V}\frac{P_1 P_2}{P_1 + P_2}$

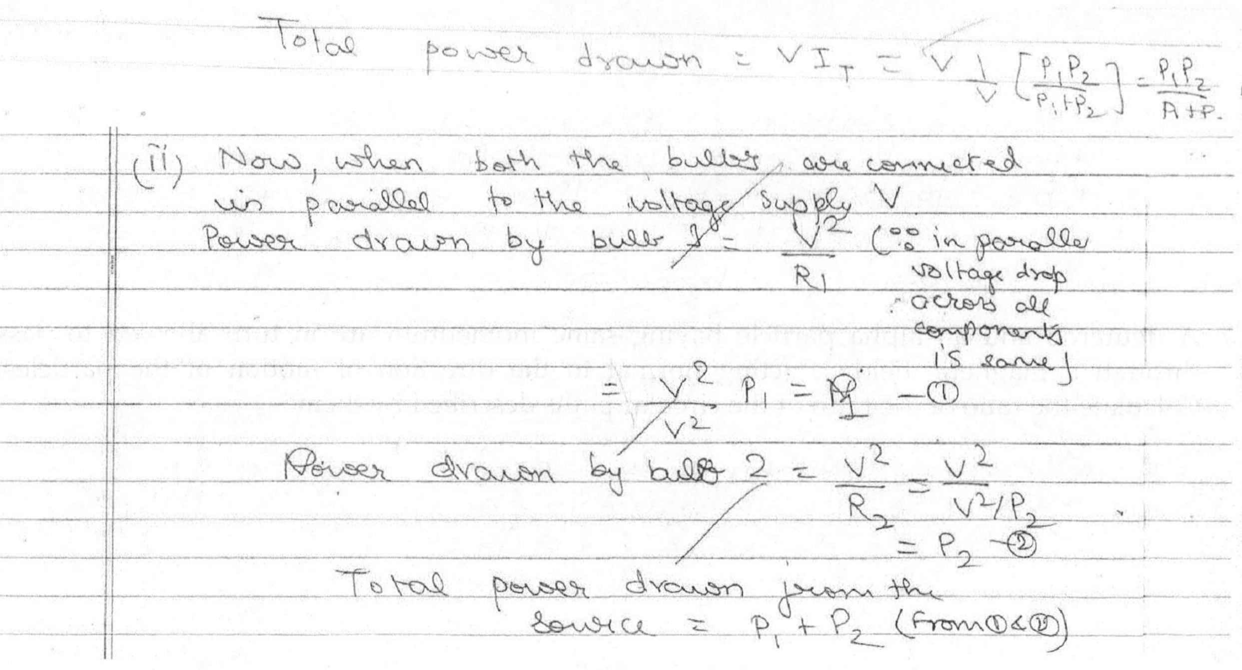
Total power drawn $= VI_T = V\frac{1}{V}\left[\frac{P_1P_2}{P_1+P_2}\right] = \frac{P_1P_2}{P_1+P_2}$.

(ii) Now, when both the bulbs are connected in parallel to the voltage supply V

Power drawn by bulb 1 $= \frac{V^2}{R_1}$ (∵ in parallel voltage drop across all components is same)

$= \frac{V^2}{V^2/P_1} = P_1$ —(1)

Power drawn by bulb 2 $= \frac{V^2}{R_2} = \frac{V^2}{V^2/P_2} = P_2$ —(2)

Total power drawn from the source $= P_1 + P_2$ (From (1) & (2))

9. Calculate the radius of curvature of an equi-concave lens of refractive index 1.5, when it is kept in a medium of refractive index 1.4, to have a power of –5D ?

OR

An equilateral glass prism has a refractive index 1.6 in air. Calculate the angle of minimum deviation of the prism, when kept in a medium of refractive index 4 /5.

9. Refractive index of prism $= 1.6 = \frac{8}{5} = n_1$

Refractive index of surrounding medium $= \frac{4\sqrt{2}}{5} = n_2$

Refractive index of prism w.r.t surroundings $= n_{12} = \frac{n_1}{n_2}$

$$\mu = \frac{8 \times 5}{5 \times 4\sqrt{2}} = \sqrt{2}$$

So, using the relation,

$$\mu = \frac{\sin\left(\frac{A+\delta_{min}}{2}\right)}{\sin\frac{A}{2}}$$

A = angle of prism = 60°, δ_{min} = angle of minimum deviation

we get,

$$\sqrt{2} = \frac{\sin\left(\frac{60+\delta_{min}}{2}\right)}{\sin 30^\circ}$$

$$\frac{1}{\sqrt{2}} = \sin\left(\frac{60+\delta_{min}}{2}\right)$$

$$\frac{60 + \delta_{min}}{2} = \sin^{-1}\frac{1}{\sqrt{2}} = 45^\circ$$

$$60^\circ + \delta_{min} = 90^\circ$$

$$\delta_{min} = 30^\circ$$

Hence, angle of minimum deviation of prism = 30°

10. State Bohr's quantization condition of angular momentum. Calculate the shortest wavelength of the Bracket series and state to which part of the electromagnetic spectrum does it belong.

OR

Calculate the orbital period of the electron in the first excited state of hydrogen atom.

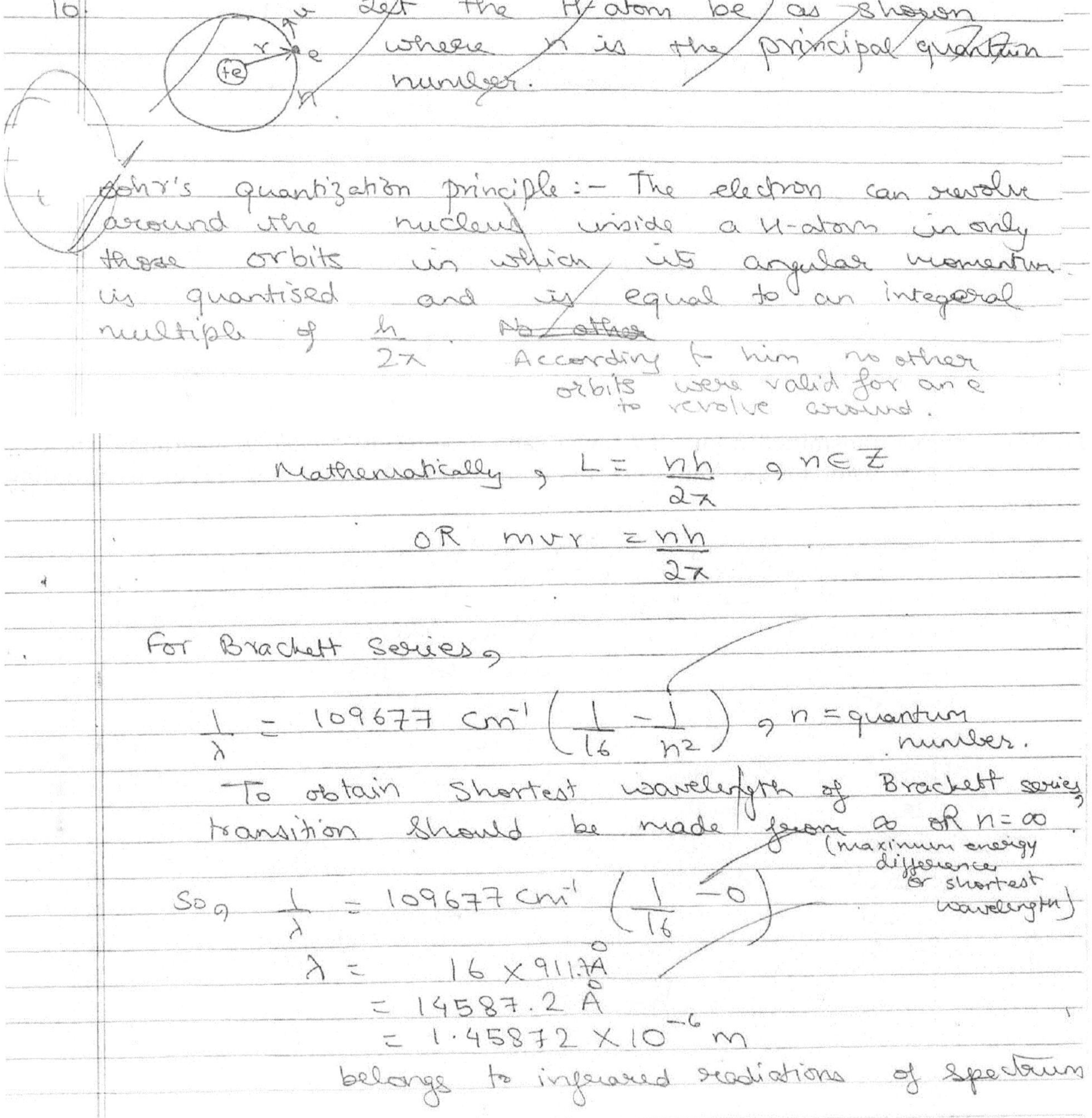

10. ~~Let the H-atom be as shown where n is the principal quantum number.~~

Bohr's quantization principle :– The electron can revolve around the nucleus inside a H-atom in only those orbits in which its angular momentum is quantised and is equal to an integral multiple of $\frac{h}{2\pi}$. According to him no other orbits were valid for an e to revolve around.

Mathematically, $L = \frac{nh}{2\pi}$, $n \in Z$

OR $mvr = \frac{nh}{2\pi}$

For Brackett Series,

$\frac{1}{\lambda} = 109677\ cm^{-1}\left(\frac{1}{16} - \frac{1}{n^2}\right)$, n = quantum number.

To obtain shortest wavelength of Brackett series, transition should be made from ∞ or n = ∞. (maximum energy difference or shortest wavelength)

So, $\frac{1}{\lambda} = 109677\ cm^{-1}\left(\frac{1}{16} - 0\right)$

$\lambda = 16 \times 911.7\ \text{Å}$

$= 14587.2\ \text{Å}$

$= 1.45872 \times 10^{-6}\ m$

belongs to infrared radiations of spectrum

11. (a) Plot a graph showing variation of deBroglie wavelength () associated with a charged particle of mass m, versus, where V is the accelerating potential.

(b) An electron, a proton and an alpha particle have the same kinetic energy. Which one has the shortest wavelength ?

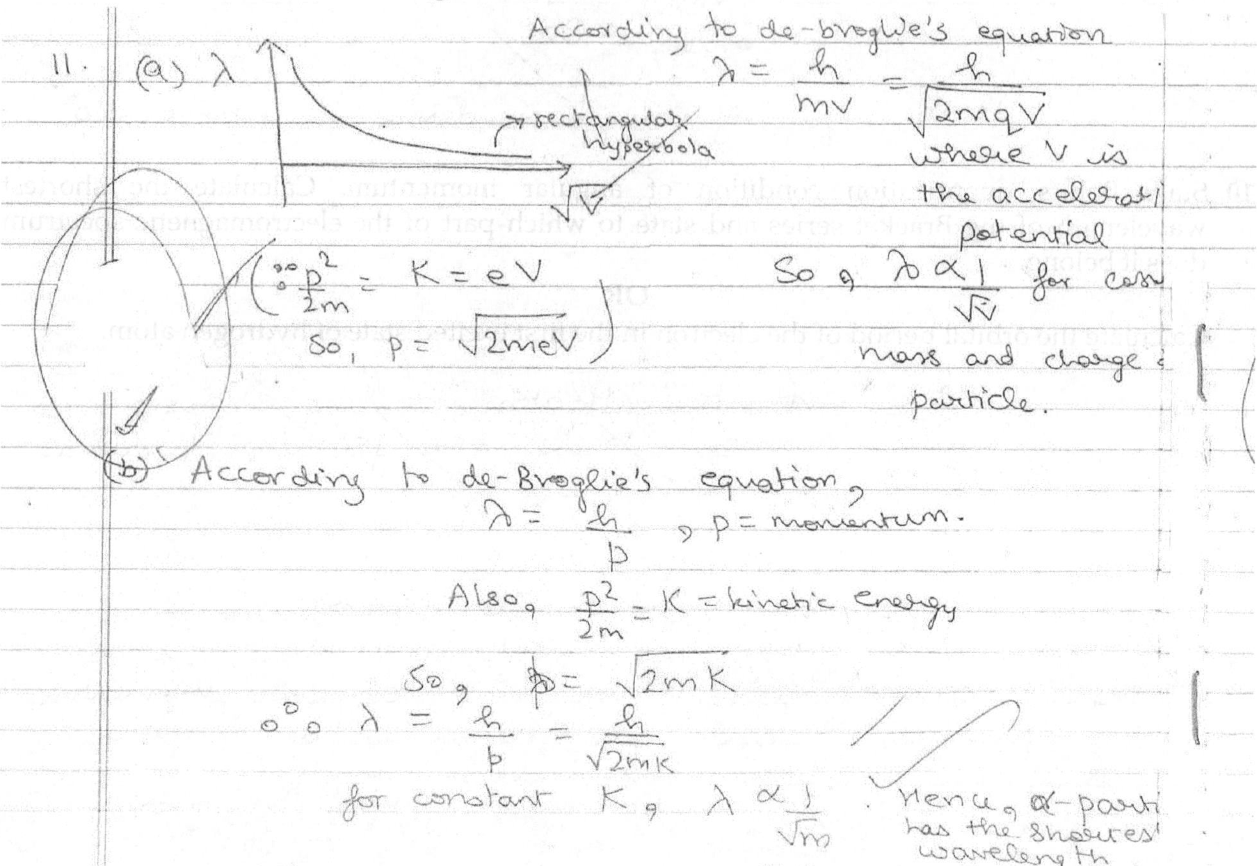

11. (a) λ

→ rectangular hyperbola

According to de-broglie's equation

$\lambda = \frac{h}{mv} = \frac{h}{\sqrt{2mqV}}$

where V is the accelerating potential

So, $\lambda \propto \frac{1}{\sqrt{V}}$ for constant mass and charge of particle.

$\left(\therefore \frac{p^2}{2m} = K = eV\right.$

$\left. So,\ p = \sqrt{2meV}\right)$

(b) According to de-Broglie's equation,

$\lambda = \frac{h}{p}$, p = momentum.

Also, $\frac{p^2}{2m} = K$ = kinetic energy

So, $p = \sqrt{2mK}$

$\therefore \lambda = \frac{h}{p} = \frac{h}{\sqrt{2mK}}$

for constant K, $\lambda \propto \frac{1}{\sqrt{m}}$

Hence, α-particle has the shortest wavelength

12. Why a signal transmitted from a TV tower cannot be received beyond a certain distance ? Write the expression for the optimum separation between the receiving and the transmitting antenna.

12. Signal transmitted from a TV tower uses space wave method for signal transmission. ∴ the transmitting antenna and the receiving antenna should lie along a straight line for effective communication. However due to curvature of earth, the signals are unable to exceed a particular distance and obstructed by the Earth's surface thus leading to a very high attenuation.

Hence, TV signals cannot be received beyond a particular distance.

Considering the transmitting tower to have a height 'h_T' and receiving tower to have a height 'h_R', the

h_T Earth R h_R

maximum range possible is equal to

$$R_{max} = \sqrt{2Rh_T} + \sqrt{2Rh_R}$$

where R = radius of earth.

So, this must be the optimum separation between the receiving and transmitting antenna.

SECTION – C

13. (a) State the underlying principle of a moving coil galvanometer.

(b) Give two reasons to explain why a galvanometer cannot as such be used to measure the value of the current in a given circuit.

(c) Define the terms :

(i) voltage sensitivity and (ii) current sensitivity of a galvanometer.

Section-C

Q.No-13

(a) Moving coil galvanometer works on the principle that a current carrying loop experiences a torque in magnetic field (∵ It has a magnetic moment associated with it). This torque produced is then balanced by the counter torque of the spring producing the necessary deflection of the needle.

OR $BINA = J\theta$, where J is the torsional constant of spring and other symbols have their usual meanings.

So, $\theta = \frac{BINA}{J}$

OR $\theta \propto I$ (current flowing through the coil)

(b) Galvanometer as such cannot be used to measure the value of current in a given circuit because :-

(i) It is a very sensitive instrument and has a very low maximum deflection current, so it will show maximum deflection most of the time and can get damaged. In such situation, it is not of much use.

(ii) Galvanometer has appreciable resistance of its own (due to windings), therefore, will alter the current flowing through the circuit and will not give accurate results (readings).

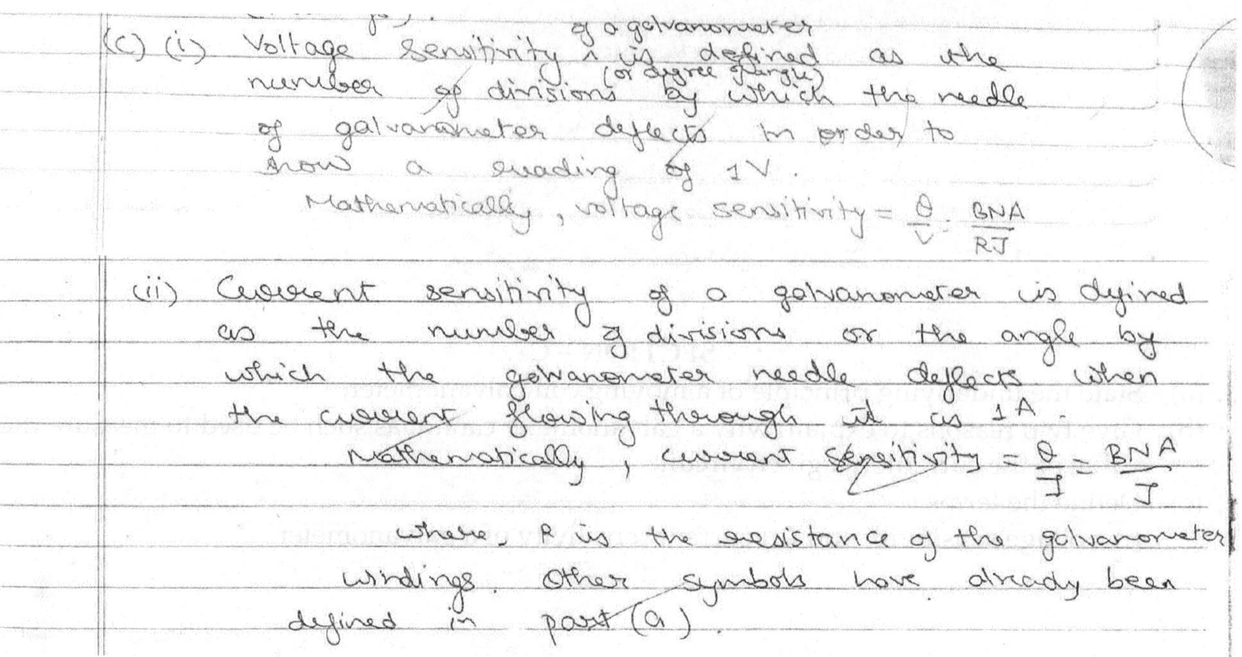

(c) (i) Voltage sensitivity of a galvanometer is defined as the number of divisions (or degree of angle) by which the needle of galvanometer deflects in order to show a reading of 1V.

Mathematically, voltage sensitivity $= \frac{\theta}{V} = \frac{BNA}{RJ}$

(ii) Current sensitivity of a galvanometer is defined as the number of divisions or the angle by which the galvanometer needle deflects when the current flowing through it is 1 A.

Mathematically, current sensitivity $= \frac{\theta}{I} = \frac{BNA}{J}$

where, R is the resistance of the galvanometer windings. Other symbols have already been defined in part (a).

14. (a) Define mutual inductance and write its S.I. unit.

(b) A square loop of side 'a' carrying a current is kept at distance x from an infinitely long straight wire carrying a current as shown in the figure. Obtain the expression for the resultant force acting on the loop.

14 (a) Mutual inductance of a coil (C_1) w.r.t another coil (C_2) is defined as the magnitude of emf induced in coil C_1 when the current through the other coil C_2 changes at the rate of 1 A s^{-1}.

Mathematically, $\mathcal{E}_1 = -M_{12}\frac{dI_2}{dt}$ (when $\frac{dI_1}{dt} = 1 As^{-1}$, $M_{12} = |\mathcal{E}_1|$)

S.I. unit of mutual inductance is henry (H)

(b) Magnetic field due to infinitely long wire at distance x,

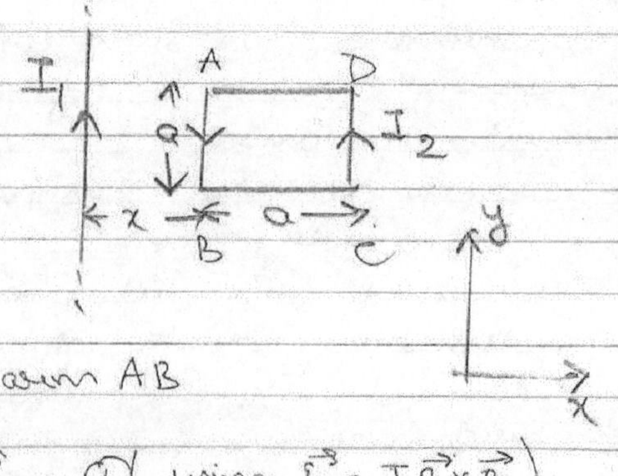

I_1, A, D, a, I_2, ← x →← a →, B, C, y, x

$B_x = \frac{\mu_0 I_1}{2\pi x}$ ⊗

Force experienced by arm AB

$\vec{F}_{AB} = \frac{\mu_0 I_1 I_2 a}{2\pi x}\hat{i}$ —(1) (using $\vec{F} = I\vec{l}\times\vec{B}$)

~~Force experienced by arm BC~~

~~$\vec{F}_{BC} = \mu$~~

Magnetic field due to infinitely long wire at a distance of x + a,

$$B_{x+a} = \frac{\mu_0 I_1}{2\pi(x+a)} \otimes$$

force experienced by ~~arm~~ DC,

$$\vec{F}_{BD} = -\frac{\mu_0 I_1}{2\pi(x+a)} I_2 a \hat{i} \quad (\text{using } \vec{F} = I\vec{l}\times\vec{B}) \quad -②$$

Clearly, force experienced by arms BC and DA will be equal in magnitude and opposite in direction, because these are placed symmetrically.

(for e.g. Force on DA will be downwards whereas force on BC will act upwards - using $\vec{F} = I\vec{l}\times\vec{B}$) —③

So, Total force on current carrying loop

$$\vec{F}_{net} = \vec{F}_{AB} + \vec{F}_{BC} + \vec{F}_{CD} + \vec{F}_{DA}$$

$$= \frac{\mu_0 I_1 I_2 a}{2\pi x}\hat{i} + (\vec{F}_{BC} + \vec{F}_{DA}) - \frac{\mu_0 I_1 I_2 a \hat{i}}{2\pi(a+x)}$$

$$\vec{F}_{net} = \frac{\mu_0 I_1 I_2 a \hat{i}}{2\pi}\left[\frac{1}{x} - \frac{1}{a+x}\right] + 0 \quad (\text{from } ①, ② \,\&\, ③)$$

$$\vec{F}_{net} = \frac{\mu_0 I_1 I_2 a \, a \hat{i}}{2\pi x(a+x)}$$

Hence, force acting on square loop is $\boxed{\frac{\mu_0 I_1 I_2 a^2}{2\pi x(a+x)}}$.

15. (a) Draw equipotential surfaces corresponding to the electric field that uniformly increases in magnitude along with the z-directions.

(b) Two charges – q and + q are located at points (0, 0, – a) and (0, 0, a). What is the electrostatic potential at the points (0, 0, z) and (x, y, 0) ?

towards right.

15 (a)

(Parallel to ~~Along~~ x-y plane)

→ Plane surface (equipotential surfaces)

Also, $d_2 < d_1$ as electric field is uniformly increasing along the z-direction (axis)

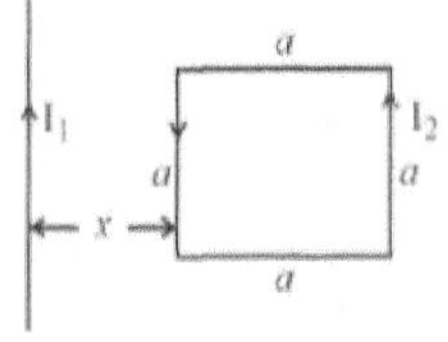

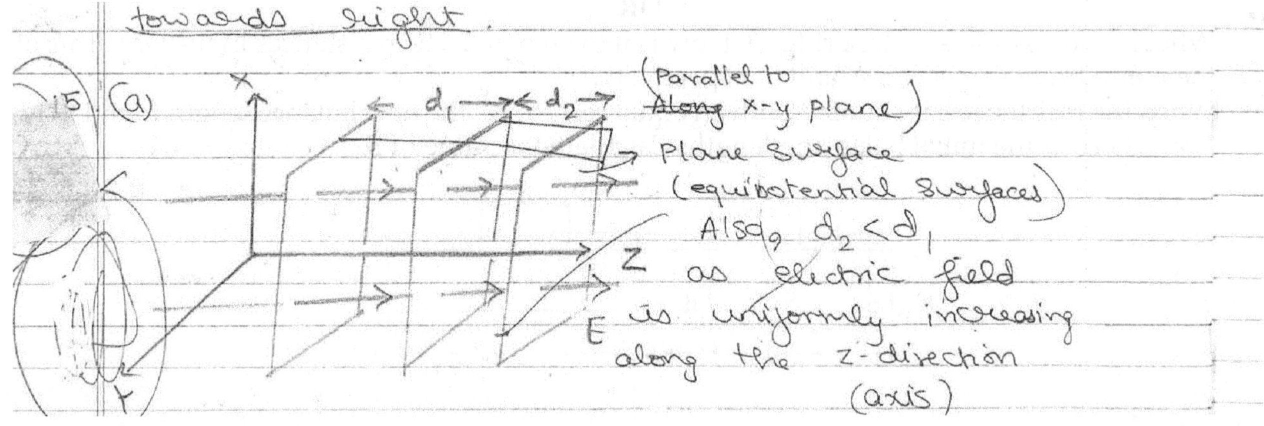

(b) Electrostatic potential at point $(x, y, 0)$ is zero because it is lying on the equitorial plane of the dipole and is equidistant from both $+q$ and $-q$.

$(0,0,-z)$ $\quad -q(0,0,-a)$ $\quad q(0,0,a)$ $\quad (0,0,+z)$

Now, potential at $(0,0,+z)$

$$V_{+z} = V_q + V_{-q} = \frac{1}{4\pi\varepsilon_0}\frac{q}{(z-a)} + \frac{1}{4\pi\varepsilon_0}\frac{(-q)}{(z+a)}$$

$$V_{+z} = \frac{q}{4\pi\varepsilon_0}\left[\frac{1}{z-a} - \frac{1}{z+a}\right]$$

$$V_{+z} = \frac{q}{4\pi\varepsilon_0}\left[\frac{z+a-z+a}{z^2-a^2}\right]$$

$$V_{+z} = \frac{2qa}{4\pi\varepsilon_0(z^2-a^2)}$$

$-q$ $\quad q$ $\quad$ z-axis

$(0,0,-z)$ $(0,0,-a)$ $\quad O \quad$ $(0,0,a)$ $(0,0,z)$

$(0,0,0)$

and
$$V_{-z} = V_{-q} + V_q$$
$$= \frac{-q}{4\pi\varepsilon_0(z-a)} + \frac{q}{4\pi\varepsilon_0}\frac{1}{(z+a)}$$
$$= \frac{q}{4\pi\varepsilon_0}\left[\frac{1}{z+a} - \frac{1}{z-a}\right]$$
$$= \frac{q\,(-2a)}{4\pi\varepsilon_0(z^2-a^2)} = \frac{-2qa}{4\pi\varepsilon_0(z^2-a^2)}$$

Hence, electrostatic potential at point $(0,0,z)$ is $\frac{1}{4\pi\varepsilon_0}\frac{2qa}{z^2-a^2}$

and at $(0,0,-z)$ is $\frac{-1}{4\pi\varepsilon_0}\frac{2qa}{z^2-a^2}$

16. Using Kirchhoff's rules, calculate the current through the 40 and 20 resistors in the following circuit :

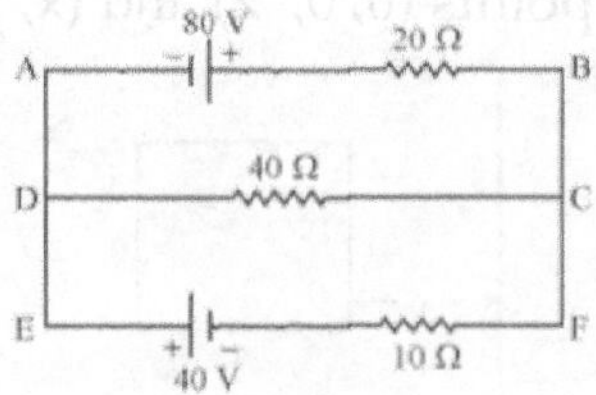

OR

What is end error in a metre bridge ? How is it overcome ? The resistances in the two arms of the metre bridge are R = 5 W and S respectively.

When the resistance S is shunted with an equal resistance, the new balance length found to be 1.5 , where is the initial balancing length. Calculate the value of S.

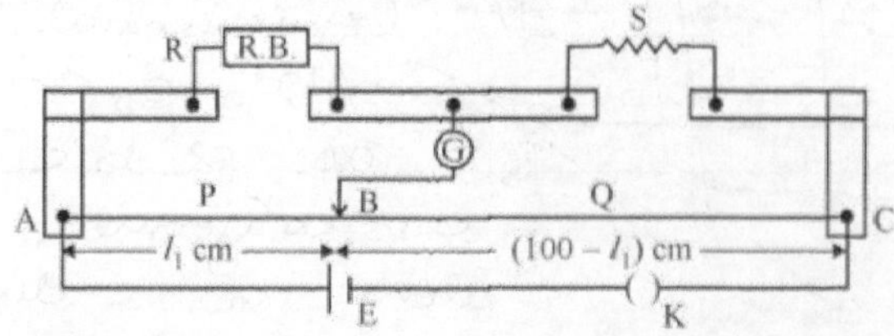

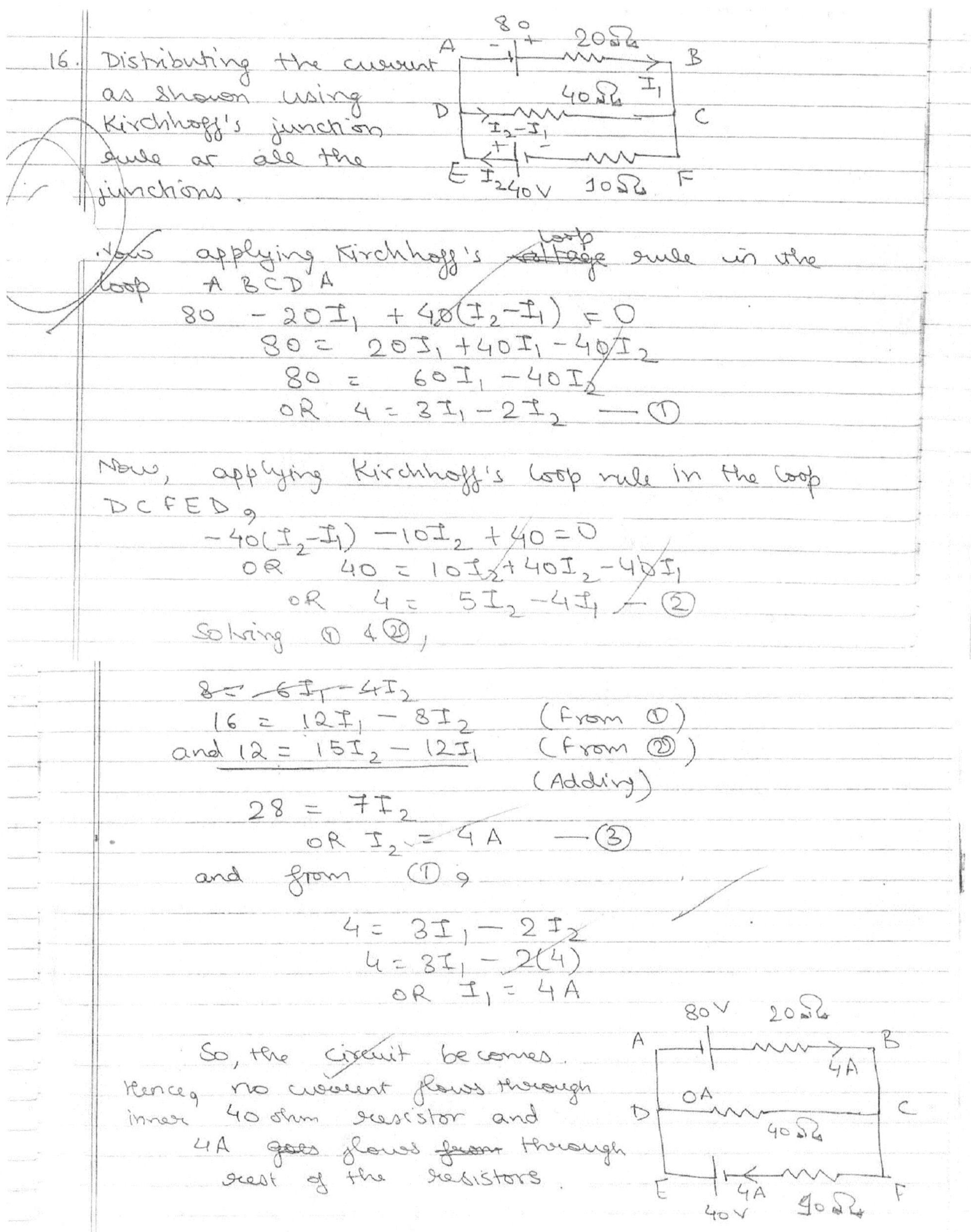

16. Distributing the current as shown using Kirchhoff's junction rule at all the junctions.

Now applying Kirchhoff's loop rule in the loop ABCDA

$80 - 20I_1 + 40(I_2 - I_1) = 0$

$80 = 20I_1 + 40I_1 - 40I_2$

$80 = 60I_1 - 40I_2$

OR $4 = 3I_1 - 2I_2$ —— (1)

Now, applying Kirchhoff's loop rule in the loop DCFED,

$-40(I_2 - I_1) - 10I_2 + 40 = 0$

OR $40 = 10I_2 + 40I_2 - 40I_1$

OR $4 = 5I_2 - 4I_1$ —— (2)

Solving (1) & (2),

$16 = 12I_1 - 8I_2$ (From (1))

and $12 = 15I_2 - 12I_1$ (From (2))

(Adding)

$28 = 7I_2$

OR $I_2 = 4\text{ A}$ —— (3)

and from (1),

$4 = 3I_1 - 2I_2$

$4 = 3I_1 - 2(4)$

OR $I_1 = 4\text{ A}$

So, the circuit becomes.

Hence, no current flows through inner 40 ohm resistor and 4A flows through rest of the resistors.

17. (a) Write the relation between half life and average life of a radioactive nucleus.

(b) In a given sample two isotopes A and B are initially present in the ratio of 1:2. Their half lives are 60 years and 30 years respectively. How long will it take so that the sample has these isotopes in the ratio of 2:1 ?

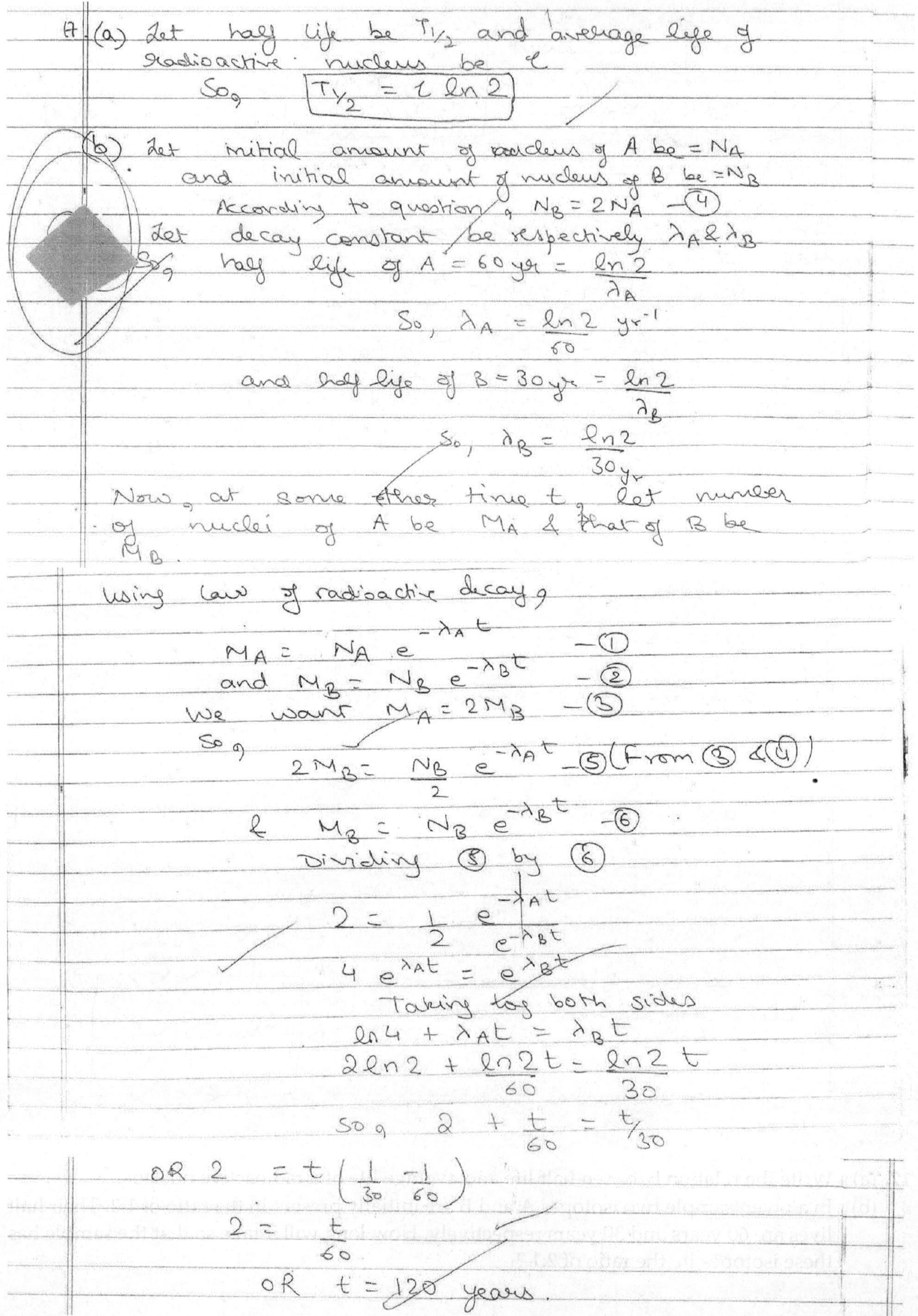

17 (a) Let half life be $T_{1/2}$ and average life of radioactive nucleus be τ.

So, $\boxed{T_{1/2} = \tau \ln 2}$

(b) Let initial amount of nucleus of A be $= N_A$

and initial amount of nucleus of B be $= N_B$

According to question, $N_B = 2N_A$ — (4)

Let decay constant be respectively λ_A & λ_B

So, half life of $A = 60\,yr = \frac{\ln 2}{\lambda_A}$

So, $\lambda_A = \frac{\ln 2}{60}\,yr^{-1}$

and half life of $B = 30\,yr = \frac{\ln 2}{\lambda_B}$

So, $\lambda_B = \frac{\ln 2}{30\,yr}$

Now, at some other time t, let number of nuclei of A be M_A & that of B be M_B.

Using law of radioactive decay,

$M_A = N_A e^{-\lambda_A t}$ — (1)

and $M_B = N_B e^{-\lambda_B t}$ — (2)

We want $M_A = 2M_B$ — (3)

So,

$2M_B = \frac{N_B}{2} e^{-\lambda_A t}$ — (5) (From (3) & (4))

& $M_B = N_B e^{-\lambda_B t}$ — (6)

Dividing (5) by (6)

$$2 = \frac{1}{2}\frac{e^{-\lambda_A t}}{e^{-\lambda_B t}}$$

$$4e^{\lambda_A t} = e^{\lambda_B t}$$

Taking log both sides

$$\ln 4 + \lambda_A t = \lambda_B t$$

$$2\ln 2 + \frac{\ln 2\, t}{60} = \frac{\ln 2}{30} t$$

So, $2 + \frac{t}{60} = t/30$

OR $2 = t\left(\frac{1}{30} - \frac{1}{60}\right)$

$2 = \frac{t}{60}$

OR $t = 120$ years.

Hence, after 120 years the required concentration of nuclei A & B will be achieved

18. Define the term wavefront. Using Huygen's wave theory, verify the law of reflection.

OR

Define the term, "refractive index" of a medium. Verify Snell's law of refraction when a plane wavefront is propagating from a denser to a rarer medium.

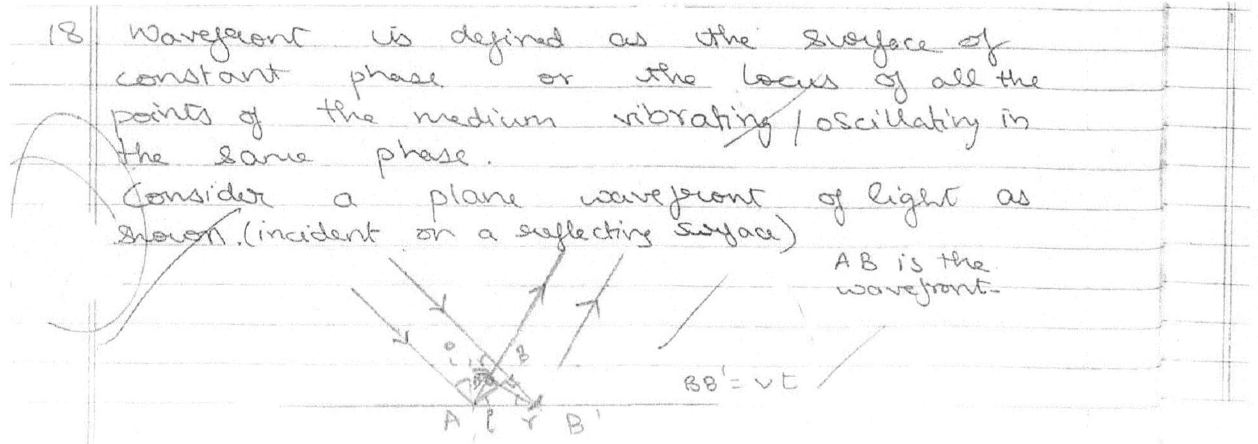

18 Wavefront is defined as the surface of constant phase or the locus of all the points of the medium vibrating / oscillating in the same phase.
Consider a plane wavefront of light as shown. (incident on a reflecting surface)

AB is the wavefront.

BB' = vt

A i r B'

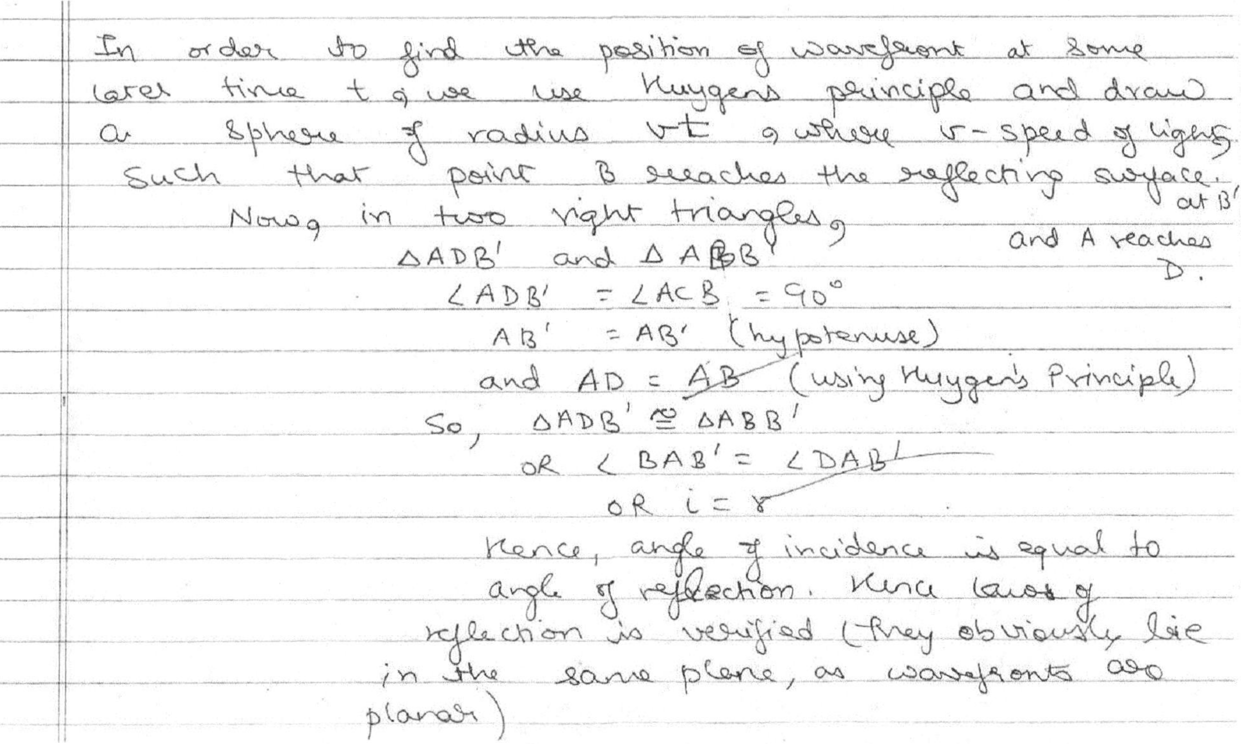

In order to find the position of wavefront at some later time t, we use Huygens principle and draw a sphere of radius vt, where v - speed of light. Such that point B reaches the reflecting surface at B' and A reaches D.
Now, in two right triangles,
ΔADB' and ΔABB'
∠ADB' = ∠ACB = 90°
AB' = AB' (hypotenuse)
and AD = AB (using Huygen's Principle)
So, ΔADB' ≅ ΔABB'
OR ∠BAB' = ∠DAB'
OR i = r
Hence, angle of incidence is equal to angle of reflection. Hence law of reflection is verified (they obviously lie in the same plane, as wavefronts are planar)

19. (a) Define the term 'self inductance' of a coil. Write its S.I. unit.

(b) A rectangular loop of sides a and b carrying current is kept at a distance 'a' from an infinitely long straight wire carrying current as shown in the figure. Obtain an expression for the resultant force acting on the loop.

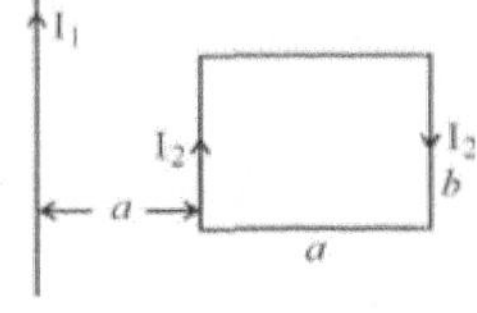

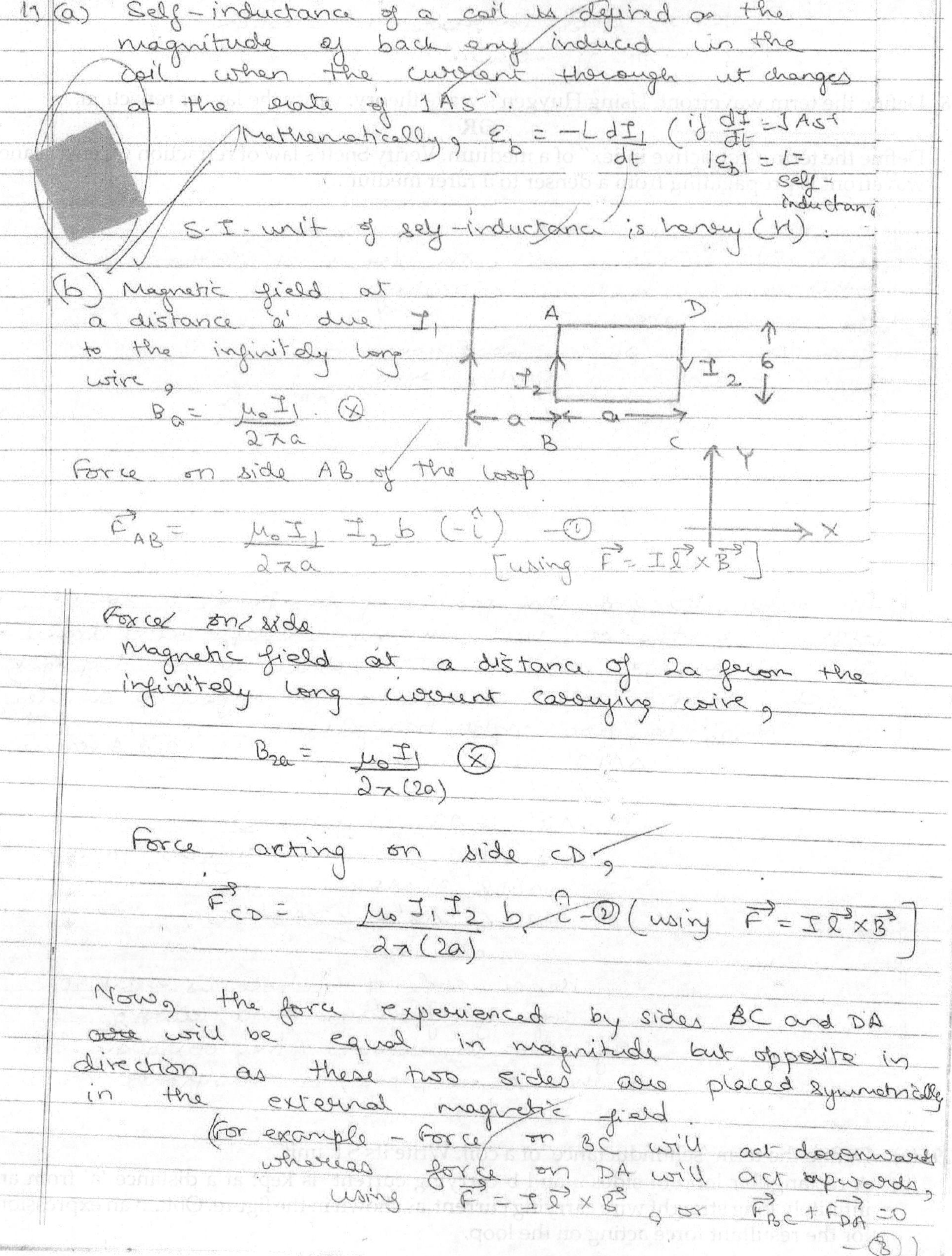

17 (a) Self-inductance of a coil is defined as the magnitude of back emf induced in the coil when the current through it changes at the rate of $1 A s^{-1}$.

Mathematically, $\mathcal{E}_b = -L \dfrac{dI_1}{dt}$ (if $\dfrac{dI}{dt} = 1 A s^{-1}$, $|\mathcal{E}_b| = L$ = self-inductance)

S.I. unit of self-inductance is henry (H).

(b) Magnetic field at a distance 'a' due to the infinitely long wire,

$$B_a = \frac{\mu_0 I_1}{2\pi a} \otimes$$

Force on side AB of the loop

$$\vec{F}_{AB} = \frac{\mu_0 I_1 I_2 b}{2\pi a}(-\hat{i}) \quad -① \quad [\text{using } \vec{F} = I\vec{l} \times \vec{B}]$$

~~Force on side~~

Magnetic field at a distance of 2a from the infinitely long current carrying wire,

$$B_{2a} = \frac{\mu_0 I_1}{2\pi(2a)} \otimes$$

Force acting on side CD,

$$\vec{F}_{CD} = \frac{\mu_0 I_1 I_2 b}{2\pi(2a)}\hat{i} \quad -② \quad \left(\text{using } \vec{F} = I\vec{l} \times \vec{B}\right)$$

Now, the force experienced by sides BC and DA ~~are~~ will be equal in magnitude but opposite in direction as these two sides are placed symmetrically in the external magnetic field.

(For example - Force on BC will act downwards whereas force on DA will act upwards, using $\vec{F} = I\vec{l} \times \vec{B}$, or $\vec{F}_{BC} + \vec{F}_{DA} = 0$ — ③)

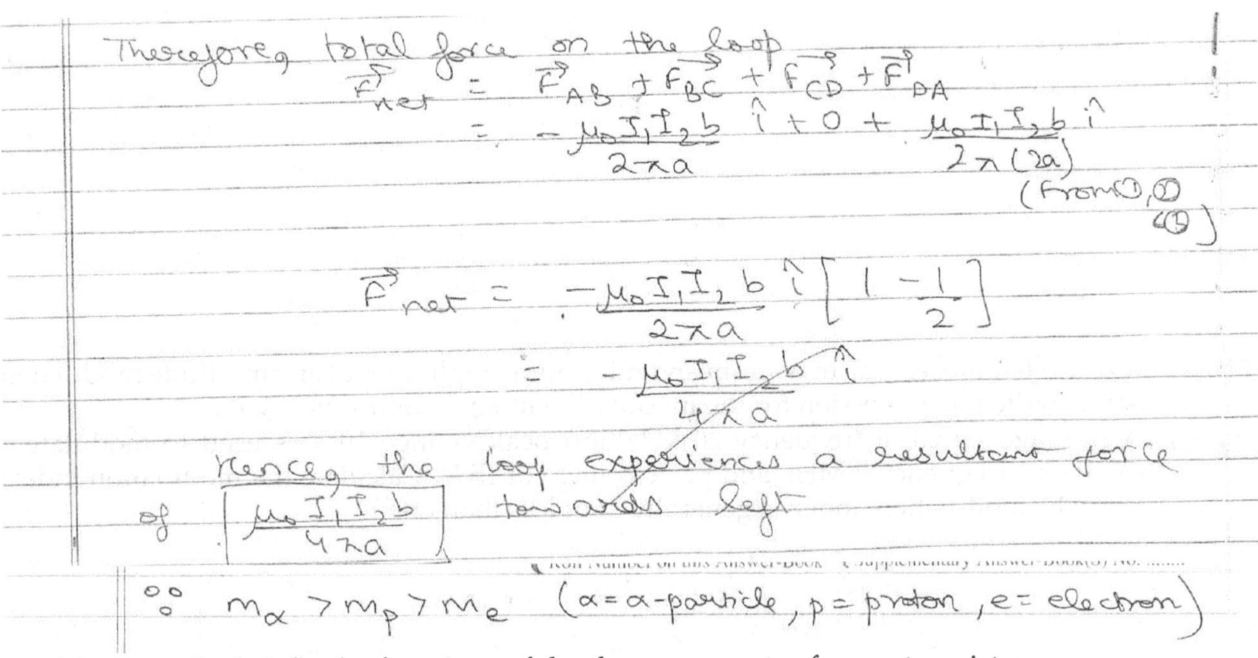

Therefore, total force on the loop

$$\vec{F}_{net} = \vec{F}_{AB} + \vec{F}_{BC} + \vec{F}_{CD} + \vec{F}_{DA}$$

$$= -\frac{\mu_0 I_1 I_2 b}{2\pi a}\hat{i} + 0 + \frac{\mu_0 I_1 I_2 b}{2\pi(2a)}\hat{i}$$

(From ①, ② & ③)

$$\vec{F}_{net} = -\frac{\mu_0 I_1 I_2 b}{2\pi a}\hat{i}\left[1 - \frac{1}{2}\right]$$

$$= -\frac{\mu_0 I_1 I_2 b}{4\pi a}\hat{i}$$

Hence, the loop experiences a resultant force of $\boxed{\frac{\mu_0 I_1 I_2 b}{4\pi a}}$ towards left

$\therefore m_\alpha > m_p > m_e$ (α = α-particle, p = proton, e = electron)

20. (a) Describe briefly the functions of the three segments of n-p-n transistor.

(b) Draw the circuit arrangement for studying the output characteristics of n-p-n transistor in CE configuration. Explain how the output characteristics is obtained.

OR

Draw the circuit diagram of a full wave rectifier and explain its working. Also, give the input and output waveforms.

20. Circuit Diagram of full wave rectifier.

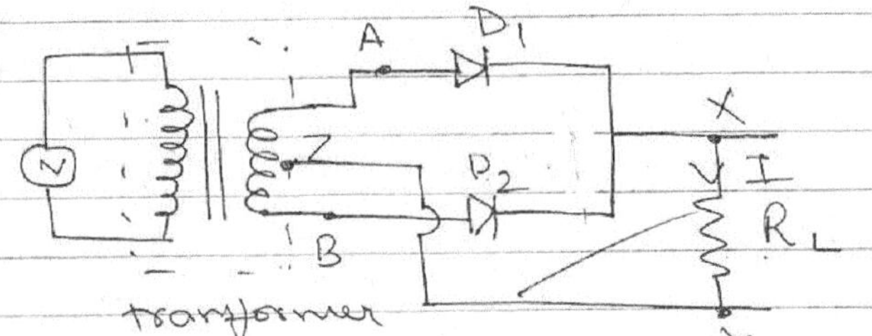

A, D_1, X, I, D_2, R_L, B, Y, transformer

Voltage forms

At A, 0, t; At B, 0, t; 0, D1, D2, D3, D4, t

Input waveforms; output waveforms.

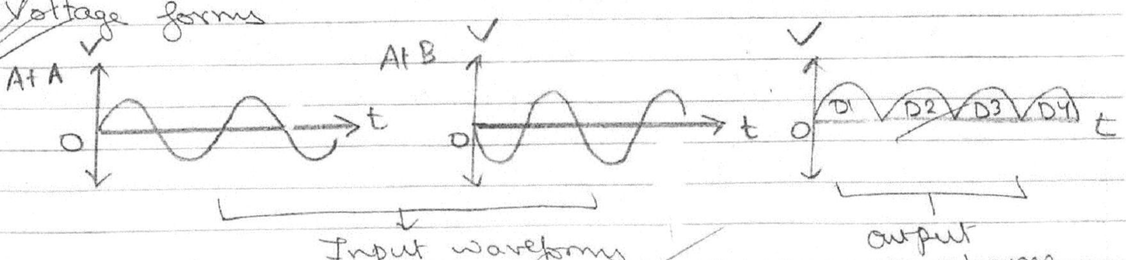

Rectifier works on the principle that a diode only conducts when it is forward biased and doesn't conduct practically when reverse biased. When voltage at A is higher than B, D1 is forward biased whereas D2 is reverse biased, therefore, current flows in the loop AXYZA and voltage appears across R_L.

When B is at a higher voltage than A, D2 is forward biased whereas D1 is reverse biased. Hence, current flows in the loop BXYZB and voltage appears across R_L. For both cases, voltage drop across R_L is unidirectional and hence, it is able to rectify the AC voltage to produce pulsating DC.

21. (a) If A and B represent the maximum and minimum amplitudes of an amplitude modulated wave, write the expression for the modulation index in terms of A & B.

(b) A message signal of frequency 20 kHz and peak voltage 10 V is used to modulate a carrier of frequency 2 MHz and peak voltage of 15 V. Calculate the modulation index. Why the modulation index is generally kept less than one ?

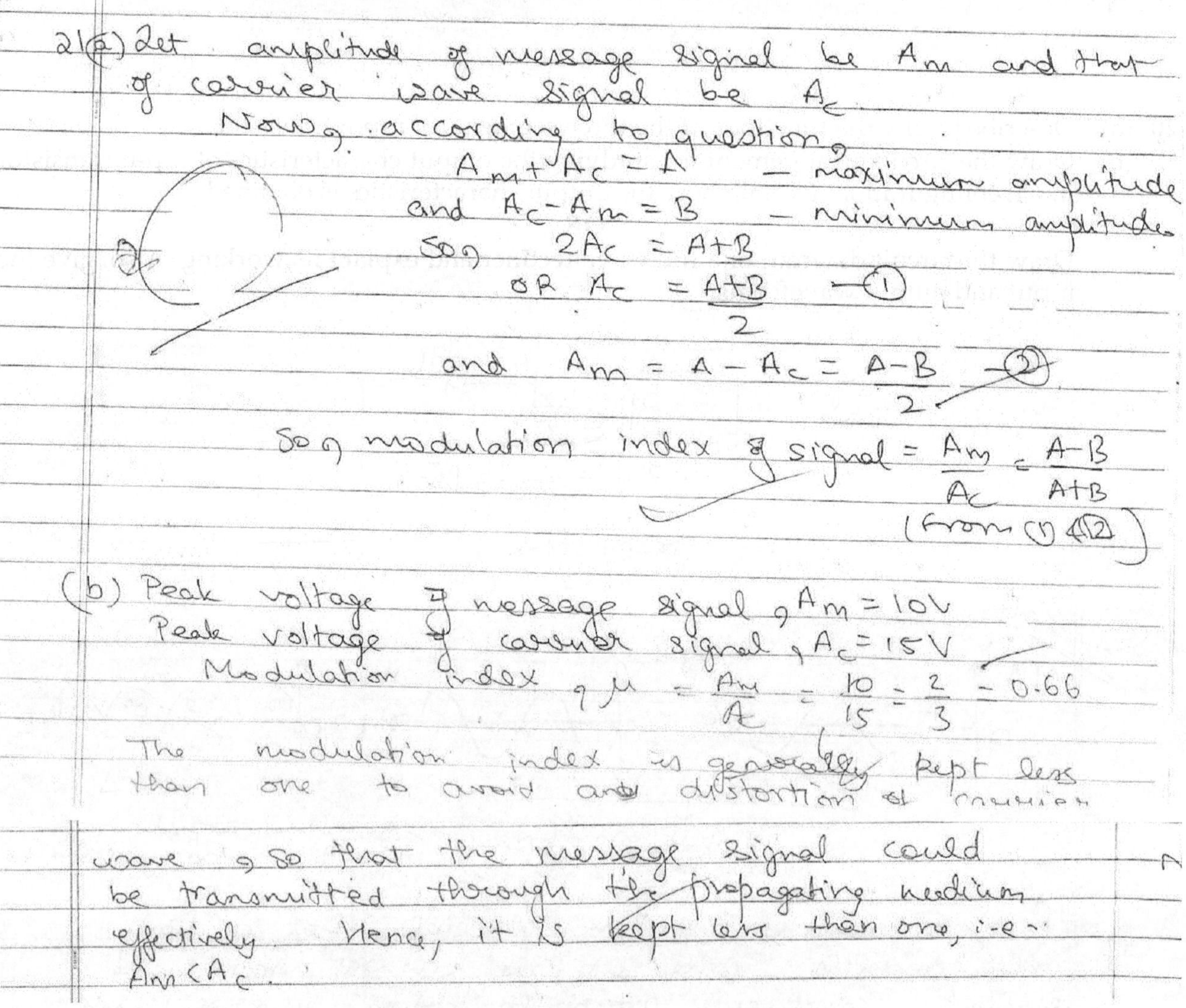

21(a) Let amplitude of message signal be A_m and that of carrier wave signal be A_c

Now, according to question,

$A_m + A_c = A$ — maximum amplitude

and $A_c - A_m = B$ — minimum amplitude

So, $2A_c = A+B$

OR $A_c = \frac{A+B}{2}$ —(1)

and $A_m = A - A_c = \frac{A-B}{2}$ —(2)

So, modulation index of signal $= \frac{A_m}{A_c} = \frac{A-B}{A+B}$ (from (1) & (2))

(b) Peak voltage of message signal, $A_m = 10V$

Peak voltage of carrier signal, $A_c = 15V$

Modulation index, $\mu = \frac{A_m}{A_c} = \frac{10}{15} = \frac{2}{3} = 0.66$

The modulation index is generally kept less than one to avoid any distortion of carrier wave, so that the message signal could be transmitted through the propagating medium effectively. Hence, it is kept less than one, i.e., $A_m < A_c$.

22. (a) State Gauss's law for magnetism. Explain its significance.

(b) Write the four important properties of the magnetic field lines due to a bar magnet.

OR

Write three points of differences between para-, dia- and ferro- magnetic materials, giving one example for each.

22.

Paramagnetic Substances	Diamagnetic Substances	Ferromagnetic Substances
(i) Are weakly attracted by the external magnetic field.	(i) Are weakly repelled by the external magnetic field	(i) Are strongly attracted by the magnetic field.
(ii) Relative magnetic permeability ≥ 1 but not very high.	(ii) Relative magnetic permeability less than 1 but >0.	(ii) Relative magnetic permeability very high >1000
(iii) Gets magnetised in same direction $\chi > 0$ and has unpaired electrons.	(iii) Gets magnetised in opposite directions, $\chi < 0$ and has paired electrons	(iii) Gets strongly magnetised in the same direction $\chi >> 1$. Consists of powerful domains

2

Behaviour in external magnetic field

N S . N S N S

e.g. Aluminium e.g. Bismuth e.g. Iron

23. (a) Identify the part of the electromagnetic spectrum used in (i) radar and (ii) eye surgery. Write their frequency range.

(b) Prove that the average energy density of the oscillating electric field is equal to that of the oscillating magnetic field.

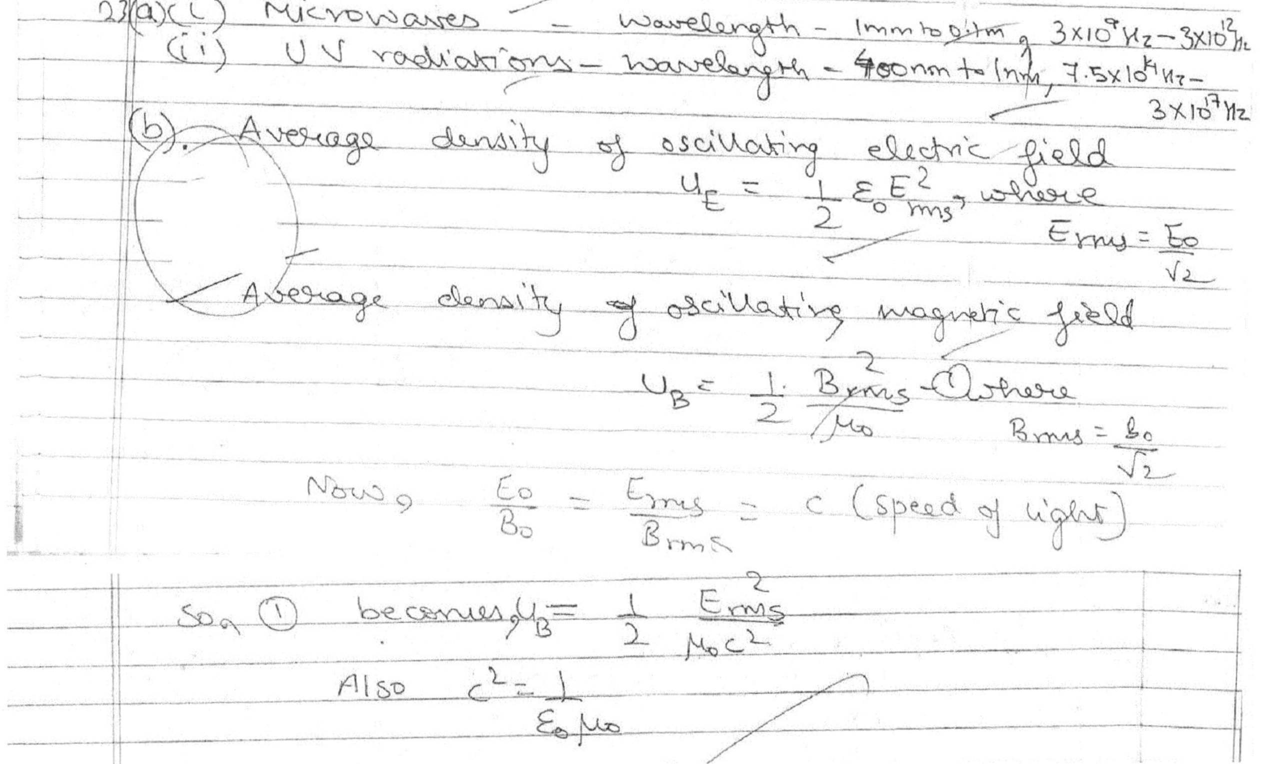
23(a)(i) Microwaves – wavelength – 1mm to 0.1m, 3×10^{9} Hz – 3×10^{12} Hz

(ii) UV radiations – wavelength – 400nm to 1nm, 7.5×10^{14} Hz – 3×10^{17} Hz

(b). Average density of oscillating electric field

$$u_E = \frac{1}{2}\varepsilon_0 E_{rms}^2, \text{ where } E_{rms} = \frac{E_0}{\sqrt{2}}$$

Average density of oscillating magnetic field

$$u_B = \frac{1}{2}\frac{B_{rms}^2}{\mu_0} \text{ ——(1) where } B_{rms} = \frac{B_0}{\sqrt{2}}$$

Now, $\frac{E_0}{B_0} = \frac{E_{rms}}{B_{rms}} = c$ (speed of light)

So, (1) becomes, $u_B = \frac{1}{2}\frac{E_{rms}^2}{\mu_0 c^2}$

Also $c^2 = \frac{1}{\varepsilon_0 \mu_0}$

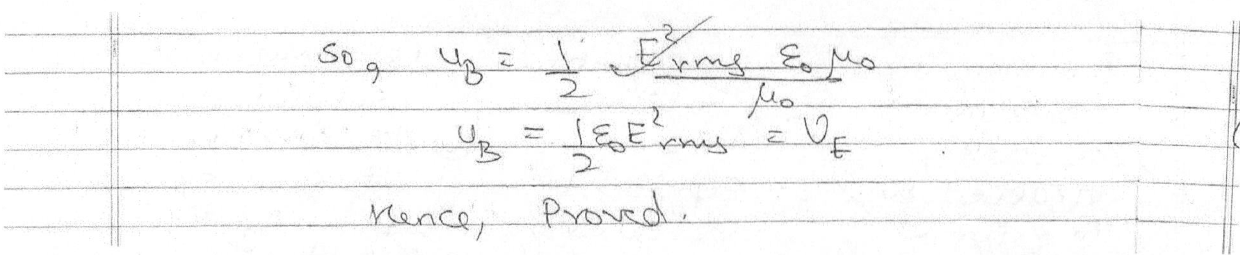

So, $u_B = \frac{1}{2} \frac{E^2_{rms} \varepsilon_0 \mu_0}{\mu_0}$

$u_B = \frac{1}{2}\varepsilon_0 E^2_{rms} = U_E$

Hence, Proved.

24. (a) Three photo diodes D1, D2 and D3 are made of semiconductors having band gaps of 2.5 eV, 2 eV and 3 eV respectively. Which of them will not be able to detect light of wavelength 600 nm ?

(b) Why photodiodes are required to operate in reverse bias ? Explain.

24. (a) Wavelength of light used = 600 nm (λ)

Energy associated with light $= \frac{hc}{\lambda}$

$= \frac{1240}{600}$ eV

$= 2.067$ eV

$\therefore$ D_1 & D_3 will not be able to detect light as energy associated with the wave is less than their band gap (2.5 eV & 3 eV respectively)

So, no e-h generation will take place.

(b) Photo-diodes are required to operate in reverse bias, because the minority carriers lead to current in this case.

For e.g. Consider an n-type semiconductor,

Let no. of electrons = n

no. of holes = p

Initially, $n >> p$. (n are majority charge carriers)

Now when light of suitable energy falls on it, no. of holes generated = Δp

no. of electrons generated = Δn

Clearly, $\Delta n = \Delta p$

So, fractional increase in holes $\frac{\Delta p}{p} >> \frac{\Delta n}{n}$

Hence, light waves of different intensities can be easily distinguished in reverse bias voltage, as fractional increase in carriers as well as current is more noticeable.

SECTION – D

25. (a) Describe any two characteristic features which distinguish between interference and diffraction phenomena. Derive the expression for the intensity at a point of the interference pattern in Young's double slit experiment.

(b) In the diffraction due to a single slit experiment, the aperture of the slit is 3 mm. If monochromatic light of wavelength 620 nm is incident normally on the slit, calculate the separation between the first order minima and the 3rd order maxima on one side of the screen. The distance between the slit and the screen is 1.5 m.

OR

(a) Under what conditions is the phenomenon of total internal reflection of light observed ? Obtain the relation between the critical angle of incidence and the refractive index of the medium.

(b) Three lenses of focal lengths +10 cm, –10 cm and +30 cm are arranged coaxially as in the figure given below. Find the position of the final image formed by the combination.

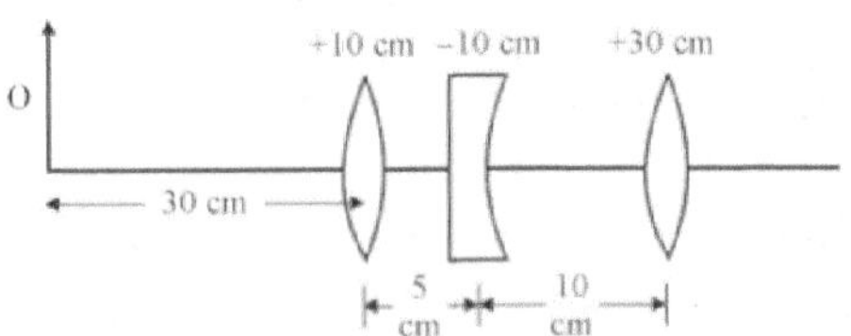

25. (a) when light travels from an optically denser medium towards an optically rarer medium, and is incident on the interface at an angle greater than the critical angle for the given pair of media, we observe total internal reflection of light.

n_1

No transmission

n_2

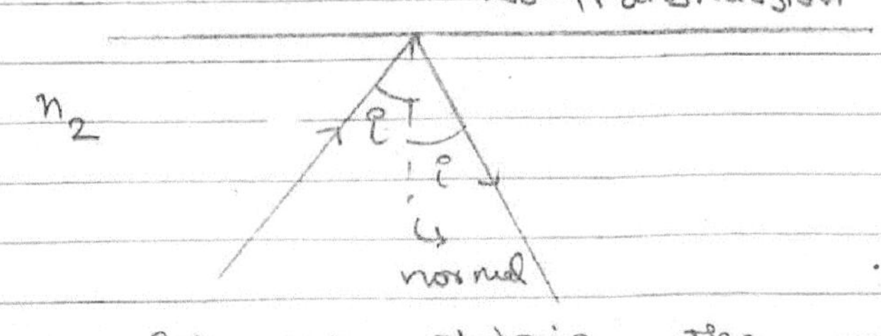

$n_2 > n_1$; and $i > \theta_c$ (critical angle)

normal

Now let us obtain the angle for which the light ray is just transmitted and grazes along the surface.

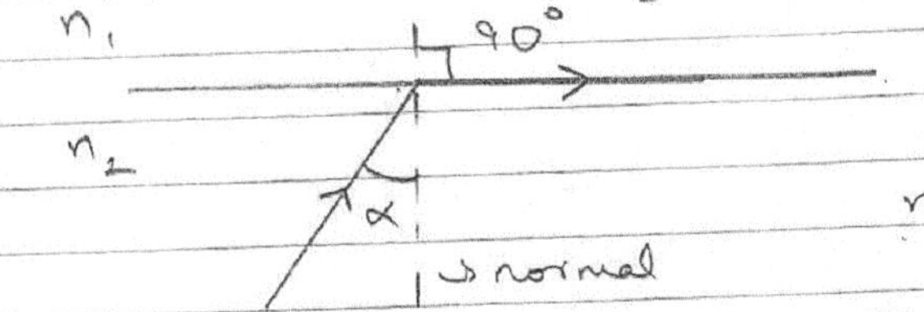

$n_2 > n_1$

n_1 90°

n_2 α normal

Applying snell's law at the interface,

$$n_2 \sin\alpha = n_1 \sin 90^\circ$$

$$\sin\alpha = \frac{n_1}{n_2}$$

[n_1 & n_2 are the refractive indices of the medium respectively]

$$\alpha = \sin^{-1}\left(\frac{n_1}{n_2}\right)$$

OR $\alpha = \sin^{-1}\left(\frac{1}{n_{21}}\right)$ [n_{21} is refractive index of denser medium w.r.t rarer medium]

Now, it is clear that if we increase the angle of incidence greater than α, no transmission of light will take place and light will be totally internally reflected. Thus, the critical angle for the pair of mediums $= \alpha = \sin^{-1}\left(\frac{1}{n_{21}}\right)$ ($\because \theta_c$ is the minimum angle above which no transmission of light takes place)

Hence, $\boxed{\theta_c = \alpha = \sin^{-1}\left(\frac{1}{n_{21}}\right)}$

(b) +10cm −10cm +30cm

← 30 cm →

← 5cm →← 10cm →

First, refraction from convex lens of focal length $f_1 = +10$ cm

Object distance, $u_1 = -30$ cm (u)

focal length, $f_1 = +10$ cm (f)

Lens formula,

$$\frac{1}{v} - \frac{1}{u} = \frac{1}{f}$$

Substituting the values,

$$\frac{1}{v} + \frac{1}{30} = \frac{1}{10}$$

$$\frac{1}{v} = \frac{1}{10} - \frac{1}{30}$$

$$\frac{1}{v} = \frac{20}{300}$$

$$v = +15\text{cm}$$

Now this image will act as a virtual object for concave lens of focal length $= -10$ cm.

So, object distance, $u_2 = (15-5)$ cm $= 10$ cm (u)

focal length, $f_2 = -10$ cm (f)

Lens formula

$$\frac{1}{v} - \frac{1}{u} = \frac{1}{f}$$

Substituting values,

$$\frac{1}{v} - \frac{1}{10} = \frac{-1}{10}$$

$$\frac{1}{v} = 0$$

OR $v \to \infty$ (image formed at infinity)

Now, the image will act as a virtual object for convex lens of focal length = +30 cm

So, object distance, $u_3 = \infty - 30$ cm $= \infty$ (u)

focal length, $f_3 = +30$ cm (f)

Lens formula:

$$\frac{1}{v} - \frac{1}{u} = \frac{1}{f}$$

Substituting values,

$$\frac{1}{v} - \frac{1}{\infty} = \frac{1}{30}$$

$$v = +30 \text{ cm}$$

Hence, the final image will be formed at 30 cm to the right of convex lens of focal length = 30 cm, or at a distance of 75 cm to the right of the object.

26. (a) Describe briefly the process of transferring the charge between the two plates of a parallel plate capacitor when connected to a battery. Derive an expression for the energy stored in a capacitor.

(b) A parallel plate capacitor is charged by a battery to a potential difference V. It is disconnected from battery and then connected to another uncharged capacitor of the same capacitance. Calculate the ratio of the energy stored in the combination to the initial energy on the single capacitor.

OR

(a) Derive an expression for the electric field at any point on the equatorial line of an electric dipole.

(b) Two identical point charges, q each, are kept 2m apart in air. A third point charge Q of unknown magnitude and sign is placed on the line joining the charges such that the system remains in equilibrium. Find the position and nature of Q.

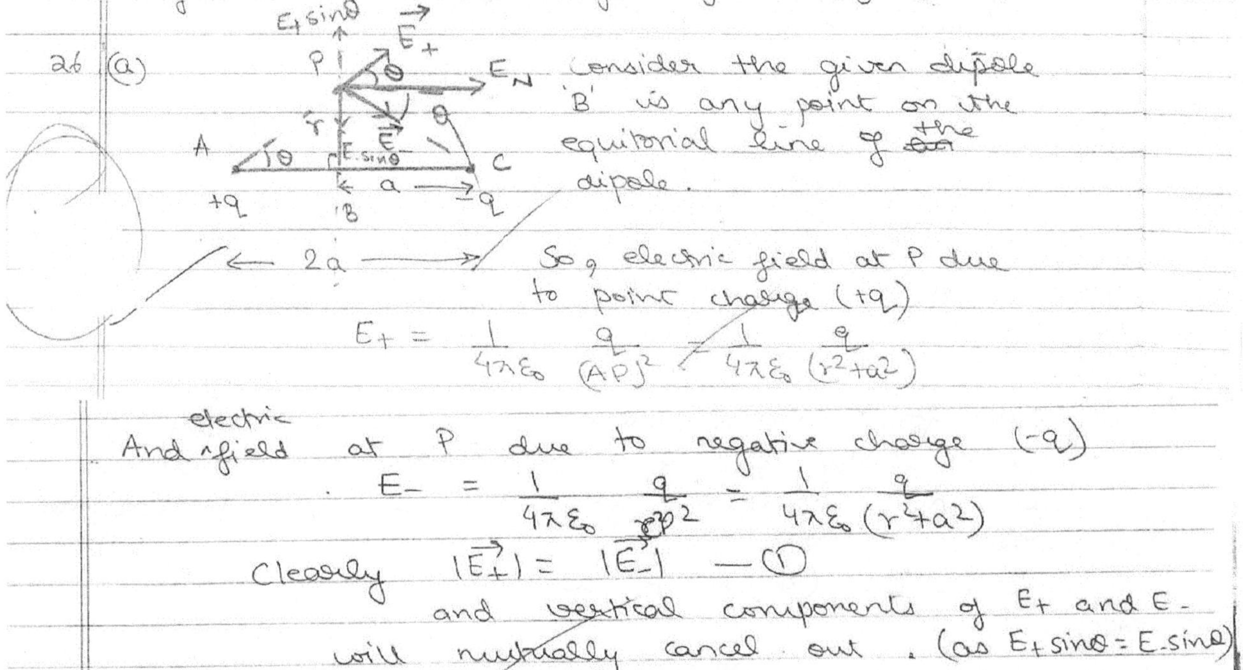

∴ net electric field is only along the horizontal as shown.

So, net electric field, $E_N = E_+ \cos\theta + E_- \cos\theta$

$E_N = 2E_+ \cos\theta$ (from ①)

Now in right ΔABP

$\cos\theta = \frac{AB}{AP} = \frac{a}{\sqrt{r^2+a^2}}$

So, $E_N = 2 \frac{1}{4\pi\varepsilon_0} \frac{q}{(r^2+a^2)} \frac{a}{(r^2+a^2)^{1/2}}$

OR $E_N = \frac{1}{4\pi\varepsilon_0} \frac{(2qa)}{(r^2+a^2)^{3/2}} = \frac{1}{4\pi\varepsilon_0} \frac{p}{(r^2+a^2)^{3/2}}$

and vectorially,

$\vec{E_N} = \frac{-1}{4\pi\varepsilon_0} \frac{\vec{p}}{(r^2+a^2)^{3/2}}$,

$\vec{p}$ is the dipole moment of the given dipole and is from right to left. ($p = 2qa$)

The negative sign indicates that $\vec{E_N}$ is opposite to $\vec{p}$ as shown.

For $r >> a$

$\vec{E_N} = \frac{-1}{4\pi\varepsilon_0} \frac{\vec{p}}{r^3}$ (ignoring the sum)

(b) B q — Q — q A ; $2-x$, x ; 2 m

Let Q be placed in between the charges as shown, at a distance of x from A as shown.

Now, for system to be in equilibrium, charges, q, q & Q should be in equilibrium.

Now total force on Q

$F_Q = \frac{-1}{4\pi\varepsilon_0} \frac{qQ}{(2-x)^2} + \frac{1}{4\pi\varepsilon_0} \frac{qQ}{x^2}$ (assuming right direction to be positive and Q to be negative)

So, $F_Q = 0 \Rightarrow (2-x)^2 = x^2$ OR $x = 1\text{ m}$

(Condition of equilibrium)

Now for any of the q to be at equilibrium it is necessary for Q to be negative to counteract for any repulsion faced by it due to the other positive charge Q.

So, F_q (q placed at A)

$$F_q = \frac{1}{4\pi\varepsilon_0}\frac{q^2}{(2)^2} - \frac{1}{4\pi\varepsilon_0}\frac{qQ}{(1)^2}$$

For equilibrium, $F_q = 0$

OR $\frac{q^2}{4} = \frac{qQ}{1}$

OR $|Q| = \frac{q}{4}$

Hence, for the entire system to be in equilibrium, Q should be placed in between the two charges (at a distance of 1m from either of them) and its value should be $-\frac{q}{4}$.

27. (a) In a series LCR circuit connected across an ac source of variable frequency, obtain the expression for its impedance and draw a plot showing its variation with frequency of the ac source.

(b) What is the phase difference between the voltages across inductor and the capacitor at resonance in the LCR circuit ?

(c) When an inductor is connected to a 200 V dc voltage, a current of 1A flows through it. When the same inductor is connected to a 200 V, 50 Hz ac source, only 0.5 A current flows. Explain, why ? Also, calculate the self inductance of the inductor.

OR

(a) Draw the diagram of a device which is used to decrease high ac voltage into a low ac voltage and state its working principle. Write four sources of energy loss in this device.

(b) A small town with a demand of 1200 kW of electric power at 220 V is situated 20 km away from an electric plant generating power at 440 V. The resistance of the two wire line carrying power is 0.5 per km. The town gets the power from the line through a 4000-220 V step-down transformer at a sub-station in the town. Estimate the line power loss in the form of heat.

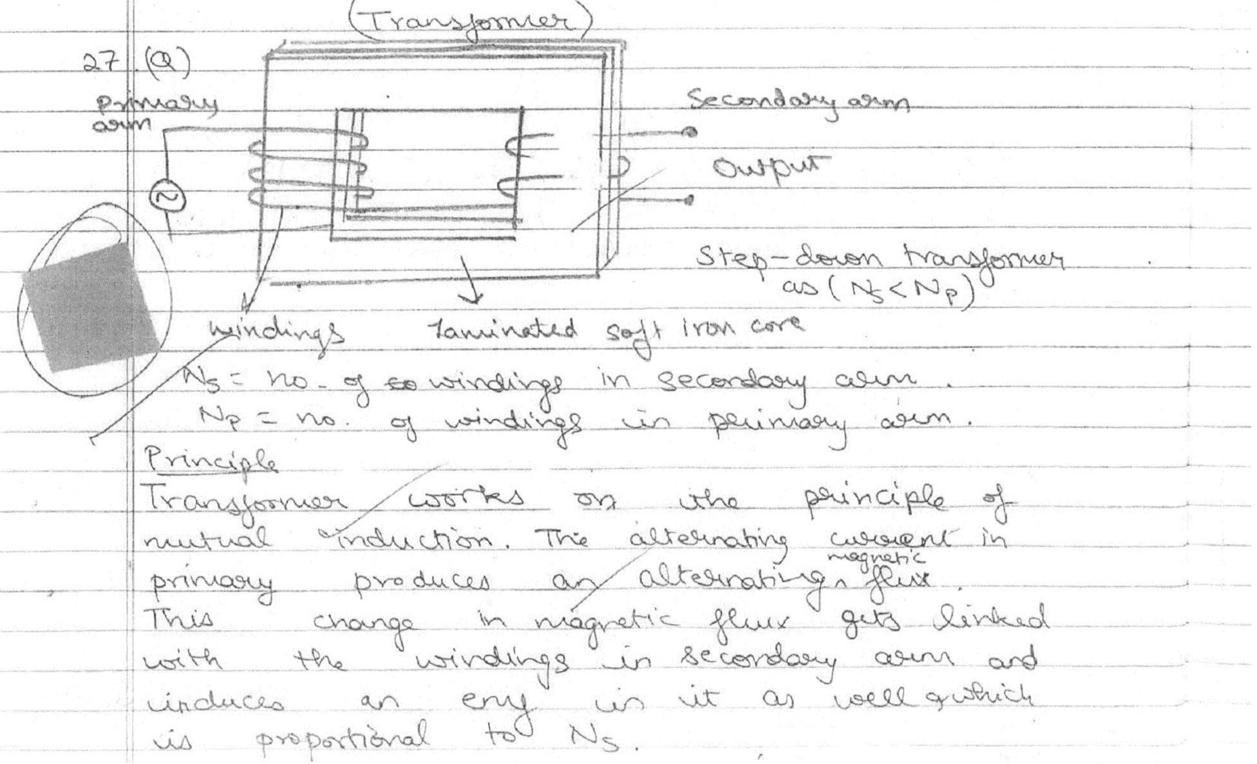

Namely, $\frac{\varepsilon_P}{\varepsilon_S} = \frac{N_P}{N_S}$ (where ε_P is the emf of primary source and ε_S is the ...

Sources of Energy loss

1) Resistive losses – Copper loss – both in wires of primary arm as well as secondary arm (if it is closed), as a wire always has finite resistance.
2) Flux leakage – The entire flux produced in primary coil may not get linked with the secondary coil and some flux may leak through the core.
3) Hysterisis loss – Due to repeated magnetisation and demagnetisation of core, some energy loss takes place which is proportional to the area of the hysteris loop.
4) Eddy current losses – Eddy currents may be induced in the core due to changing magnetic flux through it, leading to large losses of energy in the form of heat.

(b) No. of wires = 2

Resistance per km = 0.5 ohm km^{-1}

Distance between town and power plant = 20 km

Total resistance of path = $2 \times 0.5 \times 20$ = 20 ohm (R)

Now the town receives a power of 1200 kW from the power plant at voltage of 4000 V ($\because$ step down transformer is 4000–220 V, and the transformers are considered to be ideal – so no power loss).

So, current flowing = $\frac{\text{Power}}{\text{Voltage}} = \frac{1200 \times 10^3}{4000}$ A = 300 A (I)

So, line power loss in the form of heat $= I^2 R$

$= (3 \times 10^2)^2 \times 20$ W

$= 9 \times 20 \times 10^4$ W = 1800 kW

Hence, 1800 kW of energy is lost while transmitting electricity.

PHYSICS

Outside Delhi [Set-I]

2020

Time allowed : 3 hours **Maximum marks : 70**

General Instructions:

Read the following instructions very carefully and strictly follow them:

(i) *This question paper comprises **four** sections—**A, B, C** and **D**.*

(ii) *There are **37** questions in the question paper. **All** questions are compulsory.*

(iii) ***Section A: Q. no. 1 to 20** are very short-answer type questions carrying **1** mark each.*

(iv) ***Section B: Q. no. 21 to 27** are short-answer type questions carrying **2** marks each.*

(v) ***Section C: Q. no. 28 to 34** are long-answer type questions carrying **3** mark each.*

(vi) ***Section D: Q. no. 35 to 37** are also long-answer type questions carrying **5** mark each.*

(vii) *There is no overall choice in the question paper. However, an internal choice has been provided in **two** questions of **one** mark, **two** questions of **two** marks, **one** question of **three** marks and all the **three** questions of **five** marks. You have to attempt **only one** of the choices in such questions.*

(viii) *However, separate instructions are given with each section and question, wherever necessary.*

(ix) *Use of calculators and log tables is not permitted.*

(x) *You may use the following values of physical constants wherever necessary.*

$c = 3 \times 10^8$ m/s

$h = 6.63 \times 10^{-34}$ Js

$e = 1.6 \times 10^{-19}$ C

$\mu_0 = 4\pi \times 10^{-7}$ T m A^{-1}

$\varepsilon_0 = 8.854 \times 10^{-12}$ C^2 N^{-1} m^{-2}

$\frac{1}{4\pi\varepsilon_0} = 9 \times 10^9$ N m^2 C^{-2}

Mass of electron (m_e) = 9.1×10^{-31} kg

Mass of neutron = 1.673×10^{-27} kg

Mass of proton = 1.673×10^{-27} kg

Avogadro's number = 0.023×10^{23} per gram mole

Boltzmann constant = 1.38×10^{-23} JK^{-1}

SECTION-A

Note : Select the most appropriate option from those given below each question.

1. A cell of emf (E) and internal resistance r is connected across a variable external resistance R. The graph of terminal potential difference V as a function of R is :

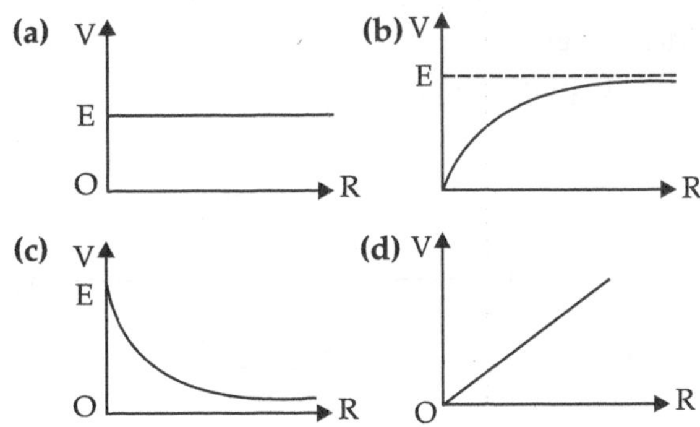

[1]

Ans. 1. **(b)**

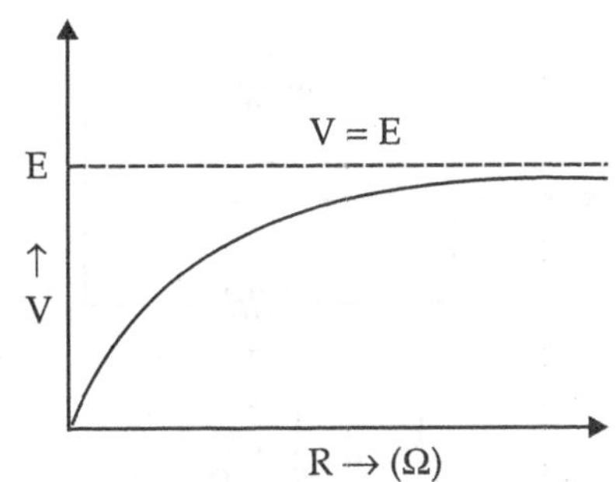

Explanation :

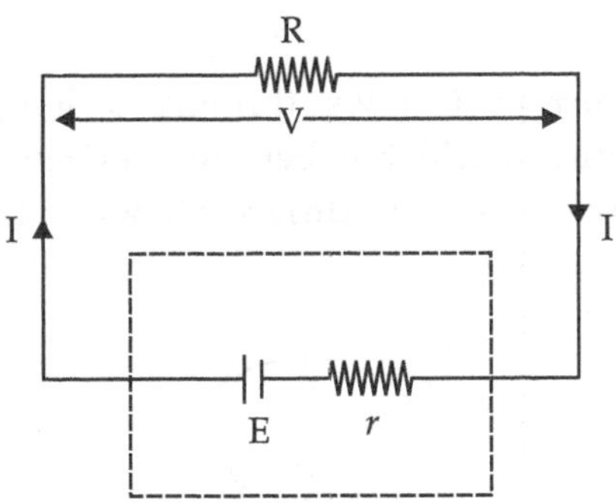

Current in circuit $I = \frac{E}{R+r}$ ...(i)

Potential difference across R will be

$$V = IR \quad ...(ii)$$

Putting the value of I from (i) in (ii)

$$V = \frac{ER}{R+r}$$

$$V = \frac{E}{\left(1+\frac{r}{R}\right)}$$

When, $R \to \infty, V = E$

$R \to 0, V = 0$

2. A uniform wire of resistance 2R is bent in the form of a circle. The effective resistance between the ends of any diameter of the circle is :

(a) 2R **(b) R**

(c) $\frac{R}{2}$ **(d) $\frac{R}{4}$** **[1]**

Ans. 2. (c)

Explanation :

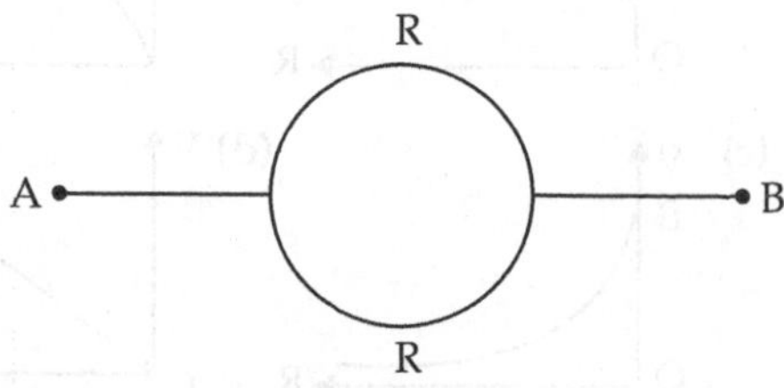

If a loop is made, then the resistances will be in parallel.

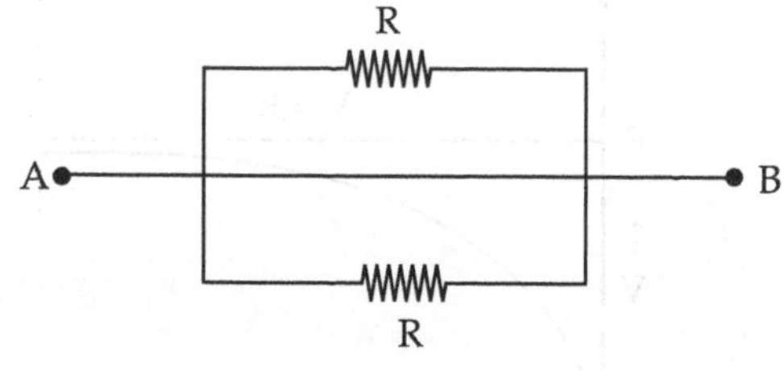

$$\frac{1}{R'} = \frac{1}{R} + \frac{1}{R} = \frac{2}{R}$$

$$R' = \frac{R}{2}$$

3. A current I flows through a long straight conductor which is bent into a circular loop of radius R in the middle as shown in the figure.

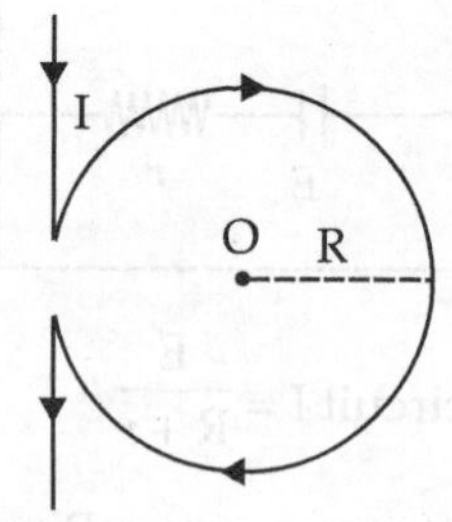

The magnitude of the net magnetic field at point O will be :

(a) Zero **(b) $\frac{\mu_0 I}{2R}(1+\pi)$**

(c) $\frac{\mu_0 I}{4\pi R}$ **(d) $\frac{\mu_0 I}{2R}\left(1-\frac{1}{\pi}\right)$** **[1]**

Ans. 3. (d)

Explanation :

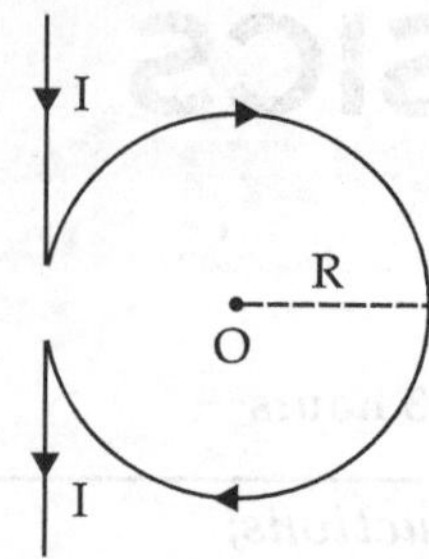

Let $\hat{n}$ be the unit vector $\perp r$ to the plane of paper.

Magnetic field $\vec{B}$ at O due to current through the entire structure is :

$$\vec{B} = \frac{\mu_0}{4\pi}\frac{2\pi I}{R}(\hat{n}) + \frac{\mu_0}{4\pi}\frac{2I}{R}(-\hat{n})$$

$$\vec{B} = \frac{\mu_0 I}{2R}\left[1 - \frac{1}{\pi}\right]\hat{n}$$

$$\vec{B} = \frac{\mu_0 I}{2R}\left[1 - \frac{1}{\pi}\right] \text{ downward}$$

4. A circular loop of radius *r*, carrying a current I lies in *y-z* plane with its centre at the origin. The net magnetic flux through the loop is :

(a) directly proportional to *r*

(b) zero

(c) inversely proportional to *r*

(d) directly proportional to I **[1]**

Ans. 4. (b)

Explanation :

Circular loop behaves as a magnetic dipole, whose one surface will be N-pole and another will be S-pole. Therefore, magnetic lines of force emerges from north and will meet at south. Hence, total magnetic flux through *y-z* plane is zero.

5. The kinetic energy of a proton and that of an α-particle are 4 eV and 1 eV, respectively. The ratio of the de-Broglie wavelengths associated with them, will be :

(a) 2 : 1 **(b) 1 : 1**

(c) 1 : 2 **(d) 4 : 1** **[1]**

Ans. 5. (b)

Explanation :

$$\lambda = \frac{h}{\sqrt{2m\,(K.E)}}$$

$$\frac{\lambda_{p^+}}{\lambda_\alpha} = \sqrt{\frac{m\alpha}{m_{p^+}}} \times \sqrt{\frac{K.E_\alpha}{K.E_{p^+}}}$$

$$\frac{\lambda_{p^+}}{\lambda_\alpha} = \sqrt{\frac{4m_{p^+}}{m_{p^+}}} \times \sqrt{\frac{1}{4}}$$

$$\frac{\lambda_{p^+}}{\lambda_\alpha} = \frac{2\times 1}{1\times 2} = \frac{1}{1} = 1:1$$

6. A photocell connected in an electrical circuit is placed at a distance 'd' from a source of light. As a result, current I flows in the circuit. What will be the current in the circuit when the distance is reduced to '$d/2$' ?

(a) I **(b) 2I**

(c) 4I **(d) 1/2** **[1]**

Ans. 6. **(c)**

Explanation :

Intensity of light $\propto \frac{1}{d^2}$

current $\propto$ intensity of light

At distance d, current is I.

At distance $d/2$, intensity of light becomes four times of its initial value.

So, current will also become 4 times.

7. A current of 10 A is flowing from east to west in a long straight wire kept on a horizontal table. The magnetic field developed at a distance of 10 cm due north on the table is :

(a) 2×10^{-5} T, acting downwards

(b) 2×10^{-5} T, acting upwards

(c) 4×10^{-5} T, acting downwards

(d) 4×10^{-5} T, acting upwards **[1]**

Ans. 7. **(a)**

Explanation :

$$\vec{B} = \frac{\mu_0 I}{2\pi r}$$

$$\vec{B} = \frac{4\pi \times 10^{-7} \times 10}{2\pi \times 10 \times 10^{-2}}$$

$$\vec{B} = 2 \times 10^{-5} \text{ T}$$

By using, right hand thumb rule it should be acting downwards.

8. When a wave undergoes reflection at an interface from rarer to denser medium, adhoc change in its phases is :

(a) $\frac{\pi}{2}$ **(b) 0**

(c) π **(d)** $\frac{\pi}{4}$ **[1]**

Ans. 8. **(c)**

Explanation :

When light is reflected from rarer to denser medium there is a phase change of π.

9. Paschen series of atomic spectrum of hydrogen gas lies in :

(a) Infrared region

(b) Ultraviolet region

(c) Visible region

(d) Partly in ultraviolet and partly in visible region **[1]**

Ans. 9. **(a)**

Explanation :

$$\bar{v} = R\left[\frac{1}{3^2} - \frac{1}{{n_2}^2}\right], n_2 = 4, 5, 6$$

Lies in the infrared region.

10. In the α-particle scattering experiment, the shape of the trajectory of the scattered α-particles depend upon :

(a) only on impact parameter

(b) only on the source of α-particles

(c) both impact parameter and source of α-particles.

(d) impact parameter and the screen material of the detector. **[1]**

Ans. 10. **(a)**

Explanation :

The shape of the trajectory of the scattered α-particles depends on the impact parameter.

$$b = \frac{kze^2 \cot \theta/2}{\frac{1}{2}mv^2}$$

Note : Fill in the blanks with appropriate answer :

11. Torque acting on an electric dipole placed in an electric field is maximum when the angle between the electric field and the dipole moment is .... **[1]**

Ans. $\theta = 90°$

Explanation :

$$\vec{\tau} = PE \sin \theta$$

$$\theta = 90°$$

$$\vec{\tau}_{max} = PE$$

12. A proton released from rest in an electric field, will start moving towards a region of .... potential in the field. **[1]**

Ans. decreasing

Explanation :

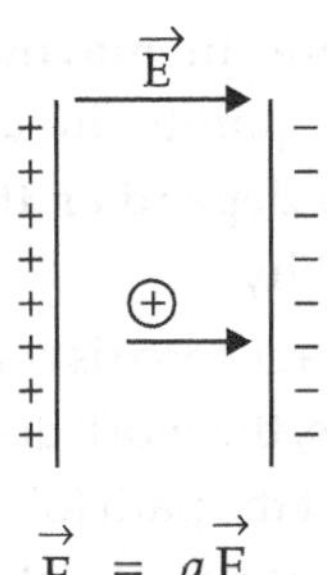

$$\vec{F} = q\vec{E}$$

Negative polarity attract positive proton.

13. To minimize the percentage error in the determination of unknown resistance of a conductor in meter bridge experiment, the balance point is adjusted near .... of the wire. **[1]**

OR

In potentiometer, a long uniform wire is used to ....potential gradient along the wire. **[1]**

Ans. mid-point (middle) or centre of wire.

Explanation :

A meter bridge is most sensitive when all four resistances are nearly equal. In that case, the null point is very close to the middle of the wire.

OR

Decrease

Explanation :

Since, $K = \frac{V}{L}$

When length increases potential gradient decreases. Hence, potentiometer becomes more sensitive.

14. Unpolarised light of intensity I_o is incident on two crossed polaroids. The intensity of light transmitted by the combination will be .... **[1]**

Ans. Zero

Explanation : As first polaroid half the light *i.e.*, $\frac{I_0}{2}$. Then the next polaroid is placed at 90° with first polaroid.

Then, $I' = \frac{I_0}{2} \cos^2 90° = 0$

15. Name the particle emitted spontaneously in the following nuclear reaction :

$$^{32}_{15}P \longrightarrow {}^{32}_{16}S + \bar{v} + \text{\_\_\_\_\_\_}$$ **[1]**

Answer the following :

Ans. β^- – particle $\left({}^{0}_{-1}e\right)$

Equation : $^{32}_{15}P \longrightarrow {}^{32}_{16}S + \bar{v} + {}^{\circ}_{-1}e$

16. The work done in moving a charge particle between two points in an uniform electric field, does not depend on the path followed by the particle. Why ? **[1]**

Ans. Since electrostatic force is conservative in nature. In other words, the work done by electric field is independent of the path followed by the particle.

17. Define 'magnetic declination' at a place earth. **[1]**

Ans. The angle between the geographical meridian and the magnetic meridian at a place is called the magnetic declination (α) at a place.

18. An A.C. source with variable frequency is connected to a parallel plate capacitor. How will the displacement current be affected with the decrease in frequency of the source ? **[1]**

Ans. When frequency of A. C. source is decreased, resistance increases so flow of current decreases. By dropping the variable frequency there will be drop or decrease in displacement current through the capacitor as it carries displacement current.

Since $I_C = \frac{V}{X_C} = \frac{V}{\frac{I}{\omega C}} = \omega CV$

19. An astronomical telescope may be a refracting type or a reflecting type. Which of the two produces image of better quality ? Justify your answer. **[1]**

Ans. The astronomical telescope of reflecting type produces better quality of image. In these telescopes, mirror of large aperture of high resolving power are used due to which image is free from chromatic and spherical aberration.

20. Can a slab of *p*-type semi-conductor be physically joined to another *n*-type semiconductor slab to form *p-n* junction ? Justify your answer. **[1]**

OR

In a *p-n* juction diode the forward bias resistance is low as compared to the reverse bias resistance. Give reason. **[1]**

Ans. No, any slab, howsoever flat will have roughness much larger than the interatomic crystal spacing (2Å to 3Å) and hence continuous contact at the atomic junction will not be possible.The junction will behave as discontinuous for the following charge carriers.

OR

When the *p-n* junction diode is forward biased, the diode's depletion layer or potential barrier shrinks. This can be considered as a practical short circuit.

Hence, resistance is negligible and current flow in forward bias is maximum.

Similarly, when it is reverse bias, depletion layer width (or potential barrier) increases. Hence, resistance increases and very small amount of current flows in reverse bias.

SECTION-B

21. Find the total charge stored in the network of capacitors connected between A and B as shown in figure :

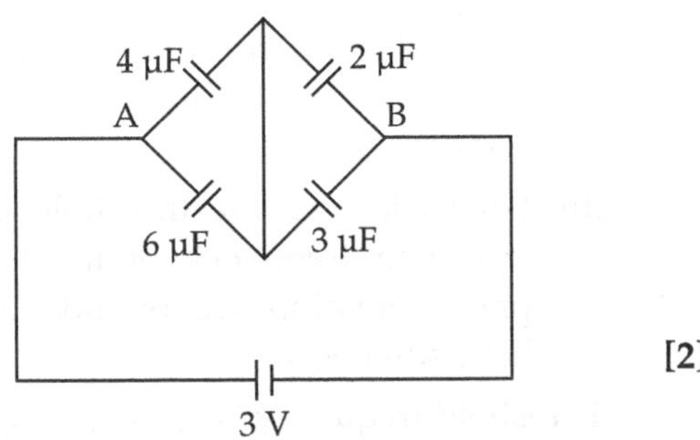

[2]

Ans. For capacitors 4μF and 2μF connected in series

$$\frac{1}{C_1} = \frac{1}{4}+\frac{1}{2} = \frac{1+2}{4} = \frac{3}{4}$$

$$C_1 = \frac{4}{3}\mu F$$

For capacitors 6μF and 3μF connected in series

$$\frac{1}{C_2} = \frac{1}{6}+\frac{1}{3} = \frac{1+2}{6} = \frac{3}{6} = \frac{1}{2}$$

$$C_2 = 2\ \mu F$$

Now, C_1 and C_2 are in parallel

$$\therefore \quad C_P = C_1 + C_2$$

$$C_P = \frac{4}{3}+2 = \frac{4+6}{3} = \frac{10}{3}$$

$$C_P = \frac{10}{3}\mu F$$

$$Q = C_P V = \frac{10}{3}\times 3$$

$$Q = 10\ \mu C$$

22. A wire of length L_o has a resistance R_o. It is gradually stretched till its length becomes $2\ L_o$.

(a) Plot a graph showing variation of its resis-tance R with its length l during stretching

(b) What will be its resistance when its length becomes $2\ L_o$? [2]

Ans. (a)

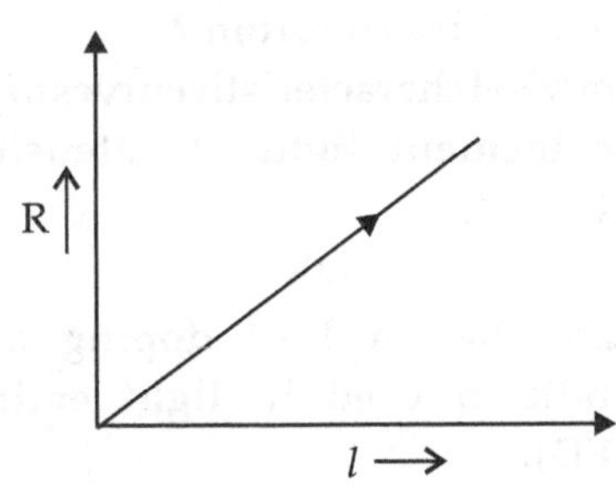

Ans. No

(b) Since, $R_o = \frac{\rho L_o}{A}$

Here, $L_{new} = 2L_o$

$$A_{new} = \frac{A}{2}$$

ρ remains same (Property of material)

$$\therefore \quad R_{new} = \frac{\rho \times 2L_o}{A/2}$$

$$R_{new} = \frac{4\rho L_o}{A} = 4R_o$$

New resistance will become 4 times of the original.

23. A resistor R and an inductor L are connected in series to a source of voltage $V = V_o \sin \omega t$. The voltage is found to lead current in phase by π/4. If the inductor is replaced by a capacitor C, the voltage lags behind current in phase by π/4. When L, C and R are connected in series with the same source,

Find the :

(i) average power dissipated and

(ii) instantaneous current in the circuit. [2]

Ans. (i)

$$P_{av} = V_{rms} \times I_{rms} \cos \phi$$

But cos ϕ is 1 at resonance

$$\therefore \quad P_{av} = V_{rms} \times I_{rms} = \frac{V^2_{rms}}{R}$$

For LCR circuit

$$\tan \theta = \frac{X_L - X_C}{R} = \frac{\omega L - \frac{1}{\omega C}}{R}$$

$$\phi = \frac{\pi}{4}$$

Initially,

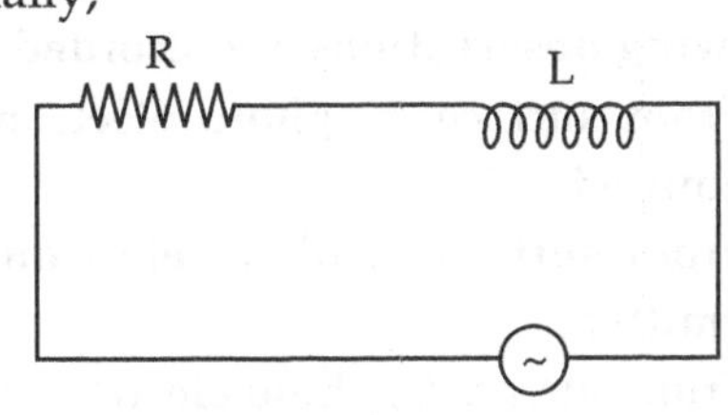

$$V = V_0 \sin \omega t$$

So, $\tan \phi = \frac{\omega L}{R}$

$$\tan \frac{\pi}{4} = \frac{\omega L}{R}$$

$$1 = \frac{\omega L}{R}$$

$$R = \omega L \quad ...(i)$$

When, inductor is removed and capacitor is placed.

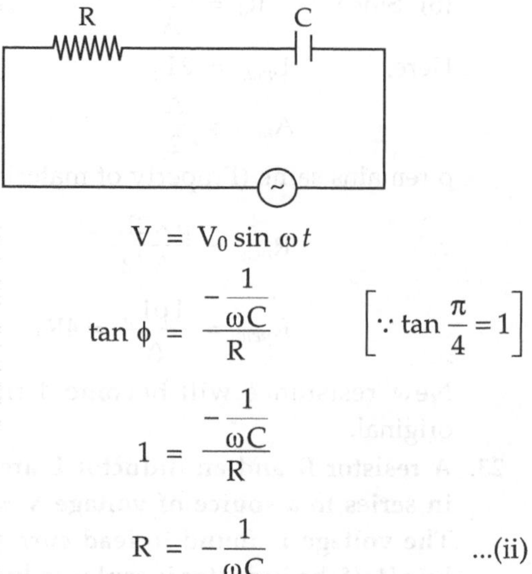

$$V = V_0 \sin \omega t$$

$$\tan \phi = \frac{-\frac{1}{\omega C}}{R} \qquad \left[\because \tan \frac{\pi}{4} = 1\right]$$

$$1 = \frac{-\frac{1}{\omega C}}{R}$$

$$R = -\frac{1}{\omega C} \qquad \text{...(ii)}$$

The – ve sign indicates that the current leads the voltage

From (i) and (ii)

$$\omega L = \frac{1}{\omega C} \text{ or } \omega = \frac{1}{\sqrt{LC}}$$

This implies that the LCR circuit is in resonance condition.

$$P_{av} = V_{rms} \times I_{rms} \cos \phi$$

But cos ϕ is 1 at resonance

$$\therefore \quad P_{av} = V_{rms} \times I_{rms} = \frac{V^2_{rms}}{R}$$

(ii) $$\text{I (Instantaneous)} = \frac{V}{R} = \frac{V_0}{R} \sin \omega t$$

$$= I_0 \sin \omega t.$$

24. Light of same wavelength is incident on three photo-sensitive surfaces A, B and C. The following observations are recorded.

(i) From surface A, photo electrons are not emitted.

(ii) From surface B, photo electrons are just emitted.

(iii) From surface C, photo electrons with some kinetic energy are emitted.

Compare the threshold frequencies of the three surfaces and justify your answer. [2]

OR

If the frequency of light incident on the cathode of a photocell is increased, how will the following be affected ? Justify your answer

(i) Energy of the photo electrons.

(ii) Photo current. [2]

Ans. From the observations made, (surface A, B and C) on the basis of photoelectric effect:

(i) For surface A, threshold frequency is more than frequency of incident light, so no emission occurs.

(ii) For surface B, threshold frequency is equal to frequency of incident light, so emission just occurs.

(iii) For surface C, the threshold frequency is less than frequency of incident light. So photo emission occurs and photo-electron have some K. E.

Threshold frequency is $\upsilon_A > \upsilon_B > \upsilon_C$.

OR

(i) The energy of the photoelectron also increases as $E = h\upsilon$.

(ii) No effect on photoelectric current. This is because of incident photon of increased energy cannot eject more than one electron from the metal surface and energy of electrons increases but no effect on photoelectric current.

25. Briefly explain how a potential barrier is set up across a *p-n* junction as a result of diffusion and drift of the charge carriers. [2]

Ans. Potential Barrier

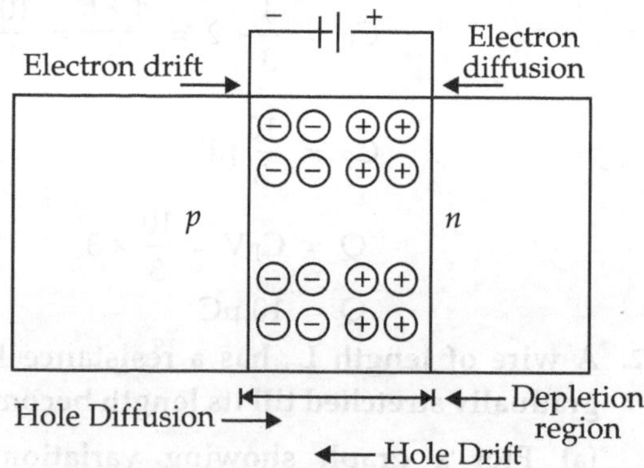

The accumulation of negative charges in the *p*-region and positive charges in the *n*-region set up a potential difference across the junction. This acts as a barrier and is called potential barrier.

26. (a) Why is a photo diode operated under reverse bias conditon ?

(b) Draw V-I characteristic curves of photodiode for incident light of intensities I_1 and I_2 ($I_1 > I_2$). [2]

OR

(a) State the level of doping and biasing condition used in light emitting diode (LED).

(b) Write any two advantages of LED over the conventional low power lamps. [2]

Ans. (a) Photodiodes convert incident light to electric current more effectively in reverse bias condition than in forward bias because the width of the depletion region increases as we increase the applied reverse bias voltage across the diode (directly proportional) in a reverse biased *p-n* junction. Hence, by applying a larger voltage, more of the incident photons get converted to electric current. So photodiodes work better in reverse biased condition.

(b)

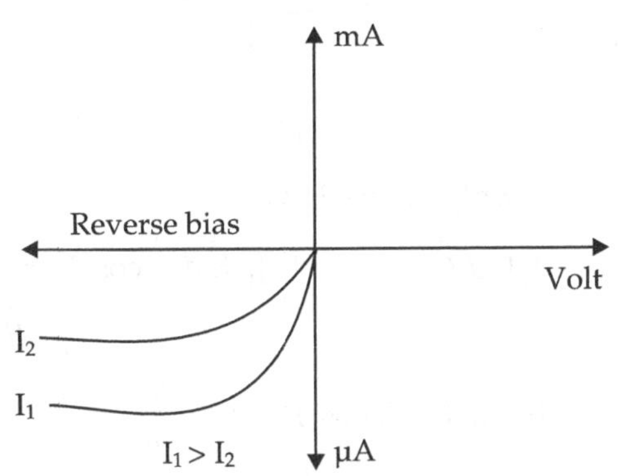

OR

(a) (LED) is a heavily doped *p-n* junction which under forward bias emits spontaneous radiations. When the diode is forward biased, electrons are sent from $n \rightarrow p$ and holes are sent from $p \rightarrow n$. At the junction boundary, the concentration of minority carriers increases as compared to the equilibrium conditions. Thus, at the junction boundary on either side of the junction, excess minority carriers are there, which recombine with majority carriers near the junction and on recombination, the energy is released in the form of photons. When the forward current of diode is small, the intensity of light emiting is small and increases when light reaches maximum. Further, increase in forward current results in decrease of light intensity. LEDs are biased such that the light emiting efficiency is maximum.

(b) LEDs are preferred because :

(i) Low operational voltage and less power

(ii) Fast action and no warmup time required.

(iii) Long life and ruggedness.

27. (a) Explain the formation of energy bands in crystalline solids.

(b) Draw the energy band diagrams of (i) a metal and (ii) a semiconductor. [2]

Ans. (a) An enormously large number of energy levels closely spaced in a very small energy range constitute an energy band.

The allowed energy bands are separated by regions in which energy levels cannot exist. These forbidden regions are called band gaps or energy gaps.

The highest energy band occupied by the valence electrons is called the valence band and the next empty allowed band is called the conduction band.

(b) (i)

(ii)

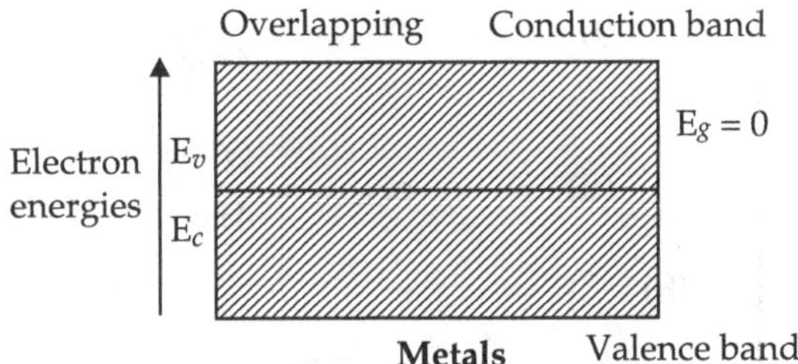

SECTION-C

28. A hollow conducting sphere of inner radius r_1 and outer radius r_2 has a charge Q on its surface. A point charge $-q$ is also placed at the centre of the sphere.

(a) What is the surface charge density on the (i) inner and (ii) outer surface of the sphere?

(b) Use Gauss' law of electrostatics to obtain the expression for the electric field at a point lying outside the sphere. [3]

OR

(a) An infinitely long thin straight wire has a uniform linear charge density λ. Obtain the expression for the electric field (E) at a

point lying at a distance x from the wire, using Gauss' law.

(b) Show graphically the variation of this electric field E as a function of distance x from the wire. [3]

Ans. (a)

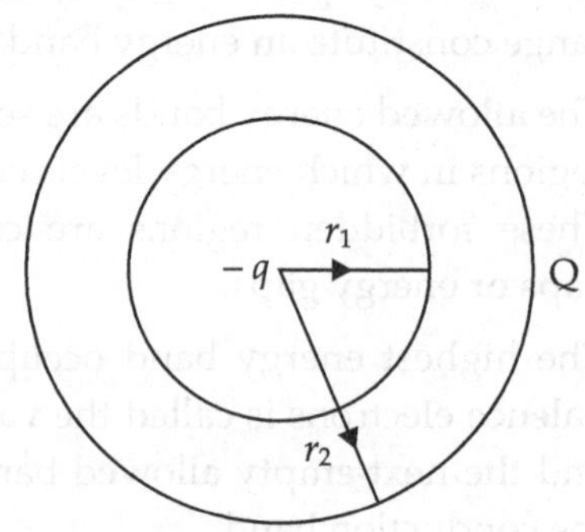

(i) $$\sigma_{Inner} = \frac{-q}{4\pi r_1^2}$$

(ii) $$\sigma_{Outer} = \frac{Q-q}{4\pi r_2^2}$$

(b)

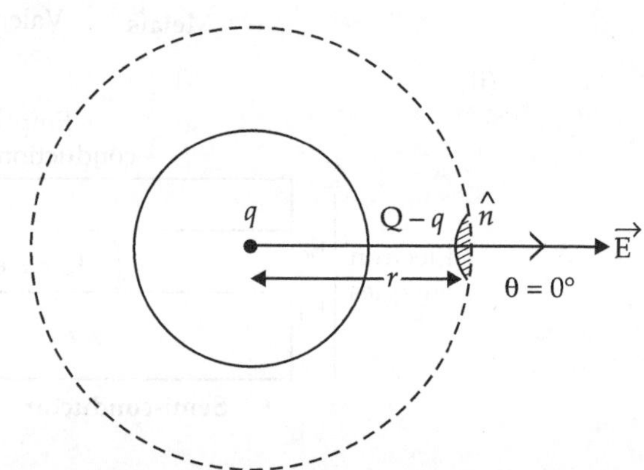

According to Gauss Law :

Electric flux, $$\phi = \frac{Q-q}{\varepsilon_o}$$

$$\oint_0^{4\pi r^2} \vec{E}\, dA \cos 0° = \frac{Q-q}{\varepsilon_o}$$

$$\vec{E}\oint_0^{4\pi r^2} dA = \frac{Q-q}{\varepsilon_o}$$

$$\vec{E}\,[A]_0^{4\pi r^2} = \frac{Q-q}{\varepsilon_o}$$

$$\vec{E} = \frac{Q-q}{4\pi\varepsilon_0 r^2} = K\left(\frac{Q-q}{r^2}\right)$$

OR

(a)

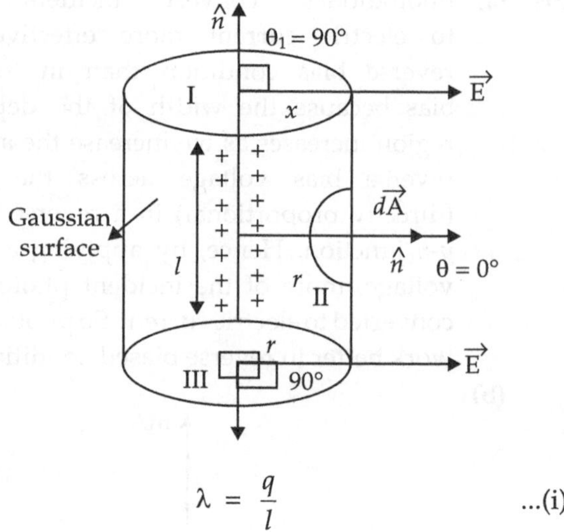

$$\lambda = \frac{q}{l} \quad \text{...(i)}$$

Apply Gauss Law :

$$\oint_I \vec{E}\, d\vec{A} \cos 90° + \oint_{II} \vec{E}\, d\vec{A} \cos 0° + \oint_{III} \vec{E}\, d\vec{A} \cos 90°$$

$$0 + \oint_{II} \vec{E}\, d\vec{A} \cos 0° + 0 = \frac{q}{\varepsilon_0}$$

$$\vec{E}\oint_0^{2\pi xl} dA = \frac{q}{\varepsilon_0}$$

$$\vec{E}\,[A]_0^{2\pi xl} = \frac{\lambda l}{\varepsilon_0} \quad \text{[From (i)]}$$

$$\vec{E} \times 2\pi xl = \frac{\lambda l}{\varepsilon_0}$$

$$\vec{E} = \frac{\lambda}{2\pi\varepsilon_0 x}$$

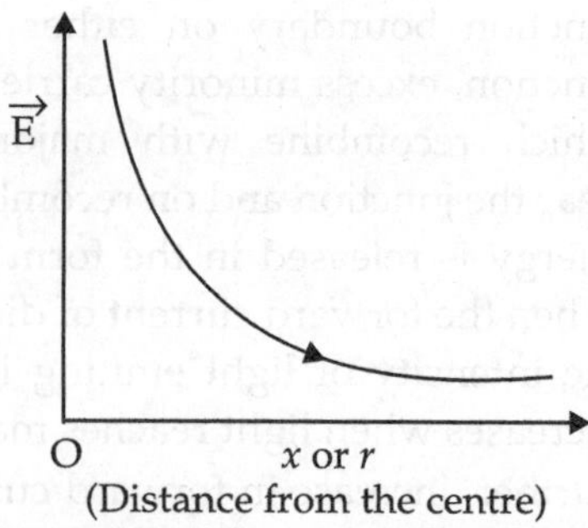

(Distance from the centre)

29. (a) Explain the principle of working of a potentiometer.

(b) In a potentiometer, a standard source of emf 5V and negligible internal resistance maintains a steady current through the potentiometer wire of length 10 m. Two primary cells of emf E_1 and E_2 are joined together in a series with (i) same polarity

and (ii) opposite polarity. The combination is connected to the potentiometer circuit in each case. The balancing length of the wire in the two cases are found to be 700 cm and 100 cm, respectively. Find the values of emf of the two cells. [3]

Ans. (a) Principle of working of potentiometer wire : The voltage drop along the wire (of uniform cross scetion) is directly proportional to the length of the wire. The potentiometer works without drawing any current from the voltage source.

(b) Emf, V = 5V

Length of potentiometer wire = 10 m = 1000 cm

Let the emf of two cells be E_1 and E_2

Balancing length in same polarity = 700 cm

Balancing length in different polarity = 100 cm

Potential gradient

$$= \frac{V}{l} = \frac{5V}{1000\text{ cm}} = \frac{1}{200}\text{ V/cm}$$

$$E_1 + E_2 = Kl_1 \quad \text{(For same polarity)}$$

$$\Rightarrow \quad E_1 + E_2 = \frac{1}{200} \times 700$$

$$\Rightarrow \quad E_1 + E_2 = 3.5\text{ cm} \quad ...(i)$$

and $$E_1 - E_2 = Kl_2 \quad \text{(For different polarity)}$$

$$\Rightarrow \quad E_1 - E_2 = \frac{1}{200} \times 100$$

$$\Rightarrow \quad E_1 - E_2 = 0.5\text{ cm} \quad ...(ii)$$

On adding (i) and (ii), we get

$$2E_1 = 4$$

$$E_1 = 2V$$

and $$E_2 = 3.5 - 2$$

$$E_2 = 1.5\text{ V}$$

Hence, the emf of cells are 2 V and 1.5 V.

30. (a) Differentiate between self inductance and mutual inductance.

(b) The mutual inductance of two coaxial coils is 2H. The current in one coil is changed uniformly from zero to 0.5A in 100 ms. Find the :

(i) change in magnetic flux through the other coil.

(ii) emf induced in the other coil during the change. [3]

Ans. (a)

Self Inductance	Mutual Inductance
1. When the current flowing through a coil changes then induced emf is produced in the coil itself. This pheno-menon is called self inductance.	1. When current in one of the two closely lying coils is changed, the emf is induced in the second coil. This phenomenon is called mutual inductance.
2. Phenomenon of self inductance is the characteristic of single coil.	2. Phenomenon of mutual inductance is the characteristic of two coils.
3. The induced current is produced in the same coil in which main current is flowing. Hence, the induced current affects the main current.	3. Main current flows in one coil and induced current is produced in the second coil. Hence, main current is not directly affected.

(b) Mutual inductance between coils M = 2H

Rate of change of current

$$\frac{dI}{dt} = \frac{0.5}{100 \times 10^{-3}}$$

$$= \frac{500}{100} = 5\text{A/s}$$

(i) Change in magnetic flux is

$$\Delta\phi = e\Delta t$$

$$= 2 \times 0.5 \qquad \left[\because e = \frac{Mdi}{dt}\right]$$

$$= 1\text{ Wb}$$

(ii) Change in magnetic flux is

$$e = \frac{-MdI}{dt}$$

$$= -2 \times 5$$

$$= -10\text{ V}$$

31. Explain with the help of a diagram, the working of a step-down transformer. Why is a laminated iron core used in a transformer ? [3]

Ans.

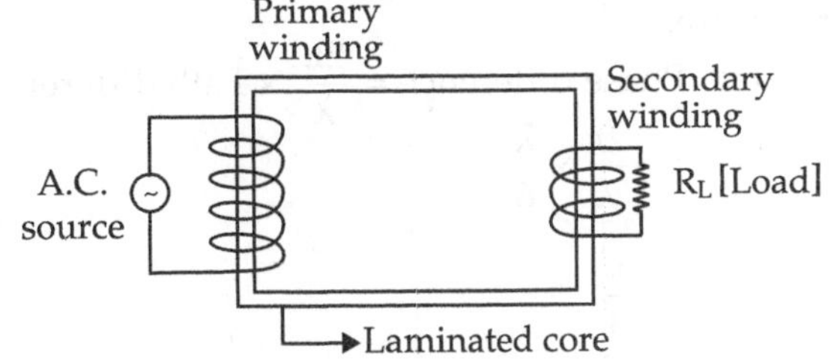

As the a.c. flows through the primary, it generates an alternating magnetic flux in the core, which also passes through the secondary. This changing flux sets up an induced emf in the secondary, also a self induced emf in the primary. If there is no leakage of magnetic flux, then flux linked with each turn of the primary will be equal to that linked with each turn of the secondary.

$$\frac{I_1}{I_2} = \frac{\varepsilon_2}{\varepsilon_1} = \frac{N_2}{N_1}$$

$$\eta = \frac{\text{Power output}}{\text{Power input}} \times 100\%$$

In step down $N_2 < N_1$

So, $\varepsilon_2 < \varepsilon_1$

Laminated iron core is used to prevent energy loss due to eddy current.

32. Name the electro-magnetic waves with their frequency range, produced in :

(a) some radioactive decay

(b) sparks during electric welding

(c) TV remote **[3]**

Ans. (a) Gamma radiations are produced during the radioactive decay of a nucleus. Frequency range is $> 3 \times 10^{17}$ Hz.

(b) Welding sparks produces ultraviolet rays. Frequency range is 7.5×10^{14} Hz-3×10^{16} Hz.

(c) T.V. remote control uses infrared radiations. Frequency range is 3×10^{12} Hz $- 4.3 \times 10^{14}$ Hz.

33. Two coherent light waves of intensity 5×10^{-2} Wm^{-2} each super-impose and produce the interference pattern on a screen. At a point where the path difference between the waves is $\lambda/6$, λ being wavelength of the wave, find the

(a) phase difference between the waves.

(b) resultant intensity at the point.

(c) resultant intensity in terms of the intensity at the maximum. **[3]**

Ans. Here,

$I_1 = I_2 = 5 \times 10^{-2}$ Wb/m^2

Here path difference $= \lambda/6$

Ans. Here,

(a) Phase difference $= \frac{2\pi}{\lambda} \times$ Path difference

$$= \frac{2\pi}{\lambda} \times \frac{\lambda}{6}$$

$$= \frac{\pi}{3}$$

(b) Resultant Intensity, $I = I_1 + I_2 + 2\sqrt{I_1 I_2} \cos\phi$

$$= 5 \times 10^{-2} + 5 \times 10^{-2} + 2\sqrt{(5 \times 10^{-2})^2} \cos\frac{\pi}{3}$$

$$= 2 \times 5 \times 10^{-2} + 2 \times 5 \times 10^{-2} \cos 60^\circ$$

$$= 10 \times 10^{-2} + 5 \times 10^{-2}$$

$$= 15 \times 10^{-2} \text{ Wb/m}^2$$

(c) The intensity is maximum when phase difference is an integral multiple of 2π

$\therefore \quad I_{max} = I_1 + I_2 + 2\sqrt{I_1 I_2}$

But $\quad I_1 = I_2 = I$

$I_{max} = 2I + 2I$

$= 4I$

$\therefore$ Resultant intensity in terms of I_{max} is

$$I' = I_{max} \cos^2 \phi/2$$

$$= I_{max} \cos^2\left(\frac{\pi/3}{2}\right) \quad \left[\because \phi = \frac{\pi}{3}\right]$$

$$= I_{max} \cos^2\left(\frac{\pi}{6}\right)$$

$$= \frac{3}{4} I_{max}$$

But $\quad I' = \frac{3}{4}[4I] = 3I$

$= 15 \times 10^{-2}$ Wb/m^2

34. Two objects P and Q when placed at different positions in front of a concave mirror of focal length 20 cm, form real images of equal size. Size of object P is three times size of object Q. If the distance of P is 50 cm from the mirror, find the distance of Q from the mirror. **[3]**

Ans. Here, focal length of concave mirror, $f = -20$ cm

Let, I_P and I_Q be the image of P and Q and O_P and O_Q be the object size of P and Q

Given,

$$I_P = I_Q$$

and $\quad O_P = 3O_Q$

$$u_P = -50 \text{ cm}$$

$$u_Q = ?$$

Now magnification of object P, $m_P = \frac{f}{f - u_P}$

$$\Rightarrow \quad \frac{I_P}{O_P} = \frac{f}{f - u_P}$$

$$\Rightarrow \quad \frac{I_P}{O_P} = \frac{-20}{-20-(-50)} = \frac{-2}{3}$$

Now, $\quad m_Q = \frac{I_Q}{O_Q} = \frac{I_P}{O_P/3}$

$$= \frac{3I_P}{O_P} = 3 \times \left(\frac{-2}{3}\right) = -2$$

But $\quad m_Q = \dfrac{f}{f - u_Q}$

$$-2 = \frac{-20}{-20 - u_Q}$$

$\Rightarrow \quad 40 + 2u_Q = -20$

$\Rightarrow \quad u_Q = -30$ cm

Hence, the distance of object Q from the mirror is 30 cm.

SECTION-D

35. (a) Show that a current carrying solenoid behaves like a small bar magnet. Obtain the expression for the magnetic field at an external point lying on its axis.

(b) A steady current of 2A flows through a circular coil having 5 turns of radius 7 cm. The coil lies in X-Y plane with its centre at the origin. Find the magnitude and direction of the magnetic dipole moment of the coil.

OR

(a) Derive the expression for the force acting between two long parallel current carrying conductors. Hence, define 1 A current.

(b) A bar magnet of dipole moment 3 Am^2 rests with its centre on a frictionless pivot. A force F is applied at right angles to the axis of the magnet, 10 cm from the pivot. It is observed that an external magnetic field of 0.25 T is required to hold the magnet in equilibrium at an angle of 30° with the field.

Calculate the value of F.

How will the equilibrium be effected if F is withdrawn ? [5]

Ans. (a) Solenoid behaves like a magnet as it has a soft iron core with insulated copper wire over it. If the north pole of a bar magnet is brought near to the negative terminal of the battery, then the solenoid repels the bar magnet as like poles repel each other. Similarly, the other end behaves as a south pole.

Consider a circular coil of radius R with centre O having 'n' turns. Suppose 'P' be any point on the axis at a distance of x from the centre O. Consider a small element of current of length dl.

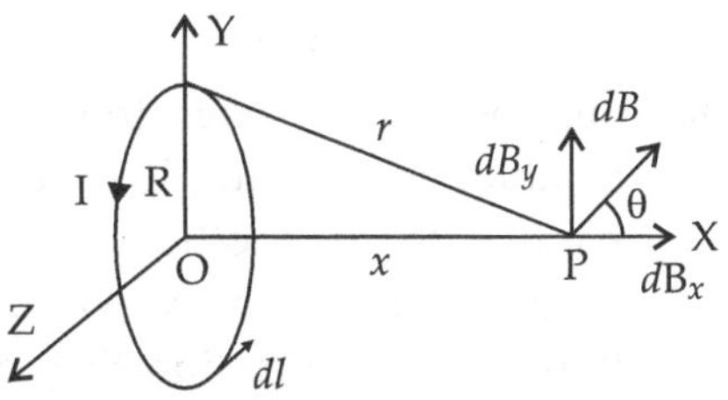

According to Biot-Savart's law, the magnetic field at point P will be

$$dB = \frac{\mu_0 I}{4\pi} \frac{|dl \times r|}{r^3}$$

Where $\quad r^2 = x^2 + R^2$

$$|dl \times r| = r\,dl$$

So, $\quad dB = \dfrac{\mu_0}{4\pi} \times \dfrac{I\,dl}{r^2}$

[$\because$ Both are perpendicular]

$$dB = \frac{\mu_0}{4\pi} \times \frac{I\,dl}{(x^2 + R^2)}$$

dB has two components *i.e.* dB_x and dB_y. dB_y is cancelled out and only the x-component remains.

$$dB_x = dB\cos\theta \quad \left[\because \cos\theta = \frac{R}{\sqrt{x^2 + R^2}}\right]$$

$$dB_x = \frac{\mu_0 I dl}{4\pi} \cdot \frac{R}{(x^2 + R^2)^{3/2}}$$

But $\quad dl = 2\pi R$

So, $\quad B = \dfrac{\mu_0\, IR \times 2\pi R}{4\pi\,(x^2 + R^2)^{3/2}}$

For n turns in the circular loop,

$$B = \frac{\mu_0\, nIR^2}{2\pi\,(x^2 + R^2)^{3/2}}\,\hat{i}$$

(b) Current in coil, $\quad I = 2A$

Number of turns, $\quad N = 5$

Radius of coil, $\quad r = 7\text{cm} = 7 \times 10^{-2}$ m

Magnetic dipole moment,

$$M = NIA$$
$$= NI(\pi r^2)$$
$$= 5 \times 2 \times \pi(7 \times 10^{-2})$$
$$= 1540 \times 10^{-4}\ \text{Am}^2$$
$$= 0.154\ \text{Am}^2$$

So, the direction of magnetic moment dipole is be perpendicular to the plane of the coil *i.e.* along Z-axis.

OR

(a) Suppose two long thin straight long conductors (or wires) PQ and RS are placed parallel to each other, carrying currents I_1 and I_2 respectively. When the current flow in same direction they experience an attractive force and when they carry current in opposite directions, they experience a repulsive force.

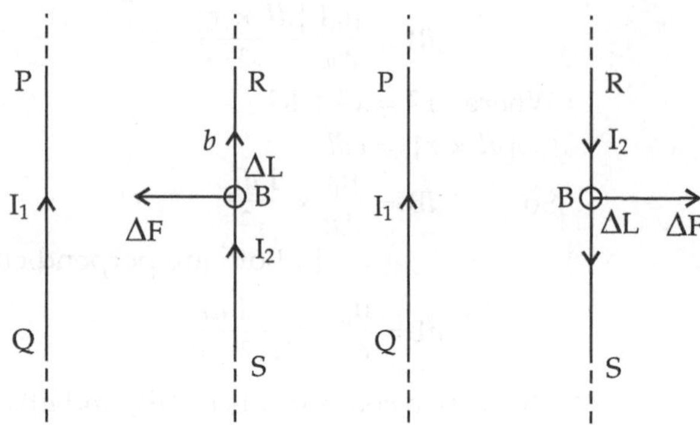

Consider a current element ΔL of wire RS. The magnetic field produced by current carrying conductor PQ at the location of other wire RS.

$$B_1 = \frac{\mu_0 I_1}{2\pi r} \quad ...(i)$$

According to Maxwell's right hand rule, the direction of magnetic field will be perpendicular to the plane of the paple.

$$\Delta F = B_1 I_2 \Delta L \sin 90°$$

$$= \frac{\mu_0 I_1}{2\pi r} I_2 \Delta L$$

∴ Total force on length L

$$F = \frac{\mu_0 I_1 I_2}{2\pi r} \Sigma \Delta L = \frac{\mu_0 I_1 I_2 L}{2\pi r}$$

∴ Force acting on per unit length of conductor

$$f = \frac{F}{L} = \frac{\mu_0 I_1 I_2}{2\pi r} \text{ N/m}$$

Thus, 1 ampere is defined as the current which when flow in each of the parallel conductors placed at separation of 1m in vacuum exerts a force of 2×10^{-1} N on 1m length of either wire.

(b) Dipole moment, $M = 3$ Am^2

External magnetic field,

$$B = .25 \text{ T}$$

Angle of field, $\theta = 30°$

Distance at which force applied from pivot,

$$r = 10 \text{ cm} = 0.10 \text{ m}$$

As magnetic bar is in equlibirium under external force.

Now, $\tau_{Force} = \tau_{M.F.}$

$\Rightarrow$ $F \times r = MB \sin\theta$

$\Rightarrow$ $F \times r \times \sin 90° = 3 \times 0.25 \times \sin 30°$

$\Rightarrow$ $F = \frac{3 \times 0.25}{0.10} \times \frac{1}{2}$

$= \frac{75}{20}$

$= 3.75$ N

If F will be removed then it will not remain in equilibrium and start rotating around its pivot.

36. (a) Draw the ray diagram showing refraction of ray of light through a glass prism. Derive the expression for the refractive index μ of the material of prism in terms of the angle of prism A and angle of minimum deviation δ_m.

(b) A ray of light PQ enters an isosceles right angled prism ABC of refractive index 1.5 as shown in figure.

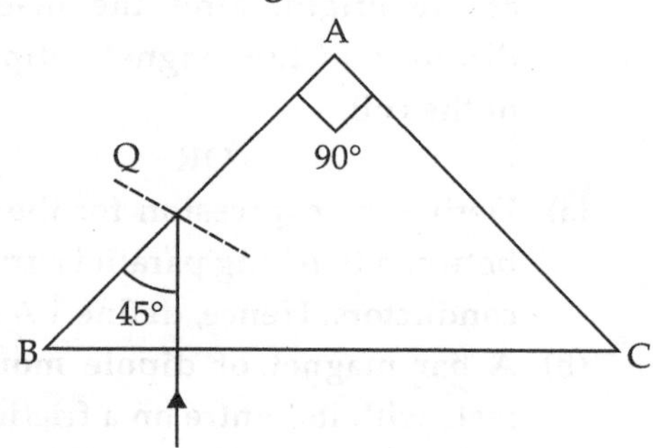

(i) Trace the path of the ray through the prism

(ii) What will be the effect on the path of the ray if the refractive index of the prism is 1.4 ?

OR

(a) Two thin lenses are placed coaxially in contact. Obtain the expression for the focal length of this combination in terms of the focal lengths of the two lenses.

(b) A converging lens of refractive index 1.5 has a power of 10 D. When it is completely immersed in a liquid, it behaves as a diverging lens of focal length 50 cm. Find the refractive index of the liquid. [5]

Ans. (a)

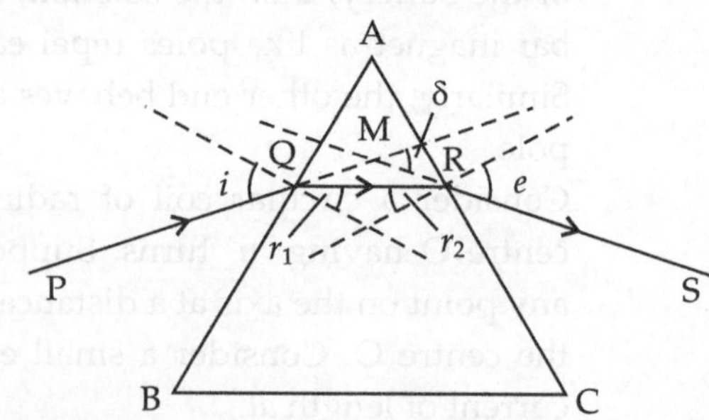

PQ = incident ray
QR = refracted ray
RS = emergent ray
$\angle A$ = angle of prism
$\angle i$ = angle of incidence
$\angle r_1$ = angle of refraction
$\angle e$ = angle of emergence
$\angle \delta$ = angle of deviation

Consider the above diagram and μ as the refractive index of the prism.

In case of minimum deviation, $\angle r_1 = \angle r_2 = \angle r$

And $A = \angle r_1 + \angle r_2$

So, $A = \angle r + \angle r = 2\angle r$

or $\angle r = \dfrac{A}{2}$

Now again,

$$A + \delta = i + e = 2i$$

or $i = \dfrac{A+\delta}{2}$

Now from Snell's Law

$$\mu = \frac{\sin i}{\sin r} = \frac{\sin\left(\frac{A+\delta}{2}\right)}{\sin\left(\frac{A}{2}\right)}$$

$$\mu = \frac{\sin\left(\frac{A+\delta}{2}\right)}{\sin\left(\frac{A}{2}\right)}$$

is the required expression

(b) (i) For critical angle,

$$\sin C = \frac{1}{\mu} = \frac{1}{1.5} = \frac{2}{3}$$

$$C = \sin^{-1}\left(\frac{2}{3}\right) = 41°8'$$

So, if incident angle is greater that 41°8', Total internal reflection will occur.

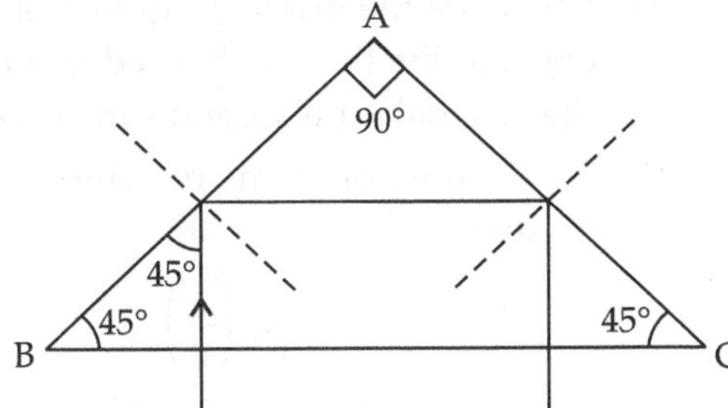

(i) If μ changes to 1.4

Then, $\sin C' = \dfrac{1}{\mu} = \dfrac{1}{1.4} = \dfrac{5}{7}$

$$C = \sin^{-1}\left(\frac{5}{7}\right) = 45°9'$$

Since, incident angle is less than the critical angle. Then total internal reflection will not occur.

OR

(a) Consider two thin lenses L_1 and L_2 of focal lenth f_1 and f_2 respectively, held coaxially in contact with each other. Let P be the point where the optical centres of the lenses coincide. (lenses being thin)

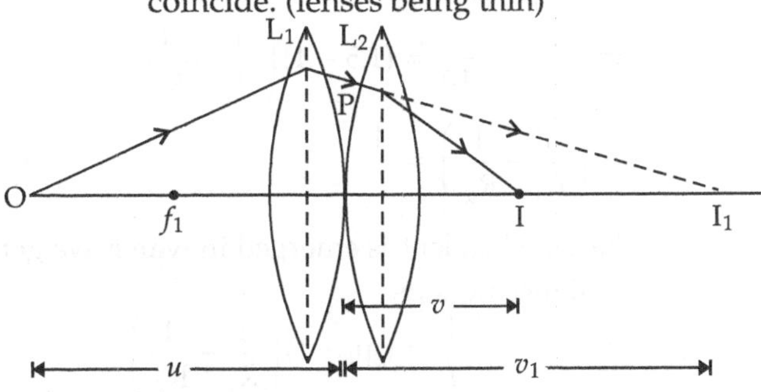

Here, OP = u (object distance)

Let L_1 alone forms the image at I_1 where, $PI_1 = v_1$ (image distance). The image I_1 serve as a virtual object for lens L_2 which forms a final image I at distance PI = v

Now using lens formula for the lens L_1, we have,

$$\frac{1}{f_1} = \frac{1}{v_1} - \frac{1}{u} \quad ...(i)$$

For the image formation by second lens L_2, we have

$$\frac{1}{f_2} = \frac{1}{v} - \frac{1}{v_1} \quad ...(ii)$$

on adding (i) and (ii), we get

$$\frac{1}{f_1} + \frac{1}{f_2} = \frac{1}{v_1} - \frac{1}{u} + \frac{1}{v} - \frac{1}{v_1}$$

$$\Rightarrow \frac{1}{f_1} + \frac{1}{f_2} = \frac{1}{v} - \frac{1}{u} \quad (iii)$$

If two lenses are considered a single lens of focal length f, which forms an image I at a distance v with an object distance being u, then we get

$$\frac{1}{v} - \frac{1}{u} = \frac{1}{f} \quad (iv)$$

On comparing (iii) and (iv), we get

$$\Rightarrow \frac{1}{f} = \frac{1}{f_1} + \frac{1}{f_2}$$

$$\Rightarrow f = \frac{f_1 f_2}{f_1 + f_2}$$

Hence, the focal length of the combined system is given by

$$f = \frac{f_1 f_2}{f_1 + f_2}$$

(b) Power of lens, P = 10D

Focal length of lens = $\frac{1}{10}$ = 10cm

Refractive index of lens $_a\mu_g = 1.5$

Focal length of diverging lens, f_1 = 50cm

According to lens formula

$$\frac{1}{f} = ({}_a\mu_g - 1)\left(\frac{1}{R_1} - \frac{1}{R_2}\right)$$

$$\Rightarrow \frac{1}{10} = (1.5 - 1)\left(\frac{1}{R_1} - \frac{1}{R_2}\right)$$

$$\Rightarrow \left(\frac{1}{R_1} - \frac{1}{R_2}\right) = \frac{1}{5} \quad ...(i)$$

Now, when lens is emerged in water, we get a diverging lens

$$\therefore \frac{1}{f} = ({}_g\mu_l - 1)\left(\frac{1}{R_1} - \frac{1}{R_2}\right)$$

$$-\frac{1}{50} = ({}_g\mu_l - 1)\left(\frac{1}{5}\right) \quad \text{[using (i)]}$$

$$\Rightarrow ({}_2\mu_l - 1) = -\frac{1}{10}$$

$$\Rightarrow {}_2\mu_l = 1 - \frac{1}{10} = \frac{9}{10}$$

$$\Rightarrow \frac{\mu_g}{\mu_l} = \frac{9}{10}$$

$$\Rightarrow \mu_l = \mu_g \times \frac{10}{9}$$

$$= 1.5 \times \frac{10}{9} = 1.67$$

Hence, refractive index of liquid is 1.67.

37. (a) Derive the law of radioactive decay

$$\mathbf{N = N_o\, e^{-\lambda t}}$$

(b) The half life of $^{238}_{92}$U undergoing α-decay is 4.5 × 10⁹ years. Find its mean life.

(c) What fraction of the initial mass of a radio-active substance will decay in five half-life periods ? [5]

OR

(a) State the postulates of Bohr's model of hydrogen atom and derive the expression for Bohr radius.

(b) Find the ratio of the longest and the shortest wavelengths amongst the spectral lines of Balmer series in the spectrum of hydrogen atom. [5]

Ans. (a) Suppose, initially the number of atom in radioactive element is N_o and N be the number of atoms after time t. After time t, let dN be the number of atoms which disintegrate in a short interval dt, then rate of disintegration will be, $\frac{dN}{dt}$, this is also called activity of element.

According to Rutherford Soddy law,

$$\frac{dN}{dt} \propto N$$

$$\frac{dN}{dt} = -\lambda N \quad ...(i)$$

Where, λ is a constant called decay or disintegration constant.

On integrating equation (i) we get

$$\log_e N = -\lambda t + c \quad ...(ii)$$

At $t = 0, N = N_o$

$$\therefore \log_e N_o = 0 + c$$

$$\Rightarrow c = \log_e N_o \quad ...(iii)$$

Put the value of c from equation (iii) in equation (ii)

$$\therefore \log_e N - \log_e N_o = -\lambda t$$

$$\Rightarrow \log_e\left(\frac{N}{N_o}\right) = -\lambda t$$

$$\Rightarrow \frac{N}{N_o} = e^{-\lambda t}$$

$$\Rightarrow N = N_o\, e^{-\lambda t}$$

(b) Mean life, $T_\alpha = \frac{1}{\lambda}$

Half life, $T_{1/2} = \frac{0.693}{\lambda}$

But $T_{1/2} = 4.5 \times 10^9$ years

$$\therefore \lambda = \frac{0.693}{4.5 \times 10^9}$$

Now $$T_\alpha = \frac{1}{\frac{0.693}{4.5 \times 10^9}} = \frac{4.5 \times 10^9}{0.693}$$

$$= 6.49 \times 10^9 \text{ years}$$

(c) Let N_o be the original number of nuclei. In one half life the number reduces to (1/2) of the original number and similar goes on.

∴ Number of atoms remained after 5 half lives

$$= N_o\left(\frac{1}{2}\right)^5$$

∴ Fraction of atoms decayed

$$= N_o - \frac{N_o}{32}$$

$$= \frac{31}{32} N_o$$

OR

(a) The main postulates of Bohr's model of hydrogen atom are as follows :

(i) The electrons in the hydrogen atom revolve around the nucleus in a circular paths of fixed radius and energy called orbits or allowed energy states.

(ii) An electron do not radiate energy if it stayed in one orbit and therefore do not fall into the nucleus.

(iii) An electron moves from a lower energy state to a higher energy state after absorbing energy but energy is emitted when electrons move from the higher energy state to lower energy state.

Suppose m be the mass of electron, v_n be its speed in n^{th} orbit of radius r_n.

$\therefore$ Centripetal force = Electrostatic force

$$\frac{mv_n^2}{r_n} = \frac{1}{4\pi\varepsilon_0}\frac{Z_e^2}{r_n^2} \quad ...(i)$$

From Bohr's postulate for quantisation of angular momentum.

$$mv_nr_n = \frac{nh}{2\pi}$$

$$\Rightarrow \quad v_n = \frac{nh}{2\pi mr_n}$$

Put this value in (i), we get

$$\frac{m}{r_n}\left(\frac{nh}{2\pi mr_n}\right)^2 = \frac{1}{4\pi\varepsilon_0}\frac{Ze^2}{r_n^2}$$

or $$r_n = \frac{\varepsilon_0 n^2h^2}{\pi mZe^2}$$

For Bohr's radius

$$n = 1$$

$$\therefore \quad r_1 = \frac{\varepsilon_0 h^2}{\pi mZe^2}$$

This is the expression for Bohr's radius.

(b) For Balmer series, wavelength is

$$\therefore \quad \frac{1}{\lambda} = k\left[\frac{1}{2^2} - \frac{1}{n^2}\right]$$

For shortest wavelength, $n = \infty$

$$\therefore \quad \frac{1}{\lambda_s} = 10^7\left[\frac{1}{2^2} - \frac{1}{\infty^2}\right]$$

$$\frac{1}{\lambda_s} = \frac{10^7}{4}$$

$$\lambda_s = 4 \times 10^{-7} \text{ m}$$

For longest wavelength, $n = 3$

$$\frac{1}{\lambda_e} = 10^7\left[\frac{1}{2^2} - \frac{1}{3^2}\right]$$

$$= 10^7\left[\frac{9-4}{36}\right]$$

$$\therefore \quad \lambda_e = \frac{36}{5}\times 10^{-7}$$

$$\frac{\lambda_e}{\lambda_s} = \frac{\frac{36}{5}\times 10^{-7}}{4\times 10^{-7}} = \frac{9}{5}$$

$$\therefore \quad \lambda_e : \lambda_s = 9 : 5$$

Outside Delhi [Set-II]

Note : Except for the following questions, all the remaining questions have been asked in previous set.

SECTION-A

Select the most appropriate option from those given below each question.

1. An electron and a proton are moving along the same direction with the same kinetic energy. They enter a uniform magnetic field acting perpendicular to their velocities. The dependence of radius of their paths on their masses is :

(a) $r \propto m$ **(b)** $r \propto \sqrt{m}$

(c) $r \propto \frac{1}{m}$ **(d)** $\frac{1}{\sqrt{m}}$ [1]

Ans. (a) $r \propto m$

Explanaion :

Since, radius of circular path

$$r = \frac{mv}{qB}$$

$$\therefore \quad r \propto m$$

6. A current of 5 A is flowing from east to west in a long straight wire kept on a horizontal table. The magnetic field developed at a distance of 10 cm due south on the table is :

(a) 1×10^{-5} **T acting downwards**

(b) 1×10^{-5} **T acting upwards**

(c) 2×10^{-5} **T acting downwards**

(d) 2×10^{-5} **T acting upwards** [1]

Ans. (a) 1×10^{-5} T downwards

5A

10 cm

$$B = \frac{\mu_0 i}{2\pi d}$$

$$= \frac{4\pi \times 10^{-7} \times 5}{2\pi \times \frac{10}{100}}$$

$$= 2 \times 10^{-7} \times 5 \times 10$$

$$B = 10^{-5} \text{ T} = 1 \times 10^{-5} \text{ T}$$

The direction is found by right hand thumb rule.

Note : Fill in the blanks with appropriate answer :

11. In β-decay, a neutron is converted into a proton, e^- and .... [1]

Ans. Electron antineutrino

$$n \longrightarrow p + e^- + \bar{v}_e$$

12. In a linearly polarized light, the electric vector oscillates along a direction perpendicular to the aligned molecules of the polaroid, called .... axis. [1]

Ans. Pass

19. Depict equipotential surface due to an electric dipole. [1]

Ans. Equipotential surface is the region in space where every point on it is same and at same electric potential. So, an electric dipole have 2 opposite charges, then a straight line passing through centre of electric dipole will be equipotential surface as shown in figure.

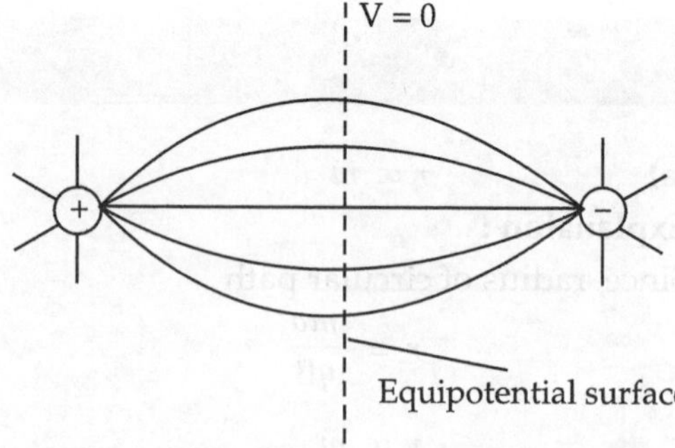

20. Define 'angle of dip' at a place on earth. [1]

Ans. The angle of dip is the angle between the direction of Earth's magnetic field and the plane of the horizon. It is also known as magnetic dip or magnetic inclination.

SECTION-B

21. A uniform wire is cut into three parts with their lengths in the ratio 2 : 3 : 6. The ends of each of these three part are connected across an ideal battery of 10 V. If a current of 5 A is drawn from the battery, find the initial resistance of the wire. [2]

Ans. Let the resistance of wire be R, its length be $11l$ (sum of all ratios) and area of cross section 'A' and its resistivity be ρ.

$$\therefore \quad R = \frac{\rho(11l)}{A}$$

Let, the length of wire cutted be $2l$, $3l$ and $6l$

$$\therefore \quad \text{Resistance of 1}^{\text{st}} \text{ part, } R_1 = \frac{\rho(2l)}{A}$$

$$\text{Resistance of 2}^{\text{nd}} \text{ part, } R_2 = \frac{\rho(3l)}{A}$$

$$\text{Resistance of 3}^{\text{rd}} \text{ part, } R_3 = \frac{\rho(6l)}{A}$$

Since, 3 resistances are connected in parallel.

$$\therefore \quad \frac{1}{R_{eq}} = \frac{1}{R_1} + \frac{1}{R_2} + \frac{1}{R_3}$$

$$= \frac{A}{\rho(2l)} + \frac{A}{\rho(3l)} + \frac{A}{\rho(6l)}$$

$$= \frac{A}{\rho l}\left[\frac{1}{2} + \frac{1}{3} + \frac{1}{6}\right]$$

$$= \frac{A}{\rho l}\left[\frac{3+2+1}{6}\right]$$

$$= \frac{A}{\rho l} \times \left(\frac{6}{6}\right)$$

$$\frac{1}{R_{eq}} = \frac{A}{\rho l}$$

$$\text{or} \quad R_{eq} = \frac{\rho l}{A}$$

$$\text{But,} \quad R_{eq} = \frac{V}{I} = \frac{10}{5} = 2\Omega$$

$$\therefore \quad \frac{\rho l}{A} = 2$$

$$\text{Now, initial resistance is } R = \frac{\rho(11l)}{A}$$

$$= 11\left(\frac{\rho l}{A}\right)$$

$$= 11 \times 2 = 22\Omega$$

Hence, the initial resistance of wire was 22 Ω.

23. Find the total charge stored in the network of capacitors connected between A and B as shown in figure.

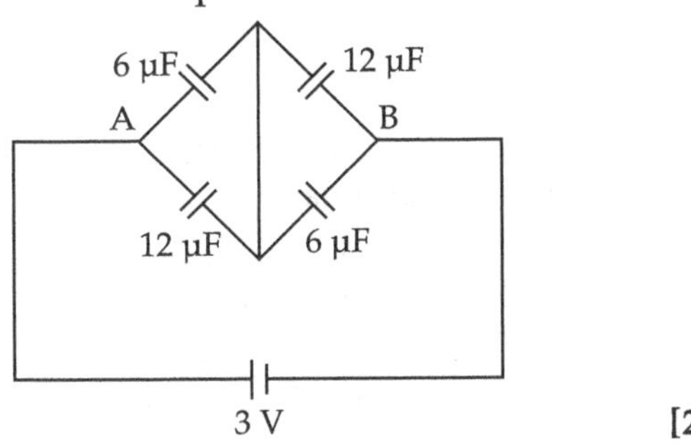

[2]

Ans.

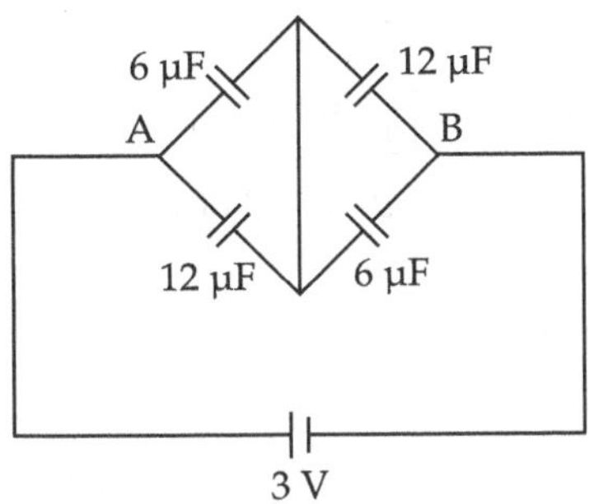

2 capacitors are connected in series between A and B.

$$\therefore \quad \frac{1}{C_1} = \frac{1}{6}+\frac{1}{12}$$

$$\frac{1}{C_1} = \frac{2+1}{12}$$

$$\Rightarrow \quad C_1 = \frac{12}{3} = 4\,\mu F$$

And 2 other capacitors are also in series between A and B.

$$\therefore \quad \frac{1}{C_2} = \frac{1}{6}+\frac{1}{12}$$

$$\frac{1}{C_2} = \frac{2+1}{12}$$

$$\Rightarrow \quad C_2 = \frac{12}{3} = 4\,\mu F$$

Now, C_1 and C_2 are in parallel

$$\therefore \quad C_{eq} = C_1 + C_2 = 4\,\mu F + 4\,\mu F = 8\,\mu F$$

$\therefore$ Total charge stored, $Q = CV = 8 \times 3\,\mu C = 24\,\mu C$

Hence, total charge stored is 2.4×10^{-7} C.

SECTION-C

30. Explain the principle of the device with diagram, which is used to provide electricity at the proper voltage for household purposes. Briefly discuss loss of energy in it due to flux leakage and its minimization. [3]

Ans. The device used to provide electricity at proper voltage for household purpose is stepdown transformer. Transformer is a static device which converts electrical power from one circuit to another without changing its frequency.

Principle : It works on the principle of mutual induction of two coils or Faraday's law of electromagnetic induction when an alternating source of voltage is introduced across a coil (which is physically coupled to another coil), the changing current through it induces an emf across the second coil.

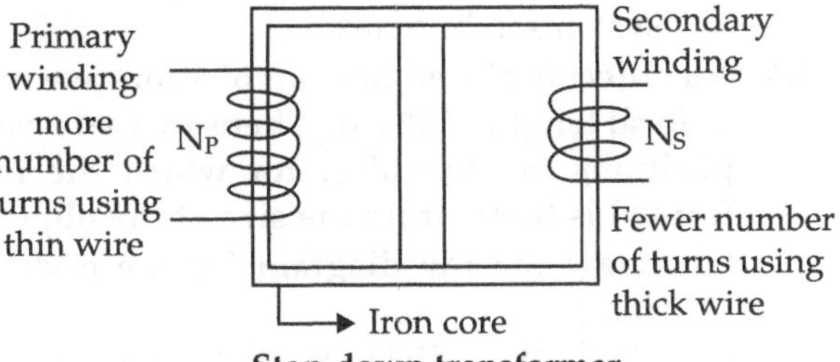

Step-down transformer

The possible sources of power losses in power transformer is :

(a) **Flux leakage** : Small amount of flux coupled from primary to secondary.
To avoid flux leakage, the coils are wound over each other again and again.

(b) **Resistance of windings** : The transformer coil wires cannot have absolutely zero resistance, thus small amount of loss due to heating arises in coil.

(c) **Core eddy currents** : Since, the core is a very good conductor itself, currents are induced in it due to changing magnetic fields, called eddy currents :
It can be reduced by making core of thin sheets by reducing the area.

(d) **Hysteresis** : Some part of energy is frozen in to the core permanently in the form of a residual magnetic field due to its ferromagnetic character.
It can be reduced by special core material which reached to zero or near zero flux density after removal of current.

31. (a) We feel the warmth of the sun light but not the pressure on our hands. Explain.

(b) Which out of wavelength, frequency and speed of an electro-magnetic wave does not change on passing from one medium to another ?

(c) A thin ozone layer in the upper atmosphere is crucial for human survival on earth, why ? [3]

Ans. (a) Because the photons of light have only energy and do not exhibit mass.

(b) When electromagnetic waves passes from one medium to another, frequency remains constant. Wavelength and speed both changes when medium of the wave changes.

(c) The thin ozone layer around the earth's atmosphere is very crucial for human survival. It protects the Earth from harmful ultraviolet (UV) rays from the sun. Without the ozone layer in the atmosphere, life on Earth would be very difficult. Plants cannot live and grow in heavy ultraviolet radiations. Ultraviolet radiations can cause skin cancer, cataracts and impaired immune systems in humans.

33. An object is placed in front of a concave mirror of focal length of 12 cm. There are two possible positions of the object for which the image formed is three times the size of the object.

(a) Draw the ray diagram for the each case, and

(b) Find the distance between the two positions of the object. [3]

Ans. Given, focal length of concave mirror, $f = -12$ cm

Magnification, $m = \pm 3$

If $m = 3$, then

$$\Rightarrow \quad 3 = \frac{v}{u}$$

$$\Rightarrow \quad v = 3u$$

By mirror formula,

$$\frac{1}{f} = \frac{1}{u} + \frac{1}{v}$$

$$\Rightarrow \quad \frac{1}{-12} = \frac{1}{u} + \frac{1}{3u}$$

$$\Rightarrow \quad \frac{1}{-12} = \frac{3+1}{3u}$$

$$\Rightarrow \quad u = \frac{4}{3} \times 12$$

$$\Rightarrow \quad u = -16 \text{ cm}$$

If $m = -3$, then

$$-3 = \frac{v}{u}$$

$$\Rightarrow \quad v = -3u$$

By mirror formula

$$\frac{1}{f} = \frac{1}{v} + \frac{1}{u}$$

$$\Rightarrow \quad \frac{1}{-12} = \frac{1}{(-3u)} + \frac{1}{u}$$

$$\Rightarrow \quad \frac{1}{-12} = \left(1 - \frac{1}{3}\right)\frac{1}{u}$$

$$\Rightarrow \quad \frac{1}{-12} = \frac{2}{3u}$$

$$\Rightarrow \quad u = -8 \text{ cm}$$

(a)

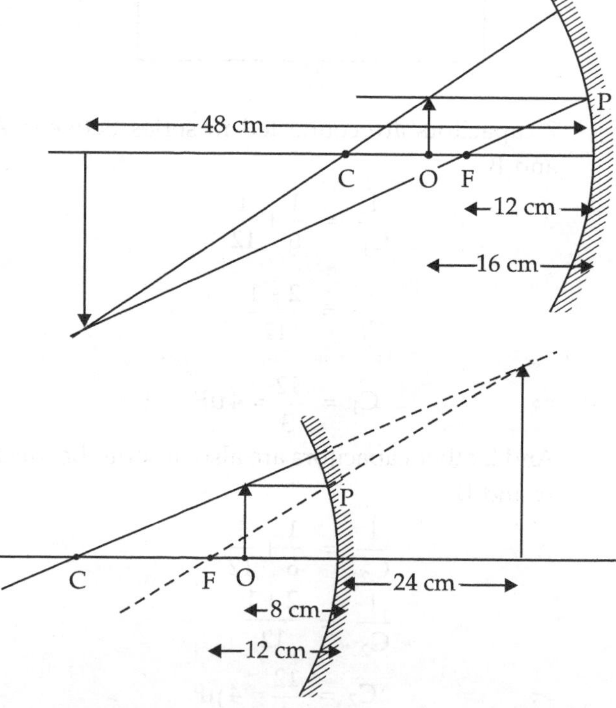

(b) The two positions of object is 16 cm and 8 cm.

$\therefore$ Difference = 16 − 8 = 8 cm

Outside Delhi [Set-III]

Note : Except for the following questions, all the remaining questions have been asked in previous sets.

SECTION-A

Select the most appropriate option from those given below each question.

2. There are uniform electric and magnetic fields in a region pointing along X-axis. An α-particle is projected along Y-axis with a velocity v. The shape of the trajectory will be :

(a) circular in XZ plane

(b) circular in YZ plane

(c) helical with its axis parallel to X-axis

(d) helical with its axis parallel to Y-axis [1]

Ans. (c) helical with its axis parallel to x-axis.

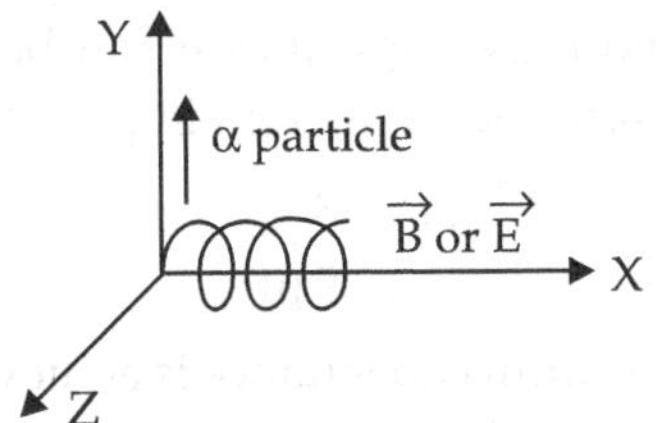

5. A current of 10 A is following from east to west in a long straight wire kept on a horizontal table. The magnetic field developed at a distance 10 cm vertically above the wire is :

(a) 1.2×10^{-5} T, acting towards south

(b) 2×10^{-5} T, acting towards north

(c) 3×10^{-5} T, acting downwards

(d) 2×10^{-5} T, acting upwards [1]

Ans. (b) 2×10^{-5} T, acting towards north

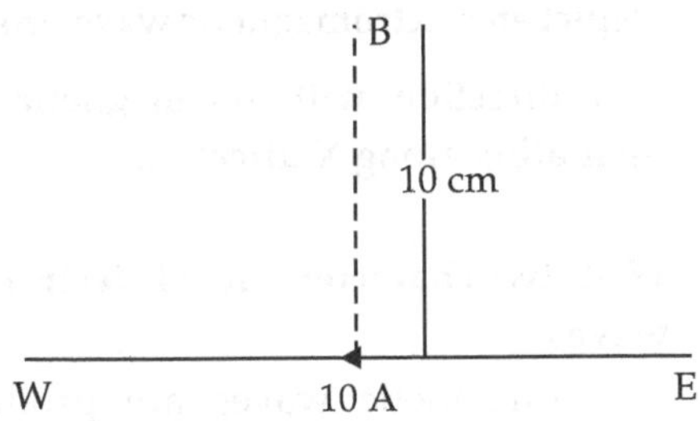

$$B = \frac{\mu_0 i}{2\pi d} = \frac{4\pi \times 10^{-7} \times 10}{2\pi \times \frac{10}{100}}$$

$$= 2 \times 10^{-5} \text{ T}$$

Direction is calculated by using right hand thumb rule.

Note : Fill in the blanks with appropriate answer :

13. A neutron is bombarded on a $^{10}_{5}B$ nucleus and an alpha particle is emitted.

The nuclear reaction involved is :

$$^{1}_{0}n + {}^{10}_{2}B \longrightarrow {}^{4}_{2}He + \text{..........}$$ [1]

Ans. $$^{10}_{5}B + {}^{1}_{0}n \longrightarrow {}^{4}_{2}He + {}^{7}_{3}Li$$

19. What is the change in the value of angle of dip when one goes from the equator to the north pole of earth ? [1]

Ans. As we move from equator to the north pole the angle of dip increases. The angle of dip is 0° at equator and 90° at the pole.

20. A charged particle (+q) moves in a uniform electric field $(\vec{E})$ in the direction opposite to $\vec{E}$. What will be the effect on its electrostatic potential energy during its motion ? [1]

Ans.

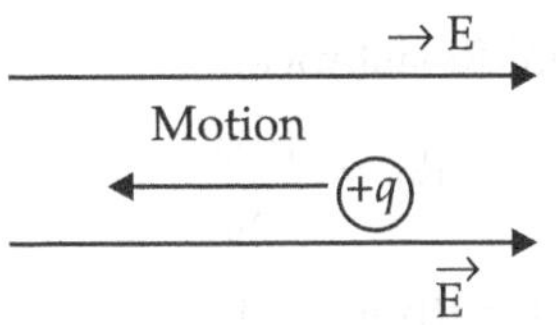

When a positive charge moves opposite to the direction of the electric field the field does negative work on it and the potential energy increases.

SECTION-B

21. Briefly explain how a potential barrier is set up across a *p-n* junction as a result of diffusion and drift of the charge carriers. [2]

Ans. During the formation of a *p-n* junction, the holes diffuse from *p*-type semiconductors to the *n*-type, and electrons diffuse from *n*-type to *p*-type. This is because of the concentration gradient across *p*-side and *n*-side.

When a hole diffuse from *p* to *n*-type, it leaves an unmovable negative charge. Similarly, when an electron diffuses from *n* to *p*-type, it leaves an unmovable positive charge. When the diffusion of holes and electrons takes place continuously across the junctions, a layer of unmovable positive and negative charges are developed on either side of the junction. This layer is called depletion layer or depletion region and the potential difference across the region is called barrier potential.

23. A wire of length L_o has a resistance R_o. It is gradually stretched till its length becomes 1.5 L_o.

(a) Plot the graph showing variation of its resistance R with its length l during stretching.

(b) What will be its resistance when its length becomes 1.5 L_o ? [2]

Ans. As the wire stretches the resistance increases.

Since, $R \propto l$

So more the distance to covered by current, more is the resistance of the wire.

(a)

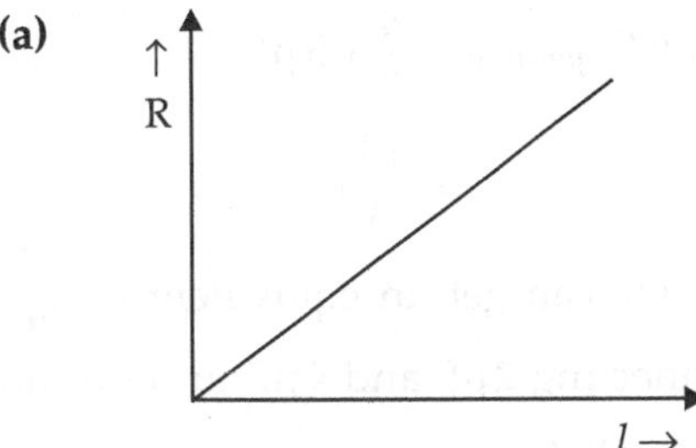

(b) Original Resistance,

$$R_0 = \frac{\rho L_0}{A_0}$$

Now, new length is 1.5 L_0

Volume of the wire remains constant.

Let, the new area be A′

$$\therefore \quad A_0L_0 = 1.5\ L_0A'$$

$$\Rightarrow \quad A' = \frac{A_0}{1.5}$$

$$\therefore \text{ New resistance, } R = \frac{\rho(1.5\ L_0)}{A'}$$

$$= \rho\frac{(1.5\ L_0)}{\frac{A_0}{1.5}}$$

$$= 2.25\frac{\rho L_0}{A_0}$$

$$= 2.25\ R_0 \quad \left[\because R_0 = \frac{\rho L_0}{A_0}\right]$$

So, new resistance becomes 2.25 times of the original resistance.

24. You are given three capacitors of 2 μF, 3 μF and 4 μF, respectively.

(a) Form a combination of all these capacitors of equivalent capacitance $\frac{13}{3}$ μF.

(b) What is the maximum and minimum value of the equivalent capacitance that can be obtained by connecting these capacitors ? **[2]**

Ans. Given, capacitor are 2 μF, 3 μF and 4 μF.

(a) The combination of 2 μF and 4 μF in series with 3 μF in parallel.

$$\frac{1}{C_s} = \frac{1}{2}+\frac{1}{4}$$

$$\Rightarrow \quad \frac{1}{C_s} = \frac{2+1}{4}$$

$$\Rightarrow \quad C_s = \frac{4}{3}\mu F$$

$$\text{Now, } C_{equivalent} = \frac{4}{3} + 3\,\mu F$$

$$= \frac{13}{3}\mu F$$

So, we can get an equivalent of $\frac{13}{3}$ μF by connecting 2 μF and 4 μF in series and 3 μF in parallel.

(b) Maximum capacitance can be achieved by joining the capacitors in parallel

i.e., $C_{max} = 2 + 3 + 4$

$= 9\,\mu F$

Minimum capacitance is achieved by joining the capacitors in series

i.e., $\frac{1}{C_{min}} = \frac{1}{2}+\frac{1}{3}+\frac{1}{4}$

$= \frac{6+4+3}{12}$

$C_{min} = \frac{12}{13}\mu F$

SECTION-C

29. (a) How are electro-magnetic waves produced ? Depict an electromagnetic wave propagating in Z-direction with its magnetic field $\vec{B}$ oscillating along X-direction.

(b) Write two characteristics of electromagnetic waves. **[3]**

Ans. (a) Electromagnetic waves are produced by accelerating or oscillating charges and do not need any material medium for propagation. The source of energy of these waves are energy of oscillating charged particles.

Mathematical expressions for electric and magnetic fields of an electromagnetic wave propagating along the *z*-axis are :

$$E_y = E_0 \sin(kz - \omega t)$$

$$B_x = B_0 \sin(kz - \omega t)$$

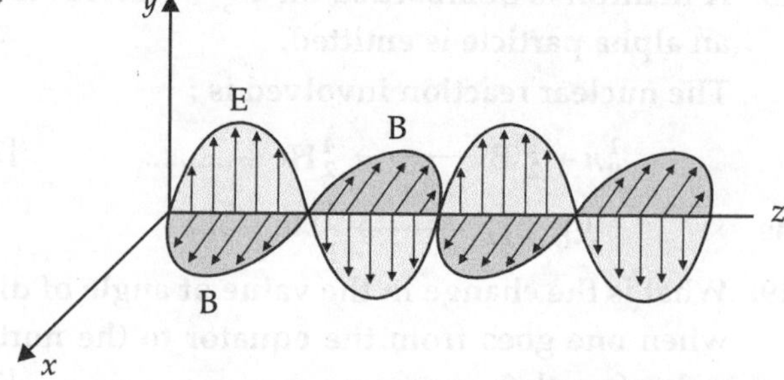

(b) Properties of electromagnetic waves are :

1. The direction of E, B and direction of propagation of waves are mutually perpendicular to one another.
2. The amplitude ratio of E and B gives velocity of light.
3. Electromagnetic waves are not deflected by electric and magnetic fields.

30. A concave mirror forms a real image of an object kept at a distance 9 cm from it. If the object is taken away from the mirror by 6 cm, the image size reduces to $\frac{1}{4}$ the of its previous size. Find the focal length of the mirror. [3]

Ans. Let, the focal length of concave mirror be $-x$.

First object distance, $u_1 = -9$ cm

Second object distance, $u_2 = -15$ cm

I case, using mirror formula

$$\frac{1}{f} = \frac{1}{u_1} + \frac{1}{v_1}$$

$$\Rightarrow \frac{1}{-x} = \frac{1}{-9} + \frac{1}{v_1}$$

$$\Rightarrow \frac{1}{v_1} = \frac{1}{9} - \frac{1}{x}$$

$$\Rightarrow \frac{1}{v_1} = \frac{x-9}{9x}$$

$$\Rightarrow v_1 = \frac{9x}{x-9} \quad ...(i)$$

II case, using mirror formula again

$$\Rightarrow \frac{1}{f} = \frac{1}{u_2} + \frac{1}{v_2}$$

$$\Rightarrow \frac{1}{-x} = \frac{1}{-15} + \frac{1}{v_2}$$

$$\Rightarrow \frac{1}{v_2} = \frac{1}{15} - \frac{1}{x} = \frac{x-15}{15x}$$

$$\Rightarrow v_2 = \frac{15x}{x-15} \quad ...(ii)$$

Now, magnification for I case,

$$m_1 = \frac{v_1}{u_1} = \frac{9x}{x-9} \times \frac{1}{(-9)} = \frac{I_1}{O}$$

Now, magnification for II case,

$$m_2 = \frac{v_2}{u_2} = \frac{15x}{x-15} \times \frac{1}{(-15)} = \frac{I_2}{O}$$

According to question,

$$I_2 = \frac{1}{4} I_1 \quad \text{[Object remains of the same size]}$$

$$\frac{x}{-(x-15)} = \frac{1}{4} \times \frac{x}{-(x-9)}$$

$$\Rightarrow \frac{1}{x-15} = \frac{1}{4(x-9)}$$

$$\Rightarrow 4x - 36 = x - 15$$

$$\Rightarrow 3x = 21$$

$$\Rightarrow x = 7 \text{ cm}$$

Hence, the focal length of mirror is 7 cm.

32. Draw the labelled diagram of an AC generator. Briefly explain its working and obtain the expression for the emf produced in the coil. [3]

Ans. Diagram of AC generator.

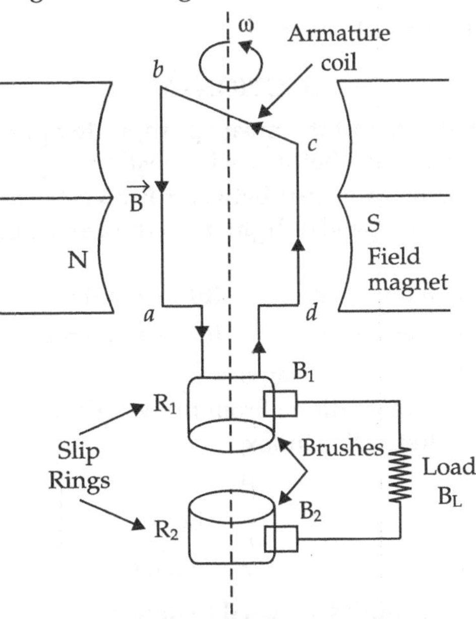

Principle and working : It works on the principle of electromagnetic induction *i.e.*, when a coil rotates continuously in a magnetic field, the effective area of the coil, linked (normally) with the magnetic field lines, changes continuously with time. The variation of a magnetic flux with time results in the production of a alternating emf in the coil.

When the armature rotates between the poles of the magnet upon an axis perpendicular to the magnetic field, the flux which links with the armature changes continuously. Due to this, an emf is induced in the armature this produces an electric current through galvanometer and the slip rings and brushes.

If N is the number of turns in coil, A be the area of coil and B be the magnetic induction, then flux ϕ is given by :

$$\phi = B.A$$

$$\phi = BA \cos \omega t$$

The emf induced in the coil is given by :

$$e = -N \frac{d\phi}{dt}$$

$$e = -N \frac{d}{dt}(BA \cos \omega t)$$

$$e = +NBA\omega \sin \omega t$$

$$e = e_0 \sin \omega t$$

Where, $e_0 = \sin \omega t$

PHYSICS
Delhi [Set-I] 2020

Time allowed : 3 hours

Maximum marks : 70

SECTION–A

Note : Select the most appropriate option from those given below each question.

1. The relationship between Brewester angle 'θ' and the speed of light 'v' in the denser medium is : **[1]**

(a) $v \tan \theta = c$ (b) $c \tan \theta = v$

(c) $v \sin \theta = c$ (d) $c \sin \theta = v$

Ans. (a) $v \tan \theta = c$

Explanation : From Brewster's law :

Refractive index,

$$\mu = \tan \theta$$

$$\frac{c}{v} = \tan \theta \qquad \left[\because \mu = \frac{c}{v}\right]$$

$$\therefore \quad c = v \tan \theta$$

2. Photo diodes are used to detect **[1]**

(a) radio waves (b) gamma rays

(c) IR rays (d) optical signals

Ans. (d) Optical signals.

Explanation : Photodiodes are used to detect the visible light.

3. The selectivity of a series LCR a.c. circuit is large, when **[1]**

(a) L is large and R is large

(b) L is small and R is small

(c) L is large and R is small

(d) L = R

Ans. (c) L is large and R is small.

Explanation : Since, selectivity depends on the quality of resonance. The quality factor is given by $Q = \omega_0 L/R$. High value of quality factor make sure that the resonance curve is sharp. Sharper the resonance curve, more selective is the LCR circuit.

4. The graph showing the correct variation of linear momentum (p) of a charge particle with its de-Broglie wavelength (λ) is : **[1]**

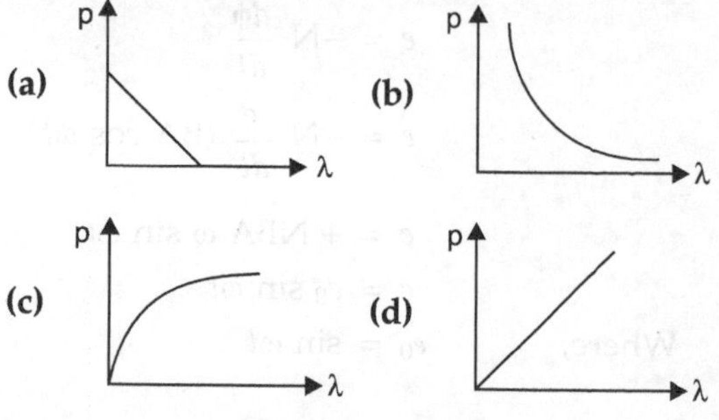

Ans. (b)

Explanation : Since $p = \frac{E}{C} = \frac{h}{\lambda}$

$$p \propto \frac{1}{\lambda},$$

By this relation we get graph between momentum and de-Broglie wavelength is a rectangular hyperbola.

5. The wavelength and intensity of light emitted by a LED depend upon : **[1]**

(a) forward bias and energy gap of the semiconductor

(b) energy gap of the semiconductor and reverse bias

(c) energy gap only

(d) forward bias only

Ans. (a) Forward bias and energy gap of the semiconductor.

Explanation : Since, the wavelength depends on energy gap as $E_g = hc/\lambda$.

And intensity will depends on the current in forward bias.

6. A charge particle after being accelerated through a potential difference 'V' enters in a uniform magnetic field and moves in a circle of radius r. If V is doubled, the radius of the circle will become : **[1]**

(a) $2r$ (b) $\sqrt{2}r$

(c) $4r$ (d) $r/\sqrt{2}$

Ans. (b) $\sqrt{2}r$

Explanation : Since, $qV = \frac{1}{2}mv^2$

$$\therefore \text{ velocity acquired, } v = \sqrt{\frac{2qV}{m}}$$

Now, radius of circular path

$$r = \frac{mv}{qB} = \frac{m}{qB}\sqrt{\frac{2qV}{m}}$$

$$= \frac{1}{B}\sqrt{\frac{2mV}{q}}$$

If V is doubled, then

$$r' = \frac{1}{B}\sqrt{\frac{2m \times (2V)}{q}}$$

$$r' = \sqrt{2}\, r$$

7. The electric flux through a closed Gaussian surface depends upon : [1]

(a) Net charge enclosed and permittivity of the medium

(b) Net charge enclosed, permittivity of the medium and the size of the Gaussian surface

(c) Net charge enclosed only

(d) Permittivity of the medium only

Ans. (a) Net charge enclosed and permittivity of medium.

Explanation : As per Gauss law,

$$\text{Electric flux } \phi = \frac{q}{\varepsilon_0}$$

8. If photons of frequency v are incident on the surfaces of metals. A & B of threshold frequencies $v/2$ and $v/3$ respectively, the ratio of the maximum kinetic energy of electrons emitted from A to that from B is : [1]

(a) 2 : 3 (b) 3 : 4

(c) 1 : 3 (d) $\sqrt{3}:\sqrt{2}$

Ans. (b) 3 : 4

Explanation : $E_k = h(v - v_0)$

$$\therefore \quad \frac{E_A}{E_B} = \frac{h(v - v/2)}{h(v - v/3)}$$

$$= \frac{v/2}{2v/3} = \frac{3}{4}$$

9. The power factor of a series LCR circuit at resonance will be : [1]

(a) 1 (b) 0

(c) 1/2 (d) $1/\sqrt{2}$

Ans. (a) 1

Explanation : Power Factor,

$$\cos\phi = \frac{R}{\sqrt{R^2 + (X_L - X_C)^2}}$$

At resonance,

$$X_C = X_L$$

$$\therefore \quad \cos\phi = \frac{R}{\sqrt{R^2}} = 1$$

10. A biconcave lens of power P vertically splits into two identical plano concave parts. The power of each part will be : [1]

(a) 2P (b) P/2

(c) P (d) $P/\sqrt{2}$

Ans. (b) P/2

Explanation: Let the focal length of biconcave lens be f and that of the planoconcave lens be f.

By Lens' maker formula, for biconcave lens

$$\frac{1}{f} = (\mu - 1)\left(\frac{1}{(-R)} - \frac{1}{R}\right) = \frac{-2(\mu - 1)}{R}$$

[where R, is the radius of curvature of biconcave lens

For planoconcave lens, the focal length

$$\frac{1}{f'} = (\mu - 1)\left(\frac{1}{(-R)}\right) = -\left(\frac{\mu - 1}{R}\right)$$

Thus, $f' = 2f$

Thus, Power $P' = \frac{P}{2}$

Note : Fill in the blanks with appropriate answer.

11. The physical quantity having SI unit NC^{-1} m is .... [1]

Ans. Electric field intensity or electric potential.

12. A copper wire of non-uniform area of crosssection is connected to a d.c. battery. The physical quantity which remains constant along the wire is .... [1]

Ans. Electric current

Explanation : Since, cross section area is nonuniform then current density electric field and drift velocity will not remain same as they vary inversely with area of cross-section.

13. A point charge is placed at the centre of a hollow conducting sphere of internal radius 'r' and outer radius '$2r$'. The ratio of the surface charge density of the inner surface to that of the outer surface will be .... [1]

Ans. –4 : 1

$$\sigma_{inside} = \frac{-q}{4\pi r^2}$$

As, $$\sigma_{outside} = \frac{+q}{4\pi(2r)^2} = \frac{q}{16\pi r^2}$$

$$\therefore \quad \frac{\sigma_{inside}}{\sigma_{outside}} = \frac{\frac{-q}{4\pi r^2}}{\frac{q}{16\pi r^2}} = \frac{-4}{1}$$

$$= -4 : 1$$

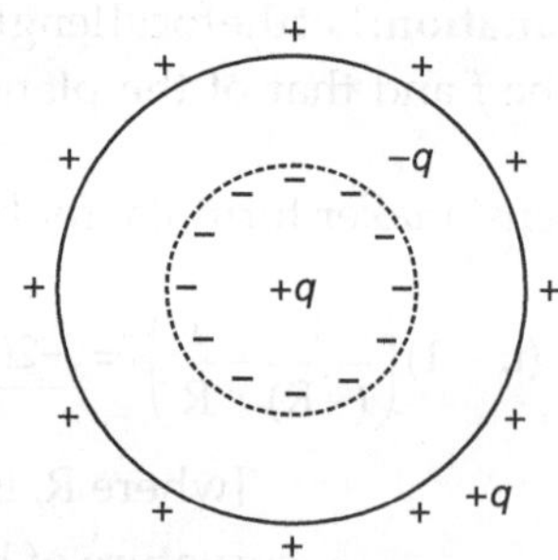

14. The .... , a property of materials C, Si and Ge depends upon the energy gap between their conduction and valence bands. [1]

Ans. Conductivity.

15. The ability of a junction diode to .... an alternating voltage, is based on the fact that it allows current to pass only when it is forward biased. [1]

Ans. rectify

Note : Answer the following :

16. Define the term 'current sensitivity' of a moving coil galvanometer. [1]

Ans. The current sensitivity of a galvano-meter is defined as the deflection produced in the galvanometer when a unit current is flowing through it.

i.e., $$I_s = \frac{\alpha}{I} = \frac{NAB}{k}$$

17. Depict the fields diagram of an electromagnetic wave propagating along positive X-axis with its electric field along Y-axis. [1]

Ans.

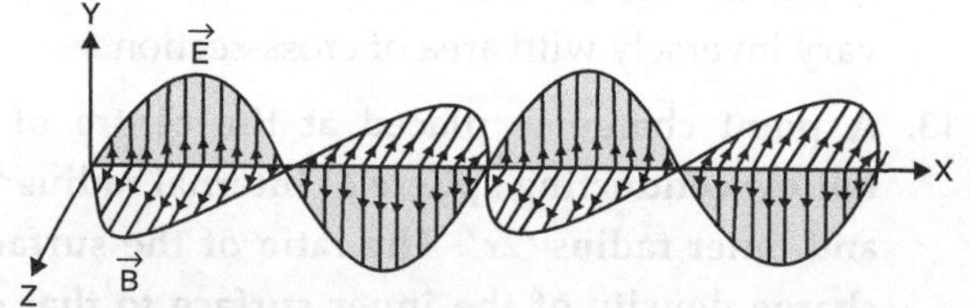

18. Write the conditions on path difference under which (i) constructive (ii) destructive interference occur in Young's double slit experiment. [1]

Ans. (i) Constructive interference :

$$\Delta x = n\lambda,$$

where, $n = 0, 1, 2$ ....

(ii) Destructive interference :

Path difference,

$$\Delta x = (2n+1)\frac{\lambda}{2}$$

where, $n = 0, 1, 2,$ ....:

19. Plot a graph showing variation of induced *e.m.f.*, with the rate of change of current flowing through a coil.

OR

A series combination of an inductor (L), capacitor (C) and a resistor (R) is connected across an ac source of emf of peak value E_0 and angular frequency (ω). Plot a graph to show variation of impedance of the circuit with angular frequency (ω). [1]

Ans.

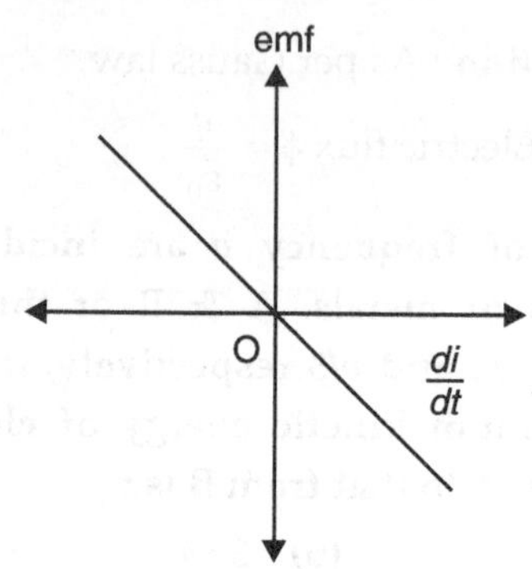

OR

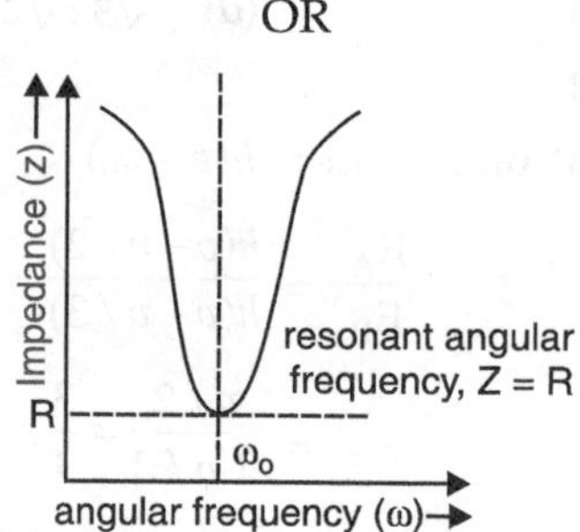

Here, ω_0 represents the resonance frequency for LCR circuit.

20. An electron moves along +*x* direction. It enters into a region of uniform magnetic field $\vec{B}$ directed along − *z* direction as shown in fig. Draw the shape of trajectory followed by the electron after entering the field. [1]

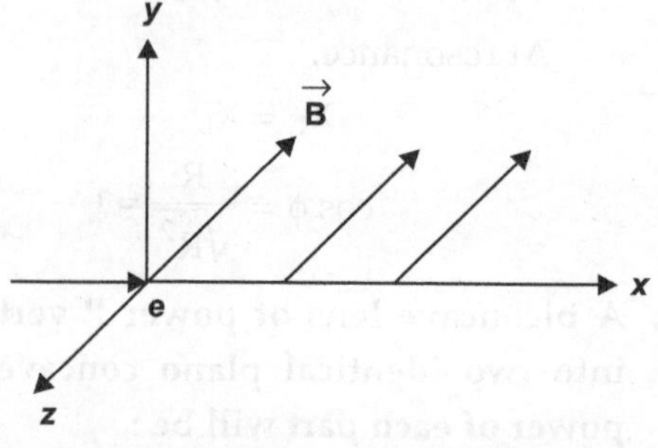

OR

A square shaped current carrying loop MNOP is placed near a straight long current carrying wire AB as shown in the fig. The wire and

the loop lie in the same plane. If the loop experiences a net force F towards the wire, find the magnitude of the force on the side 'NO' of the loop.

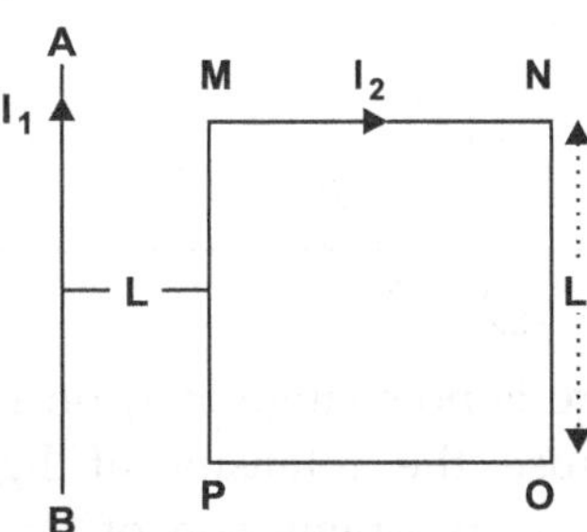

Ans. According to Fleming's left hand rule, the electron will move downward *i.e.*, in negative Y direction. It will follow a semi-circular path.

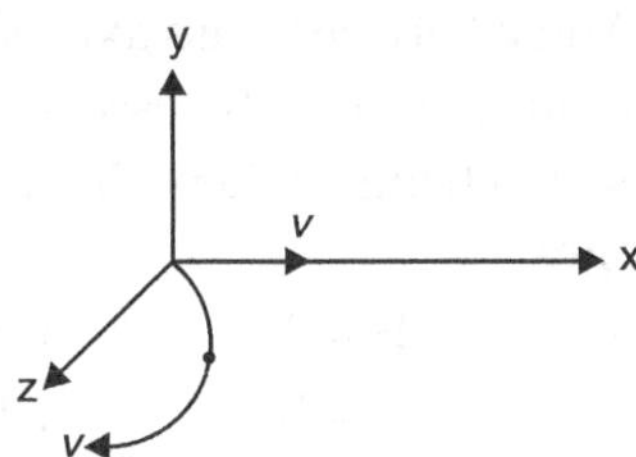

OR

The force on side NO will be repulsive, as current in side NO and wire AB is in opposite direction.

$$\therefore \quad F_{NO} = \frac{\mu_0 I_1 I_2 L}{2\pi r}$$

$$= \frac{\mu_0 I_1 I_2 L}{2\pi(L+L)}$$

$$= \frac{\mu_0 I_1 I_2 L}{2\pi \times 2L}$$

$$= \frac{\mu_0 I_1 I_2}{4\pi} \quad \text{(away from the wire)}$$

SECTION–B

21. Derive the expression for the torque acting on an electric dipole, when it is held in a uniform electric field. Identify the orientation of the dipole in the electric field, in which it attains a stable equilibrium. [2]

OR

Obtain the expression for the energy stored in a capacitor connected across a dc battery. Hence define energy density of the capacitor. [2]

Ans. Dipole in a Uniform External Field :

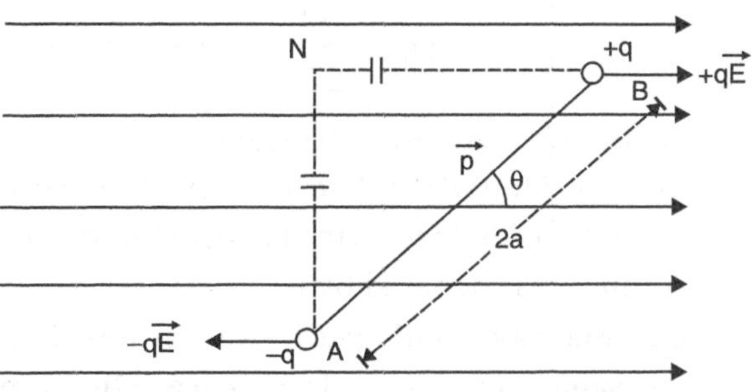

Consider an electric dipole consisting of charges $-q$ and $+q$ and of length $2a$ placed in a uniform electric field $\vec{E}$ making an angle θ with electric field.

Force on charge $-q$ at A $= -q\vec{E}$ (opposite to $\vec{E}$)

Force on charge $+q$ at B $= +q\vec{E}$ (along $\vec{E}$)

Electric dipole is under the action of two equal and unlike parallel force, which give rise to a torque on the dipole

τ = Force × Perpendicular distance between the two forces

$$\tau = qE(AN) = qE \times (2a \sin\theta)$$

$$\tau = q(2a)\, E \sin\theta$$

$$\tau = pE \sin\theta \qquad (\because p = q \times 2a)$$

$$\therefore \quad \vec{\tau} = \vec{p} \times \vec{E}$$

The dipole will be in stable equilibrium when $\theta = 0°$, then the net torque acting on it will be zero.

OR

The work done in charging the capacitor gets stored in it in the form of potential energy. The electric field on one plate due to charge Q and charge density σ on it $= \frac{\sigma}{2\varepsilon_0}$

The potential energy of one plate at a distance d from the other plate.

U_E = (electric p.d. between them) × (electric charge on the other plate)

$$= \frac{\sigma d}{2\varepsilon_0} \times Q$$

(taking reference potential on the first plate as zero).

Putting $\sigma = \frac{Q}{A}$

$$U_E = \frac{dQ}{2\varepsilon_0 A} = \frac{Q^2}{2C}$$

[as $C = \frac{\varepsilon_0 A}{d}$ for parallel plate capacitor]

$$U_E = \frac{Q^2}{2C} = \frac{1}{2}CV^2 = \frac{1}{2}QV$$

This is the required expression.

Energy Density : The energy density of a capacitor is defined as the energy stored per unit volume. It is denoted by *u*.

22. Gamma rays and radio waves travel with the same velocity in free space. Distinguish between them in terms of their origin and the main application. **[2]**

Ans.

Radio waves	Gamma rays
1. They are originated from rapid acceleration and deceleration of electrons. They are generated artificially by transmeters.	1. They are originated from the setting process of an excited nucleus of a radionuclide after it undergoes radioactive decay.
2. The wavelengths of radio waves is greater than gamma rays .	2. The wavelength of gamma rays are smaller than radio waves.
3. Radio waves are used in communication and broadcasting system.	3. They are used in medicine (radiotherapy), industry (sterilization and disinfection) and the nuclear industry.

23. Light from a sodium lamp (S) passes through two polaroid sheets P_1 and P_2 as shown in fig. What will be the effect on the intensity of the light transmitted (i) by P_1 and (ii) by P_2 on rotating polaroid P_1 about the direction of propagation of light ? Justify your answer in both cases. **[2]**

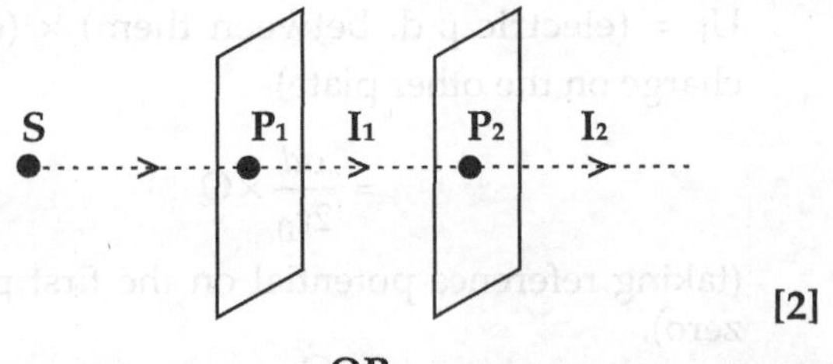

[2]

OR

Define the term 'wave front of light'. A plane wave front AB propagating from denser medium (1) into a rarer medium (2) is incident on the surface P_1P_2 separating the two media as shown in fig. Using Huygen's principle, draw the secondary wavelets and obtain the refracted wave front in the diagram.

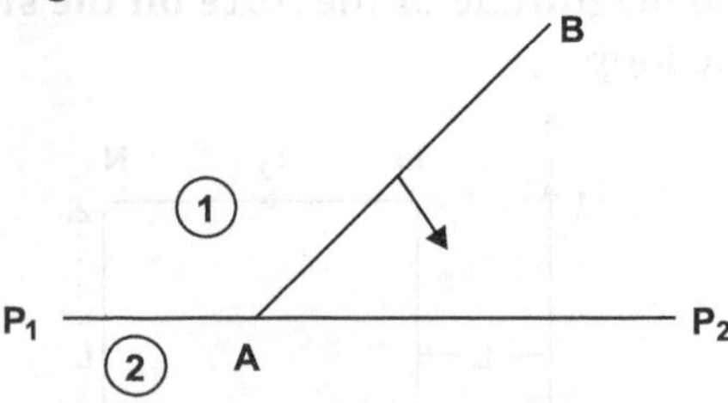

Ans. (i) There is no change in I_1 on rotation of P_1, because the intensity of light does not change by orientation of pass axis of the polaroid.

(ii) I_1 = Intensity of polarised light from P_1

I_2 = Intensity of polarised light from P_2

θ = Angle between pass axis of P_1 and P_2

By Malus law $I_2 = I_1 \cos^2\theta$

Thus, I_2 change when P_1 is rotated as θ changes

$I_2 = 0$ when, $\theta = 90°$

$I_2 = I_1$ when, $\theta = 0°$

OR

Wavefront of light : A wavefront is the locus of all the points in space that reach a particular distance by a propagating wave in same phase at any instant.

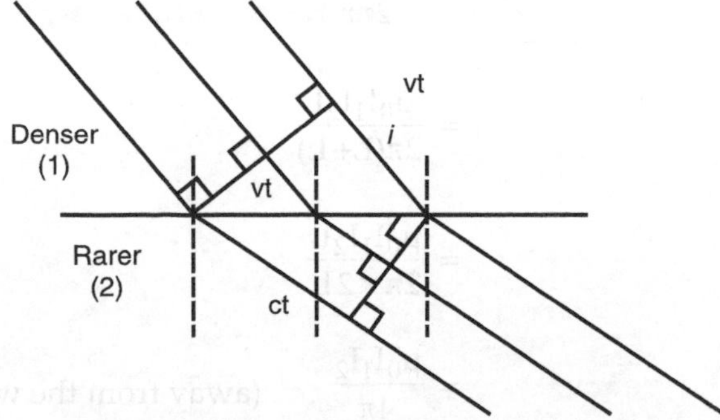

24. A heavy nucleus P of mass number 240 and binding energy 7.6 MeV per nucleon splits in to two nuclei Q and R of mass numbers 110, 130 and binding energy per nucleon 8.5 MeV and 8.4 MeV, respectively. Calculate the energy released in the fission. **[2]**

Ans. Given : Mass number of P = 240

Binding energy per nucleon = 7.6 MeV

Mass number of Q = 110

Binding energy per nucleon = 8.5 MeV

Mass number of R = 130

Binding energy per nucleon = 8.4 MeV

Total B.E. of P $= 240 \times 7.6$

$= 1824$ MeV

Total B.E. of Q and R $= 110 \times 8.5 + 130 \times 8.4$
$= 935 + 1092 = 2027$ MeV

$\therefore$ Energy released in the fission $= 2027 - 1824$
$= 203$ MeV

25. Figure shows the stopping potential (V_0) for the photo electron versus ($1/\lambda$) graph, for two metals A and B, λ being the wavelength of incident light.

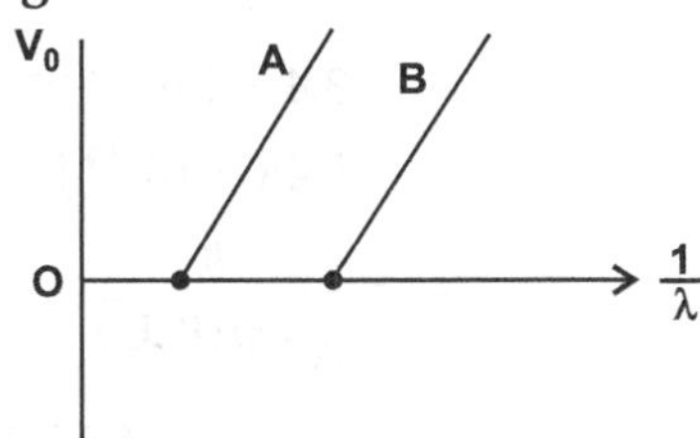

(a) How is the value of Planck's constant determined from the graph?

(b) If the distance between the light source and the surface of metal A is increased, how will the stopping potential for the electrons emitted from it be effected? Justify your answer. [2]

Ans. (a) We know that

$$eV_0 = \frac{hc}{\lambda} - \phi$$

or $$V_0 = \frac{hc}{e}\left(\frac{1}{\lambda}\right) - \frac{\phi}{e} \quad ...(i)$$

where e = Charge on electron
h = Planck's constant,
ϕ = work function of metal surface

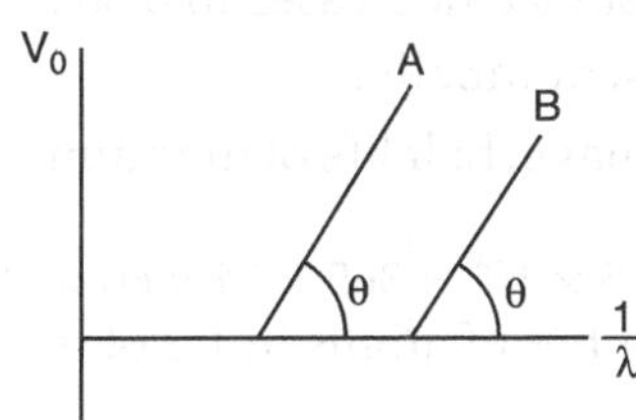

Equation (i) is a straight line as shown in the graph.

Here, slope of line $= \tan\theta = \frac{hc}{e}$

or $$h = \frac{e \tan\theta}{c} \quad ...(ii)$$

Planck's constant can easily be determined by substituting the values of the slope of graph, speed of light and electronic charge in equation (ii)

(b) Stopping potential only depends on frequency of incident light. If the distance is increased, intensity will decrease but no change takes place in stopping potential.

26. Use Bohr's model of hydrogen atom to obtain the relationship between the angular momentum and the magnetic moment of the revolving electron. [2]

Ans. Circulating electron around a heavy nucleous constitutes a current given by :

$$I = \frac{e}{T} = \frac{e}{\frac{2\pi R}{v}} = \frac{ev}{2\pi R}$$

So, there will be magnetic moment associated with this circulating current is given by :

$$\mu_e = I.A$$
$$= \frac{ev}{2\pi R} \times \pi R^2$$
$$= \frac{evR}{2} \quad ...(i)$$

Angular momentum of circulating electron is given by

$$l = mvR$$

so $$vR = \frac{l}{m} \quad ...(ii)$$

Put the value of equation (ii) in (i), we get

$$\mu_e = \frac{el}{2m}$$

which is the required expression.

27. In a single slit diffraction experiment, the width of the slit is increased. How will the (i) size and (ii) intensity of central bright band be affected? Justify your answer. [2]

Ans. In a single slit diffraction experiment, if the width of the slit is increased then :

(a) the size of the central bright band reduces to half. As width of central maxima is $\frac{2\lambda}{a}$.

If 'a' increases, then width decreases.

(b) Intensity of the central bright band increases and more brighter fringe is obtained on the screen, because more light will be allowed to pass. And the light energy is now squeezed into a smaller area on the screen because the size of the central maxima is reduced. These factors make the intensity increase manyfold.

SECTION–C

28. (a) Differentiate between electrical resistance and resistivity of a conductor.

(b) Two metallic rods, each of length L, area of cross A_1 and A_2, having resistivities ρ_1 and ρ_2 are connected in parallel across a d.c. battery. Obtain the expression for the effective resistivity of this combination. [3]

Ans. (a)

Electrical Resistance	Resistivity
1. Resistance opposes the flow of free electrons charges in the conductor	1. Resistivity is the property of the material which defines the resistance of the material having specific dimension.
2. It is measured by the ratio of the p.d.(V) across it to the current (I) flowing through it. $R = \frac{V}{I}$	2. It is numerically equal to the resistance of a sample of unit length and unit area of cross-section at certain temperature.
3. Its S.I. unit is Ω	3. Its. S.I. unit is Ω mv

(b) Here, length of each rod is L.

Area of cross sections are A_1 and A_2 and resistivities are ρ_1 and ρ_2 respectively

Resistance of 1st metallic rod, $R_1 = \frac{\rho_1 L}{A_1}$

Resistance of 2nd metallic rod, $R_2 = \frac{\rho_2 L}{A_2}$

∴ Effective resistance when connected in parallel.

$$R_{eq} = \frac{R_1 R_2}{R_1 + R_2}$$

$$\frac{\rho_{eq} L}{A_1 + A_2} = \frac{\rho_1 \rho_2 L^2 A_1 A_2}{A_1 A_2 L(\rho_1 A_2 + \rho_2 A_1)}$$

$$\rho_{eq} = \frac{\rho_1 \rho_2 (A_1 + A_2)}{\rho_1 A_2 + \rho_2 A_1}$$

29. Calculate the de-Broglie wavelength associated with the electron revolving in the first excited state of hydrogen atom. The ground state energy of the hydrogen atom is – 13.6 eV. [3]

Ans. De Broglie wavelength,

$$\lambda = \frac{h}{mv} = \frac{h}{p}$$

where, p is momentum of electron

Kinetic energy (KE) and momentum (p) are related as

$$KE = \frac{p^2}{2m} \quad \text{(where, } m = \text{mass of electron)}$$

$$\Rightarrow \quad p = \sqrt{2m(KE)}$$

$$\Rightarrow \quad \lambda = \frac{h}{\sqrt{2m(KE)}}$$

According to Bohr's model,

Kinetic energy of $e^- = |\text{T.E. of } e^-|$

$$= \left| -\frac{13.6 \times Z^2}{n^2} \right| eV$$

for hydrogen, Z = 1 and first excited state implies $n = 2$.

$$\text{K.E.} = \frac{13.6 \times 1^2}{2^2} = 3.4 eV$$

$$= 3.4 \times 1.6 \times 10^{-19} \text{ J}$$

$$= 5.44 \times 10^{-19} \text{ J}$$

$$\therefore \quad \lambda = \frac{h}{\sqrt{2m(\text{K.E.})}}$$

$$= \frac{6.63 \times 10^{-34}}{\sqrt{2 \times 9.1 \times 10^{-31} \times 5.44 \times 10^{-19}}}$$

$$= 6.66 \times 10^{-10} m$$

$$= 6.66 Å$$

30. (a) Define the term decay constant of a radioactive substance. [3]

(b) The half life of $^{238}_{92}U$ undergoing α decay is 4.5 × 10⁹ years. Calculate the activity of 10g sample of $^{238}_{92}U$. [3]

Ans. (a) Decay constant of a radioactive substance is defined as the ratio of its instantaneous rate of disintegration to the number of atoms present at that time.

i.e., $N = N_0 e^{-\lambda t}$

It can also be defined as the reciprocal of time when the number of atoms in a radioactive substance decreases to 0.308 of the number present intially.

(b) We have, half life of uranium $T_{1/2} = 4.5 \times 10^9$ years

$= 4.5 \times 10^9 \times 365 \times 24 \times 60 \times 60$ seconds.

Number of atoms in 1 g of uranium,

$$N = \frac{6.023 \times 10^{23}}{238} \text{ atoms}$$

Number of atoms in 10g of uranium,

$$N_1 = \frac{6.023 \times 10^{23} \times 10}{238} \text{ atoms}$$

Activity of sample = λN

$$= \frac{0.693}{T_{1/2}} \times N_1$$

$$= \frac{0.693 \times 6.023 \times 10^{23} \times 10}{4.5 \times 10^9 \times 3.156 \times 10^7 \times 238}$$

$= 1.235 \times 10^5$ disintegration/second.

31. What is a solar cell? Draw its V-I characteristics. Explain the three processes involved in its working. [3]

OR

Draw the circuit diagram of a full wave rectifier. Explain its working showing its input and output waveforms. [3]

Ans. Solar Cell : It is a semiconductor device used to convert photons of solar light into electricity.

V.I. characteristic of solar cell :

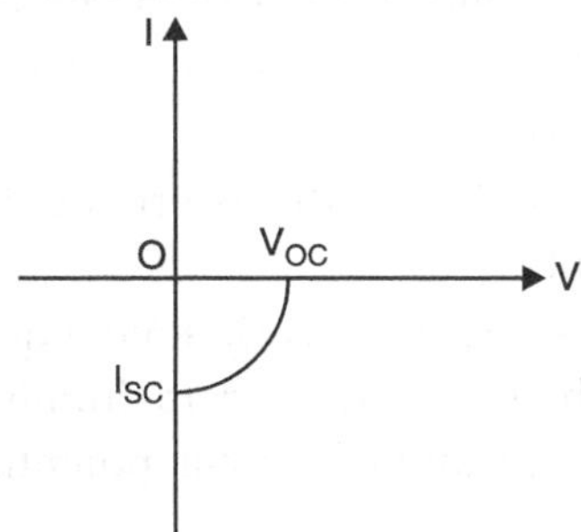

1. I_{SC} is the short circuit current when the load resistance R_L is zero.
2. V_{OC} is the open circuit voltage when R_L is infinity.

The generation of emf by a solar cell, when light falls on it, is due to the following 3 basic process : generation, separation and collection.

(i) Generation of e-h pairs due to light (with $h\nu > E_g$) close to the junction.

(ii) Separation of electrons and holes due to electric field of the depletion region. Electrons are swept to *n*-side and holes to *p*-side.

(iii) The electrons reaching the *n*-side are collected by the front contact and holes reaching *p*-side are collected by the back contact. Thus, *p*-side becomes positive and *n*-side becomes negative giving rise to photo voltage.

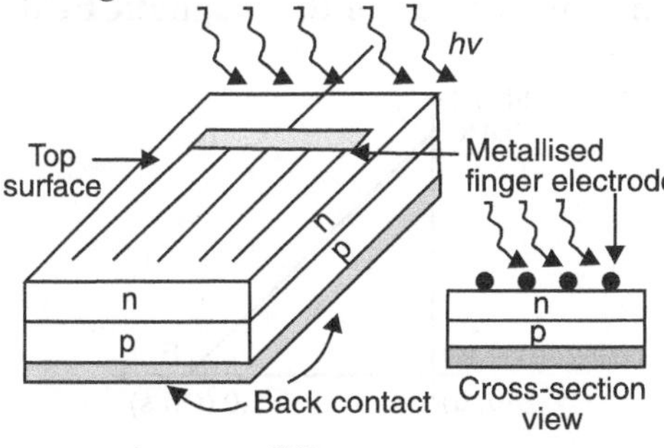

OR

Full wave rectifier :

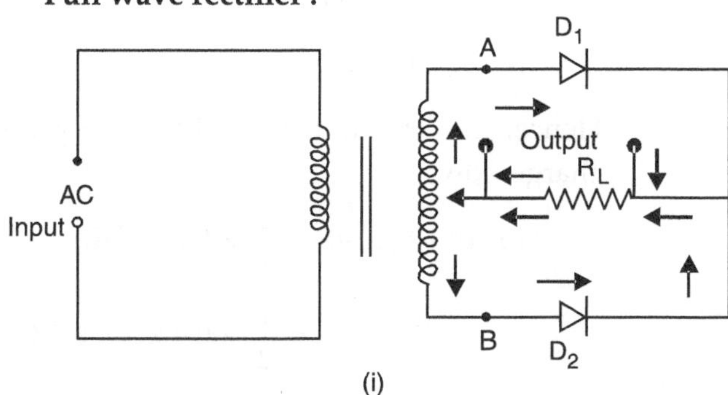

(i)

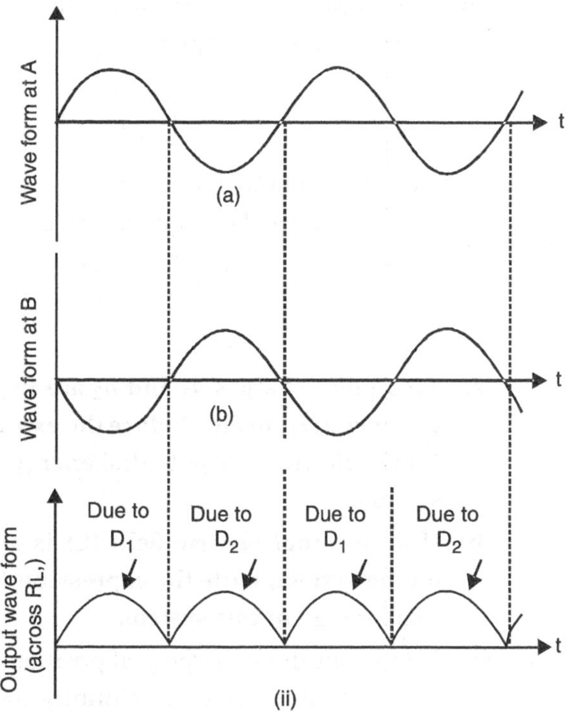

(ii)

Explanation : In positive half cycle of AC, end A becomes positive and D_1 becomes forward biased and D_2 is reverse biased, so D_1 conducts and D_2 doesn't. So conventional current flows through D_1, R_L and upper half of secondary winding . Similarly, during negative half cycle of AC diode D_2 becomes forward biased and D_1 is reverse biased, current flows through D_2, R_L and lower half of secondary winding. Thus, current flows in same direction in both half cycle input AC voltage.

32. An optical instrument uses a lens of power 100 D for objective lens and 50 D for its eyepiece. When the tube length is kept at 25cm, the final image is formed at infinity. [3]

(a) Identify the optical instrument.

(b) Calculate the magnification produced by the instrument

Ans. (a) Power of lens = $\frac{1}{\text{focal length}}$

So, focal length of objective lens = $\frac{1}{100}m$ = 1cm

Focal length of eye piece = $\frac{1}{50}m$ = 2cm

Since, the objective has a smaller focal length than the eye piece, the instrument is a compound microscope.

(b) Magnification produced when the image is formed at infinity is given by,

$$m = \left(\frac{L}{f_o}\right)\left(\frac{D}{f_e}\right)$$

Here, tube length, L = 25cm

Least Distance of distinct vision, D = 25cm

$$\therefore \quad m = \frac{25 \times 25}{1 \times 2} = \frac{625}{2}$$

$$= 312.5$$

33. (a) Two point charges q_1 and q_2 are kept at a distance of r_{12} in air. Deduce the expression for the electrostatic potential energy of this system. [3]

(b) If an external electric field (E) is applied on the system, write the expression for the total energy of this system. [3]

Ans. (a) Let us consider a charge q_1 at position vector r_1 and charge q_2 which is at infinity and to be brought at point P_2 having position vector r_2 and dW is the small amount of work done in moving a charge through a small distance dx.

+q_1 ← r_{12} → q_2 .... q_2
$P_1(r_1)$ $P_2(r_2)$ (at ∞)

$$dW = F.dx$$

$$= -k\frac{q_1q_2}{r^2}dx$$

$$W = \int_{\infty}^{r_{12}} \frac{kq_1q_2}{r^2}dx$$

$$= -kq_1q_2\int_{\infty}^{r_{12}} \frac{1}{r^2}dx$$

$$= -kq_1q_2\left[\frac{-1}{r}\right]_{\infty}^{r_{12}}$$

$$= kq_1q_2\left[\frac{1}{r_{12}}\right]$$

$$W = \frac{kq_1q_2}{r_{12}}$$

(b) Let us consider a system of two charges q_1 and q_2 from the origin. Let the external electric field applied be E.

Let the work done in bringing the charge q_1 from infinity to r_1 is given as $qV(r_1)$ and the work done in bringing charge q_2 from infinity to r_2 be $q_2V(r_2)$. But the work required to be done on q_2 will include the field of q_1 along the electric field E is given as, $\frac{kq_1q_2}{r_{12}}$.

The potential energy of the charge q at a distance r from the origin in an external electric field is $qV(r)$.

Where, $V(r)$ is the external potential at that point.

Thus, the total work done required to bring both the charges from infinity to present configuration or total potential energy of system is = $q_1V(r_1) + q_2V(r_2) + \frac{kq_1q_2}{r_{12}}$

34. When a conducting loop of resistance 10Ω and area 10 cm² is removed from an external magnetic field acting normally, the variation of induced current in the loop with time is shown in the figure.

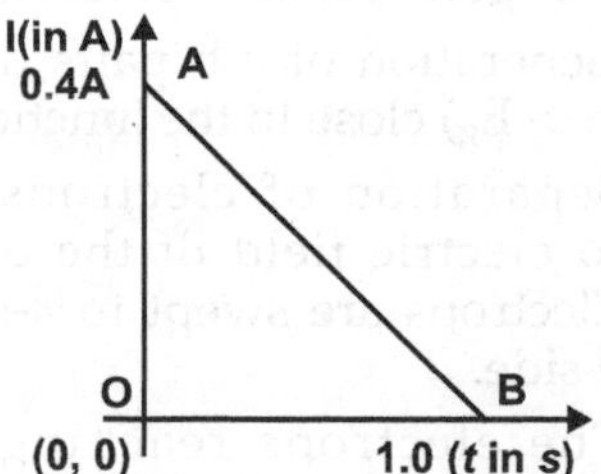

Find the :

(i) total charge passed through the loop.

(ii) change in magnetic flux through the loop.

(iii) magnitude of the magnetic field applied. [3]

Ans.

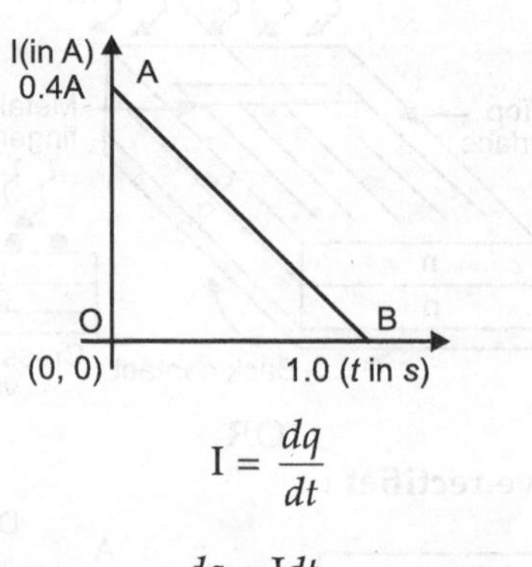

(i) $$I = \frac{dq}{dt}$$

$$dq = Idt$$

Hence, the area under the I-t curve gives charge flow.

$$\text{Area of I-}t\text{ curve} = \frac{1}{2} \times OA \times OB$$

$$= \frac{1}{2} \times 0.4 \times 1 = 0.2\ C$$

Total charge passed through loop (Δq) = 0.2 C

(ii) Now, we know

$$\Delta q = \frac{\Delta\phi}{R}$$

$$\therefore \quad \Delta\phi = \Delta q \times R$$

$$= 0.2 \times 10 = 2\ Wb$$

(iii) Now, change in flux, $\Delta\phi = B\,(\Delta A)$

$$2 = B\left(\frac{10}{100\times100}m\right)$$

$$\therefore \quad B = \frac{2\times100\times100}{10}$$

$$B = 2000T$$

Hence, the magnitude of field applied = 2000T

SECTION–D

35. (a) Define the term 'focal length of a mirror'. With the help of a ray diagram, obtain the relation between its focal length and radius of curvature. [5]

(b) Calculate the angle of emergence (e) of the ray of light incident normally on the face AC of a glass prism ABC of refractive index $\sqrt{3}$. How will the angle of emergence change qualitatively, if the ray of light emerges from the prism into a liquid of refractive index 1.3 instead of air?

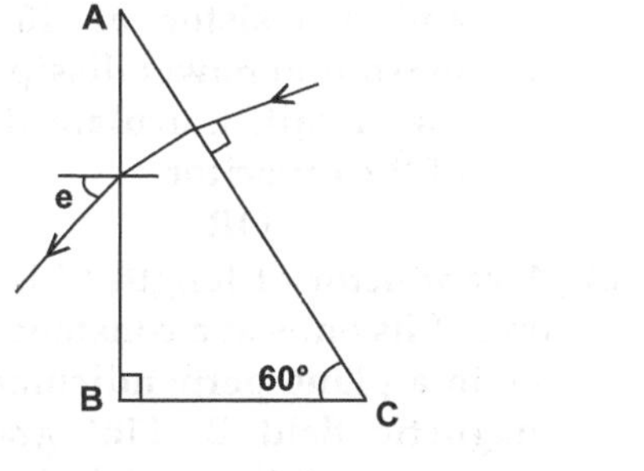

[5]

OR

(a) Define the term 'resolving power of a telescope'. How will the resolving power be effected with the increase in :

(i) Wavelength of light used.

(ii) Diameter of the objective lens.

Justify your answers.

(b) A screen is placed 80 cm from an object. The image of the object on the screen is formed by a convex lens placed between them at two different locations separated by a distance 20 cm. Determine the focal length of the lens. [5]

Ans. (a) Focal length of a mirror : It is defined as the distance between the centre of a lens or curved mirror and its focus.

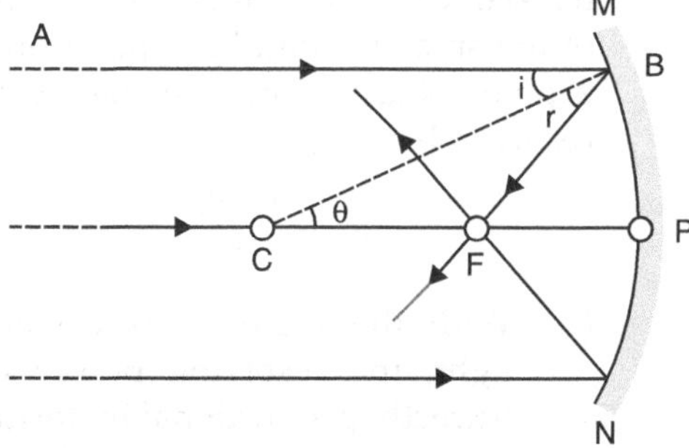

Consider a ray of light AB, parallel to the principal axis and incident on a spherical mirror at point B. The normal to the surface at point B is CB and CP = CB = R, radius of curvature. The ray AB after reflection from mirror passes through the focus F (concave mirror) or will appear to diverge from F (convex mirror) and obeys the law of reflection, $i = r$

Now, $\angle BCP = \theta = i$

In ΔCBF, $\theta = r$

$\therefore$ BF = FC

If the aperture of mirror is small, B lies close to P and therefore BF = PF.

or FC = FP = PF

or PC = PF + FC = PF + PF

or $R = 2PF = 2f$

or $f = \frac{R}{2}$

This is the required expression.

(b) By Snell's law, $\mu_1 \sin i = \mu_2 \sin r$

$$\mu_1 = \mu_{prism} = \sqrt{3}, \mu_2 = 1$$

$$\therefore \quad \mu_{prism} \sin 30° = \sin e$$

$$\sqrt{3}\times\frac{1}{2} = \sin e$$

$$\Rightarrow \quad e = 60°$$

Now, when external medium is changed to liquid of $\mu_2 = 1.3$, then

$$\mu_{prism} \sin 30° = \mu_2 \sin e$$

$$\sqrt{3}\times\frac{1}{2} = 1.3 \sin e$$

$$\therefore \quad e = \sin^{-1}\left(\frac{\sqrt{3}}{2\times1.3}\right)$$

$$= 41.77°$$

Hence, the angle of emergence reduces to 41.77° from 60°.

OR

(a) Resolving power of an astronomical telescope: It is defined as the reciprocal of the smallest angular separation between 2 point objects whose images can be just resolved by the telescope :

$$\text{R.P.} = \frac{1.22\lambda}{D}$$

(i) With the increase in wavelength of light, the resolving power increases. (directly proportional by formula)

(ii) With the increase of diameter of objective lens the resolving power decreases. (inversely proportional by formula)

(b) Case 1

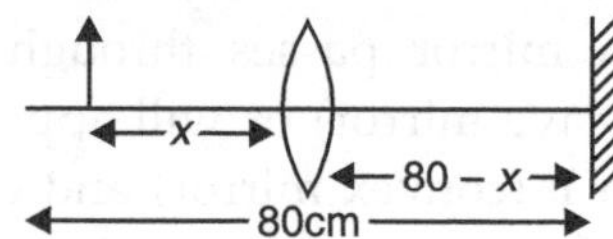

Let

Object distance, $u = x$

image distance, $v = 80 - x$

focal length $= f$

According to lens formula,

$$\frac{1}{f} = \frac{1}{v} + \frac{1}{u}$$

$$\frac{1}{f} = \frac{1}{80-x} + \frac{1}{x} \quad ...(i)$$

Similarly,

Case 2

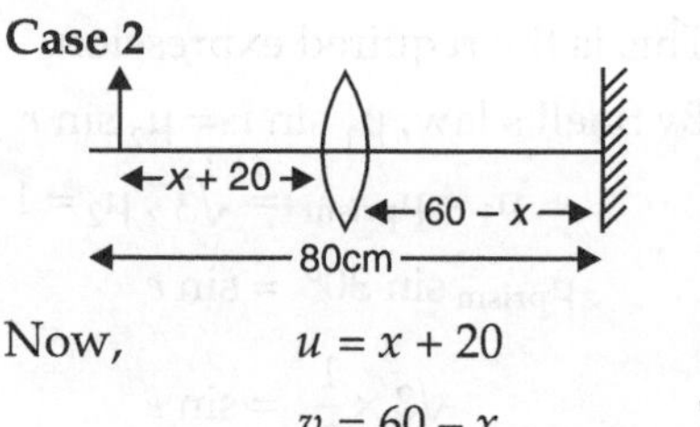

Now, $u = x + 20$

$v = 60 - x$

$$\therefore \quad \frac{1}{f} = \frac{1}{60-x} + \frac{1}{20+x} \quad ...(ii)$$

on comparing (i) and (ii) we get

$$\frac{1}{80-x} + \frac{1}{x} = \frac{1}{60-x} + \frac{1}{20+x}$$

$$\Rightarrow \quad \frac{80}{(80-x)x} = \frac{80}{(60-x)(20+x)}$$

$$80x - x^2 = 1200 + 40x - x^2$$

$$\Rightarrow \quad 40x = 1200$$

$$\Rightarrow \quad x = 30 \text{ cm}$$

Put the value of x in equation (i)

$$\therefore \quad \frac{1}{f} = \frac{1}{80-30} + \frac{1}{30}$$

$$\frac{1}{f} = \frac{1}{50} + \frac{1}{30}$$

$$\Rightarrow \quad f = 18.75 \text{ cm}$$

Hence, the focal length of lens is 18.75 cm.

36. (a) Show that an ideal inductor does not dissipate power in an a.c. circuit.

(b) The variation of inductive reactance (X_L) of an inductor with the frequency (*f*) of the ac source of 100 V and variable frequency is shown in the figure given.

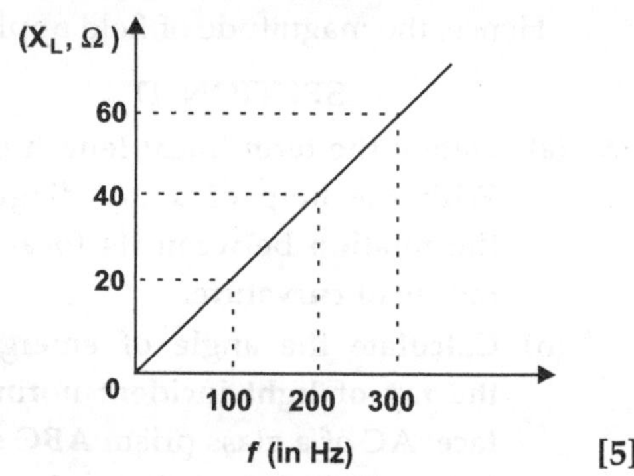

[5]

(i) Calculate the self-inductance of the inductor.

(ii) When this inductor is used in series with a capacitor of unknown value and a resistor of 10 Ω at 300s^{-1}, maximum power dissipation occurs in the circuit. Calculate the capacitance of the capacitor.

OR

(a) A conductor of length '*l*' is rotated about one of its ends at a constant angular speed 'ω' in a plane perpendicular to a uniform magnetic field B. Plot graphs to show variations of the emf induced across the ends of the conductor with (i) angular speed ω and (ii) length of the conductor *l*.

(b) Two concentric circular loops of radius 1 cm and 20 cm are placed coaxially.

(i) Find mutual inductance of the arrangement.

(ii) If the current passed through the outer loop is changed at a rate of 5 A/ms, find the emf induced in the inner loop. Assume the magnetic field on the inner loop to be uniform. [5]

Ans. (a) Power in a.c., circuit,

$$P = VI \cos \phi$$

For pure inductive circuit, the phase difference between current and voltage is $\pi/2$

$\therefore \quad \phi = \pi/2$

$$\cos\phi = \cos\pi/2 = 0$$

Therefore, zero power is dissipated. This current is sometimes referred to as watt-less current.

(b) (i) Since, $X_L = \omega L$

and $\omega = 2\pi f$, f is the frequency in Hz.

So, $L = \frac{X_L}{\omega} = \frac{X_L}{2\pi f}$

$$= \frac{20}{2\pi \times 100} = \frac{40}{2\pi \times 200} = \frac{60}{2\pi \times 300}$$

$$= 31.84 \times 10^{-3} \approx 32 \text{ mH.}$$

(ii) Since, power dissipation is maximum when

$$X_L = X_C$$

or $\omega L = \frac{1}{\omega C}$

or $C = \frac{1}{\omega^2 L}$

$$= \frac{1}{4\neq^2 f^2 L}$$

$$= \frac{1}{4 \times 3.14 \times 3.14 \times 300 \times 300 \times 32 \times 10^{-3}}$$

$= 8.8\,\mu F$

OR

(a) (i)

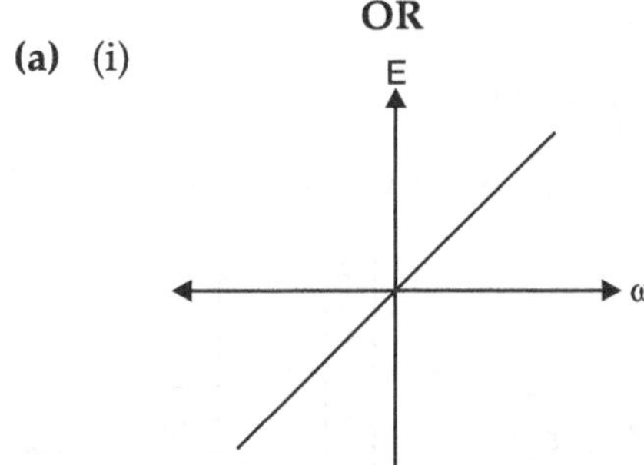

Induced Emf $= E = \frac{B\omega L^2}{2}$

(ii)

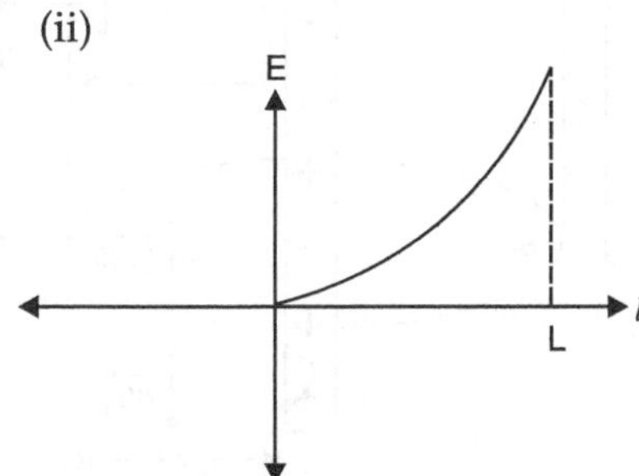

(b) (i) Since, magnetic flux, $\phi = MI$

And magnetic field at the centre of bigger loop, $\vec{B}$.

$$\frac{\mu_0 I}{2R} = \frac{4\pi \times 10^{-7}}{2 \times 20 \times 10^{-2}}$$

$$= \pi \times 10^{-6} I$$

Flux through the smaller loop

$$\phi = BA_s$$

$$= \frac{4\pi \times 10^{-5} I}{40} \times \pi(0.01)^2$$

$$= \pi^2 \times 10^{-10} \times I$$

$\therefore \quad M = \frac{\phi}{I}$

$$= \frac{\pi^2 \times 10^{-10} \times I}{I}$$

$$= 9.86 \times 10^{-10} H$$

(ii) Emf included

$$= \frac{-d\phi}{dt}$$

$$= -9.86 \times 10^{-10} \times \frac{dI}{dt}$$

$$= -9.86 \times 10^{-10} \times 5$$

$$= -4.93 \times 10^{-9} V$$

37. (a) Write two important characteristics of equipotential surfaces.

(b) A thin circular ring of radius r is charged uniformly so that its linear charge density becomes λ. Derive an expression for the electric field at a point P at a distance x from it along the axis of the ring. Hence, prove that at large distances ($x >> r$), the ring behaves as a point charge. [5]

OR

(a) State Gauss's law on electrostatics and derive an expression for the electric field due to a long straight thin uniformly charged wire (linear charge density λ) at a point lying at a distance r from the wire.

(b) The magnitude of electric field (in NC^{-1}) in a region varies with the distance r (in m) as

$$E = 10r + 5$$

By how much does the electric potential increase in moving from point at r = 1m to a point at r = 10m. [5]

Ans. (a) Characteristics of equipotential surfaces are :

(i) Potential remains same at all the points on the equipotential surface.

(ii) No work is required to move a charge on an equipotential surface.

(b) Let us consider a thin circular ring of radius r with charge density as λ.

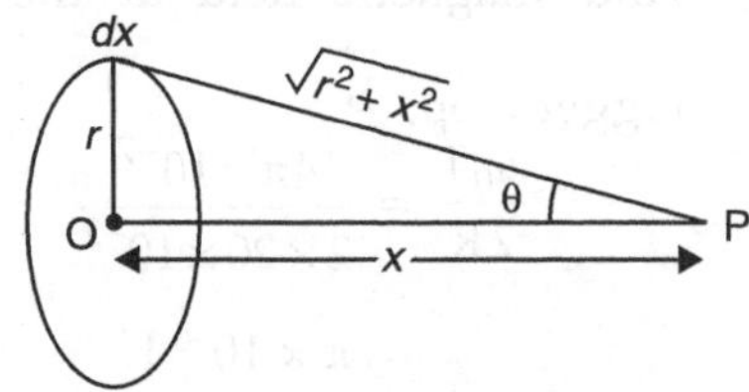

Let us consider a small charge element (dx) on the ring having small charge (dq)

$$dq = \lambda dx$$

The electric field due to this charge element at point P is given as :

$$dE = \frac{1}{4\pi\varepsilon_0}\frac{dq}{(r^2+x^2)}$$

$$dE = \frac{1}{4\pi\varepsilon_0}\frac{\lambda dx}{(r^2+x^2)}$$

Electric field at point P will have 2 components $d\vec{E}\cos\theta$ in horizontal direction and $d\vec{E}\sin\theta$ in vertical direction. The vertical component will cancel out the effect of each other due to the presence of the diametrically opposite element and horizontal component will survive at point P.

$$\therefore \quad \cos\theta = \frac{x}{\sqrt{r^2+x^2}}$$

Now, $$dE\cos\theta = \frac{\lambda x dx}{4\pi\varepsilon_0(r^2+x^2)^{3/2}}$$

$$\therefore \quad dq = \lambda dx$$

$$\therefore \quad dE\cos\theta = \frac{x dq}{4\pi\varepsilon_0(r^2+x^2)^{3/2}}$$

On integrating, we get

$$E_x = \int dE\cos\theta$$

$$= \int\frac{x dq}{4\pi\varepsilon_0(r^2+x^2)^{3/2}}$$

$$= \frac{x.Q}{4\pi\varepsilon_0(r^2+x^2)^{3/2}}$$

Where, Q is the total charge on ring

When $x >> r$, then $r^2 + x^2 \approx x^2$

$$\therefore \quad E_x = \frac{xQ}{4\pi\varepsilon_0 x^3} = \frac{Q}{4\pi\varepsilon_0 x^2}$$

So, at large distance, the ring behaves like a point charge.

OR

(a) Electric Field ($\vec{E}$) due to a long straight uniformly charged wire of charge density λ : Consider a cylindrical Gaussian surface of radius r and length l coaxial with line charge. The cylindrical Gaussian surface may be divided into three parts : (1) curved surface S_1 (2) flat surface S_2 and (3) flat surface S_3.

By symmetry, the electric field has the same magnitude E at each point of curved surface S_1 and is directed radially outward. We consider small elements of surfaces S_1, S_2 and S_3. The surface element vector $d\vec{S_1}$ is directed along the direction of electric field (*i.e.*, angle between $\vec{E}$ and $d\vec{S_1}$ is 0°); the elements $d\vec{S_2}$ and $d\vec{S_3}$ are directed perpendicular to field vector $\vec{E}$ (*i.e.*, angle between $d\vec{S_2}$ and $\vec{E}$ and $d\vec{S_3}$ and $\vec{E}$ is 90°).

Electric flux through the cylindrical surface.

$$\oint_S \vec{E}\,d\vec{S} = \oint_{S_1}\vec{E}\,d\vec{S_1} + \oint_{S_2}\vec{E}\,d\vec{S_2} + \oint_{S_3}\vec{E}\,d\vec{S_3}$$

$$= \oint_{S_1} E dS_1 \cos 0° + \oint_{S_2} E dS_2 \cos 90° + \oint_{S_3} E dS_3 \cos 90°$$

$$= E\int dS_1 + 0 + 0$$

$= E \times$ area of curved surface $= E \times 2\pi rl$

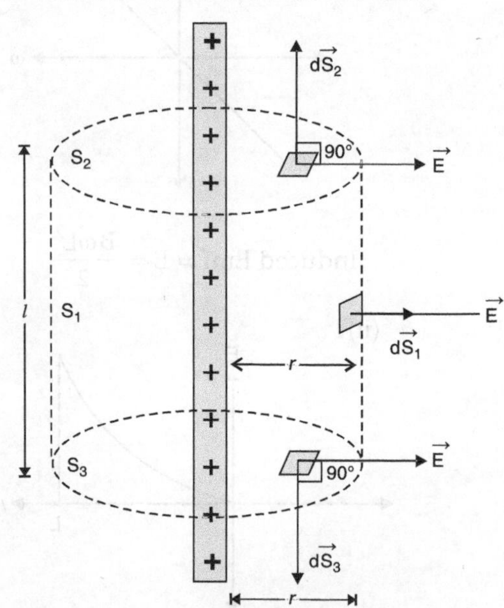

Charge enclosed,

$$q = \lambda l$$

By Gauss' theorem

$$\phi_E = \frac{q}{\varepsilon_0}$$

$$E.2\pi rl = \frac{\lambda l}{\varepsilon_0}$$

$$E = \frac{\lambda}{2\pi\varepsilon_0 r}$$

(b) Let point A be at $r = 1$ m and point B be given at $r = 10$m

V_A = potential at A

V_B = potential at B

$$\therefore \quad V_B - V_A = \int_A^B \vec{E}.d\vec{r}$$

$$= -\int_1^{10} (10r + 5)dr$$

$$= -\left[\frac{10r^2}{2} + 5r\right]_1^{10}$$

$$= -\left[\left(\frac{10(10)^2}{2} + 5(10)\right) - \left(\frac{10(1)^2}{2} + 5 \times 1\right)\right]$$

$$= -[550 - 10] = -540 \text{ V}$$

Delhi [Set-II]

Note : Except for the following questions, all the remaining questions have been asked in previous set.

SECTION–A

Note : Select the most appropriate option from those given below each question.

6. A biconvex lens of focal length f is cut into two idendical plano-convex lenses. The focal length of each part will be : [1]

(a) f (b) $f/2$

(c) $2f$ (d) $4f$

Ans. (c) $2f$

Explanation : Let the focal length of biconvex lens be f and plano-convex lens be f'.

Then by lens maker's formula

For biconvex lens

$$\frac{1}{f} = (\mu - 1)\left(\frac{1}{R} + \frac{1}{R}\right) = \frac{2(\mu - 1)}{R}$$

Where, R is the radius of curvature of biconvex lens.

For plano-convex lens

$$\frac{1}{f'} = (\mu - 1)\left(\frac{1}{R}\right) = \frac{\mu - 1}{R}$$

Thus $$\frac{1}{f} = \frac{2}{f'}$$

or $$f' = 2f$$

7. The phase difference between the current and the voltage in series LCR circuit at resonance is : [1]

(a) π (b) $\pi/2$

(c) $\pi/3$ (d) zero

Ans. (d) zero

Explanation : At resonance, the circuit is purely resistive and there is no phase difference between current and voltage.

8. Photons of frequency v are incident on the surfaces of metals A and B of threshold frequencies 3/4 v and 2/3 v, respectively. The ratio of the maximum kinetic energy of electrons emitted from A to that from B is : [1]

(a) 2 : 3 (b) 4 : 3

(c) 3 : 4 (d) 3 : 2

Ans. (c) 3 : 4

Explanation : According to Einstein photoelectric equation

$$hv = hv_0 + K_{max}$$

For first metal

$$hv = h\left(\frac{3}{4}v\right) + K_1$$

$$\Rightarrow \quad K_1 = \frac{hv}{4}$$

For second metal

$$hv = h\left(\frac{2}{3}v\right) + K_2$$

$$\Rightarrow \quad K_2 = \frac{hv}{3}$$

Thus, the ratio is

$$\frac{K_1}{K_2} = \frac{hv/4}{hv/3} = \frac{3}{4}$$

SECTION–B

21. (a) Define one Becquerel.

(b) A radioactive substance disintegrates into two types of daughter nuclei, one type with disintegration constant λ_1 and the other type with disintegration constant λ_2. Determine the half-life of the radioactive substance.

Ans. (a) One Becquerel (B*q*) is defined as the activity of a quantity of radioactive sample in which one nucleus decays per second.

(b) $X \xrightarrow{\lambda_1} D_1$

$X \xrightarrow{\lambda_2} D_2$

(Where, X is a radioactive nuclei and D_1, D_2 are the daughter nuclei)

$$\lambda_{effective} = \lambda_1 + \lambda_2$$

$$t_{1/2} = \frac{\log 2}{\lambda_{effective}} = \frac{0.693}{\lambda_1 + \lambda_2}$$

23. In case of photo electric effect experiment, explain the following facts giving reasons.

(a) The wave theory of light could not explain the existence of the threshold frequency.

(b) The photo electric current increases with increase of intensity of incident light.

Ans. (a) If light would only be a wave its energy would be represented by the amplitude of the light wave. A more intense light source, even if it will be of lower frequency, would be able to eject out photoelectrons from the metal surface and photoelectric current would be generated.

But actually a light below a certain threshold frequency could not generate any current, no matter how intense the light is.

(b) With the increased intensity of light the more photons will be incident on metal surface. If the frequency of incident light is more than the threshold frequency, then the more number of photons means more photoelectrons will be emitted. This result in more photoelectric current.

SECTION–C

30. (a) Define internal resistance of a cell.

(b) A cell of emf E and internal resistance *r* is connected across a variable resistor R. Plot the shape of graphs showing variation of terminal voltage V with (i) R and (ii) circuit current I. **[3]**

Ans. (a) Internal resistance is the resistance which is present within the battery that resists the current flow when connected to a circuit.

(b) (i) Since, $$V = \frac{ER}{R+r}$$

$$\Rightarrow \quad V = \frac{E}{1+\frac{r}{R}}$$

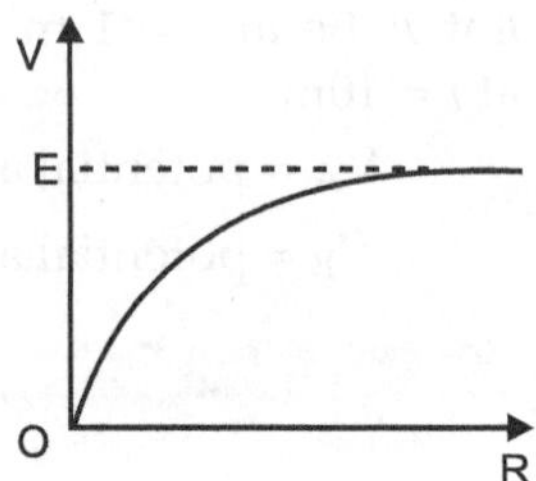

(ii) Since $V = E - IR$

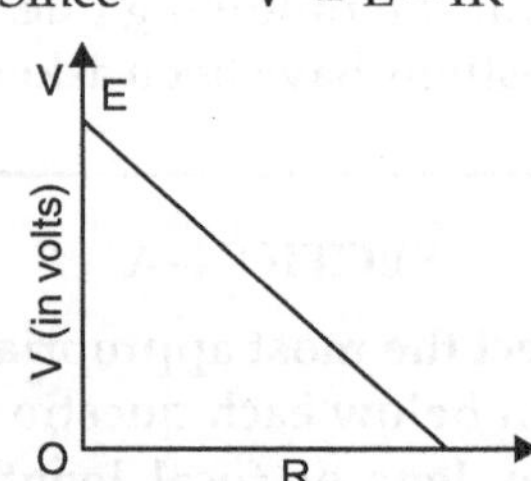

32. When a conducting loop of resistance 10 Ω and area 10 cm^2 is removed from an external magnetic field acting normally, the variation of induced current I in the loop with times is as shown in the figure.

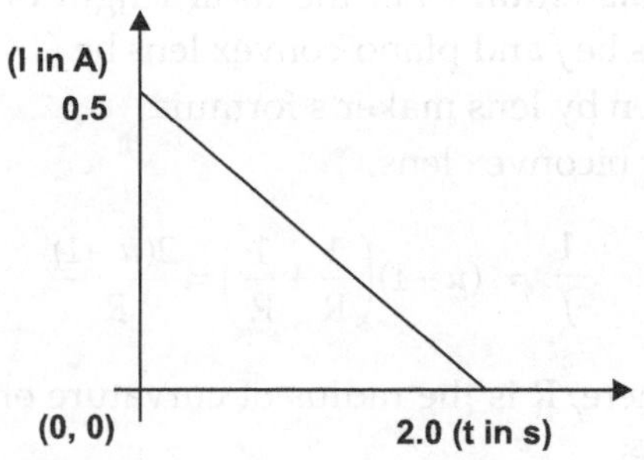

Find the :

(i) total charge passed through the loop.

(ii) change in magnetic flux through the loop.

(iii) magnitude of the magnetic field applied. **[3]**

Ans. (a) Since, $$I = \frac{dq}{dt}$$

$$\Rightarrow \quad dq = Idt$$

Area of I – *t* curve = Total charge flows

$$\therefore \quad Q = \frac{1}{2} \times 2 \times 0.5 = 0.5C.$$

Hence, total charge that flow through loop is 0.5C

(b) Since, $\Delta Q = \frac{\Delta\phi}{k}$

$$\Rightarrow \quad \Delta\phi = \Delta Q \times k$$

$$= \frac{1}{2} \times 10\Omega$$

$$= 5Wb$$

Hence, change in magnetic flux is 5Wb.

(c) Now, $\Delta\phi = B(\Delta A)$

$$B = \frac{\Delta\phi}{\Delta A}$$

$$= \frac{5}{\frac{10}{100 \times 100}} = 5000 \text{ T}$$

Hence, the magnetic field applied is 5000 T

33. Draw the curve showing the variations of binding energy per nucleon with the mass number of nuclei. Using it explain the fusion of nuclei lying on ascending part and fission of nuclei lying on descending part of this curve.

Ans.

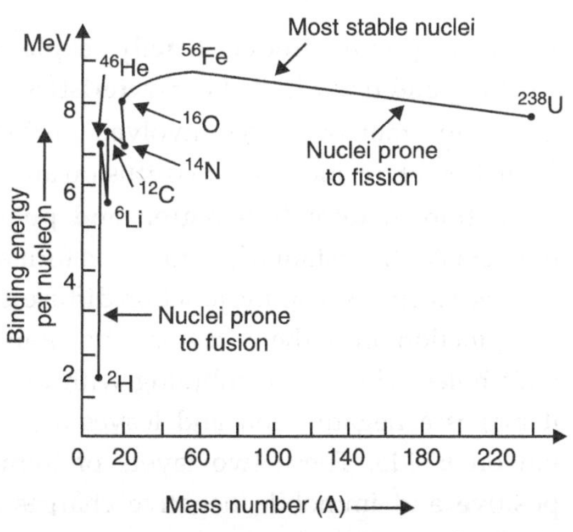

The graph above tells us that binding energy per nucleon is smaller for heavier nuclei as well as for lighter nuclei than for middle order nuclei (with mass no. between 30 to 170). Heavier nuclei are unstable and they undergo fission and lighter nuclei which are unstable undergo fusion in order to form a stable nuclei lying in the range of mass number 30 to 170.

34. An optical instrument uses a lens of power 100D for objective lens and 50 D for its eyepiece. When the tube length is kept at 20cm, the final image is formed at infinity. [3]

(a) Identify the optical instrument.

(b) Calculate the magnification produced by the instrument.

Ans. (a) Power of lens = $\frac{1}{\text{Focal length}}$

So, focal length of objective lens

$$= \frac{1}{100} m = 1 \text{cm}$$

Focal length of eye piece = $\frac{1}{50} m = 2 \text{cm}$

Since, the objective has a smaller focal length than the eye piece, the instrument is a compound microscope.

(b) Magnification produced is formed at infinity is given by,

$$m = \left(\frac{L}{f_o}\right)\left(\frac{D}{fe}\right)$$

where, tube length L = 20cm and least distance of distinct vision, D = 25cm

$$m = \frac{20}{1} \times \frac{25}{2} = 250$$

Delhi [Set-III]

Note : Except for the following questions, all the remaining questions have been asked in previous sets.

SECTION–B

21 Write shortcomings of Rutherford atomic model. Explain how these were overcome by the postulates of Bohr's atomic model. [2]

Ans. According to Rutherford's model, electrons revolve around the nucleus in a circular path. But particles that are in motion on a circular path would undergo acceleration and it causes radiation of energy by charged particles. Eventually, electrons will loose energy and fall on to the nucleus. Thus, making nucleus unstable. Rutherford's model cannot explain the stability of the atom.

Bohr modified this model by proposing that the electrons move in orbits of fixed size and energy. The energy of an electrons depends on the size of the orbit and is lower for smaller orbits. Radiation occur only when electron jump from one orbit to another.

24. Calculate for how many years will the fusion of 2.0 kg deuterium keep 800 W electric lamp glowing. Take the fusion reaction as :

$$^2_1H + ^2_1H \longrightarrow ^3_1He + ^1_0n + 3.27 \text{ MeV}$$

Ans. 2 deuterium nuclei fuse to produce 3.27 MeV

$$= 3.27 \times 1.6 \times 10^{-13} \text{ J} = 5.238 \times 10^{-13} \text{ J}$$

Energy produced per nuclei

$$= \frac{5.238 \times 10^{-13}}{2} = 2.61 \times 10^{-13} \text{ J}$$

2 kg deuterium atoms will have

$$= \frac{6.023 \times 10^{23}}{2} \times 2000$$

$$= 6.023 \times 10^{26} \text{ deuterium atom}$$

6.022×10^{26} deuterium atoms will produce

$$= 2.61 \times 10^{-13} \times 6.023 \times 10^{26} \text{ J}$$

$$= 15.77 \times 10^{13} \text{ J}$$

$$\because \quad \text{Power} = \frac{E}{t}$$

$$800 = \frac{15.77 \times 10^{13}}{t}$$

$$\Rightarrow \quad t = \frac{15.77 \times 10^{13}}{800}$$

$$= 1.971 \times 10^{11} \text{ seconds}$$

$$t = \frac{1.971 \times 10^{11}}{365 \times 24 \times 60 \times 60}$$

$$= 6.24 \times 10^3 \text{ years.}$$

Hence, electric lamp will glow for 6.24×10^3 years.

SECTION–C

28. (a) Define the term 'half-life' of a radioactive substance.

(b) The half life of $^{238}_{92}U$ undergoing α decay is 4.5×10^9 years. Calculate the activity of 5g sample of $^{238}_{92}U$. **[3]**

Ans. (a) Half life of a radioactive element is a time in which the given amount of the substance reduces to half of the original value as a consequence of decay.

(b) We have

$$t_{1/2} = 4.5 \times 10^9 \text{ years}$$

$$= \frac{0.693}{\lambda}$$

$$\lambda = \frac{0.693}{4.5 \times 10^9 \times 365 \times 24 \times 60 \times 60}$$

Number of nuclei present in sample,

$$N_0 = \frac{5 \times 6.022 \times 10^{23}}{238}$$

$$\text{Activity} = A = \frac{dN}{dt} = \lambda N_0$$

$$= \left[\frac{0.693}{4.5 \times 10^9 \times 365 \times 24 \times 3600}\right] \times \frac{5 \times 6.022 \times 10^{23}}{238}$$

$$\cong 0.61 \times 10^5 \text{ disintegration/sec.}$$

29. Explain the formation of potential barrier and depletion region in a *p-n* junction diode. What is the effect of applying forward bias on the width of depletion region?

OR

What is photo diode? Briefly explain its working and draw its V-I characteristics. **[3]**

Ans. Two important processes involved during the formation of *p-n* junction are diffusion and hole.

Formation of depletion region and potential barrier : At the instant of *p-n* junction formation, the free electrons near the junction diffuse across the junction into the *p*-region and combines with holes. Thus, on combining with the hole, it makes a negative ion and leaves a positive ion on *n*-side. These two layers of immobile positive and immobile negative charges form the depletion charge.

Further, as electrons diffuse across the junction, a point is reached where the negative charge repels only further diffusion of electron. This depletion region now acts as a barrier. Now the external energy is supplied to get electrons to move across the barrier of electric field.

The potential difference required to move the electron through the electric field is called barrier potential.

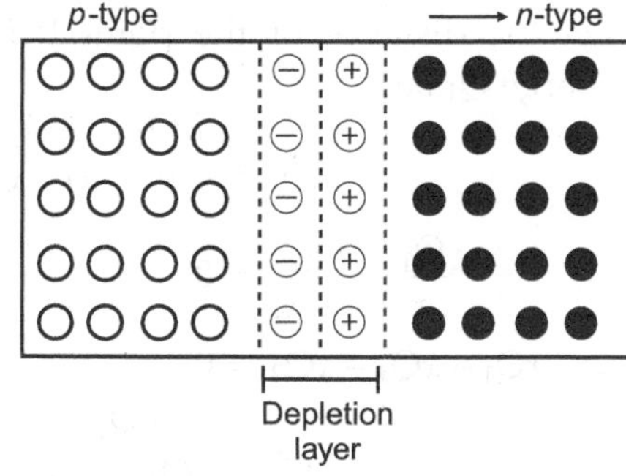

The effect of forward bias on *p-n* region is that the depletion layer will be reduced.

OR

Photodiode : It is an optoelectronic device in which current carriers are generated by photons through photo-excitation *i.e.,* photo conduction by light.

A photodiode is a special *p-n* junction diode made of photosensitive semiconducting material. It is operated under reverse bias below the breakdown voltage. The conductivity of *p-n* junction photodiode is modulated by the absorption of incident light in or near the depletion layer which exists at the *p-n* junction. The conductivity of *p-n* junction photodiode increases with the intensity of light falling on it.

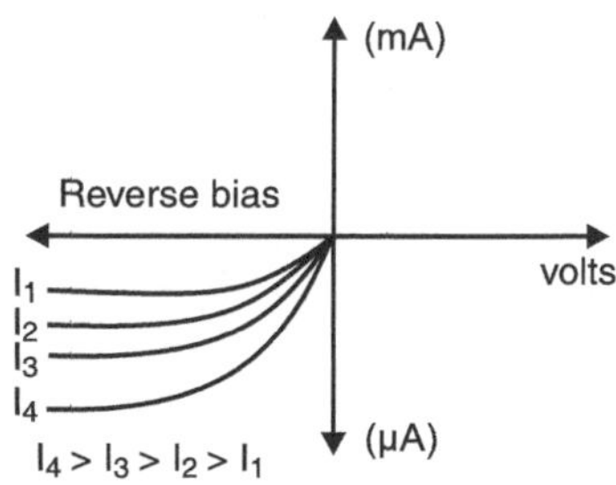

When the photodiode is reverse biased with a voltage less than its breakdown voltage and no light is incident on its junction, the reverss current is extremely small. This current is called dark current.

30. Calculate the de-Broglie wavelength associated with the electron in the first 2nd excited state of hydrogen atom. The ground state energy of the hydrogen atom is 13.6 eV. [3]

Ans. De broglie wavelength is, $\lambda = \frac{h}{mv} = \frac{h}{p}$

where, p is the momentum of an electron

K.E. and momentum are related by,

$$\text{K.E.} = \frac{p^2}{2m}$$

$$\Rightarrow \quad p = \sqrt{2m\text{K.E.}}$$

$$\Rightarrow \quad \lambda = \frac{h}{\sqrt{2m\text{K.E.}}}$$

Then, $\quad \text{K.E of } e^- = |\text{T.E of } e^-|$

$$= \left| -\frac{13.6 \times Z^2}{n^2} \right| eV$$

For Hydrogen (Z = 1) and second excited state implies $n = 3$

$$\text{K.E.} = \frac{13.6 \times 1^2}{3^2}$$

$$= 1.51 \text{ eV}$$

$$= 1.51 \times 1.6 \times 10^{-19} \text{ J}$$

$$= 2.41 \times 10^{-19} \text{J}$$

$$\therefore \quad \lambda = \frac{h}{\sqrt{2m(\text{KE})}}$$

$$= \frac{6.63 \times 10^{-34}}{\sqrt{2 \times 9.1 \times 10^{-31} \times 2.41 \times 10^{-19}}}$$

$$= 1.001 \times 10^{-9} \text{ m}$$

$$= 10 \text{Å}$$

32. (a) Two point charges $+Q_1$ and $-Q_2$ are placed r distance apart. Obtain the expression for the amount of work done to place a third charge Q_3 at the midpoint of the line joining two charges.

(b) At what distance from charge $+Q_1$ on the line joining two charges (in terms of Q_1, Q_2 and r) will this work done be zero. [3]

Ans. According to the question,

Q_1 ← r → Q_2
← r/2 → ← r/2 →
Q_3 (at ∞)

(a) Work done :

$$W = \frac{kQ_1Q_3}{r/2} - \frac{kQ_3Q_2}{r/2}$$

$$W = \frac{2rQ_1Q_3}{r} - \frac{2rQ_3Q_2}{r}$$

$$= \frac{2rQ_3(Q_1 - Q_2)}{r}$$

$$= \frac{2Q_3(Q_1 - Q_2)}{4\pi\varepsilon_o r}$$

(b) For work done to be zero : Work done will be zero for the point where the electric potential is zero consider a point P to be lying in between the two charges, where electric potential is zero.

Let the distance of the point P from the charge Q_1 be x.

$$V = \frac{kQ_1}{x} - \frac{kQ_2}{(r-x)} = 0$$

$$\Rightarrow \frac{Q_1}{x} - \frac{Q_2}{r-x} = 0$$

$$\Rightarrow rQ_1 - xQ_1 - xQ_2 = 0$$

$$\Rightarrow x = \frac{rQ_1}{(Q_1 + Q_2)}$$

●●